The Mathematics *of* Egypt, Mesopotamia, China, India, and Islam

A Sourcebook

The Mathematics *of* Egypt, Mesopotamia, China, India, and Islam

A Sourcebook

Victor Katz, *Editor*

Annette Imhausen
Eleanor Robson
Joseph Dauben
Kim Plofker
J. Lennart Berggren

PRINCETON UNIVERSITY PRESS • PRINCETON AND OXFORD

Published by Princeton University Press, 41 William Street,
Princeton, New Jersey 08540

In the United Kingdom: Princeton University Press, 3 Market Place,
Woodstock, Oxfordshire OX20 1SY

Library of Congress Cataloging-in-Publication Data

The mathematics of Egypt, Mesopotamia, China, India,
and Islam: a sourcebook/Victor Katz, editor.
p. cm.
Includes bibliographical references and index.

ISBN-13: 978-0-691-11485-9 (hardcover: alk. paper)
ISBN-10: 0-691-11485-4 (hardcover: alk. paper)

1. Mathematics, Ancient–Sources. 2. Mathematics–History–Sources.
I. Katz, Victor J.
QA22.M3735 2007
510.9–dc22 2006030851

British Library Cataloging-in-Publication Data is available

This book has been composed in Minion and Univers
Printed on acid-free paper. ∞
press.princeton.edu

Printed in the United States of America

3 5 7 9 10 8 6 4 2

Contents

Chapter 2
Mesopotamian Mathematics
Eleanor Robson

Chapter 3
Chinese Mathematics
Joseph W. Dauben

Preface

Kim Plofker and I conceived of this book at a meeting of the Mathematical Association of America several years ago. We agreed that, since there was now available a fairly large collection of English translations of mathematical texts from Egypt, Mesopotamia, China, India, and Islam, the time was ripe to put together an English sourcebook. This book would have texts gathered together so that they could easily be studied by all those interested in the mathematics of ancient and medieval times. I secured commitments to edit the five sections from outstanding scholars with whom I was familiar, scholars who had already made significant contributions in their fields, and were fluent in both the original languages of the texts and English. There are a growing number of scholars investigating the works of these civilizations, but I believe that this group was the appropriate one to bring this project to fruition.

The editors decided that to the extent possible we would use already existing English translations with the consent of the original publisher, but would, where necessary, produce new ones. In certain cases, we decided that an existing translation from the original language into French or German could be retranslated into English, but the retranslation was always made with reference to the original language. As it turned out, the section editors of both the Egyptian and Mesopotamian sections decided to produce virtually all new translations, because they felt that many previous translations had been somewhat inadequate. The other sections have a mix of original translations and previously translated material.

Each of the five sections of the Sourcebook has a preface, written by the section editor, giving an overview of the sources as well as detailing the historical setting of the mathematics in that civilization. The individual sources themselves also have introductions. In addition, many of the sources contain explanations to help the reader understand the sometimes fairly cryptic texts. In particular, the Chinese and Indian sections have considerably more detailed explanations than the Islamic section, because Islamic mathematicians, being well schooled in Greek mathematics, use "our" techniques of mathematical analysis and proof. Mathematicians from China and India, and from Egypt and Mesopotamia as well, come from traditions far different from ours. So the editors of those sections have spent considerable effort in guiding the reader through this "different" mathematics.

The book is aimed at those having knowledge of mathematics at least equivalent to a U.S. mathematics major. Thus, the intended audience of the book includes students studying mathematics and the history of mathematics, mathematics teachers at all levels, mathematicians, and historians of mathematics.

The editors of this Sourcebook wish to thank our editors at Princeton University Press, David Ireland and Vickie Kearn, who have been extremely supportive in guiding this book from concept to production. We also wish to thank Dale Cotton, our production editor, Dimitri Karetnikov, the illustration specialist, and Alison Anderson, the copy editor, for their friendly and efficient handling of a difficult manuscript. Finally, as always, I want to thank my wife Phyllis for her encouragement and for everything else.

Victor J. Katz
Silver Spring, MD
December, 2005

Permissions

Chapter 1

SECTION IV: The material from the Demotic papyri is reprinted from Richard Parker, *Demotic Mathematical Papyri*, © 1972 by Brown University Press, by permission of University Press of New England, Hanover, NH.

Chapter 3

SECTION IV: The material from the *Zhou bi suan jing* is reprinted from *Astronomy and Mathematics in Ancient China: The Zhou Bi Suan Jing*, by Christopher Cullen, © 1996, with the permission of Cambridge University Press.

SECTION V: The material from the *Nine Chapters on the Mathematical Art* and the *Sea Island Mathematical Manual* is reprinted from *The Nine Chapters on the Mathematical Art: Companion and Commentary*, by Shen Kangshen, John Crossley, and Anthony W.-C. Lun, © 1999, with the permission of Oxford University Press.

SECTION VI.A.: The material from the *Sunzi suan jing* is reprinted from *Fleeting Footsteps: Tracing the Conception of Arithmetic and Algebra in Ancient China*, by Lam Lay Yong and Ang Tian Se, © 1992, with the permission of World Scientific Publishing Co. Pte. Ltd.

SECTION VI.B.: The material from the *Zhang qiujian suan jing* is reprinted from "Zhang Qiujian Suanjing (The Mathematical Classic of Zhang Qiujian): An Overview," by Lam Lay Yong, *Archive for History of Exact Sciences* 50 (1997), 201–240, with the kind permission of Springer Science and Business Media.

SECTION VII.A: The material from chapters 1, 2, 4 of the *Shu shu jiu zhang* is reprinted from *Chinese Mathematics in the Thirteenth Century: The Shu-shu chiu-chang of Ch'in Chiu-shao*, by Ulrich Libbrecht, © 1973, with the permission of MIT Press. The material from chapter 3 of the *Shu shu jiu zhang* is reprinted from *A History of Chinese Mathematics*, by Jean-Claude Martzloff, Stephen Wilson, trans., © 1997, with the kind permission of Springer Science and Business Media.

SECTION VII.B: The material from Li Zhi's *Ce yuan hai jing* is reprinted from "Chinese Polynomial Equations in the Thirteenth Century," in Li Guihao *et al.*, eds., *Explorations in the History of Science and Technology in China*, Chinese Classics Publishing House, Shanghai, © 1982. The material from Li Zhi's *Yigu yanduan* is reprinted from "Li Ye and his Yi Gu Yan Duan (Old Mathematics in Expanded Sections)," by Lam Lay Yong and Ang Tian Se, *Archive for History of Exact Sciences* 29 (1984), 237–265 with the kind permission of Springer Science and Business Media.

SECTION VII.C.: The material from Yang Hui's *Xiangjie Jiuzhang Suanfa* is reprinted from "On The Existing Fragments of Yang Hui's Hsiang Chieh Suan Fa," by Lam Lay Yong, *Archive for History of Exact Sciences* 6 (1969), 82–88 with the kind permission of Springer Science and Business Media. The material from Yang Hui's *Yang Hui Suan Fa* is reprinted from *A Critical Study of the Yang Hui Suan Fa*, by Lam Lay Yong, © 1977, with the permission of Singapore University Press.

SECTION VIII: The material from Matteo Ricci and Xu Guangxi's, "Prefaces" to the First Chinese Edition of Euclid's *Elements* is reprinted from *Euclid in China*, by Peter M. Engelfriet, © 1998 by permission of Brill Academic Publishers.

Chapter 4

SECTION II.B: The material from the *Śulbasūtras* is reprinted from *The Sulbasutras*, by S. N. Sen and A. K. Bag, © 1983 by permission of the Indian National Science Academy.

SECTION II.D.: Figure 2.5 is reprinted from *The Concept of Śūnya*, by A. K. Bag and S. R. Sarma, eds., © 2003 by permission of the Indian National Science Academy.

SECTION III.A.: The material from the *Āryabhaṭīya* is reprinted from *Expounding the Mathematical Seed: a Translation of Bhāskara I on the Mathematical Chapter of the Āryabhaṭīya*, by Agathe Keller, © 2005 by permission of Birkhäuser Verlag AG.

SECTION III.C.: The material from the *Bakhshālī Manuscript* is reprinted from *The Bakhshali Manuscript: An Ancient Indian Mathematical Treatise*, © 1995 by permission of Egbert Forsten Publishing. The material from the *Bījagaṇita* of Bhāskara II with commentary of Sūryadāsa is reprinted from *The Sūryaprakāśa of Sūryadāsa: A Commentary on Bhāskaracarya's Bījagaṇita*, by Pushpa Kumari Jain, © 2001 by permission of the Oriental Institute in Vadodara.

SECTION III.E.: The material from the *Śiṣyadhīvṛddhidatantra* of Lalla is reprinted from *Śiṣyadhivṛddhida Tantra of Lalla*, by Bina Chatterjee, © 1981 by permission of the Indian National Science Academy.

SECTION IV.B.: The material from the *Kriyākramakarī* of Śaṅkara is reprinted from *Studies in Indian Mathematics: Series, Pi and Trigonometry*, by T. Hayashi, T. Kusuba, and M. Yano, Setsuro Ikeyama, trans., by permission of the authors.

SECTION V.A.: The material from the *Gaṇitakaumudī* of Nārāyana Paṇḍita is reprinted from *Combinatorics and Magic Squares in India: A Study of Nārāyana Paṇḍita's "Gaṇitakaumudi", chapters 13-14*, a Brown University dissertation by Takanori Kusuba and is used by permission of the author.

Chapter 5

SECTION II: The material from the Banū Mūsā *On Conics* is reprinted from *Apollonius' Conics Books V-VII: The Arabic Translation of the Lost Greek Original in the Version of the Banu Musa*, by G. J. Toomer, © 1990 with the kind permission of Springer Science and Business Media. The material from ibn al-Haytham's *Completion of the Conics* is reprinted from *Ibn al-Haytham's Completion of the Conics*, by J. P. Hogendijk, © 1985 with the kind permission of Springer Science and Business Media.

SECTION III: The material from al-Khwārizmī's *Treatise on Hindu Calculation* is reprinted from "Thus Spake al-Khwārizmī: A Translation of the Text of Cambridge University Library Ms. Ik.vi.5," by John Crossley and Alan Henry, *Historia Mathematica* 17 (1990), 103–131 with permission from Elsevier Ltd. The material from al-Uqlīdisī's *Chapters on Hindu Arithmetic* is reprinted from *The Arithmetic of al-Uqlidisi*, by A. Saidan, © 1978 by permission of Reidel Publishers. The material from Kūshyār ibn Labān's *Principles of Hindu Reckoning* is reprinted from *Kushyar ibn Laban Principles of Hindu Reckoning*, by M. Levey and M. Petruck, © 1966 by permission of The University of Wisconsin Press.

SECTION IV: The material from al-Samaw'al on al-Karajī and binomial coefficients is reprinted from *The Development of Arabic Mathematics: Between Arithmetic and Algebra*, by Roshdi Rashed, © 1994 with the kind permission of Springer Science and Business Media. The material from the algebra of Omar Khayyam is reprinted from *The Algebra of Omar Khayyam*, by D. Kasir, © 1931 with the permission of Columbia Teachers College Press.

SECTION V: The material from al-Bāghdadī on balanced numbers is reprinted from "Two Problems of Number Theory in Islamic Times," by J. Sesiano, *Archive for History of Exact Sciences* 41, (1991), 235–238 with the kind permission of Springer Science and Business Media.

SECTION VI.A.: The material from al-Kūhī and al-Sābī on centers of gravity is reprinted from "The Correspondence of Abū Sahl al-Kūhī and Abū Isḥāq al-Sābī," by J. Lennart Berggren, *Journal for the History of Arabic Science* 7, (1983), 39–124 with the permission of the *Journal*. The material from al-Kūhī on the astrolabe is reprinted from "Abū Sahl al-Kūhī's Treatise on the Construction of the Astrolabe with Proof: Text, Translation and Commentary," by J. Lennart Berggren, *Physis* 31 (1994), 141–252 with the permission of Leo S. Olschki, publisher. The material from al-Sijzī on the construction of the heptagon is reprinted from "Greek and Arabic Constructions of the Regular Heptagon," by J. P. Hogendijk, *Archive for History of Exact Sciences* 30 (1984), 197–330 with the kind permission of Springer Science and Business Media. The material on the anonymous treatise on the nine-sided figure is reprinted from "An Anonymous Treatise on the Regular Nonagon," by J. L. Berggren, *Journal for the History of Arabic Science* 5 (1981), 37–41 with the permission of the *Journal*. The material from al-Jazarī on constructing geometrical instruments is reprinted from *The Book of Knowledge of Ingenious Mechanical Devices by Ibn al-Razzāz al-Jazarī*, by Donald Hill, © 1974 by permission of Reidel Publishers. The material from al-Tūsī on the theory of parallels is reprinted from *A History of Non-Euclidean Geometry: Evolution of the Concept of a Geometric Space*, by B. A. Rosenfeld, © 1988 with the kind permission of Springer Science and Business Media.

SECTION VI.B.: The material from Abū al-Wafā' on the geometry of artisans is reprinted from "Mathematics and Arts: Connections between Theory and Practice in the Medieval Islamic

World", by Alpay Özdural, *Historia Mathematica* 27, (2000), 171–201 with permission from Elsevier, Ltd. The two patterns for muqarnas are reprinted from *Topkapi Scroll-Geometry and Ornament in Islamic Architecture*, by G. Necipoglu, © 1995 with permission of the Getty Research Institute.

SECTION VII: The material from al-Bīrūnī on geodesy is reprinted from *The Determination of the Coordinates of Cities: Al-Bīrūnī's Taḥdīd al-Amākin*, by Jamil Ali, © 1967 with permission of the American University of Beirut Press, Beirut, Lebanon. The material from al-Kāshī on interpolation using second order differences is reprinted from "A Medieval Interpolation Scheme Using Second Order Differences," by E. S. Kennedy, in *Studies in the Islamic Exact Sciences by E. S. Kennedy, Colleagues, and Former Students*, edited by D. A. King and M. H. Kennedy, © 1983 with permission of the American University of Beirut Press, Beirut, Lebanon.

SECTION VIII: The material from al-Bīrūnī on the number of possibilities for eclipse data is reprinted from *The Determination of the Coordinates of Cities: Al-Bīrūnī's Taḥdīd al-Amākin*, by Jamil Ali, © 1967 with permission of the American University of Beirut Press, Beirut, Lebanon.

SECTION IX: The material from al-Kūhī on the certainty deriving from mathematics is reprinted from "The Correspondence of Abū Sahl al-Kūhī and Abū Isḥāq al-Sābī: A Translation with commentaries," by J. Lennart Berggren, *Journal for the History of Arabic Science* 7, (1983), 39–124 with the permission of the *Journal*. The material from al-Sijzī on heuristics is reprinted from *Al-Sijzī's Treatise on Geometrical Problem Solving*, by J. P. Hogendijk, with the permission of the author.

The Mathematics *of* Egypt, Mesopotamia, China, India, and Islam

A Sourcebook

Introduction

A century ago, mathematics history began with the Greeks, then skipped a thousand years and continued with developments in the European Renaissance. There was sometimes a brief mention that the "Arabs" preserved Greek knowledge during the dark ages so that it was available for translation into Latin beginning in the twelfth century, and perhaps even a note that algebra was initially developed in the lands of Islam before being transmitted to Europe. Indian and Chinese mathematics barely rated a footnote.

Since that time, however, we have learned much. First of all, it turned out that the Greeks had predecessors. There was mathematics both in ancient Egypt and in ancient Mesopotamia. Archaeologists discovered original material from these civilizations and deciphered the ancient texts. In addition, the mathematical ideas stemming from China and India gradually came to the attention of historians. In the nineteenth century, there had been occasional mention of these ideas in fairly obscure sources in the West, and there had even been translations into English or other western languages of certain mathematical texts. But it was only in the late twentieth century that major attempts began to be made to understand the mathematical ideas of these two great civilizations and to try to integrate them into a worldwide history of mathematics. Similarly, the nineteenth century saw numerous translations of Islamic mathematical sources from the Arabic, primarily into French and German. But it was only in the last half of the twentieth century that historians began to put together these mathematical ideas and attempted to develop an accurate history of the mathematics of Islam, a history beyond the long-known preservation of Greek texts and the algebra of al-Khwarizmi. Yet, even as late as 1972, Morris Kline's monumental work *Mathematical Thought from Ancient to Modern Times* contained but 12 pages on Mesopotamia, 9 pages on Egypt, and 17 pages combined on India and the Islamic world (with nothing at all on China) in its total of 1211 pages.

It will be useful, then, to give a brief review of the study of the mathematics of Egypt, Mesopotamia, China, India, and Islam to help put this *Sourcebook* in context.

To begin with, our most important source on Egyptian mathematics, the Rhind Mathematical Papyrus, was discovered, probably in the ruins of a building in Thebes, in the middle of the nineteenth century and bought in Luxor by Alexander Henry Rhind in 1856. Rhind died in 1863 and his executor sold the papyrus, in two pieces, to the British Museum in 1865. Meanwhile, some fragments from the break turned up in New York, having been acquired also in Luxor by the American dealer Edwin Smith in 1862. These are now in the

Brooklyn Museum. The first translation of the Rhind Papyrus was into German in 1877. The first English translation, with commentary, was made in 1923 by Thomas Peet of the University of Liverpool. Similarly, the Moscow Mathematical Papyrus was purchased around 1893 by V. S. Golenishchev and acquired about twenty years later by the Moscow Museum of Fine Arts. The first notice of its contents appeared in a brief discussion by B. A. Turaev, conservator of the Egyptian section of the museum, in 1917. He wrote chiefly about problem 14, the determination of the volume of a frustum of a square pyramid, noting that this showed "the presence in Egyptian mathematics of a problem that is not to be found in Euclid." The first complete edition of the papyrus was published in 1930 in German by W. W. Struve. The first complete English translation was published by Marshall Clagett in 1999.

Thus, by early in the twentieth century, the basic outlines of Egyptian mathematics were well understood—at least the outlines as they could be inferred from these two papyri. And gradually the knowledge of Egyptian mathematics embedded in these papyri and other sources became part of the global story of mathematics, with one of the earliest discussions being in Otto Neugebauer's *Vorlesungen über Geschichte der antiken Mathematischen Wissenschaften* (more usually known as *Vorgriechische Mathematik*) of 1934, and further discussions and analysis by B. L. Van der Waerden in his *Science Awakening* of 1954. A more recent survey is by James Ritter in *Mathematics Across Cultures.*

A similar story can be told about Mesopotamian mathematics. Archaeologists had begun to unearth the clay tablets of Mesopotamia beginning in the middle of the nineteenth century, and it was soon realized that some of the tablets contained mathematical tables or problems. But it was not until 1906 that Hermann Hilprecht, director of the University of Pennsylvania's excavations in what is now Iraq, published a book discussing tablets containing multiplication and reciprocal tables and reviewed the additional sources that had been published earlier, if without much understanding. In 1907, David Eugene Smith brought some of Hilprecht's work to the attention of the mathematical world in an article in the *Bulletin of the American Mathematical Society*, and then incorporated some of these ideas into his 1923 *History of Mathematics.*

Meanwhile, other archaeologists were adding to Hilprecht's work and began publishing some of the Mesopotamian mathematical problems. The study of cuneiform mathematics changed dramatically, however, in the late 1920s, when François Thureau-Dangin and Otto Neugebauer independently began systematic programs of deciphering and publishing these tablets. In particular, Neugebauer published two large collections: *Mathematische Keilschrift-Texte* in 1935–37 and (with Abraham Sachs) *Mathematical Cuneiform Texts* in 1945. He then summarized his work for the more general mathematical public in his 1951 classic, *The Exact Sciences in Antiquity.* Van der Waerden's *Science Awakening* was also influential in publicizing Mesopotamian mathematics. Jens Høyrup's survey of the historiography of Mesopotamian mathematics provides further details.

Virtually the first mention of Chinese mathematics in a European language was in several articles in 1852 by Alexander Wylie entitled "Jottings on the Science of the Chinese: Arithmetic," appearing in the *North China Herald*, a rather obscure Shanghai journal. However, they were translated in part into German by Karl L. Biernatzki and published in *Crelle's Journal* in 1856. Six years later they also appeared in French. It was through these articles that Westerners learned of what is now called the Chinese Remainder problem and how it was initially solved in fourth-century China, as well as about the ten Chinese classics and the Chinese algebra of the thirteenth century. Thus, by the end of the nineteenth century,

European historians of mathematics could write about Chinese mathematics, although, since they did not have access to the original material, their works often contained errors.

The first detailed study of Chinese mathematics written in English by a scholar who could read Chinese was *Mathematics in China and Japan*, published in 1913 by the Japanese scholar Yoshio Mikami. Thus David Eugene Smith, who co-authored a work solely on Japanese mathematics with Mikami, could include substantial sections on Chinese mathematics in his own *History* of 1923. Although other historians contributed some material on China during the first half of the twentieth century, it was not until 1959 that a significant new historical study appeared, volume 3 of Joseph Needham's *Science and Civilization in China*, entitled *Mathematics and the Sciences of the Heavens and the Earth.* One of Needham's chief collaborators on this work was Wang Ling, a Chinese researcher who had written a dissertation on the *Nine Chapters* at Cambridge University. Needham's work was followed by the section on China in A. P. Yushkevich's history of medieval mathematics (1961) in Russian, a book that was in turn translated into German in 1964. Since that time, there has been a concerted effort by both Chinese and Western historian of mathematics to make available translations of the major Chinese texts into Western languages.

The knowledge in the West of Indian mathematics occurred much earlier than that of Chinese mathematics, in part because the British ruled much of India from the eighteenth century on. For example, Henry Thomas Colebrooke collected Sanskrit mathematical and astronomical texts in the early nineteenth century and published, in 1817, his *Algebra with Arithmetic and Mensuration from the Sanscrit of Brahmegupta and Bhascara.* Thus parts of the major texts of two of the most important medieval Indian mathematicians were available in English, along with excerpts from Sanskrit commentaries on these works. Then in 1835, Charles Whish published a paper dealing with the fifteenth-century work in Kerala on infinite series, and Ebenezer Burgess in 1860 published a translation of the *Sūrya-siddhānta*, a major early Indian work on mathematical astronomy. Hendrik Kern in 1874 produced an edition of the *Āryabhaṭīya* of Aryabhata, while George Thibaut wrote a detailed essay on the *Śulbasūtras*, which was published, along with his translation of the *Baudhāyana Śulbasūtra*, in the late 1870s. The research on medieval Indian mathematics by Indian researchers around the same time, including Bāpu Deva Sāstrī, Sudhākara Dvivedī, and S. B. Dikshit, although originally published in Sanskrit or Hindi, paved the way for additional translations into English.

Despite the availability of some Sanskrit mathematical texts in English, it still took many years before Indian contributions to the world of mathematics were recognized in major European historical works. Of course, European scholars knew about the Indian origins of the decimal place-value system. But in part because of a tendency in Europe to attribute Indian mathematical ideas to the Greeks and also because of the sometimes exaggerated claims by Indian historians about Indian accomplishments, a balanced treatment of the history of mathematics in India was difficult to achieve. Probably the best of such works was the *History of Indian Mathematics: A Source Book*, published in two volumes by the Indian mathematicians Bibhutibhusan Datta and Avadhesh Narayan Singh in 1935 and 1938. In recent years, numerous Indian scholars have produced new Sanskrit editions of ancient texts, some of which have never before been published. And new translations, generally into English, are also being produced regularly, both in India and elsewhere.

As to the mathematics of Islam, from the time of the Renaissance Europeans were aware that algebra was not only an Arabic word, but also essentially an Islamic creation. Most early algebra works in Europe in fact recognized that the first algebra works in that continent were translations of the work of al-Khwārizmī and other Islamic authors. There was also some awareness that much of plane and spherical trigonometry could be attributed to Islamic authors. Thus, although the first pure trigonometrical work in Europe, *On Triangles* by Regiomontanus, written around 1463, did not cite Islamic sources, Gerolamo Cardano noted a century later that much of the material there on spherical trigonometry was taken from the twelfth-century work of the Spanish Islamic scholar Jābir ibn Aflaḥ.

By the seventeenth century, European mathematics had in many areas reached, and in some areas surpassed, the level of its Greek and Arabic sources. Nevertheless, given the continuous contact of Europe with Islamic countries, a steady stream of Arabic manuscripts, including mathematical ones, began to arrive in Europe. Leading universities appointed professors of Arabic, and among the sources they read were mathematical works. For example, the work of Ṣadr al-Ṭūsī (the son of Naṣīr al-Dīn al-Ṭūsī) on the parallel postulate, written originally in 1298, was published in Rome in 1594 with a Latin title page. This work was studied by John Wallis in England, who then wrote about its ideas as he developed his own thoughts on the postulate. Still later, Newton's friend, Edmond Halley, translated into Latin Apollonius's *Cutting-off of a Ratio*, a work that had been lost in Greek but had been preserved via an Arabic translation.

Yet in the seventeenth and eighteenth centuries, when Islamic contributions to mathematics may well have helped Europeans develop their own mathematics, most Arabic manuscripts lay unread in libraries around the world. It was not until the mid-nineteenth century that European scholars began an extensive program of translating these mathematical manuscripts. Among those who produced a large number of translations, the names of Heinrich Suter in Switzerland and Franz Woepcke in France stand out. (Their works have recently been collected and republished by the Institut für Geschichte der arabisch-islamischen Wissenschaften.) In the twentieth century, Soviet historians of mathematics began a major program of translations from the Arabic as well. Until the middle of the twentieth century, however, no one in the West had pulled together these translations to try to give a fuller picture of Islamic mathematics. Probably the first serious history of Islamic mathematics was a section of the general history of medieval mathematics written in 1961 by A. P. Yushkevich, already mentioned earlier. This section was translated into French in 1976 and published as a separate work, *Les mathématiques arabes (VIIIe–XVe siècles)*. Meanwhile, the translation program continues, and many new works are translated each year from the Arabic, mostly into English or French.

By the end of the twentieth century, all of these scholarly studies and translations of the mathematics of these various civilizations had an impact on the general history of mathematics. Virtually all recent general history textbooks contain significant sections on the mathematics of these five civilizations. As this sourcebook demonstrates, there are many ideas that were developed in these five civilizations that later reappeared elsewhere. The question that then arises is how much effect the mathematics of these civilizations had on what is now world mathematics of the twenty first-century. The answer to this question is very much under debate. We know of many confirmed instances of transmission of mathematical ideas from one of these cultures to Europe or from one of these cultures to another, but there are numerous instances where, although there is circumstantial evidence of transmission, there is no definitive documentary evidence. Whether such will be found as more translations are made and more documents are uncovered in libraries and other institutions around the world is a question for the future to answer.

References

Biernatzki, Karl L. 1856. "Die Arithmetic der Chinesen." *Crelle's Journal* 52: 59–94.

Burgess, Ebenezer. 1860. "The Sūryasiddhānta." *Journal of the American Oriental Society*, 6.

Clagett, Marshall. 1999. *Ancient Egyptian Science: A Source Book.* Vol. 3, *Ancient Egyptian Mathematics.* Philadelphia: American Philosophical Society.

Colebrooke, Henry Thomas. 1817. *Algebra with Arithmetic and Mensuration from the Sanscrit of Brahmegupta and Bhascara.* London: John Murray.

Datta, B., and A. N. Singh., 1935/38. *History of Hindu Mathematics: A Source Book.* 2 vols. Bombay: Asia Publishing House.

Hilprecht, Hermann V. 1906. *Mathematical, Metrological and Chronological Tablets from the Temple Library of Nippur* Philadelphia: University Museum of Pennsylvania.

Høyrup, Jens. 1996. "Changing Trends in the Historiography of Mesopotamian Mathematics: An Insider's View." *History of Science* 34: 1–32.

Kern, Hendrik. 1874. *The Aryabhatiya, with the Commentary Bhatadipika of Paramadicyara.* Leiden: Brill.

Kline, Morris. 1972. *Mathematical Thought from Ancient to Modern Times.* New York: Oxford University Press.

Mikami, Yoshio. 1913. *The Development of Mathematics in China and Japan.* Leipzig: Teubner.

Needham, Joseph. 1959. *Science and Civilization in China.* Vol. 3, *Mathematics and the Sciences of the Heavens and the Earth.* Cambridge: Cambridge University Press.

Neugebauer, Otto. 1934. *Vorlesungen über Geschichte der antiken Mathematischen Wissenschaften. I. Vorgriechische Mathematik.* Berlin: Springer-Verlag.

———. 1935–37. *Mathematische Keilschrift-Texte.* Berlin: Springer-Verlag.

———. 1945. *Mathematical Cuneiform Texts* (with Abraham Sachs). New Haven: American Oriental Society.

———. 1951. *The Exact Sciences in Antiquity.* Princeton: Princeton University Press.

Peet, Thomas. 1923. *The Rhind Mathematical Papyrus, British Museum 10057 and 10058. Introduction, Transcription, Translation and Commentary.* London.

Ritter, James. 2000. "Egyptian Mathematics." In Helaine Selin, ed., *Mathematics Across Cultures: The History of Non-Western Mathematics.* Dordrecht: Kluwer.

Smith, David Eugene. 1907. "The Mathematical Tablets of Nippur." *Bulletin of the American Mathematical Society* 13: 392–398.

———. 1923. *History of Mathematics.* 2 vols. Boston: Ginn.

Struve, W. W. 1930. *Mathematische Papyrus des Staatlichen Museums der Schönen Künste in Moskau.* Quellen und Studien zur Geschichte der Mathematik A1. Berlin: Springer-Verlag.

Suter, Heinrich. 1986. *Beiträge zur Geschichte der Mathematik und Astronomie im Islam.* 2 vols. Frankfurt: Institut für Geschichte der arabisch-islamischen Wissenschaften.

Thibaut, George. 1875. "On the Sulba-sutra." *Journal of the Asiatic Society of Bengal.* 44: 227–275.

Thureau-Dangin, François. 1983. *Textes mathématiques babyloniens* (Ex Oriente Lux 1). Leiden: Brill.

Turaev, Boris. A. 1917. "The Volume of the Truncated Pyramid in Egyptian Mathematics." *Ancient Egypt* 3: 100–102.

Van der Waerden, B. L. 1954. *Science Awakening.* New York: Oxford University Press.

Whish, Charles. 1835. "On the Hindu quadrature of a circle." *Transactions of the Royal Asiatic Society of Great Britain and Ireland* 3: 509–23.

Woepcke, Franz. 1986. *Etudes sur les mathématiques arabo-islamiques.* 2 vols. Frankfurt: Institut für Geschichte der arabisch-islamischen Wissenschaften.

Wylie, Alexander. 1852. "Jottings on the Science of the Chinese Arithmetic." *North China Herald*, Aug.–Nov. nos. 108–13, 116–17, 119–21.

Yushkevich, A. P. 1961. *The History of Mathematics in the Middle Ages* (in Russian). Moscow: Fizmatgiz.

———. 1964. *Geschichte der Mathematik in Mittelalter*, Leipzig: Teubner.

———. 1976. *Les mathématiques arabes* (*VIIIe –XVe siècles*). Paris: Vrin.

1 Egyptian Mathematics

Annette Imhausen

Preliminary Remarks

The study of Egyptian mathematics is as fascinating as it can be frustrating. The preserved sources are enough to give us glimpses of a mathematical system that is both similar to some of our school mathematics, and yet in some respects completely different. It is partly this similarity that caused early scholars to interpret Egyptian mathematical texts as a lower level of Western mathematics and, subsequently, to "translate" or rather transform the ancient text into a modern equivalent. This approach has now been widely recognized as unhistorical and mostly an obstacle to deeper insights. Current research attempts to follow a path that is sounder historically and methodologically. Furthermore, writers of new works can rely on progress that has been made in Egyptology (helping us understand the language and context of our texts) as well as in the history of mathematics.

However, learning about Egyptian mathematics will never be an easy task. The main obstacle is the shortage of sources. It has been over 70 years since a substantial new Egyptian mathematical text was discovered. Consequently, we must be extremely careful with our general evaluation of Egyptian mathematics. If we arbitrarily chose six mathematical publications of the past 300 years, what would we be able to say about mathematical achievements between 1700 and 2000 CE? This is exactly our situation for the mathematical texts of the Middle Kingdom (2119–1794/93 BCE). On the positive side, it must be said that the available source material is as yet far from being exhaustively studied, and significant and fascinating new insights are still likely to be gained. Also, the integration of other texts that contain mathematical information helps to fill out the picture. The understanding of Egyptian mathematics depends on our knowledge of the social and cultural context in which it was created, used, and developed. In recent years, the use of other source material, which contains direct or indirect information about Egyptian mathematics, has helped us better understand the extant mathematical texts.

This chapter presents a selection of sources and introduces the characteristic features of Egyptian mathematics. The selection is taken from over 3000 years of history. Consequently, the individual examples have to be taken within their specific context. The introduction following begins with a text *about* mathematics from the New Kingdom (1550–1070/69 BCE) to illustrate the general context of mathematics within Egyptian culture.

To introduce this text, we need to bear in mind that the development and use of mathematical techniques began around 1500 years earlier with the invention of writing and number systems. The available evidence points to administrative needs as the motivation for this development. Quantification and recording of goods also necessitated the development of metrological systems, which can be attested as early as the Old Kingdom and possibly earlier. Metrological systems and mathematical techniques were used and developed by the scribes, that is, the officials working in the administration of Egypt. Scribes were crucial to ensuring the smooth collection and distribution of available goods, thus providing the material basis for a prospering government under the pharaoh. Evidence for mathematical techniques comes from the education and daily work life of these scribes. The most detailed information can be gained from the so-called mathematical texts, papyri that were used in the education of junior scribes. These papyri contain collections of problems and their solutions to prepare the scribes for situations they were likely to face in their later work.

The mathematical texts inform us first of all about different types of mathematical problems. Several groups of problems can be distinguished according to their subject. The majority are concerned with topics from an administrative background. Most scribes were probably occupied with tasks of this kind. This conclusion is supported by illustrations found on the walls of private tombs. Very often, in tombs of high officials, the tomb owner is shown as an inspector in scenes of accounting of cattle or produce, and sometimes several scribes are depicted working together as a group. It is in this practical context that mathematics was developed and practised. Further evidence can be found in three-dimensional models representing scenes of daily life, which regularly include the figure of one or more scribes. Several models depict the filling of granaries, and a scribe is always present to record the respective quantities.

While mathematical papyri are extant from two separate periods only, depictions of scribes as accountants (and therefore using mathematics) are evident from all periods beginning with the Old Kingdom. Additional evidence for the same type of context for mathematics appears during the New Kingdom in the form of literary texts about the scribal profession. These texts include comparisons of a scribe's duties to duties of other professions (soldier, cobbler, farmer, etc.). It is clear that many of the scribes' duties involve mathematical knowledge. The introduction begins with a prominent example from this genre.

Another (and possibly the only other) area in which mathematics played an important role was architecture. Numerous extant remains of buildings demonstrate a level of design and construction that could only have been achieved with the use of mathematics. However, which instruments and techniques were used is not known nor always easy to discern. Past historiography has tended to impose modern concepts on the available material, and it is only recently that a serious reassessment of this subject has been published.[1] Again, detailed accounts of mathematical techniques related to architecture are only extant from the Middle Kingdom on. However, a few sketches from the Old Kingdom have survived as well, which indicate that certain mathematical concepts were present or being developed. These concepts then appeared fully formed in the mathematical texts.

Throughout this chapter Egyptian words appear in what Egyptologists call "transcription." The Egyptian script noted only consonants (although we pronounce some of them as vowels today). For this reason, transcribing hieratic or hieroglyphic texts means to transform the

[1]See [Rossi 2004].

text into letters which are mostly taken from our alphabet and seven additional letters (*ꜣ*, *ꜥ*, *ḥ*, *ḫ*, *ẖ*, *ṯ*, *ḏ*). In order to be able to read Egyptian, Egyptologists therefore agreed on the convention to insert (in speaking) short "e" sounds where necessary. The pronunciation of the Egyptian transcription alphabet is given below. This is a purely modern convention—how Egyptian was pronounced originally is not known. The Appendix contains a glossary of all Egyptian words in this chapter and their (modern) pronunciation.

Letter	Pronunciation
ꜣ	ă
j	i or j
ꜥ	ā
w	ŭ, ū, or w
b	b
p	p
f	f
m	m
n	n
r	r
h	h
ḥ	like Arabic ḥ
ḫ	like ch in "loch"
ẖ	like German ch in "ich"
z	voiced s
s	unvoiced s
š	sh
q	emphatic k
k	k
g	g
t	t
ṯ	like ch in "touch"
d	d
ḏ	dg as in "judge"

I. Introduction

The passage below is taken from Papyrus Anastasi I,[2] an Egyptian literary text of the New Kingdom (1550–1070/69 BCE). This composition is a fictional letter, which forms part of a debate between two scribes. The letter begins, as is customary for Egyptian letters, with the writer Hori introducing himself and then addressing the scribe Amenemope (by the shortened form Mapu). After listing the necessary epithets and wishing the addressee well, Hori recounts receiving a letter of Amenemope, which Hori describes as confused and insulting. He then

[2]The hieroglyphic transcription of the various extant sources can be found in [Fischer-Elfert 1983]. An English translation of the complete text is [Gardiner 1911]; this however is based on only ten of the extant 80 sources. The *editio princeps* is [Fischer-Elfert 1986]. The translation given here is my own, which is based on the work by Fischer-Elfert.

proposes a scholarly competition covering various aspects of scribal knowledge. The letter ends with Hori criticizing the letter of his colleague again and suggesting to him that he should sit down and think about the questions of the competition before trying to answer them.

The mathematical section of the letter, translated below, comprises several problems similar to the collections of problems found in mathematical papyri. However, in this letter, the problems are framed by Hori's comments (and sometimes insults), addressed to his colleague Amenemope (Mapu). Hori points out several times the official position which Amenemope claims for himself ("commanding scribe of the soldiers," "royal scribe") and teases him by calling him ironically "vigilant scribe," "scribe keen of wit," "sapient scribe," directly followed by a description of Amenemope's ineptness. In between, Hori describes several situations in which Amenemope is required to use his mathematical knowledge.

Note that while in each case the general problem is easy to grasp, there is not enough information, in fact, for a modern reader to solve these mathematical problems. This is partly due to philological difficulties: even after two editions the text is still far from fully understood. The choice of this extract as the first source text is mainly meant to illustrate the social and cultural context of mathematics in ancient Egypt.

Papyrus Anastasi I, 13,4–18,2

Another topic

Look, you come here and fill me with (the importance of) your office. I will let you know your condition when you say: "I am the commanding scribe of the soldiers." It has been given to you to dig a lake. You come to me to ask about the rations of the soldiers. You say to me: "Calculate it!" I am thrown into your office. Teaching you to do it has fallen upon my shoulders.

I will cause you to be embarrassed, I will explain to you the command of your master—may he live, prosper, and be healthy. Since you are his royal scribe, you are sent under the royal balcony for all kinds of great monuments of Horus, the lord of the two lands. Look, you are the clever scribe who is at the head of the soldiers.

A ramp shall be made of (length) 730 cubits, width 55 cubits, with 120 compartments, filled with reeds and beams. For height: 60 cubits at its top to 30 cubits in its middle, and an inclination (*sqd*) of 15 cubits, its base 5 cubits. Its amount of bricks needed shall be asked from the overseer of the troops. All the assembled scribes lack someone (i.e., a scribe) who knows them (i.e., the number of bricks). They trust in you, saying: "You are a clever scribe my friend. Decide for us quickly. Look, your name has come forward. One shall find someone in this place to magnify the other thirty. Let it not be said of you: there is something that you don't know. Answer for us the number (lit. its need) of bricks." Look, its measurements are before you. Each one of its compartments is of 30 cubits (in length) and a width of 7 cubits.

Hey Mapu, vigilant scribe, who is at the head of the soldiers, distinguished when you stand at the great gates, bowing beautifully under the balcony. A dispatch has come from the crown prince to the area of Ka to please the heart of the Horus of Gold, to calm the raging lion. An obelisk has been newly made, graven with the name of his majesty—may he live, prosper, and be healthy—of 110 cubits in the length of its shaft, its pedestal of 10 cubits, the circumference of its base makes 7 cubits on all its sides, its narrowing towards the summit 1 cubit, its pyramidion 1 cubit in height, its point 2 digits.

Add them up in order to make it from parts. You shall give every man to its transport, those who shall be sent to the Red Mountain. Look, they are waiting for them. Prepare the way for the crown prince Mes-Iten. Approach; decide for us the amount of men who will be in front of it. Do not let them repeat writing while the monument is in the quarry. Answer quickly, do not hesitate! Look, it is you who is looking for them for yourself. Get going! Look, if you hurry, I will cause your heart to rejoice.

I used to [...] under the top like you. Let us fight together. My heart is apt, my fingers listen. They are clever, when you go astray. Go, don't cry, your helper is behind you. I let you say: "There is a royal scribe with Horus, the mighty bull." And you shall order men to make chests into which to put letters that I will have written you secretly. Look, it is you who shall take them for yourself. You have caused my hands and fingers to be trained like a bull at a feast until every feast in eternity.

You are told: "Empty the magazine that has been loaded with sand under the monument for your lord—may he live, prosper, and be healthy—which has been brought from the Red Mountain. It makes 30 cubits stretched upon the ground with a width of 20 cubits, passing chambers filled with sand from the riverbank. The walls of its chambers have a breadth of 4 to 4 to 4 cubits. It has a height of 50 cubits in total. [...] You are commanded to find out what is before it. How many men will it take to remove it in 6 hours if their minds are apt? Their desire to remove it will be small if (a break at) noon does not come. You shall give the troops a break to receive their cakes, in order to establish the monument in its place. One wishes to see it beautiful.

O scribe, keen of wit, to whom nothing whatsoever is unknown. Flame in the darkness before the soldiers, you are the light for them. You are sent on an expedition to Phoenicia at the head of the victorious army to smite those rebels called Nearin. The bow-troops who are before you amount to 1900, Sherden 520, Kehek 1600, Meshwesh <100>, Tehesi 880, sum 5000 in all, not counting their officers. A complimentary gift has been brought to you and placed before you: bread, cattle, and wine. The number of men is too great for you: the provision is too small for them. Sweet Kemeh bread: 300, cakes: 1800, goats of various sorts: 120, wine: 30. The troops are too numerous; the provisions are underrated like this what you take from them (i.e., the inhabitants). You receive (it); it is placed in the camp. The soldiers are prepared and ready. Register it quickly, the share of every man to his hand. The Bedouins look on in secret. O learned scribe, midday has come, the camp is hot. They say: 'It is time to start'. Do not make the commander angry! Long is the march before us. What is it, that there is no bread at all? Our night quarters are far off. What is it, Mapu, this beating we are receiving (lit. of us)? Nay, but you are a clever scribe. You cease to give (us) food when only one hour of the day has passed? The scribe of the ruler—may he live, prosper, and be healthy—is lacking. Were you brought to punish us? This is not good. If Pa-Mose hears of it, he will write to degrade you."

The extract above shows that mathematics constituted an important part in a scribe's education and daily life. Furthermore, it illustrates the kind of mathematics that was practiced in Egypt. The passages cited refer to mathematical knowledge that a scribe should have in order to handle his daily work: accounting of grain, land, and labor in pharaonic Egypt. There have been several attempts to reconstruct actual mathematical exercises from the examples referred to in this source. All of them have met difficulties, which are caused not only by the numerous philological problems but also by the fact that the problems are deliberately "underdetermined." These examples were not intended to be actual mathematical problems that the Egyptian reader (i.e., scribe) should solve, but they were meant to remind him of types of mathematical problems he encountered in his own education.

Timeline [a]	Extant Mathematical Texts [b]	Script
Archaic Period Dyn. 1–2 (c. 300–2686 BCE)		
Old Kingdom Dyn. 3–8 (2686–2160 BCE)		HIERATIC
First Intermediate Period Dyn. 9–10 (2160–2025 BCE)		
Middle Kingdom Dyn. 11–12 (2025–1773 BCE)	***pMoscow (E4676)*** ***Math. Leather Roll (BM10250)*** ***Lahun Fragments*** ***pBerlin 6619*** ***Cairo Wooden Boards*** ***pRhind (BM10057–8)***	
Second Intermediate Period Dyn. 13–17 (1773–550 BCE		
New Kingdom Dyn. 18–20 (1550–1069 BCE)	***Ostracon Senmut 153*** ***Ostracon Turin 57170***	
Third Intermediate Period Dyn. 21-25 (1069–656 BCE)		
Late Period Dyn. 26–31 (664–332 BCE)		
Greek/Roman Period (332 BCE–395 CE)	*pCairo JE 89127–30* *PCairo JE 89137–43* *pBM 10399* *pBM10520* *pBM10794* *pCarlsberg 30*	DEMOTIC

[a]Dates according to [Shaw 2000].
[b]In this column Hieratic texts are listed in bold and italic, while Demotic texts are listed in italic.

Educational texts are the main source of our knowledge today about Egyptian mathematics. As already mentioned, there are very few sources available. These are listed in the table above. (Note that only mathematical texts, i.e., texts which teach mathematics, are included here, and therefore *pAnastasi I* is not listed.) Egyptian mathematical texts belong to two distinct groups: table texts and problem texts. Examples of both groups will be presented in this chapter. These are complemented by administrative texts that show mathematical practices in daily life.

The following paragraphs present the Egyptian number system, arithmetical techniques, Egyptian fraction reckoning, and metrology, in order to make the sources more easily accessible.

I.a. Invention of writing and number systems

The earliest evidence of written texts in Egypt at the end of the fourth millennium BCE consists of records of names (persons and places) as well as commodities and their quantities. They show the same number system as is used in later times in Egypt, a decimal system without positional notation, i.e. with a new sign for every power of 10:

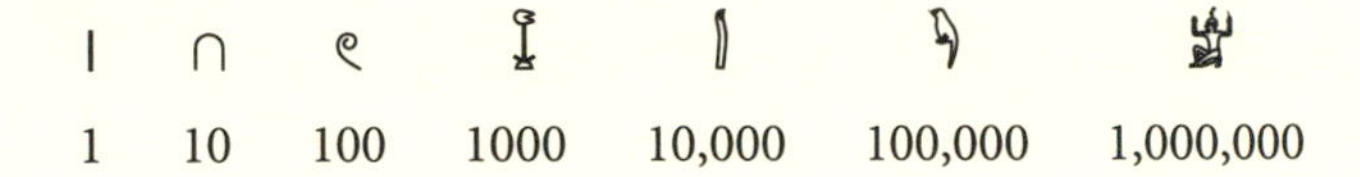

Naqada tablets CG 14101, 14102, 14103

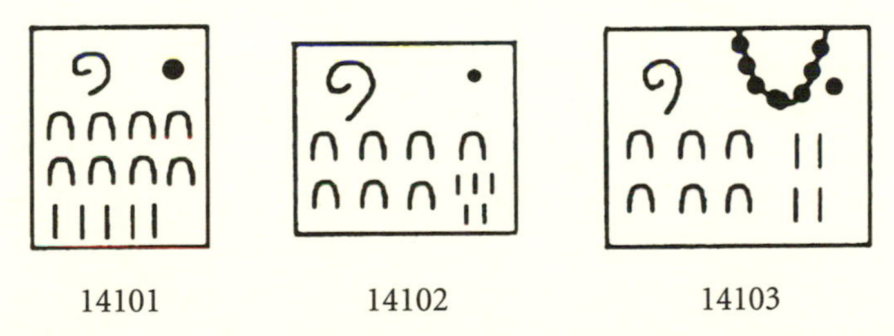

These predynastic tablets were probably attached to some commodity (there is a hole in each of the tablets), and represented a numeric quantity related to this commodity. The number written on the first tablet is 185; the sign for 100 is written once, followed by the sign for 10 eight times, and the sign for 1 five times. The second tablet shows the number 175, and the third tablet 164. In addition, a necklace is drawn on the third tablet. This is interpreted as a tablet attached to a necklace of 164 pearls.

Parallel with the hieroglyphic script, which throughout Egyptian history was mainly used on stone monuments, a second, simplified script evolved, written with ink and a reed pen on papyrus, ostraca, leather, or wood. This cursive form of writing is known as "hieratic script." The individual signs often resemble their hieroglyphic counterparts. Over time the hieratic script became more and more cursive, and groups of signs were combined into so-called ligatures.

Hieroglyphic script could be written in any direction suitable to the purpose of the inscription, although the normal direction of writing is from right to left. Thus, the orientation of the individual symbols, such as the glyph for 100, varies. Compare the glyph for 100 in the table above and in the illustrated tablets. Hieratic, however, is always written from right to left. While hieroglyphic script is highly standardized, hieratic varies widely depending on the handwriting of the individual scribe. Therefore, it is customary in Egyptology to provide a hieroglyphic transcription of the hieratic source text.

Notation for fractions: Egyptian mathematics used unit fractions (i.e., $\frac{1}{2}$, $\frac{1}{3}$, $\frac{1}{4}$, etc.) almost exclusively; the single exception is $\frac{2}{3}$. In hieratic, the number that is the denominator is written with a dot above it to mark it as a fraction. In hieroglyphic writing the dot is replaced by the hieroglyph ⊂⊃ ("part"). The most commonly used fractions $\frac{2}{3}$, $\frac{1}{2}$, $\frac{1}{3}$, and $\frac{1}{4}$ were written by special signs:

hieratic	hieroglyphic	Value
		$\frac{2}{3}$
		$\frac{1}{2}$
		$\frac{1}{3}$
	×	$\frac{1}{4}$

More difficult fractions like $\frac{3}{4}$ or $\frac{5}{6}$ were represented by sums of unit fractions written in direct juxtaposition, e.g., $\frac{3}{4} = \frac{1}{2}\frac{1}{4}$ (hieroglyphic ⊂ ×); $\frac{5}{6} = \frac{2}{3}\frac{1}{6}$ (hieroglyphic). In transcription, fractions are rendered by the denominator with an overbar, e.g., $\frac{1}{2}$ is written as $\bar{2}$. The fraction $\frac{2}{3}$ is written as $\bar{\bar{3}}$.

I.b. Arithmetic

Calculation with integers: the mathematical texts contain terms for addition, subtraction, multiplication, division, halving, squaring, and the extraction of a square root. Only multiplication and division were performed as written calculations. Both of these were carried out using a variety of techniques the choice of which depended on the numerical values involved. The following example of the multiplication of 2000 and 5 is taken from a problem of the *Rhind Mathematical Papyrus* (remember that the hieratic original, and therefore this hieroglyphic transcription, are read from right to left):

Rhind Mathematical Papyrus, problem 52

\\.	2000		./
2	4000		||
\\4	8000		|||| /
Total	10,000		

The text is written in two columns. It starts with a dot in the first column and the number that shall be multiplied in the second column. The first line is doubled in the second line.

Therefore we see "2" in the first column and "4000" in the second column, the third line is twice the second ("4" in the first column, "8000" in the second column).

The first column is then searched for numbers that add up to the multiplicative factor 5 (the dot in the first line counts as "1"). This can be achieved in this example by adding the first and third lines. These lines are marked with a checkmark (\). The result of the multiplication is obtained by adding the marked lines of the second column. If the multiplicative factor exceeds 10, the procedure is slightly modified, as can be followed in the example below from problem 69 of the *Rhind Mathematical Papyrus*. The multiplication of 80 and 14 is performed as follows:

Rhind Mathematical Papyrus, problem 69

·	80	∩∩∩∩ ∩∩∩∩	·
\ 10	800	9999 9999	∩ /
2	160	∩∩∩ ∩∩∩9	II
\ 4	320	∩∩999	IIII /
Total	1120	∩∩9 [illegible]	[illegible]

After the initial line, we move directly to 10; then the remaining lines are carried out in the usual way, starting with double of the first line.

Divisions are performed in exactly the same way, with the roles of first and second column switched. The following example is taken from problem 76 of the *Rhind Mathematical Papyrus*. The division that is performed is $30 \div 2\frac{1}{2}$.

Rhind Mathematical Papyrus, problem 76

·	$2\bar{2}$	⊃II	·
\ 10	25	III II∩∩	∩ /
\ 2	5	III II	II /
Total	12	II∩	[illegible]

Again we find two columns. This time the divisor is subsequently either doubled or multiplied by 10. Then the second column is searched for numbers that add up to the dividend 30. The respective lines are marked. The addition of the first column of these lines leads to the result of the division.

Calculation with fractions: the last example of the division included a fraction ($2\bar{2}$); however, in this example it had little effect on the performance of the operation. From previous examples of multiplication (and division) it is obvious that doubling is an operation which has to be performed frequently. If fractions are involved, the fraction has to be doubled. If the fraction is a single unit fraction with an even denominator, halving the denominator easily does this, e.g., double of $\overline{64}$ is $\overline{32}$ ($\frac{2}{64} = \frac{1}{32}$). If, however, the denominator is odd (or a series of unit fractions is to be doubled), the result is not as easily found. For this reason the so-called $2 \div N$ table was created. This table lists the doubles of odd unit fractions. Examples of this table can be found in the section of table texts following this introduction. Obviously, the level of difficulty in carrying out these operations usually rises considerably as soon as fractions are involved.

The layout of a multiplication with fractions is the same as the layout of the multiplication of integers. The following example—the result of which is unfortunately partly destroyed—is taken from problem 6 of the *Rhind Mathematical Papyrus*, the multiplication of $\overline{\overline{3}}\ \overline{5}\ \overline{30}$ with 10.

Rhind Mathematical Papyrus, problem 6

.	$\overline{\overline{3}}\ \overline{5}\ \overline{30}$
\2	$1\ \overline{\overline{3}}\ \overline{10}\ \overline{30}$
4	$3\ \overline{2}\ \overline{10}$
\8	$7\ \overline{5}$

Total 9 loaves of bread. This is it.

Note that the multiplication with ten in this case is not performed directly, but explicitly carried out through doubling and addition.

Divisions with a divisor greater than the dividend use a series of halvings starting either with $\frac{2}{3}$ ($\frac{2}{3}, \frac{1}{3}, \frac{1}{6}, \ldots$) or with $\frac{1}{2}$ ($\frac{1}{2}, \frac{1}{4}, \frac{1}{8}, \ldots$). For instance the division $70 \div 93\frac{1}{3}$ in problem 58 of the *Rhind Mathematical Papyrus* is performed as follows:

Rhind Mathematical Papyrus, problem 58

.	$93\ \overline{3}$
\\$\overline{2}$	$46\ \overline{\overline{3}}$
\\$\overline{4}$	$23\ \overline{3}$
[Total	$\overline{2}\ \overline{4}$]

In more difficult numerical cases the division is first carried out as a division with remainder. The remainder is then handled separately.

I.c. Metrology

Note that the following overview is by no means a complete survey of Egyptian metrology, but includes only those units which are used in the sources of this chapter.

The approximate values given here are derived from the approximation that 1 cubit ≈ 52.5 cm, which is used in standard textbooks. It must be noted however, that this was determined as an average of cubit rods of so-called votive cubits, that is, cubits that have been placed in a tomb or temple as ritual objects. (Some other cubit rods have been unearthed that bear signs of actually having been used by architects and workers.) A valid "standard cubit" throughout Egypt did not exist. Naturally, the same holds for area and volume measures.

Length measures

1 *ḫt*	= 100 cubits	≈ 52.5 m
1 cubit	= 7 palms	≈ 52.5 cm
1 palm	= 4 digits	≈ 7.5 cm
1 digit		≈ 19 mm

Area measures

1 *ḫ3-t3*	= 10 *sṯ3.t*	≈ 27562.5 m^2
1 *sṯ3.t*	= (1 *ḫt*)^2	≈ 2756.25 m^2
1 area-cubit	= 1 cubit × 100 cubit	≈ 27.56 m^2

Volume measures

1 *ḫ3r*	= 16 *ḥq3.t*	≈ 76.8 l[3]
1 *ḫ3r*	= 20 *ḥq3.t* = 2/3 cubic-cubit	≈ 96.5 l[4]
1 *ḥq3.t*	= 10 *hnw*	≈ 4.8 l
1 *hnw*	= 32 *r3*	≈ 0.48 l
1 *r3*		≈ 15 ml

II. Hieratic Mathematical Texts

Egyptian mathematical texts can be assigned to two groups: table texts and problem texts. Table texts include tables for fraction reckoning (e.g., the 2 ÷ *N* table, which will be the first source text below, and the table found on the *Mathematical Leather Roll*) as well as tables for the conversion of measures (e.g., *Rhind Mathematical Papyrus*, Nos. 47, 80, and 81). Problem texts state a mathematical problem and then indicate its solution by means of step-by-step instructions. For this reason, they are also called procedure texts.

---

[3]This is the New Kingdom value, from Dynasty 20 onward. In the Old Kingdom and Middle Kingdom, the value was 1 *ḫ3r* = 10 *ḥq3.t*.

[4]This is the value found in the *Rhind Mathematical Papyrus*.

The extant hieratic source texts (in order of their publication) are

- *Rhind Mathematical Papyrus* (BM 10057–10058)
- *Lahun Mathematical Fragments* (7 fragments: UC32114, UC32118B, UC32134, UC32159–32162)
- *Papyrus Berlin* 6619 (2 fragments)
- *Cairo Wooden Boards* (CG 25367 and 25368)
- *Mathematical Leather Roll* (BM 10250)
- *Moscow Mathematical Papyrus* (E4674)
- *Ostracon Senmut* 153
- *Ostracon Turin* 57170

Most of these texts were bought on the antiquities market, and therefore we do not know their exact provenance. An exception is the group of mathematical fragments from Lahun, which were discovered by William Matthew Flinders Petrie when he excavated the Middle Kingdom pyramid town of Lahun.

II.a. Table texts

UC 32159

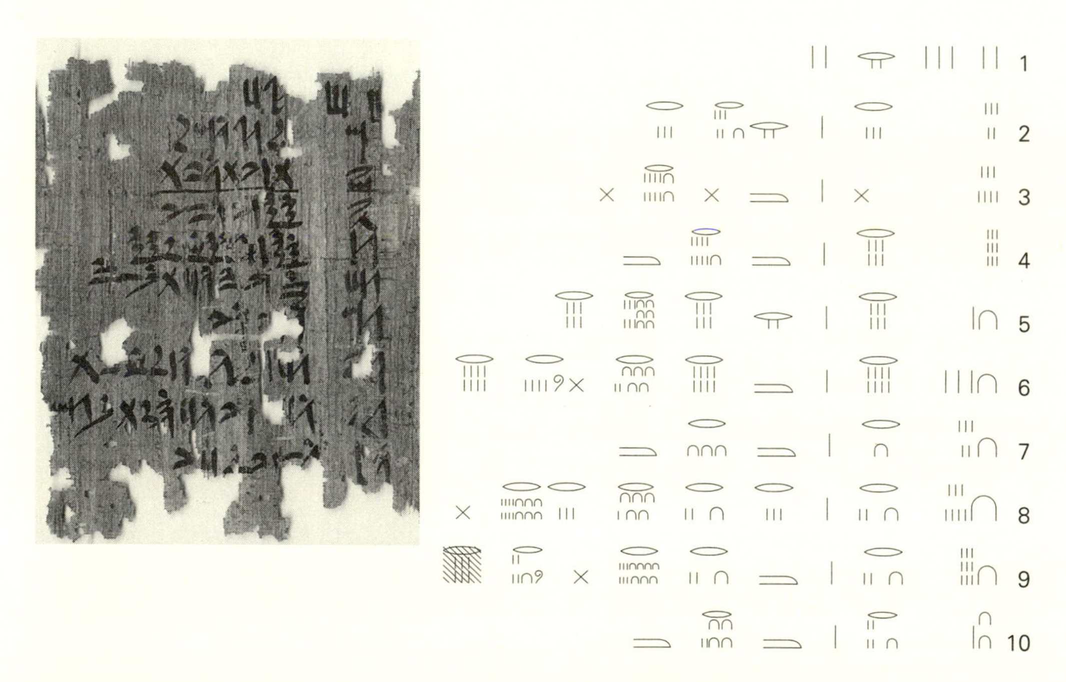

Reprinted by permission of Petrie Museum of Eqyptian Archaeology, University College, London.

The photograph shows a part of the so-called $2 \div N$ table from one of the Lahun fragments. The hieroglyphic transcription of the fragment on the photo is given next to it. This table was used to aid fraction reckoning. Remember that Egyptian fraction reckoning used only unit fractions and the fraction $\frac{2}{3}$. As multiplication consisted of repeated doubling, multiplication of fractions often involved the doubling of fractions. This can easily be done if the denominator is even. To double a unit fraction with an even denominator, its denominator has to be halved, e.g., $2 \times \frac{1}{8} = \frac{1}{4}$.

However the doubling of a fraction with an odd denominator always consists of a series of two or more unit fractions, which are not self-evident. Furthermore, there are often several possible representations; however, Egyptian mathematical texts consistently used only one, which can be found in the $2 \div N$ table. Below is the transcription of our example into numbers:

	Column I	*Column II*		
1	2 3	$\overline{\overline{3}}$ 2		
2	5	$\overline{3}$ $1\overline{\overline{3}}$	$\overline{15}$ $\overline{3}$	
3	7	$\overline{4}$ $1\overline{2}\overline{4}$	$\overline{28}$ $\overline{4}$	
4	9	$\overline{6}$ $1\overline{2}$	$\overline{18}$ $\overline{2}$	
5	11	$\overline{6}$ $1\overline{\overline{3}}\overline{6}$	$\overline{66}$ $\overline{6}$	
6	13	$\overline{8}$ $1\overline{2}\overline{8}$	$\overline{52}$ $\overline{4}$	$\overline{104}$ $\overline{8}$
7	15	$\overline{10}$ $1\overline{2}$	$\overline{30}$ $\overline{2}$	
8	17	$\overline{12}$ $1\overline{3}\overline{12}$	$\overline{51}$ $\overline{3}$	$\overline{68}$ $\overline{4}$
9	19	$\overline{12}$ $1\overline{2}\overline{12}$	$\overline{76}$ $\overline{4}$	$\overline{114}$ [$\overline{6}$]
10	21	$\overline{14}$ $1\overline{2}$	$\overline{42}$ $\overline{2}$	

The numbers are grouped in two columns. The first column contains the divisor N (in the first line only it shows both dividend 2 and divisor 3). The second column shows alternatingly fractions of the divisor and their value (as a series of unit fractions). For example, the second line starts with the divisor 5 in the first column; therefore it is $2 \div 5$ that is expressed as a series of unit fractions. It is followed in the second column by $\overline{3}$, $1\overline{\overline{3}}$, $\overline{15}$, and $\overline{3}$. This has to be read as $\overline{3}$ of 5 is $1\overline{\overline{3}}$ and $\overline{15}$ of 5 is $\overline{3}$. Since $1\overline{\overline{3}}$ plus $\overline{3}$ equals 2, the series of unit fractions to represent $2 \div 5$ is $\overline{3}$ $\overline{15}$.

The Recto of the *Rhind Mathematical Papyrus* contains the $2 \div N$ table for $N = 3$ to $N = 101$. Here, the solutions are marked in red ink, rendered as **bold** in the transcription below. There have been several attempts to explain the choices of representations in the $2 \div N$ table. These attempts were mostly based on modern mathematical formulas, and none of them gives a convincing explication of the values we find in the table. It is probable that the table was constructed based on experiences in handling fractions. Several "guidelines" for the selection of suitable fractions can be discerned. The author tried to keep the number of fractions to represent $2 \div N$ small; we generally find representations composed of two or three fractions only. Another guiding rule seems to be the choice of fractions with a small denominator over a bigger denominator, and the choice of denominators that can be decomposed into several components.

Rhind Mathematical Papyrus, 2 ÷ N Table

N	*2 ÷ N*	*N*	*2 ÷ N*
3	$\overline{\overline{\mathbf{3}}}$ 2	53	$\overline{\mathbf{30}}$ 1 $\overline{\overline{3}}$ $\overline{10}$ $\overline{\mathbf{318}}$ $\overline{6}$ $\overline{\mathbf{795}}$ $\overline{15}$
5	$\overline{\mathbf{3}}$ 1 $\overline{\overline{3}}$ $\overline{\mathbf{15}}$ $\overline{3}$	55	$\overline{\mathbf{30}}$ 1 $\overline{\overline{3}}$ $\overline{6}$ $\overline{\mathbf{330}}$ $\overline{6}$
7	$\overline{\mathbf{4}}$ 1 $\overline{2}$ $\overline{4}$ $\overline{\mathbf{28}}$ $\overline{4}$	57	$\overline{\mathbf{38}}$ 1 $\overline{2}$ $\overline{\mathbf{114}}$ $\overline{2}$
9	$\overline{\mathbf{6}}$ 1 $\overline{2}$ $\overline{\mathbf{18}}$ $\overline{2}$	59	$\overline{\mathbf{36}}$ 1 $\overline{2}$ $\overline{12}$ $\overline{18}$ $\overline{\mathbf{236}}$ $\overline{4}$ $\overline{\mathbf{531}}$ $\overline{9}$
11	$\overline{\mathbf{6}}$ 1 $\overline{\overline{3}}$ $\overline{6}$ $\overline{\mathbf{66}}$ $\overline{6}$	61	$\overline{\mathbf{40}}$ 1 $\overline{2}$ $\overline{40}$ $\overline{\mathbf{244}}$ $\overline{4}$ $\overline{\mathbf{488}}$ $\overline{8}$ $\overline{\mathbf{610}}$ $\overline{10}$
13	$\overline{\mathbf{8}}$ 1 $\overline{2}$ $\overline{8}$ $\overline{\mathbf{52}}$ $\overline{4}$ $\overline{\mathbf{104}}$ $\overline{8}$	63	$\overline{\mathbf{42}}$ 1 $\overline{2}$ $\overline{\mathbf{126}}$ $\overline{2}$
15	$\overline{\mathbf{10}}$ 1 $\overline{2}$ $\overline{\mathbf{30}}$ $\overline{2}$	65	$\overline{\mathbf{39}}$ 1 $\overline{\overline{3}}$ $\overline{\mathbf{195}}$ $\overline{3}$
17	$\overline{\mathbf{12}}$ 1 $\overline{3}$ $\overline{12}$ $\overline{\mathbf{51}}$ $\overline{3}$ $\overline{\mathbf{68}}$ $\overline{4}$	67	$\overline{\mathbf{40}}$ 1 $\overline{2}$ $\overline{8}$ $\overline{20}$ $\overline{\mathbf{335}}$ $\overline{5}$ $\overline{\mathbf{536}}$ $\overline{8}$
19	$\overline{\mathbf{12}}$ 1 $\overline{2}$ $\overline{12}$ $\overline{\mathbf{76}}$ $\overline{4}$ $\overline{\mathbf{114}}$ $\overline{6}$	69	$\overline{\mathbf{46}}$ 1 $\overline{2}$ $\overline{\mathbf{138}}$ $\overline{2}$
21	$\overline{\mathbf{14}}$ 1 $\overline{2}$ $\overline{\mathbf{42}}$ $\overline{2}$	71	$\overline{\mathbf{40}}$ 1 $\overline{2}$ $\overline{4}$ $\overline{40}$ $\overline{\mathbf{568}}$ $\overline{8}$ $\overline{\mathbf{710}}$ $\overline{10}$
23	$\overline{\mathbf{12}}$ 1 $\overline{\overline{3}}$ $\overline{4}$ $\overline{\mathbf{276}}$ $\overline{12}$	73	$\overline{\mathbf{60}}$ 1 $\overline{6}$ $\overline{20}$ $\overline{\mathbf{219}}$ $\overline{3}$ $\overline{\mathbf{292}}$ $\overline{4}$ $\overline{\mathbf{365}}$ $\overline{5}$
25	$\overline{\mathbf{15}}$ 1 $\overline{\overline{3}}$ $\overline{\mathbf{75}}$ $\overline{3}$	75	$\overline{\mathbf{50}}$ 1 $\overline{2}$ $\overline{\mathbf{150}}$ $\overline{2}$
27	$\overline{\mathbf{18}}$ 1 $\overline{2}$ $\overline{\mathbf{54}}$ $\overline{2}$	77	$\overline{\mathbf{44}}$ 1 $\overline{2}$ $\overline{4}$ $\overline{\mathbf{308}}$ $\overline{4}$
29	$\overline{\mathbf{24}}$ 1 $\overline{6}$ $\overline{24}$ $\overline{\mathbf{58}}$ $\overline{2}$ $\overline{\mathbf{174}}$ $\overline{6}$ $\overline{\mathbf{232}}$ $\overline{8}$	79	$\overline{\mathbf{60}}$ 1 $\overline{4}$ $\overline{15}$ $\overline{\mathbf{237}}$ $\overline{3}$ $\overline{\mathbf{316}}$ $\overline{4}$ $\overline{\mathbf{790}}$ $\overline{10}$
31	$\overline{\mathbf{20}}$ 1 $\overline{2}$ $\overline{20}$ $\overline{\mathbf{124}}$ $\overline{4}$ $\overline{155}$ $\overline{5}$	81	$\overline{\mathbf{54}}$ 1 $\overline{2}$ $\overline{\mathbf{162}}$ $\overline{2}$
33	$\overline{\mathbf{22}}$ 1 $\overline{2}$ $\overline{\mathbf{66}}$ $\overline{2}$	83	$\overline{\mathbf{60}}$ 1 $\overline{3}$ $\overline{20}$ $\overline{\mathbf{332}}$ $\overline{4}$ $\overline{\mathbf{415}}$ $\overline{5}$ $\overline{\mathbf{498}}$ $\overline{6}$
35	$\overline{\mathbf{30}}$ 1 $\overline{6}$ $\overline{\mathbf{42}}$ $\overline{\overline{3}}$ $\overline{6}$	85	$\overline{\mathbf{51}}$ 1 $\overline{\overline{3}}$ $\overline{\mathbf{255}}$ $\overline{3}$
37	$\overline{\mathbf{24}}$ 1 $\overline{2}$ $\overline{24}$ $\overline{\mathbf{111}}$ $\overline{3}$ $\overline{\mathbf{296}}$ $\overline{8}$	87	$\overline{\mathbf{58}}$ 1 $\overline{2}$ $\overline{\mathbf{174}}$ $\overline{2}$
39	$\overline{\mathbf{26}}$ 1 $\overline{2}$ $\overline{\mathbf{78}}$ $\overline{2}$	89	$\overline{\mathbf{60}}$ 1 $\overline{3}$ $\overline{10}$ $\overline{20}$ $\overline{\mathbf{356}}$ $\overline{4}$ $\overline{\mathbf{534}}$ $\overline{6}$ $\overline{\mathbf{890}}$ $\overline{10}$
41	$\overline{\mathbf{24}}$ 1 $\overline{\overline{3}}$ $\overline{24}$ $\overline{\mathbf{246}}$ $\overline{6}$ $\overline{\mathbf{328}}$ $\overline{8}$	91	$\overline{\mathbf{70}}$ 1 $\overline{5}$ $\overline{10}$ $\overline{\mathbf{130}}$ $\overline{\overline{3}}$ $\overline{30}$
43	$\overline{\mathbf{42}}$ 1 $\overline{42}$ $\overline{\mathbf{86}}$ $\overline{2}$ $\overline{\mathbf{129}}$ $\overline{3}$ $\overline{\mathbf{301}}$ $\overline{7}$	93	$\overline{\mathbf{62}}$ 1 $\overline{2}$ $\overline{\mathbf{186}}$ $\overline{2}$
45	$\overline{\mathbf{30}}$ 1 $\overline{2}$ $\overline{\mathbf{90}}$ $\overline{2}$	95	$\overline{\mathbf{60}}$ 1 $\overline{2}$ $\overline{12}$ $\overline{\mathbf{380}}$ $\overline{4}$ $\overline{\mathbf{570}}$ $\overline{6}$
47	$\overline{\mathbf{30}}$ 1 $\overline{2}$ $\overline{15}$ $\overline{\mathbf{141}}$ $\overline{3}$ $\overline{\mathbf{470}}$ $\overline{10}$	97	$\overline{\mathbf{56}}$ 1 $\overline{2}$ $\overline{8}$ $\overline{14}$ $\overline{28}$ $\overline{\mathbf{679}}$ $\overline{7}$ $\overline{\mathbf{776}}$ $\overline{8}$
49	$\overline{\mathbf{28}}$ 1 $\overline{2}$ $\overline{4}$ $\overline{\mathbf{196}}$ $\overline{4}$	99	$\overline{\mathbf{66}}$ 1 $\overline{2}$ $\overline{\mathbf{198}}$ $\overline{2}$
51	$\overline{\mathbf{34}}$ 1 $\overline{2}$ $\overline{\mathbf{102}}$ $\overline{2}$	101	$\overline{\mathbf{101}}$ 1 $\overline{\mathbf{202}}$ $\overline{2}$ $\overline{\mathbf{303}}$ $\overline{3}$ $\overline{\mathbf{606}}$ $\overline{6}$

Mathematical Leather Roll

Column 1		Column 2		Column 3		Column 4	
$\overline{10}\ \overline{40}$	it is $\overline{8}$	$\overline{30}\ \overline{45}\ \overline{90}$	it is $\overline{15}$	$\overline{10}\ \overline{40}$	it is $\overline{8}$	$\overline{18}\ \overline{36}$	it is $\overline{12}$
$\overline{5}\ \overline{20}$	it is $\overline{4}$	$\overline{24}\ \overline{48}$	it is $\overline{16}$	$\overline{5}\ \overline{20}$	it is $\overline{4}$	$\overline{21}\ \overline{42}$	it is $\overline{14}$
$\overline{4}\ \overline{12}$	it is $\overline{3}$	$\overline{18}\ \overline{36}$	it is $\overline{12}$	$\overline{4}\ \overline{12}$	it is $\overline{3}$	$\overline{45}\ \overline{90}$	it is $\overline{30}$
$[\overline{10}]\ \overline{10}$	it is $\overline{5}$	$\overline{21}\ \overline{42}$	it is $\overline{14}$	$\overline{10}\ \overline{10}$	it is $\overline{5}$	$\overline{30}\ \overline{60}$	it is $\overline{20}$
$[\overline{6}\ \overline{6}]$	it is $\overline{3}$	$\overline{45}\ \overline{90}$	it is $\overline{30}$	$\overline{6}\ \overline{6}$	it is $\overline{3}$	$\overline{15}\ \overline{30}$	it is $\overline{10}$
$[\overline{6}\ \overline{6}\ \overline{6}]$	it is $\overline{2}$	$\overline{30}\ \overline{60}$	it is $\overline{20}$	$\overline{6}\ \overline{6}\ \overline{6}$	it is $\overline{2}$	$\overline{48}\ \overline{96}$	it is $\overline{32}$
$[\overline{3}\ \overline{3}]$	it is $\overline{\overline{3}}$	$\overline{15}\ \overline{30}$	it is $\overline{10}$	$\overline{3}\ \overline{3}$	it is $\overline{\overline{3}}$	$\overline{96}\ \overline{192}$	$\overline{64}$
$[\overline{25}]\ \overline{15}\ \overline{75}\ \overline{200}$	it is $\overline{8}$	$\overline{48}\ \overline{96}$	it is $\overline{32}$	$\overline{25}\ \overline{15}\ \overline{75}\ \overline{200}$	it is $\overline{8}$		
$[\overline{50}]\ \overline{30}\ \overline{150}\ \overline{400}$	it is $\overline{16}$	$\overline{96}\ \overline{192}$	$\overline{64}$	$\overline{50}\ \overline{30}\ \overline{150}\ \overline{400}$	it is $\overline{16}$		
$\overline{25}\ \overline{50}\ \overline{150}$	it is $\overline{6}$			$\overline{25}\ \overline{50}\ \overline{150}$	it is $\overline{6}$		
$[\overline{9}\ \overline{18}]$	it is $\overline{6}$			$\overline{9}\ \overline{18}$	it is $\overline{6}$		
$[\overline{7}\ \overline{14}]\ \overline{28}$	it is $\overline{4}$			$\overline{7}\ \overline{14}\ \overline{28}$	it is $\overline{4}$		
$[\overline{12}\ \overline{24}]$	it is $[\overline{8}]$			$\overline{12}\ \overline{24}$	it is $\overline{8}$		
$\overline{14}\ \overline{21}\ \overline{42}$	it is $[\overline{7}]$			$\overline{14}\ \overline{21}\ \overline{42}$	it is $\overline{7}$		
$[\overline{18}\ \overline{27}]\ \overline{54}$	it is $[\overline{9}]$			$\overline{18}\ \overline{27}\ \overline{54}$	it is $\overline{9}$		
$[\overline{12}\ \overline{33}]\ \overline{66}$	it is $[\overline{11}]$			$\overline{12}\ \overline{33}\ \overline{66}$	it is $\overline{11}$		
$[\overline{28}\ \overline{49}]\ \overline{196}$	it is $[\overline{13}]$			$\overline{28}\ \overline{49}\ \overline{196}$	it is $\overline{13}$		
				$\overline{30}\ \overline{45}\ \overline{90}$	it is $\overline{15}$		
				$\overline{2}[\overline{4}]\overline{4}[\overline{8}]$	it is $[\overline{1}]\overline{6}$		

The *Mathematical Leather Roll* is another aid for fraction reckoning. It contains 26 sums of unit fractions which equal a single unit fraction. The 26 sums have been noted in two columns, followed by another two columns with the same 26 sums. The numeric transcription given above shows the arrangement of the sums of the source.

Apart from fraction reckoning, tables were also needed for the conversion of different measuring units. An example of these tables can be found in the *Rhind Mathematical Papyrus*, No. 81. Here, two systems of volume measures, *ḥqꜣ.t* and *hnw*, are compared. *ḥqꜣ.t* is the basic measuring unit for grain, with 1 *ḥqꜣ.t* equaling 10 *hnw*. The *ḥqꜣ.t* was used with a system of submultiples, which were written by distinctive signs:

$\frac{1}{2}$ *ḥqꜣ.t*

$\frac{1}{4}$ *ḥqꜣ.t*

$\frac{1}{8}$ *ḥqꜣ.t*

$\frac{1}{16}$ *ḥqꜣ.t*

$\frac{1}{32}$ *ḥqꜣ.t*

$\frac{1}{64}$ *ḥqꜣ.t*

In older literature about Egyptian mathematics these signs are often interpreted as hieratic versions of the hieroglyphic parts of the eye of the Egyptian god Horus. However, texts from the early third millennium as well as depictions in tombs of the Old Kingdom, which show the same signs prove that the eye of Horus was not connected to the origins of the hieratic signs.[5] 1 *ḥqꜣ.t* also equals 32 *rꜣ*, the smallest unit for measuring volumes.

The table found in No. 81 of the *Rhind Mathematical Papyrus* is divided into three parts. Each part is introduced by an Egyptian particle (in the translation rendered as "now"). The first section of the table, arranged in two columns, lists the submultiples of the *ḥqꜣ.t* as *hnw*. Due to the values of the submultiples, each line is half of its predecessor. The following two sections are both laid out in three columns. The first column gives combinations of the submultiples of the *ḥqꜣ.t* and *rꜣ.w*. The second column lists the respective volume in *hnw*. The last column contains the volumes as fractions of the *ḥqꜣ.t*, this time not written in the style of submultiples but as a pure numeric fraction of the unit *ḥqꜣ.t*.

The source text of these last two sections shows a rather large number of errors. Out of 82 entries 11 are wrong. Some of these errors seem to be simple writing errors; some follow from using a faulty entry in a previous line or column to calculate the new entry. The table is given here with all the original (sometimes wrong) values followed by footnotes that give the correct value and—if possible—an explanation for the error. It is difficult to account for the large number of mistakes in this table. The *Rhind Mathematical Papyrus* (of which this table is a part) is a collection of tables and problems, mostly organized in a carefully thought out sequence. It was presumably the manual of a teacher.

[5] See [Ritter 2002] for a detailed discussion.

Rhind Mathematical Papyrus, No. 81

Another reckoning of the hnw

Now	$\overline{2}$	*ḥqꜣ.t*	5
	$\overline{4}$	*ḥqꜣ.t*	2 $\overline{2}$
	$\overline{8}$	*ḥqꜣ.t*	1 $\overline{4}$
	$\overline{16}$	*ḥqꜣ.t*	$\overline{2}\ \overline{8}$
	$\overline{32}$	*ḥqꜣ.t*	$\overline{4}\ \overline{16}$
	$\overline{64}$	*ḥqꜣ.t*	$\overline{8}\ \overline{32}$

Now	$\overline{2}\ \overline{4}\ \overline{8}$	*ḥqꜣ.t*		as *hnw*, it is 8 $\overline{2}\ \overline{4}$	
	$\overline{2}\ \overline{4}$	*ḥqꜣ.t*		· it is 7 $\overline{2}$	
	$\overline{2}\ \overline{8}\ \overline{32}$	*ḥqꜣ.t*	3 $\overline{3}$ *rꜣ.w*	· 6 $\overline{2}\ \overline{16}$[6]	it is $\overline{\overline{3}}$ of a *ḥqꜣ.t*
	$\overline{2}\ \overline{8}$	*ḥqꜣ.t*		· 6 $\overline{4}$	it is $\overline{5}$ of a *ḥqꜣ.t*[7]
	$\overline{4}\ \overline{8}$	*ḥqꜣ.t*		· 3 $\overline{2}\ \overline{4}$	it is 3 of a *ḥqꜣ.t*[8]
	$\overline{4}\ \overline{32}\ \overline{64}$	*ḥqꜣ.t*	1 $\overline{\overline{3}}$ *rꜣ.w*	· 3 $\overline{4}\ \overline{8}$[9]	it is $\overline{7}$ of a *ḥqꜣ.t*[10]
	$\overline{4}$	*ḥqꜣ.t*		· 2 $\overline{2}$	it is $\overline{4}$ of a *ḥqꜣ.t*
	$\overline{8}\ \overline{16}$	*ḥqꜣ.t*	4 *rꜣ.w*	· 2	it is $\overline{5}$ of a *ḥqꜣ.t*
	[$\overline{8}\ \overline{32}$]	*ḥqꜣ.t*	3 $\overline{3}$ *rꜣ.w*	· [1 $\overline{\overline{3}}$]	it is [$\overline{6}$ of] a *ḥqꜣ.t*

Now	$\overline{8}\ \overline{16}$	*ḥqꜣ.t*	4 *rꜣ.w*	it is 2 *hnw*	**it is $\overline{5}$ of a *ḥqꜣ.t***
	$\overline{16}\ \overline{32}$	*ḥqꜣ.t*	2 *rꜣ.w*	it is 1 *hnw*	**it is $\overline{10}$ of a *ḥqꜣ.t***
	$\overline{32}\ \overline{64}$	*ḥqꜣ.t*	1 *rꜣ.w*	it is $\overline{2}$ *hnw*	**it is $\overline{20}$ of a *ḥqꜣ.t***
	$\overline{64}$	*ḥqꜣ.t*	3 *rꜣ.w*	it is $\overline{4}$ *hnw*	**it is $\overline{40}$ of a *ḥqꜣ.t***
	$\overline{16}$	*ḥqꜣ.t*	1 $\overline{3}$ *rꜣ.w*	it is $\overline{\overline{3}}$ *hnw*	**it is $\overline{30}$ of a *ḥqꜣ.t***[11]
	$\overline{32}$	*ḥqꜣ.t*	$\overline{\overline{3}}$ *rꜣ.w*	it is 3 *hnw*	**it is $\overline{60}$ of a *ḥqꜣ.t***[12]

---

[6]Correct value: 6 $\overline{\overline{3}}$ *hnw*. Possible explanation for the mistake: $\overline{2}\ \overline{8}\ \overline{32}$ *ḥqꜣ.t* = 6 $\overline{2}\ \overline{16}$ *hnw*, therefore it is likely that 3 $\overline{3}$ *rꜣ.w* of the first column were forgotten when the second column was determined.

[7]Correct value: $\overline{2}\ \overline{8}$ of a *ḥqꜣ.t*.

[8]Correct value: $\overline{4}\ \overline{8}$ of a *ḥqꜣ.t*.

[9]Correct value: 3 $\overline{48}$ *hnw*. Possible explanation for the mistake: What I read as $\overline{4}$ may be a very badly written $\overline{40}$, but it seems more probable to read $\overline{4}$.

[10]Correct value: $\overline{4}\ \overline{32}\ \overline{48}$ of a *ḥqꜣ.t*.

[11]Correct value: $\overline{15}$ of a *ḥqꜣ.t*.

[12]Correct value: $\overline{30}$ of a *ḥqꜣ.t*. Possible explanation for the mistake: Calculation based on wrong entry in previous line of this column.

64	*ḥqꜣ.t*	3 *rꜣ.w*	it is 5 *hnw*[13]	**it is 50 of a *ḥqꜣ.t***[14]
$\overline{2}$	*ḥqꜣ.t*		it is 5 *hnw*	**it is $\overline{2}$ of a *ḥqꜣ.t***
$\overline{4}$	*ḥqꜣ.t*		it is 2 $\overline{2}$ *hnw*	**it is $\overline{4}$ of a *ḥqꜣ.t***
$\overline{2}\ \overline{4}$	*ḥqꜣ.t*		it is 7 $\overline{2}$ *hnw*	**it is $\overline{2}\ \overline{4}$ of a *ḥqꜣ.t***
$\overline{2}\ \overline{4}\ \overline{8}$	*ḥqꜣ.t*		it is 8 $\overline{2}$ *hnw*[15]	**it is $\overline{2}\ \overline{4}\ \overline{8}$ of a *ḥqꜣ.t***
$\overline{2}\ \overline{8}$	*ḥqꜣ.t*		it is 6 $\overline{4}$ *hnw*	**it is $\overline{2}\ \overline{8}$ of a *ḥqꜣ.t***
$\overline{4}\ \overline{8}$	*ḥqꜣ.t*		it is $\overline{2}\ \overline{4}$ *hnw*[16]	**it is $\overline{4}\ \overline{8}$ of a *ḥqꜣ.t***
$\overline{2}\ \overline{8}\ \overline{32}$	*ḥqꜣ.t*	3 $\overline{3}$ *rꜣ.w*	it is 6 $\overline{\overline{3}}$ *hnw*	**it is $\overline{\overline{3}}$ of a *ḥqꜣ.t***
$\overline{4}\ \overline{16}\ \overline{64}$	*ḥqꜣ.t*	1 $\overline{\overline{3}}$ *rꜣ.w*	it is 3 $\overline{3}$ *hnw*	**it is $\overline{3}$ of a *ḥqꜣ.t***
$\overline{8}$	*ḥqꜣ.t*		it is 1 $\overline{4}$ *hnw*	**it is $\overline{8}$ of a *ḥqꜣ.t***

II.b. Problem texts

The extant hieratic mathematical texts contain approximately 100 problems, most of which come from the *Rhind* and *Moscow Mathematical Papyri.* The problems can generally be assigned to three groups:

- pure mathematical problems teaching basic techniques
- practical problems, which contain an additional layer of knowledge from their respective practical setting
- non-utilitarian problems, which are phrased with a pseudo-daily life setting without having a practical application (only very few examples extant)

The following sections present selected problems of all three groups. Because problems are often phrased elliptically, occasionally other examples from the same problem type must be read in order to understand the problem. This will be seen from the first two examples (*Rhind Mathematical Papyrus*, problems 26 and 27). Unfortunately, due to the scarcity of source material, many problem types exist only in a few examples or even only in one.

The individual sources share a number of common features. They can generally be described as rhetorical, numeric, and algorithmic. "Rhetoric" refers to the texts being written without the use of any symbolism (like +, –, √). The complete procedure is written as a prose text, in which all mathematical operations are expressed verbally. "Numeric" describes the absence of variables (like x and y). The individual problems always use concrete numbers. Nevertheless, it is quite obvious that general procedures were taught through these concrete examples without being limited to specific numeric values. "Algorithmic" refers to the way mathematical knowledge was taught in Egypt—by means of procedures. The solutions to the problems are given as step by step instructions which lead to the numeric result of the given problem.

---

[13]Correct value: $\overline{6}$ *hnw*.

[14]Correct value: $\overline{60}$ of a *ḥqꜣ.t*.

[15]Correct value: 8 $\overline{2}\ \overline{4}$ *hnw*. Possible explanation for the mistake: the author forgot to write $\overline{4}$.

[16]Correct value 3 $\overline{2}\overline{4}$ *hnw*. Possible explanation for the mistake: the author forgot to write 3.

While the problem texts show these similarities, each source also shows some characteristics which make it distinct from the others. For instance, the examples from the *Rhind Mathematical Papyrus* usually include problems, the instructions for their solution, verification of results, and calculations related to the instructions or calculations as part of the verification. The examples from the *Moscow Mathematical Papyrus* only note the problem and the instructions for its solution. Furthermore, the two texts show slightly different ways of expressing these instructions. Since it is by no means self-evident to a modern reader how to read (and understand) these texts, the first example will be discussed in full detail. For the examples of practical problems, a basic knowledge of their respective backgrounds is often essential to understand the mathematical procedure. Therefore the commentary to those problems may contain an overview of their setting.

A note on language and translations

The problem texts show a high level of uniformity in grammar and wording. The individual parts of a problem, that is, title, announcement of its data, instructions for its solution, announcement and verification of the result are clearly marked through different formalisms. This will be mirrored in the translations given in this chapter. The individual termini for mathematical objects and operations were developed from daily life language. Thus the Egyptian *w3ḥ* ("to put down") became the terminus for "to add." In my translations I have used modern mathematical expressions wherever it is clear that the same concept is expressed. This can be assumed for all of the basic arithmetic operations. However, scholars have not yet determined if there are, as in the Mesopotamian case, subtle differences between apparent synonyms. The use of different grammatical structures to distinguish individual parts of a problem text can be summarized as shown in the following table.

Section of the problem text	*Grammatical markers*
Title	infinitive construction
Announcement of given data	2nd person construction, directly addressing the pupil
Instructions for solution	2nd person, imperative or *sḏm.ḫr.f* (see below)
Announcement of intermediate results	*sḏm.ḫr.f* (3rd person)
Announcement of final result	nominal constructions
Working	purely numerical

The instructions use a special verb form called the *sḏm.ḫr.f*, which indicates a necessary consequence from a previously stated condition. It is found not only in mathematical texts but also in medical texts. In mathematical texts it is used in the instructions as well as in announcing intermediate results. In translations this was traditionally rendered by "you are to..." in instructions and by the present tense "it becomes" in the announcement of intermediate results. This practice ignores the fact that the verb form used in both cases is the same, and should, consequently, be translated as such. In my translations I have used "shall" to express *sḏm.ḫr.f.*

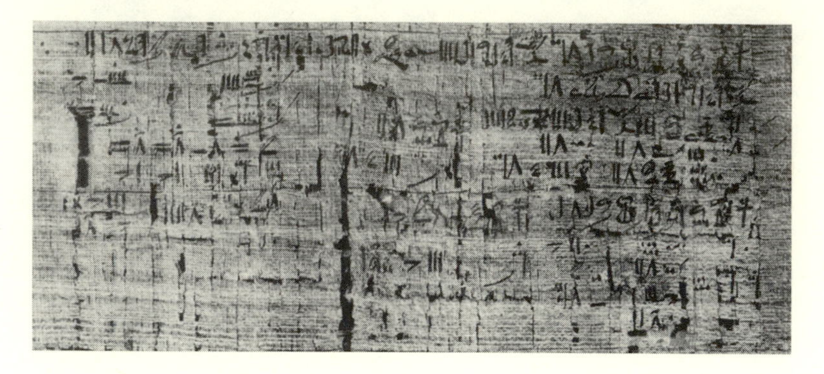

Rhind Mathematical Papyrus Problems 26 and 27.
Reprinted by permission of The British Museum.

Rhind Mathematical Papyrus, Problem 26

A quantity, its $\overline{4}$ (is added) to it *so that 15 results*

Calculate with 4.
You shall calculate its $\overline{4}$ as 1. Total 5.
Divide 15 by 5.

\ .	5
\2	10

3 shall result.
Multiply 3 times 4.

.	3
2	6
\4	12

12 shall result.

.	12
$\overline{4}$	3 Total 15.

The quantity 12
its $\overline{4}$ 3, **total 15.**

This problem belongs to the group of ʿḥʿ-problems, named after the characteristic term used in the title of each of these problems. ʿḥʿ is the Egyptian word for "quantity" or "number." The ʿḥʿ-problems, as can be seen from the example above, teach the procedure for determining an unknown quantity (ʿḥʿ) from a given relation with a known result. This example presents a quantity to be determined, which becomes 15 if its fourth is added to it. The text of the problem can be divided into three sections:

- title and given data
- procedure to solve the problem
- verification

The beginning of the problem is marked by the use of red ink (rendered as bold print in the transliteration). The procedure is then given as a sequence of instructions, sometimes followed by their respective calculations. For example, after the instruction "divide 15 by 5" we see the actual operation carried out. Once the result is obtained a verification is executed, first in the form of a calculation and then indicated by the use of red ink, as a complete statement.

In order to achieve a close reading of the source text, the individual steps of the solution have to be followed as such. We can make this procedure clearer if we rewrite the given instructions using our basic mathematical symbolism (+, –, ×, ÷). The procedure stated in the problem looks as follows after this rewriting ([ ] indicate ellipses in the text):

data		$\overline{4}$
		15
sequence of instructions	**1**	$[1 \div \overline{4}] = 4$
	2	$4 \times \overline{4} = 1$
	3	$4 + 1 = 5$
	4	$15 \div 5 = 3$
	5	$3 \times 4 = 12$
verification	$\mathbf{v_1}$	$12 \times \overline{4} = 3$
	$\mathbf{v_2}$	$12 + 3 = 15$

The text starts by announcing the given data of the problem: $\overline{4}$ and 15. In the rewritten form they are noted above the sequence of instructions. The instructions begin with "Calculate with 4." Since 4 is the inverse of the first datum ($\overline{4}$), there must have been one step in the calculation that has not been noted in the source text, namely the calculation of the inverse of $\overline{4}$. In the rewritten procedure above, we include this as step **1**. To indicate that it was not noted in the source text, we use square brackets ($[1 \div \overline{4}]$). Step **2** is the multiplication of the result of step **1** with the first datum ($4 \times \overline{4}$). Step **3** adds the result of steps **1** and **2:** 4 + 1. Step **4** uses the second datum (15) and the result of step **3:** 15 ÷ 5. Step **5** finally is the multiplication of the results of steps **1** and **4:** 3 × 4.

By following the procedure in this rewritten form several observations can be made. The basic structure of the text is sequential; results obtained in one step may be used in later step(s). Thus the result of **1** is used in **2**, **3**, and **5**; the result of **2** is used in **3**, the result of **3** is used in **4**, and the result of **4** is used in **5**. Data can be used at any time in the procedure. In this example the first datum ($\overline{4}$) appears in steps **1** and **2**; the second datum (15) in step **4**. Other numbers appearing in the instructions are either inherent to the specific mathematical operation carried out (e.g., the number 1 in the calculation of the inverse), or to the procedure itself (we will see an example of this later). The scribe must have known these numbers; they were learned with the sequence of operations of the procedure.

The different categories of "numbers" can be made even more obvious by rewriting the procedure again, this time indicating the data as D_1 ($=\overline{4}$) and D_2 (=15), and the result of step number *n* by ***n***, and the constants as before by their numerical value:

	D_1
	D_2
1	$[1 \div D_1]$
2	$\mathbf{1} \times D_1$
3	$\mathbf{1} + \mathbf{2}$
4	$D_2 \div \mathbf{3}$
5	$\mathbf{4} \times \mathbf{1}$
$\mathbf{v_1}$	$\mathbf{5} \times D_1$
$\mathbf{v_2}$	$\mathbf{5} + \mathbf{v_1} = D_2$

Again the sequential character is obvious. Rewriting procedure texts in this way enables a modern reader to compare the procedure of different problems more easily, as well as to see similarities between individual examples.

The solution of this example uses the so-called method of false position. A wrong solution (= 4) is assumed. In order to make this wrong solution suitable for the following calculations, it is determined here as the inverse of the first datum. The unknown (false solution) and its fractional part are then added (= 5). This is compared to the given (correct) result (= 15). Since the result obtained with the assumed number is three times smaller than the given result, the assumed number has to be multiplied by 3 to obtain the correct solution.

Rhind Mathematical Papyrus, Problem 27

A quantity, its $\overline{5}$ (is added) to it *so that 21 results*

.	5	
$\overline{5}$	1	Total 6.
\ .	6	
\ 2	12	
\ $\overline{2}$	3	Total 21.
\ .	3 $\overline{2}$	
2	7	
\ 4	14 (sic! source text 15)	

The quantity 17 $\overline{2}$,

its $\overline{5}$ .3 $\overline{2}$ **Total 21.**

Problem 27 also belongs to the group of ʿḥʿ-problems. Indeed, it is very similar to its predecessor, problem 26. However, after the title, which again includes the given data, only three calculations are noted, and not a single instruction. A comparison with the calculations of problem 26 reveals that the procedure of solving this problem is identical. This can best be seen if we rewrite the procedure in the same way as we have done in problem 26. The rewritten procedure shows the similarity (operations are reconstructed based on the calculations):

	No. 27				*No. 26*
	$\overline{5}$		D_1		$\overline{4}$
	21		D_2		15
1	$[1 \div \overline{5}] = 5$	**1**	$[1 \div D_1]$	**1**	$[1 \div \overline{4}] = 4$
2	$5 \times \overline{5} = 1$	**2**	$\mathbf{1} \times D_1$	**2**	$4 \times \overline{4} = 1$
3	$5 + 1 = 6$	**3**	$\mathbf{1} + \mathbf{2}$	**3**	$4 + 1 = 5$
4	$21 \div 6 = 3\overline{2}$	**4**	$D_2 \div \mathbf{3}$	**4**	$15 \div 5 = 3$
5	$3\overline{2} \times 5 = 17\overline{2}$	**5**	$\mathbf{4} \times \mathbf{1}$	**5**	$3 \times 4 = 12$
$\mathbf{v_1}$	$17\overline{2} \times \overline{5} = 3\overline{2}$	$\mathbf{v_1}$	$\mathbf{5} \times D_1$	$\mathbf{v_1}$	$12 \times \overline{4} = 3$
$\mathbf{v_2}$	$17\overline{2} + 3\overline{2} = 21$	$\mathbf{v_2}$	$\mathbf{5} + \mathbf{v_1} = D_2$	$\mathbf{v_2}$	$12 + 3 = 15$

Moscow Mathematical Papyrus, Problem 25

Method of calculating a quantity calculated times 2
together with (it, i.e., the quantity), it has come to 9.
Which is the quantity that was asked for?
You shall calculate the sum of this quantity and this 2.
3 shall result.
You shall divide 9 by this 3.
3 times shall result.
Look, 3 is that which was asked for.
What has been found by you is correct.

This example from the *Moscow Mathematical Papyrus* shows several differences to the style of the *Rhind Mathematical Papyrus.* Only the instructions were noted, no calculation was written down. Also, after the statement of the solution, no verification is carried out; instead we find a note stating that the solution is correct.

The title indicates that it is another example of an *ꜥḥꜥ*-problem. However, in this example, instead of adding a fractional part of the unknown quantity to itself, a multiple of it must be added. Consequently, the procedure to solve this problem differs from the two previous examples.

	2		D_1
	[1]		D_2
	9		D_3
1	$1 + 2 = 3$	**1**	$D_1 + D_2$
2	$9 \div 3 = 3$	**2**	$D_3 \div \mathbf{1}$

Rhind Mathematical Papyrus, Problem 50

Method of calculating *a circular* **area** *of 9 ḫt*
What is its amount as area?
You shall subtract its (i.e., the diameter's) $\overline{9}$ as 1,
while the remainder is 8.
You shall multiply 8 times 8.
It shall result as 64.
It is its amount as area: 64 *sṯꜣ.t.*

Calculation how it results: (9 *ḫt*)

·	9
its $\overline{9}$	1

subtraction from it, remainder: 8

·	8
2	16
4	32
\8	64

Its amount as area: 64 *sṯꜣ.t.*

This problem teaches the Egyptian algorithm to calculate the area of the circle of diameter 9 *ẖt*: one ninth of the diameter is subtracted from it, and the remainder is squared. The procedure uses the diameter (given in this example as 9 *ẖt*) and the constant $\bar{9}$. The source text of problem 50 shows another feature found in some of the Egyptian problem texts. There is a drawing of the calculated object, a little bigger than the column breadth in which it is written. The drawing of the circle has its characteristic dimension, its diameter, written inside it. This type of drawing has been named an *in-line-drawing* by Jim Ritter. As in other drawings of Egyptian mathematical texts, they are sufficiently accurate to show the "idea" of the represented object. However, they are not technical drawings and the information we can gain from them is limited.

Rhind Mathematical Papyrus, Problem 48

.	8 *sṯꜣ.t*
2	16 *sṯꜣ.t*
4	32 *sṯꜣ.t*
\8	64 *sṯꜣ.t*

\ .	9 *sṯꜣ.t*
2	18 *sṯꜣ.t*
4	36 *sṯꜣ.t*
\8	72 *sṯꜣ.t*
Total:	81 *sṯꜣ.t.*

Rhind Mathematical Papyrus, Problem 48. Reprinted by permission of The British Museum.

The text of this problem comprises a drawing (into which the number 9 is inscribed) and two calculations. The calculations can easily be identified as two multiplications, namely 8 times 8 *sṯꜣ.t* and 9 times 9 *sṯꜣ.t*. The drawing shows a square of base 9 (9 is the number written inside it) with another geometric figure inscribed into it. The second calculation (9 times 9 *sṯꜣ.t*) determines the area of the square. The first calculation can be interpreted as the calculation of the area of a circle of diameter 9, as in the previous example of problem 50. Only the last of the three steps of the algorithm was written down in the form of its working. This suits the drawing which we can identify as a circle inscribed into a square. Again, as in the case of the *in-line-drawing* of the previous problem, the sketch is sufficiently accurate to give an idea of the objects; however, it is far from being a technical drawing.[17]

---

[17]Previous interpretations of this problem, which tried to use this drawing alone to establish how the Egyptian method to calculate the area of a circle was developed, ignored this (as well as the two calculations referring to the drawing). It is not possible from the extant sources to follow the development of mathematical techniques. What we see are techniques presented in a form suitable for teaching junior scribes, and not the research notes of advanced scribes.

> MODERN EXCURSION The Egyptian procedure does not involve π. It is not based on a dependence of circumference on radius or diameter. However, it is of course possible to transform this procedure (from a modern point of view) into a formula and compare this to our modern formula $A = \frac{1}{4}\pi d^2$.
>
> $$A = \left(\tfrac{8}{9}d\right)^2 = \tfrac{64}{81}d^2 = \tfrac{1}{4}\left(\tfrac{256}{81}\right)d^2$$
>
> $$\tfrac{256}{81} \approx 3.16$$
>
> While the concept of π as the ratio of circumference to diameter is absent from the Egyptian procedure, its exactness, compared to the modern formula, is as if π were approximated by 256/81 ≈ 3.16.

Moscow Mathematical Papyrus, Problem 10

*Method of calculating a **nb.t***

If you are told, a *nb.t* with diameter $4\overline{2}$ and [?] as *ꜥḏ*. Let me know its area.
You shall calculate $\overline{9}$ of 9, because as for the *nb.t*, it is $\overline{2}$ of [...].
1 shall result.
You shall calculate the remainder as 8.
You shall calculate $\overline{9}$ of 8.
$\overline{\overline{3}}\ \overline{6}\ \overline{18}$ shall result.
You shall calculate the remainder of these 8 after these $\overline{\overline{3}}\ \overline{6}\ \overline{18}$.
$7\overline{9}$ shall result.
You shall calculate $7\overline{9}$ times $4\overline{2}$.
32 shall result.
Behold it is its area.
What has been found by you is correct.

Problem 10 of the *Moscow Mathematical Papyrus*, like the previous examples, teaches the calculation of an area. However, in contrast to every other example of area or volume calculation, it contains neither an *in-line-drawing* nor a sketch. Which type of area is calculated in this problem? By reading the translation of the source text, it becomes obvious that this question is not easily answered. (What is a *nb.t*?) There has been some scholarly discussion about this problem, and we still do not know for sure which object we face here.

A first clue should be given by the Egyptian designation of the object '*nb.t*'. This has been discussed in detail by the Egyptologist Friedhelm Hoffmann.[18] Unfortunately the word '*nb.t*' appears only in this problem. By comparison with later texts and similar words, it can be concluded that it can refer to either a three-dimensional object like a hill (mathematically idealized this would be a half-sphere or a half cylinder) or a two-dimensional segment of a circle. Are there any further clues in the source text? We learn about a diameter of $4\overline{2}$. Then there seems to be a second dimension, indicated by the Egyptian word *ꜥḏ*. This is not followed by a numerical value.

---

[18]See detailed discussion in [Hoffmann 1996].

Let us now look at the procedure, to see if we can get some clarification from the instructions given to solve the problem. Also, you may wonder by now how two areas as different as a half-sphere and a half-cylinder can both be contenders for the object of one problem.

The numeric procedure looks as follows (the question mark is put in for the value of the possible second datum):

$4\overline{2}$

?

1	$[2 \times 4\overline{2}] = 9$
2	$\overline{9} \times 9 = 1$
3	$9 - 1 = 8$
4	$\overline{9} \times 8 = \overline{\overline{3}}\,\overline{6}\,\overline{18}$
5	$8 - \overline{\overline{3}}\,\overline{6}\,\overline{18} = 7\overline{9}$
6	$7\overline{9} \times 4\overline{2} = 32$

Looking at the series of calculations, we can make a few initial observations. The instructions start with the calculation of $\overline{9}$ of 9. In order to obtain the "9" of step **2**, the datum $4\overline{2}$ has to be doubled, which was put in as step **1** in the rewritten procedure. Steps **2** and **3** resemble the procedure of calculating the area of a circle. Thus $\overline{9}$ is presumably a constant. Another look at the calculations reveals that there are five instances where numbers appear that are not results of a previous step. Step **1** includes 2 (constant) and $4\overline{2}$ (datum). Step **2** has $\overline{9}$ (constant). Step **4** shows another occurrence of $\overline{9}$ which is likely to be a constant. And finally, step **6** has another $4\overline{2}$. There are two possible interpretations of this last $4\overline{2}$:

- It refers again to the diameter. Consequently we assume that there is only one datum ($= 4\overline{2}$) and the mentioning of *ʿd̲* is erroneous.
- It refers to the second datum *ʿd̲*, which is of the same value ($4\overline{2}$) as the diameter.

Before we go on working with the source text, let us compare our modern calculation of the area of a hemisphere and the area of a semicylinder.

Modern excursion The surface A_s of a hemisphere of radius r is calculated as $A_s = 2\pi r^2$.
The surface A_c of a semi-cylinder of radius r and height h is calculated as $A_c = \pi rh$.
If we put the diameter d instead of the radius r, the two formulas become $A_s = \frac{1}{2}\pi d^2$ and $A_c = \frac{1}{2}\pi dh$.
As we have seen earlier, if we assume the existence of two data (*tp-rꜣ* and *ʿd̲*), both with the same value in this problem ($4\overline{2}$),then the two formulas yield indeed the same result with the data given in this problem.

Let's have another look at the procedure, this time rewritten in its most general form:

D_1

$D_2(?)$

1	$2 \times D_1$	**4**	$\overline{9} \times$ **3**
2	$\overline{9} \times$ **1**	**5**	**3** – **4**
3	**1**–**2**	**6**	**5** $\times D_2$ (or D_1)

The resemblance to the procedure of calculating the area of a circle has already been mentioned. Note, however, that if D_1 is the diameter of the object (which it must be according to the result of this problem), then the first step is the calculation of double the diameter. If we accept the second datum *ʿḏ* as being $4\bar{2}$, we can interpret the first five steps of the algorithm as the calculation of half the circumference of a circle of diameter (*tp-rꜣ*) $4\bar{2}$. The last step is then the multiplication of base and height to obtain the area.

Moscow Mathematical Papyrus, Problem 14

Method of calculating a ⏢.
If you are told ⏢ of 6 as height, of 4 as lower side, and of 2 as upper side.
You shall square these 4. 16 shall result.
You shall double 4. 8 shall result.
You shall square these 2. 4 shall result.
You shall add the 16 and the 8 and the 4. 28 shall result.
You shall calculate $\bar{3}$ of 6. 2 shall result.
You shall calculate 28 times 2. 56 shall result.
Look, belonging to it is 56.
What has been found by you is correct.

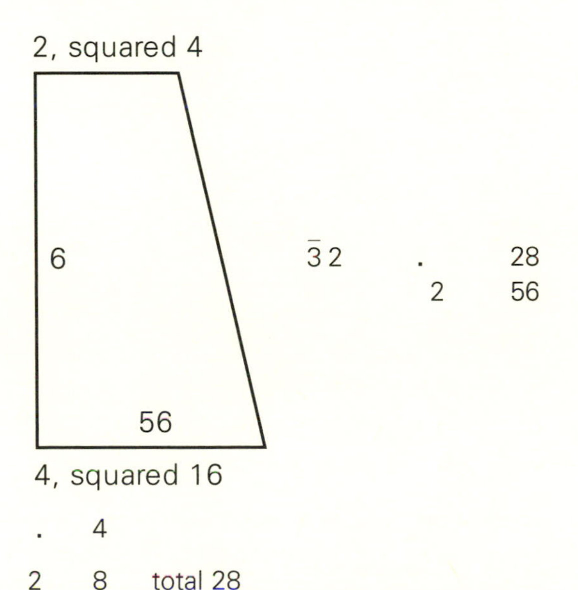

$\bar{3}$ 2 . 28
2 56

This problem, again from the *Moscow Mathematical Papyrus*, teaches the method for calculating the volume of a truncated pyramid. The truncated pyramid is not designated by an Egyptian term, but rather by its *in-line-drawing* ⏢. After the instructions, a sketch drawing is made and, exceptional for the *Moscow Papyrus*, calculations are noted. The sketch includes the data of the object, and the results of operations performed with these data. Thus, the lower side is indicated as 4, followed by its square 16 (used in the calculation). The same is done for the upper side (2, and its square 4) and the height (6, and its third 2). The multiplication 2 × 4 is indicated below the drawing, followed by the total of 16, 4, and 8 (28). Next to the result of the calculation of a third of the height (2), the final multiplication carried out in this problem (2 × 28) is noted. The result of the problem, the volume of the truncated pyramid (56), is indicated inside the drawing.

> If one transforms this procedure into a modern formula, the result is
>
> $$V = \tfrac{1}{3}(a^2 + 2ab + b^2),$$
>
> which is the correct formula for the calculation of the volume of a truncated square pyramid of upper base *a*, lower base *b*, and height *h* (not an approximation).

There have been several attempts to determine how this procedure was discovered by the Egyptians. However, these are all only more or less likely speculations.[19] The mathematical texts themselves give no indications how the procedures taught in them were found, nor do the administrative texts.

Rhind Mathematical Papyrus, Problem 56

***Method of reckoning** a pyramid: 360 as base,*
250 as height of it.
Let me know its *sqd*.
You shall calculate the half of 360. It shall result as 180.
You shall divide 180 by 250. $\overline{2}\,\overline{5}\,\overline{50}$ of one cubit shall result.
One cubit is 7 palms. You shall multiply by 7.

·	7
$\overline{2}$	3 $\overline{2}$
$\overline{5}$	1 $\overline{3}\,\overline{15}$
$\overline{50}$	$\overline{10}\,\overline{25}$

Its *sqd* is 5 $\overline{25}$ palms.

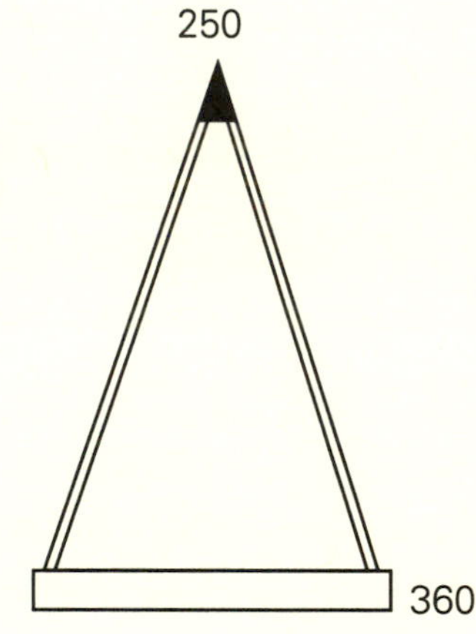

Problem 56 is one example of the six problems from the *Rhind Papyrus* relating to pyramids. All six of the problems teach the relation between base, height, and the slope of the sides. The Egyptians used the term *sqd* to describe the slope of the walls. The *sqd* measures how many palms an inclined plane retreats on a vertical height of one cubit. It is always measured in palms or palms and digits. Consequently, the *sqd* of a pyramid can be calculated as

$$sqd \text{ [palms]} = 7 \text{ [palms]} \cdot \frac{\frac{1}{2} base \text{ [cubits]}}{height \text{ [cubits]}}$$

[19]See, for example [Gillings 1964], [Gunn/Peet 1929, 178–185], [Thomas 1931], and [Vetter 1933].

This problem first gives the base and height of a pyramid. Its *sqd* is to be determined. The text of the problem is accompanied by a sketch of a pyramid. The numerical values of base and height are written next to this drawing.

Rhind Mathematical Papyrus, Problem 41

***Method of calculating a circular granary of 9,** 10.*
You shall subtract $\overline{9}$ of 9 as 1, remainder 8.
Multiply 8 times 8. 64 shall result.
You shall multiply 64 times 10. It shall result as 640.
Add its half to it. It shall result as 960.

It is its amount in ẖꜣr.
You shall calculate $\overline{20}$ of 960 as 48.
This is its content in quadruple *ḥqꜣ.t*: grain 48 *ḥqꜣ.t*
Method of its procedure.

.	8	.	64
2	16	\10	640
4	32	\\$\overline{2}$	320
\8	64	total	960
		$\overline{10}$	96
		$\overline{20}$	48

Rhind Papyrus problem 41 and the following example (problem 42) teach the calculation of the volume of a granary with a circular base. Egyptian tomb decorations as well as archaeological finds describe two types of granaries, those with a circular base that look like cones, and those with a rectangular base. The conic granaries are treated as cylinders in the examples of the mathematical problems. Consequently, the calculation of their volume consists of the Egyptian procedure for determining the area of the circle and the multiplication of this area by the given height. However, this is not the end of the procedure of this problem, as rewriting the algorithm shows:

	9
	10
1	$\overline{9} \times 9 = 1$
2	$9 - 1 = 8$
3	$8 \times 8 = 64$
4	$64 \times 10 = 640$
5	$\overline{2} \times 640 = 320$
6	$640 + 320 = 960$
7	$\overline{20} \times 960 = 48$

The dimensions of the granary are given in cubits (not explicitly stated in this problem). Therefore the resulting volume in **4** is obtained in cubic cubits. This needs to be transferred into the volume units usually used with large amounts of grain, *ẖꜣr* (obtained in step **6** as 960), and hundreds of *ḥqꜣ.t* (obtained in step **7** as 48). As in previous examples of the *Rhind Papyrus*, we find the actual calculations carried out at the end of the problem.

Lahun Fragment UC 32160 (Griffith, Petrie Papyri IV.3), column 1–2

12			
	1365$\bar{3}$ (circled)	8	
$\bar{\bar{3}}$	8	\\.	256
$\bar{3}$	4	2	512
total	16	\\4	1024
\\.	16	[\\]$\bar{3}$	85 $\bar{3}$
\\10	160		total 1365 $\bar{3}$
\\ 5	80		
total	256		

The text of this fragment does not have a problem and instructions for its solution. Instead, we find a drawing and several calculations. They belonged to a problem and its procedure—which were written either on a separate papyrus or on a now lost part of this papyrus. Three calculations are associated with the drawing. They are written in two columns under and next to it. The first calculation (ll. 1–3) is the multiplication $1\bar{3} \times 12 = 16$; the second calculation (ll. 4–7) the multiplication $16 \times 16 = 256$; and the second column holds the calculation $5\bar{3} \times 256 = 1365\bar{3}$.

From these calculations and the numerical values given in the drawing (without knowing anything else about the problem) we can reconstruct the following procedure (steps are indicated as ***n′***, since there may be steps before the ones reconstructed here):

	12		D_1
	8		D_2
1′	$1\bar{3} \times 12 = 16$	**1′**	$1\bar{3} \times D_1$
2′	$16 \times 16 = 256$	**2′**	**1′** × **1′**
3′	$256 \times 5\bar{3} = 1365\bar{3}$	**3′**	**2′** × $5\bar{3}$

At this point it is unclear if $1\bar{3}$ and $5\bar{3}$ are further data, derived from the given data, or constants inherent to the problem. Also, the second datum (8), which is known from the *in-line-drawing*, does not appear in this procedure.

A comparison with the problems of the *Rhind Mathematical Papyrus* brings us to problem 43 where similar multiplications are carried out: $1\bar{3} \times 8 = 10\bar{\bar{3}}$ and subsequently $10\bar{\bar{3}} \times 10\bar{\bar{3}} = 113\bar{\bar{3}}\,\bar{9}$. Following this is the calculation of $113\bar{\bar{3}}\,\bar{9} \times 4$, 4 being $\bar{\bar{3}}$ of another datum of problem 43. This also fits our procedure, for the $5\bar{3}$—which we meet in the last step of our procedure—is indeed $\bar{\bar{3}}$ of 8. Therefore we can reconstruct the following procedure:

	12		D_1
	8		D_2
1	$1\bar{3} \times 12 = 16$	**1**	$1\bar{3} \times D_1$
2	$16 \times 16 = 256$	**2**	**1** × **1**
3	$\bar{\bar{3}} \times 8 = 5\bar{3}$	**3**	$\bar{\bar{3}} \times D_2$
4	$256 \times 5\bar{3} = 1365\bar{3}$	**4**	**2** × **3**

Problem 43 of the *Rhind Mathematical Papyrus* is the calculation of the volume of a granary with a circular base from its given diameter and height. Unfortunately the text of this problem is corrupt; therefore it was not chosen as an example in this selection. It teaches an alternative method of determining the volume of a circular based granary, the result of which is expressed directly as an amount in *ẖ3r*.

The drawing found at the beginning of the calculations of the Lahun fragment suits this interpretation: indicated are the circular shape of the base, the diameter (12), which is written above it, the height (8), which is written on its side, and its content in *ẖ3r* ($1365\bar{3}$), which is written inside it. Thus the author appears to be calculating the volume in *ẖ3r* of a cylinder of diameter d and height h in the form $\left(\frac{4}{3}d\right)^2\left(\frac{2}{3}h\right)=\frac{32}{27}d^2h$. This is evidently equivalent to using the algorithm of problems 48 and 50 of the *Rhind Mathematical Papyrus* for calculating the area of a circle of diameter d, which we have expressed in the modern formula $A=\frac{1}{4}\left(\frac{256}{81}\right)d^2$.

Rhind Mathematical Papyrus, Problem 65

Method of calculating 100 loaves of bread for 10 men

a sailor, a commander, and a watchman as doubles.
Its procedure:
You shall add these beneficiaries (ration receivers). 13 shall result.
Divide the 100 loaves by 13. $7\,\bar{\bar{3}}\,\overline{39}$ shall result.
You shall say:
This is the ration for these 7 men, and the sailor, commander, and watchman as doubles.

. $7\,\bar{\bar{3}}\,\overline{39}$	the sailor	$15\,\bar{3}\,\overline{26}\,\overline{78}$
. $7\,\bar{\bar{3}}\,\overline{39}$	the commander	$15\,\bar{3}\,\overline{26}\,\overline{78}$
. $7\,\bar{\bar{3}}\,\overline{39}$	the watchman	$15\,\bar{3}\,\overline{26}\,\overline{78}$
. $7\,\bar{\bar{3}}\,\overline{39}$		
. $7\,\bar{\bar{3}}\,\overline{39}$	total: 100.	
. $7\,\bar{\bar{3}}\,\overline{39}$		
. $7\,\bar{\bar{3}}\,\overline{39}$		

The Egyptian ration system was based on the distribution of quantities of grain, and of bread and beer. It constituted the core of Egyptian administration, and it must have been a frequent task for a scribe to calculate amounts of food for various beneficiaries (see also the respective section of the *Anastasi I Papyrus* in the introduction of this chapter).

This problem has 100 loaves of bread to be distributed among 10 men. Three of them shall receive double the amount the others do. The solution determines a "corresponding" number of recipients who would all get the same ration. That is, the three persons receiving the double amount are counted twice. In the Egyptian text, this procedure is called "adding the beneficiaries." The basic ration is then calculated by the division of the given loaves by the number of "recipients." For those who get the double share, the basic ration must be doubled. The verification at the end adds up the individual rations.

Moscow Mathematical Papyrus, Problem 15

Method of Calculating Barley

If you are told: 10 *ḥq3.t* of barley made as beer of *psw* 2.
Indeed, let me know the (amount of) beer.
You shall calculate this 10 times 2. 20 shall result.
Look, 20 jars of beer.
What has been found by you is correct.

Within the Egyptian mathematical problem texts, the so-called bread-and-beer problems hold an important position due to their frequency.[20] In order to understand the relevance of these problems, we need to consider their practical background. The production of bread and beer in Egypt can be traced by archaeological artifacts as well as by representations on tomb walls decorated with scenes from daily life. There is a certain variety in details depicted in these scenes. For instance, not every depiction shows the same sequence of steps, and some of them are represented only in a few instances.[21] The most detailed tomb scenes are found during the Old Kingdom, i.e. prior to the mathematical texts. During the Middle Kingdom, models representing scenes from daily life become more prominent. While these are often less detailed than the Old Kingdom tomb scenes, they show no significant change in the representation of baking and brewing.

The procedure starts with the depiction of workers taking grain out of the granary and handing it to the bakers. An official notes the amounts that are given out. The actual production of bread begins with purifying and grinding the grain. Bakers then make the flour into dough and distribute it into pre-heated bread molds, in which it is baked. The preparation of beer uses bread that is mixed with dates and water in a large bin. The dates serve to fortify the taste as well as to bring yeast to the mixture for fermentation. The yeast organisms are found on the skins of the dates. Before further processing, the mixture is probably left alone for fermentation. Next the mixture is run through a strainer. The beer is finally poured into vessels. The last scenes of these representations show the delivery of produced goods. A scribe is shown measuring the delivered quantities.

A connection between the production of bread and beer and the mathematical bread and beer problems can be traced by means of the terminology that is used. The representations of baking and brewing show the distribution of grain and the delivery of bread and beer. To be able to control the production, the distributed grain must have a specific equivalent in bread and beer, that is, a certain quantity of bread and beer of a known quality. It must be possible therefore to calculate how many loaves of bread can be baked with a given amount of grain. The entities and formulations used to describe this are found in the bread and beer problems of the mathematical texts.

The most important *terminus technicus* here is the Egyptian word *psw*. The *psw* denotes how many bread loaves or vessels of beer are prepared from one *ḥq3.t* of grain. Therefore it can be calculated as the quotient of the number of bread loaves or vessels of beer and the amount of grain that has been used to produce them.

---

[20]Nine problems in the *Rhind Papyrus* (problems 69–78) and eleven problems in the *Moscow Papyrus* (problems 5, 8, 9, 12, 13, 15, 16, 20, 21, 22, and 24) belong to this group.

[21]The following description is based on the work of Dina Faltings, who examined the available material from the Old Kingdom [Faltings 1998].

$$psw_{bread} \equiv \frac{\text{number of bread loaves (made from 1 } \textit{ḥqꜣ.t} \text{ of grain)}}{\text{1 } \textit{ḥqꜣ.t} \text{ of grain}}$$

$$= \frac{\text{number of bread loaves}}{\text{amount of grain used for their production}}$$

$$psw_{beer} \equiv \frac{\text{number of vessels of beer (made from 1 } \textit{ḥqꜣ.t} \text{ of grain)}}{\text{1 } \textit{ḥqꜣ.t} \text{ of grain}}$$

$$= \frac{\text{number of vessels of beer}}{\text{amount of grain used for their production}}$$

The higher the *psw* of a bread or beer, the lower is its actual content of grain. In the problems of the *Moscow Papyrus* the *psw* of bread is constantly 20 (20 breads were made from 1 *ḥqꜣ.t* of grain). The *psw* of beer varies between 2 and 6.

With this information, the mention of *psw* as well as the calculation in our problem become understandable. Given are 10 *ḥqꜣ.t* of barley, which shall be made into beer of *psw* 2. According to the definition of the *psw*, the amount of beer is calculated by the multiplication of the two data. This is the easiest example of a bread and beer problem. Two problems of a similar kind are found at the beginning of the bread and beer problems from the *Rhind Mathematical Papyrus.* These are followed by problems in which a given amount of bread of a given first *psw* value shall be exchanged for bread of a second *psw* value, the amount of which has to be determined. Yet another variation can be found in many of the bread and beer problems of the *Moscow Mathematical Papyrus.* There, different types of ingredients, namely barley, emmer, dates, and *bšꜣ* (probably malt), are shown to have different values. Barley is used as a kind of reference grain. As a consequence, the quantities of barley mentioned in the more complex problems of the *Moscow Papyrus* have two possible meanings. They denote either a quantity used for the actual baking/brewing, or the value of (another) quantity of a different type of grain that is used for baking or brewing.

Moscow Mathematical Papyrus, Problem 23

Method of Calculating the work-rate of a cobbler.
If you are told the work rate of a cobbler:
If he (only) cuts (leather for sandals) it is 10 (pairs of sandals) per day;
if he (only) finishes (the sandals) it is 5 (pairs of sandals) per day.
How much is (his work-rate) per day if he cuts (the leather) and finishes (the sandals)?
You shall calculate the parts of this 10 and this 5. The total shall result as 3.
You shall divide 10 by it. $3\overline{3}$ times shall result.
Look, it is $3\overline{3}$ per day.
What has been found by you is correct.

The handicraft context of problem 23 of the *Moscow Mathematical Papyrus* is named in its setting: it is the work of a cobbler. We have access to this context through artefacts from settlement sites as well as depictions of handicraft processes in tombs. The basic product of an Egyptian cobbler, as can be seen from many depictions found in tombs, is sandals. The production of sandals consists of the preparation of the leather, its cutting, and finally the finishing process of the shoe, which incorporates putting together the leather pieces and, eventually, applying some kind of decoration to the leather.

The title of the problem provides the information that the work-rate of a cobbler must be calculated. The following lines state that the cobbler has to cut leather for ten pairs of sandals as his work-rate per day if he only cuts the leather; but if the leather has been already cut, he has to finish five pairs per day. What is looked for is the amount of his contribution, given the condition that the cobbler does both parts of the sandal production, cutting and finishing the shoe. The following instructions given to solve the problem use the fact that the contribution of the cobbler, if he only cuts leather, is twice the contribution if he finishes the sandals. If he has to cut ten pairs or finish five as his daily work-rate, it will take three days work to amount to ten pairs of sandals. What was wanted was the number of sandals per day if he does both. This is obtained by dividing 10 by 3, which results in $3\frac{1}{3}$ pairs of sandals.

III. Mathematics in Administrative Texts

The following section introduces several examples of administrative documents, another genre of texts that incorporate mathematical knowledge. Contrary to the mathematical problem texts, in this type of text only data and results are noted. The mathematical operations that were executed to obtain the results from the data are "missing." Consequently, it is not always possible to reconstruct the way the results were obtained. Furthermore, many of these texts are badly preserved, so that some parts of data or results are no longer extant. Finally, not every administrative text with numerical information is mathematically interesting; many of them are simple lists of data.

III.a. Middle Kingdom texts: The Reisner papyri

The *Reisner Papyri* provide us with the exceptional case of an example of a reasonably well-preserved text including some mathematically interesting passages, of which two will be discussed here. These papyri were found during excavations at Naga ed-Deir conducted by the Egyptologist George A. Reisner in 1901–1904: "Among the later tombs on the slope below the cliff, one contained four rolls of hieratic papyri, badly worm eaten; and another yielded a set of poisoned arrows."[22]

The *Reisner I Papyrus*, from which the following two examples are taken, measures approximately 3.50 meters, and it has a height of 31.6 cm. As in other accounting papyri of the Old and Middle Kingdoms the papyrus is divided by several horizontal lines which were meant to help the scribe in aligning his entries. Based on the palaeography, the regnal years found in the text (without the name of the ruler), and personal names, the text has been assigned to the reign of Senusret I (1956–1911 BCE).

The documents are accounts of building construction and carpentry workshops, including lists of workmen arranged in groups, as well as calculations related to construction projects and the necessary workers to perform them. Several accounts refer to a dockyard workshop,

[22] See [Reisner 1904, p. 108]—A photo of the papyri in situ can be found in [Simpson 1963, frontispiece].

including lists of copper tools (axes, adzes, saws, etc.) by units of copperweight, presumably for recasting. For the purpose of editing and reference, the editor W. K. Simpson divided the document into several sections designated by letters of the alphabet. The chosen division reflects the scribes' own divisions to a major extent. A number of these sections belong together due to their subject. Five sections of the *Reisner I Papyrus* constitute records of the construction of a building, presumably a temple (sections G, H, I, J, and K, of which section I is the first of our examples here).[23] They contain four accounts followed by a summary. The layout of these accounts is quite clear. On the left side of the text (= the right side of the translation) we find six columns that list lengths, widths, depths, units, the product of the four previous columns, and the respective number of workers (which is left blank in Section I). The right side of the text (= the left side of the translation) indicates dates and circumstances of the work recorded otherwise by numeric entries alone. In section I this contains seven dates, as well as various details about the place of work and the material involved. The dates are given in the usual way as *x*. (month) of a season, *y* day. The Egyptian calendar distinguished three seasons *ꜣḥ.t* (inundation), *prt* (emergence of the crops), and *šmw* (harvest). Each season had four months of 30 days. Furthermore, there were five additional days (epagomenals) at the end of the year.

Unfortunately, some of the terms for materials are not known from any other source, and the occurrences in this text are not sufficient to establish their meaning. Therefore, the translation given here is only a provisional one.[24] It is interesting to see that all of the activities are characterized by lengths, widths, depths, and a subsequent multiplier (units), because these structure the necessary data to calculate the total volume. The individual actions described seem to indicate that this section of the text is concerned with the production of bricks.

The account lists three dimensions, a number of units, and the product of these. The last column (enlistees) is left blank. Almost all measurements are given as cubits, except for three, explicitly marked as palms. In three instances the products have been miscalculated (marked in my translation by sic! with the correct value given in brackets). In addition there are three numbers in red (indicated in bold in the translation) written in between the columns. Their meaning and relation to the other entries in this section are not clear.

The second example, *Papyrus Reisner I*, section O, records workers' compensation over a period of 72 days. The heading of the first column gives a date, presumably the beginning of the period covered in the table, as well as the information where the work took place. The first column consists of a list of 20 names. Following this are six columns, each of which includes a 12-day period. The entries in these columns are divided into the record of days the individual worker worked in the respective 12 days and the amount of *trsst*-bread[25] he received as compensation for this work. Within the entries of the record of days worked, black numbers indicate actual days worked, red numbers indicate absence. In the first column we find 12, 12, and 2 in red, in the second column 3, 2, and 3, and in the third column 8, 7, 11, and 11. These days of absence are added, and the totals are noted in black at the bottom of the text: 26 for the first column, 8 for the second column and 47 (miscalculated for 37) for the third column.

---

[23]A useful introduction into the subject is [Arnold 1991].

[24]For further information see the extensive lexicographical commentary in [Simpson 1963].

[25]The Egyptians had a variety of bread and cake types. What kind of bread/cake the individual names designate is not know. The bread used in this text is called *trsst* in Egyptian. It is attested in only one other source, which also deals with rations.

A Construction Account: Papyrus Reisner I, Section I

	4. *prt*, **day 15**	lengths	widths	depths	<units>	product	equaling enlistees
	given [to him] in molding the ground: the great chamber	12	5	$\overline{2}$	1	30	
	[given to him in] the august [chamber]	[1]5	5	$\overline{2}$	1	$37\overline{2}$	
ḫ3[26]	[..........] given to him in the eastern chapel of the glorious chambers	8	5	[$\overline{2}$]	1	20	
	[.........given] to him in ... for ...	[1]8	1[1]	[$\overline{3}$]	1	132	
	the western *mḫ3w*[27]	32	4	[$\overline{4}$]	1	32	
ḫ3	the eastern *mḫ3w*	52	3 **298[..]**	[$\overline{4}$]	1	39	
	[1. *šmw*], day 28, given to him as fill: the great chamber	24	5 palms	$\overline{2}$	1	8; 4 palms	
	[given to] him carrying *srft*[28]	26	6	5 palms	1	111; 3 palms	
	[..]	20	5	5 palms	1	71; 3 palms	
	given to him in loosening brick clay	27	7	2	1	378	
	2. *šmw*, day 1, given to him in removing water from the field	8	7	2	1	112	
	2. *šmw*, day 2, given to him as builders in the tower	$1\overline{2}$	$1\overline{2}$	2	2	9	
		$2\overline{2}$	$1\overline{2}$	$1\overline{2}$	2	$11\overline{4}$	
		$3\overline{2}$	$2\overline{2}$	$1\overline{2}$	2	$25\overline{4}$ (sic! $26\overline{4}$)	
		4	$2\overline{2}$ **$81\overline{2}$**	$1\overline{2}$	2	36 (sic! 30)	
	2. *šmw*, day [..], completed for him in brick-clay of the fields	10	$5\overline{2}$	<..>$\overline{4}$	1	55 (sic! $13\overline{2}\overline{4}$)	
		16	$5\overline{2}$	[...]	1	$75\overline{4}$	
		8	6 **$556\overline{4}$**	**[1]**	1	48	
	completed for him in large size brick						

[26] *ḫ3*: mark used to call attention to related items.

[27] *mḫ3w*: not known from any other text.

[28] *srft*: material used in construction; lighter in weight than stone or sand. There is a detailed discussion in [Simpson, 1963, pp. 74–75].

Recompensation of Workers: Papyrus Reisner I, Section O

	12 days	*trsst*	12 days	*trsst*		12 days	*trsst*	12 days	*trsst*	12 days	*trsst*	12 days	*trsst*
*Year 24, **2nd month of šmw, day 21***													
List of enlistees who are in This:													
Nakhti's son Se-ankhi-nedjes	12	94	12	80		11	80	12	88	10	80	15	120
Senmutef's son Zi-n-Wosret **Redi-wi-Sobek**	10	**7** 56	12	96		11	244			10	80	15	
Senbebu's son Zi-n-Min	**12**	\	12	88		11	80	12	94	10	80	15	120
Se-ankhi's son Anhur-nakhte	12	**8** 64	12	88		11	80	12	62	10	80	15	80
Wosre's son Ikeki	12	**2** 80	9 **3**	72		11	88	12	48	10	80	15	120
Senet's son Kemni	12	94	12	96		11	88	12	80	10	80	15	120
Iri's son Mentu-hotep	12	90	12	88		11	88	12	88	10	80	15	120
Renef-ankhu's son Si-Anhur[29]	**12**	\	12	88		11	56	12	88	10	80	15	120
Hedjenenu's son Gem-mutef	12	94	10 **2**	80		11	88	12	88	10	80	5	40
Ankhu's son Anhur-nakhte	12	94	12	84		11	88	12	92	10	80	15	120
Inyotef's son Sefkhy	12	94	12	96		11	88	12	80	10	80	15	120
Sobek-wosre's son Sobek-nakhte	12	94	12	96	60	11	88	12	88	10	80	15	130
Sobeknakhte's son Sobek-nofre	12	94	12	96		11	88	12	92	10	80	15	120
Nakht-aa's son Nakht-tjen	12	94	12	96		11	88	12	92	10	80	15	120
Sobek-nakhte's son Shemai	12	94	9 **3**	72		11	88	12	88	10	80	15	120
Seankhi's son Sobek-hotep	10 **2**	80	12	96		3 **8**	24	\	\				
Irne's son Yu	12	94	12	96		11	88	12	84	10	80	15	120
Zi-n-Wosret's son Sehetep-ib	12	94	12	96		4 **7**	32	\	\				
Ameny's son Neferkhau	12	64	5	40		\ **11**		\	\				
his brother Sefkhy	12	94	12	96		\ **11**		\	\				
total	26	**1610**	8	**1546**		47	**1564**	**1530**	**100**		***1280***		***1710***

[29]Note that the original source text is "messier" than this translation: In this line, as well as in the next, and in the third line from the bottom further names (Nakhti in this line, Wehemy and Si-nefer in the next, and Merernedjes further below) have been inserted in between the columns that list the days worked and respective *trsst* loaves.

Since the number of days worked and the compensation in bread are both noted, the number of *trsst* loaves divided by the number of days worked results in the number of *trsst* loaves given to a worker per day. We would expect this to be constant; however, the number of *trsst* loaves per day for the individual persons varies. Sometimes it even varies for one individual on different days. It is hard to say what might have caused this variation, since the only information we get here is the actual numbers of *trsst* loaves given to the workers. It is noteworthy, however, that the most frequent ratio is that of 8 *trsst* loaves per day.

The totals given in red at the bottom of the sheet (1610, 1546, 1564, 1530/100, 1280, and 1710) equal only in one instance (1280) the sum of values given in the respective columns above. Therefore—if we don't assume that the scribe simply didn't know how to add—it seems that values from another source were also used in establishing the totals.

III.b. New Kingdom texts: Ostraca from Deir el Medina[30]

No mathematical texts from the New Kingdom are extant. Instead, there is a significant number of administrative and other texts which show the application of mathematical knowledge. The majority of these documents come from the area of Thebes.

The tombs of the pharaohs of the New Kingdom are located in a desert valley, on the western bank of the Nile River opposite Luxor. This valley is today known as the Valley of the Kings. The workers (quarrymen, plasterers, draftsmen, and painters) who constructed and decorated these tombs lived in a village located in another nearby desert valley. This site is today known as Deir el-Medina. The village was founded in the eighteenth dynasty and was inhabited until the end of the twentieth dynasty. Apart from the remains of this settlement, tens of thousands of documents have been discovered there, including (among others) letters, administrative and legal documents, magic and religious texts, as well as texts and sketches relating to the construction of tombs. These sources give us some insight into daily life at a workers' settlement during the New Kingdom. The government provided the workers at Deir el-Medina with all necessary commodities for their life as well as equipment for their work. Wages consisted mostly of grain. They were higher than what could have been consumed, and thus supposedly included a surplus used for exchange.

The Egyptian week had ten days, the last two of which were free. In addition there seem to have been long weekends and afternoons that were free as well. The working day lasted for approximately four hours in the morning and four hours in the afternoon with a break around lunch. The crews were divided into two groups, one for the right side of the tomb and one for the left side. Each group had its own foreman and assistant. The work on the tomb began with the selection of a site in the valley, probably under the eyes of a royal commission. Then a plan of the tomb was drawn, presumably on papyrus. The typical tomb consisted of several descending corridors and a number of rooms, the last of which usually contained the sarcophagus. The construction of the tomb usually began with quarrying the corridors and rooms. The plasterers worked behind the quarrymen and covered the uneven walls with a layer of gypsum and whitewash to enable their further decoration. Then the proposed texts and designs would be drafted and checked by a master. Finally, these texts were painted or

[30] A more detailed overview of daily life in Deir el Medina and the construction of tombs in the New Kingdom can be found in [McDowell 1999] and [Bierbrier 1982].

sculpted. Documents, mostly ostraca, of each of these stages in the work have been preserved and provide further insights.

The following example is a calculation of volumes presumably from the quarrying of a tomb. It is a so-called palimpsest; that is, another text had been written on the ostracon and erased before this text was written on it. The text begins immediately—without any heading—with the enumerations of dimensions. Above the beginning are a few sign rests; however, they do not seem to belong to this text.

Ostracon IFAO 1206[31]

front

1) its length 18 cubits, width 5, height 8: 180 cubic cubits[32]
2) its length 20 cubits, width 6, height 6: 720 cubic cubits
3) its length 15 cubits, width 15, height 6: 1350 cubic cubits
4) the room which is on the side, its length 15 cubits, width 5, height 5 [: 350 cubic cubits]
5) Another:
6) its length 16 cubits, width 4, 3 palms, height 5: 350[33]
7) its length 20 cubits, width 5, height 5: 500
8) its length 15 cubits, width 5, 4 palms, height 6: 504[34]
9) makes as cubic cubits 4204
10) makes 5 months, each of 720 cubic cubits
11) (and) 1 month of 604 cubic cubits, remainder 116.
12) Three months of cubic cubits
13) each of 720, 2160

back

1) 6964 cubic-cubits[35]
2) makes 9 months, that which is lacking (to use up the 9 months) is 116.
3) When the vizier put (it) into my hands:
4) 4116 cubic cubits and 2160,[36]
5) makes 4 months lacking
6) in the year 29: 8 months fulfilled.
7) Remainder 604 cubic cubits for year 30,
8) its remainder is 8400.
9) Year 31: 604 (each) in 7 months,
10) summer till inundation,
11) its remainder 3596.
12) Total of its remainders from year 30 till 31, 3rd month of inundation
13) 11996

---

[31]The Ostracon was published in [Wimmer 2000].

[32]Error in calculation of cubic cubits. The correct result is 720 cubic cubits.

[33]The result in cubic cubits is only an approximation. The exact result is 354.3 cubic cubits.

[34]The correct result is 501.4 cubic cubits.

[35]Error in calculation. The correct result is 6364.

[36]Error in copying from the other side. Correct value: 4204 (instead of 4116).

sides

1) day 21 lacking, day 22 sealing
2) 3 decades of 240 cubic cubits: 720 per month

Commentary

Front and back of this ostracon bear 13 lines of text each. In addition there are two lines of text noted on the sides. The first four lines of text seem to belong to the construction of one tomb, then, after the heading "another" in line 5 another three lines of dimensions follow. All but one (line 4) of these lines also give the respective volumes after listing the three dimensions. Line 9 represents presumably the sum of all previous volumes; however the result given (if the 350 cubic cubits from line 4 are included) differs by 250 cubic cubits. The total of cubic cubits is then divided by the monthly amount of cubic cubits which normally would be quarried (720) which results (line 10) in 5 months and (line 11) a sixth month of 604 cubic cubits. The remainder is the difference of 604 from a regular month of 720. The next two lines calculate the possible amount of quarrying for three months.

The back of the ostracon begins with the addition of the two previous volumes (4204 and 2160) resulting in 6364, which the scribe again divides by the monthly volume of 720, with the result of 9 months (and a surplus of 116). The remaining text (lines 3–13) determine the distribution of this (and at least one other volume to be quarried) over the period of 3 regnal years (year 29, 30, 31, presumably of king Ramses III). The work of the 6364 cubic cubits is given to the scribe of this ostracon after 4 months of year 29 had already passed. The remaining 8 months of that year are fully worked (with 720 cubic cubits each), which leaves a remainder of 604 from cubic cubits for year 30. Obviously there are another 8400 cubic cubits after that to remove (line 8), and some work seems to be planned for year 31 (line 9) However the information given is not detailed enough to reconstruct exactly what was going on in the years 30 and 31. The final line gives the total of remainders from year 30 till 31, 3rd month of inundation as 11996 (which is the sum of 8400 and 3596).

The inscription on the sides states that 720 cubic cubits were divided into 3 decades (= 3 weeks = 1 month) each of 240 cubic cubits.

IV. Mathematics in the Graeco-Roman Period

IV.a. Context

Roughly 1500 years later than their hieratic predecessors, the demotic mathematical texts provide the second corpus of mathematical documents from Ancient Egypt. They were written after Egypt had been under Persian rule twice (525–404 and 343–332 BCE), which already suggests the possibility of contacts between Egyptian and Mesopotamian mathematics. Furthermore, at least since the foundation of Naukratis (630/620 BCE), Greeks had permanently been in Egypt. Then, in 332 BCE, with the conquest of Egypt by Alexander, Egypt became part of the Hellenistic world, and in 30 BCE Egypt became a province of the Roman Empire. This is the historical and political background to the demotic sources.

"Demotic" designates the script as well as the grammar of the Egyptian language at that time. The demotic script developed from the hieratic script. It is more cursive than its predecessor and also contains more ligatures. It is therefore no longer possible to transcribe demotic texts (easily) into their hieroglyphic counterparts. The grammar is the intermediate stage between Late Egyptian and Coptic.

Compared to earlier mathematical texts, the demotic sources show some features which are a continuation of Egyptian tradition, but also several changes, some of which may be due to a Mesopotamian influence. As in the earlier hieratic corpus, we can distinguish between table texts and problem texts. The problem texts can still be characterized formally as numeric, rhetorical, and algorithmic. However, the verb-form *sḏm.ḫr.f*, characteristic of the hieratic mathematical texts, is no longer used (in fact, it is no longer existent in demotic at all). Furthermore the terminology for individual mathematical operations has sometimes changed. And, as will become obvious from the following examples, the problem types as well as the algorithms for their solution have been modified.

The extant demotic mathematical papyri comprise eight sources.[37] Five of them, which contain 72 short problems and tables, were published in [Parker 1972]. Two more were published by the same author in two articles, and another, unidentified example was edited by Eugene Revillout. All of these publications are more than 30 years old and, due to development in the study of demotic, are in need of reworking. However, for the time being, Parker's translations, based on the knowledge of the entire corpus, are the best available, and so are used in the following few examples. The problem numbers given are those of [Parker 1972].

IV.b. Table texts

BM 10520 (No. 54)

64
128
192
256
320
384
448
512
576
640
704
768
832
896
960
1024

Written as one column of numbers only, this "table" comprises the multiples of 64 from 64 (= 1 × 64) up to 1024 (= 16 × 64).

[37]Detailed references for the individual papyri and their publications can be found in [Fowler 1999, 258 and note 80].

BM 10794 (No. 67)

The method of taking [$\overline{150}$ to 10]

1 to $\overline{150}$
2 to $\overline{90}$ $\overline{450}$
3 to $\overline{60}$ $\overline{300}$
4 to $\overline{45}$ $\overline{225}$
5 to $\overline{30}$
6 to $\overline{30}$ [$\overline{150}$]
7 to $\overline{30}$ $\overline{90}$ [$\overline{450}$]
8 to $\overline{20}$ $\overline{300}$
9 to $\overline{30}$ [$\overline{45}$ $\overline{225}$]
10 to $\overline{15}$

One can follow the building of this table according to certain groups; for example, from the calculation of $2 \div 150 = \overline{90}\ \overline{450}$ it is easy to get to $4 \div 150$ by simply halving the respective denominators $\overline{90}$ and $\overline{450}$. This is also possible for the pairs $3 \div 150$ and $6 \div 150$ as well as $5 \div 150$ and $10 \div 150$. A second technique, which can be observed from $6 \div 90$ onward, uses the addition of previous solutions, $7 \div 150 = 5 \div 150 + 2 \div 150 = \overline{30} + \overline{90}\ \overline{450} = \overline{30}\,\overline{90}\ \overline{450}$. This technique may have been used for $6 \div 150$, $7 \div 150$, $8 \div 150$, $9 \div 150$, and $10 \div 150$ (in the cases of $8 \div 150$ and $10 \div 150$ with a further step, combining two of the unit fractions). Note that $6 \div 150$ and $10 \div 150$ could have been found by both techniques alike.

IV.c. Problem texts

pCairo JE 89127–30, 89137–43 (No. 7)[38]

The things you (should) know about the articles of cloth. Viz.
If it is said to you: "Have sailcloth made for the ships,"
and it is said to you: "Give 1000 cloth-cubits to one sail;
have the height of the sail be (in the ratio) 1 to 1 $\overline{2}$ the width,"
(here is) the way of doing it. Viz.
Find its half, when it happens that the ratio is 1 to 1 $\overline{2}$: result 1500.
Cause that it reduce to its square root: result 38 $\overline{\overline{3}}$ $\overline{20}$.
You shall say: "The height of the sail is 3[8] $\overline{\overline{3}}$ $\overline{20}$ cubits."
You shall take to it $\overline{\overline{3}}$—since it happens that it is [to 1 $\overline{2}$] that 1 makes [a ratio]:
result 25 $\overline{\overline{3}}$ $\overline{10}$ $\overline{90}$.
It is the width.

The problem calculates length and width of a rectangular sail from its given area and ratio of length and width. As we have done in the earlier examples, it is possible here, too, to rewrite the algorithm in a more symbolic form.

---

[38]Translation of [Parker 1972, 19].

	1000		D_1
	$1\bar{2}$		D_2
1	$1\bar{2} \times 1000 = 1500$	**1**	$D_2 \times D_1$
2	$\sqrt{1500} = 38\ \bar{\bar{3}}\ \overline{20}$	**2**	$\sqrt{\mathbf{1}}$
3	$\bar{\bar{3}} \times 38\ \bar{\bar{3}}\ \overline{20} = 25\ \bar{\bar{3}}\ \overline{10}\ \overline{90}$	**3**	$\overline{D_2} \times \mathbf{2}$

The solution determines the height from the given area (D_1) and ratio (D_2) in steps (1) and (2); then the width is calculated using the height and the given ratio (3).

To understand these instructions, a look at a modern solution may help. It must be stressed, however, that this *is not a translation of the source into modern terminology*, nor is it an explanation how the Egyptian mathematician arrived at this particular algorithm. It is merely meant as a help for the modern reader to understand what is going on in the algorithm.

Modern solution If we designate the height as x and the width as y, the information given in the data can be expressed as follows:

(I) $x \times y = 1000$

(II) $x \div y = \frac{3}{2}$ or (II*) $y = x \times \frac{2}{3}$

(II*) in (I) : $x^2 \times \frac{2}{3} = 1000$ and thus $x^2 = \frac{3}{2} \times 1000$

(remember that $\frac{3}{2} \times 1000$ is what is calculated as the first step of the algorithm above). From this we obtain $x = \sqrt{\frac{3}{2} \times 1000}$ (which is calculated in the second step of the algorithm); and finally $y = \frac{2}{3} \times x$ (the last step of the algorithm).

The following problem is well known from Mesopotamian sources. We are given the length of an erect pole (leaning against a wall). The pole's foot is then moved outward a given distance, and it has to be determined how far the top of the pole has been lowered. It is generally assumed that a transmission of mathematical and astronomical knowledge from Mesopotamia to Egypt occurred, possibly in the times of Persian rule of Egypt.[39] However, a detailed study of this assumed transmission has yet to be done.

pCairo JE 89127–30, 89137–43 (No. 26)[40]

A pole which is 10 cubits [when erect to (the) top].
[If the number] of its foot (moved) outward is 8 cubits,
wh[at is the lowering of its top from it?]
You shall reckon 10, 10 times: result 100.
You shall reckon 8, 8 times: result 6[4].
Take it [from 1]00: [result] 36.
Cause that it reduce to its square root: result 6.
Ta[ke it from 10: remainder 4].
You shall say:
"Four cubits is [the] number [of the lowering of its top from it]"

[39] See [Høyrup 2002, 405–6] and [Parker 1972, 5–6].
[40] Translation of [Parker 1972, 35–37].

As in the previous problem we can easily follow the algorithm, which we can also compare to the modern solution:

	10		D_1
	8		D_2
1	$10 \times 10 = 100$	**1**	$D_1 \times D_1$
2	$8 \times 8 = 64$	**2**	$D_2 \times D_2$
3	$100 - 64 = 36$	**3**	**1 – 2**
4	$\sqrt{36} = 6$	**4**	$\sqrt{\mathbf{3}}$
5	$10 - 6 = 4$	**5**	D_1 **– 4**

Modern solution The pole (of length x), the height which it measures against the wall once the foot is moved out (z), and the distance which the foot was moved out (y) constitute a right triangle.

Therefore

$$z^2 = x^2 - y^2.$$

(The algorithm calculates x^2 and y^2 in the first two steps, and then the difference (z^2) in the third step.)

We are looking for "the lowering of its top from it," which is in our modern terminology $x - z$. (The algorithm moves on to extract the square root (= z), and finally calculates the difference $x - z$.)

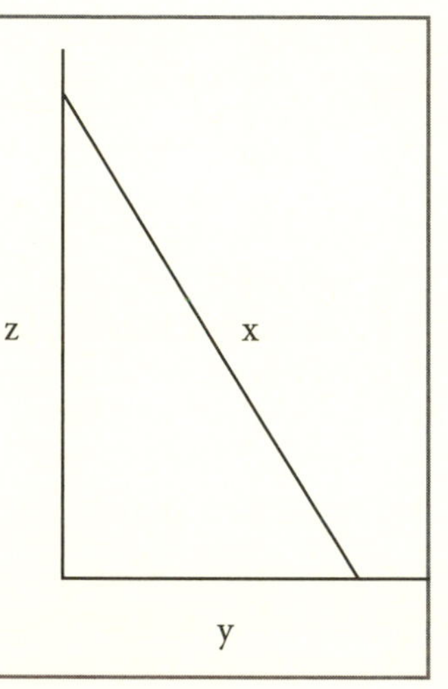

pBM 10520 (No. 64)[41]

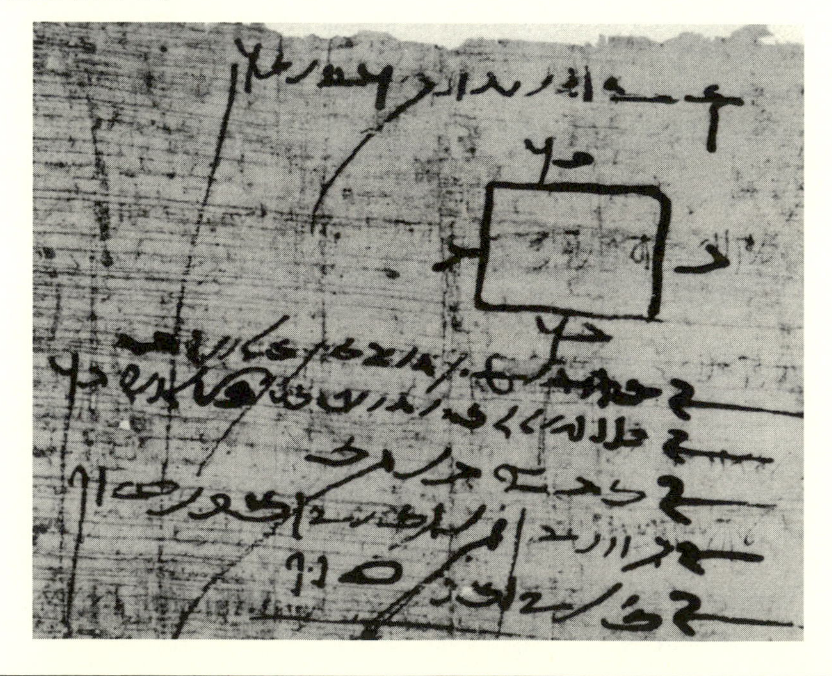

[41]Translation of [Parker 1972, 71].

A piece (of land).
Its plan.

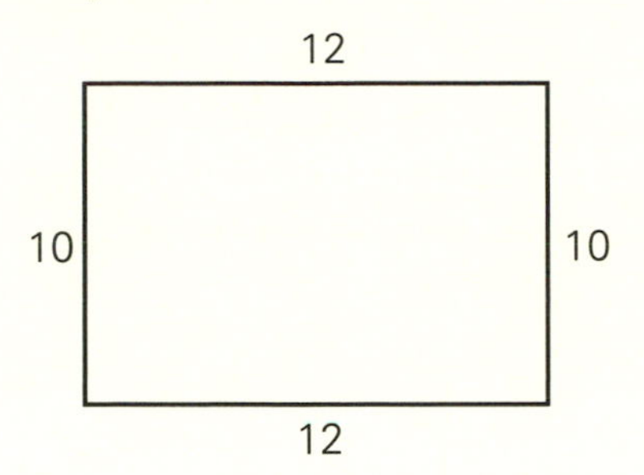

You shall add the south and the north: (result) 20. Viz.
Its half, 10.
You shall add the east and the west: (result) 24. Viz.
Its half, 12.
You shall reckon 10 <12> times: result: 120.
You shall carry 100 into 120 in order to bring another (formulation):
result $1\bar{5}$ (land) cubits.
You shall say:
"In order to bring another (formulation), $1\bar{5}$ (land) cubits."

This problem calculates the area of a rectangular field of sides 10 and 12. We already know an earlier Egyptian calculation of the area of a rectangle from the *Rhind Mathematical Papyrus* (problem 49). The algorithm of that problem multiplies the two sides (as we would expect) to obtain the area. The algorithm of our problem, however, begins with calculating the sum of the opposing sides, which is then halved. The results of this, since the geometrical shape is a rectangle, is of course again the (given) length of the sides. These are then multiplied and finally the result is calculated in a typical unit for measuring fields, the (land)-cubit, a strip of 1 cubit × 100 cubits. An advantage of this method is its applicability to approximate areas of all types of quadrilaterals, especially irregular ones, for which it is generally used.

V. Appendices

V.a. Glossary of Egyptian terms

As already mentioned in the introduction, the Egyptian script wrote only consonants. In order to be able to "speak Egyptian," Egyptologists conventionally insert short "e" sounds where necessary. The list below gives the Egyptian word in transcription, its (modern) pronunciation in brackets, and then its meaning.

ꜣḫ.t (achet) Egyptian season of inundation
ꜥḥꜥ (acha) "quantity"
ꜥḏ (adge) dimension of *nbt*, maybe its height
wꜣḥ (wach) "to add"
bšꜣ (besha) grain product
prt (peret) Egyptian season of the emergence of the crops
psw (pesu) unit to characterize grain content of bread or beer
mḫꜣw (mechau) unknown word
nb.t (nebet) object which is calculated in problem 10 of the Moscow Papyrus
rꜣ (ra) "part" (?), smallest measuring unit for volumes
hnw (henu) unit to measure volumes
ḥqꜣ.t (hekat) unit to measure volumes of grain
ḫꜣ (cha) mark used to call attention to related items
ḫꜣ-tꜣ (cha-ta) measuring unit for areas
ḫt (chet) measuring unit for lengths
ḫꜣr (char) measuring unit for volumes of grain
srft (serfet) material used in construction
sqd (seked) "inclination"
sṯꜣt (setshat) measuring unit for areas
sḏm.ḫr.f (sedgem cher ef) "he shall hear"
sḏm.ḫr.k (sedgem cher ek) "you shall hear"
šmw (shemu) Egyptian season of harvest
tp-rꜣ (tep-ra) diameter
trsst (terseset) type of bread

V.b. Sources

Introduction

pAnastasi I, 13,4–18,2
Hieratic Text: British Museum, 1842, pl. XLVII–LII

Invention of writing and number system

Naqada Tablets
Quibell, 1904, p. 216 and pl. XLIII

Arithmetic

Multiplication RMP, No. 52

Hieratic Text: Robins/Shute, 1987, pl. 16
Multiplication RMP, No. 69
Hieratic Text: Robins/Shute, 1987, pl. 19
Division RMP, No. 76
Hieratic Text: Robins/Shute, 1987, pl. 20
Multiplication RMP, No. 6
Hieratic Text: Robins/Shute, 1987, pl. 9
Division RMP, No. 58
Hieratic Text: Robins/Shute, 1987, pl. 17

Hieratic mathematical texts

Table texts

UC32159 (2:n table)
Copyright: Petrie Museum of Egyptian Archaeology, University College London
Reprinted with permission
RMP, 2:N table
Hieratic Text: Robins/Shute, 1987, pl. 1–8
Mathematical Leather Roll
Hieratic Text: Glanville, 1927
RMP, No. 81
Hieratic Text: Robins/Shute, 1987, pl. 20

Problem texts

RMP (BM EA 10057 and 10058), No. 26 and 27
Copyright: The British Museum
Reprinted with permission
MMP, No. 25
Hieratic Text: Struve, 1930, Col. XLV
RMP, No. 50
Hieratic Text: Robins/Shute, 1987, pl. 16
RMP (BM EA 10057 and 10058), No. 48
Copyright: The British Museum
Reprinted with permission
MMP, No. 10
Hieratic Text: Struve, 1930, Col. XVIII–XX
MMP, No. 14
Hieratic Text: Struve, 1930, Col. XXVII–XXIX
RMP, No. 56
Hieratic Text: Robins/Shute, 1987, pl. 17
RMP, No. 41
Hieratic Text: Robins/Shute, 1987, pl. 14
UC 32160
Hieratic Text: Griffith, 1898, pl. VIII or
Collier/Quirke 2002, CD files UC 32160-f
RMP, No. 65

Hieratic Text: Robins/Shute, 1987, pl. 19–20
MMP, No. 15
Hieratic Text: Struve, 1930, Col. XXX
MMP, No. 23
Hieratic Text: Struve, 1930, Col. XLII

Mathematics in administrative texts

Middle Kingdom texts: the Reisner papyri

Reisner I, Section I
Hieratic Text: Simpson, 1963, pl. 15
Reisner I, Section O
Hieratic Text: Simpson, 1963, pl. 21

New Kingdom texts: Ostraca from Deir el Medina

Ostracon IFAO 1206
Hieratic Text: Wimmer, 2000, pl. XLVI and XLVIII

Mathematics in the late period

Table texts

BM 10520, No. 54
Demotic Text: Parker, 1972, pl. 19
BM 10794, No. 67
Demotic Text: Parker, 1972, pl. 24

Problem texts

Cairo JE 89127–30, 89137–43, No. 7
Demotic Text: Parker, 1972, pl. 2
Cairo JE 89127–30, 89137–43, No. 26
Demotic Text: Parker, 1972, pl. 9

V.c. References

Arnold, Dieter. 1991. *Building in Egypt.* New York, Oxford: Oxford University Press.

Bierbrier, Morris L. 1982. *The Tomb-builders of the Pharaohs.* London: British Museum Publications.

British Museum. 1842. *Select papyri in the hieratic character from the collections of the British Museum.* London: British Museum.

Collier, Mark, and Stephen Quirke. 2002. *The UCL Lahun papyri: letters.* Oxford: Archaeopress.

Faltings, Dina. 1998. *Die Keramik der Lebensmittelproduktion im Alten Reich: Ikonographie und Archäologie eines Gebrauchsartikels.* Studien zur Archäologie und Geschichte Altägyptens 14. Heidelberg: Heidelberger Orientverlag.

Fischer-Elfert, Hans Werner. 1983. *Die satirische Streitschrift des Papyrus Anastasi I.* 2nd ed. Kleine Ägyptische Texte. Wiesbaden: Otto Harrassowitz.

Fischer-Elfert, Hans Werner. 1986. *Die satirische Streitschrift des Papyrus Anastasi: Übersetzung und Kommentar.* Ägyptologische Abhandlungen 44. Wiesbaden: Otto Harrassowitz.

Fowler, David H. 1999. *The mathematics of Plato's Academy: a new reconstruction.* Oxford: Oxford University Press.

Gardiner, Alan H. 1911. *Egyptian hieratic texts. series I: literary texts of the New Kingdom.* Leipzig: J.C. Hinrichs.

Gillings, Richard J. 1964. "The volume of a truncated pyramid in Ancient Egypt." *Mathematics Teacher* 57: 552–55.

Glanville, Stephen R. K. 1927. "The mathematical leather roll in the British Museum." *Journal of Egyptian Archaeology* 13: 232–239.

Griffith, Francis L. 1898. *The Petrie papyri: Hieratic papyri from Kahun and Gurob.* London: Quaritch.

Gunn, Battiscomb, and Thomas E. Peet. 1929. "Four geometrical problems from the Moscow Mathematical Papyrus." *Journal of Egyptian Archaeology* 15: 167–85.

Hoffmann, Friedhelm. 1996. "Die Aufgabe 10 des Moskauer mathematischen Papyrus." *Zeitschrift für Ägyptische Sprache und Altertumskunde* 123: 19–26.

Høyrup, Jens. 2002. *Lengths, widths, surfaces: a portrait of Old Babylonian algebra and its kin.* Berlin: Springer.

McDowell, Andrea G. (1999). *Village life in ancient Egypt: Laundry lists and love songs.* Oxford: Oxford University Press.

Quibell, James E. 1904. *Archaic objects.* Catalogue général des antiquités égyptiennes du Musée du Caire. nos. 11001–12000 et 14001–14754. Cairo: Institut Français d'Archéologie Orientale.

Parker, Richard A. 1972. *Demotic mathematical papyri.* Providence: Brown University Press.

Peet, Thomas E. 1923. *The Rhind Mathematical Papyrus, British Museum 10057 and 10058.* London: Hodder & Stoughton.

Reisner, George A. 1904. "Work of the expedition of the University of California at Naga-Ed-Der." *Annales du Service des Antiquités de l'Égypte* 5(1904).

Ritter, James. 2002. "Closing the Eye of Horus." In John M. Steele and Annette Imhausen (eds.), *Under One Sky: Astronomy and Mathematics in the Ancient Near East.* Alter Orient und Altes Testament 297. Münster: Ugarit Verlag.

Robins, Gay, and Charles Shute. 1987. *The Rhind Mathematical Papyrus: An Ancient Egyptian Text.* London: British Museum Publications.

Rossi, Corinna. 2004. *Architecture and Mathematics in Ancient Egypt.* Cambridge: Cambridge University Press.

Simpson, William Kelly. 1963. *Papyrus Reisner I: The Records of a Building Project in the Reign of Sesostris I.* Boston: Museum of Fine Arts.

Shaw, Ian (ed.) 2000. *The Oxford History of Ancient Egypt.* Oxford: Oxford University Press.

Struve, Wasili W. 1930. *Mathematischer Papyrus des staatlichen Museums der schönen Künste in Moskau.* Otto Neugebauer, Julius Stenzel, Otto Toeplitz (eds.), Quellen und Studien zur Geschichte der Mathematik, Astronomie und Physik, Abteilung A: Quellen 1. Berlin: Julius Springer.

Thomas, W. R. 1931. "Moscow Mathematical Papyrus, No. 14." *Journal of Egyptian Archaeology* 17: 50–52.

Vetter, Quido. 1933. "Problem 14 of the Moscow Mathematical Papyrus." *Journal of Egyptian Archaeology* 19: 16–18.

Wimmer, Stefan. 2000. "Welches Jahr 29?". In R. J. Demarée and A. Egberts (eds.), *Deir el-Medina in the Third Millennium AD* (Egyptologische Uitgaven 14). Leiden: Nederlands Instituut Voor Het Nabije Oosten.

2 Mesopotamian Mathematics

Eleanor Robson

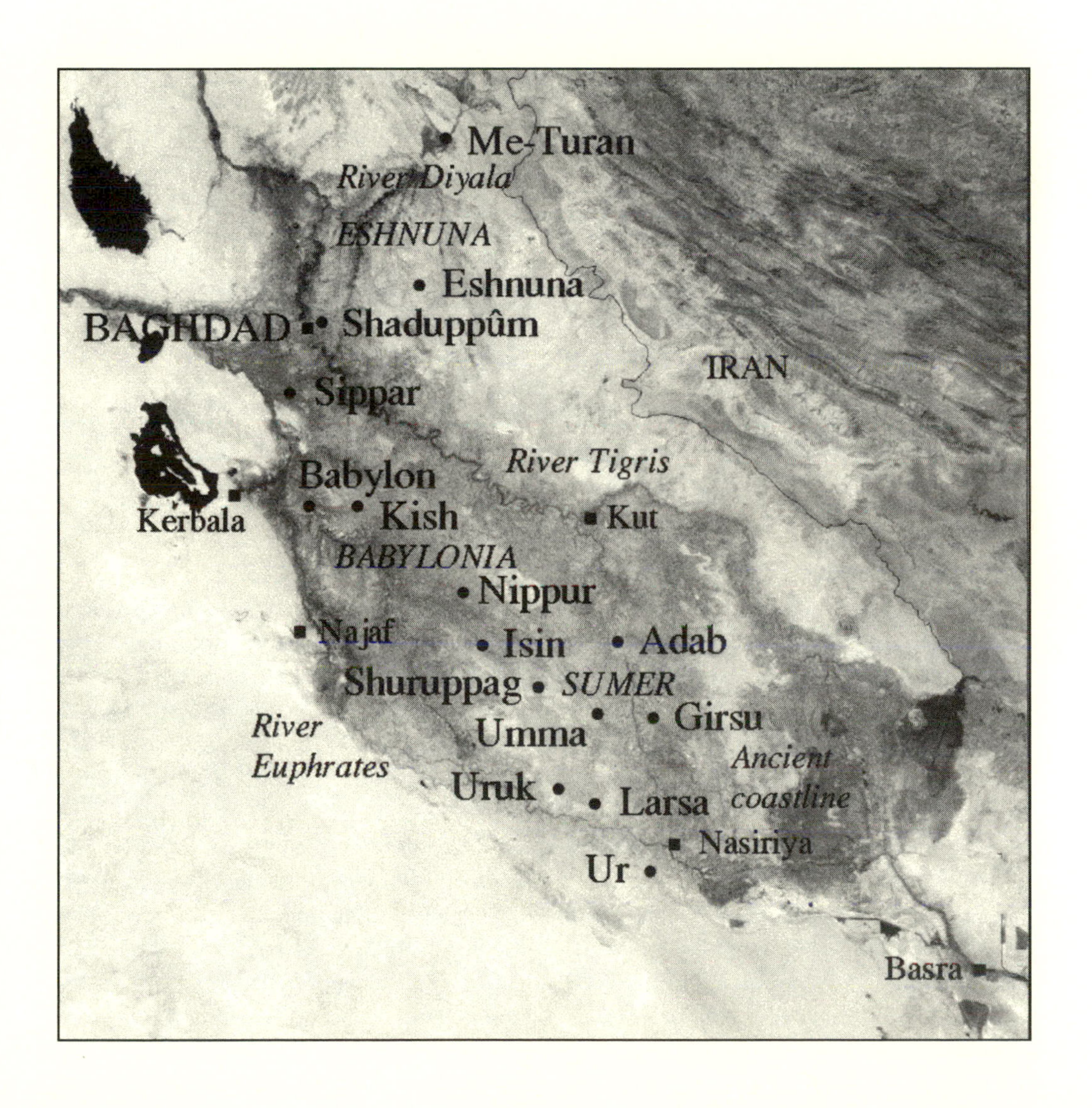

I. Introduction

It is tempting to think that, because it all happened such a long time ago, there is little new to say about mathematical developments in ancient Mesopotamia (southern Iraq and neighbouring areas). The standard histories of mathematics all tell much the same story: of the sexagesimal place value system, approximations to $\sqrt{2}$, Pythagorean triples, and a little algebra. Those achievements, impressive as they are, are but a tiny fraction of what could and should be told. Many thousands of Mesopotamian mathematical documents survive, written on clay tablets in the cuneiform script; hundreds have been edited and published since the beginning of the twentieth century. New translations of over sixty of them are presented here, spanning some three thousand years and thousands of square kilometers. Some are just a few lines of calculation or a roughly sketched diagram; others are long compilations of mathematical problems or highly sophisticated arithmetical tables.

In fact, as we shall see in this introduction, modern scholarly understanding of Mesopotamian mathematics is changing and improving at a rate unparalleled since the 1930s. Not only are dozens of new sources published every year, causing constant reevaluation of the historical corpus, but recent developments in the scripts, languages, history and archaeology of the region have stimulated exciting new lines of research that contextualise ancient mathematical practice in a way that was unimaginable even ten years ago. This introduction briefly sketches answers to three questions, showing why and how this section of the Sourcebook has been compiled:

- How have Western views of Mesopotamian mathematics changed over the last two millennia?
- Who wrote mathematics in ancient Mesopotamia, why, and how?
- How have the translations presented here been chosen and produced?

At the end of the introduction there are also practical notes on sources, metrology, and bibliography.

I.a. Mesopotamian mathematics through Western eyes

Myths and rumors

The mathematical achievements and interests of ancient Iraq have undergone periodic reassessments over the last two thousand years, ever since writers in the Greek tradition began to make comments and observations. But the most thorough overhauls have come since the rediscovery of cuneiform culture in the second half of the nineteenth century and again in the last decades of the twentieth.

For the most part Greek writers took Egypt to be the birthplace of mathematics, but they credited the Babylonian, or Chaldaean, priests of southern Iraq with astrology and the ability to make predictions from the stars. Strabo, for instance, names three Babylonian astronomers, two of whom have since been identified in cuneiform sources too. The transmission to Classical science of Babylonian observational data, values of periodicities, and even the sexagesimal place value system for fractional values in astronomical calculations, all confirm that there were direct contacts during the Persian and Hellenistic periods (c. 550–150 BCE). There are no early Greek traditions about Babylonian mathematics, however. It was only around 300 CE that Iamblichus claimed that Pythagoras had spent time in Babylon—nearly a millennium earlier, in the sixth century BCE. Without corroboration from earlier or more reliable sources, however, there is little reason to believe this report: rather, it is part of the fabrication of Pythagorean tradition in Late Antiquity.[1]

Thereafter, various confused traditions circulated about the ancient Chaldaeans. Isidore of Seville, Bede, Bacon, Recorde, and Wallis all associated them vaguely with mathematics and astronomy. The entire tradition was grounded solely on the repetition and elaboration of a common core of myths and stories: there was not a trace of Babylonian-Chaldaean mathematical writing to base it on until the dramatic finds of the mid-nineteenth century.

In our own image

In the 1840s rival British and French teams began to uncover and document the remains of vast stone palaces near Mosul, now in northern Iraq but then part of the Ottoman Empire whose capital was Istanbul. The adventurers quickly identified the ruins they were digging as the ancient Assyrian city of Nineveh, known to them through the stories of the Old Testament and Classical authors. They thus claimed it as part of their own, European heritage, and were little interested in its place in Middle Eastern history and tradition per se. Thus unwittingly the tone was set for interpreting ancient Assyrian—and, later, Babylonian and Sumerian—remains. On the one hand the finds represented the "cradle of civilization," mankind's first tottering steps toward European sophistication. At the same time they embodied the birth of the exotic, decadent Orient—a view corroborated by partisan Classical historians like Herodotus, who had written pejoratively of Mesopotamia in the fifth century BCE at the height of the Greek wars against Persia and its subject peoples. Third, they were potential witnesses to events described in the Old Testament, appearing at a crucial juncture in Western European intellectual history when geology and evolution were beginning to enter the public consciousness to challenge the literal veracity of the Bible.

The decipherment of cuneiform throughout the 1840s and '50s was an international effort, spearheaded by Classical scholars and clergymen. Babylonian and Sumerian remains were discovered en masse in the region between Baghdad and Basra from the 1870s onward, ancient sites to be plundered for inscribed and monumental objects for museum display and philological study. The Sumerian language and culture created a particular philosophical problem for cuneiform scholars: here was a major civilization, clearly older than Assyria or Babylonia yet not attested anywhere in the Bible. Gradually the study of ancient Mesopotamia was forced to break free of its Classico-Biblical origins, in Europe if not in America. The advent of stratigraphic, recorded archaeology in the first years of the twentieth century brought with it the potential for integrating the history witnessed on the tablets with the material circumstances in which they were written—a potential that even now is not always fully realized.[2]

Rational reconstruction

Mathematical cuneiform tablets were first published in the late nineteenth century, but they made little sense. In 1906 Hermann Hilprecht, director of the University of Pennsylvania's long-running excavations at Nippur, published a book in which he edited and discussed multiplication and reciprocal tables, as well as metrological lists and tables, and reviewed the totality of sources published so far. Embroiled in a scandal over the veracity of his claims of

---

[1]The relationship between Babylonian and classical astronomy has been much studied in recent years; see for instance [Jones 1996]. [Burkert 1972] is the classic account of the late antique construction of Pythagoreanism.

[2]See [Larsen 1996] for a detailed history of decipherment.

a "temple library," Hilprecht and his work were not well received in the Assyriological community. However, the historian of mathematics David Eugene Smith reviewed the book enthusiastically and incorporated the less fantastical parts into his influential *History of Mathematics*.[3]

Meanwhile, Vincent Scheil and François Thureau-Dangin were making significant corrections and additions to Hilprecht's work on mathematical tables. Ernst Weidner published the first edition of an Old Babylonian mathematical problem in 1916. But the study of cuneiform mathematics changed forever when famously the young Otto Neugebauer began his program of decipherment and publication in the late 1920s. By 1945 Neugebauer had systematically trawled all the great cuneiform collections of Western Europe and North America for mathematical tablets. So monumental were his publications, so thorough his mathematical analysis, that it was genuinely thought for many decades that there was little more to be said. A series of public lectures in the late 1940s, published as *The Exact Sciences in Antiquity*, has remained the public face of cuneiform mathematics ever since.[4]

New sources continued to appear, particularly from Iraqi excavations in the area around Baghdad, and from French digs in southwest Iran. But Neugebauer's interpretative paradigm remained paramount: analysis in mathematical terms only, highlighting features such as the use of the Pythagorean theorem that could be taken as an index of Babylonian progression toward modernity. Questions of authorship, context, and function were systematically overlooked; textuality and materiality played no part in the academic discourse of the mid-twentieth century. The focus remained on the mathematics of the early second millennium BCE, the so-called Old Babylonian period. When put into international context, it was presented either as "pre-Greek"—post-Euclidean Late Babylonian mathematics being late, corrupt, and irrelevant in this view—or even (in Van Der Waerden's eyes at least) an oriental degeneration of a mythical proto-European mathematics that had somehow spread eastward across Asia in the Neolithic.[5]

In its own right

Attitudes changed slowly from the 1970s onward. Marvin Powell recognized the importance of Sumerian mathematics, problematizing the evolution (not invention) of the sexagesimal place value system in the third millennium BCE. Jens Høyrup reevaluated Old Babylonian "algebra" by analysing its language rather than forcing it to fit modern algebraic models. Jöran Friberg was the first to study seriously the mathematics of the sixth century BCE and later. My own contribution, perhaps an inevitable outcome of a training in Middle Eastern studies rather than history of science, has been to reach out into archaeological context and social history.[6]

At the same time, David Fowler has led a radical reinterpretation of the origins of Greek mathematics. He has thrown out the old assumption of Babylonian influence on the early development of the Classical tradition, according to which Babylonian methods were abandoned in the supposed "incommensurability crisis"—leaving no trace in the historical record—in favor of a move toward the abstractions of Euclideanism. In this light Babylonian mathematics can no longer be portrayed as the infancy of the "Western tradition." Instead, the

---

[3][Hilprecht 1906; Smith 1907, 1923–25]. See [Robson 2002a] for more on the relationship between Hilprecht, Smith, and Plimpton.

[4]See e.g., [Scheil 1915; Thureau-Dangin 1934, 1936, 1938; Weidner 1916; Neugebauer 1935–37; Neugebuaer and Sachs 1945; Neugebauer 1951].

[5]See e.g., [Baqir 1950, 1962; Bruins and Rutten 1961; Van Der Waerden 1954, 1983].

[6]See e.g., [Powell 1976; Høyrup 2001, 2002; Friberg 1993, 1999; Friberg et al. 1990].

TABLE 2.1: The Rediscovery of Mesopotamian Mathematics since 1800

Date	*Scholarly developments*[7]	*Political events*[8]
pre-1800	Traditional Biblical narratives, Greek stories, and travellers' tales of ancient Mesopotamia	1638–1918: Iraq under Ottoman rule
1820	1819: Tiny display of undeciphered cuneiform tablets at the British Museum	
1840	1842: Anglo-French archaeological rediscovery of ancient Assyria (northern Iraq)	
1860	1857: Akkadian cuneiform officially deciphered at the Royal Asiatic Society, London	
1880	1880s–: Mass recovery of cuneiform tablets in Babylonia (southern Iraq)	1881: Museum of the Ancient Orient founded in Istanbul, the Ottoman capital
	1900: First drawings of Old Babylonian mathematical tablets published	
1900	1903–: Progress in understanding sexagesimal numeration and tables	
	1910s–: Widespread adoption of scientific excavation methods	1914–18: First World War stops excavation
1920	1916: First decipherment of an Old Babylonian mathematical problem	1920: Formation of modern Iraqi state
	1922: First publication of Late Babylonian mathematics	1932: Iraqi independence from British Mandate
1940	1930–45: "Golden age" of translation; major editions by Neugebauer, Sachs, and Thureau-Dangin	1940s: Iraqi excavations begin
	1956: First volume of the *Chicago Assyrian Dictionary* published	
1960	1961–: Publication of mathematics from Susa (southwest Iran) and western peripheries	1968: Ba'athist coup in Iraq
	1976–: Increased study of third-millennium mathematics in Sumerian	1970s–80s: Large-scale damming projects on the Euphrates, Tigris and Diyala rivers drive rescue excavations in archaeologically unexplored areas of Iraq
1980	1984: First volume of the *Pennsylvania Sumerian Dictionary* published	
	1990–: Høyrup develops discourse analysis of Old Babylonian mathematics	1990–: Gulf War ends excavations; UN sanctions result in widespread archaeological looting
2000	1997–: Study of archaeological and social context of Mesopotamian mathematics	
		2003: Iraq War results in end of Ba'athist regime; pillage of museums and archaeological sites

[7]For more details of the mathematical historiography see [Høyrup 1996]; for archaeology the best account is still [Lloyd 1980].

[8]For the political background see [Tripp 2002].

picture is much more complicated, inviting us to study it on its own terms, not simply as a precursor to something else. Farther afield, there has been widespread critical examination of academia's imperialist legacy. Influenced by Edward Said's seminal *Orientalism*, postcolonialist historians have highlighted and challenged European and American scholarship's appropriation of the Middle East's past for the West. Similarly, an influential anthropological critique of the categories "us/them" and "sameness/otherness" has drawn attention to the familiarizing strategies used by nineteenth- and twentieth-century scholars to domesticate mainstream historical interpretations of ancient Mesopotamian culture toward the West and away from the Middle East.[9]

It is hoped that the wide range of mathematical sources translated in this chapter reflects these changing attitudes: they have been chosen not as extraordinary or surprising examples of modernity with which you are invited to identify. Rather they represent the typical products of scribal culture, reflecting a wide variety of textual genres from the simplistic to the sophisticated, from rote-learned tables and rough calculations to carefully constructed word problems. But inevitably, as research methodologies develop and multiply, and ever more sources are discovered and deciphered, the selections and interpretations given here will come to seem as dated as those of a hundred years ago. Scholarly fashions come and go, but the tablets themselves endure.

I.b. Mathematics and scribal culture in ancient Iraq

Mathematics is not created out of nothing—it is written by individuals operating within the social and intellectual norms and conventions of the societies in which they dwell. Thus coming to grips with another culture's mathematics is not simply a matter of translating one notation into another. Instead we need to explore the personal, intellectual, and social circumstances under which it was written. Paradoxically, of all ancient and "other" mathematics, Mesopotamia provides us with the most potential for contextualising interpretation. First, relatively imperishable clay tablets leave us with a written legacy of primary sources many times greater than other ancient societies such as Egypt (or compare the Greek tradition, where original documents are almost nonexistent). Second, Iraq has some incredibly well preserved and carefully recorded urban archaeology, including many sites where mathematical tablets have been found. To study its mathematics, then, only as mathematics and not as the product of a person's body, brain, and culture, we willfully ignore a historical source of unparalleled richness that has the potential to help us understand the interconnections between mathematics and other aspects of culture and society that no other ancient civilization can match.

Rather than outlining here such a history of Mesopotamian mathematics (Table 2), we shall look instead at three different archaeological contexts in which mathematics has been found. As we shall see, we have to abandon modern notions of universal literacy and numeracy; in this world reading, writing, and calculating were almost exclusively professional activities to which even the wealthy and powerful did not necessarily have access. Writing was used only by temples, palaces, and affluent families, primarily to record property ownership and the rights to income. It will thus be important for us to distinguish between numeracy, which uses mathematics for the accounting and administration of assets, and mathematics, as an intellectual activity for teaching, learning, and creating new mathematical skills and ideas.[10]

---

[9]See, e.g., [Fowler 1999; Said 1978; Bahrani 1998].

[10]See [Robson 2001a] for a discussion of contexualization; and [Robson forthcoming] for a social history of mathematics in ancient Iraq.

TABLE 2.2: **Overview of Mathematical Developments in, Ancient Mesopotamia**

Date	*Mathematical developments*	*Socio-political background*[11]
4000 BCE	Pre-3200: Preliterate token-based accounting	Increasing urbanization in southern Iraq
3500 BCE	3200: Literate numeracy; the first school mathematics	Uruk period/Early Bronze Age Sumerian language
3000 BCE	Sophisticated accounting and quantitative planning	Early Dynastic period: city states
2500 BCE	School mathematics; c. 2050: first attestation of the mature sexagismal place value system	Akkadian language Territorial empires of Akkad and Ur
2000 BCE	c. 1850–1650: widespread evidence of "pure" mathematics in scribal training: line geometry, concrete algebra, quantity surveying	City states; empire of Babylon: Middle Bronze Age, or Old Babylonian period
1500 BCE	Cuneiform culture and sexagesimal numeracy spread from southern Iraq	"Amarna age" of international diplomatic contact across the Middle East; Late Bronze Age
1000 BCE	800 BCE–: quantitive methods in Assyrian scholarship	Assyrian empire; Aramaic language and the alphabet; Iron Age
500 BCE	400 BCE–: mathematics in the temples of Uruk and Babylon	Persian and Seleucid empires: Late Babylonian period
0 BCE/CE	75 CE: the last known datable cuneiform tablet; transmission of mathematical knowledge and practice to other languages	Parthian empire

Inana's temple in Uruk, 3200 BCE

By the late fourth millennium, the flat marshlands of the south of modern-day Iraq were teeming not only with wildlife but also with people. The region's population centers, which were the largest the world had yet known, were sustained through a sophisticated network of socioeconomic interactions. In earlier societies most fit and healthy individuals had been economically active providers and producers as well as consumers. Now new social relations of unprecedented complexity relied also on managers, administrators, and organizers, who earned their living and prestige not through production but through the oversight and control of the community as a whole. This new social class dispensed justice, managed communal building and agricultural projects, and took the lead in religious life, and it did so through the institution of the temple.

The temple was at the literal and metaphorical heart of every large Sumerian settlement. Made of mud brick and whitewashed in brilliant limestone plaster, it dominated the flat marshland landscape as a conspicuous emblem of the city's wealth, prestige, and functionality. At one level the home of the city's patron god or goddess, it was also the economic powerhouse of the city and its hinterland. Through offerings and large-scale use of labor, whether forced or voluntary, temples came to own vast tracts of arable land, where barley was grown and huge herds of sheep and goats were tended. All of these assets had to be managed in order to provide for the god and his or her followers; this is the first known context in which

[11]For a general overview of Mesopotamian history, see [Kuhrt 1995; Roaf 1990; or Van De Mieroop 2004].

writing was used. Managers at the goddess Inana's temple E-ana in Uruk, some time in the thirty-third century BCE, adapted a long-used system of accounting with clay tokens in order to record, manage, and predict the wealth of their employer. Onto flattened clay surfaces they impressed stylized outlines of accounting tokens to represent numbers, and scratched pictograms and other symbols to stand for the objects they were counting and accounting for. The accountants used a dozen or so different numeration and metrological systems, depending on the type of commodity they were dealing with. As trainees they thus had to learn how to deal with various number bases and conversions between different measurement systems as well as how to write some 1200 symbols representing all the different categories and subtypes of objects, people, animals, land, and other assets they managed. The world's earliest known piece of school mathematics is an exercise in calculating the areas of two fields, yielding conspicuously round answers (see pp. 73–74). We do not know exactly who wrote it, or exactly when, but it was found along with about 5,000 other tablets, mostly temple accounts but including several hundred other school exercises, mostly vocabulary lists. The tablets had been thrown away when they were no longer needed, along with other rubbish discarded by the administrators, and then reused as building rubble when Inana's temple was rebuilt and refurbished some time before 3000 BCE.[12]

A scribal school in Nippur, 1740 BCE

A few hundred miles north of Uruk, and about a millennium and a half later, the city of Nippur was an important religious center of the kingdom of Babylonia, which covered most of south and central Iraq. Home of the great god Enlil, it was where kings traditionally came to receive Enlil's blessings and permission to rule. Some 100 meters south of his great temple complex E-kur was an unassuming small house in a block of other small houses, which had been built out of mud brick in the late nineteenth century BCE. Now some eighty years later it was occupied by a priest who ran a small school in his tiny front courtyard, where he had built a bench and a bitumen-lined bin in which to soak and recycle the tablets that his students wrote.

Here he taught one or two students, perhaps his own sons. For their elementary education he followed a system used by other teachers in the city: first the basics of impressing wedge-shaped marks in the clay with a reed stylus, then the careful copying, repetition, and memorization of a standard sequence of visually simple cuneiform signs. Next his young charges learned how to write the Sumerian words for various objects, grouped according to the materials they were made of, just as their long-ago predecessors had in Uruk. Then came a series of more abstract exercises in understanding the complexities of Sumerian and the cuneiform script, including a set series of multiplication tables and metrological lists (cf. pp. 82–90). Only then were the students ready to write whole sentences of Sumerian, and to take their first steps in Sumerian literature, a subject in which this particular teacher had a special interest. The literary compositions were not simply stories; they educated the young men in the myths and belief system of the temple, and inculcated in them a strong self-identity as professionally literate and numerate scribes. (Translations of some of these literary works are scattered throughout this chapter.) The seemingly endless rote-learning was interrupted occasionally by the opportunity to practice mathematical calculations—but the students' results were not always correct.

[12]See [Nissen et al. 1993, 1–46] for more detail on the Uruk accounting system.

Sometime after 1739 BCE, when the house needed to be repaired, nearly 1500 school tablets were used as bricks and building material, becoming embedded in the fabric of the house for the rest of its life.[13]

A scholarly household in Uruk, 420 BCE

By the late fifth century BCE Babylonia was a province of the enormous Persian Empire, though so far the end of native rule after millennia had made little impact on the day-to-day lives of the inhabitants. On the eastern outskirts of Uruk, inside the city wall, was a large mud-brick house arranged around a central courtyard. It was occupied by a family of scholars with close associations to the sky god Anu's temple Resh in the city center, who traced their descent from an ancestor called Shangû-Ninurta, "Chief administrator of the god Ninurta." The men of the family made their living as incantation-priests and healers, using a mixture of herbal remedies, incantations, and supernatural diagnostic and prognostic techniques to care for their patients.

The family owned a library of nearly two hundred scholarly and professional works on clay tablets and waxed wooden writing boards, stored in large terracotta jars in a special room off the courtyard. The older men taught the younger male members of the family to write Sumerian and Akkadian and to calculate. The boys' early efforts were recycled in a bituminized area of the courtyard near the tablet room. By their late teens they were ready to copy the sophisticated reference materials of their trade, signing and dating their tablets and dedicating them to an older family member. Part of their profession involved a deep understanding of the complex ominous calendar of auspicious, inauspicious, and evil days so that their medication and ritual could be as effective as possible. This led to an interest in astronomy and mathematics, which involved not only copying arithmetical tables and collections of word problems from earlier originals but also checking that the tables were indeed correct (see pp. 167–170).[14] When the family vacated the house, for reasons unknown, they left the library *in situ*. The abandoned house eventually collapsed or was demolished, crushing the storage jars and their contents, and a new house built over the ruins of the old.

These three examples show, then, that mathematics was not a leisure pursuit in ancient Iraq; nor was it an exclusive professional activity supported by institutional patronage. Rather, it was a fundamental part of the process of becoming professionally literate and numerate, whether as accountant, priest, or scholar. All the mathematics presented here should be understood in such a pedagogical, and usually domestic, context.

I.c. From tablet to translation

Abandonment and discovery

After their useful life was over, most clay tablets were dunked into water and recycled without ever being baked for posterity. This is true not only of trainee scribes' school exercises, but also of archival documents belonging to temples, palaces, and families. With the development of libraries in the mid-second millennium, reference copies began to be kept of important scholarly works, both in repositories attached to temples or palaces, and in smaller family

---

[13]See [Robson 2001, 2002] for more on the school known as House F in Nippur.

[14]See [Robson forthcoming, chap. 8] for more on the Shangû-Ninurta family, and other mathematically inclined occupants of their house.

libraries. Thus archaeologists tend to find mathematical tablets only in very particular circumstances: when individual tablets were lost, or when large quantities were re-used as building materials (as in Inana's temple at Uruk or the school house in Nippur), or when the building in which they were stored was abandoned suddenly without the opportunity to clear it out (as in the scholarly family's house in Uruk). Frustratingly, then, what remains is not representative of what was written, and is rarely found exactly in the context in which it was used. Even more frustratingly, the bulk of the mathematical tablets now available for study in museum collections were not in fact carefully excavated, but simply dug out and sold, with no record of their findspots, to eager collectors and museum agents in the days before the advent of scientific archaeology, when the importance of context could hardly be guessed at.

Since the beginning of the twentieth century, however, archaeologists have recognized that archaeology is an inherently destructive process, and have thus put increasing emphasis on recording as much as possible of an excavation in the knowledge that what they are digging can never otherwise be recreated. Thus in the excavations of the three places discussed above—from the 1920s, 1950s, and 1970s, respectively—the position of each cuneiform tablet (and every other artefact) was recorded as it was found, in relation to the building that housed it and with respect to a three-dimensional coordinate grid imposed across the whole archaeological site. Although for much of the twentieth century cuneiform tablets have been published very separately from the other results of the same archaeological digs, it is nevertheless possible, with a lot of painstaking work, to reunite the tablets with their find context and thus to recreate the circumstances in which they were abandoned at the end of their useful life.[15] Tablets are often excavated in a fragile state, often in several pieces. Those fragments need to be cleaned of their surface dirt, especially from the cuneiform wedges of the writing itself. They are baked, desalinated, and joined to the other surviving pieces of the tablet. Each excavated piece is given a unique identifier by which it is referred to in the excavation records; often each tablet is assigned a further identifying number when it enters a museum collection, based either on the day it was catalogued for the museum, or a running number based on the total inventory of the collection. Tablets are stored in drawers, often in their own individual boxes, and are monitored by conservators for signs of deterioration.

Publication and interpretation

Publication of cuneiform tablets is a multistage operation. The first task is usually to make an accurate drawing of the tablet, in order to carry out as close a reading of it as possible. Detailed scrutiny enables the scholar to decide, wedge by wedge, which marks on the tablet are writing and which are simply surface damage. As she makes her preliminary sketch she also writes out a first transliteration—that is, a sign by sign alphabetic representation of the cuneiform writing. Cuneiform has almost no punctuation marks, and words are usually written together without spaces in between. There are about 600 different signs in its repertory, most of which have several different context-dependent meanings. They might be numerals, syllables, logograms representing whole words, determinatives indicating the type of word that follows, or phonetic indicators designed to help the interpretation of logograms. The sign for 4, for instance, can also be the syllable /sha/, the logogram GAR standing for the verb "to put," or the logogram NINDAN, which means either the length unit "rod" or "bread." Thus identifying the

[15]See [Matthews 2003] on the principles and theory of Mesopotamian archaeology.

signs the scribe wrote is not enough—appropriate values have to be found in order to recover a meaningful text.[16]

Translation of cuneiform tablets, even mathematical ones with their formulaic language and repetitive structure, thus requires a thorough knowledge of both Akkadian and Sumerian, especially as Sumerian lexical bases, written with one cuneiform sign, were often used as logograms standing for a multisyllable Akkadian word. The same script is used for both the Sumerian and Akkadian languages. Sumerian, one of the first languages in the world to be written down, has no surviving linguistic relatives. People probably stopped speaking Sumerian some time in the early second millennium BCE, but it continued to be written, especially for scholarly and religious purposes, right until the last few centuries BCE. Structurally, it is a little like German or Turkish, in that it constructs words through agglutination or stickiness: long chains of grammatical particles attach to a lexical base, but the words and the particles themselves, and the way that they combine, are unique to Sumerian. This makes it very difficult to understand, and it is a standing joke that there are as many radically different interpretations of Sumerian grammar as there are Sumerologists. Akkadian, on the other hand, is very well understood, as it is a Semitic language closely related to Hebrew and Arabic. Like other Semitic languages, the core meanings of words are held in three root consonants, and their grammatical function with a sentence expressed by different combinations of vowels and consonants around them. The vocabulary of Akkadian is often very close to Hebrew and Arabic, although it also borrowed and adapted words from Sumerian too. Akkadian was written from the mid-third millennium onward and died out (as a written language) at about the same time as Sumerian, though its relative Aramaic had mostly replaced it as a spoken and written language by the seventh century BCE.[17]

Publication does not stop at drawing and photographing, editing and translating. It also involves interpretation, relating the source to the rest of the contemporary mathematical corpus as well as to the other tablets with which it was found. As we have seen with our three brief examples, no scribe ever wrote nothing but mathematics, so to understand their mathematics and how they thought about it we must look at the other things they wrote and thought about too. That is beyond the scope of this book, but English translations of other genres of cuneiform tablets are now widely available.[18]

On translation

Translation is never a straightforward task. Jens Høyrup and Jöran Friberg advocate "conformal" translations, in which the English is made to conform as closely as possible to the original Akkadian semantics and syntax. While, in conjunction with a transliteration, this translation style enables the reader to get as close as possible to the scribe's view of the text, it produces almost incomprehensible stand-alone English prose. Here I have tried to strike a middle ground between "conformality" and the modernizing translation style of Neugebauer and his successors. I have kept natural English word order throughout, and while I have followed Høyrup and Friberg's ideal of employing standard translations for technical terms I have avoided neologisms wherever possible.

[16]See [Walker 1987] for a fuller account of the cuneiform writing system.

[17]See [Michalowski 2004; Huehnergard and Woods 2004] for overviews of the Sumerian and Akkadian languages.

[18]E.g., [Michalowski 1993; Foster 1996; Dalley 2000; Black et al. 2004].

Thus I have distinguished consistently between, for instance, symmetrical and asymmetrical addition: X *u* Y *kamārum* "to sum X and Y" versus X *ana* Y *wasābum* "to add X to Y." The latter is a very physical operation; its subtractive counterpart is X *ina* Y *nasāhum* "to take away X from Y." Similarly several sorts of multiplication are distinguished within the Old Babylonian mathematical corpus. Simple numerical multiplication of the sort found in tables is represented by the logogram A.RÁ "times." Geometrical multiplication, X *u* Y *šutakūlum*, in which lines are multiplied to form areas, literally "to make X and Y hold each other," I have translated as "to combine X and Y." Repetition of the same entity N times, X *ana* N *esēpum*, has become "to copy X N times." Finally, there is a more general verb, X *ana* Y *našûm*, literally "to raise X to Y," for which I have retained the phrase "to multiply X by Y."

I have also chosen to translate the names of some geometrical figures more literally, as their semantic range is not identical to ours. Thus *mithartum*, usually translated "square," may mean the area or side of a square, or even square root. I have thus chosen the term "square-side." Similarly *kippatum*, "circle," may stand for the area of a circle or its circumference; I have not distinguished them in the translations. On the other hand I have not (yet) been able to find a satisfactory single translation for *siliptum*, both "rectangle" and "diagonal," so I have translated it with one or the other term as appropriate. Then there is a range of geometrical figures which have no direct counterpart in modern mathematics: *santakkum* "wedge" is a three-sided figure, not strictly a triangle as one or more sides may be curved. A "crescent moon," *uskarum*, is a circle cut by a chord; it usually means a semicircle but may be a larger or smaller segment than that. Irregular quadrilaterals that may have one or more curved sides are called *pūt alpim*, "ox's brow," while the *apsamikkum*, a concave square figure composed of four inverted quarter-segments of the circumference of a circle, I have more loosely rendered as "cow's nose." The shape delimited by two one-third segments of a circle is an "ox's eye," *īn alpim*, while that composed of two one-quarter segments is a "barge," *makurrum*. A length running across the interior of a geometrical figure is called *tallum*; most often it means a diameter, but it can have other meanings too, so I translate it here as "dividing line." Finally, in geometrical algebra a line may be given an extra dimension, always of length 1, in order to convert it to an area: this is called a *wasītum*, or "projection."

In any event, it is important to remember that, however faithfully I have attempted to render the original sources into English, what you will be reading are *my translations*, and thus interpretations which are open to doubt and challenge. They are not the primary sources themselves, but an early twenty-first-century representation of them which is necessarily far removed from the originals. You will be reading alphabetic texts from a printed book in a familiar language, perhaps seated at a desk in a library or an office, a physical experience far removed from squatting on the ground in bright sunlight to pore over a clay tablet held in your hand. It will never be possible to fully comprehend this mathematics as it was meant to be read, for we cannot entirely escape our own twenty-first-century lives and brains and training, however hard we try (nor, perhaps, should we want to). But even if the enterprise is ultimately doomed to failure, that does not mean it is not rewarding and satisfying to try.

I.d. Explananda

Sources

The tablets translated here come from the following collections. For the most part I have based my translations on personal inspection of the tablets themselves, or of good photographs. This was not possible, for obvious reasons, in the case of the tablets from the Iraq Museum (IM), Baghdad, where published scale drawings (hand copies) have had to suffice.

A	The Oriental Institute of the University of Chicago, Chicago.
AO, AOT	The Louvre, Paris; including excavations at Telloh (AOT). Tablets studied courtesy of Béatrice André-Salvini of the Département des Antiquités Orientales.
Ash	The Ashmolean Museum, Oxford. Tablets studied courtesy of the late Roger Moorey, former Keeper of Antiquities.
BM, *UET*	The British Museum, London; including the excavations at Ur (*UET*). Tablets studied courtesy of Christopher Walker, formerly Senior Assistant Keeper of the Ancient Near East, and the Trustees of the British Museum.
CBS, UM	Collection of the Babylonian Section, University of Pennsylvania Museum of Archaeology and Anthropology, Philadelphia. Tablets studied courtesy of Steve Tinney, Curator of the Babylonian Section.
HS	The Hilprecht Collection, University of Jena.
IM, Db, Haddad, W	The Iraq Museum, Baghdad; including excavations at Tell Dhiba'i (Db); Tell Haddad, ancient Me-Turan (Haddad); and Uruk (W).
Plimpton	The George A Plimpton Collection, Columbia University. Tablets studied courtesy of Jane Siegel of the Rare Book and Manuscript Library.
Strasbourg	Cuneiform collection of the Bibliothèque Nationale et Universitaire, Strasbourg.
TSS	Istanbul Arkeoloji Müzelerinde, Istanbul; excavations at Shuruppag.
VAT	Vorderasiatisches Museum, Berlin. Tablets studied by Jeremy Black courtesy of Joachim Marzahn, curator.
YBC	Yale Babylonian Collection, New Haven. Tablets studied courtesy of Ulla Jeyes, curator; and with the generous assistance of Paul-Alain Beaulieu.

Many cuneiform tablets are also in the hands of private collectors. However, as the export of Iraqi antiquities has been banned for many decades, and the market in them illegal in all member states of the United Nations since May 2003, the study or purchase of privately owned material is tantamount to handling stolen goods, except in the extremely rare cases where a long pedigree of ownership can be proved. Furthermore, the illicit market in cuneiform tablets encourages the continued plundering of badly protected, fragile archaeological sites and the consequent destruction of vital historical data for future generations. There is expected to be an upsurge in the underground international trade in cuneiform tablets as a result of the devastating archaeological looting in the aftermath of the 2003 Iraq War. Tablets for sale in shops, at auction, or on the web should be reported to the relevant national art theft police authority or to Interpol so that they may be confiscated and repatriated.

Metrology[19]

Early metrology The accountants of late fourth-millennium Uruk used at least twelve different metrological systems, depending on what they were measuring or counting. For instance, when counting discrete objects, their notation distinguished between the living and the dead, and between fish and cheese. However, identical symbols were used in different systems with different meanings. Although these systems were reformed and simplified over the coming centuries, some of the notational ambiguity remained in Early Dynastic metrology, as did the bundling of number and unit into a single sign.

Area units: ≈ 3.9 km^2 ≈ 64.8 ha ≈ 6.48 ha ≈ 2.16 ha ≈ 3600 m^2

Length units: ≈ 3.6 km ≈ 360 m ≈ 60 m ≈ 6 m

Length measures

1 rod	= 2 reeds	≈ 6 meters
1 reed	= 3 seed-cubits	≈ 3 meters
1 seed-cubit	= 2 cubits	≈ 1 meter
1 cubit	= 2 half-cubits	≈ 50 cm
1 cubit	= 3 double-hands	
1 double-hand	= 10 fingers	≈ 17 cm
1 finger		≈ 17 mm

Area measures

1 rod × 1 rod	= 1 sar	≈ 36 m^2
100 sar	= 1 iku	≈ 3600 m^2
1 eshe	= 6 iku	≈ 2.16 ha
1 bur	= 3 eshe = 18 iku	≈ 6.5 ha
1 shekel	= 1/60 sar	≈ 0.6 m^2

Capacity measures

1 **sila**	≈ 1 liter
1 ban	= 10 sila (10 liters)
1 bariga	= 6 ban (60 liters)
1 lidga	= 4 bariga (240 liters)
1 (great) gur	= 2 lidga (480 liters)
1 granary	= 2400 (great) gur (1,152,000 liters)

Metrology from the twenty-first to the sixteenth centuries Standard units of calculation (which are often implicit within Old Babylonian problems) are shown in **bold**.

[19]Following [Nissen et al. 1993, 30–31; Powell 1990; George 1993: 119].

Length measure

1 finger	≈ 17 mm
1 cubit	= 30 fingers (0.5 m)
1 **rod**	= 12 cubits (6 m)
1 chain	= 1 00 cubits or 5 rods (30 m)
1 cable	= 1 00 rods (360 m)
1 league	= 30 00 rods or 30 cables (10.8 km)

(1 **cubit** is the standard unit of height.)

Area and volume measure

1 area **sar**	= 1 rod square (36 m^2)
1 volume **sar**	= 1 area sar × 1 cubit (18 m^3)
1 ubu	= 50 sar (1800 m^2 or 900 m^3)
1 iku	= 2 ubu = 1 40 sar (3600 m^2 or 1800 m^3)
1 eshe	= 6 iku (2.16 ha or 108,000 m^3)
1 bur	= 3 eshe (6.48 ha or 324,000 m^3)

Units from the iku upward were not written explicitly, but used special unit-specific notation. These writings are indicated in the translations by putting the units in brackets, thus: "2 (bur) 1 (eshe) 3 (iku) area." Special, absolute value signs were used to write multiples of the bur: 10, 60, 600, 3600, and occasionally the "big 3600," or 3600 × 3600.

Capacity

1 **sila**	≈ 1 liter. The sila may be divided into 60 shekels.
1 ban	= 10 sila (10 liters)
1 bariga	= 6 ban (60 liters)
1 gur	= 5 bariga (300 liters)

Ban and usually bariga units were not written explicitly, but used special unit-specific notation. These writings are indicated in the translations by putting the units in brackets, thus: "1 gur 2 (bariga) 3 (ban) 4 sila." Multiples of the gur were written with the sexagesimal place value system.

Weight

1 grain	≈ 0.05 g
1 shekel	= 3 00 grains (8.3 g)
1 **mina**	= 1 00 shekels (0.5 kg)
1 talent	= 1 00 minas (30 kg)

Multiples of the talent were written with the sexagesimal place value system.

Brick measure

1 brick **sar**	= 720 bricks
Size of a small, unbaked brick: 15 × 10 × 5 fingers	
Number of small bricks in 1 volume sar: 1 26 24 bricks	= 7;12 brick sar
Size of a square, baked brick: 20 × 20 × 5 fingers	
Number of square bricks in 1 volume sar: 32 24 bricks	= 2;42 brick sar

First-millennium metrologies

Arû measure for lengths and areas

1 (big) finger	≈ 3 cm		
1 *arû* cubit	= 24 (big) fingers (0.75 m)		
1 rod	= 12 cubits (9 m)	1 rod × 1 rod	= 1 sar (≈ 81 m^2)
		1 ubu	= 50 sar (0.4 ha)
		1 iku	= 2 ubu (0.81 ha)
		1 eshe	= 6 iku
		1 bur	= 3 eshe = 18 iku
		1 shar	= 60 bur

Arû "seed measure" for areas

1 sila	≈ 270 m^2	
1 ban	= 10 sila (0.27 ha)	
1 bariga	= 6 ban (1.62 ha)	= 2 iku
1 gur	= 5 bariga (8.1 ha)	= 10 iku

Cable "reed measure" for lengths and areas

1 finger	≈ 2 cm	
1 cable-cubit	= 24 fingers (0.5 m)	
1 rod	= 12 cubits (6 m)	1 rod × 1 rod = 1 sar (36 m^2)
1 chain	= 5 rods or 50 cubits (30 m)	
1 cable	= 2 chains or 10 rods (60 m)	1 cable × 1 cable = 25 sar (900 m^2)
		1 iku = 1 40 sar (0.36 ha)

Cable "seed measure" for area

1 grain	= 70 cm^2
1 nindan	≈ 1800 grams (7.5 m^2)
1 sila	= 10 nindan (75 m^2)
1 ban	= 6 sila (450 m^2)
1 bariga	= 6 ban (0.27 ha)
1 gur	= 5 bariga (1.35 ha)

"Reed measure" for lengths and areas

1 finger	≈ 2 cm	1 finger × 1 finger	= 1 small finger (4 cm^2)
		1 cubit × 1 finger	= 1 grain (100 cm^2)
		1 reed × 1 finger	= 1 (area) finger (730 cm^2)
1 cubit	= 24 fingers (0.5 m)	1 cubit × 1 cubit	= 1 small cubit (0.25 m^2)
		1 reed × 1 cubit	= 1 (area) cubit (1.75 m^2)
1 reed	= 7 cubits (3.5 m)	1 reed × 1 reed	= 1 (area) reed (12.25 m^2)
1 rod	= 2 reeds or 14 cubits (7 m)		

Bibliography

I have in general restricted bibliographical references to the place of first publication of each tablet, and the most recent comprehensive studies as of late 2003. Duncan Melville's wide-ranging bibliography of Mesopotamian mathematics and my shorter, annotated one can be found at <http://it.stlawu.edu/~dmelvill/mesomath/index.html>, along with many other useful introductory resources.

Editorial conventions

Words missing from the original tablets are restored within square brackets, [thus], or where not restorable as [......]. Uncertain translations are marked with a question mark: (?); badly written signs are followed by an exclamation mark: !. Accidental omissions are restored <within angle brackets>, accidental inclusions within double angle brackets «within double angle brackets». Occasional editorial glosses are shown in round brackets, (thus); editorial comments are shown in italics in square brackets, [*thus*]. Errors of vocabulary and calculation are noted and corrected; errors of spelling and grammar are not. I have represented numerals by numerals and syllabically written numbers by words. Sexagesimal number notation follows [Friberg 1987–90, 534]: the "sexagesimal point" is marked throughout by a semicolon, but its placing is often conjectural; space is left between sexagesimal places.

II. The Long Third Millennium, c. 3200–2000 BCE

II.a. Uruk in the late fourth millennium

Some five thousand clay tablets, reused as building rubble in the central temple precinct of the city of Uruk, constitute the world's oldest assemblage of written records, dating from some time in the last quarter of the fourth millennium BCE. While the large majority of the tablets record transactions within the domestic economy of the city, some 10 percent were written by trainee administrators as they learned to write. Most of those exercises are standardized lists of the nouns used within the book-keeping system, but one tablet, now known as W 19408,76, contains two exercises on calculating the areas of fields. It is thus the world's oldest piece of recorded mathematics [Nissen et al. 1993, fig. 50; Damerow 2001, 258].

Different commodities were counted with different number bases and used the accounting tokens and their numeral successors in different ways. W 19408,76 uses just three different number signs, all within the length measurement system. Small circles represent 10 rods each, c. 60 meters, while the large D-shaped impressions represent units of 60 rods, c. 360 meters long. The combination of the two—a small circle within a large D—represents 60 × 10 = 600 rods, or c 3.6 km. There are just two other signs on the tablet: a horizontal line representing length and a vertical line representing width. Each side of the tablet thus records two lengths and two widths; no areas are marked. But if, in each case, the average width is multiplied by the average length—the so-called agrimensors' method for finding the area of a field, $A = (l_1 + l_2)(w_1 + w_2)/4$—the area found is a conspicuously round number in the contemporaneous area system (1 rod × 1 rod = 1 sar, 1800 sar = 1 bur). As we shall see, both the agrimensors' method and conspicuously round numbers are prominent features of school mathematics later in the third millennium too.

W 19408,76

Obverse

— 600 600	600 60 60 10 10 I 60 60 10 60
600 600 —	600 60 60 10 10 I 60 60 10

$$\{(2 \times 600) + (2 \times 600)\}/2 \times \{(600 + 5 \times 60 + 3 \times 10) + (600 + 4 \times 60 + 3 \times 10)\}/2$$
$$= (1200 + 1200)/2 \times (930 + 870)/2$$
$$= 1200 \times 900 \text{ (rods)} = 1{,}080{,}000 \text{ (area sar)}$$
$$= 600 \text{ (area bur)}$$

Reverse

600 60 60 60 60 — 60 10 10 60	600 60 600 10 10 I
600 60 10 10 600 60 10 10 — 60	60 60 10 10 60 60 10 10 60 60 I 60 60

$$\{(600 + 6 \times 60 + 2 \times 10) + (2 \times 600 + 3 \times 60 + 4 \times 10)\}/2 \times \{(2 \times 600 + 60 + 2 \times 10) + (8 \times 60 + 4 \times 10)\}/2$$
$$= (980 + 1420)/2 \times (1280 + 520)/2$$
$$= 1200 \times 900 \text{ (rods)} = 1{,}080{,}000 \text{ (area sar)}$$
$$= 600 \text{ (area bur)}$$

II.b. Shuruppag in the mid-third millennium

If W 19408,76 is the world's oldest known mathematical exercise, then VAT 12593 is the first securely datable mathematical table in world history. It comes from the Sumerian city of Shuruppag to the north of Uruk, c. 2600 BCE. It is ruled into three columns on each side with ten rows on the front or obverse side. The first two columns of the obverse list length measures from c. 3.6 km to 360 m in descending units of 360 m, followed by the Sumerian word sá ("equal" and/or "opposite"), while the final column gives their products in area measure. Only six rows are extant or partially preserved on the reverse. They continue the table in smaller units, from 300 to 60 m in 60 m steps, and then perhaps (in the damaged and missing lower half) from 56 to 6 m in 6 m steps [Deimel 1923, no. 82; Robson 2003, 27–30].

VAT 12593

[1 × 600] rods	[1 × 600 (rods) square	3 × 60 + 2 × 10 (bur)]
9 × 60 (rods)	9 × 60 (rods) square	2 × 60 + 4 × 10 + 2 × 1 (bur)
8 × 60 (rods)	8 × 60 (rods) square	2 × 60 + 8 × 1 (bur)
7 × 60 (rods)	7 × 60 (rods) square	[1 × 60] + 3 × 10 + 8 × 1 (bur)
6 × 60 (rods)	[6 × 60 (rods) square]	1 × 60 + 1 × 10 + 2 × 1 (bur)
5 × 60 (rods)	5 × 60 (rods) square	5 × 10 (bur)
4 × 60 (rods)	4 × 60 (rods) square	3 × 10 + 2 × 1 (bur)
[3 × 60 (rods)]	3 × 60 (rods) square	1 × 10 + 8 × 1 (bur)
2 × 60 (rods)	2 × 60 (rods) square	8 × 1 (bur)
1 × 60 (rods)	1 × 60 (rods) square	2 × 1 (bur)
5 × 10 (rods)	5 × 10 (rods) square	1 (bur) + 1 (eshe) + 1 (iku)
4 × 10 (rods)	4 × 10 (rods) square	2 (eshe) + 4 (iku)
3 × 10 (rods)	3 × 10 (rods) square	1 (eshe) + 3 (iku)
2 × 10 (rods)	2 × 10 (rods) square	[4 (iku)]
1 × 10 (rods)	[1 × 10 (rods) square	1 (iku)]

VAT 12593 Obverse

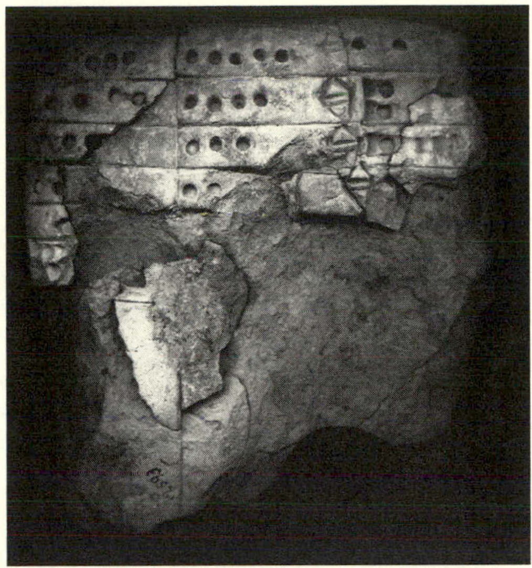

VAT 12593 Reverse

While at first sight this table appears simply to list the areas of square fields, in fact it is akin to much more abstract multiplication tables from later periods. Prior to the invention of the sexagesimal place value system, there was no concept of abstract numeration: numbers were thought of not as independent entities but as attributes of concrete objects—the length of a line, for instance, or the quantity of sheep in a flock. Therefore in the third millennium BCE to square a number meant, at some level, to construct a square area from two equal lengths.

But VAT 12593 is not the only mathematical tablet from Shuruppag. Among the others are a fascinating pair of exercises, now called TSS 50 and TSS 671, which are two different attempts at the same division problem [Jestin 1937; Høyrup 1982; Melville 2002]. The aim is to calculate the number of men who can be fed by a large quantity of grain if each is to receive 7 sila—not an amount of rations that ever appears in the administrative record. We can therefore see it as an exploration of the relationship between the sexagesimal system and 7, its lowest coprime number.

TSS 50

A granary of barley. 1 man received 7 sila. (What are) its men? 4 × 36,000 + 5 × 3600 + 4 × 600 + 2 × 60 + 51. 3 sila of grain remains.

TSS 671

A granary of barley. 1 man received 7 sila. (What are) the workers? 4 × 36,000 + 5 × 3600 + 3 × 600 + 6 × 60.

The answers show different strategies: one scribe chooses to give an answer with a remainder; the other to give an approximation, or round number, which is significantly smaller than the first. The difficulty in interpretation rests on knowing exactly how large a granary was supposed to be—for which there is no independent evidence. If the answer to TSS 50 is correct, then the granary would have been equal to 2400 gur, where 1 gur = 480 sila. [Høyrup 1982] has suggested that TSS 671 gives an incomplete answer; more recently [Melville 2002] has convincingly explained the answer on TSS 671 as arising through an error in solving the problem through repeated addition.

Sadly, we have no detailed context for these three tablets because their excavation numbers were lost or never recorded. Nevertheless we can make some general remarks based on the findspots of other school tablets in Shuruppag [Martin 1988, 82–103]. Around a thousand tablets were excavated there, almost all of them from houses and buildings that burned down in a city-wide fire in about 2600 BCE. The 60-odd school tablets come, in the main, from two widely separated buildings: an office in the south of the city from which donkeys appear to have been disbursed but which also housed a "library"; and a house to the northwest that has been tentatively characterized as a scribal center, as only school tablets have been found in and around it.

II.c. Nippur and Girsu in the twenty-fourth century BCE

The earliest known mathematical diagram containing quantitative textual data is on a small round tablet found in a secondary archaeological context in a shrine of the god Enlil's temple in Nippur.[20] To judge from its findspot and handwriting, it almost certainly dates to the Old Akkadian period, c. 2350–2250 BCE. IM 58045, also known by its excavation number 2NT 600, is a round tablet with a diagram of a bisected trapezoid [Friberg 1990, 541; Damerow 2001, 263]. The diagram presents a topological view of the shape under consideration, with all quantitative information contained in the surrounding text. Differences in length are minimized in the diagram, which shows a symmetrical trapezoidal area with a dividing line. If we assume that the round form of the tablet signals a student's exercise, then we have to decide what the original problem might have been. If it were simply to find the area of the whole figure using the agrimensors' formula, then the central dividing line might have been simply to mark the mean length, 12 cubits or 2 reeds, of the two vertical sides:

12 cubits × (17 + 7)/2 cubits = 144 square cubits = 1 sar.

IM 58045

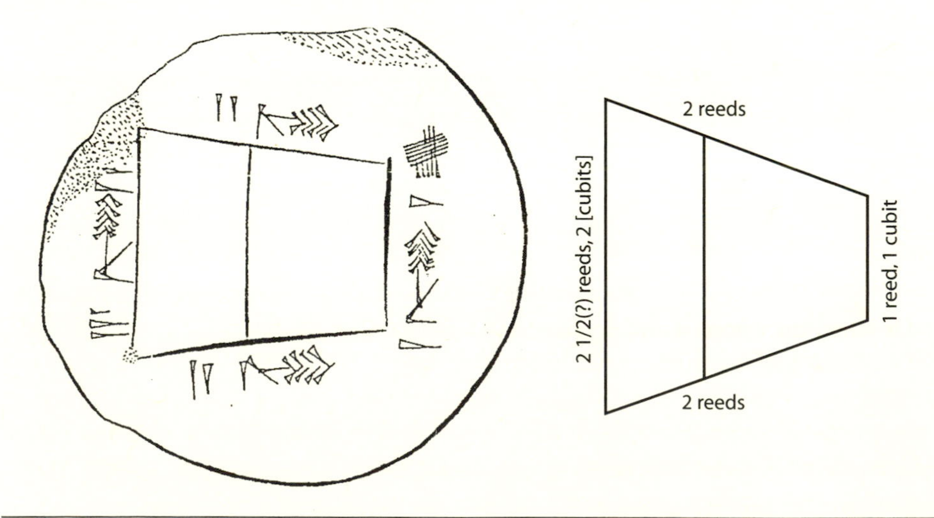

[20]Findspot information from the Nippur excavation notebooks held by the Babylonian Section of the University of Pennsylvania Museum of Anthropology and Archaeology.

As in the area problems from late fourth-millennium Uruk (above), we see (if the restoration of the longer side is correct) that the problem has been set to yield a simple unit answer.

It may be that the problem was a more complex one—perhaps to find the length of the middle line such that it bisects the area into two equal halves of 1/2 sar each. In later, Old Babylonian mathematics it was commonly known that triples of integers obeying the rule $a^2 + b^2 = 2c^2$ yielded the opposite sides, a and b, and bisecting line c of a trapezoid (e.g., YBC 4675 on page 101). It may be that such a rule was known already in the Old Akkadian period, for $(17^2 + 7^2) / 2 = (289 + 49) / 2 = 169 = 13^2$, a satisfyingly integer solution to the putative problem [Damerow 2001, 263–64].

Strikingly, all other surviving mathematical problems from the same period are also concerned with the properties of length and area [Westenholz 1975; Powell 1976; Whiting 1984; Foster and Robson 2004]. Three typical examples are given below.

HS 815

The long side is 60 + 7 1/2 rods.
<What is> the short side? <The area is> 1 (iku).
<Its short side is> 1 rod, 5 cubits, 2 double-hands, 3 fingers, and 1/3 finger.

In this problem from Nippur the aim is to find the short side of a rectangle of unit area, whose long side is known [Pohl 1935, no. 65; Foster and Robson 2004, no. 8]. The answer is correct, but no-one has yet put forward a convincing explanation of how the calculation was carried out. The scribe may have used repeated addition, as suggested by [Melville 2002] for the division problems from Shuruppag above. On the other hand, this type of problem may presage ideas of reciprocity, which developed with the sexagesimal place value system; in the Old Babylonian period we certainly see mutually reciprocal numbers treated as the sides of a rectangle of unit area (e.g., YBC 6967, on page 102).

Ash 1924.689

Square 1 × 216,000 + 4 × 60 rods 4 seed-cubits
(*1 blank line*)
Square 1 × 3600 + 1 × 60 + 32 rods 1 seed-cubits
Ur-Ishtaran
7 × 216,000 + 4 × 36,000 + 7 × 60 + 17 (bur) 1 (eshe) 3 1/2 (iku), 10 sar 16 2/3 shekels is found.

There are two problems on this tablet from the region of Girsu, both assigned to a student named Ur-Ishtaran, whose name is also found on another contemporary mathematical problem and at least one working administrative record [Gelb 1970, no. 112; Foster and Robson 2004, no. 3]. In each case he has to find the area of the squares defined by the lengths given. He (or the person who set the problem) has written a lengthy answer to one of the problems on the back of the tablet, but it bears no obvious relationship to either of them. [Friberg 1990, 540–41] suggests that the scribe intentionally or accidentally understood these traditionally sexagesimal signs to have decimal values, but it is unclear to me how this would work.

AO 11409

The long side is 1 03 1/2 rods.
The short side by the watercourse is 22 1/2 (rods).
The lower long side is 1 30 (rods).
The short side (reached by) irrigation is 22 1/2 (rods)
The harrowed area is 2 (eshe) 1 1/2 (iku).
Foxy the felter.

On this tablet from the Girsu region the problem is to find the area of an irregular quadrilateral, dressed up to seem like a real surveyor's record of a field [Foster and Robson 2004, no. 12]. As we have seen, the standard Mesopotamian method for finding the area of an irregular quadrilateral involves taking the averages of opposite sides, and multiplying. That would result in an area of approximately 2 eshe 5 1/4 iku. Instead, the scribe appears to have multiplied the length of the short side by 60 (rods) to get an area of 1350 sar, or 13 1/2 iku, namely the 2 eshe 1 1/2 iku given on the tablet. Although this problem mentions a "harrowed" area, it is probably not about work rates, but rather uses the harrowed field as a quasi-realistic pretext for calculating the area of an irregular quadrilateral. "Foxy the felter" may have been the student assigned the problem to solve (as in Ash 1924.689 mentioned above), or perhaps the person to whom the field was assigned in the problem-setting scenario.

II.d. Umma and Girsu in the twenty-first century BCE

The sexagesimal place value system was in place by the twenty-first century BCE, as attested by reciprocal tables (or rather, lists), known in increasing numbers in recent years. BM 106444, probably from the city of Umma, gives integer fractions of 60 expressed in the sexagesimal place value system: that is, without using fractional notation [Robson 2003–4].

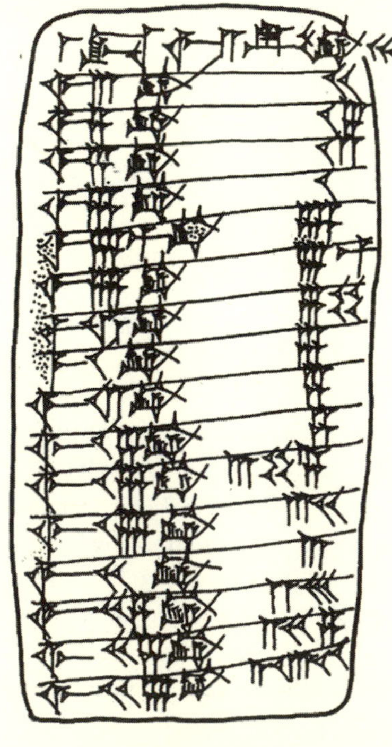

BM 106444 Obverse

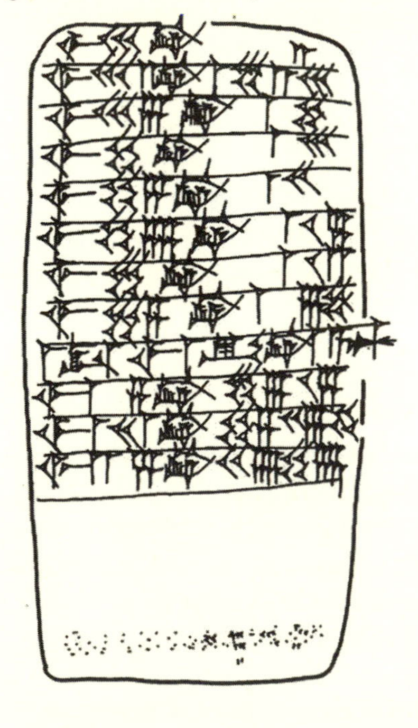

BM 106444 Reverse

BM 106444

Sixty: its 2nd part is	30
its 3rd part is	20
its 4th part is	15
its 5th part is	12
its 6th part is	10
its 7 1/2-th part is	8
its 8th part is	7;30
its 10-minus-1th part is	6;40
its 12th part is	5
its 15th part is	4
its 16th part is	3;45
its 18th part is	3;20
its 20th part is	3
its 24th part is	2;30
its 25th part is	2;24
its 27th part is	2;13 20
its 30th part is	2
its 32nd part is	1;52 30
its 36th part is	1;40
its 40th part is	1;30
its 45th part is	1;20
its 48th part is	1;15
its 50th part is	1;12
its 54th part is	1;06 40

Evidence for the use of the sexagesimal place value system (SPVS) at this period, as in all periods following, is scanty, because it was meant to serve solely as a calculating tool, not as a notation for permanent record. A clear example of the function of the SPVS is provided by AOT 304, a round school tablet from Girsu that may date back as far as the Ur III period [Thureau-Dangin 1903, no. 413; Robson 1999, 66]. On the obverse the calculation is laid out in the form of a model document, imitating the layout of administrative texts of the time, except that the height and width are given in reverse order to the usual practice. On the reverse the three measurements are written in sexagesimal place value notation.

AOT 304

Obverse	Length: 6 rods, 4 1/3 cubits.	Reverse	6 31 50
	Height: 1/2 rod.		3
	Width: 2 cubits, 5 fingers.		10 50
	Its volume: 3 1/2 sar, 2 1/2 shekels.		
	Its bricks: 25 1/2 sar.		

Although the scribe has mistranscribed the length and height as 6;31 50 and 3 instead of 6;21 40 rods and 6 cubits respectively, the width is correctly given as 0;10 50 rods. These produce a volume of 3;32 14 35 volume sar, rounded up on the obverse to 3;32 30. This volume is multiplied by 7;12, a standard constant (see YBC 7243 on page 149), to give 25;30 brick sar, or 18,360 bricks (a brick sar was considered to comprise 720 bricks, regardless of their individual size). This last calculation is not recorded on the tablet. The name of the object is not given, but the constant used suggests that it concerns a pile of standard unbaked bricks, measuring 15 × 10 × 5 fingers, or 0;00 00 41 40 volume sar. Using the dimensions given on the obverse, instead of the erroneous transcriptions on the reverse, the correct volume and bricks would be 6;53 28 20 volume sar and 49;37 brick sar, or 35,724 bricks. [Friberg 2001, 93] gives an alternative explanation of the calculation on the reverse of the tablet.

Mathematics in Sumerian literature (i)

This humorously testy dialogue between two colleagues was popular in the scribal schools of Old Babylonian Nippur [ETCSL 5.1.3; Black et al. 2004, 277–80]. The older man begins by reminiscing idealistically about his days as a student, exhorting the younger man to pay attention and model himself on the strictures of his old teacher. The young man robustly responds that he has no need of such lecturing for he is now a fully competent scribe and administrator. The older man finally retracts his words, acknowledging that his younger colleague is now ready to become a teacher himself. They end by praising Nisaba, the patron goddess of scribes. Although neither mathematics nor any other school subject is explicitly mentioned here, we see the wide range of numerate tasks the scribe of a large household might have carried out, from allocating rations to the domestic staff to managing farm land and the people and animals working on it.

A Supervisor's Advice to a Younger Scribe

1–8[*The supervisor speaks:*] 'One-time member of the school, come here to me, and let me explain to you what my teacher revealed. Like you, I was once a youth and had a mentor. The teacher assigned a task to me—it was man's work. Like a springing reed, I leapt up and put myself to work. I did not depart from my teacher's instructions, and I did not start doing things on my own initiative. My mentor was delighted with my work on the assignment. He rejoiced that I was humble before him and he spoke in my favor.

9–15'I just did whatever he outlined for me—everything was always in its place. Only a fool would have deviated from his instructions. He guided my hand on the clay and kept me on the right path. He made me eloquent with words and gave me advice. He focused my eyes on the rules which guide a man with a task: zeal is proper for a task, time-wasting is taboo; anyone who wastes time on his task is neglecting his task.

16–20'He did not vaunt his knowledge: his words were modest. If he had vaunted his knowledge, people would have frowned. Do not waste time, do not rest at night—get on with that work! Do not reject the pleasurable company of a mentor or his assistant: once you have come into contact with such great brains, you will make your own words more worthy.

21–26'And another thing: you will never return to your blinkered vision; that would be greatly to demean due deference, the decency of mankind. The heart is calm in

……, and sins are absolved. An empty-handed man's gifts are respected as such. Even a poor man clutches a kid to his chest as he kneels. You should defer to the powers that be and …… —that will calm you.

27–28'There, I have recited to you what my teacher revealed, and you will not neglect it. You should pay attention—taking it to heart will be to your benefit!'

29–35The learned scribe humbly answered his supervisor. 'I shall give you a response to what you have just recited like a magic spell, and a rebuttal to your charming ditty delivered in a bellow. Don't make me out to be an ignoramus—I will answer you once and for all! You opened my eyes like a puppy's and you made me into a human being. But why do you go on outlining rules for me as if I were a shirker? Anyone hearing your words would feel insulted!

36–41'Whatever you revealed of the scribal art has been repaid to you. You put me in charge of your household and I have never served you by shirking. I have assigned duties to the slave girls, slaves, and subordinates in your household. I have kept them happy with rations, clothing, and oil rations, and I have assigned the order of their duties to them, so that you do not have to follow the slaves around in the house of their master. I do this as soon as I wake up, and I chivvy them around like sheep.

42–49'When you have ordered offerings to be prepared, I have performed them for you on the appropriate days. I have made the sheep and banquets attractive, so that your god is overjoyed. When the boat of your god arrives, people should greet it with respect. When you have ordered me to the edge of the fields, I have made the men work there. It is challenging work which permits no sleep either at night or in the heat of day, if the cultivators are to do their best at the field-borders. I have restored quality to your fields, so people admire you. Whatever your task for the oxen, I have exceeded it and have fully completed their loads for you.

50–53'Since my childhood you have scrutinized me and kept an eye on my behavior, inspecting it like fine silver—there is no limit to it! Without speaking grandly—as is your shortcoming—I serve before you. But those who undervalue themselves are ignored by you—know that I want to make this clear to you.'

54–55[*The supervisor answers:*] 'Raise your head now, you who were formerly a youth. You can turn your hand against any man, so act as is befitting.'

56–59[*The scribe speaks:*] 'Through you who offered prayers and so blessed me, who instilled instruction into my body as if I were consuming milk and butter, who showed his service to have been unceasing, I have experienced success and suffered no evil.'

60–61[*The supervisor answers:*] 'The teachers, those learned men, should value you highly. You who as a youth sat at my words have pleased my heart.

62–72'Nisaba has placed in your hand the honour of being a teacher. For her, the fate determined for you will be changed and so you will be generously blessed. May she bless you with a joyous heart and free you from all despondency …… at whatever is in the school, the place of learning. The majesty of Nisaba …… silence.

For your sweet songs even the cowherds will strive gloriously. For your sweet songs I too shall strive and shall ...... They should recognise that you are a practitioner (?) of wisdom. The little fellows should enjoy like beer the sweetness of decorous words: experts bring light to dark places, they bring it to closes and streets.'

[73–74]Praise Nisaba who has brought order to ...... and fixed districts in their boundaries, the lady whose divine powers are divine powers that have no rival!

III. The Old Babylonian Period, c. 2000–1600 BCE

III.a. Arithmetical and metrological tables

Trainee scribes memorized a large number of reciprocal and multiplicative relationships as a normal part of their elementary education, writing them down as part of the memorization process [Robson 2002]. Thousands of surviving tablets show there to have been a fixed order and more or less fixed content to the lists of number facts the students learned. BM 80150 from Sippar is one of the few containing a large portion of the whole series. It starts with a reciprocal table descended from the twenty-first century exemplars (see BM 106444, page 79), continues with multiplication tables for 17 head numbers from 50 to 9 (the entire series runs to 1;15), and finishes with another standard list of mutually reciprocal pairs of integers [Pinches 1963, no. 42; Nissen et al. 1993, fig. 124]. All known school mathematical tables are in Sumerian, or are at least written exclusively with logograms.

BM 80150

Its 2/3	40	45	1 20	[16	13 20]	[15	11 15]
Its [half]	30	48	1 15	[17]	1[4 10]	[16	12]
———		50	1 12	[18]	15	[17	12 45]
[2]	30	54	1 06 40	[19]	15 [50]	[18	13 30]
[3]	20	1	1	[20]	16 40	19	[14 15]
[4]	15	1 04	56 15	[30]	25	20	[15]
[5]	12	1 21	44 26 40	40	33 20	30	22 30
6	10	———		50 times 50 41 40		40	30
8	7 30	1	50	———		50	37 30
9	6 40	2	1 40	1	45	45 times 45 33 45	
10	6	3	2 30	2	1 30	———	
12	5	4	3 20	3	[2 15]	44 26 40 times 1	
15	4	5	4 10	4	[3]		44 26 40
16	3 45	6	5	5	3 45	2	1 28 53 20
1[8]	3 20	7	5 50	6	4 30	3	2 13 20
20	3	8	6 40	7	5 15	4	2 57 46 40
2[4]	2 30	9	7 30	8	6	5	3 42 13 20
[25]	2 24	[10]	8 20	9	6 45	6	4 26 40
[27	2 13 20]	[11]	9 10	10	7 30	7	5 11 06 40
30	2	[12]	10	[11	8 15]	8	5 55 33 20
32	1 52 [30]	[13]	10 50	[12	9]	9	6 40
36	1 40	[14]	11 40	[13	9 45]	10	7 24 26 40
40	1 30	[15]	12 30	[14	10 30]	11	8 08 53 20

12	8 53 20
13	9 37 46 40
14	[10 22] 13 20
[15	11] 06 40
[16	11 51] 06 40
[17	12 35] 33 20
[18	1]3 20
19	14 04 26 40
20	14 48 53 20
30	22 13 20
40	29 37 46 40
50	36 53 13 20

44 26 40 [times 44 26 40]
[32 55 1]8 31 06 40

1	40
2	1 20
3	2
4	2 40
5	3 20
6	4
7	4 40
8	5 20
[9]	6
[10]	6 40
[11]	7 20
[12]	8
[13]	8 40
[14]	9 20
[15]	10
[16]	10 40
[17]	11 20
[18]	12
19	[12 40]
20	[13 20]
30	[20]
40	26 40
50	33 20

1	36
2	1 [1]2
3	1 [4]8
4	224
5	3
6	3 36
7	4 12
8	4 48
9	5 24
10	6
11	6 36
12	7 12
13	7 48
14	8 24
15	9
16	9 36
17	10 12
18	10 48
19	11 24
20	12
30	18
40	24
50	30

36 times 36 21 36

1	30
2	1
3	1 30
4	2
5	2 30
[6]	3
[7]	3 30
[8]	4
[9]	4 30
10	5
11	5 30
12	6
13	6 30
14	7
15	7 30
16	8
17	8 30
18	9
19	9 30
20	10
30	15
40	20
50	25

1	25
2	50
3	1 15
4	1 40
5	2 05
6	2 30
[7]	2 55
[8]	3 20
[9]	3 45
[10]	4 10
11	4 35
12	5
13	5 [25]
14	[5 50]
15	6 15
16	6 40
17	7 05
18	7 [30]
19	7 55
20	8 20
30	12 30
40	16 40
50	20 50

25 times 25 10 25

1	24
2	48
3	1 12
4	1 36
5	2
6	2 24
7	2 48
8	[3 12]
9	[3 36]
10	[4]
11	[4 24]
12	[4 48]
13	5 [12]
14	5 36
15	6
16	6 24
17	6 48
18	7 12
19	7 36
20	8
30	12
40	16
50	20

24 times 24 9 [36]

1	22 30
2	45
3	1 07 30
4	1 30
5	1 52 [30]
6	2 1[5]
7	2 3[7 30]
8	3
9	3 2[2 30]
10	3 45
11	4 07 [30]
12	[4 30]
13	[4 52 30]
14	[5 15]
15	5 37 30
16	6
17	6 22 30
18	6 45
19	7 07 [30]
20	7 30
30	11 15
40	15
50	18 4[5]

22 30 times 22 30 [8 26 15]

1	20
2	40
3	1
4	1 20
5	1 40
6	2
7	2 20
8	2 40
9	3
10	3 20
11	3 40
12	4
13	4 20
14	4 40
15	5
16	5 20
17	5 40
18	6
19	6 20
20	6 40
30	10
40	13 20
50	16 40

1	18
2	36
3	54
4	1 12
5	1 30
6	1 48
7	2 06
8	2 24
9	2 42
10	3
11	3 18

12 3 36
13 3 54
14 4 12
15 4 30
16 4 48
17 5 06
18 5 24
19 5 42
20 6
30 9
40 12
50 15

---

1 1[6 40]
2 [33 20]
3 [50]
4 1 06 40
5 1 23 20
6 1 40
[7] 1 56 40
[8] 2 13 20
[9] 2 30
[10] 2 46 40
[11] 3 03 20
[12] 3 20
[13] 3 36 40
[14] 3 53 20
[15] 4 10
[16] 4 26 40
[17] 4 43 20
[1]8 5
[1]9 5 16 40
20 5 33 20
30 8 20
40 11 06 40
50 13 53 20
16 40 times 16 40
4 37 46 40

---

1 16
2 32
3 48
4 1 04
5 1 20
6 1 36
7 1 52
8 2 08
9 [2 24]
[10 2 40]
[11 2 56]
[1]2 3 [12]
13 3 28
14 3 44
15 4
16 4 16
17 4 32
18 4 «52»<48>
19 5 04
20 5 20
30 8
40 «12» <10 40>
50 «15» <13 20>

---

1 15
2 30
3 45
4 1
5 1 15
6 1 30
7 1 45
8 2
9 2 15
10 2 30
11 2 45
12 3
13 3 15
14 3 30
15 3 45
16 4
17 4 15
18 4 30
19 4 [45]
20 [5]
30 7 30
40 10
50 12 30

---

1 12 30
2 25
3 37 30
4 50
5 [1 02 30]
6 [1 15]
7 [1 27 30]
8 [1 40]
9 [1 52 30]
10 [2 05]
11 [2 17 30]
12 [2 30]
13 [2 42 30]
14 [2 55]
15 [3 07 30]
16 [3 20]
17 [3 32 30]
18 [3 45]
19 [3 57 30]
20 [4 10]
30 [6 1]5
40 [8 20]
50 10 25
12 30 times 12 30
2 36 15

---

1 12
2 24
3 36
4 48
5 1
[6] 1 12
[7] 1 24
[8] 1 36
9 [1 48]
10 [2]
11 [2 12]
12 [2 24]
13 [2 36]
14 [2 4]8
15 3
16 3 12
17 3 24
18 3 36
19 3 48
20 4
30 6
40 8
50 10

---

1 10
2 20
3 30
4 40
5 50
6 1
[7] 1 10
[8] 1 20
9 1 30
10 1 40
11 1 50
12 2
13 2 10
14 2 20
15 2 30
16 2 40
17 2 50
18 3
19 3 10
20 3 20
30 5
40 6 40
50 8 20

---

1 9
2 18
3 27
4 36
5 45
6 54
7 1 03
8 1 12
9 1 21
10 1 30
11 1 39
12 1 48
13 1 [5]7
14 [2] 06
15 [2 1]5
16 [2 2]4
17 2 33
18 2 42
19 2 51
20 3
30 4 30
40 6
50 7 30

---

2 05	28 48 is its inverse	[4 26 40]	3 30 is its inverse
[8 53 20]	6 45 is its inverse	[8 53 20]	6 45 is its inverse
4 10	14 24 is its inverse	[17 46 40 3]	22 30 is its inverse
8 20	7 12 is its inverse	[35 33 201 4]	1 15 is its inverse
16 40	3 36 is its inverse	[1 1]1 06 40 «its inverse»	50 37 30 is its inverse
33 20	1 48 is its inverse	2 22 13 20 [...]	25 18 45 [is its inverse]
1 06 40	54 is its inverse		
[2 13 20]	27 is its inverse		

It is possible to get a sense of how long it took trainee scribes to learn the complete series of multiplication tables from three dated tablets from Larsa, Ash 1924.447, 1924.451, and YBC 11924, written by a single scribe [Robson 2004, 15; MCT, 23]. The first two, both multiplication tables for 24, were written within four days of each other, the second time more legibly than the first. Six months, it is assumed, separates them from the third tablet, a table for 4, written in 1815 BCE. As the 24 times table came a quarter of the way through the entire series and the 4 times table three-quarters of the way through, it seems safe to deduce that the latter was written later and that the whole sequence took a year to master in the context of the Larsa scribal curriculum.

Ash 1924.447		*Ash 1924.451*		*YBC 11924*	
[24] times 1	24	24 times 1	24	[4] times 1	4
[times] 2	48	times 2	48	[times] 2	8
[times 3]	1 12	times 3	1 12	[times 3]	12
times 4	1 36	times 4	1 36	[times 4]	16
times 5	2	[times] 5	2	times 5	[20]
times 6	2 24	[times] 6	2 24	times 6	[24]
times 7	2 48	[times] 7	2 48	times 7	28
times 8	3 12	times [8]	3 12	times 8	[32]
times 9	3 36	times 9	3 36	times 9	[36]
times 10	4	times 10	4	times 10	[40]
times 11	4 24	times 11	4 24	times 11	[44]
times 12	4 48	times 12	4 48	times 12	[48]
times 13	5 12	times 13	5 12	[times 13	52]
times 14	5 36	times 14	5 36	[times 14	56]
times 15	6	times 15	6	times 15	1
times 16	6 24	times 16	6 24	times 16	1 04
times 17	6 48	times 17	6 48	times [17]	1 08
times 18	7 12	times 18	7 12	times [18]	1 12
times 19	7 36	times 19	7 36	[times 19]	1 16
times 20	8	times 20	8	[times 20]	1 20
times 30	12	times 30	12	[times] 30	2
times 40	16	times 40	16	[times] 40	2 40
times 50	20	times 50	20	[times] 50	3 20
<Long> tablet of Suen-apil-Urim. Harvest Month (XII), day 9. Praise Nisaba (and) Ea!		Finished. Harvest Month (XII), day 13. Long tablet of Suen-Apil-Urim. Praise Nisaba!		Long tablet of Suen-apil-Urim. Festival of Inana (VI), day 11. Year of Enki's temple in Urim.	

Old Babylonian metrological lists and tables have been little studied since their basic relationships were determined in the 1940s. However, it is now clear that they too were memorized as part of the elementary scribal school curriculum, both as lists and as tables, though the pedagogical purpose of each has yet to be distinguished [Robson 2002]. Ash 1931.137 from Kish is a metrological list of the measures in the capacity, weight, and area systems [Robson 2004, no. 19]. Some lines are omitted, others erroneously included, and the scribe did not understand the notation for the large units of capacity or area. Long lists like these were presumably written at the end of a learning phase to demonstrate that the whole metrological series had been mastered. (See metrology summary in the Introduction.)

Ash 1931.137

1/3 sila grain
1/2 sila grain
2/3 sila grain
5/6 sila grain
1 sila grain
1 1/3 sila grain
1 1/2 sila grain
1 2/3 sila grain
1 5/6 sila grain
2 sila grain
3 sila grain
4 sila grain
5 sila grain
6 sila grain
7 sila grain
8 sila grain
9 sila grain
1 (ban) grain
1 (ban) 1 sila grain
1 (ban) 2 sila grain
1 (ban) 3 sila grain
1 (ban) 4 sila grain
1 (ban) 5 sila grain
1 (ban) 6 sila grain
1 (ban) 7 sila grain
1 (ban) 8 sila grain
1 (ban) 9 sila grain
2 (ban) grain
3 (ban) grain
4 (ban) grain
5 (ban) [grain]
1(bariga) [grain]
1(bariga) 1 (ban) [grain]
1(bariga) 2 (ban) [grain]
1(bariga) 3 (ban) grain
1(bariga) 4 (ban) grain
1(bariga) 5 (ban) grain
2 (bariga) grain
2 (bariga) 1 (ban) grain
2 (bariga) 2 (ban) grain
2 (bariga) 3 (ban) grain
2 (bariga) 4 (ban) grain
2 (bariga) 5 (ban) grain
«2 (bariga) grain»
3 (bariga) grain
3 (bariga) 1 (ban) grain
3 (bariga) 2 (ban) grain
3 (bariga) 3 (ban) grain
3 (bariga) 4 (ban) grain
3 (bariga) 5 (ban) grain
4 (bariga) grain
4 (bariga) 1 (ban) grain
4 (bariga) 2 (ban) grain
4 (bariga) 3 (ban) grain
4 (bariga) 4 (ban) grain
4 (bariga) 5 (ban) grain
1 (gur) grain
[1 (gur) 1 (bariga)] grain
[1 (gur) 2 (bariga) grain]
[1 (gur) 3 (bariga)] grain
[1 (gur) 4 (bariga) grain]
[2 (gur)] grain
[3 (gur) grain]
[4 (gur) grain]
[5 (gur)] grain]
[6 (gur)] grain
[7 (gur)] grain
[8 (gur) grain]
[9 (gur) grain]
[10 (gur) grain]
[11 (gur) grain]
[12 (gur) grain]
13 (gur) grain
14 (gur) grain
15 (gur) grain
16 (gur) grain
17 (gur) grain
18 (gur) grain
19 (gur) grain
20 (gur) grain
30 (gur) grain
40 (gur) grain
50 (gur) grain
60 (gur) grain
1 10 (gur) grain
1 20 (gur) grain
1 30 (gur) grain
1 40 (gur) grain
1 50 (gur) grain
2 00 (gur) grain
3 00 (gur) grain
4 00 (gur) grain
5 00 (gur) grain
6 00 (gur) grain
7 00 (gur) [grain]
8 00 (gur) [grain]
9 00 (gur) grain.
600 (gur) grain
600 + 1 00 (gur) grain
600 + 2 00 (gur) grain
600 + 3 00 (gur) grain
600 + 4 00 (gur) grain
600 + 5 00 (gur) grain
600 + 6 00 (gur) grain
600 + 7 00 (gur) grain
600 + 8 00 (gur) grain
600 + 9 00 (gur) grain
[2 × 600 (gur) grain]
[3 × 600 (gur) grain]
[4 × 600 (gur) grain]
[5 × 600 (gur) grain]
3600 (gur) grain
3600 + 600 (gur) grain
3600 + 2 × 600 (gur) grain
3600 + 3 × 600 (gur) grain
3600 + 4 × 600 (gur) grain
3600 + 5 × 600 (gur) grain
3600 (gur) grain
2 × 3600 (gur) grain
3 × 3600 (gur) grain
4 × 3600 (gur) grain
5 × 3600 (gur) grain
6 × 3600 (gur) grain
7 × 3600 (gur) grain
8 × 3600 (gur) grain
9 × 3600 (gur) grain
«3600 (gur) grain»
«2 × 3600 (gur) grain»
«3 × 3600 (gur) grain»
«4 × 3600 (gur) grain»
«5 × 3600 (gur) grain»
«6 × 3600 (gur) grain»
«7 × 3600 (gur) grain»
«8 × 3600 (gur) grain»
«9 × 3600 (gur) grain»
36,000 (gur) grain
72,000! (gur) grain
108,000! (gur) grain
144,000! (gur) grain
3,600 (gur) big grain

1/2 grain silver
1 [grain] silver
1 1/2 grain silver
2 grains silver
2 1/2 grains silver

3 grains silver
4 grains silver
5 grains silver
6 grains silver
7 grains silver
8 grains silver
9 grains silver
10 grains silver
11 grains silver
12 grains silver
13 grains silver
14 grains silver
15 grains silver
16 grains silver
17 grains silver
18 grains silver
19 grains silver
20 [grains silver]
[21 grains] silver
[22 grains] silver
[22] 1/2 grains silver
23 grains silver
24 grains silver
25 grains silver
26 [grains] silver
27 [grains] silver
28 grains silver
29 grains silver
6th (shekel) silver
6th (shekel)
5 grains silver
6th (shekel)
10 (gr) silver
[4th] (shekel) silver
[4th (shekel)
5 grains silver]
[4th] (shekel)
<10> grains silver
[1/3] shekel silver
[1/3] shekel 5 (gr) silver
[1/3] shekel 10 (gr) silver
1/3 shekel 15 (gr) silver
1/3 shekel 20 (gr) silver
1/2 shekel silver
1/2 shekel 5 (grains) silver
1/2 shekel 10 (gr) silver
1/2 shekel 15 grains silver
1/2 shekel 20 (gr) silver
2/3 shekel silver
2/3 shekel
5 gr silver
2/3 shekel
10 (gr) silver
2/3 shekel
15 (gr) silver
[2/3 shekel]
20 (gr) silver
[5/6 shekel] silver
[5/6 shekel 5 (gr)] silver
[5/6 shekel 10 (gr) silver]
[5/6 shekel 15 (gr) silver]
[5/6 shekel 20 (gr) silver]
1 sixty … [silver]
1 shekel silver
1 1/3 shekel silver
[1 1/2] shekel silver
<1> 2/3 shekel silver
<1> 5/6 shekel silver
2 shekels silver
3 shekels silver
4 shekels silver
5 shekels silver
6 shekels silver
7 shekels silver
8 shekels silver
8 shekels silver
9 shekels silver
10 shekels silver
[11] shekels silver
[12 shekels] silver
13 [shekels silver]
14 [shekels silver]
15 [shekels] silver
16 shekels silver
17 shekels silver
[18] shekels silver
[19] shekels [silver]
[1/3 mina silver]
[1/2] mina silver
[2/3] mina silver
5/6 mina silver
1 mina silver
1 1/3 mina silver
1 1/2 mina silver
1 2/3 mina silver
1 5/6 mina silver
2 minas [silver]
3 minas [silver]
4 minas silver
5 minas silver
6 minas silver
7 minas silver
8 minas silver
9 minas silver
10 minas silver
11 minas silver
12 minas silver
13 minas silver
14 minas silver
15 minas silver
16 minas silver
17 minas silver
18 minas silver
19 minas silver
[20] minas silver
[30] minas silver
[40] minas silver
[50] minas silver
1 talent [silver]
1 talent 10 minas [silver]
1 talent 20 minas silver
1 talent 30 minas silver
1 [talent] 40 minas silver
1 talent 50 minas [silver]
2 talents [silver]
3 talents [silver]
4 talents silver
[5 talents silver]
[6 talents silver]
[7 talents silver]
[8 talents silver]
[9 talents] silver
[10] talents silver
[11] talents silver
[12] talents silver
[13] talents silver
14 talents silver
15 talents silver
16 talents silver
17 talents silver
18 talents silver
19 talents silver
10 talents silver
20 talents silver
30 talents silver
40 talents silver
50 talents silver
1 00 talents silver

1/3 sar [area]
1/2 sar area
2/3 sar area
5/6 sar area
[1] sar area
1 1/3 sar area
1 1/2 sar area
1 2/3 sar area
1 5/6 sar area
2 sar area
3 sar area
4 sar area
5 sar area
6 sar area
7 sar area
8 sar area
<9 sar area>
10 sar area
11 sar area
12 sar area
[13 sar area]
[14 sar area]
[15 sar area]
[16 sar area]
[17 sar area]
[18 sar area]
[19 sar area]
[20 sar area]
[30] sar area
40 sar area
«50 sar area»
1 (ubu) area
1 (ubu) 10 sar area
1 (ubu) 20 sar area
1 (ubu) 30 sar area
1 (ubu) 40 sar area
<1 (iku) area>
1 (iku) 1 (ubu) area
2 (iku) area
2 (iku) 1 (ubu) area
3 (iku) area
3 (iku) 1 (ubu) area
4 (iku) area
4 (iku) 1 (ubu) area
5 (iku) area
5 (iku) 1 (ubu) area
1 (eshe) area
1 (eshe) 1 (iku) area
1 (eshe) 2 (iku) area
1 (eshe) 3 (iku) area
1 (eshe) 4 (iku) area
1 (eshe) 5 (iku) area
2 (eshe) area
2 (eshe) 1 (iku) area
2 (eshe) 2 (iku) area
2 (eshe) 3 (iku) area
2 (eshe) 4 (iku) area
2 (eshe) 5 (iku) area
1 (bur) area
1 (bur) 1 (eshe) [area]
1 (bur) 2 (eshe) [area]
2 (bur) [area]
3 (bur) [area]
4 (bur) [area]
5 (bur) [area]
6 (bur) [area]

7 (bur) area	18 (bur) area	6 × 60 (bur) area	«8 × 60 (bur) area»
8 (bur) area	19 (bur) area	7 × 60 (bur) area	«9 × 60 (bur) area»
9 (bur) area	20 (bur) [area]	8 × 60 (bur) area	600 (bur) area
[10 (bur) area]	30 (bur) area	9 × 60 (bur) area	1200! (bur) area
11 (bur) area	40 (bur) area	«60 (bur) area»	1800! (bur) area
12 (bur) area	50 (bur) area	«2 × 60 (bur) area»	2400! (bur) area
13 (bur) area	<60 (bur) area>	«3 × 60 (bur) area»	<3,000 (bur) area>
14 (bur) area	2 × 60 (bur) area	«4 × 60 (bur) area»	600 (bur) big area
15 (bur) area	3 × 60 (bur) area	«5 × 60 (bur) area»	
16 (bur) area	4 × 60 (bur) area	«6 × 60 (bur) area»	
17 (bur) area	5 × 60 (bur) area	«7 × 60 (bur) area»	

Ash 1923.366 is a six-sided prism from Larsa containing two metrological tables of length and height, a table of inverse squares, and a table of inverse cubes [Robson 2004, no. 9]. Metrological tables give the equivalent sexagesimal writings for each unit, expressed in terms of the standard unit: here rods for horizontal length and cubits for vertical height. They are the crucial link between the absolute notations of the various metrological systems and the sexagesimal place value system, enabling scribes to convert between the two notations for recording, calculating, and back again. The tables following were less widely copied; they give the squares and cubes of integers from 1 to 60 and 1 to probably 30.

Ash 1923.366

[1] finger	10	[1/2 rod 5] cubits	[55]	[40] rods	40	17 cables	17
[2] fingers	20	[1 rod]	1	[45] rods	45	18 cables	18
[3] fingers	30	[1] 1/2 rods	1 [30]	[50] rods	50	19 cables	19
[4] fingers	40	[2] rods	[2]	[55] rods	55	2/3 league	20
[5] fingers	50	[2] 1/2 rods	[2 30]	1 cable	1	1 league	30
[6] fingers	1	[3] rods	[3]	1 cable 10 rods	1 10	1 1/2 leagues	45
[7] fingers	1 10	[3] 1/2 rods	[3 30]	1 cable 20 rods	1 20	1 2/3 leagues	50
[8] fingers	1 20	[4] rods	4	1 cable 30 rods	1 30	2 leagues	1
[9] fingers	1 30	[4] 1/2 rods	4 30	1 cable 40 rods	1 40	3 leagues	1 30
[1/3] cubit	1 40	[5] rods	5	1 cable 50 rods	1 50	4 leagues	2
[1/2] cubit	2 30	[5] 1/2 rods	5 30	2 cables	2	5 leagues	2 30
[2/3] cubit	3 20	[6] rods	6	3 cables	3	6 leagues	3
[1] cubit	5	[6] 1/2 rods	6 30	4 cables	4	7 leagues	3 30
[1 1/3] cubits	6 [40]	7 rods	7	5 cables	5	8 leagues	4
[1 1/2] cubits	[7] 30	7 1/2 rods	7 30	6 cables	6	9 leagues	4 30
[1 2/3] cubits	8 20	8 rods	8	7 cables	7	10 leagues	5
[2] cubits	10	8 1/2 rods	8 30	8 cables	8	11 leagues	5 30
[3] cubits	15	9 rods	9	9 cables	9	12 leagues	6
[4] cubits	[20]	9 1/2 rods	9 30	10 cables	10	13 leagues	6 30
[5] cubits	[25]	[10] rods	10	11 cables	11	14 leagues	7
[1/2 rod]	30]	15 rods	15	12 cables	12	15 leagues	7 30
[1/2 rod 1] cubit	[35]	[20] rods	20	13 cables	13	[16] leagues	8
[1/2 rod 2] cubits	[40]	25 rods	25	14 cables	14	17 leagues	8 30
[1/2 rod 3] cubits	[45]	[30] rods	30	1/2 league	15	18 leagues	9
[1/2 rod 4] cubits	[50]	[35] rods	35	16 cables	16	19 leagues	9 30

20 leagues	10
30 leagues	15
40 leagues	20
50 leagues	25
[1 00] leagues	30
[...] leagues	[...]

1 finger	2
2 fingers	4
3 fingers	6
4 fingers	8
5 fingers	10
6 fingers	12
7 fingers	14
8 fingers	16
9 fingers	18
1/3 cubit	20
1/2 cubit	30
2/3 cubit	40
1 cubit	1
1 1/3 cubits	1 20
1 1/2 cubits	1 30
[1] 2/3 cubits	1 40
2 cubits	2
3 cubits	3
[4] cubits	4
[5] cubits	5
1/2 rod	6
1/2 rod 1 cubit	7
1/2 rod 2 cubits	8
1/2 rod 3 cubits	9
1/2 rod 4 cubits	10
1/2 rod 5 cubits	11
1 rod	12
1 1/2 rods	18
2 rods	24
2 [1/2] rods	30
[3] rods	36
[3] 1/2 rods	42
4 rods	48
4 1/2 rods	54
5 rods	1
5 1/2 rods	1 06
6 rods	1 12
6 1/2 rods	1 [18]
7 rods	[1 24]
7 1/2 rods	[1 30]
8 rods	1 36
8 1/2 rods	1 42
9 rods	1 48
[9] 1/2 rods	1 54
10 rods	2

1 squares 1
4 squares 2
[9] squares 3
16 squares 4
[25 squares 5]
[3]6 [squares] 6
[4]9 [squares] 7
[1] 04 [squares] 8
[1] 21 [squares] 9
[1] 40 squares 10
2 01 squares 11
2 24 squares 12
2 49 squares 13
3 16 squares 14
3 45 squares 15
4 16 squares 16
4 49 squares 17
5 24 squares 18
6 01 squares 19
6 40 squares 20
7 21 squares 21
8 04 squares 22
8 49 squares 23
9 36 squares 24
10 25 squares 25
11 16 squares 26
12 09 squares 27
[13] 04 squares 28
14 01 squares 29
15 squares 30
16 01 squares 31
17 04 squares 32
18 09 squares 33
19 16 squares 34
20 25 squares [35]
21 36 squares [36]
22 49 squares 37
2[4] 04 squares 38
25 21 squares 39
26 40 squares 40
28 01 squares 41
29 24 squares 42
30 49 squares 43
32 16 squares 44
33 45 squares 45
35 16 squares 46
36 49 squares 47
38 24 squares 48
40 01 squares 49
41 40 squares 50
[43 21 squares 51]
[45 04 squares 52]
[46 49 squares 53]
[48 36 squares 54]
[50 25 squares 55]
[52 16 squares 56]
[54 09 squares 57]
[56 04 squares 58]
58 01 [squares 59]
1 00 00 [squares 1 00]

1 [cubes 1]
8 [cubes 2]
2[7 cubes 3]
1 04 [cubes 4]
2 [05 cubes 5]
3 [36 cubes 6]
5 [43 cubes 7]
8 [32 cubes 8]
12 [09 cubes 9]
16 [40 cubes 10]
22 1[1 cubes 11]
28 [48 cubes 12]
36 [37 cubes 13]
45 [44 cubes 14]
[...

About half a dozen known tablets list powers of the integers 5, 9, 16, 1 40, and 3 45 to ten powers, all but the first of which are themselves squares [MKT: I 77–78]. The purpose of these tables has never been satisfactorily elucidated. The scribe of the unprovenanced tablet BM 22706 lost track of an empty sexagesimal place in the sixth power of 1 40, leading to erroneous values in the rest of the table [Nissen et al. 1993, fig. 128]. On the other hand, the writer of IM 73355, from Larsa, made copying or memorization errors in the higher reaches of the tables for 16 and 3 45 (which, incidentally, are mutually reciprocal) that did not affect the values given in the following lines [Arnaud 1994, no. 55]. Like many Larsa school tablets, IM 73355 finishes with a line of praise to the deities of scribalism and the student's name. Note that even on the original tablet of IM 73355, the tabulation breaks down in the final few lines (just as in the translation) because the entries in the first column are too long for the tablet.

BM 22706

		(*correct values:*)
[1 40] times	1 40 2 46 40	
times 1 40	4 37 46 40	
times 1 40	7 42 57 46 [40]	
times 1 40	12 51 36 17 46 [40]	
times 1 40	21 26 <00> 29 37 46 40	
times 1 40	35 44 09 22 57 46 40	35 43 20 49 22 57 46 40
times 1 40	59 33 35 38 [16 17 46 40]	59 32 14 43 18 16 17 46 40
times 1 40	1 3[9] 15 [59 40 27 08] 37 [46 40]	1 39 13 44 32 10 27 09 37 46 40
times 1 40	2 45 2[6] 39 [28 25 14 22 57 46 40]	2 38 42 54 13 57 25 16 03 03 46 40

5 times 5	25
times 5	2 05
times 5	10 25
times 5	52 05
times 5	4 20 25
times 5	21 42 05
times 5	1 48 30 25
times 5	9 02 32 05
times 5	45 12 40 25

IM 73355

3 45 times	3 45
14	03 45
times 52 44	03 45
times 3 17 45 14	03 45
times 12 21 34 <37> 44	03 45
times 46 20 [5]4 51 30 14	03 45
times 2 53 [48 2]5 43 «06» <08> 22 44	03 45
times 10 51 4[6 3]6 26 46 25 15 14	03 45
times 40 44 09 46 40 24 04 42 07 <4>4	03 45
times 2 [3]2 45 36 40 01 30 17 «27 59» <37 58> 00 14	03 45

16 times	16
[4]	16
[times] 1 08	16
[times] 18 12	16
[times] 4 51 [16]	16
[times] 1 17 [40 20]	16
[times] 20 42 45 24	16
[times 5 3]1 24 «24» 06 28	16
[times 1] 28 «24 26» <22 25> 43 «22» <32>	16
[times] 23 33 58 51 36 36	16

Finished. (Praise to) Nisaba and Haia.
Nawir-[...] wrote it.

Mathematics in Sumerian literature (ii)

Lipit-Eshtar ruled the city and kingdom of Isin from 1934 to 1924 BCE. Like many early Mesopotamian kings he commissioned Sumerian poems in his honor, two of which later became standards in the scribal schools because of the simplicity of their language and clarity of their imagery. Here literacy and numeracy, both in the goddess Nisaba's gift, are closely tied to royal justice and fairness, which in turn generates the support and trust of the gods in the king's rule [ETCSL 2.5.5.2; Black et al. 2004, 308–11].

Praise Poem of King Lipit-Eshtar

[1–14]Lipit-Eshtar, proud king, enthroned prince, most seemly offshoot of kingship, who walks like Utu, brilliant light of the Land, lofty in nobility, riding on the great divine powers; who settles the people in the four quarters; favored by Enlil, beloved by Ninlil, trustworthy youth with shining eyes, worthy of the throne-dais, whose seemly head is adorned with the tiara, the good headdress, who holds in his hand the sceptre over the black-headed, prince Lipit-Eshtar, son of Enlil, wise shepherd, who leads the people to let them relax ...... in pleasant shade, lord, great bison, beloved by An! Your trust is put in mother Ninlil; Lipit-Eshtar, you exert great power.

[15–24]You, who speak as sweet as honey, whose name suits the mouth, longed-for husband of Inana, to whom Enki gave broad wisdom as a gift! Nisaba, the woman radiant with joy, the true woman, the scribe, the lady who knows everything, guides your fingers on the clay: she makes them put beautiful wedges on the tablets and adorns them with a golden stylus. Nisaba generously bestowed upon you the measuring rod, the surveyor's gleaming line, the yardstick, and the tablets which confer wisdom.

[25–39]Lipit-Eshtar, Enlil's son, you have realized justice and righteousness. Lord, your goodness covers everything as far as the horizon. King Lipit-Eshtar, counsellor with huge intelligence, who never tires of discussion, wise one whose decisions guide the people, amply wise, knowing everything in great detail! To decide justly the lawsuits of foreign countries, you recognize true and false even in people's thoughts. Lipit-Eshtar, you ...... the wicked, but you also know how to save some-one by commuting his death sentence; you know how to free someone from the severe punishment, from the jaws of destruction. The mighty do not commit rob-bery and the strong do not abuse the weak anymore: you have established justice in Sumer and Akkad and made the Land feel content.

[40–50]Lipit-Eshtar, king of Isin, king of Sumer and Akkad, you are the tablet writer of Nibru; Lipit-Eshtar, you are the constant attendant of the E-kur, Enlil's house. You are the beloved one of Enlil's and Ninlil's hearts. Hero Ninurta is your mighty commissioner. Chief minister Nuska is your aid in all matters. You have been rightly chosen by Nintud as the purification priest of Kesh. When in Urim, you are the youth who has the attention of Suen. You are the one to whom Enki gave the good headdress in Eridug. In Unug, Lipit-Eshtar, you are the delight of holy Inana's heart. In Isin, Ninisina set up your lofty throne-dais.

51–56 Among joyful songs of the heart, in an auspicious regnal year, the prince, the powerful prince surpassing in greatness and majesty, your father Ishme-Dagan, king of the Land, made the foundations of his throne firm for you. On the orders of An and Enlil, he silenced the loud (?) strife of the foreign countries.

57–63 Lipit-Eshtar, Enlil's son, you have made every mouth speak of your righteousness. The tablets will forever speak your praise in the school. May the scribes ...... and glorify you greatly! May eulogies of you never cease in the school! Perfect shepherd, youthful son of Enlil, Lipit-Eshtar, be praised!

III.b. Mathematical problems

Old Babylonian mathematical problems can be divided roughly into three types, whose boundaries are very blurred: problems about shape, area, and volume (line geometry); problems about finding unknowns by means of techniques such as completing the square (geometrical algebra); and problems about quasi-realistic labour scenarios that require knowledge of socially determined constants to solve them (quantity surveying). Tablets may simply list problems to be solved, with or without a numerical answer; or they may also give worked solutions in the form of algorithmic instructions. Problems are usually grouped thematically within a tablet and occasionally collected into series of tablets too. Some worked solutions also show the calculations to be performed. Examples of each are given below.

Most known tablets of mathematical problems were excavated, stored, and published without regard to their city or date of origin, at a time when tablets were considered not as archaeological artefacts but rather as bearers of text. It is almost impossible accurately to date or provenance mathematical tablets (or any other sort) from internal evidence alone; it is safest simply to assert that they were written in southern Iraq in the early second millennium BCE, most likely between about 1820 and 1740. Two notable exceptions are the tablets excavated from ancient towns along the Diyala Valley (the ancient kingdom of Eshnuna) to the northeast of Baghdad; and those that can be ascribed to a scribe named Sîn-iqisham, active in the city of Sippar in the later Old Babylonian period.

All the examples translated here were written in the Akkadian language, with more or less dense use of Sumerian logograms. Some of the technical terminology of Old Babylonian mathematics is explained in the note on translations on pp. 67–68. Diagrams are transcriptions of original illustrations on the tablets unless otherwise noted.

Line geometry

One tablet above all others has been instrumental to modern understanding of Old Babylonian line geometry: an unprovenanced collection of forty plane geometrical problems, with diagrams but no answers, now known as BM 15285 [Gadd 1922; Saggs 1960; Robson 1999, 208–17]. About a third of it is now missing, but the complete tablet must have measured rather more than 30 by 50 cm. Shadings on the diagrams mark my restorations of damage on the tablet.

All the problems concern plane figures drawn inside a square of 60 rods, on a four by four grid of smaller squares. They are organized in increasing order of complexity, with no solutions or numerical answers, giving the impression of a textbook. The first six problems concern just squares and circles, the following six only squares and right triangles; but problems (xxiv) onwards use many figures which are alien to modern mathematics, but which can be identified by relating diagram and text [Robson 1999, 34–56]. The absence of numerical answers on the tablet means that it is often impossible to know what the intended solutions were.

BM 15285

(i)

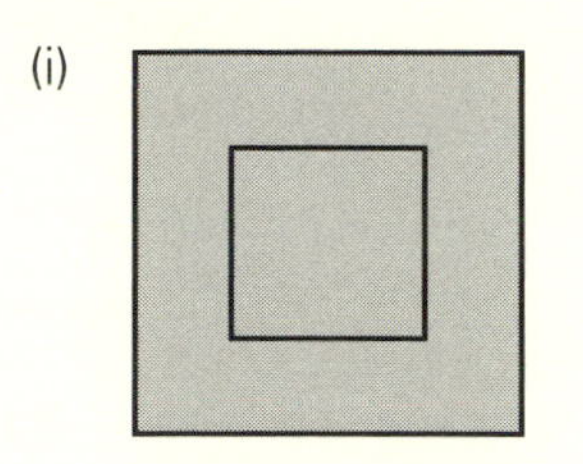

[The square-side is 1 cable. I extended a border on each side and then I drew a second square-side. What is its area?]

(ii)

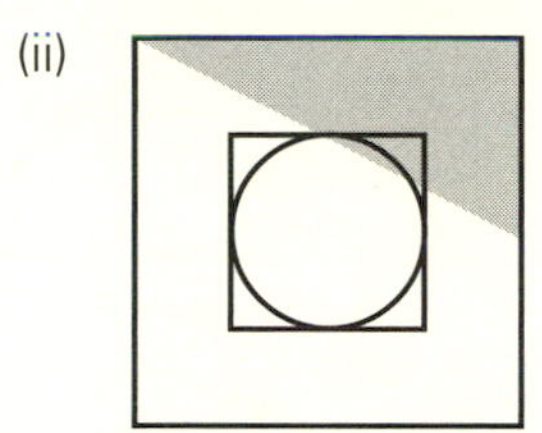

The square-side is 1 cable. I extended a border on each side and then I drew a second square-side. Inside the square-side I drew a circle. What are their areas?

(iii)

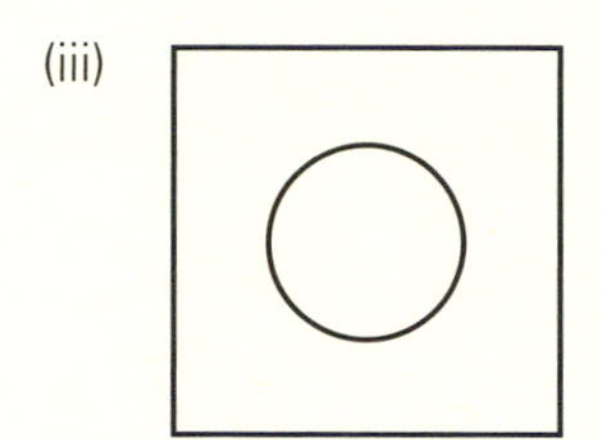

The square-side is 1 cable. I extended a border on each side and I drew a circle. What is its area?

(iv)

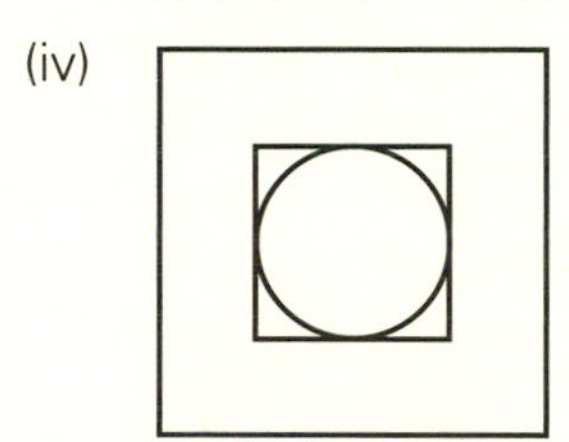

The square-side is 1 cable. Inside it I drew a square-side and a circle. The circle that I drew touched the square-side. What are their areas?

(v)

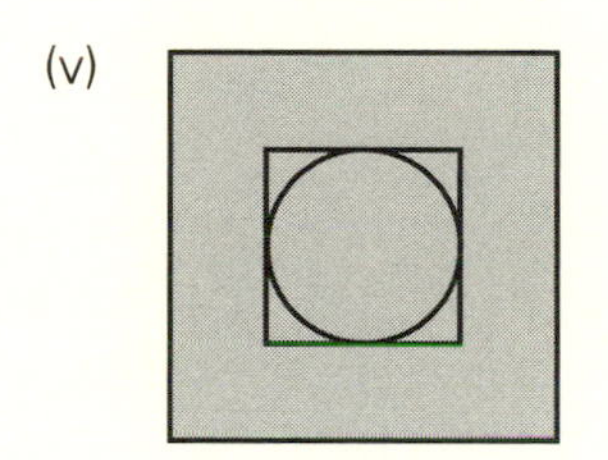

[The square-side is 1 cable.] <I drew> a second [square-side. Inside the second square-side I drew] 4 wedges and 1 circle. What are their areas?

(vi)

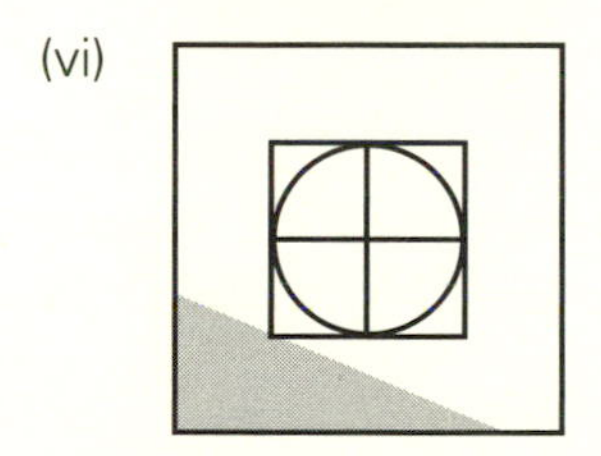

[The square-side is 1 cable. <I drew> a second square-side. Inside the second square-side] I drew [4 squares and 1 circle.] What are their areas?

(vii)

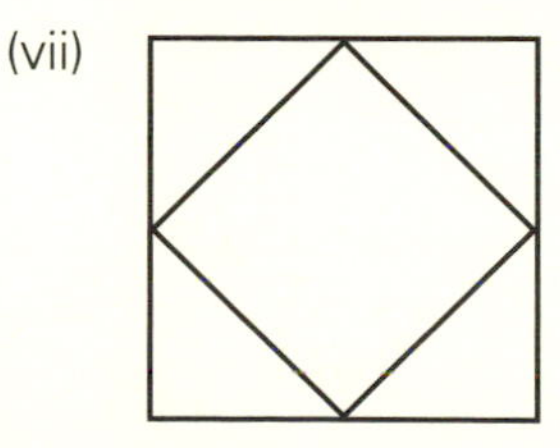

The square-side is 1 cable. Inside it I drew a second square-side. The square-side that I drew touches the outer square-side. What is its area?

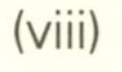

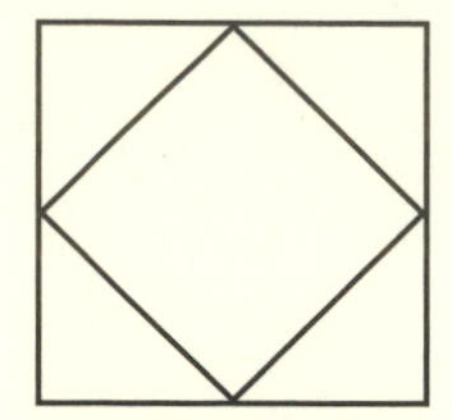

The square-side is 1 cable. Inside it <I drew> 4 wedges and 1 square-side. The square-side that I drew touches the second square-side. What is its area?

(ix)

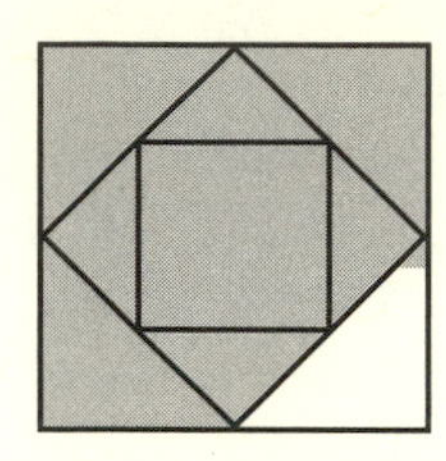

The square-side is 1 cable. Inside it I drew a square-side. The square-side that I drew touches the <first> square-side. Inside the second square-side I drew a third square-side. <The square-side> that I drew touches the <second> square-side. What is its area?

(x)

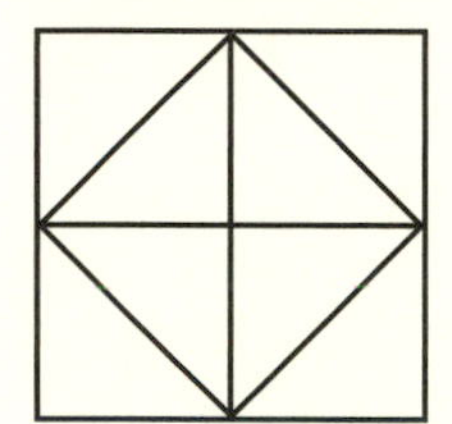

The square-side is 1 cable. Inside it I drew 8 wedges. What are their [areas]?

(xi)

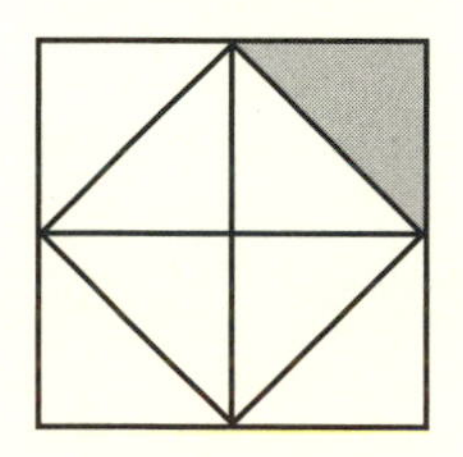

The square-side is 1 cable. Inside it I drew a square-side. The square-side that I drew touches the <first> square-side. Inside the <second> square-side I drew 4 wedges. <What are their areas?>

(xii)

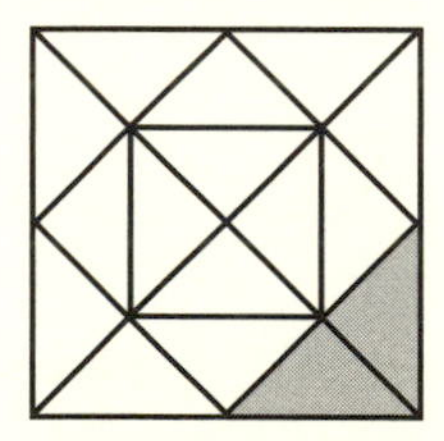

The square-side is 1 cable. Inside it [I drew] 16 wedges. What are their areas?

(xiii)

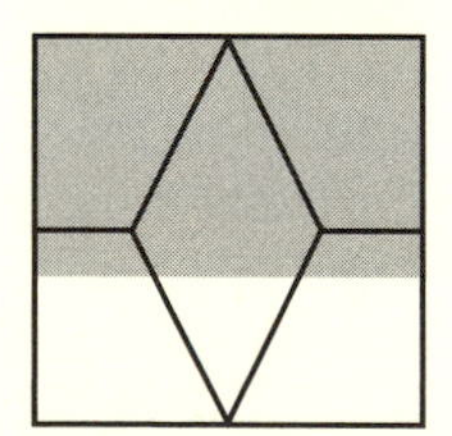

The square-side is 1 cable. Inside it I drew 4 ox's brows and 2 wedges. What are their areas?

(xiv)

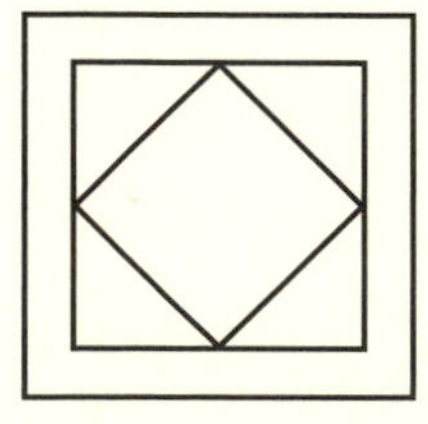

The square-side is 1 cable. I extended half on each side and I drew a square-side. Inside the second square-side I drew a third square-side. What is its area?

(xv) [*Diagram and text missing.*]

(xvi) [*Diagram and text missing.*]

(xvii)

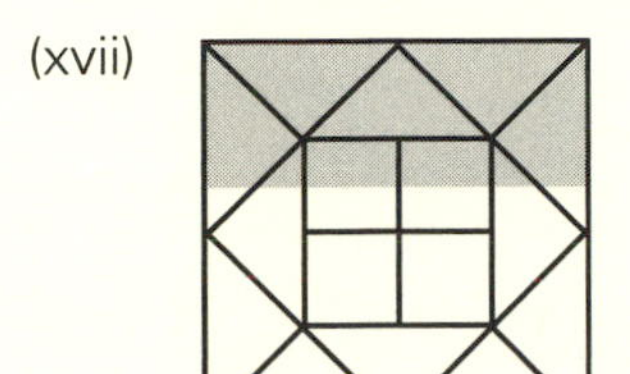

The square-side is 1 cable. <Inside it> I drew 12 wedges and 4 squares. What are their areas?

(xviii)

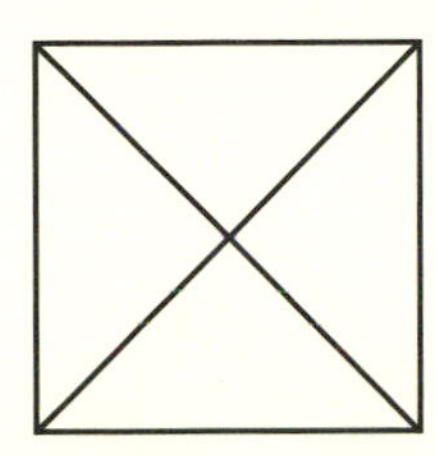

The square-side is 1 cable. Inside it I drew 4 wedges. What are their areas?

(xix) [*Diagram and text missing.*]

(xx) [*Diagram and text missing.*]

(xxi) [*Diagram and text missing.*]

(xxii) [*Diagram and text missing.*]

(xxiii)

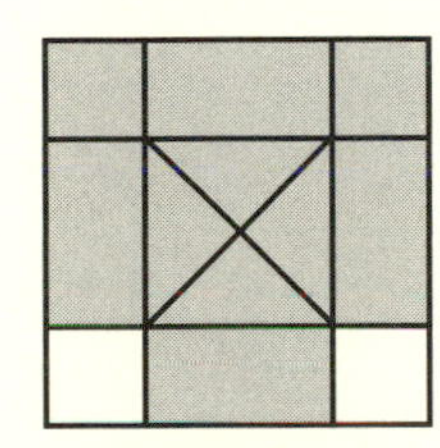

The square-side is 1 cable. Inside it <I drew> 4 squares, 4 rectangles and 4 wedges. What are their areas?

(xxiv)

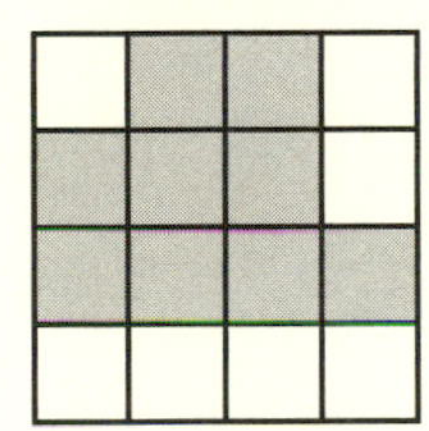

The square-side [is 1 cable]. Inside it I drew 16 squares. What are their areas?

(xxv)

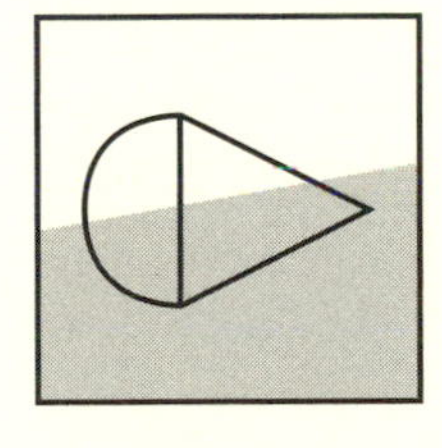

The … is 1 cable. … the width.

(xxvi) [*Diagram and text missing.*]

(xxvii) [*Diagram and text missing.*]

BM 15285

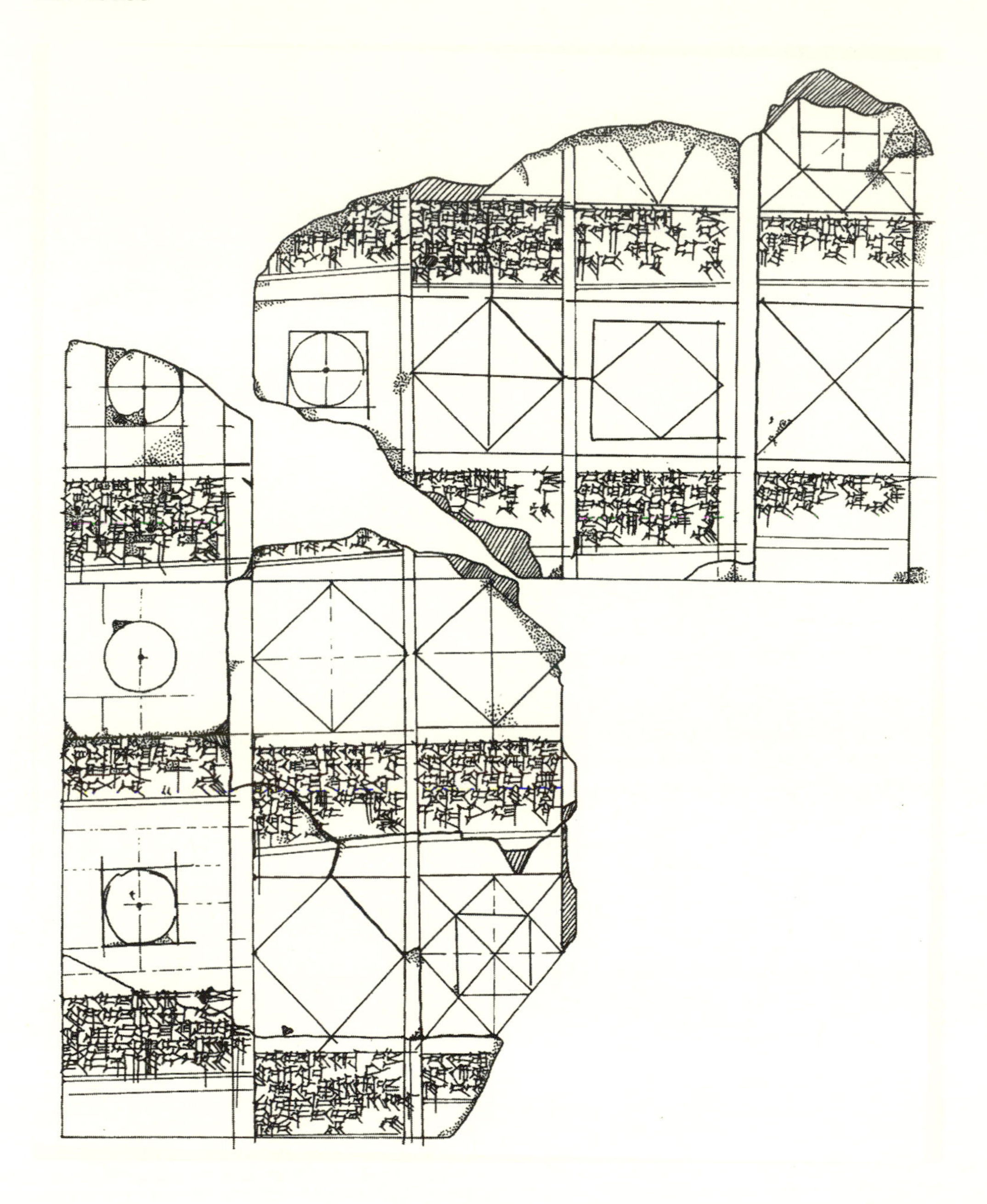

Obverse

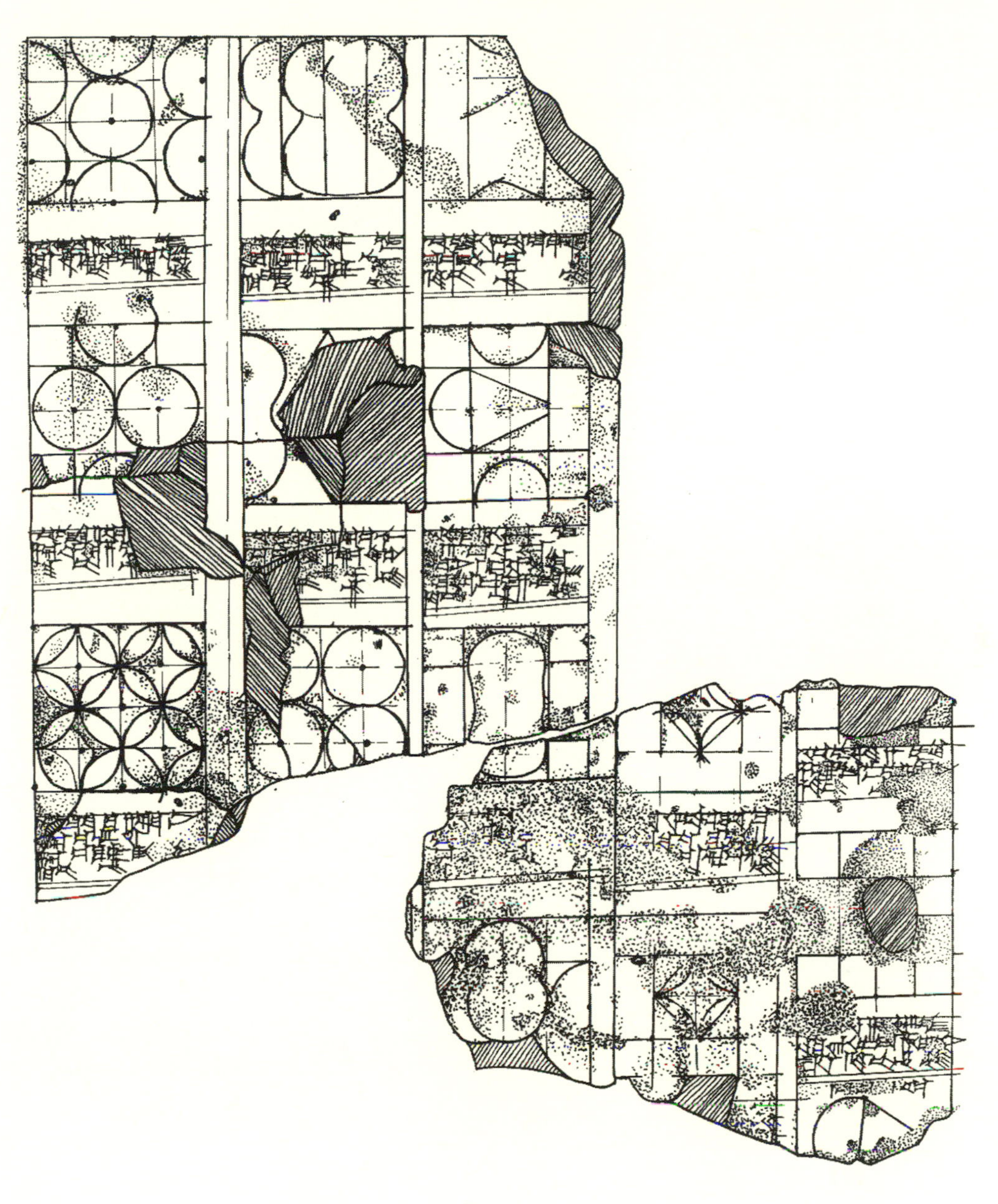

Reverse

(xxviii)

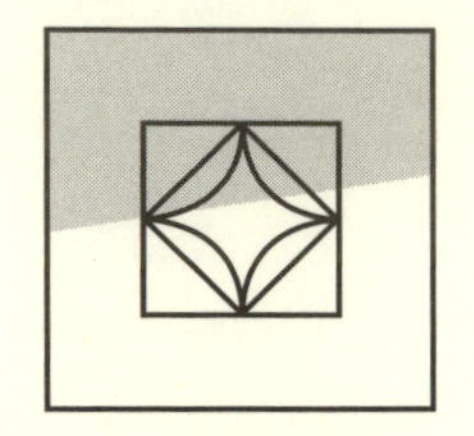

The square-side is 1 [cable]. I extended a border on each side and I drew a square-side. Inside the square-side that [I drew is] 1 cow's nose. What is [its area]?

(xxix)

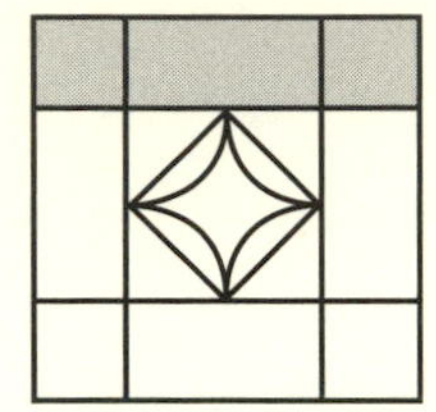

[*Text missing.*]

(xxx)

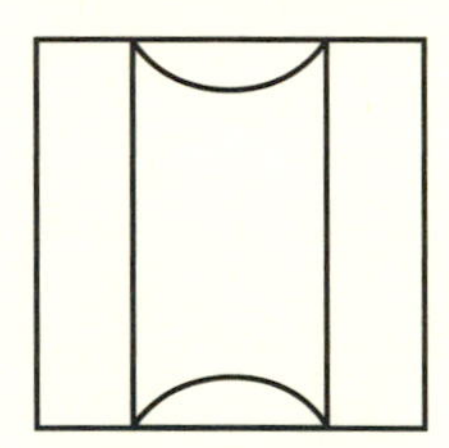

The square-side is 1 cable. I extended a border on each side and I drew the shape of a lyre. What is its area?

(xxxi)

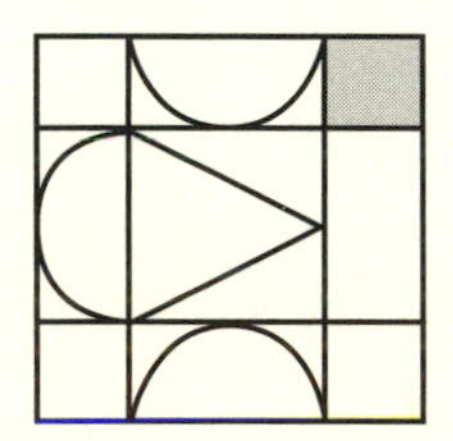

The square-side is 1 cable. [Inside it] are 2 crescent moons, [1] wedge, 1 cone (?), 1 rectangle and 4 squares. What are their areas?

(xxxii)

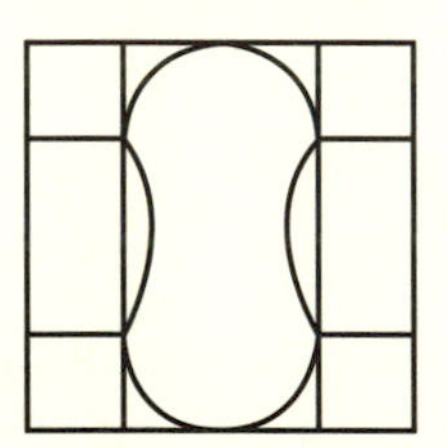

The square-side is [1] cable. Inside [it are 2] rectangles, [1] oval (?) and 4 squares. What are their areas?

(xxxiii)

[*Text missing.*]

(xxxiv)

The square-side is 1 cable. Inside it are 3 bows and 1 rectangle. What are their areas?

(xxxv)

The square-side is 1 cable. Inside it are 2 bows, 1 barge (?) and 4 ox's brows. [......] What are their [areas]?

(xxxvi)

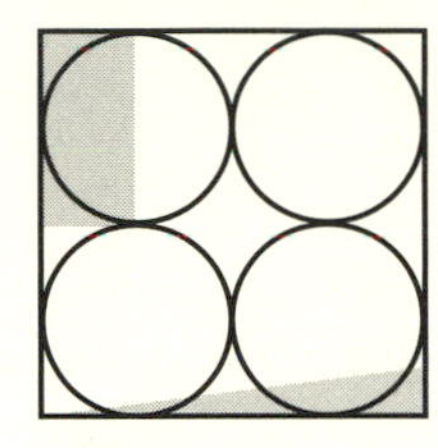

[*Text missing.*]

(xxxvii) [*Diagram and text missing.*]

(xxxviii)

The square-side is 1 cable. Inside it are 1 circle and 6 crescent moons. What are their areas?

(xxxix)

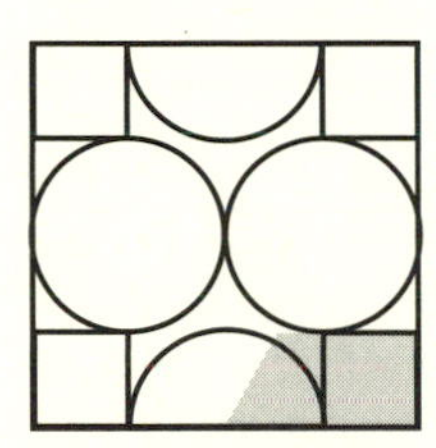

The square-side is 1 cable. <Inside it are> 2 circles, 2 crescent moons and 4 squares. What are their areas?

(xl)

The square-side is 1 cable. <Inside it are> 4 wedges, 16 barges, 5 cow's noses. What are their areas?

The properties of right triangles are explored in two model solutions of problems, both with diagrams, both from the ancient kingdom of Eshnuna (modern Diyala Valley [Baqir 1950, 1962; Høyrup 2002, 231–34, 257–61].

IM 55357

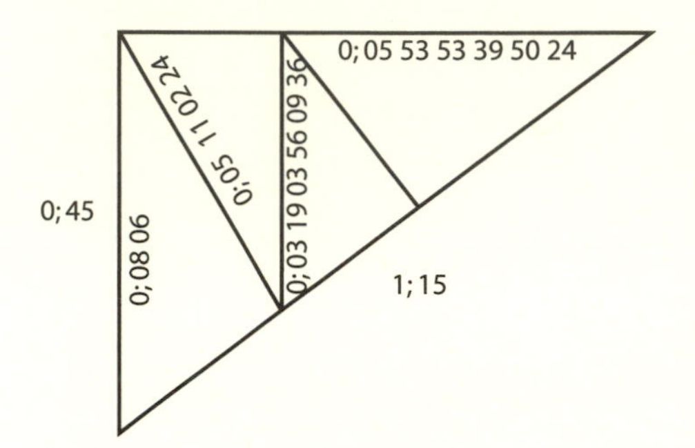

[1–5]A wedge. The length is 1, the long length 1;15, the upper width 0;45, the complete area 0;22 30. Within 0;22 30, the complete area, the upper area is 0;08 06, the next area 0;05 11 02 24, the third area 0;03 19 03 56 09 36, the lower area 0;05 53 53 39 50 24. What are the upper length, the middle length, the lower length, and the vertical?

[6–10]You, when you proceed: solve the reciprocal of 1, the length. Multiply by 0;45. You will see 0;45. Multiply 0;45 by 2. You will see 1;30. Multiply 1;30 by 0;08 06, the upper area. You will see 0;12 09. What squares 0;12 09? 0;27, the <length of the upper> wedge, squares (it). Break off <half of> 0;27. You will see 0;13 30. Solve the reciprocal of 0;13 30. Multiply by 0;08 06, the upper area. You will see 0;36, the dividing (?) length of 0;45, the width.

[11–17]Turn back. Take away 0;27, the length of the upper wedge, from 1;15. You leave behind 0;48. Solve the reciprocal of 0;48. You will see 1;15. Multiply 1;15 by 0;36. You will see 0;45. Multiply 0;45 by 2. You will see 1;30. Multiply 1;30 by 0;05 11 02 24. You will see 0;07 46 33 36. What squares 0;07 46 33 36? 0;21 36 squares (it). The width of the second triangle is 0;21 35. Break off half of 0;21 36. You will see 0;10 48. Solve the reciprocal of 0;10 48. <Multiply> by ... [*Here the text stops, unfinished.*]

Db$_2$-146

[1–16]If they ask you about a diagonal, as follows, "The diagonal is 1;15, the area 0;45. How much are the length and width?" you, when you proceed, draw 1;15, the diagonal, (and) its counterpart and then combine them so that 1;33 45 will come up. Its foundation (?) is 1;33 45. Copy your area times 2 so that 1;30 will come up. Take (it) away from 1;33 45 so that «1 30» 0;03 45 is the remainder. Take the square-side of 0;03 45 so that 0;15 will come up. <Break off> its half <so that> 0;07 30 will come up. Multiply by 0;07 30 so that 0;00 56 15 will come up. The ...... is 0;00 56 15. <Sum> 0;45, your area, over the ...... <so that> 0;45 56 15 will come up. Take the square-side of 0;45 56 15 so that 0;52 30 will come up. Draw 0;52 30 (and) its counterpart and then add the 0;07 30 that you combined to one, take away from one. Your length is 1, the width 0;45.

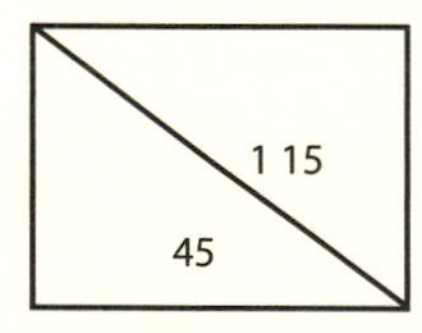

[16–25]If the length is 1, the width 0;45, how much are the area and diagonal? [You, when] you [proceed], combine the length so that [1 will come up]. Let your head hold [1]. Return and then combine 0;45, the width so that 0;33 45 will come up. Add it to your length so that 1;33 45 will come up. Take the square-side of 1;33 45 so that 1;15 will come up. Your diagonal is 1;15. Multiply your length by the width so that 0;45, your area, <will come up>. That is the procedure.

Mathematical problems were often constructed artificially. The trapezoid in YBC 4675 cannot actually exist; it has been created simply as a vehicle for exploring the trapezoidal triple 7, 13, 17 first seen in the Old Akkadian period (IM 58045, page 76) [MCT, B; Høyrup 2002, 244–49]. As diagrams were never drawn it scale, it is unlikely that anyone ever noticed that the trapezoid is not constructible. The function of the two long sides is simply to average out to 5 00.

YBC 4675

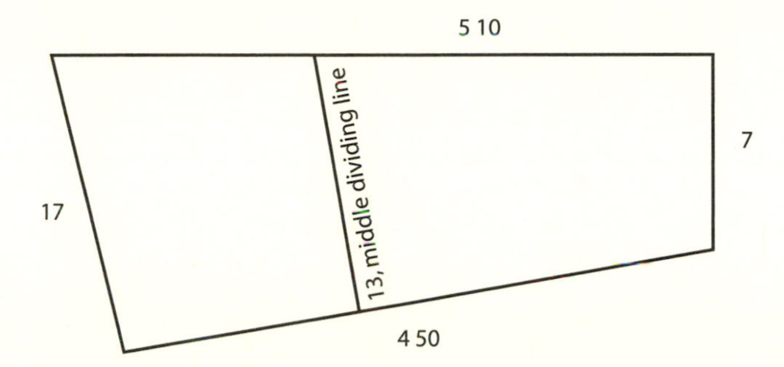

1–7 If an area with sloping lengths—the first length is 5 10, the second length 4 50, the upper width 17, the lower width 7, its area is 2 (bur). The area is divided into two, 1 (bur) each. How much is my middle dividing line? How much should I put for the long length and the short length so that 1 (bur) is correct; and for the second 1 (bur) how much should I put for the long length and the short length so that 1 (bur) may be correct?

7–12 You sum both the complete lengths and then you break off their halves, so that 5 00 will come up for you. You solve the reciprocal of 5 00 that came up for you, and then—as for the upper width which exceeded the lower width by 10—you multiply by 10, the excess, so that it will give you 0;02. You turn around.

12–16 You combine 17, the upper width, so that 4 49 will come up for you. You take away 2 00 from the middle of 4 49 so that 2 49 is the remainder. You take its square-side so that 13, the middle dividing line, will come up for you.

17–R7 You sum 13, the middle dividing line that came up for you, and 17, the upper width, and then you break of their halves so that 15 will come up for you. You solve the reciprocal of 15, and then you multiply by 1 (bur), the area, so that it will give you 2 00. You multiply the 2 00 which came up for you by 0;02, the rising-factor (?), so that 4 will come up for you. You add 4 that came up to you to 2 cables, so that 2 04 is the long length. You take away 4 from the second 2 cables so that 1 56 is the short length. You do (the necessary calculations) so that (you will see that) 1 (bur) is correct. You turn around.

R7–R16 You sum 13, the middle dividing line that came up for you, and 7, the lower width, and then you break their halves so that 10 will come up for you. You solve the reciprocal of 10, and you multiply by 1 (bur), the area, so that 3 cables will come up for you. You multiply the 3 cables which came up for you by 0;02, the rising-factor (?), and the 6 will come up for you. You add 6 to 3 cables so that 3 07 is the long length. You take away 6 from 3 cables so that 2 54 is the short length. You combine (what is necessary) so that (you will see that) 1 (bur) is correct.

Geometrical algebra

In recent years Jens Høyrup has written extensively and authoritatively on Old Babylonian geometrical algebra, which comprises about half of the Old Babylonian corpus of mathematical problems [Høyrup 2002]. He has shown that underlying the apparently arithmetized texts are fundamentally concrete formulations that describe how to manipulate areas and lines in order to find unknowns. YBC 6967 is one of the simplest examples [MCT, text Ua; Høyrup 2002, 55–58]. The problem is to find a pair of mutual reciprocals knowing only the difference between them (and, by definition, that their product is 60: see BM 106444, on page 79). By visualizing the unknowns as the sides of rectangle of area 60 the rectangle can be manipulated into a gnomon and the original lengths found by completing the square:

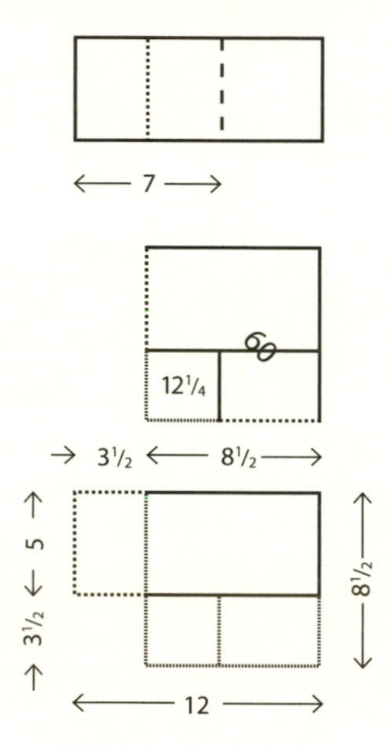

No such diagrams survive on the tablets themselves, but their use is inferred through a literal interpretation of the operations described. They explain, for instance, how two halves of a quantity can be manipulated independently of each other—whereas in symbolic algebra, a halved quantity is simply shrunk, rather than cut in two—and why each of four different verbs of multiplication is used in any particular context, as each has a specific geometrical or arithmetical function.

YBC 6967

A reciprocal exceeds its reciprocal by 7. What are the reciprocal and its reciprocal? You: break in two the 7 by which the reciprocal exceeds its reciprocal so that 3;30 (will come up). Combine 3;30 and 3;30 so that 12;15 (will come up). Add 1 00, the area, to the 12;15 which came up for you so that 1 12;15 (will come up). What squares 1 12;15? 8;30. Draw 8;30 and 8;30, its counterpart, and then take away 3;30, the holding-square, from one; add to one. One is 12, the other is 5. The reciprocal is 12, its reciprocal is 5.

BM 13901, an unprovenanced set of twenty-four model solutions of problems in quadratic geometrical algebra, progresses from very simple scenarios about single squares to complex situations involving two squares or more [Thureau-Dangin 1936; Høyrup 2002, 50–77 passim]. It is analyzed in detail by [Høyrup 2001].

The diagrams of problems (i), (ii), and (xxxiii) below are modern attempts to interpret the scribe's words. In problem (i), the assignment is to find the side of a square given the sum of its area and side. In order to conceptualize this sum as a rectangle, the scribe converts the side into an area by multiplying it by 1, the "projection," and appending it to the square. As in YBC 6967, the solution is found through breaking the resultant rectangle at the midway point of the difference between the two sides, and rearranging the whole to create an L-shaped figure in order to generate a large square whose area can be calculated. The procedure is shown in this modern reconstruction:

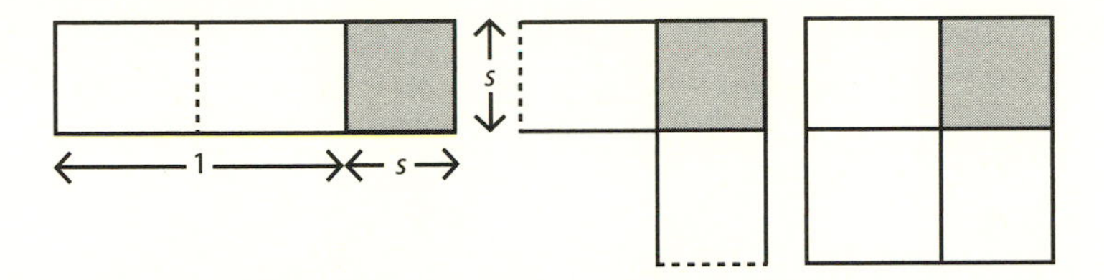

Problem (ii), on the other hand, provides the difference between the area and side of the square before asking for the side. In this case, a "projection" is removed from the square to produce a rectangle. Again the difference between the two sides is cut in half and rearranged in a by now familiar configuration:

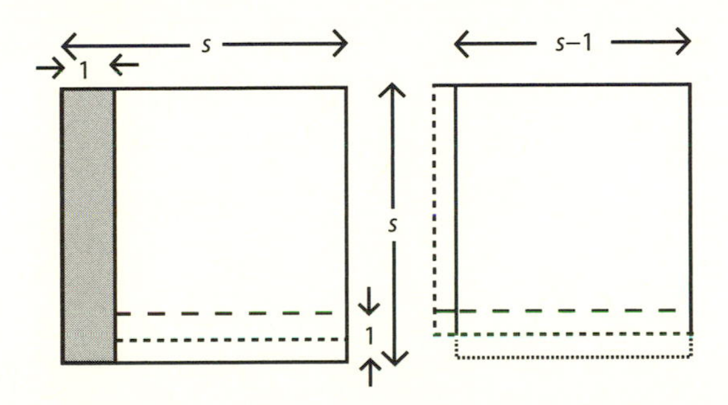

The following problems use similar cut and paste methods. In problem (xxiii), for instance, the four sides of the square are "projected" on to its area to make a cross-shaped figure of known area. Dividing this area by four produces an L-shaped figure that will then yield the answer in the usual way:

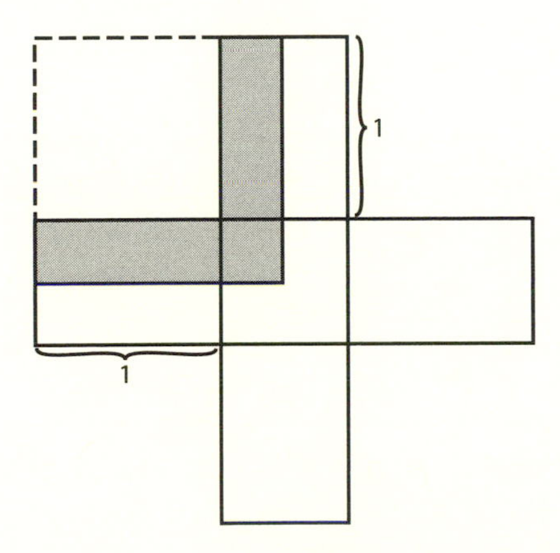

Interestingly, the final problem, (xxxiv), contains several calculation errors, perhaps because the scribe was losing concentration. The mistakes are eventually made to cancel each other out, presumably to produce the correct result without the bother of rewriting the procedure. It is in essence the same problem as (xiv) and those on Strasbourg 363 (page 110).

BM 13901

(i) I summed the area and my square-side so that it was 0;45. You put down 1, the projection. You break off half of 1. You combine 0;30 and 0;30. You add 0;15 to 0;45. 1 squares 1. You take away 0;30 which you combined from inside 1 so that the square-side is 0;30.

(ii) I took away my square-side from inside the area so that it was 14 30. You put down 1, the projection. You break off half of 1. You combine 0;30 and 0;30. You add 0;15 to 14 30. 14 30;15 squares 29;30. You add 0;30 which you combined to 29;30 so that the square-side is 30.

(iii) I took away a third of the area. I added a third of the square-side to inside the area so that it was 0;20. You put down 1, the projection. [You take away] a third of 1, the projection, [and] you multiply 0;40 by 0;20. You write down 0;13 20. You break [half of 0;20], the third which you added. You combine 0;10 and 0;10. You add 0;01 40 to 0;13 20. 0;15 [squares] 0;30. You take away [0;10 which you combined from inside 0;30] so that (it is) 0;20. The reciprocal of 0;40 [is 1;30. You multiply by 0;20 so that] the square-side is [0;30].

(iv) [I took away] a third [of the area]. I summed [the area and] my square-side (so that) it was 4 46;40. You put down [1, the projection]. You take away 0;20, a third of 1, the projection, and you multiply 0;40 by 4 46;40, and [you write down] 3 11;[06 40]. You break [in half] 1, the projection. You [combine] 0;30 and 0;30. You add [0;15 to 3 11;06 40]. 3 11;21 40 squares 13;50. [You take away 0;30] which you combined from [inside] 13;40 and <(it is) 13;20. The reciprocal of 0;40 is> 1;30. You multiply [by] 13;20 so that the square-side is 20.

(v) [I summed the area and my square-side and a third] of my square-side [so that it was 0;55]. You put down [1, the projection]. You add a third of [1, the projection to 1]: 1;20. [You combine] its half, 0;40, [and 0;40]. You add 0;26 40 to 0;55 and [1;21 40 squares 1;10. Take away 0;40 that you] combined from the middle of 1;10 so that the square-side is [0;30].

(vi) [I summed the area and two-thirds] of my square-side [so that it was 0;35]. You put down [1, the projection]. Two-thirds of [1, the projection] (is) 0;40. You combine its half, 0;20 and 0;20. [You add 0;06 40 to 0;35 and then] 0;41 40 squares 0;50. You] take away 0;20 that you combined from the middle of 0;50] so that the square-side is 0;30.

(vii) [I summed my square-side seven times and the area] eleven times [so that it was 6;15]. You write down [7 and 11. You multiply] 11 by 6;15 [so that (it is) 1 08;45]. You break off [half of 7. You combine 3;30 and 3;30.] You add [12;15 to 1 08;45] and [1 21 squares 9. You take away 3;30 that you] combined from the middle of 9 [and you write down 5;30. The reciprocal of] 11 cannot be solved. [What should I put down by 11 that] will give me [5;30]? [Its quotient(?) is 0;30. The square-side is 0;30.]

(viii) [I summed the areas of my two square-sides so that] (it was) 0;21 40. [And I summed my square-sides so that it was 0;50]. You break [off half of 0;21] 40. [You write down 0;10 50. You break off half of 0;50.] You combine [0;25 and 0;25. You take away 0;10 25 from the middle of 0;10 50 and then 0;00 25] squares [0;05]. You add 0;05 to the first 0;25 [so that the first square-side is 0;30]. You take away 0;05 from the second 0;25 so that [the second square-side is 0;20].

(ix) I summed the areas of my two square-sides so that it was 0;21 40. A square-side exceeds the (other) square-side by 0;10. You break off half of 0;21 40 and you write down 0;10 50. You break off half of 0;10 and then you combine 0;05 and 0;05. You take away 0;00 25 from the middle of 0;10 50 and then 0;10 25 squares 0;25. You write down 0;25 twice. You add 0;05 that you combined to the first 0;25 so that the square-side is 0;30. You take away 0;05 from the middle of the second 0;25 so that the second square-side is 0;20.

(x) I summed the areas of my two square-sides so that (it was) 21;15. A square-side is less than the (other) square-side by a seventh. You put down 7 and 6. You combine 7 and 7: 49. You combine 6 and 6. You sum 36 and 49 so that (it is) 1 25. The reciprocal of 1 25 cannot be solved. What should I put down by 1 25 that will give me 21;15? 0;15 squares 0;30. You multiply 0;30 by 7 so that the first square-side is 3;30. You multiply 0;30 by 6 so that the second square-side is 3.

(xi) I summed the areas of my two square-sides so that (it was) 28;15. A square-side exceeds the (other) square-side by a seventh. You write down 8 and 7. You combine 8 and 8: 1 04. You combine 7 and 7. You sum 49 and 1 04: 1 53. The reciprocal of 1 53 cannot be solved. What should I put down by 1 53 that will give me 28;15? 0;15 squares 0;30. You multiply 0;30 by 8 so that the first square-side is 4. You multiply 0;30 by 7 so that the second square-side is 3;30.

(xii) I summed the areas of my two square-sides so that (it was) 0;21 40. I combined my two square-sides so that (it was) 0;10. You break off half of 0;21 40 and then you combine 0;10 50 and 0;10 50. It is 0;01 57 «46» <21> 40. You combine 0;10 and 0;10. You take away 0;01 40 from 0;01 57 «46» <21> 40 and 0;00 17 «46» <21> 40 squares 0;04 10. You add 0;04 10 to the first 0;10 50 and then 0;15 squares 0;30. The first square-side is 0;30. You take away 0;04 10 from the middle of the second 0;10 50 and then 0;06 40 squares 0;20. The second square-side is 0;20.

(xiii) I summed the areas of my two square-sides so that (it was) 0;28 20. A square-side was a quarter of the (other) square-side. You write down 4 and 1. You combine 4 and 4: 16. You combine 1 and 1. You sum 1 and 16 so that (it is) «16» <17>. The reciprocal of 17 cannot be solved. What should I put down by 17 that will give me 0;28 20? 0;01 40 squares 0;10. You multiply 0;10 by 4 so that the first square-side is 0;40. You multiply 0;10 by 1 so that the second square-side is 0;10.

(xiv) I summed the areas of my two square-sides so that (it was) 0;25 25. A square-side was two-thirds of the (other) square-side [and 0;05] rods. You write down 1 and 0;40 and 0;05 over 0;40. [You combine] 0;05 and 0;05. [You take away 0;00 25 from the middle of 0;25 25 and then you write down 0;25. You combine 1 and 1: 1. You combine 0;40 and 0;40. You sum 0;26 40 and 1 and then you multiply 1;26 40 by 0;25 and then you write down 0;36 06 40.] You [multiply 0;05 by] 0;40 and then you multiply 0;03 20 by 0;03 20. You add 0;00 11 06 40] to 0;36 06 40 [and then 0;36 17 46 40 squares 0;46 40. You take away 0;03] 20 that you combined [from the middle of 0;46 40] and then you write down 0;43 20. [The reciprocal of 1;26 40 cannot] be solved. What [should I put down] by 1;26 40 [that] will give me [0;43 20]? Its quotient (?) is 0;30. [You multiply 0;30 by 1 so that] the first square-side is [0;30. You multiply 0;30 by 0;40 and then] you sum [0;20] and [0;05] so that the second square-side [is 0;25].

(xv) I summed [the areas of my four] square-sides so that (it was) 0;27 05. [A square-side was two-thirds], a half, a third of the (other) square-side(s). You write down [1 and 0;40 and 0;30 and 0;20]. You combine 1 and 1: <it is> 1. [You combine 0;40 and 0;40]: it is [0;26] 40. You combine 0;30 and 0;30: it is 0;15. [You combine 0;20 and 0;20. You sum 0;06] 40 and 0;15 and 0;26 40 and 1. [The reciprocal of 1;48] 20 cannot be solved. [What] should I put down [by 1;48 20] that will give me 0;27 05? [0;15 squares 0;30.] You multiply [0;30 by 1] so that the first square-side is 0;30. [You multiply 0;30 by 0;40] so that the second square side is 0;20. [You multiply 0;30 by 0;30] so that the third square-side is 0;15. [You multiply 0;30 by 0;20 so that] the fourth square-side is 0;10.

(xvi) I took away [a third of the square-side] from inside the area so that (it was) 0;05. [You put down 1, the projection.] A third of 1, the projection is 0;20. You break off [half of 1, the projection]. You multiply 0;30 by 0;20 so that (it is) 0;10. [You] combine [0;10 and 0;10]. You add 0;01 40 to 0;05 and then [0;06 40 squares 0;20]. You add 0;10 that you combined to 0;20 so that the square-side is 0;30.

(xvii) I summed [the areas of] my three square-sides so that (it was) 10 12;45. A square-side was a seventh of the (other) square-side. You write down 49 and 7 and 1. [You] combine 49 and 49: <it is> 40 01. You combine 7 and 7: it is 49. You combine 1 and 1: <it is> 1. You sum 40 01 and 49 and 1 so that (it is) 40 51. The reciprocal of 40 51 cannot be solved. What should I put down by 40 51 that will give me 10 12;45? Its quotient (?) is 0;15. 0;15 squares 0;30. You multiply 0;30 by 49 so that the first square-side is 24;30. You multiply 0;30 by 7 so that the second square-side is 3;30. You multiply 0;30 by 1 so that the third square-side is 0;30.

(xviii) I summed the areas of my three square-sides so that (it was) 0;23 20. A square-side exceeds the (other) square-side by 0;10. You multiply the 0;10 which exceeds by 1. You multiply 0;10 by 2. <You combine> 0;20 and 0;20: it is 0;06 40. You combine 0;10 and 0;10. You add 0;01 40 to 0;06 40. You take away 0;08 20 from the middle of 0;23 20 and then you multiply 0;15 by 3, the square-sides. You write down 0;45. You sum 0;10 and 0;20 and then you combine 0;30 and 0;30 You add 0;15 to 0;45 and then 1 squares 1. You take away 0;30 that you combined and then you write down 0;30. The reciprocal of 3, the square-sides, is 0;20. You multiply by 0;30. The square-side is 0;10. You add 0;10 to 0;10 so that the second square-side is 0;20. You add 0;10 to 0;20, so that the third square-side is 0;30.

(xix) I combined the square-sides and then I summed (it and) the area. I combined as much as a square-side exceeds the (other) square-side with itself. I took it away from [the middle of the area] so that (it was) 0;23 20. I [summed] my square-sides [so that (it was) 0;50]. You copy 0;23 20 twice. You write down 0;46 40. [You combine 0;50 and 0;50. You take away 0;41 40 from the middle of 0;46 40 and then the reciprocal of 0;05 is 12. You multiply 0;05 by 0;05. 0;00 25 squares 0;05. You break off half of 0;50. You add 0;25 to 0;05, so that the first square-side is 0;30. You take away 0;05 from the middle of 0;25, so that the second square-side is 0;20.]

(xx)–(xxii) [*Missing.*]

(xxiii) An area. I added four widths and the area so that (it was) 0;41 40. You write down 4, the four sides. The reciprocal of 4 is 0;15. You raise 0;15 by 0;41 40, so that (it is) 0;10 25. You write it down. You add 1, the projection, and then 1;10 25 squares 1;05. You take away 1, the projection, which you added and then you copy 0;05 twice and then 0;10 rods squares itself.

(xxiv) I summed the areas of my three square-sides so that (it was) 0;29 10. A square-side is (equal to) two-thirds of the (other) square-side and 0;05 rods, (which is equal to) half the (third) square-side and 0;02 [30] rods. You write down 1 and 0;40 and 0;20 <and> 0;05 over 0;40. You write down 0;02 30 over 0;20. You break off half of 0;05. You add 0;02 30 to 0;02 30. You combine 0;05 and 0;05. You write down 0;00 25. You combine 0;05 and 0;05. You add 0;00 25 to 0;00 25 and then you take away «0;25 25» <0;00 50> from the middle of 0;29 10. You write down 0;03 45. You combine 1 and 1: 1. You combine 0;40 and 0;40: 0;26 40. You combine 0;20 and 0;20. You sum 0;06 40 and 0;26 40 and 1 and then you multiply 1;22 30 by 0;03 45 so that (it is) 0;05 50. You multiply 0;40 by 0;05: 0;03 20. You multiply [0;20] by 0;02 30 <added to half of 0;05>: 0;00 50. You sum 0;03 20 and 0;50 and then you combine 0;04 10 and 0;04 10. You add 0;00 17 21 40 to 0;05 50 and then 0;06 07 21 40 squares 0;19 10. You take away 0;04 10 from the middle of 0;19 10 and then you «copy 0;15 twice» <multiply by the reciprocal of 1;33 40>. You multiply 0;30 by 1 so that the first square-side is 0;30. You multiply 0;30 by 0;40 and then you add 0;20 to 0;05 so that the second square-side is 0;25. You break off half of 0;25 and then you add 0;12 30 to 0;02 30 so that the third [square-side] is 0;15.

Some tablets contain long, densely formulated lists of problems whose methods of solution are not known. A 24194 gives a particularly complex sequence of algebraic problems, with seven basic scenarios each followed by a long sequence of variations [MCT, text T]. In each case there are a rectangle of known area (1 eshe = 600 sar) and a complicated relationship between the length and width of that rectangle; the goal is to find the length and width individually. In this extract the first expression is equivalent to $l \times [2\ 29 + \{l + w + 2(l - w)\}/14]/(7 \times 11) = 32$ (rods). For each of the seven scenarios a set of problems is generated by adding or subtracting the complicated expression (CE) to or from a shorter expression (SE) about the length and/or width of the rectangle; for instance, the first problems below ask for $2\text{CE} + l = 34$, $l - \text{CE} = 28$, where the SE is simply the length, l. A statement such as “I copied 20 times; it exceeded by 10” should be understood as $20\text{CE} + 10 = \text{SE}$, as here “it” refers to the SE.

The colophon gives the total number of problems as 240 and states that it is the tenth tablet in a series. Two long extracts (obverse II 9218–34 and reverse III 2–end) are presented here. Curly brackets have been added to the first to help group mathematical expressions correctly.

A 24194

An area of 1 (eshe). A 14th part of {length (and) width and 2 times the excess of [length over] width}; [I added 2 29]. I added a 7th part of an 11th part (of all this) to the length. (It was) 32.

I copied 2 times, I added. (It was) 34.
I took away. (It was) 28.
I copied 2 times, I took away. [(It was) 26.]
[I copied 1]5 times. [(It was) the length.]
I copied 20 times. It exceeded by 10.
I copied 10 times. It fell short by 10.
I added to the width. (It was) 22.
I copied 2 times, I added. (It was) 24.
I took away. (It was) 18.
I copied 2 times, I took away. (It was) 16.
I copied 10 times. (It was) the width.

An area of 1 (eshe). Length and excess of length over width and a 4th part (of all this) and then I added 6. Its 8th part (of all this); I added 15. Its 11th part (of all this); I added the length. Its 8th part (of all this) I added [to the length. (It was) 3]4.

I copied 2 times, I added. (It was) 38.
I took away. (It was) 26.
I copied 2 times, I took away. (It was) 22.
I copied 7;30 times. (It was) the length.
I copied 10 times. It exceeded by 10.
I copied 5 times. It fell short by 10.
I added the width. [(It was) 24.]
I copied 2 times, I added. [(It was) 28.]
I took away. (It was) 16.
[I copied 2] times, I took away. (It was) 12.
[I copied 5] times. (It was) the width.

I copied 5 times. It exceeded by 4.
I added [length and width]. (It was) 54.
I copied 2 times, I added. [(It was) 58.]
I took away. (It was) 46.
I copied 2 times, I took away. (It was) 42.
I copied [12];30 [times]. (It was) length, width.
I copied 15 [times]. It exceeded by 10.
I copied 10 [times]. It fell short by 10.

[I added to the length and the excess of length over width.] (It was) 44.
[I copied 2] times, I added. (It was) 48.
I took away. (It was) 36.
I copied 2 times, I took away. (It was) 32.
[I copied 10] times. (It was) the length.
I copied 12 times. It exceeded by 8.
I copied 8 times. It fell short by [8].

I added to the length and a 3rd part of the length [to] the width [and] a 4th part of the width so that (it was) 1 09.
I copied 2 [times], I added. (It was) 1 13.
I took away, so that (it was) 1 01.
I copied 2 times, I took away. (It was) 57.
I copied 20 times. It exceeded by 15.
I copied 15 times. It fell short by 5.

I added [a 3rd] part of the length, a 4th part of the width and 3 times the excess of the length over the width, so that (it was) 49.
I copied «3» <2> times, I added. (It was) 53.
I took away, so that (it was) 41.
I copied 2 times, I took away. [(It was) 37]
I copied 10 times. It exceeded by 5.
I copied 15 times. It fell short by 15.

I added 3 times the length, 2 times [the width], so that (it was) 2 14.
I copied 2 times, I added. (It was) 2 18.
I took away. (It was) 2 06.
I copied 2 times, I took away. (It was) 2 02.
I copied 30 times. It fell short by 10.
I copied 40 times. It exceeded by 30.
I added [....... (It was) 2 54.]
I copied 2 times, I added. [(It was) 2 58.]
[I took away.] (It was) 2 46.
I copied 2 times, I took away. (It was) 2 42.
I copied 40 times. It exceeded by 10.
I copied 45 times. It fell short by 10.

4 sixties of exercises, tablet 10.

A variety of scenarios was used to set up problems in finding unknowns, as shown by a group of small tablets now in Strasbourg Museum [Frank 1928 nos., 7, 8, 10, 11; MKT: 243–56, 259–63, 311–313; TMB: 84–92; Høyrup 2001, 239–44 (Str 367)].

The three problems of Strasbourg 363 ask for the sides of two different squares, given their sums and various relationships between them and the side of an auxiliary square, which is used in the solution to all three problems. If for convenience we denote the sides of the unknown squares as u and v, and that of the auxiliary square as s, the first problem states that $v = 2/3s - 10$ (and implicitly $u = s$), the second $v = 2/3s + 5$ and $u = s + 10$, and the third $v = 2/3s + 5$, $u = s + 20$. The procedure for solving them is essentially the same in each case.

First, the combined area of the squares is transformed into a sum of auxiliary squares, and then scaled and manipulated into an L-shape in the usual way. Diagrams for the first problem are reconstructed below:

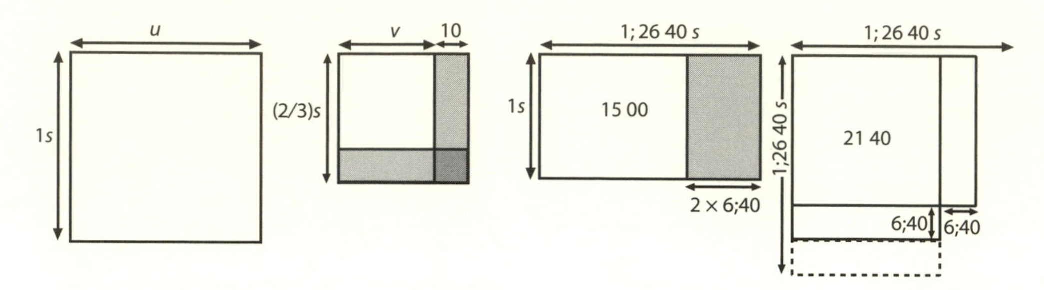

Strasbourg 363

(i) Sum the areas of 2 square-sides so that (it will give) 16 40. (One) square-side is 2/3 of the (other) square-side. Take away 10 from the small square-side. What are the square-sides?

You, when you proceed: combine 10. It will give 1 40. Take away 1 40 from 16 40 so that it will give 15 00. Combine 1 so that it will give 1. Combine 0;40 so that (it will give) 0;26 40. Sum 1 and 0;26 40 so that it will give 1;26 40. Multiply 1;26 40 by 15 00. It will give 21 40. Multiply 0;40 by 10 so that it will give 6;40. Combine 6;40 so that it will give 44;26 40. Add 44;26 40 to 21 40 so that it will give 22 24;26 40. The square-side of 22 24;26 40 is 36;40. Add the 6;40 that you combined to 36;40 so that it will give 43;20. What should I put to 1;26 40 so that it will give 43;20? Put 30. Multiply 30 by 1 so that the large square-side is 30. Multiply 30 by 0;40 so that it will give 20. Take away 10 from 20 so that the small square-side is 10.

(ii) Sum the areas of 2 square-sides so that (it will give) 37 05. (One) square-side is 2/3 of the (other) square-side. Add 10 to the large square-side, add 5 to the small square-side. What are the square-sides?

You, when you proceed: combine 10 so that it will give 1 40. Combine 5. It will give 25. Sum 1 40 and 25 so that it will give 2 05. Take away 2 05 from 37 05 so that it will give 35 00. Combine 1 so that it will give 1. Combine 0;40 so that (it will give) 0;26 40. Sum 1 and 0;26 40 so that it will give 1;26 40. Multiply 1;26 40 by 35 00 so that it will give 50 33;20. Multiply 10 by 1. [It will give] 10. Multiply 0;40 by 5 so that it will give 3;20. Sum 10 and 3;20, so that it will give 13;20. Combine 13;20. It will give 2 57;46 40. Add 2 57;46 40 to 50 33;20 so that it will give 53 31;06 40. The square-side of 55 31;06 40 is 56;40. Take away 13;20 from 56;40 so that it will give 43;20. What should I put to 1;26 40 so that it will give 43;20? Put 30. Multiply 30 by 1 so that it will give 30. Add 10 to 30 so that the large square-side is 40. Multiply 30 by 0;40 so that it will give 20. Add 5 to 20 so that it will give 25, the small square-side.

(iii) Sum the areas of 2 square-sides so that (it will give) 52 05. (One) square-side is 2/3 of the (other) square-side. Add 20 to the large square-side, add 5 to the small square-side. What are the square-sides?

You, when you proceed: combine 20 so that it will give 6 40. Combine 5 so that it will give 25. Sum 6 40 and 25 so that it will give 7 05. Take away 7 05 from 52 05 so that it will give 45 00. Combine 1 so that it will give 1. Combine 0;40. It will give 0;26 40. Sum 1 and 0;26 40 so that it will give 1;26 40. Multiply 1;26 40 by 45 00 so that it will give 1 05 00. Combine 20 and 1 so that <it will give 20>. Combine 0;40 and 5 so that it will give 3;20. Sum 20 and 3;20 so that it will give 23;20. Combine 23;20 so that it will give 9 04;26 40. Add 9 04;46 40 to 1 05 00 so that it will give 1 14 04;46 40. The square-side of 1 14 04;46 40 is 1 06;40. Take away 23;20 from 1 06;40 so that it will give [43;20]. What should I put to 1;26 40 so that it will give 43;20? Put 30. Multiply 30 by 1. It will give 30. Add 20 to 30 so that the large square-side is 50. Multiply 30 by 0;40. It will give 20. Add 5 to 20 so that it will give 25, the small square-side.

Strasbourg 364

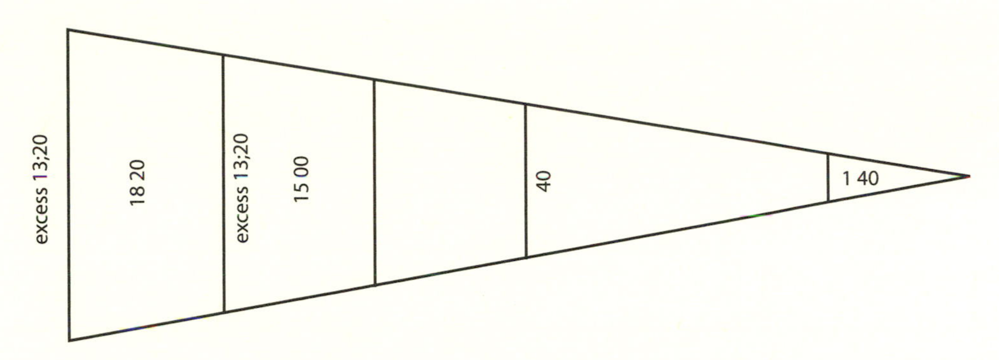

(i) A wedge with 5 rivers inside. The area of the upper river is 18 20, the (area of the) [2nd] river [15 00]. The upper width exceeds the dividing line by 13;20. [Dividing line exceeds] dividing line by 13;20. I do not know the length and area of the 3rd river. The 4th dividing line is 40, the [5]th area is 1 40.

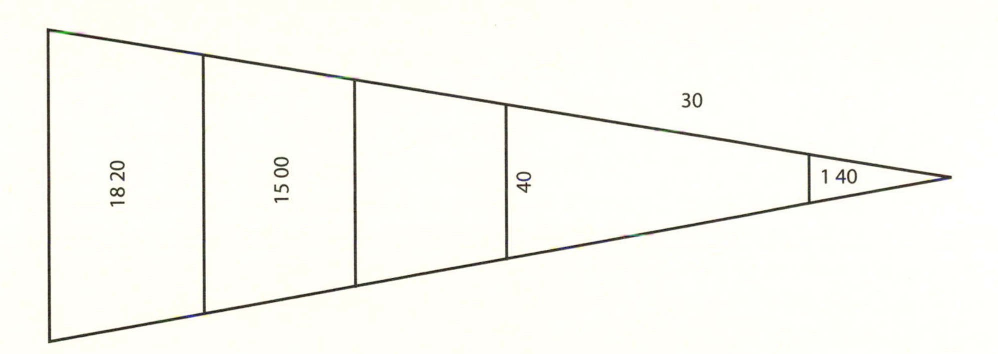

(ii) A wedge with 5 rivers inside. The upper area is 18 20, the 2nd area 15 00. I do not know the length and area of the third river. The 4th dividing line is 40, the «5th» <4th> length is 30, the <5th> area 1 40. What are the dividing lines and the upper width?

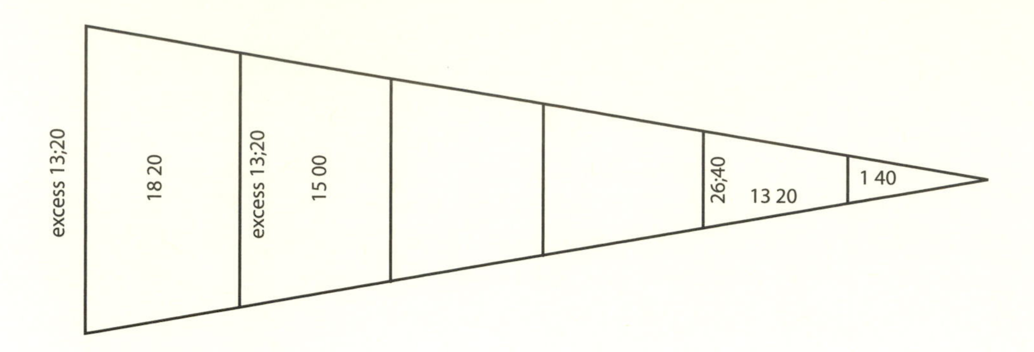

(iii) A wedge with 5 rivers inside. The upper area is 18 20, the second area 15 00. I do not know the third area. The 4th area is 13 20; ½ (across) it is 26;40 (wide). I do not know the 5th area. The upper width exceeds the dividing line by 13;20. Dividing line exceeds dividing line by 13;20. What are the area, lengths, and dividing lines?

(iv) and (v) [*Missing.*]

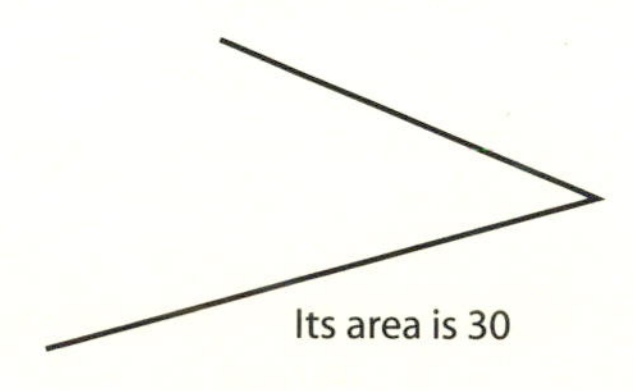

(vi) A wedge of area 1. [I went down] from the upper length [and] I lay the dividing line across [......] but I do not [know] how much I laid the dividing line across. [......] I went down 3 rods 4 cubits, and then I laid across a ...... I went [...] and then I installed a dyke. From the dyke that I installed I laid across a [......] but I do not know how much I went (along) the dividing line. The area is 5 16 40. «Earth» How much area did I take and how much did I leave behind?

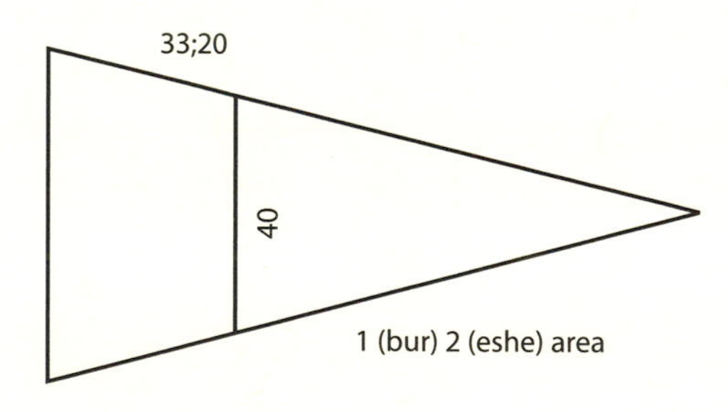

(vii) A wedge. I do not know the length and upper width. The area is 1 (bur) 2 (eshe). From the upper width I went down 33;20, so that the dividing line was 40. What are the length and (upper) width?

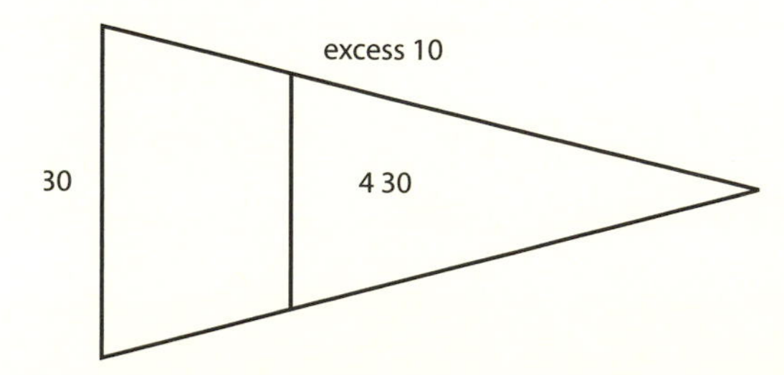

(viii) A wedge with 2 rivers inside. The upper width is 30, the lower area is 4 30. The lower length exceeds the upper length by 10. <What are the lengths?>

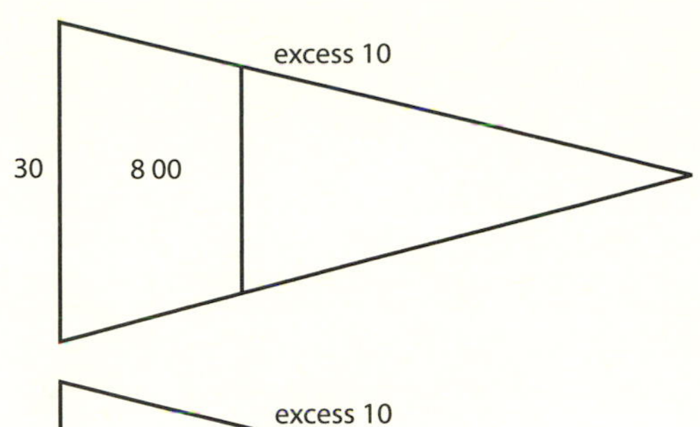

(ix) A wedge with 2 rivers inside. The upper width is 30, the upper area 8 00. The lower length exceeds the upper length by 10. What are the lengths?

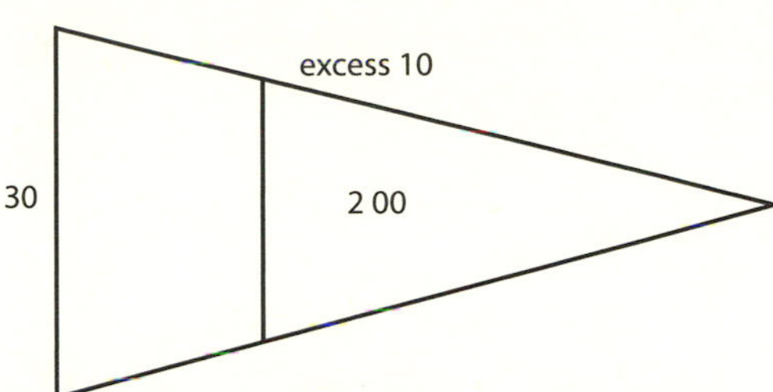

(x) A wedge with 2 rivers inside. The upper width is 30, [the lower area 2 00]. The lower length [exceeds] the upper length by 10. [What are the lengths?]

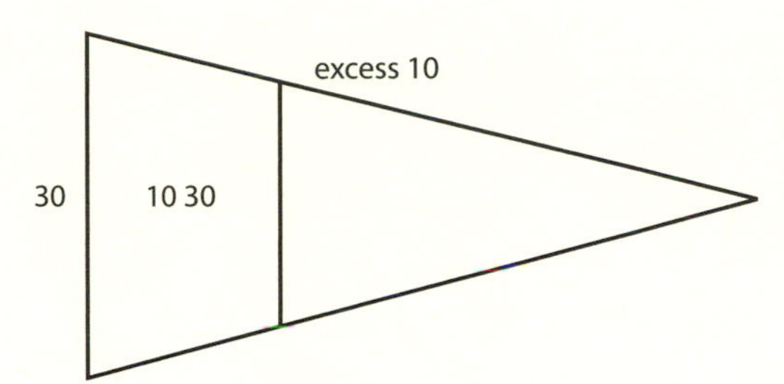

(xi) A wedge with 2 rivers inside. [The upper width is 30], the upper area 10 30. The [upper] length [exceeds the lower length by 10. What are the lengths?]

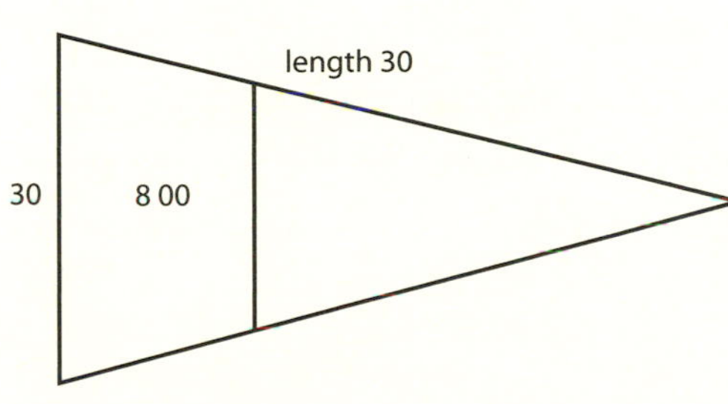

(xii) A wedge [with 2 rivers inside. The upper width is 30, the upper area] 8 00. [The lower length is 30. ……]

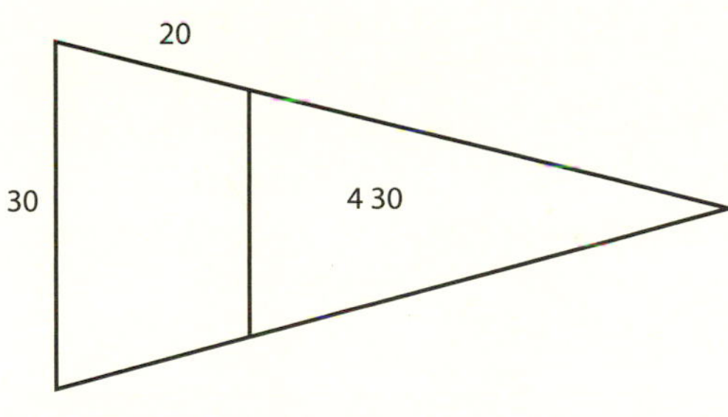

(xiii) A wedge [with 2 rivers inside. The upper width is 30, the upper length] 20. [The area is 4 30. ……]

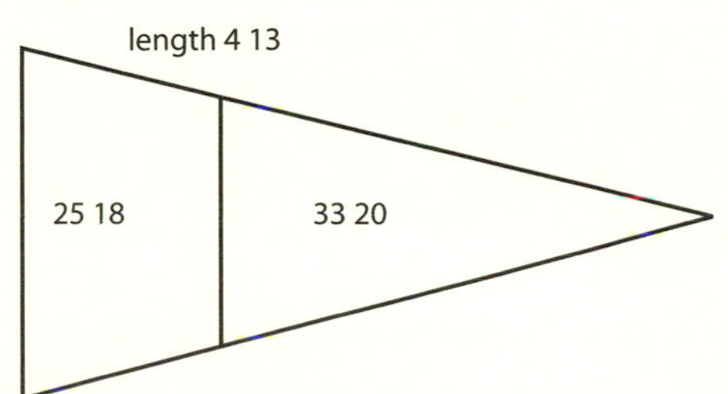

(xiv) A wedge [with 2 rivers inside. The upper length is 4 13, the upper area] 25 18. [The lower area is 33 20. ……]

Strasbourg 367

An ox's brow with 2 rivers inside. The upper area is 13 03, the 2<nd> area is 22 57. A 3rd part of the lower length is in the upper length. Sum that by which the upper width exceeds the dividing line and (that by which) the dividing line exceeds the lower width. (It will give) 36. What are their lengths, the widths, and the dividing line?

You, when you proceed: put down 1 and 3. Sum 1 and 3. (It will give) 4. Solve the reciprocal of 4 so that (it will give) 0;15. Multiply 0;15 by 36. It will give 9. Multiply 9 by 1. It will give 9. Multiply 9 by 3. (It will give) 27. That by which the upper width exceeds the dividing line is 9. That by which the dividing line exceeds the lower width is 27.

Solve the reciprocal of 1. Multiply 1 by 13 03. It will give 13 03. Solve the reciprocal of 3. Multiply 0;20 by 22 57. It will give 7 39. By what does 13 03 exceed 7 39? It exceeds by 5 24. Sum 1 and 3. (It will give) 4. Break 1/2 of 4. (It will give) 2. Solve the reciprocal of 2. <Multiply> 0;30 by 5 24. It <will give> 2 42, the falsely counted (?). <The reciprocal of> 2 42 cannot be solved. What should I put to 2 42 so that it will give 9? Put 0;03 20. Solve the reciprocal of 0;03 20. It will give 18. Multiply 18 by 1. The upper length is 18. Multiply 18 by 3. The lower length «the lower length» is 54.

Break 1/2 of 36. Multiply «17» <18> by 1 12. (It will give) 21 36. Solve (*sic*) 21 36 from 36 00, the area. (It will give) 14 24. Solve the reciprocal of 1 12, the length. Multiply 0;00 50 by 14 24. It will give 12. Add 12 to 36 so that (it will give) 48. The upper width is 48. Add 12 to 27. The dividing line is 39. It will give 12, the lower width.

Strasbourg 368

I took a reed but I did not know its measurement. I broke off 1 cubit from it and then I went a length of 1 sixty. I returned to it what I had broken off from it and then I went a width of its 30. The area is 6 15. What is the width (*sic*) of the reed?

You, when you proceed: put down 1 00 and 30. Put down 1, the reed that you do not know. You multiply (it) by 1 00, its sixty that you went, so that the false length is 1 00. Multiply 30 by that 1, (so that) the false width is 30. Multiply 30, the false width, by 1 00 «30», the false length. The false area is 30 00. Multiply «1» 30 00 by 6 15, the true area, so that it will give you 3 07 30 00. Multiply the 0;05 that was broken off by the false length. It will give 5. Multiply 5 by the false width. It will give 2 30. Break 1/2 of 2 30. (It will give) 1 15. Combine 1 15. (It will give) 1 33 45. Add to 3 07 30 00. (It will give) 3 [09 03 4]5. What is the square-side? [The square-side] is 13 45. Add the 1 15 that you combined to it. It will give 15 00. Solve the reciprocal of 30 00, the false area. (It will give) 0;00 02. Multiply 0;00 02 by 15 00. The width (*sic*) of the reed is 30.

Quantity surveying

Fully half the known corpus of Old Babylonian mathematical problems uses the management of building work and agricultural labor as a pretext for setting exercises in line geometry or geometric algebra. Although many problems use terminology and technical constants that are also known from earlier administrative practice (see YBC 7243 on page 150) the majority of the problems are highly unrealistic, both in the measurements of the objects described and in the nature of the questions posed.

Carrying bricks, building earthen walls, and repairing canals and associated earth works are among the commonest problem-setting scenarios. YBC 4673, which sets the problems and occasionally gives numerical answers, gives examples of each [MKT, III 29–34, pl. 3; Friberg 2001, passim]. The colophon counts 23 exercises on the tablet. In fact there are only 21 sections, two of which run over column boundaries; and the last five sections in fact comprise an out of place procedure for calculating the number of workers needed to make a dam [Robson 1999, 102–105].

YBC 4673

(i) A brick. The length is 1/2 cubit, the width 1/3 cubit, its height 5 fingers. What are its area and volume?

(ii) 1 man carried 9 sixties of bricks for a length of 30 rods. They gave him 1 (ban) of grain. Now, he carried 5 sixties of bricks and then finished the bricks. How much grain did they give him? 5 1/2 sila 3 1/3 shekels of grain.

(iii) A builder carried 9 sixties of bricks for a length of 30 rods. They gave him 1 (ban) of grain. Now, he carried 6 sixties of bricks and then finished the bricks. [......] the reciprocal of the work rate for carrying bricks [......] If 2 15, the bricks [......] the reciprocal of the work rate [...... What are] the length of carrying and the grain [given]?

(iv) 1/2 (iku) 4 sar of bricks were put down (to be carried) for [a length] of 30 rods. What laborers should I put down to finish the bricks in 1 day? 1 12 laborers.

(v) One man carried earth over a distance of 30 rods, and then he built a brick-pile. For what proportion of the day did he carry earth; for what proportion of the day did he build the brick-pile; and how many bricks were there? 2 40.

(vi)–(viii) [*These problems are too damaged for translation.*]

(ix) An earth wall. The [breadth] is 1 cubit, the height [1] cubit, the work rate [0;03 4]5. What length does 1 man take? The length is 1/2 rod 3 cubits.

(x) The volume of an earth wall is 5 shekels, the breadth 1 cubit, the height 1 cubit, the work rate 0;03 45. What length does 1 man [take]? The length is 1 rod.

(xi) The volume of an earth wall is 5 shekels, the breadth 2 cubits, its height 1 cubit, the work rate 0;03 45. What length does 1 man <take>? The length is 1/2 rod.

(xii) An earth wall. Its length is 5 cables, the breadth 2 cubits, its height 1/2 rod. In 1 cubit (height) it decreases 1/3 cubit in width and then a man demolishes so that a height of 1 1/2 cubits is left. How long is the length (demolished in 1 day)? ......

(xiii) A dike. The breadth is 1 cubit, its height 1 cubit. In 1 cubit (height) it slopes 2 cubit. What earth is packed down in a length of 1/2 rod?

(xiv) An old dike. The breadth is 1 cubit, the height 1 cubit. In 1 cubit it slopes <1 cubit>. Now, I added a breadth of 1 cubit [......], a height of 1 cubit. What earth is packed down in a length of 1/2 rod? [The volume is 10 shekels]. What are the new volume [and the old volume]? [The old volume is] 2 [1/2 shekels, the new volume] 7 [1/2 shekels].

(xv) An old dike. The breadth is 2 cubits, its height 2 cubits. Now, I added a breadth of 1 cubit, a height of 1 cubit. In 1 cubit (height) it sloped 1 cubit. What earth is packed down in a length of 1/2 rod? The volume is 1/3 sar and 2 1/2 shekels. What are the old volume and the new volume? The old volume is 10 shekels, the new volume 12 1/2 shekels.

(xvi) 1/3 mina of washed (?) wool, 1/2 mina of carded wool, 1 1/2 shekels of finished (?) wool. 6 shekels are diminished from 1 mina. 6 shekels are taken away from 1 mina of wool so that (there is) 5/6 mina 4 shekels of washed (?) wool. 1/2 mina 6 shekels of carded wool are taken away from 5/6 mina 4 shekels of washed (?) wool so that (there are) 18 shekels of «carded» <finished (?)> wool. What wool did I give to 1 woman? Washed (?) wool, carded wool, finished (?) wool.

(xvii) Multiply 2/3, the reeds, by 7 1/2 sar, the volume, so that (you will see) 5 sar, the reeds. Multiply 1/3 sar, the earth, by 7 1/2 sar, so that (you will see) 2 1/2 sar, the earth.

(xviii) Solve the reciprocal of 5 shekels, cutting (?) reeds. Multiply by 5, the volume of the reeds, so that you will see 1 sixty, the workers.

(xix) Solve the reciprocal of 6 shekels, carrying reeds. Multiply by 5 sar, the volume, so that you will see 50, the workers.

(xx) Solve the reciprocal of 2/3, the reeds. Multiply by 5 sar, the volume, then and you will see 7;30, the workers.

(xxi) Solve the reciprocal of 0;10, the work rate. Multiply by 2 1/2 sar, the volume, and you will see 15, the workers.

23 exercises, second tablet.

Quantity surveying could also be used as a pretext for developing complex problems on geometrical algebra. The problems about trenches on YBC 4657 are initially simple, but quickly move away from realistic scenarios as sums and differences of parameters are given. YBC 4662 and YBC 4663 give worked solutions to all but ten of the thirty-one problems on YBC 4657; presumably a fourth tablet originally had those on too [MCT, texts G, H, J].

YBC 4657

(i) A trench. [The length is 5 rods], the width is [1 1/2] rods, its depth is 1/2 rod. The work rate is 10 shekels of earth, its wages are 6 grains for a hired man. What are the area, volume, [laborers], and silver? The area is 7 1/2, the volume 45, [the laborers] 4 30, the silver 9 shekels.

(ii) The silver for a trench is [9 shekels], the width 1 1/2 rods, <its depth> 1/2 rod. The work rate is 10 shekels of earth, its wages are 6 grains for a hired man. [What is] its length? The length is 5 rods «rods».

(iii) The silver for a trench is [9 shekels], the length 5 rods, its depth 1/2 rod. The work rate is 10 shekels of earth, its wages are 6 grains for a hired man. [What is] its width? The width is 1 1/2 rods.

(iv) The silver for a trench is [9 shekels], its length [5] rods, the width 1 1/2 rods. The work rate is 10 shekels of earth, its wages are 6 grains for a hired man. What is its depth? Its depth is [1/2 rod].

(v) The silver for a trench is 9 shekels, the length 5 rods, the width 1 1/2 rods, its [depth 1/2 rod]. Its wages are 6 grains for a hired man. What is the earth of the work rate? The work rate is 10 shekels of earth.

(vi) The silver for a trench is [9 shekels, the length] 5 [rods, the width 1 1/2] rods, its depth 1/2 rod. The work rate is 10 shekels of earth. [How much are] its wages for a [hired] man? Its wages are [6 grains] for a hired man.

(vii) The silver for a trench is [9 shekels], its depth [1/2 rod. The work rate is 10 shekels.] Its wages are [6] grains for a hired man. Add the length [and width] so that [(it is) 6 1/2 rods]. What are the [length and] width?

(viii) The silver [for a trench is 9 shekels. Its depth is 1/2 rod], the work rate [10 shekels], its wages 6 grains for a hired man. [The length exceeds the width by 3 1/2 rods.] What are the length and width? The width is 5 rods, the length 1 1/2 <rods>.

(ix) A trench. [The length is 5 rods, the width 1 1/2 rods,] its [depth 1/2 rod.] What are the area and volume? The area is 7 1/2 sar, the volume 45.

(x) The volume of a trench is [45 sar, the width 1 1/2 rods], its depth 1/2 rod. What is its length? Its length is 5 rods.

(xi) The volume of a trench is 45 [sar], the length [5] rods, its depth 1/2 rod. What is its width? Its width is 1 1/2 rods.

(xii) The volume of a trench is 45 [sar], the length [5] rods, its width 1 1/2 rods. What is its depth? Its depth is 1/2 rod.

(xiii) The volume of a trench is 45 [sar], its depth 1/2 rod. Add the length and the width so that (it is) 6 1/2 rods. What are the length and the width? The length is 5 rods, the width 1 1/2 rods.

(xiv) The volume of a trench is 45 sar, its depth 1/2 rod. The length exceeds the width by 3 1/2 rods. What are the length and width? The length is 5 rods, the width 1 1/2 rods.

(xv) A trench. Add the area and volume so that (it is) 52;30. The width is 1 1/2 rods, its depth 1/2 rod. What is its length? Its length is 5 rods.

(xvi) A trench. Add the area and volume so that (it is) 52;30. The length is 5 rods, its depth 1/2 rod. What is [its width]? [The width] is 1 1/2 rods.

(xvii) [A trench.] Add [the area and volume] so that (it is) <52;30>. Its depth is 1/2 rod. [Add] the length and [width so that (it is)] 6 [1/2 rods]. What are [the length and width]?

(xviii) [A trench.] Add [the area and volume] so that (it is) 52;30. Its depth is 1/2 rod. The length [exceeds] the width by 3 [1/2 rods]. What are [the length and width]?

(xix) A trench in an area of 7 1/2 sar; the volume is 45. Add the length and width so that (it is) 6 1/2 rods. What are the length, width, and its depth?

(xx) A trench in an area of 7 1/2 sar; the volume is 45. The length exceeds the width by 3 1/2 rods. What are the length, width, and its depth? The length is 5 rods, the width 1 1/2 rods, its depth 1/2 rod.

(xxi) A trench in an area of 7 1/2 sar; the volume is 45. Its depth is a 7th <part> of that by which the length exceeds the width. What are the length and width? The length is 5 rods, the width 1 1/2 rods.

(xxii) A trench. The length is 5 rods, the width 1 1/2 rods, its depth 1/2 rod. The work rate is 10 shekels of earth. What length does 1 man take? He takes a length of 6 2/3 fingers.

(xxiii) A trench. The length is 5 rods, the width 1 1/2 rods, its depth 1/2 rod. The work rate is 10 shekels of earth. What length do 30 laborers take? They take a length of 1/2 rod 2/3 cubit.

(xxiv) A trench. The length is 5 rods, the width 1 1/2 rods, its depth 1/2 rod. The work rate is 10 shekels of earth. In how many days do 30 laborers finish? They finish in 9 days.

(xxv) A trench. The width is 1 1/2 rods, its depth 1/2 rod. The work rate is 10 shekels of earth. 30 laborers finish it in 9 days. What is its length? Its length is 5 rods.

(xxvi) A trench. The length is 5 rods, its depth 1/2 rod. The work rate is 10 shekels of earth. 30 laborers finish it in 9 days. What is its width? The width is 1 1/2 rods.

(xxvii) A trench. The length is 5 rods, the width 1 1/2 rods. The work rate is 10 shekels of earth. 30 laborers finish it in 9 days. What is its depth? Its depth is 1/2 rod.

(xxviii) A trench. The length is 5 rods, the width 1 1/2 rods, its depth 1/2 rod. 30 laborers finish it in 9 days. What is the earth of the work rate? The work rate is 10 shekels.

(xxix) A trench. 30 laborers finish it in 9 days. Its depth is 1/2 rod, the work rate is 10 shekels. Add the length and width and (it is) 6 1/2 rods. What are the length and width? The length is 5 rods, its width 1 1/2 rods.

(xxx) A trench. 30 laborers finish it in 9 days. Its depth is 1/2 rod, the work rate is 10 shekels. The length exceeds the width by 3 1/2 rods. What are the length and width? The length is 5 rods, its width 1 1/2 rods.

(xxxi) A trench. Its square-side is 2 1/2 rods, its depth 3 1/3 cubits. The work rate is 10 shekels. Its wages are 1 (ban) of grain for each hired man. What are the area, volume, laborers, and grain? The area is 6 sar and a 4th part, the volume 20 5/6 sar, the laborers 2 05, the grain 4 (gur) 5 (ban).

31 exercises about trenches.

YBC 4663

(i) A trench. The length is 5 rods, <the width> 1 1/2 rods, its depth 1/2 rod. The work rate is 10 <shekels of> earth, [its wages] 6 [grains]. What are the area, volume, laborers, and silver? You, when you proceed: combine the length and the width. It will give you 7;30. Multiply 7;30 by the depth. It will give you 45. Solve the reciprocal of the work rate. It will give you 6. Multiply by 45. It will give you 4 30. Multiply 4 30 by the wages. It will give you 0;09. That is the procedure.

(ii) 9 shekels of silver for a trench. <The width> is 1 1/2 rods, its depth 1/2 rod. The work rate is 10 <shekels of> earth, its [wages] 6 grains. What is its length? You, when you proceed: combine length and its depth. It will give you 9. Solve the reciprocal of the work rate, multiply by 9. It will give you 54. Multiply 54 by the wages. It will give you 0;01 48. <Solve> the reciprocal of 1 48. It will give you 33;20. Multiply 33;20 by 0;09, the silver. It will give you 5. Its length is 5 rods. That is the procedure.

(iii) 9 <shekels of> silver for a trench. The length is 5 rods, its depth 1/2 rod. The work rate is 0;10, its wages 6 grains. What is its width? You, when you proceed: combine the length and its depth. It will give you 30. Solve the reciprocal of the work rate, multiply by 30. It will give you 3 00. Multiply 3 00 by the wages. It will give you 0;06. Solve the reciprocal of 0;06, multiply by 0;09, the silver. It will give you its width. The width is 1 1/2 rods. That is the procedure.

(iv) 9 shekels of silver for a trench. The length is 5 rods, the width 1 1/2 rods. The work rate is 10 <shekels of> earth, its wages 6 grains. What is its depth? You, when you proceed: combine the length and width. It will give you 7;30. Solve the reciprocal of the work rate, multiply by 7;30. It will give you 45. Multiply 45 by the wages. It will give you 0;01 30. Solve the reciprocal of 0;01 30. It will give you 40. Multiply 40 by 0;09, the silver. It will give you 6, its depth. Its depth is 1/2 rod.

(v) 9 <shekels of> silver for a trench. The length is 5 rods, the width 1 1/2 rods, its depth 1/2 rod. The wages are 6 grains. What is the work rate? You, when you proceed: combine the length and width. It will give you 7;30. Multiply 7;30 by its depth. It will give you 45. Multiply 45 by the wages. It will give you 0;01 30. Solve the reciprocal of 0;09, the silver. It will give you 6;40. Multiply 6;40 by 0;01 30. It will give you the work rate. The work rate is 10 shekels.

(vi) 9 shekels of silver for a trench. The length is 5 rods, the width 1 1/2 rods, its depth 1/2 rod. The work rate is 10 shekels. What are its wages? You, when you proceed: combine the length and width. It will give you 7;30. Multiply 7;30 by its depth. It will give you 45. Solve the reciprocal of the work rate, multiply by 45. It will give you 4 30. Solve the reciprocal of 4 30. It will give you 0;00 13 20. Multiply 0;00 13 20 by 0;09, the silver. It will give you the wages. Its wages are 6 grains.

(vii) «9 shekels of silver for a trench.» 9 <shekels of> silver for a trench. I summed the length and width so that (it was) 6;30. [Its depth] is 1/2 rod, the work rate 10 shekels, its wages 6 grains. What are the length and its width? You, when you proceed: solve the reciprocal of its wages, multiply by 9 shekels, the silver. It will give you 4 30. Multiply 4 30 by the work rate. It will give you 45. Solve the reciprocal of its depth, multiply by 45. It will give you 7;30. Break off 1/2 of the length and width that were summed. It will give you 3;15. Combine 3;15. It will give you 10;33 45. Take away 7;30 from the middle of 10;33 45. It will give you 3; 03 45. Take its square-side. It will give you 1;45. Add to 1 (copy of 3;15), take away from 1 (copy of 3;15). It will give you length and width. The length is 5, the width 1 1/2 rods.

(viii) 9 <shekels of> silver for a trench. The length exceeds the width by 3;30. Its depth is 1/2 rod, the work rate 10 shekels. Its wages are 6 grains. What are the length and width? You, when you proceed: solve the reciprocal of the wages, multiply by 0;09, the silver. It will give you 4 30. Multiply 4 30 by the work rate. It will give you 45. Solve the reciprocal of 1/2 rod, multiply by 45. It will give you 7;30. Break off 1/2 of that by which the length exceeds the width. It will give you 1;45. Combine 1;45. It will give you 3;03 45. Add 7;30 to 3;03 45. It will give <you> 10;33 45. Take its square-side. It will give you 3;15. Put down 3;15 twice. Add 1;45 to 1 (copy of 3;15), take away 1;45 from 1 (copy of 3;15). It will give you length and width. The length is 5 rods, the width 1 1/2 rods. That is the procedure.

YBC 4662

(xix) A trench in an area of 7 1/2 sar. The volume is 45 sar. I summed the length and width, so that (it was) 6;30. [What are] the length, width, and [its depth]? You, when you proceed: [solve] the reciprocal of 7 1/2, the area, and multiply by 45 sar, the volume. It will give [you] 6, its depth. Solve the reciprocal of its depth. It will give you 10. [Multiply] 10 by 45, the volume. It will give you 7;30. [Break] off 1/2 of the length and width that were summed. It will give you 3;15. Combine 3;15 «by» <and> 3;15. It will give you 10;33 45. Take away 7;30 from inside [10;33 45]. It will give you 3;03 45. [Take its square-side.] It will give you 1;45. Add [1];45 to 1 (copy of 3;15), [take away 1;45] from [1 (copy of 3;15)]. It will give you the length and width. [The length is] 5 rods, [the width 1 1/2].

(xx) A trench in an area of 7 1/2 sar. The volume is 45 sar. The length <exceeds> the width by 3;30. What are the length, width, and its depth? You, when you proceed: solve the reciprocal of 7;30, the area. It will give you 0;08. Multiply 0;08 by 45, the volume. It will give you 6, its depth. Solve the reciprocal of 6, its depth. It will give you 0;10. Multiply 0;10 by 45, the volume. It will give you 7;30. Break off [1/2 of that by which the length] exceeds the width. It will give you [1;45]. Combine [1]; 45 «by» <and> 1;45. It will give you [3;03 4]5. [Add] 7;30 to the middle of [3;03 45]. It will give you 10;33 45. Take its square-side. It will give you 3;15. Put down 3;15 twice. Add 1;45 to 1 (copy of 3;15), take away 1;45 from 1 (copy of 3;15). It will give you the length [and] width. The length is 5 rods, the width 1 1/2.

(xxi) A trench <in> an area of 7 1/2 sar. The volume is 45 sar. Its depth is a 7th part of that by which the length exceeds the width. What are the length, width, and its depth? You, when you proceed: solve the reciprocal of 7 1/2 sar, the area, [multiply] by 45, [the volume]. It will give you its depth. Break off 1/2 of the seven that was put down. It will give you 3;30. Solve the reciprocal of its depth. It will give you 0;10. Multiply 0;10 by 45, the volume. It will give you 7;30. Break off 1/2 of 3;30. It will give you 1;45. Combine 1;45 «by» <and> 1;45. It will give you 3;03 45. Add 7;30 to the middle of 3;03 54. It will give you 10;33 45. [Take] the square-side of 10;33 45. It will give you 3;15. Put down 3;15 <twice> and then add 1;45 to 1 (copy of 3;15), take away 1;45 from 1 (copy of 3;15). It will give you the length and width. The length is 5 rods, [the width 1 1/2 rods].

(xxii) A trench. [The length is 5 rods, the width 1 1/2 rods,] its depth 1/2 rod. [The work rate is] 10 [shekels. How much length does 1 man take?] You, [when you] proceed: [combine width and its depth]. It will give [you] 9. [Solve the reciprocal of 9. It will give] you [0;06 40. Multiply] 0;06 40 by the work rate. [It will give] you [0;01 06 40]. 1 man takes 0;01 06 40.

(xxiii) [A trench. The length is 5 rods, the width 1 1/2 rods], its depth [1/2 rod]. The work rate is 10 shekels. [How much length do 30 men take? You], when you [proceed: ......]

(xxiv) [A trench. The length is 5 rods, the width 1 1/2 rods, its depth 1/2 rod. The work rate is] 10 shekels. In how many [days do 30 laborers] finish it? [You], when you proceed: combine the length and width. It will give you [7];30. Multiply 7;30 by its depth. It will give you 45. Solve the reciprocal of the work rate. It will give you 6. Multiply 45 by 6. It will give you 4 30. Solve the reciprocal of 30, the laborers. It will give you 0;02. Multiply by 4 30. It will give you 9. 30 laborers finish it in 9 days.

(xxv) A trench. The width is 1 1/2 rods, its depth 1/2 rod, the work rate 10 shekels. 30 laborers finish it in 9 days. What is its length? You, when you proceed: combine the width and its depth. It will give you 9. Solve the reciprocal of the work rate. [It will give you 6]. Multiply 9 by 6. It will give you 54. Solve the reciprocal of 54. It will give you 0;01 06 40. Multiply 30 and 9. It will give you 4 30. Multiply 4 30 by 0;01 06 40. It will give you the length. The length is 5 rods.

(xxvi) A trench. The length is 5 rods, its depth 1/2 rod, the work rate 10 shekels. 30 laborers finish it in 9 days. What is its width? You, when you proceed: combine the length and its depth. It will give you 30. Solve the reciprocal of the work rate. It will give you 6. Multiply [30] by 6. <It will give you 3 00>. Solve <the reciprocal of> 3 00. You will see 0;00 20. [Combine] 30, the laborers, and 9. It will give you 4 30. You multiply 4 30 by 0;00 20 so that it will give you the width. The width is 1 1/2 rods.

(xxvii) A trench. [The length is 5 rods, the width 1 1/2 rods], the work rate 10 shekels. <30 laborers finish it in 9 days>. [What is] its depth? [You], when you [proceed]: combine the length and width. It will give you [7;30]. Solve the reciprocal of the work rate, <multiply by 7;30>. It will give you 45. Solve the reciprocal of 45. It will give you 0;01 20. Combine 30, the laborers <and> the 9 days. It will give you [4 30]. Multiply 4 30 by 0;01 20. [It will give you 6. Its depth is 1/2 rod.]

(xxviii) A trench. The length is 5 rods, [the width 1 1/2 rods], its depth 1/2 rod. <30 laborers finish it in 9 days>. What is the work rate? You, when you proceed: combine the length and width. You will see 7;30. Multiply 7;30 by its depth. You will see 45. Combine 30, the laborers, and the 9 days. You will see 4 30. Solve the reciprocal of 4 30. You will see 0;00 13 20. [Multiply] 0;00 13 20 by [45]. It will give you the work rate. The work rate is 10 shekels.

YBC 4662 obverse, right edge, reverse (upside-down)

The following small problem, YBC 7997, has created immense difficulties of interpretation since it was published [MCT, text Pa; Friberg 2001, 97–98]. As I understand it, the aim is to calculate the number of standard bricks that can fit into a cylindrical kiln of circumference 1 1/2 rods and perhaps of height 3/4 rod. The attempt to find the height in cubits is not successful, but the rest of the procedure is followed correctly. When the volume of the kiln is found it is multiplied by the constant 7;12, the number of standard bricks, counted in brick-sar, found in a volume sar (see already AOT 304 on page 79).

YBC 7997

A kiln. The circle is 1;30 rods. You divide half of the dividing line into four and then «you combine 0;15, the quarter.»[21] 0;03 45 will come up. Then you multiply (it) by 12, of the depth, so that 0;45, the depth, will come up. You return. You combine 1;30, the circle, multiply (it) by 0;05 so that 0;11 15 will come up. Then you combine 0;45 and 0;11 15 and then you multiply as much as comes up by 7;12 so that it gives you the (number of) bricks.

Another popular mathematical topic was the grain capacity of cylindrical storage vessels. A variety of conversion constants was used, the most common being 6 00 00 sila per volume sar. The first six problems of Haddad 104 [Al-Rawi and Roaf 1984] deal with this topic; the rest concern work rates for plastering and for making bricks [Robson 1999, 75–82]. Problems (i)–(vi) all concern the volumes of cylindrical vessels with vertical or tapering sides. Their cross-sectional area is found using a method equivalent to the formulae $c = 3d$ and $A = 0;05c^2$; note that there is no all-purpose equivalent to π. Then if necessary the upper and lower surfaces are averaged, and then multiplied by the height. The volume in sar is multiplied by 6 00 00 to find the capacity in sila, which is then re-expressed in larger capacity units if appropriate. The tablet is one of the few collections of Old Babylonian mathematical problems with a good archaeological context: it was found in the same house as a collection of Sumerian literature and Akkadian magical incantations in the ancient town of Me-Turan in the kingdom of Eshnuna, to the northeast of Baghdad [Cavigneaux 1999].

Haddad 104

(i) The procedure of a log. Its dividing line is 0;05 (rods, or) a cubit. How much is it suitable for storing? When you proceed: put down the depth equal to the dividing line. Make 0;05 into a depth so that 1 will come up. Triple 0;05, the dividing line, so that 0;15 will come up. The circle of the log is 0;15. Combine 0;15 so that 0;03 45 will come up. Multiply 0;03 45 by 0;05, the constant of a circle, so that 0;00 18 45, the area, will come up. Multiply 0;00 18 45 by 1, the depth, so that 0;00 18 45, the volume, will come up. Multiply 0;00 18 45 by 6 00 00, (the constant) of storage, so that 1 52;30 will come up. The log contains 1 (bariga) 5 (ban) 2 1/2 sila of grain. That is the procedure.

(ii) If a log, whose bottom is 0;05, its top 0;01 40, is 0;30 (rods, or) a reed, long, how much grain does it contain? When you proceed: make 0;30, the length of the log, into a depth so that 6 will come up. Return. Sum and break 0;05, the bottom, and 0;01 40,

---

[21]The second half of this sentence is clearly superfluous. The dividing line is 0;30 rods long, so an eighth of it would be 0;03 45 rods, with or without combining the quarter into a square area. But it is not clear why the depth of a kiln should be an eighth of its dividing line. In this case it would be 3 meters across by some 375 mm deep.

the top, so that 0;03 20 will come up. Triple 0;03 20 so that 0;10, the circle of the log, will come up. Combine 0;10 so that 0;01 40 will come up. Multiply 0;01 40 by 0;05, the constant, so that 0;00 08 20, the area, will come up. Multiply 0;00 08 20 by 6, the length of the log, so that 0;00 50, the volume, will come up. Multiply by 0;00 50 by 6 00 00, (the constant) of storage and 5 00 will come up. The log contains 1 gur of grain.

(iii) If a log, whose bottom is 0;05, its top 0;01 40, is 5 (rods or), a half-rope, long, it price is 1 talent of silver. Now, I am carrying 1 mina of silver. Remove (grain to the value of) 1 mina of silver from the log, either from its base or from its top, and then give it to me. When you proceed: make 5, the length of the log, into a depth so that 1 00 will come up. Return. Sum and break 0;05, the bottom, and 0;01 40, the top, so that 0;03 20, the thickness of the log, will come up. Triple 0;03 20 so that 0;10, the circle of the log, will come up. Combine 0;10 so that 0;01 40 will come up. Multiply 0;01 40 by 0;05, the constant, so that 0;00 08 20, the area, will come up. Multiply 0;00 08 20 by 1 00, the length of the log, so that 0;08 20, the volume, will come up. Multiply 0;08 20 by 6 00 00, (the constant) of storage, and 50 00 (sila) will come up. The log contains 10 gur of grain.

10 gur grain is the storage (equivalent) of 1 talent of silver. If 50 00 (sila) of grain is the storage of 1 talent of silver, how much is the storage of 1 mina of silver? Solve the reciprocal of 1 mina of silver so that 1 will come up. Multiply 1 by 1 00, a talent of silver, so that 1 00 will come up. Solve the reciprocal of 1 00 so that 0;01 will come up. Multiply 0;01 by 50 00, the grain, so that 50 will come up. 5 (ban) of grain is the storage of 1 mina silver. Return.

Ask: "if a log whose bottom is 0;05 (contains) 5 (ban) of grain, how much is its length?" Solve the reciprocal of 6 00 00, (the constant) of storage, so that 0;00 10 will come up. Multiply 0;00 10 by 50, the grain, so that 0;08 20 will come up. Let your head keep 0;08 20. Return. Triple 0;05, the bottom, so that 0;15, the circle of the log, will come up. Combine 0;15 so that 0;03 45 will come up. Solve the reciprocal of 0;03 45 so that 16 will come up. Multiply 16 by the 0;08 20 that your head is keeping, so that 0;02 13 20 will come up. You go up 0; 02 13 20, thirteen fingers and 1/3 finger, the length of the log, and then you trim it off and then you give it for 1 mina of silver. Return.

Work out its storage. Make 0;02 13 20 into a depth so that 0; 26 40 will come up. Triple 0;05, the bottom, so that 0;15, the circle of the log, will come up. Combine 0;15 so that 0;03 45 will come up. Multiply 0;03 45 by 0;05, the constant of a circle, so that 0;00 18 45, the area, will come up. Multiply 0;00 18 45 by 0;26 40, the length of the log, so that 0;00 08 20, the volume, will come up. Multiply 0;00 08 20 by 6 00 00, (the constant) of storage, so that 50 will come up. The storage of 1 mina of silver is 5 (ban) of grain. Return.

Trim from its top. Solve the reciprocal of 1 mina of silver so that 1 will come up. Multiply 1 by 1 00, a talent of silver, so that 1 00 will come up. Solve the reciprocal of 1 00 so that 0;01 will come up. Multiply 0;01 by 50 00, the grain of the whole log, and 50 will come up. The storage of 1 mina of silver is 5 (ban) of grain.

If 50 (sila of) grain is the storage of 1 mina of silver and 0;01 40 the top of the log, how much should I descend so that I may remove (grain to the value of) 1 mina of silver? Solve the reciprocal of 6 00 00, (the constant) of storage, so that

0;00 00 10 will come up. Multiply 0;00 00 10 by 50, the grain, so that 0;00 08 20 will come up. Let your head keep 0;00 08 20. Return. Triple 0;01 40, the top of the log, so that 0;05, the circle of the log, will come up. Combine 0;05 so that 0;00 25 will come up. Solve the reciprocal of 0;00 25 so that 2 24 will come up. Multiply 2 24 by the 0;00 08 20 that your head is keeping so that 0;20 will come up. You descend 0;20, four cubits, and then you trim it off and then you give it for 1 mina of silver. Return.

Work out its storage (equivalent). Make 0;20, the length of the log, into a depth so that 4 will come up. Triple 0;01 40, the top of the log, so that 0;05, the circle of the log, will come up. Combine 0;05 so that 0;00 25 will come up. Multiply 0;00 25 by 0;05, the constant, so that 0;00 02 05, the area, will come up. Multiply 0;00 02 05 by 4, the length of the log, so that 0;00 08 20, the volume, will come up. Multiply 0;00 08 20 by 6 00 00, (the constant) of storage, so that 50 will come up. The log holds 5 (ban) of grain. That is the procedure.

(iv) The procedure of a sila (measuring vessel). The dividing line of my sila (vessel) is 0;01 (rods, or) six fingers. What should I make deep so that it amounts to 1 sila? When you proceed: solve the reciprocal of 6 00 00, (the constant) of storage, so that 0;00 00 10 will come up. Return. Triple 0;01, the dividing lines, so that 0;03, the circle of the sila (vessel), will come up. Combine 0;03 so that 0;00 09 will come up. Solve the reciprocal of 0;00 09 so that 6 40 will come up. Multiply 6 40 by 0;00 00 10 so that 0;01 06 40 will come up. The depth is 0;01 06 40 (rods, or) six fingers and two-thirds of a finger.

If the depth is 0;01 06 40, the dividing line 0;01, how much grain does my sila (vessel) hold? Make 0;01 06 40 into a depth so that 0;13 20 (cubits) will come up. Triple 0;01, the dividing line, so that 0;03, the circle of the sila (vessel), will come up. Combine 0;03 so that 0;00 09 will come up. Multiply 0;00 09 by 0;05, (the constant) of a circle, so that 0;00 00 45, the area, will come up. Multiply 0;00 00 45 by 0;13 20, the depth, so that 0;00 00 10, the volume, will come up. Multiply 0;00 00 10 by 6 00 00, (the constant) of storage, so that 1 sila of grain will come up.

If the grain is 1 sila, my depth 0;01 06 40, what are my dividing line and my circle? Make 0;01 06 40 into a depth so that 0;13 20 (cubits) will come up. Solve the reciprocal of 0;13 20 so that 4;30 will come up. Return. Solve the reciprocal of 0;05, (the constant) of a circle, so that 12 will come up. Solve the reciprocal of 6 00 00, (the constant) of storage, so that 0;00 00 10 will come up. Multiply 0;00 00 10 by 12 so that 0;00 02 will come up. Multiply 0;00 02 by 4;30 so that 0;00 09 will come up. Have its square-side come up so that 0;03 will come up. The circle of the sila (vessel) is 0;03. Take a third of 0;03 so that 0;01, the dividing line, will come up. That is the procedure.

(v) If the dividing line of a ban (measuring vessel) is 0;02, (and) it is filled by a ban and 2/3 sila of grain, what should I trim off so that it amounts to 1 (ban)? When you proceed: have the depth come up. Solve the reciprocal of 6 00 00, (the constant) of storage, so that 0;00 00 10 will come up. Multiply 0;00 00 10 by 1 (ban) 2/3 sila of grain so that 0;00 01 06 40 will come up. Return and then triple 0;02, the dividing line, so that 0;06, the circle of the ban (vessel), will come up. Combine 0;06 so that 0;00 36 will come up. Solve the reciprocal of 0;00 36 so that 1 40 will come up. Multiply 1 40 by 0;00 01 06 40 so that 0;02 57 46 40, the depth, will come up. Return.

The dividing line of 2/3 sila of grain is 0;02. How deep should I make it so that it amounts to 2/3 sila? Solve the reciprocal of 6 00 00, (the constant) of storage, so that 0;00 00 10 will come up. Multiply 0;00 00 10 by 0;40, the grain, so that 0;00 00 06 40 will come up. Return. Triple 0;02, the dividing lines, so that 0;06, the circle, will come up. Combine 0;06 so that 0;00 36 will come up. Solve the reciprocal of 0;00 36 so that 1 40 will come up. Multiply 1 40 by the 0;00 00 06 40 that your head was keeping, so that 0;00 11 06 40, the depth, will come up. You go down and then you trim off 0;00 11 06 40.

Take away 0;00 11 06 40 from the middle of 0;02 57 46 40, the former depth, <so that> 0;02 46 40, the depth of 1 (ban) of grain, will come up. Make 0;02 46 40 into a depth so that 0;33 20 (cubits) will come up. Triple 0;02, the dividing line, so that 0;06, the circle of the ban (vessel), will come up. Combine 0;06 so that 0;00 36 will come up. Multiply 0;00 36 [by] 0;05, (the constant) of a circle, so that 0;00 03, the area, will come up. Multiply 0;00 03 by 0;33 20, the depth, so that 0;00 01 40, the volume, will come up. Multiply 0;00 01 40 by 6 00 00, (the constant) of storage, so that 10 will come up. It amounts to 1 (ban) of grain. That is the procedure.

(vi) Three bariga (measuring vessels) and 2 (gur, or 10) sixties (sila) of grain. The first is a bariga, the second 5 (ban), the third 4 (ban). What grain do they issue? When you proceed: sum 1 (bariga), 5 (ban), and 4 (ban), so that 2 30 (sila) will come up. Solve the reciprocal of 2 30 so that 0;00 24 will come up. Multiply 0;00 24 by 10 00, the grain, so that 4 will come up. Multiply 4 by 1 00 so that the big one issues 4 sixties, and then multiply 4 by 50 so that the second one issues 3;20 sixties. Multiply 4 by 40 so that the third issues 2;40 sixties. <That is the procedure.>

(vii) The work rate of plastering. He puts down 1 (rod, or) two reeds square and a thickness of 0;00 10 (rods, or) a finger and then he plasters for the whole day. If it is 1 (rod, or) two reeds square and he makes it 0;00 10 (rods, or) a finger thick, how much mud is there? When you proceed: make the thickness of the plastering into a depth so that 0;02 will come up. Return. Combine 1, the square-side of the plastering, so that 1, the area, will come up. Multiply 1 by 0;02, the thickness, so that 0;02, the volume, will come up. Multiply 0;02 by 5 00 00, (the constant) of the measured amount (?), so that 10 00 will come up. 1 man plasters 2 gur of mud for a whole day. That is the procedure.

(viii) If height of the plastering is 0;40 (rods, or) eight cubits, the thickness of my plastering 0;00 10 (rods, or) a finger, what are the length that I plaster and the mud? When you proceed: write down 1 sar, the mud that 1 man plasters in a whole day. Solve the reciprocal of 0;40, the height, so that 1;30 will come up. Multiply 1;30 by 1 sar of mud so that 1;30 will come up. For a whole day you plaster a length of 1;30 (rods, or) three reeds. Multiply 1;30, the length, by 0;40, the height of the plastering, so that 1 sar, the area, will come up. Multiply 1 by 0;02, the thickness of the plastering, so that 0;02, the volume, will come up. Multiply 2 by 5 00 00, (the constant) of the measured amount (?), so that 10 00 will come up. 1 man plasters 10 00 (sila, or) <2> gur of mud in a whole day.

(ix) Brickage, combined constant. What is the work rate of brick-making and what is the output of 1 man? When you proceed: write down 0;20, the work rate of destroying (?), 0;20, the work rate of brick-making, 0;10, the work rate of mixing. Solve the reciprocal of 0;20 so that 3 will come up. Solve the reciprocal of 0;20 so that 3 will come up. Solve the reciprocal of 0;10 so that 6 will come up. Sum them so that 12 will come up. Solve the reciprocal of 12 so that 0;05 will come up. The combined work rate is 0;05. Return. Combine 0;03 20, the square-side of a brick, so that 0;00 11 06 40, the area of a brick, will come up. Return. Make 0;01, the thickness of the brick, a depth so that 0;12 will come up. Multiply 0;12 by 0;00 11 06 40 so that 0;00 02 13 20, the volume of a brick, will come up. Solve the reciprocal of 0;00 02 13 20 so that 27 00 will come up. Multiply 27 00 by 0;05, the work rate, so that 2;15 sixties of bricks, the output of 1 man, will come up. Triple 2;15. 6;45 sixties, the output of the work rate, will come up.

(x) The work rate of brick-making. I carry for 5 (rods, or) a half-rope and then I make bricks. What is the output of 1 man? When you proceed: write down 0;20, the destroying(?), 0;20, the brick-making, 0;10, the mixing. Return and then solve the reciprocal of 5, the distance, so that 0;12 will come up. Multiply 0;12 by 45 00, the going, so that 9 00 will come up. Multiply 9 00 by 0;00 02 13 20, the basket, so that 0;20 is the volume you will carry here for 5 (rods, or) a half-rope. Write down 0;20, the volume, by the side of 0;10, the mixing. Return. Solve the reciprocal of 0;20 so that 3 will come up. Solve the reciprocal of 0;20 so that 3 will come up. Solve the reciprocal of 0;10 so that «solve so that» 6 will come up. Solve the reciprocal of 0;20, the volume, so that 3 will come up. Sum (them) so that 15 will come up. Solve the reciprocal of 15 so that 0;04 will come up. Multiply 0;04 by 1, the day, so that 0;04, the work rate, will come up. Return. Combine 0;03 20, the square-side of a brick, so that 0;00 11 06 40, the area of a brick, will come up. Multiply 0;00 11 06 40 by 0;12, the thickness of a brick, so that 0; 00 02 13 20, the volume of a brick, will come up. Solve the reciprocal of 0;00 02 13 20, the volume of a brick, so that 27 00 will come up. Multiply by 0;04, the work rate, so that 1 48 bricks, the output of 1 man, will come up. Triple 1 48 so that 5 24, the output of the work rate, will come up. That is the procedure.

Mixed problems

We have already seen that, while problems are often grouped thematically on tablets, it is difficult to impose modern boundaries between different topics of Old Babylonian mathematics. The following compilations confirm that apparent eclecticism. Strasbourg 362, for instance, has problems on arithmetic progressions, geometrical algebra, and quantity surveying [Frank 1928, 6; MKT, I 239–43; TMB, 82–84].

Strasbourg 362

(i) 10 brothers (inherited) 1 2/3 minas of silver. Brother exceeded brother (but) I do not know by how much they exceeded. The 8th share was 6 shekels. By how much did brother exceed brother?

You, when you proceed: solve the reciprocal of 10, the workers (*sic*), so that it will give 0;06. You multiply 0;06 by 1 2/3 minas, the silver, so that it will give 0;10. You copy two times so that it will give 0;20. Copy 0;06, the eighth share, two times so that it will give 0;12. Take away 0;12 from 0;20 so that it will give 0;08. Let your head keep 0;08.

Sum 1 and 1, the lower part (?), so that it will give 2. Copy 2 two times so that it will give 4. You add 1 to 4 so that it will give 5. You solve 5 from 10, the workers, so that it will give 5. You solve the reciprocal of a 5th part so that it will give 0;12. You multiply 0;12 by 0;08 so that it will give 0;01 36, the 0;01 36 (minas) that brother exceeded brother.

(ii) [*Too broken to translate.*]

(iii) Sum a 7th part of the length, a 7th part of the width, and a 7th part of the area, so that (it is) 2. Sum [the length] and the width, so that (it is) 5;50. What are the length and width? The length is 3;30, the width 2;20.

(iv) Sum a 7th part of the length and the area, so that (it is) 27. The width is 0;30. What are the length and area? The length is 42, the area 21.

(v) A (measuring) reed of 1 cubit. 1 finger kept falling off for me until it was finished. What length did I go? I went a length 1 rod 3 1/2 cubits.

(vi) A siege ramp, of length 10 rods, width 1 1/2 rods. 3 governors (each) took on a length of 3 rods 4 cubits. One (had) 1 sixty workers, the 2nd (had) 1 20 workers, the 3rd (had) 1 40 workers. The earth was transported [......]. [What are] my siege ramp and my depth, and how much earth [was transported]?

Scribal students were expected not simply to memorize mathematical procedures but also to practice carrying them out. AO 8862 is a four-sided prism from mid-eighteenth-century Larsa with eight problems on geometrical algebra and quantity surveying, giving both worked solutions and rough calculations [MKT, I 108–23; II pls. 35–38; TMB, 64–71; Friberg 2001, passim]. It is thought that prisms were written by students to show that they had successfully memorized an elementary exercise or mastered a literary work; in this case the student shows that he can lay out calculations appropriately as well as recall the correct steps to solve the problems set. Occasionally the arithmetical operations are written within the model procedure and occasionally too the student writes out the procedure in the first person ("I") instead of the more usual second person ("you"). The prism is dedicated to the goddess Nisaba.

AO 8862

O Nisaba!

(i) Length and width. I combined length and width so that I made an area. I turned around. I added as much as the length exceeded the width to the middle of the area so that (it was) 3 03. I returned. I summed the length and width so that (it was) 27. What are the length, width, and area?

27	3 03, sums	
15	length	
		3 00, area
12	width	

You, when you proceed: add 27, the sum of the length and width, to the middle of [3 03] so that (it is) 3 30. Add 2 to 27 so that (it is) 29. You break off half of 29 and then 14;30 times 14;30 is 3 30;15. You take away 3 30 from the middle of 3 30;25 so that the remainder is 0;15. 0;15 squares 0;30. Add 0;30 to one 14;30 so that the length is 15. You take away 0;30 from the second 14;30 so that the width is 14. You take away the 2 that you added to 27 from 14, the width, so that the true width is 12.

I combined 15, the length, and 12, the width and then 15 times 12 is 3 00, the area. By what does 15, the length exceed 12, the width? It exceeds by 3. Add 3 to the middle of 3 00, the area. The area is 3 03.

(ii) Length and width. I combined length and width so that I built an area. I turned around. I added half of the length and a third of the width to the middle of my area so that (it was) 15. I returned. I summed the length and width so that (it was) 7. What are the length and width?

You, when you proceed: you write down [2], the writing of a half, [and] 3, the writing of a third, and then you solve the reciprocal of 2, 0;30.—0;30 times 7 is 3;30—I multiply <0;30> by 7, the sums (*sic*) of length and width, and then take away 3;30 from 15, my sums, so that the remainder is 11;30. Go no further. I combine 2 and 3—3 times 2 is 6—and then the reciprocal of 6 will give you 0;10. I take away 0;10 from [7], your sums of the length and width, so that the remainder is 6;50. I (?) break off half of 6;50 so that it will give you 3;25. You write down 3;25 twice and then—3;25 times 3;25—I take away 11;40 15 from the middle of 11;30 (*sic*) so that the remainder is 0;10 25. <0;10 25 squares 0;25.> You add 0;25 to one 3;25 so that (it is) 3;50. And you add that which I took away from the sum of the length and width to 3;[50] so that the length is 4. I take away 0;25 from the second 3;25 so that the width is 3.

<15> 7, sums
4, length
12, area
3, width

(iii) Length and width. I combined length and width so that I built an area. I turned around and then I combined as much as the length exceeded the width with the sum of the length and my [width] and then I added it to the middle of my area so that (it was) 1;13 20. I returned. I summed the length and width [so that] it was 1;40.

1;40 1;13 20, sums
1, length
0;40, area
0;40, width

You, when you proceed: <combine> 1;40, the sum of the length and width—1;40 times 1;40 is 2;46 40. You take away 1;13 20, the area, from 2;46 40 so that (it is) 1;33 20. Go no further. You break off half of 1;40 and then—0;50 times 0;50—you add 0;41 40 to 1;33 20 and then 2;15 squares 1;30. By what does 1;40 exceed 1;30? It exceeds by 0;10. Add 0;10 to 0;50. The length is 1. Take away 0;10 from 0;50 so that the width is 0;40.

(iv) Length and width. I combined the length and width «the length and width» so that I built an area. I returned. I summed the length and width so that it was square with the area. I summed the length, width, and area so that (it was) 9. What are the length, width, and area?

(v) For a length of three ropes one man carried 9 sixties of bricks here so that I gave him 2 (ban) grain. Now, the builder has had me provide (for them) and so I called for 5 laborers. And then one carried one part of it to me, the second twice it, the third three times it, the fourth four times it, the fifth five times it. How many of the bricks the first one carried to me did he entrust to me, and then how much grain did I give him?

30	36	1 20	9
1	1 12	2 40	
2	1 48	4	
3	2 24	5 20	
4	3	6 40	
5	<3 36>	<8>	
20 grains			

(vi) For three ropes one man carried 9 sixties of bricks here so that I gave him 2 (ban) grain. Now, the builder has had me provide (for them) and so I called for 4 laborers. And then the first one carried a seventh part of it, the second an eleventh of it, the third a thirteenth of it, the fourth a fourteenth of it. How many bricks did he entrust to me, and then how much grain did I give him?

30	9	
7	1 24	3 06 40
11	2 12	<4> 53 20
13	2 36	5 46 40
14	2 48	6 [1]3 20
1 30	20	grain

(vii) If (they ask you), "For a length of three ropes I added the bricks, the laborers, [and] my days so that (it was) 2 20. My days were (equal to) two-thirds of my workers." Select the bricks, the laborers, and my days for me.

[6]	2	2 20
		work rate of a day
1	30	6
		one laborer, 30 work rates, 30 laborers
		1 30 bricks
40	20	20 days.

(viii) I summed the areas, my volume, [and] the bricks, so that (it was) 12;30 02. I converted [...] into a depth. [......] my earth.

BM 85194, with BM 85210, was the first mathematical cuneiform tablet to be published, as early as 1900. It appeared in a volume of hand copies (drawings) of British Museum tablets [King 1902]. Unfortunately for the early decipherers the procedures it outlines are riddled with accidental omissions and additions as well as numerical errors and rare words. Neugebauer [MKT, I 142–93] and Thureau-Dangin [TMB, 21–39] made brave attempts at translation but it has not been tackled in its entirety again since then.

The thirty-five problems are grouped roughly thematically. Some are more difficult to interpret than others. Problem (iv) concerns a circular city surrounded by a moat with a rampart outside it, built from the earth excavated in the construction of the moat. Problems (vi)–(viii) concern waterclocks which drip water into large vessels of constant cross-section, thereby changing the heights of water in them [Fermor and Steele 2000]. Problems (x)–(xiii) have so far resisted satisfactory interpretation, as the meaning of their subject matter, *imlû*, is unknown and it is unclear whether to translate a key term as "circle" or "depth." Problems (xiv)–(xv) deal with reed bundles in the shape of tapering cylinders by finding their average cross-sections.

Problems (xvi)–(xix) concern trapezoidal figures, whether bricks for a cylindrical well or the cross-sections of sloping walls. Problem (xxix), about the area of a circular segment, has not yet been satisfactorily interpreted. Problem (xxxiii) is an apparently unnecessary complication of a very simple problem. Problems (xxxiv) and (xxxv) deal with cylindrical baskets, using the metrological conversion constant 1 sar volume = 6 40 00 sila. Two lines of problem (xxxiv) are badly damaged, but it is possible to restore the gist of the text here, if not its exact wording.

BM 85194

(i) A ramp. At the base of the volume the lower breadth is 1 rod, the top 0;30 rods, the height 4. In front of the city gate the lower breadth is 1;20, the upper breadth 1 rod, the height 6. What is the volume? Demarcate the length for 1 man. You: sum <1 and> 1;30. You will see 2;30. Break off 1/2 of 2;30. Put down 1;15. Sum 1 and 0;30. You will see 1;30. Break off 1/2 of 1;30. You will see 0;45. Sum 1;15 and 0;45. You will see 2. <Break off> its 1/2. You will see 1. Put (it) down. Sum 6, the height, and 4, the height <at> the base of the volume. You will see 10. Break its 1/2. You will see 5. Multiply 10, the length, by 5. You will see 50, the volume. Solve the reciprocal of 0;10, the work rate. You will see 6. Multiply 50, the volume, by 6. You will see 5 00, the laborers. Solve the reciprocal of the laborers. You will see 0;00 12. Multiply 0;00 12 by 10, the length. You will see 0;02. 1 man takes 0;02 (rods). That is the procedure.

(ii) Temple foundations. The length is 0;30, the width 0;20. On each side 0;10 squares the support. The depth is 3 cubits. <What is> the volume? Demarcate the length for 1 man. You: square 0;10. You will see 0;01 40. Multiply 0;01 40 by 3 cubits, the depth. You will see 0;05, the volume of the support. Return. See the volume of the length. «Square-side.» <Multiply> 0;20 <the length> by 10, the width. You will see 0;03 20. Multiply 0;03 20 by 3 cubits, the depth. You will see 0;10. The volume of the length is 0;10. The (total) volume <is 0;20>. Solve the reciprocal of 0;10, the work rate. You will see 6. Raise 0;20 by 6. You will see 2. 1 man takes a length of 2. That is the procedure.

(iii) Temple foundations. The length is 0;30, the width 0;20, the depth 3 cubits. <On each side 0;10 squares the two supports.> Demarcate the length for 1 man in 1/2 a day. You: square 0;10. You will see 0;01 40. Multiply 0;01 40 by [3 cubits], the depth. You will see 0;05, the volume. Return. See the volume of the second support. Square 0;10. Multiply 0;01 40 by 3 cubits. You will see 0;05, the second volume. Multiply 0;20, «the square-side of» the length by 0;10, the width. You will see 0;03 20. Multiply 0;03 20 by 3 cubits, the depth. «Multiply 0;03 20 by 3 cubits, the depth.» You will see 0;10, the volume. Return. Demarcate the length assigned to 1 man in 1/2 <a day>. Break off 1/2 of 0;05, the volume. You will see 0;02 30. Solve the reciprocal of 0;10, the work rate. You will see 6. Multiply 0;02 30 by 6. You will see 15. The laborers are 15 strong. Solve «solve» the reciprocal of 15. You will see 0;04. Multiply 0;10 by 0;04. You will see 0;40. The length demarcated to 1 man is 0;40. Return. See 0;10, the work rate. Break off 1/2 of 0;10. You will see 0;05. Multiply 0;40 by 0;05. You will see 0;03 20. Multiply 0;03 20 by 3 cubits, the depth. You will see 0;10, the work rate. That is the procedure.

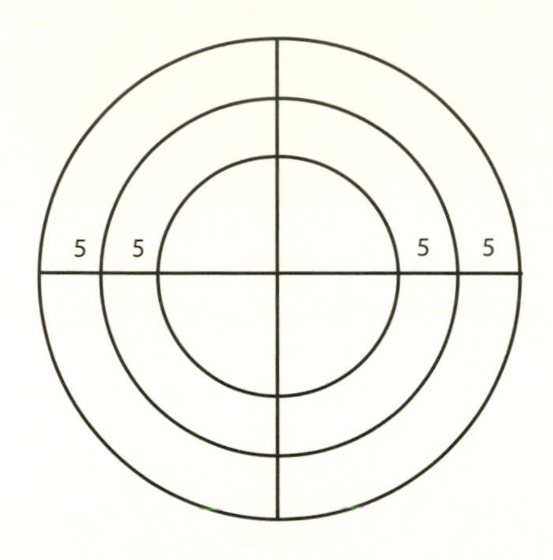

(iv) A city. I encircled it with a circle of sixty (rods). It projected 5 on each side and then I built a moat. The depth was 6. I took away a volume of 1 07 30. <It projected> 5 on each side. Above the moat I built a dyke. That dyke sloped 1 cubit in 1 cubit. What are the base, top, and height? And what is the circle of the dyke? You: as the circle is sixty, what is the dividing line? Take away a third part of sixty, the circle. You will see 20. The dividing line is 20. Double 5, the border. You will see 10. Add 10 to 20, the dividing line. You will see 30. Triple the dividing line. You will see 1 30. The circle of the moat is 1 30.

Return. Square 1 30. You will see 2 15 00. Multiply 2 15 00 by 0;05, (the constant of) the circle. You will see 11 15, the area. Multiply 11 15 by 6, the depth. <You will see 1 07 30, the volume.> Solve the reciprocal of 0;10, the work rate. You will see 6. Multiply 6 by 1 07 30, the volume. You will see 6 45 00. The laborers are 6 45 00 strong. Solve the reciprocal of 6 45 00. You will see 0;00 08 53 20. Multiply 0;00 08 53 20 by 1 30, the circle. You will see 0;13 20 (the length for 1 man) demarcated in the moat.

Return. See 0;10 the work rate. Solve the reciprocal of 1 30, the circle. You will see 0;00 40. Multiply 0;00 40 by 0;13 20. You will see 0;00 08 52 30. Multiply 0;00 08 53 20 by 1 07 30, the volume. You will see 0;10, the work rate.

Return. See the dyke. Copy 0;05, the slope. You will see 0;10. Copy 0;10. You will see 0;20. Multiply 0;20 by 1 07;30. You will see 22;30. What should I add to 22;30 so that the addition may satisfy a «remainder» <square-side> and that which is added may satisfy a square-side? Add 5 03;45. You will see 27 33;45. The square-side of the base is 5;15. What is the square-side of 5 03;45? The square-side is 2;15, the top. Solve the reciprocal of 2 sixties, the circle of the dyke. You will see 0;00 30. <Multiply> 0;00 30 by 1 07;30. You will see 33;45. Return. Sum the base and the top. You will see 7;30. Break off 1/2 of 7;30. You will see 3;45. Solve the reciprocal of 3;45. You will see 0;16. «Sum» <multiply> 0;16 by 33;45. You will see 9. The height of the dyke is 9. That is the procedure.

(v) A wall. The length is sixty, the top 0;30, the base 1, the height 6. <What is> the volume? Demarcate the length for 1 man. You: sum 0;30 and 1. You will see 1;30. Break off 1/2 of 1;30. You will see 0;45. Multiply 0;45 by 6, the height. You will see 4;30, the volume. Solve the reciprocal of 0;10, the work rate. You will see 6. Multiply 4;30 by 6. You will see 27, the laborers. Solve the reciprocal of 27, the laborers. You will see 0;02 13 20. Multiply 0;02 13 20 by 1 cable (the length of the wall). You will see 2;13 20. 1 man takes (it). That is the procedure.

(vi) An outflow water clock. I opened (it) so that the outflow was 1/2 sila. It did not reach the 1 sila mark by a 4th part of 0;00 10, a finger. How high is the surface over the surface? You: solve the reciprocal of 0;01 40, the height of the water clock. You will see 36. Multiply 36 by 0;30. You will see 18. Multiply 18 by 0;00 02 30. You will see 0;00 45, by which the surface exceeds the surface. That is the procedure.

(vii) A water clock. I opened (it) so that <the outflow was> 1/2 sila. The surface area exceeded the surface by 0;00 45. What did I take out <to reach> the 1 sila mark? You: solve the reciprocal of 0;30, the «1» <1/2> sila outflow. You will see 2. Multiply 2 by 0;01 40. You will see 0;03 20. Multiply 0;03 20 by 0;00 45. You will see 0;00 02 30, by which it did not reach the 1 sila mark. That is the procedure.

(viii) An outflow water clock. I opened 0;03 20. The outflow was a broken sila. It did not reach the 1 sila mark by 0;00 00 44 26 40, a 9th part of 2/3 of 0;00 10, a finger. By what did the surface exceed the surface? Solve the reciprocal of 0;01 40. You will see 36. Multiply 36 by 0;03 20. You will see 3. Multiply 2 by 0;00 00 44 26 40. The surface is higher than the surface by 0;00 01 28 53 20. The procedure.

(ix) A wedge (?). 20 squares it. What is its area? You: square 20. You will see 6 40. Take away a «4th» <1/2> of 6 40. You will see 3 20. The area is 2 iku. That is the procedure.

(x) The *imlû* is 1. What is the circle (*or:* depth)? You: put down 4 and 3, the seed. Multiply 4 by 3. You will see 12. <Solve> the reciprocal of 18. You will see 0;03 20. Multiply 0;03 20 by 12. You will see 0;40. The circle (*or:* depth) is 0;40. That is the procedure.

(xi) The circle (*or:* depth) is 0;40. <What is> its «weight» <*imlû*>? <You: put down 4 and 3, the seed.> Multiply 4 by 3. You will see 12. Solve the reciprocal of 12. You will see 0;05. Multiply 18, its clay (?), by 0;05. You will see 1;30. Multiply 1;30 by 0;40, the circle (*or:* depth). You will see 1. That is the procedure.

(xii) Its *imlû* is 0;30. What is the circle (*or:* depth)? You: multiply 4 by 3. You will see 12. Solve the reciprocal of 18. You will see 0;03 20. Multiply 0;03 20 by 12. You will see 0;40. Multiply 0;40 by 0;30. You will see 0;20, the circle (*or:* depth). That is the procedure.

(xiii) The circle (*or:* depth) is 0;20. What is the *imlû*? You: multiply 4 by 3. You will see 12. Solve the reciprocal of 12. You will see «you will see» 0;05. Multiply 18, the *imlû*, by 0;05. «Multiply» <You will see> 1;30. Multiply 1;30 by 0;20, the circle (*or:* depth). You will see 0;30. The *imlû* is 0;30. That is the procedure.

(xiv) A reed bundle. The lower circle is 0;04, the upper circle «0;01» <0;02>, the height 6. What are the volume and the dividing line of the upper volume and lower volume? You: square 0;04. You will see 0;00 16. Multiply 0;00 16 by 0;05, (the constant) of the circle. You will see 0;00 01 20. Square 0;02. You will see 0;00 04. Multiply 0;00 04 by 0;05, (the constant) of the circle. You will see 0;00 00 20. Sum 0;00 01 20 and 0;00 00 20. You will see 0;00 01 40. Break off 1/2 of 0;00 01 40. You will see 0;00 00 50. Multiply 0;00 00 50 by 6, the height. You will see 0;00 05, the volume of reed bundles. That is the procedure.

(xv) A reed bundle. The base is 0;04, the top 0;02, the height 6, the volume 0;00 05. It went up 3 cubits. What are the dividing line and volume? You: by what does 0;04, the base, «go up over» <exceed> 0;02, the top? It exceeds by 0;02. Solve the reciprocal of 6, the height. You will see 0;10. Multiply 0;10 by [0;02]. You will see 0;00 20. Multiply 0;00 20 by 3 that you went up. You will see 0;01. Take away 0;01 from 0;04, the base. [You will see 0;03.] The dividing line is 0;03. That is the procedure.

(xvi) Baked bricks for a well. The length is 0;03 20, the upper [width] 0;02 [30], the lower width 0;01 40. [How many bricks] are put down for the well? You: by [what] does 0;02 30, the upper width, exceed [0;01 40, the lower width]? It exceeds by 0;00 50. Solve the reciprocal of 0;00 50. [You] will see [1 12]. Multiply 1 12 by 0;01 40. You [will see] 2. [Square 2.] You will see 4. Multiply 0;03 20, the length, by 4. [You] will see 0;1[3 20], the large dividing line. Copy 0;13 20 times 3. You [will see] 0;40. The circle is [0;40].

[Return. What is] one layer (of bricks)? Solve [the reciprocal of 0;01 40, the upper width]. You will see 36. [Multiply] 36 [by 0;40, the circle. You will see 24, one] layer (of bricks).

[Return. What is] the central circle? Copy [0;03 20, the length] of a brick. <You will see 0;06 40.> [Add] 0;06 40 to 0;13 [20. You will see 0;20.] Copy 0;20 times 3. You will see 1. [The central circle is] 1. That is the procedure.

(xvii) A wall. The height is 36, the top 1/2 [rod, 3] cubits. I descended from the top by as much as the base. What are what I descended, the base, the dividing line, and the volume? You: in 1 cubit the slope is 0;00 50. Copy 0;00 50. You will see 0;01 40. Multiply 0;01 40 by 36, the height. You will see 1. Put it down. Square 0;45. You will see 0;33 45. Sum 1 and 0;33 45, so that you will see 1;33 45. What is the square-side? The square-side is 1;15. The base is 1;15. Multiply 1;15 by 12, the ratio (?) of height. You will see 15, <that I descended>. Put down 0;25, the slope.

Return. By what does 1;15, the base, exceed 0;45 <, the top>? It exceeds by 0;30. Multiply 0;25, the slope, by 0;30, the excess. You will see 0;12 30. Add 0;12 30 to 0;45, the top, so that you will see 0;57 30, the dividing line.

Return. See the <lower> volume. Add 0;57 30 and 0;45 the top. You will see 1;42 30. Break 1/2 of 1;42 30. You will see 0;51 15. <Raise> 0;51 15 by the 15 that you descended. You will see 12;48 45, the upper volume.

Return. See the lower volume. Sum 0;57 30 and 1;15. <You will see> 2;12 30. Break off 1/2 of 2;12 30. You will see 1;06 «40» <15>. Multiply 1;06 40 by 21, the (remaining) height. You will see 23;11 15, the lower volume. That is the procedure.

(xviii) A wall. The volume is 36, the height 36. In 1 cubit (height) the (total) slope is 0;00 50. <What are> the base and the top? You: solve the reciprocal of 36, «the volume» <the height>. You will see 0;01 40. Multiply 0;01 40 by 36. You will see 1. Put down «2» <1>. Multiply 0;00 25, the slope (of one side), by 36. You will see 0;15. Take away 0;15 from 1. You will see 0;45, the top. Add 0;15 to 1. You will see 1;15, the base. That is the procedure.

(xix) Baked bricks for a well. The length is 0;03 20. The upper width <exceeds> the lower width by 0;00 50. The circle is 0;40 «exceeds». What are the upper width and the lower width? You: take away the 0;20th part (*sic*) of 0;40. You will see 0;13 20.

Solve the reciprocal of 0;03 20. You will see 18. Multiply 18 by 0;13 20. You will see 4. Break off 1/2 of 4. You will see 2. Multiply 0;00 50, the excess, by 2. You will see 0;01 40. Put down 0;01 40 as the ratio (?) of the upper width and lower width. Add 0;00 50 to 0;01 40. You will see 0;02 30. The upper width is 0;02 30. That is the procedure.

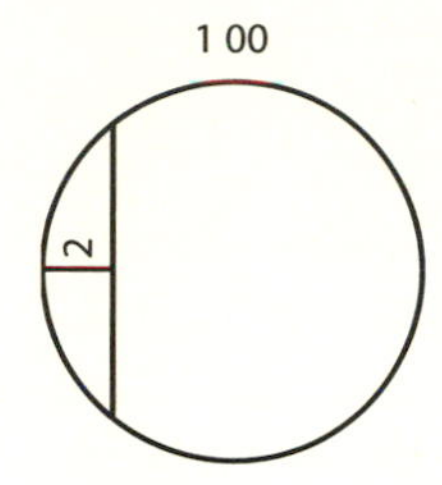

(xx) A circle was 1 00. I descended 2 rods. What was the dividing line (that I reached)? You: «you:» «Square» <double> 2. You will see 4. Take away «you will see» 4 from 20, the dividing line. You will see 16. Square 20, the dividing line. You [will see] 6 40. Square 16. You will see 4 16. Take away 4 16 from 6 40. You will see 2 24. What squares 2 24? 12 squares it, the dividing line. That is the procedure.

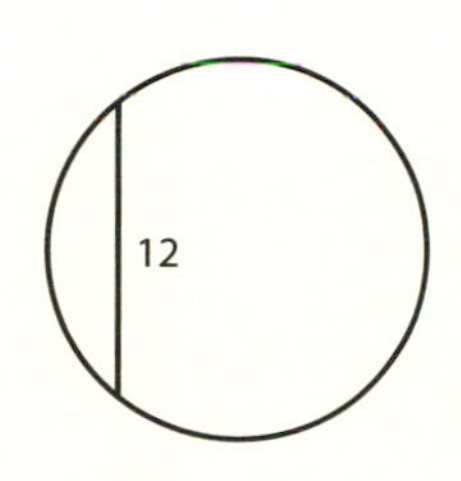

(xxi) If I circled a circle of 1 00 (so that) the dividing line was 12, <what> was it that I descended? You: square 20, the dividing line. You will see 6 40. Square 12. <You will see 2 24.> Take away 2 24 from 6 40. You will see 4 16. What squares <4> 16? «4» <16> squares it. <Take away 16 from 20. You will see 4.> Break off 1/2 of 4. You will see 2. What you descended is 2. The procedure.

(xxii) A (measuring vessel of 1) bariga. The dividing line is 0;04, the grain 1 00. What are the depth and circle? You: square the dividing line. You will see 0;00 16. Solve the reciprocal of 0;00 16. You will see 3 45. The depth is 3 45. The procedure.

(xxiii) A textile. The length is 48 (rods). In 1 day she weaves 0;20 (rods). In how many days will she cut (it) off (the loom)? You: solve the reciprocal of 0;20. You will see 3. Multiply 48 by 3. You will see 2 24. She will cut (it) off after 4 months, 24 days.

(xxiv) A (measuring vessel of 1) ban is full of grain. I received [1 sila] of grain from inside it. <Its dividing line is> 0;02. What depth should I descend so that it is 1 sila? [You:] divide 0;02 30, the depth, by 10. Solve the reciprocal of 10. You will see 0;06. Multiply [0;02] 30 by 0;06. You will see 0;00 15. The depth is 0;10 fingers (and) 1/2 of 10 fingers. May you see 1 sila. Square 0;02, the dividing line. You will see 0;00 04. Multiply 0;00 15, the depth, by 0;00 04. You will see 0;00 00 01. The grain is 1 sila. That is the procedure.

(xxv) I shall capture a city hostile to Marduk with a volume of 1 30 00. I established the foundations of the volume as 6 (and) 8 (rods) to reach the wall. The vertical of the volume is 36. How much length should I trample down so that I may capture [the city], and [what is] the length behind the hole (?)? [You:] solve [the reciprocal] of 6, the foundations of the volume. You will see 0;10. [Multiply] 0;10 by [1 30 00, the volume]. You will see [15 00]. Solve the reciprocal of 8. You will see 0;07 30. Multiply [0;07 30 by 15 00]. You will see 1 52;30. Copy 1 52;30. [You will see 3 45.] Multiply 3 «44» <45> by 36. You will see 2 15. [Square] 1 52;«20» <30>. You will see [3 30];56 15. [Take away] 2 15 from 3 30;56 15. What squares [1 1]5;56 «14» <15>? You will see 1 07;30. Take away [1 07;30 from] 1 52;30. You will see 45, the height of the wall. [Break off 1/2 of 45.] You will see [2]2;[30]. Solve the reciprocal of 22;30. You will see 0;02 40. Multiply [15 00 by] 0;02 40. The length is 40. Return. See 1 30 00, the wall. Multiply 22;30, [1/2 (?) the height] by 40, the length. You will see 15 00. Multiply 15 00 by 6. You will see 1 30 00. The volume is 1 30 00. That is the method.

(xxvi) I shall capture a city hostile to Marduk with a volume of 1 30 00. I shall walk from the base of the volume a length of 32 (rods) in front of me. The height of the volume is 36. What length should I trample so that I may capture the city? What is the length that is in front of the hole (?)? You: solve the reciprocal of 32. You will see 0;01 52 30. Multiply 0;01 52 30 by 36, the height. You will see 1;07 30. Solve the reciprocal of 6, the foundations of the volume. You will see 0;10. Multiply 1 30 00, the volume, by 0;10. you will see 15 00. Copy 15 00 twice. You will see 30 00. Multiply 30 00 by 1;07 30. You will see 33 45. What squares 33 «42» <45>? 45 squares (it). The height of the wall is 45. By how much does «44» <45>, the height of the wall, exceed 36, the height of the volume? It exceeds by 9. Solve the reciprocal of 1;07 30. You will see 0;53 20. Multiply 0;53 20 by 9. You will see 8. You tread a length of 8 (rods) in front of you.

(xxvii) From 4 furrows 0;30 fell off. I [harvested] 1 sila of grain from 1/2 rod. What is the grain of 1 (bur) area? You: [solve] the reciprocal of 4, the furrows. You will see 0;15. Multiply 0;15 by 0;30. You will see 0;07 30. Solve the reciprocal of 0;07 30. You will see 8. Multiply 8 by 30 00, the area. You will see 4 00 00. The length of 1 (bur) of area is 8 leagues. Solve the reciprocal of 30 00, the area. You will see 0;00 02. Multiply 4 00 00 by 0;00 02. You will see 8. There are 8 (00 00 sila, or) «in» 1 36 gur of grain.

(xxviii) A moat, 10 (rods) on each side. The height (*sic*) is 18. In 1 cubit (height) the slope is 1. <What are> the base and volume? You: sum 0;05 and 0;05. You will see 0;10. Multiply [0;10] by 18, the height. You will see 3. Take away 3 from 10. [You will see] 7, the base. Return. Sum <7>, the base and <10>, the top «10». You will see 17. [Break off 1/2 of 17.] You will see 8;30. Square. You will see 1 12;15. [Put down] 1 12;[15]. Take away a «second» <fourth> part of 3, the excess of the top over the base. Add 0;45 to 1 12;15 so that you will see 1 13. Multiply 18 by 1 13. You will see 22 30. The volume is 2 (eshe) 1 1/2 (iku). That is the procedure.

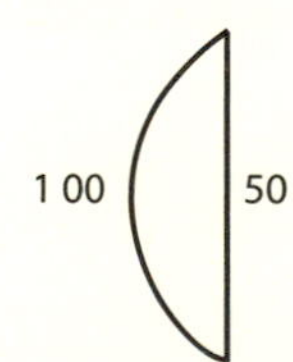

(xxix) A crescent moon. The (arc of) circle is 1 00, the dividing line 50. <What is> the area? You: by what does 1 sixty, the circle, exceed 50? The excess is 10. Multiply 50 by 10, the excess. You will see 8 20. Square 10, the dividing line. You will see 1 40. Take away 1 40 from 8 20. You will see 7 30. The area is [7 30]. The procedure.

(xxx) If a boat carries 1 sar of bricks, what grain does it carry? You: 0;41 40 is the volume of [1 00] bricks. Multiply 0;41 40 by 5, (the constant) of storage. The «volume» <grain> is 3;28 00. The «brick» <grain> of 1 brick is 3 1/3 sila 8 1/3 shekels. Multiply 3;28 20 by 12 «fingers» <sixties>. You will see 41 40. You will see 8 (gur) 1 40 (sila) grain. The procedure.

(xxxi)–(xxxii) [*These problems are too broken for translation.*]

(xxxiii) The length is [0;30], the area 0;10. What is the width? You: [break off 1/2 of 0;30.] You [will see] 0;15. Solve the reciprocal of 0;15. [You] will see 4. [Multiply] 0;10 by [4]. You will see [0;40]. Multiply 0;40 by 0;30 «0;30», the length. [You will see 0;20], the width.

(xxxiv) A basket. [It] circled [a circle of] 0;03 [fingers]. (From it I removed 1 sila of grain.) What height did it (the level of the grain) descend? You: [square 0;03. You] will see [0;00 09]. Multiply 0;00 09 by 0;05, (the constant) of a circle. You will see 0;00 00 45. [Solve the reciprocal of 0;00 00 45.] You will see 1 20. Multiply the reciprocal of 6 40 00 by «0;00 09» <1 20>. [You will see 0;00 12.] It descended a height of 0;00 12 rods. The procedure.

(xxxv) I enclosed a basket of 1 sila (with) 0;10 fingers. What length did I go? You: square 0;10. You will see 0;01 40. Multiply 0;01 40 by 0;05, (the constant) of a circle. You will see 0;00 08 20. Solve the reciprocal of 0;00 08 20. You will see 7 12. Multiply the reciprocal of 6 40 00 by 7 12. You will see 0;01 04 48, the length. The procedure.

Total 35 trails of calculation.

The word "trails" in the colophon of BM 85194 normally means "animal tracks" in nonmathematical contexts. It is particular to a group of eight mathematical tablets including BM 96957+VAT 6598, whose colophon—uniquely for a collection of problems—identifies it as the property of one Ishkur-mansum, son of Sin-iqisham [Robson 1996, 1997, 1999, 231–45]. He is probably to be identified with an Ishkur-mansum who was active as a teacher in the city of Sippar in around 1630 BCE—but on the evidence of these tablets he cannot have been a particularly competent teacher of mathematics [Robson forthcoming chap. 4].

Problems (i)–(xvii) concern the number of bricks needed to build walls and houses. It is assumed that 1/6 of each wall comprises mortar, and 1/3 of the total floor area of each house comprises the walls. The final group of problems gives three different approximations to the Pythagorean rule; problems (xxiii)–(xxv) are heavily damaged, and their reconstruction is conjectural.

The larger fragment, VAT 6598, was published in part by [Weidner 1916] and thus problems (xviii) and (xxi) enjoy the privilege of being the first mathematical problems to be deciphered. Explanatory glosses written in tiny signs below the main line of writing are shown here in subscript characters.

BM 96957+VAT 6598

(i) [A wall.] The width is [2 cubits], the length 2 1/2 rods, the height 1 1/2 rods. [How many bricks? You]: multiply 2 cubits, the width, by 2 1/2 rods, the length. You will see 0;25$_{\text{the ground}}$. Multiply [0;25] by 18, the height. You will see 7;30$_{\text{7 1/2 sar, the volume}}$. Multiply 7;30 by 6, the constant of a wall. You will see 45. The bricks are 45 sar. The procedure.

(ii) A wall. The length is 2 1/2 rods, the height 1 1/2 rods, the bricks 45 sar. What is the width of my wall? You: multiply 2;30, the length, by 18, the height, so that you will see [45]. Keep (it). Solve$_{\text{the reciprocal}}$ the reciprocal of 6, the constant of a wall. [Multiply] by [45 sar, the bricks]. You will see 7;30$_{\text{the volume}}$. Keep (it). [Solve] the reciprocal of the 45 that you are keeping. [You will see 0;01 20.] Multiply 0;01 20 by the $7;30_{\text{the volume}}$ that you are keeping and [you will see 0;10. The width of the wall is 2 cubits.] The procedure.

(iii) [*Missing. Presumably on finding the bricks needed for the wall; cf. (xiv).*]

(iv) [*Missing. Presumably on finding the length and width given their sum; cf. (xv).*]

(v) [A wall. The height is 1 1/2 rods, the bricks 45 sar. The length exceeds the width of the wall by 2;20 rods. What are the length and width of my wall? You: solve the reciprocal of 6, the constant of a wall. You will see 0;10. Multiply 0;10 by 45 sar, the bricks. You will see 7;30. The volume is 7 1/2 sar. Solve the reciprocal of 18, the height. You will see 0;03 20. Multiply 0;03 20 by 7;30. You will see $0;25_{\text{the ground}}$. Keep (it). Break off 1/2 of 2;20, by which the length exceeds the width of the wall. You will see 1;10. Put (it) down 1;10 twice. Square (it).] You will see 1;21 40. Add the 0;25 that you are keeping to 1;21 40. You will see 1;46 40. What is the square-side of 1;46 40? 1;20 is the square-side. Put (it) down twice. Add 1;10 to 1;20. You will see 2;30. The length of the wall is 2;30. Take away 1;10 from 1;20. The remainder is 0;10. The width of the wall is 2 cubits. The procedure.

(vi) The <area of a> house is 5 sar. For a height of 2 1/2 rods how many bricks should I get made? You: take a 3rd part of 5 sar. You will see $1;40_{\text{the walls}}$. Multiply 1;40 by 2 1/2 rods, the height. You will see 4;10. You will get 2 1/2 iku of bricks made and then you will pile up (a) 5 sar (area of) house to a height of 2 1/2 rods. The procedure.

(vii) If the bricks are 2 1/2 iku, the height 2 1/2 rods, what (area of) house should I build? You: solve the reciprocal of 2 1/2 rods, the height. Multiply by 4;10, the 2 1/2 iku of bricks. You will see 1;40. [Keep (it).] Solve [the reciprocal of 0;20, the constant of a built house]. You will see 3. [Multiply] 3 by [the 1;40 that you are keeping. You will see 5.] The (area of the) house is 5 sar. [The procedure.]

(viii)–(xi) [*The rest of the column is missing. It may have contained up to four problems of approximately the same length as (vi) and (vii), the first of which was almost certainly about finding the height of the house given the area and bricks.*]

(xii) A wall of baked bricks. The width is 2 cubits, the height 1 rod, the baked bricks 9 sar. What is the length of my wall? You: solve the reciprocal of 2;15, the constant of a wall. You will see 0;26 40. Multiply 0;26 40 by 9 sar, the baked bricks. You will see $4_{\text{the volume}}$. Keep (it). Solve the reciprocal of 12, (or) 1 rod, the height of the wall. You will see 0;05. Multiply 0;05 by the 4 that you are keeping. You will see $0;20_{\text{the position}}$. Solve the reciprocal of 0;10, the width; multiply by the 0;20 that you saw. You will see 2. The length is 2 rods. The procedure.

(xiii) The length of a wall of baked bricks is 2 rods, the height 1 rod, the baked bricks 9 sar. What is the width of my wall? You: take the reciprocal of 2;15, the constant of a wall. You will see 0;26 40. Multiply 0;26 40 by 9 sar, the baked bricks <that> you spread out for your wall. You will see $4_{\text{the volume}}$. Solve the reciprocal of 1 rod, the height of your wall. Multiply by the 4 that you saw. You will see 0;20. The area of your wall is 1/3 sar. Solve the reciprocal of 2 rods, the length. You will see 0;30. Multiply 0;30 by 0;20. You will see 0;10. The width of the wall is 2 cubits. The procedure.

(xiv) The length of a wall is 2 rods, the width of the wall 2 cubits. 9 sar of baked bricks are put down for the wall. How high can it be with 9 sar of baked bricks? You: solve the reciprocal of 2;15, the constant of baked [bricks]. You will see 0;26 40. Multiply 0;26 40 by 9 [sar, the baked bricks that] are put down [for] the wall. You will see $4_{\text{the volume}}$. [Keep (it). Multiply 2 rods, the length, by 0;10. You will see 0;20.] Solve the reciprocal of 0;20. You will see 3. Multiply 3 by the 4 [that] you are keeping. [You will see 12.] You will elevate (it) [to a height of 1 rod] with 9 sar of baked bricks. The procedure.

(xv) [A wall of] baked [bricks]. The height of the wall is 1 rod, the baked bricks 9 sar. I summed the length and thickness [of the wall] so that (it was) 2;10. What are the length and thickness of my wall? [You:] solve the [reciprocal of 2;15], the constant of baked bricks. You will see 0;26 40. Multiply 0;26 40 [by] 9 sar, the baked bricks. You will see $4_{\text{the volume}}$. Keep 4, the volume of your baked bricks. Solve [the reciprocal of 12], 1 rod, the height of your baked bricks. You will see 0;05. Multiply 0;05 by the 4 that you are keeping. You will see 0;20. The area is $0;20_{\text{the position}}$. Keep (it). Break off 1/2 of 2;10, [the sum] of the length and thickness of the wall. You will see 1;05. Square 1;05. [You will see] 1;10 25. Take away [0;20], the area [that] you are keeping, from 1;10 25. You will see 0;50 25. What is the square-side of 0;50 25? [The square-side is 0;55.] Add [0;55] to the 1;05 that you squared. You will see 2. The length is 2 rods. Take away [0;55 from 1;05 that] you squared. The remainder is 0;10. The thickness is 2 cubits. The procedure.

(xvi) [A wall of baked bricks.] I laid 9 sar of [baked] bricks. The length [exceeds the thickness of the wall] by [1;50]. The height is 1 rod. [What are] the length of the wall and the thickness of my wall? [You:] solve [the reciprocal of 2];15, the constant of a wall of baked bricks. You will see 0;26 40. Multiply 0;26 40 [by 9 sar], the baked [bricks]. You will see $4_{\text{the volume}}$. The volume of the baked bricks is 4 sar. Solve [the reciprocal] of 12, the height of the wall. You will see 0;05. Multiply 0;05 by 4, the volume of baked bricks. You will see 0;20. Keep 0;20, the area. Break off 1/2 of 1;50, by which the length exceeds the thickness of the wall. You will see 0;55. Put down [0;55 twice]. Square 0;55. You will see 0;50 25. Add 0;20, the area, to 0;50 25. You will see 1;10 25. What is the square-side of 1;10 25? [The square-side] is 1;05. Put (it) down twice. Add the [0;55 that] you squared to 1;05. You will see 2. [The length] of the wall is [2 rods. Take away 0;55 from 1;05.] The remainder is 0;10. The thickness [of the wall] is 2 cubits. [The procedure.]

(xvii) A wall of baked bricks. [The length of the wall is 2 rods.] It is [2 cubits] thick at the bottom, [1 cubit thick at the top]. The height is 2 rods. What heaping (?) does my wall [heap up, and] in 1 cubit how much does it slope? You: sum 2 cubits, the lower base, [and 1 cubit], that it tapers to at the top. You will see 0;15. Break off 1/2 of 0;15. You will see $0;07\ 30_{\text{the position}}$. Multiply 0;07 30 by 2 rods, the length. You will see 0;15. Multiply [0;15 by] 24, the height. You will see 6. Multiply 6 by [2;1]5, the constant of baked bricks. You will see 13;30. The baked bricks are 13 1/2 sar. Break off 1/2 of 13;30. You will see 6;45. A heaping (?) of 6 2/3 <sar> 5 shekels will be heaped up for you with your wall. Return. See how much it slopes in 1 cubit. By how much does 2 cubits, the lower base, exceed 0;05 that is on the top? It exceeds by 0;05. Keep (it). Because they said, "What slope does it slope in 1 cubit?," [put down 0;05]. Solve the reciprocal of 0;05. You will see 12. [Multiply] by 2 rods, the height.

[You will see 24.] Solve the reciprocal of 24. You will see 0;02 30. Multiply [0;02 30 by] 0;05, (which is) 1 cubit, the ratio. You will see 0;00 12 30. [Multiply 0;00 12 30 by 6 00.] You will see [1;15]. In 1 cubit the wall slopes a slope of 1 finger and a quarter of 1 finger. The procedure.

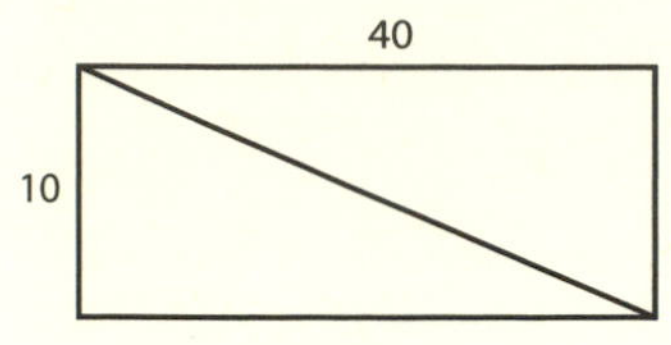

(xviii) A gate. The height is 1/2 <rod>, 2 cubits, the breadth 2 cubits. What is its diagonal? You: square 0;10, the breadth. You will see 0;01 40, the ground. Take the reciprocal of 0;40 (rods, or 8) cubits, the height. Multiply by 0;01 40, the ground. You will see 0;02 30. Break off 1/2 of 0;02 30. You will see 0;01 15. [Add] 0;01 15 [to 0;40, the height]. You will see 0;41 15. [The diagonal] is 0;41 15. The procedure.

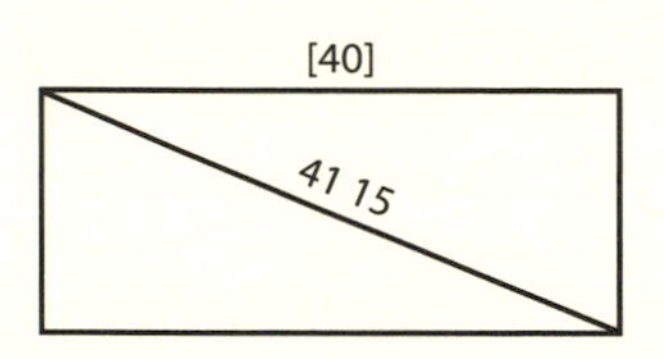

(xix) If the gate has height 0;40 (rods, or 8) cubits and diagonal 0;41 15, what is the breadth? You: take away 0;40, the height, from 0;41 15, the diagonal. The remainder is 0;01 15. Copy 0;01 15. You will see 0;02 30. Multiply 0;40, the length, by 0;02 30, the product (?) that you saw. You will see 0;01 40.What is the square-side? 0;10 is the square-side. The breadth is 0;10. The procedure.

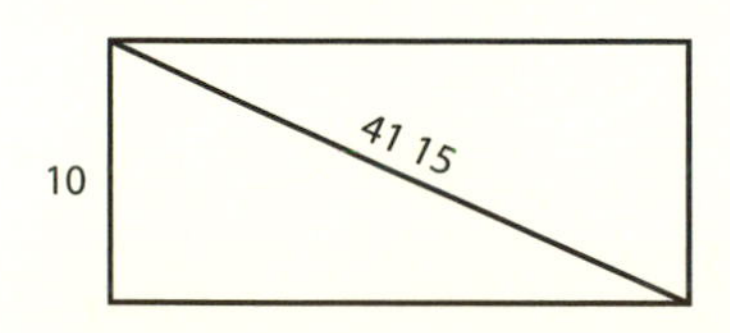

(xx) The breadth is 2 cubits, the diagonal 0;41 15. What is the height? You: no (solution). «One.»

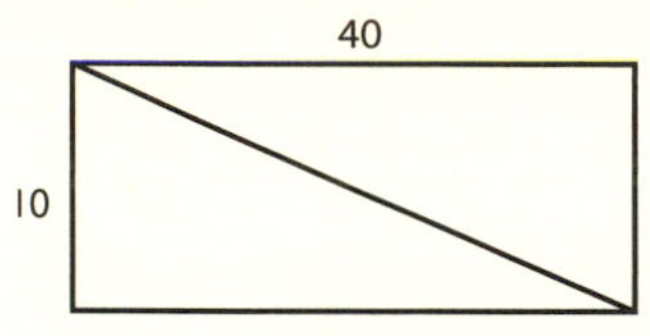

(xxi) The breadth is 2 cubits, the height 0;40 (rods, or 8) cubits. What is its diagonal? You: square 0;10, the width. You will see 0;01 40, the ground. Multiply 0;01 40 by 0;40 (rods, or 8) cubits, the height, so that you will see 0;01 06 40. Copy (it). You will see 0;02 13 20. Add it to 0;40 (rods, which is 8) cubits, the height. You will see 0;42 13 20, the diagonal. The procedure.

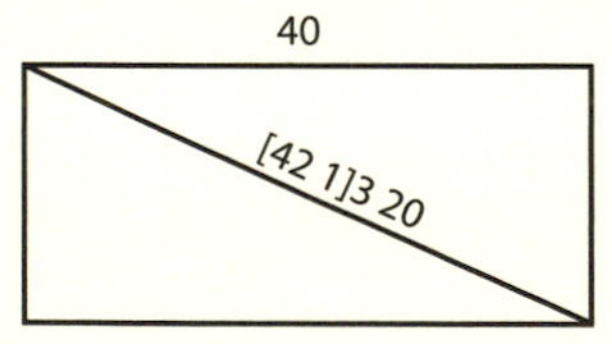

(xxii) [*Text missing. The following reconstruction is conjectural:* The height is 0;40 (rods, or 8) cubits, the diagonal 0;42 13 20. What is the breadth? You: take away 0;40, the height, from 0;42 13 20. You will see 0;02 13 20. Copy 0;40. You will see 1;20. Solve the reciprocal. Multiply it by 0;02 13 20 so that you will see 0;01 40. What is the square-side? The square-side is 0;10, the breadth. The procedure.]

(xxiii) [*Diagram and much text missing. The following reconstruction is conjectural:* The breadth is 2 cubits, the height 0;40 (rods, or 8) cubits. What is its diagonal? You: square 0;10, the width. You will see 0;01 40. Square 0;40, the length. You will see 0;26 40. Sum 0;26 40 and 0;01 40. You will see 0;28 20. What is the square-side? The square-side is 0;41 13 51 48 (?), the diagonal.] The procedure.

(xxiv) [*Diagram missing.* The height is 0;40 (rods, or 8) cubits, the diagonal 0;41 13 51 48 (?).] What is the breadth? You: [square 0;41 13 51 48 (?).] You will see [0;28] 20. [Square 0;40, the height. You will see 0;26 40. Take away 0;26 40] from 0;28 20. [You will see 0;01 40. What is the square-side? The square-side is 0;10, the breadth.] The procedure.

(xxv) [*Diagram missing.* The breadth is 2 cubits, the diagonal 0;41 13 51 48 (?). What is the height? You:] square [0;41 13 51 48 (?)], the diagonal. The squared number (?) is 0;28 20. [Square 0;10, the breadth.] You will see 0;01 40. [Take away] 0;01 40 from 0;28 20. [The remainder is 0;26 40.] What is the square-side? The square-side is 0;40. The procedure.

One. Total 25 (?) trails. Ishkur-mansum, son of Sîn-iqisham.

Mathematics in Sumerian literature (iii)

This fictionalized letter, from Ishbi-Erra (first king of the Isin dynasty, c. 2017–1985 BCE) to Ibbi-Suen, last king of Ur (c. 2028–2004 BCE), describes how he, while still in the latter king's service, has been sent north to buy grain in order to alleviate the famine in the south, but is held back by incursions of nomadic Martu people. It reads suspiciously like an OB school mathematics problem: the first paragraph gives the silver-grain exchange rate and the total amount of silver available (72,000 shekels); in the second the silver has been correctly converted into grain. Next that huge capacity measure is divided equally among large boats. The numbers are conspicuously round and easy to calculate with; those in the final, damaged part of the section quoted are reminiscent of the final multiplicands of a standard multiplication table or the sexagesimal fractions 1/3, 1/2, [2/3], 5/6. At one level, its is no more than a pretext to show simple mathematics and metrology at work in a quasi-realistic context [ETCSL 3.1.17].

Letter from Ishbi-Erra to King Ibbi-Suen about the Purchase of Grain

[1–2]Say to Ibbi-Suen, my lord: this is what Ishbi-Erra, your servant, says:

[3–6]You ordered me to travel to Isin and Kazallu to purchase grain. With grain reaching the exchange rate of one shekel of silver per gur, 20 talents of silver have been invested for the purchase of grain.

[7–12]I heard news that the hostile Martu have entered inside your territories. I entered with 72,000 gur of grain—the entire amount of grain—inside Isin. Now I have let the Martu, all of them, penetrate inside the Land, and one by one I have seized all the fortifications therein. Because of the Martu, I am unable to hand over this grain for threshing. They are stronger than me, while I am condemned to sitting around.

[13–16]Let my lord repair 600 barges of 120 gur draught each; 72 solid boats, 20 ……, 30 ……, placing (?) 50 …… and 60 (?) boat doors on the boats (?), may he also …… all the boats.

[17–21]Let them bring it up by water, along the Kura and the Palishtum water-courses, to the grain heaps (?) that are spread out. And I myself intend to go (?) and meet them (?). The place there where the boats moor will be under my responsibility. Let them load up huge amounts of grain (?), the entire amount of grain; it should reach (?) you.

[22–23]If you have not got enough grain, I myself shall have grain brought in to you.

[24–28]My lord has become distressed about the battles in Elam. But the Elamites' grain rations have quickly been exhausted, so do not slacken your forces! Do not fall head first into their slavery, nor follow at their heels!

[29–30]I have at my disposal enough grain to meet the needs of your palace and of all the cities for 15 years.

[31–33]Let it be my responsibility to guard for you Isin, and Nippur. My lord should know this!

III.c. Rough work and reference lists

The Old Babylonian mathematical corpus has long been categorized into tables and problems. However, in the last few years two or three other types of mathematical output have begun to be recognized too [Robson 1999, 9–15]. First, diagrams and calculations were drawn up by students in the process of doing arithmetical exercises and solving mathematical problems, often on roughly made, frequently reused round or square tablets ("rough work"). Second, lists of suitable numerical parameters were drawn up by teachers to assist in setting particular types of problems, as were the technical constants so frequently used in the mathematical problems (together "reference lists"). A third type of mathematical writing, the "model document," is more problematic by its very nature, as instances are often difficult to distinguish from the real, working administrative records they are designed to mimic. Examples include AO 11409 (page 78) and AOT 304 (page 79).

Geometrical diagrams

The problem behind the diagram on YBC 7302 is to find the area of circle of circumference 3 [MCT, 44]. The student squares the circumference and multiplies by the constant 0;05 ≈ $1/4\pi$ (see YBC 7243, p. 149).

YBC 7302 Obverse (Reverse blank)

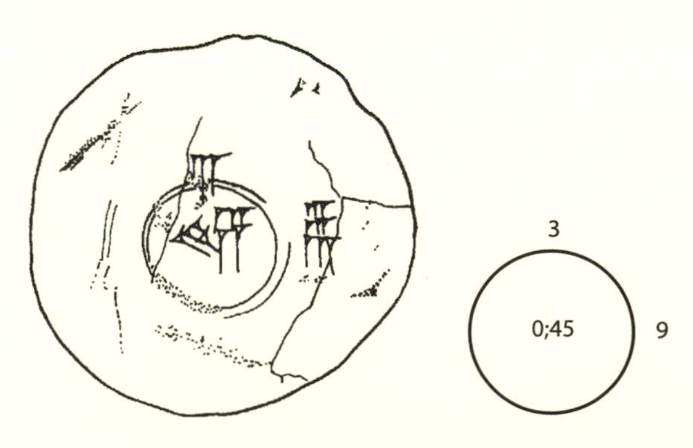

The student who made Ashmolean 1931.91 had the task of finding the area of a right triangle given two lengths [Robson 2004]. The correct answer to the problem should be 3;45 × 1;52 30 × 0;30 = 3;30 56 15, but the student misplaced a sexagesimal place in an intermediate calculation as the following reconstructions show:

Ash 1931.91

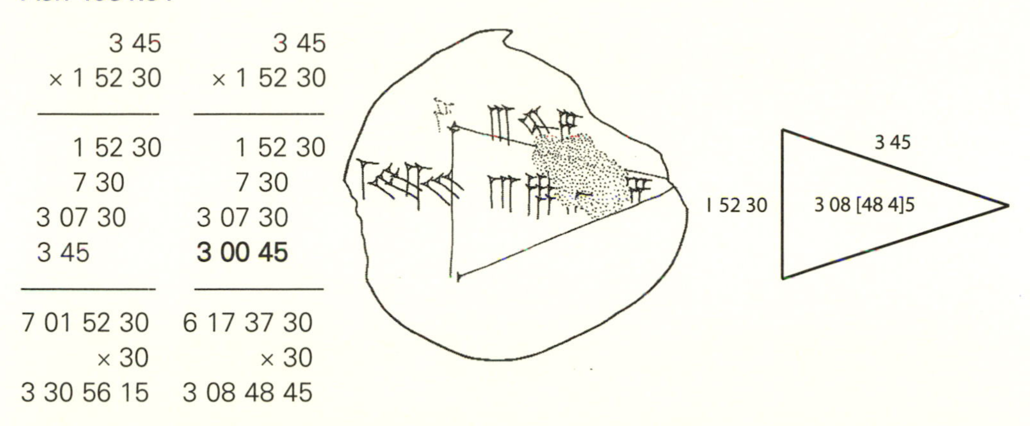

The problem behind the diagram on the obverse of YBC 7289 was probably to calculate the diagonal of a square of side 30 (see too the third line of BM 80209, page 150) [MCT, 42; Fowler and Robson 1998]. The famous approximation to $\sqrt{2}$, namely 1;24 51 10, was probably simply copied (or memorized) from a list of constants like YBC 7243, page 149. The previously unpublished reverse shows the much-erased remains of a solution to an analogous problem about the diagonal of a 3-4-5 rectangle. The calculations below it may belong to another problem entirely.

YBC 7289 Obverse (left) and Reverse (right)

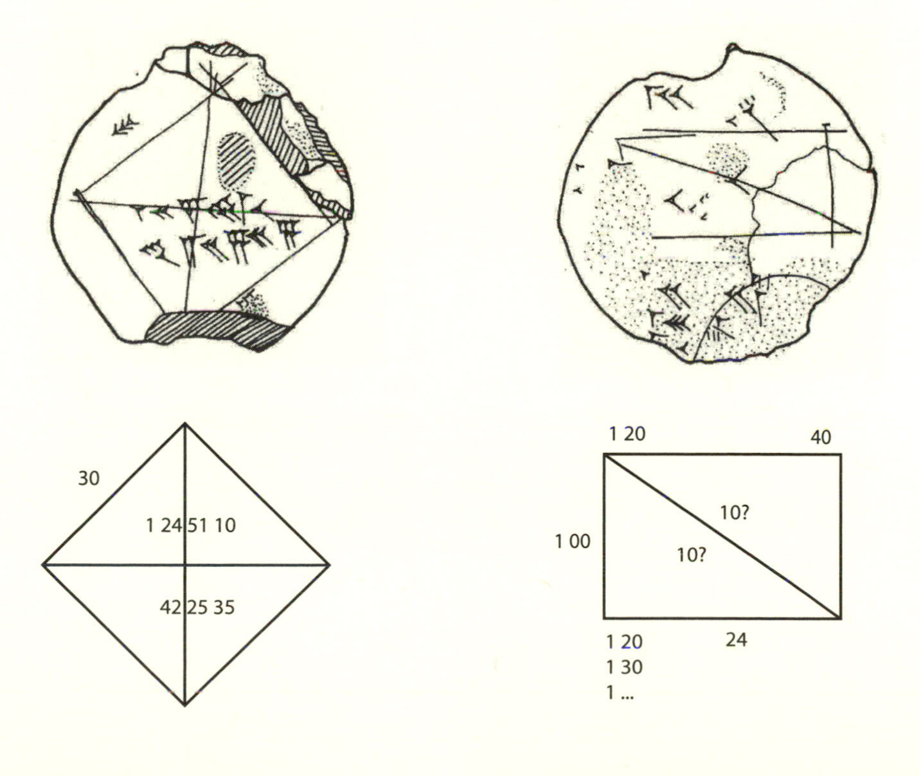

Both the obverse and the reverse of YBC 7290 carry diagrams of trapezoids, only one of which is numbered [MCT, 44; Nemet-Nejat 2002, no. 15]. The area of the trapezoid has been calculated using the so-called agrimensors' formula, known in Mesopotamia since at least the late fourth millennium BCE (see W 19408, 76 on page 74).

YBC 7290 Obverse (left) and Reverse (right)

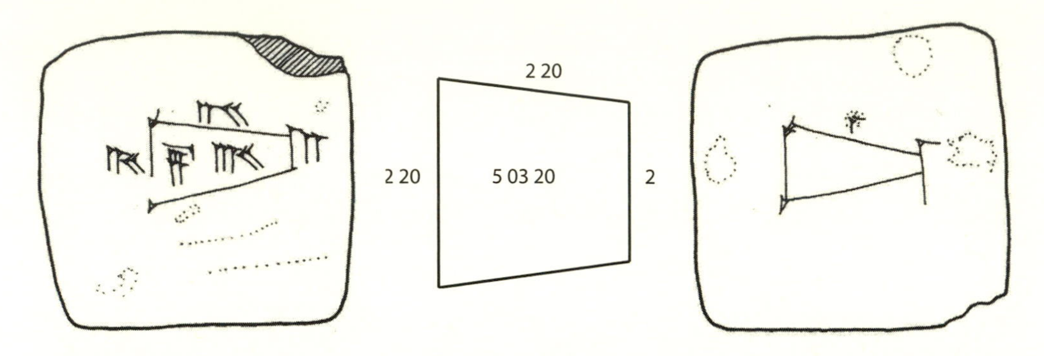

The diagram on YBC 11126 shows another trapezoid and accompanying calculations [MCT, 44]. In the bottom right corner the two short sides, 45 and 22 30, are added, halved, and multiplied by the length, 3, to produce the area, 1 41 15. The import of the other numerals is not clear.

YBC 11126 Obverse (Reverse blank)

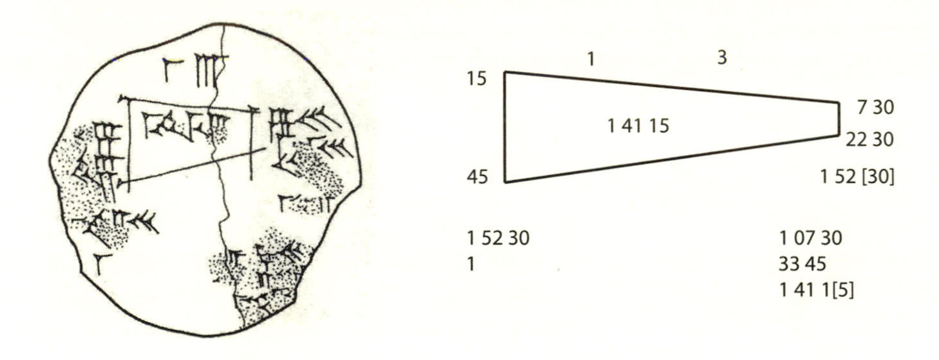

The trapezoid on the obverse of Ash 1922.168 is partitioned into three sub-areas with short sides in the ratio 1 : 2 : 3 [Robson 1999, 273–74]. It is not clear whether the student doing the calculation was given the values of the intermediate short sides or expected to find them himself. At any rate, he used the agrimensors' formula to find the area of each sub-area, writing the constant 0;30 three times to mark the fact that the opposite lengths must be added and multiplied by 0;30 (or divided by two). The reverse (not shown here) is covered in tiny semi-erased numerals.

Ash 1922.168 Obverse

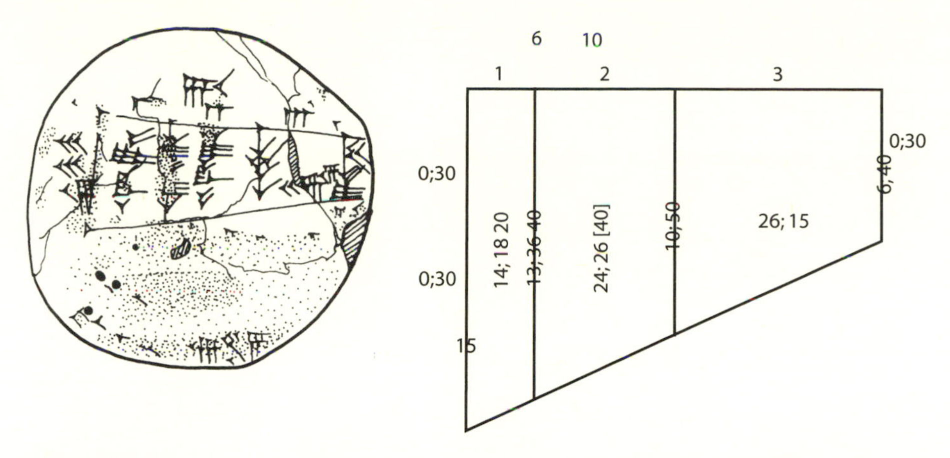

Calculations

The examples of rough work presented so far are all unprovenanced, but there is one group of forty-six tablets from a single house in the Old Babylonian city of Ur which reveal a great deal about how calculations were laid out [Robson 1999, 245–72; Friberg 2000, 101–36]. They all concern multiplication and division, none addition or subtraction; and are all written on round tablets whose obverse sides have been used to record Sumerian proverbs, which are also translated here. If the teaching sequence at Ur was similar to that known from Nippur (and that is a supposition that is far from proven) then this suggests that students were first given active practice in mathematical calculation (rather than the rote memorization of lists and tables) toward the end of their elementary schooling. Two groups of about half a dozen tablets have the same type of exercises but with different numerical values, suggesting that this particular school had between one and six pupils. In the following, the tilde symbol (~) indicates a reciprocal relationship.

Squaring calculations were laid out so that the multiplicands were in vertical alignment. The surface of *UET* 6/2 211 is badly eroded, but the calculation is still clearly visible.

UET 6/2 211

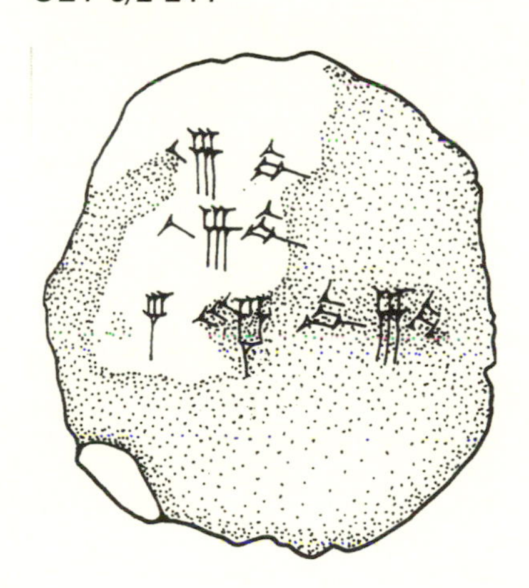

16 40

16 40

4 37 46 40

A lion caught a wild boar. He roared: "Your flesh has not yet filled my mouth, but your squeals have deafened my ears!" [ETCSL, 6.2.3 66].

UET 6/2 222 is more complicated. It shows a repeated pattern of squares in a computation which comes down to: $(0;15^2 \times 17)^2 \times 16^2 = 17^2$. Because 15^2 and 16 are mutually reciprocal, the relationship is trivially obvious, but is nevertheless appealing in its use of consecutive integers. While the squarings are laid out vertically, other multiplications are recorded with the multiplicands in horizontal alignment. There are no rulings on the tablet. This type of circular calculation—which seems to be an exercise in its own right, rather than the rough work for a problem text—is otherwise unknown.

UET 6/2 222

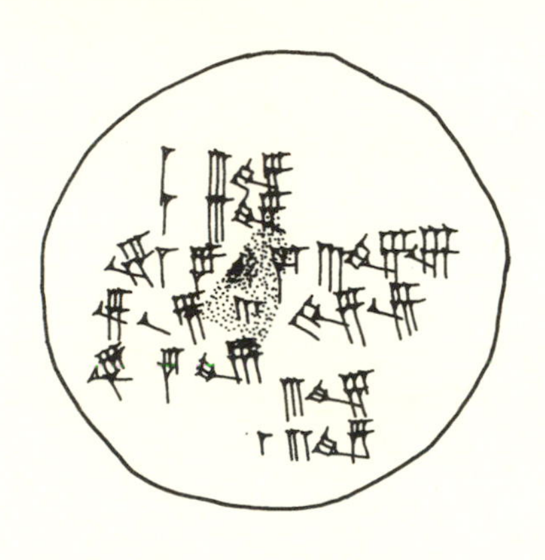

1 03 45
(×) 1 03 45
(=) 1 07 44 03 45 (×) 16
(=) 18 03 45 (×) 16
(=) 449 (= 17 × 17)
15
(×) 15
(=) 3 45
(×) 17
(=) 1 03 45

When a dog snarls, throw a morsel into his mouth [ETCSL, 6.2.3 78].

The largest group of calculations from Ur consists of continued multiplications. Multiplicands are written to the left of a vertical line, products to the right. When a division is necessary, the reciprocal of the divisor is written immediately to the right of it and a multiplication is carried out.

On both *UET* 6/2 236 and 293 (and three other tablets from Ur) the third operation is a division by 10, followed by a multiplication by 2. It may be that the students are working out the labor needed to dig canals in the shape of rectangular prisms using the daily work rate of 0;10 volume-sar and sometimes converting the depth into cubits [Robson 1999, 255]. Then the laborers' wages are calculated at 6 grains or 0;02 shekels a day [Friberg 2000, 124]. See, for instance, the problems on YBC 4657, page 116.

UET 6/2 293

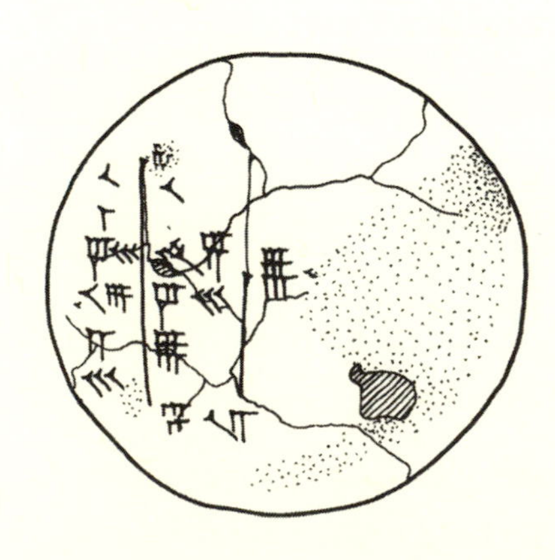

10
(×) 1 (=) 10
(×) 4 30 (=) 45
(÷) 10 (~) 6 (=) 4 30
(÷) 30 (~) 2 (=) 9
5 (~) 12

He who shaves his head gets more hair. And he who gathers the barley gains more and more grain [ETCSL, 6.2.3 42].

UET 6/2 236

Make the donkey sit like this! I am making it lift its shrivelled penis! [ETCSL, 6.2.3 63].

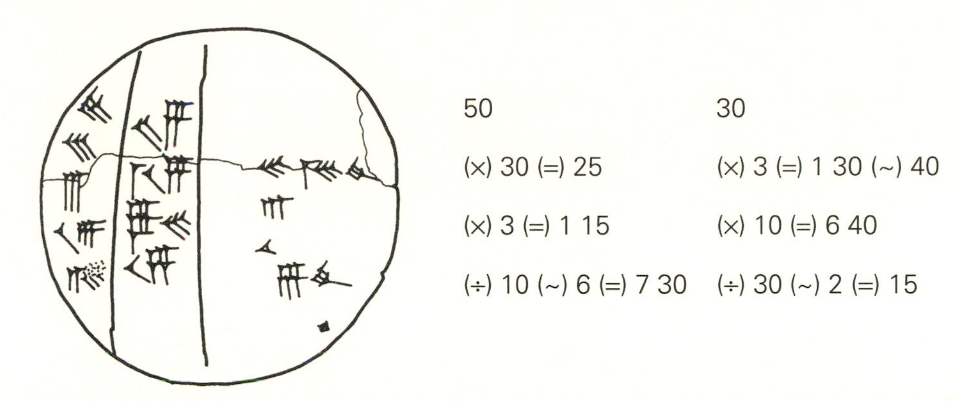

50
(×) 30 (=) 25
(×) 3 (=) 1 15
(÷) 10 (~) 6 (=) 7 30
30
(×) 3 (=) 1 30 (~) 40
(×) 10 (=) 6 40
(÷) 30 (~) 2 (=) 15

UET 6/2 295 shows a simple but messily written example of the method used to find sexagesimally regular reciprocals. Although it works with any regular number, in fact one only ever finds it used on the number 2 05 or any value $2^n \times 2\ 05$ (see the list at the end of BM 80150 on page 82) [Sachs 1947; Robson 2002, 353–54].

The product of any reciprocal pair is, by definition, 1. We can therefore imagine 2;05 as the side of a rectangle whose area is 1 (fig. a); the task is to find the length of the other side. We can measure off a part of the first side, so that it has a length that is in the standard reciprocal table—in this case 0;05, whose reciprocal is 12. We can thus draw another rectangle with lines

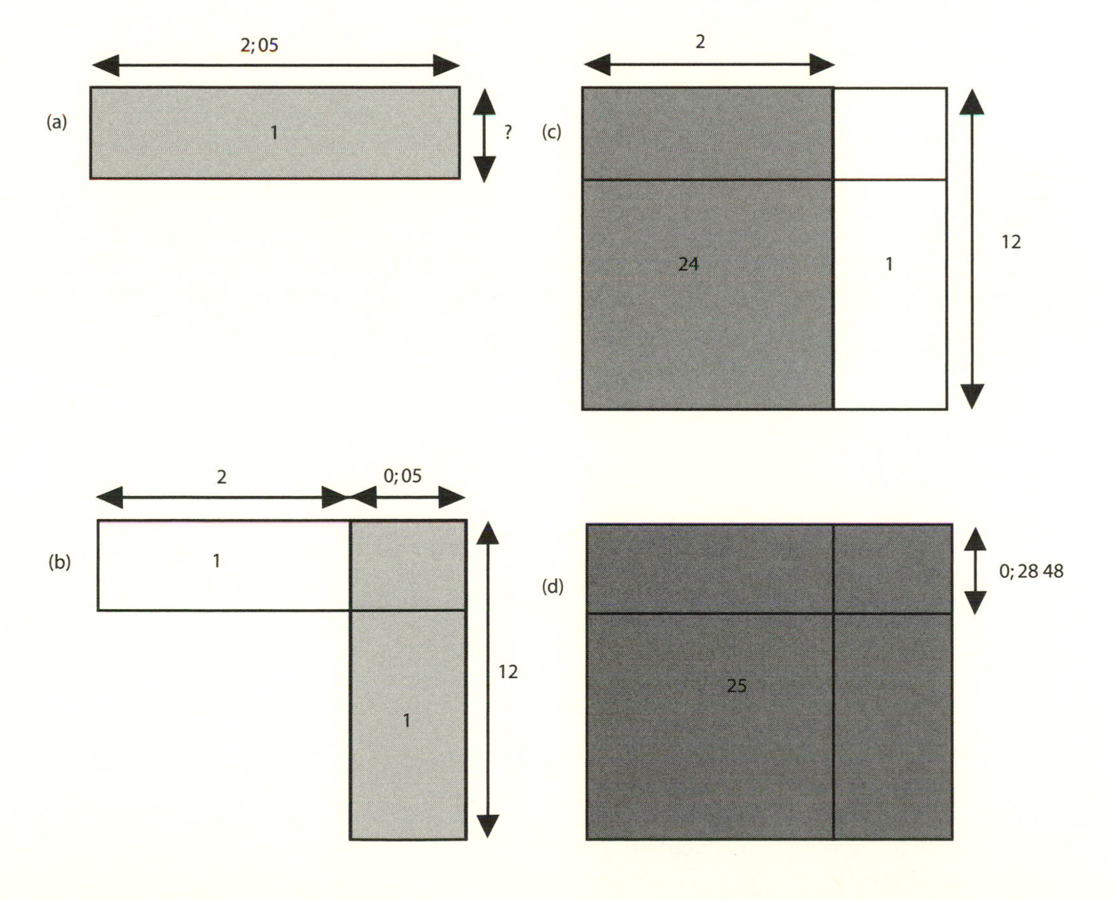

of these lengths, whose area will also be 1 (fig. b). This gives us an L-shaped figure. We can fill it in to make a rectangle by multiplying the 12 by 2, the part of the original length that we haven't used yet—24 (fig. c). Add 1, the area of the 12 by 0;05 rectangle. The total area is 25. This new large rectangle is 25 times bigger than the original rectangle, with area 1. Therefore 12 is 25 times bigger than our mystery reciprocal. We divide 12 by 25 by finding the inverse of 25—0;02 24—and multiplying. The reciprocal we wanted to find is thus 12 × 0;02 24 = 0;28 48 (fig. d). This is the number in the middle of the calculation. The scribe then checks his result by working backwards from 0;28 48 to 2;05 again.

UET 6/2 295

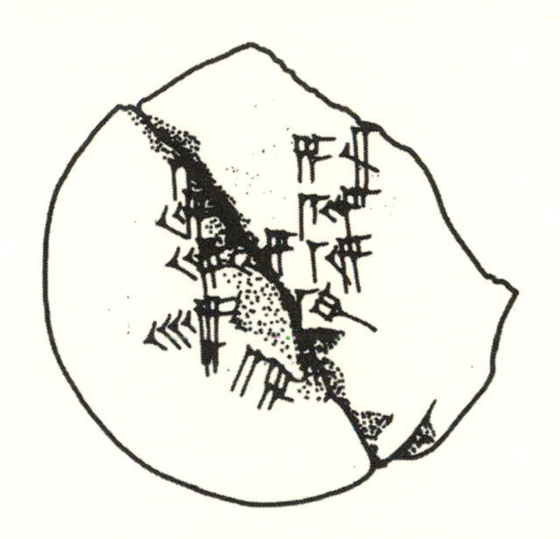

2 05 (~) 12

25 (= 1 + 2 × 12 ~) 2 24

28 48 (= 12 × 2 24)

28 48 (~) 1 15

36 (= 1 + 28 × 1 15 ~) 1 40

2 05 (= 1 15 × 1 40)

He who can say, "Let him hurry, let him run, let him be strong, and he will carry it!" is a lucky man [ETCSL, 6.2.3 165].

Some calculations are laid out in a tabular format. [Friberg 2000, 113–6] suggests that *UET* 6/2 274 lays out the calculations for a problem in geometric algebra, in which the sum of two squares is known to be 12;30, and that the square-side of one is 7 times the square-side of the other. The problem is solved using "false position"; see for example BM 13901 (xiii) on page 105.

UET 6/2 274

When the young scribe is absent, it is a bad thing. No rushes are torn up to make his bedding [ETCSL, 6.2.3 186].

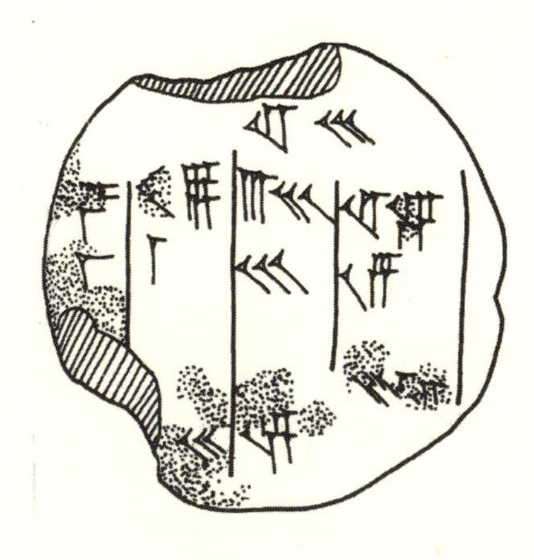

			12 30
7	49	3 30	12 15
1	1	30	15
			50 (~) 1 12
	30	15	

Reference lists

YBC 7243, from Larsa, is a list of coefficients that gives many of the technical constants used in the problems and rough work translated here [MCT, Text Ue; Robson 1999, list E]. While many of the constants used in line geometry and in quantity surveying (especially the mathematics of bricks and of labor) are well understood, others will remain little more than empty names and numbers until problems are discovered that use and name them. The contents of this list and about a dozen others are discussed in detail by [Robson 1999].

YBC 7243

0;05	a circle
0;15	the area of a [semicircle]
0;10	the area of a semi[circle, 2nd one]
0;26 40	the area of a cow's nose
0;12 30	the area of a bow
0;13 20	the area of a barge
5 24	the area of [......]
0;15 36 (and) 0;10	the area of a semicircle ......
0;45	the area of ......
1;24 51 10	the diagonal of a square
1	a square
45	a water-dipper
0;03 45	a mud wall
18	an excavation
11 08	white clay
54	... clay
6 40	clods
4 16	clods
0;00 05	a boat
0;10	the work rate of an excavation
0;20	material thrown up from a canal
0;03 36	straw
1 15	a warehouse for straw
10, 45 (and) 6	bitumen that is from the oven
0;15	destruction
0;48	the waters of a field
2 13 20	black dye
20	[......]
20	a hod (?)
15	water
0;15	bitumen
6 40	carriage of earth
9 00	the weight of copper
1 20	pipe/mold (?) of gold
4 48 00	the thickness of a log
1 12	pipe/mold (?) of copper
1 48	pipe/mold (?) of gold
2 24	(a kind of mineral)
1;12	a 1-cubit brick
7;12	a brick pile

4 30 00	carriage
2;30 (and) 5;20	......
48, 1 40 (and) 1 20	......
2;42	baked brick
1 41 15	of square brick
45	going
3	of a 1/2 cubit brick
5 20	reed work
1 15	carriage
12	water clock
26 40	thinness of silver
20	sheep fat.

About half a dozen tablets are known which list different sets of compatible parameters for school problems, enabling a teacher to set different exercises on the same theme (see, for instance, *UET* 6/2 236 and 293 on page 146–147). BM 80209, from Sippar, gives values for simple problems on squares (compare YBC 7289 on page 143) and circles (compare YBC 7302 on page 142) [Pinches 1963, no. 39; Friberg 1981]. Then more difficult algebraic exercises are planned, involving the circumference, area, and sometimes diameter of the circle. The teacher must have been working out the values as he went along, as the numbers scribbled at the end of the list are closely related to the last two entries in the list [Friberg 1981, 62].

BM 80209

[If] it squares [... on each side], what is the area?
[If] it squares [... on each side], what is the area?

[If] it squares 20 on each side, what is the dividing line?

If it squares 10 on each side, what is the extension?

If the area is 8 20, what is the circle?
If the area is 2 13 20, what is the circle?
If the area is 3 28 20, what is the circle?
If the area is 5, what is the circle?

Add 1/2 a length to the area of a circle: 8 25
«Add» <Take away> 1/2 a length from the area of a circle: 8 15
Add 1 length to the area of a circle: 8 30
Take away 1 length from the area of a circle: 8 10
Add 1 1/3 lengths to the area of a circle: 8 33 20
Take away 1 1/3 lengths from the area of a circle: 8 06 40
Add 1 1/2 lengths to the area of a circle: 8 35
Take away 1 1/2 lengths [from] the area of a circle: 8 05
Add 1 2/3 lengths [to] the area of a circle: 8 36 [40]
Take away 1 2/3 lengths [from the area] of a circle: 8 03 [20]
Add 1 1/4 lengths [to the area of a circle]: 8 32 [30]
Take away [1 1/4] lengths from [the area of a circle]: 8 07 30

[Add 1] 1/5 lengths to the area [of a circle: 8] 32
[Take away 1] 1/5 lengths from the area [of a circle]: 8 08
Sum the area of 2 circles: 41 40. Circle exceeds circle by 10
Sum the area of 2 circles: 3 28 20. Circle exceeds circle by 10
Sum the area of 2 circles: 5 39 40. Circle exceeds circle by 10
Sum the area of 2 circles: 8 28 20. Circle exceeds circle by 10
Sum the area of the circle, the dividing line of the circle and the perimeter of the circle, so that (it is) 8 33 20
Sum the area of the circle, the dividing line of the circle and the perimeter of the circle, so that (it is) 1
Sum the area of the circle, the dividing line of the circle and the perimeter of the circle, so that (it is) 1 55
Sum the area of the circle, the dividing line of the circle and the perimeter of the circle, so that (it is) 3 06 40

[3 06 40] 15 33 20 40
[......] «2» 26 40 «2»
[......16] 4 3 20 40
[1 55] 9 35 33 40 4 40

Plimpton 322 from Larsa, undoubtedly the most famous piece of Old Babylonian mathematics, is also a teacher's list of parameters, probably for problems on right triangles or rectangles; see YBC 7289 on page 143, and BM 96957+VAT 6598 on page 137 [MCT, text A; Robson 2001]. One or two more columns have broken off the left-hand side of the tablet. Their contents have been the subject of much speculation. However, the terminology used in the headings undoubtedly links the table to Old Babylonian geometrical algebra (see YBC 6967 on page 102).

Plimpton 322

The holding-square of the diagonal [from which 1] is taken away, so that the short side comes up	Square-side of the width	Square-side of the diagonal	Its name
[1 59] 00 15	1 59	2 49	1st
[1 56 56] 58 14 «56» <50 06> 15	56 07	«3 12 01» <1 20 25>	2nd
[1 55 07] 41 15 33 45	1 16 41	1 50 49	3rd
1 53 10 29 32 52 16	3 31 49	5 09 01	4th
1 48 54 01 40	1 05	1 37	[5]th
1 47 06 41 40	5 19	8 01	[6th]
1 43 11 56 28 26 40	38 11	59 01	7th
1 41 33 «59» <45 14> 3 45	13 19	20 49	8th
1 38 33 36 36	«9» <8> 01	12 49	9th
1 35 10 02 28 27 24 26 40	1 22 41	2 16 01	10th
1 33 45	45	1 15	11th
1 29 21 54 2 15	27 59	48 49	12th
1 27 <00> 03 45	«7 12 01» <2 41>	4 49	13th
1 25 48 51 35 6 40	29 31	53 49	14th
1 23 13 46 40	«56» <28>	53	[15]th

Mathematics in Sumerian literature (iv)

Although at first sight *The Farmer's Instructions* is a bucolic stroll through the agricultural year, in fact parts of it are deeply heavily mathematized [ETCSL, 5.6.3]. Work rates, sowing rates, and grain yields are all quantified, consistent with the practices of institutional land management in early Mesopotamia [Robson 1999, 157–65]. The composition was widely copied in Old Babylonian scribal schools; see too the description of agricultural overseeing in *A Supervisor's Advice to a Younger Scribe* on pages 80–81.

The Farmer's Instructions

[1]Old man cultivator gave advice to his son:

[2–7]When you have to prepare a field, inspect the levees, canals and mounds that have to be opened. When you let the flood water into the field, this water should not rise too high in it. At the time that the field emerges from the water, watch its area with standing water; it should be fenced. Do not let cattle herds trample there.

[8–13]After you cut the weeds and establish the limits of the field, level it repeatedly with a thin hoe weighing two-thirds of a mina. Let a flat hoe erase the oxen tracks, let the field be swept clean. A maul should flatten the furrow bottoms of the area. A hoe should go round the four edges of the field. Until the field is dry it should be smoothed out.

[14–22]Your implements should be ready. The parts of your yoke should be assembled. Your new whip should hang from a nail—the bindings of the handle of your old whip should be repaired by artisans. The adze, drill, and saw, your tools and your strength, should be in good order. Let braided thongs, straps, leather wrappings, and whips be attached securely. Let your sowing basket be checked, and its sides made strong. What you need for the field should be at hand. Inspect your work carefully.

[23–29]The plow oxen will have back-up oxen. The attachments of ox to ox should be loose. Each plow will have a back-up plow. The assigned task for one plow is 10 (bur), but if you build the implement at 8 (bur), the work will be pleasantly performed for you. 3 (?) bariga of grain will be spent on each 1 (bur) area.

[30–34]After working one plow's area with a *bardil* plow, and after working the *bardil* plow's area with a *tugsig* plow, till it with the *tuggur* plow. Harrow once, twice, three times. When you flatten the stubborn spots with a heavy maul, the handle of your maul should be securely attached, otherwise it will not perform as needed.

[35–40]When your field work becomes excessive, you should not neglect your work; no one should have to tell anyone else: "Do your field work!" When the constellations in the sky are right, do not be reluctant to take the oxen force to the field many times. The hoe should work everything.

[41–45]When you have to work the field with the seeder-plow, your plow should be properly adjusted. Put a leather sealing on the *kashu* of your plow. Provide your beam with narrow pegs. Your boards should be spread. Make your furrows.

[46–54]Make eight furrows per rod of width; the barley will lodge in more closely spaced furrows. When you have to work the field with the seeder-plow, keep your eye on the man who drops the seed. The grain should fall two fingers deep. You should put one shekel of seed per rod. If the barley seed is not being inserted into the hollow of the furrow, change the wedge of your plow share. If the bindings become loose, tighten them.

55–63Where you have made vertical furrows, make slanted furrows, and where you have made slanted furrows, make vertical furrows. Straight furrows will give you edges that are wide enough and nice (?). Your crooked furrows should be straightened out. Make the furrows clear. Plow your portion of field. The clods should be picked out. The furrows should be made wide where the soil is open, and the furrows should be narrower where the soil is clogged: it is good for the seedlings.

64–73After the seedlings break open the ground, perform the rites against mice. Turn away the beaks of small birds. When the plants overflow the narrow bottoms of the furrows, water them with the water of the first seed. When the plants resemble a ...... reed mat, water them. Water the plants when they are heading. When the plants are fully leafed out, do not water them or they will become infected by leaf rust. When the barley is right for husking, water it. It will provide a yield increase of one sila per ban.

74–80When you have to reap the barley, do not let the plants become overripe. Harvest at the right time. One man is to cut the barley, and one to tie the sheaves; and one before him should apportion the sheaves: three men should harvest for you. Do not let those who gather the barley bruise the grain. They should not scatter the grain when it is in the stacks.

81–90Your daily work starts at daybreak. Gather your force of helpers and grain gatherers in sufficient number and lay down the sheaves. Your work should be carefully done. Although they have been having stale coarse flour, do not let anyone thresh for your new bread—let the sheaves have a rest. The rites for the sheaves should be performed daily. When you transport your barley, your barley carriers should handle small amounts (?).

91–95Mark the limits of a vacant lot of yours. Establish properly your access paths to it. Your wagons should be in working order. Feed the wagon's oxen. Your implements should be ....

96–103Let your prepared threshing floor rest for a few days. When you open the threshing floor, smooth its surface (?). When you thresh, the teeth of your threshing sledge and its leather straps should be secured with bitumen. When you make the oxen trample the grain, your threshers should be strong. When your grain is spread on the ground, perform the rites of the grain not yet clean.

104–106When you winnow, put an intelligent person as your second winnower. Two people should work at moving the grain around.

107–109When the grain is clean, lay it under the measuring stick. Perform the rites in the evening and at night. Release the grain at midday.

110–111Instructions of the god Ninurta, son of Enlil—Ninurta, faithful farmer of Enlil, your praise be good!

IV. Later Mesopotamia, c. 1400–150 BCE

The late second and early first millennia BCE

The lack in our knowledge of post-Old Babylonian mathematics is in part a reflection of the fact that late second-millennium Babylonia as a whole is little studied. What does survive shows a continuity of approach and terminology with the Old Babylonian period, which means that sources without archaeological context have to be dated according to their handwriting and spelling conventions, neither of which can yet be used accurately.

AO 17264 dates to some time in the later second millennium BCE [Thureau-Dangin 1934; MKT, I 126–34; TMB, 74–76; Brack-Bernsen and Schmidt 1990]. The problem is to divide a quadrilateral field into six so that pairs of brothers get equal shares; but the method of solution offered is almost entirely fanciful. The configuration of the field can be reconstructed as follows:

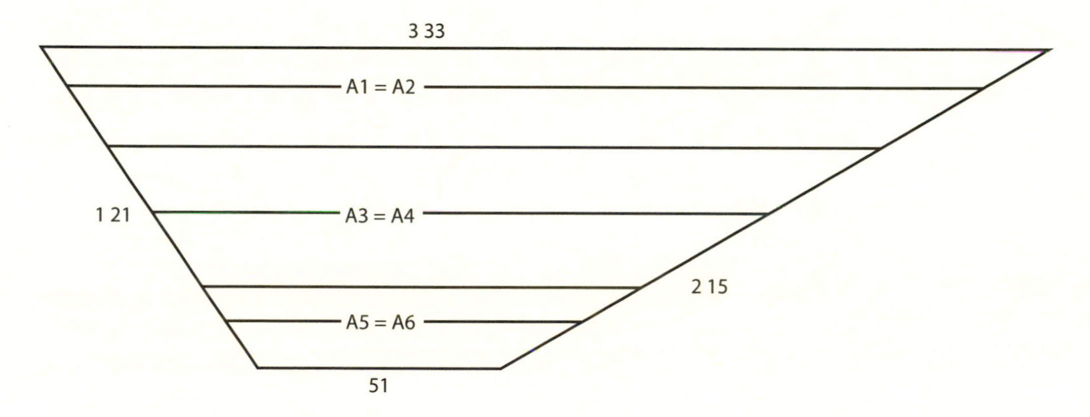

AO 17264

1–4(A field.) ...... «length». The upper <length> is 2 15, the lower length 1 21, [the upper width] 3 33, the lower width 51. There are 6 brothers. The oldest and the following are equal; 3 and 4 are equal; 5 and 6 are equal. What are the boundaries: the dividing lines and the descenders?

4–10You, when you proceed: sum 3 33, the upper width, and 51, the lower width, so that it makes 4 24. Return, and then solve the reciprocal of 2 15, the <upper> length, so that it makes 0;00 26 40. Multiply 0;00 26 40 by 1 21, the <lower> length, so that it makes 0;36. Add 0;36 to 4 24 «lower», so that it makes 4 24;36. Return, and then sum 2 15, the upper length and 1 21, the lower length, so that it makes 3 36. Break 3 36, so that it makes 1 48. Solve the reciprocal of 1 48, so that it makes 0;00 33 20. Multiply 0;00 33 20 by 4 24;36, so that it makes 2;27. The 2nd dividing line is 2 27 (*sic*).

10–12Return, and then by what does 2 15, the upper length, exceed 1 31, the lower length? It exceeds by 54. Take away 54 from 2 27, the 2nd dividing line, so that the remainder is 1 33. «The remainder» The 4th dividing line is 1 33.

[12–17]Return, and then combine 3 33, the upper width. It makes 12 36 09. Return, and then combine 2 27, the second dividing line, so that it makes 6 00 09. Sum 6 00 09 and 12 36 09, so that it makes 18 36 18. Break 18 36 18, so that it makes 9 18 09. Make its square-side come up, so that it makes 3 03. The upper dividing line is «2» <3> 03.

[17–21]Return, and then [combine] 2 27, [the 2nd] dividing line, [so that] it makes 6 00 09. Return, and then [combine] 1 33, the 4th dividing line, [so that] it makes 2 24 09. [Sum] 2 24 09 and 6 00 09, [so that] it makes 8 24 18. Break 8 24 18, so that it makes 4 12 09. Make its square-side come up, so that it makes 2 03. The 3rd dividing line is 2 03.

[21–25]Return, and then combine 1 33, the 4th dividing line, so that it makes 2 24 09. Return, and then combine 51, the lower width, so that it makes 43 21. sum 43 21 and 2 24 09, so that <it makes> 3 07 30. Break 3 07 30, so that it makes 1 33 45. Make its square-side come up, so that it makes 1 15. The 5th dividing line is 1 15.

[25–30]Return, and then by what does 3 33, the <upper> width, exceed 51, the lower width? It exceeds by 2 42. Solve the reciprocal of 2 42, so that it makes 0;00 22 13 20. Return, and then by what does 3 33 exceed 3 03, the <upper> diagonal? It exceeds by 30. Multiply 30 by 0;00 22 13 20, so that it makes 0;11 06 40. Multiply 0;11 06 40 by 2 15, the upper length, and 1 21, the lower length, so that it makes 25 and 15. The upper desc<ender> is 25, the lower desc<ender> is 15.

[30–33]Return, and by how much does 3 33, the upper dividing line, exceed 2 27, the 2nd dividing line? It exceeds by 36. Multiply 36 by 0;00 22 13 20, so that it makes 0;13 20. Multiply 0;13 20 by 2 15, the <upper length>, and 1 21 the <lower> length, so that it makes 30 and 18. The 2nd upper desc<ender> is 30, the 2nd lower desc<ender> is 18.

[33–37]Return, and by how much does 2 27, the second dividing line, exceed 2 03, the 3rd dividing line? [It exceeds by] 24. Multiply 24 by 0;00 22 13 20, so that [it makes] 0;08 53 20. Multiply [0;08 53 20] by 2 15, the <upper> length and 1 [21, the lower length], so that it makes 20 and 12. The [upper] desc<ender> is 20, the lower [descender is 12].

[37–38]Return. You make the 3 remaining desc<enders> like [the previous ones]. That is the procedure.

Another late second-millennium problem, HS 245 from Nippur, is known in two further versions from Nineveh, the seventh-century BCE capital of Assyria [Horowitz 1998, 179–82]. It uses the measurement of distances between a group of stars which rise heliacally in sequence as a pretext for exploring division by sexagesimally irregular numbers. One of the Assyrian versions (Sm 162) was written on the back of a circular star map, "extracted for reading" from an unknown scholarly series by Nabû-zuqup-kena, a renowned expert in omens and portents. This suggests that the text had lost its scholastic function by then.

HS 245

I summed 19 from the moon to the Pleiades, 17 from the Pleiades to Orion, 14 from Orion to the Arrow, 11 from the Arrow to the Bow, 9 from the Bow to the Yoke, 7 from the Yoke to Scorpio, 4 from Scorpio to Anta-gub, so that it was 2 sixties leagues. How far is a god above (another) god? You, when you proceed: sum 19, 17, 14, 13, 11, 9, 7, (and) 4, so that you will see 1 21. The reciprocal of 1 21 is 0;00 44 26 40. [*Here the text breaks off.*]

Sm 162 Reverse

[......]
[......] to the Arrow. 13 (?) from the Arrow to the Yoke
[......] put. Sum them so that it is 1 21.
The moon is placed [......] above the Pleiades;
[the Pleiades] is placed [...... above] Taurus;
[Taurus] is placed [...... above] the Arrow;
[the Arrow] is placed [...... above] the Yoke;
[the Yoke] is placed [...... above] Scorpio;
[Scorpio] is placed [...... above] Centaurus.
[......] they are held
[......] how far are they?

[...] elevated king of the Abzu, great gods. [......] Extracted for reading. [Tablet of Nabû-zuqup-kena son of] Marduk-shuma-iqisha, scribe.

Tables of squares and inverse squares of integers and half-integers are known from the Old Babylonian period onwards. Ash 1924.796, from sixth-century Kish, is a beautifully written library copy, with a decorative pattern of firing holes on its surface [MKT, II pl. 34 (there called 1931–38); Robson 2004, no. 28]. It uses three vertically aligned diagonal wedges to indicate empty sexagesimal places, here marked by a colon : .

Ash 1924.796

[1 times 1	1	the square-side is 1]
[1 30 times 1 30	2 15	the square-side is 1 30]
[2 times 2	4	the square-side is 2]
[2 30 times 2 30	6 15	the square-side is 2 30]
[3 times 3	9	the square-side is 3]
[3 30 times 3 30	12 15	the square-side is 3 30]
[4 times 4	16	the square-side is 4]
[4 30 times 4 30	20 15	the square-side is 4 30]
[5 times 5	25	the square-side is 5]
[5 30 times 5 30	30 15	the square-side is 5 30]
[6 times 6	36	the square-side is 6]
[6 30 times 6 30	42 15	the square-side is 6 30]
[7 times 7	49]	the square-side is 7
[7 30 times 7 30	56 15]	the square-side is 7 30

[8 times 8	1 04]	the square-side is 8
[8 30 times 8 30	1 12] 15	the square-side is 8 30
[9 times 9	1 21]	the square-side is 9
[9 30 times 9 30	1 30] 15	the square-side is 9 30
[10 times 10	1] 40	the square-side is 10
[10 30 times 10 30]	1 50 15	the square-side is 10 30
[11 times 11]	2 01	the square-side is 11
[11 30 times] 11 30	2 12 15	the square-side is 11 30
[12] times 12	2 24	the square-side is 12
[12] 30 times 12 30	2 36 15	the square-side is 12 30
[13] times 13	2 49	the square-side is 13
[13] 30 times 13 30	3 02 15	the square-side is 13 30
[14] times 14	3 16	the square-side is 14
[14] 30 times 14 30	3 30 15	the square-side is 14 30
[15] times 15	3 45	the square-side is 15
15 30 times 15 30	4 : 15	the square-side is 15 30
16 times 16	4 16	the square-side is 16
16 30 times 16 30	4 32 15	the square-side is 16 30
17 times 17	4 49	the square-side is 17
17 30 times 17 30	5 06 15	the square-side is 17 30
18 times 18	5 24	the square-side is 18
18 30 [times] 18 30	5 42 15	the square-side is 18 30
19 [times] 19	6 01	the square-side is 19
19 30 [times] 19 30	6 20 15	the square-side is 19 30
20 [times 20]	6 40	the square-side is 20
20 30 [times 20] 30	7 «50» 15	the square-side is 20 30
[21 times] 21	7 21	the square-side is 21
[21 30 times] 21 30	7 42 15	the square-side is 21 30
[22 times] 22	8 04	the square-side is 22
[22 30] times 22 30	8 26 15	the square-side is 22 30
[23] times 23	8 49	[the square-side is] 23
[23 30] times 23 30	9 12 15	[the square-side is] 23 30
[24] times 24	9 36	[the square-side is] 24
[24 30] times 24 30	10 : 15	the square-side [is] 24 30
[25 times] 25	10 25	the square-side [is] 25
[25 30 times] 25 30	10 50 15	the square-side [is] 25 30
[26 times] 26	11 16	[the square-side is] 26
[26 30 times 26] 30	11 42 [15	the square-side is] 26 30
[27 times 27	12] 09	the square-side is 27
[27 30 times 27 30]	12 36 15	the square-side is 27 30
[28 times 28]	13 04	the square-side is 28
[28 30 times 28 30]	13 32 15	the square-side is 28 30
[29 times 29	14] 01	the square-side is 29
[29 30 times 29 30	14 30] 15	the square-side is 29 30
[30 times 30	15	the square-side is] 30
[30 30 times 30 30	15 30 15	the square-side is] 30 30

[31 times 31	16 01	the square-side is] 31
[31 30 times 31 30	16 32 15	the square-side is] 31 30
[32 times 32	17 04	the square-side is] 32
[32 30 times 32 30	17 36 15	the square-side is] 32 30
[33 times 33	18 09	the square-side is] 33
[33 30 times 33 30	18 42 15	the square-side is 33 30]
[34 times 34	19 16	the square-side is 34]
[34 30 times 34 30	19 50 15	the square-side] is [34 30]
[35 times 35	20 25]	the square-side is 35
[35 30 times 35 30	21 : 15]	the square-side is 35 [30]
[36 times 36	21 36]	the square-side is 36
[36 30 times 36 30	22 12 15]	the square-side is 36 [30]
[37 times] 37	22 49	[the square-side is 37]
[37 30] times 37 30	23 26 15	the square-side [is 37 30]
[38] times 38	24 «1» 4	the square-side [is 38]
[38] 30 times 38 [30]	24 42 15	[the square-side is 38 30]
[39] times 39	25 21	the square-side is 39
[39] 30 times 39 [30]	26 : 15	the square-side is 39 [30]
[40] times 40	26 40	the square-side is 40
[40] 30 times 40 30	27 20 15	the square-side is [40 30]
[41] times 41	28 01	the square-side is [41]
[41] 30 times 41 30	28 42 15	the square-side [is 41 30]
[42] times 42	29 24	the square-side is [42]
[42] 30 times 42 30	30 : 06 15	the square-side is [42 30]
[43] times 43	30 [4]9	the square-side is [43]
[43] 30 times 43 30	31 32 [15]	the square-side is 43 [30]
[44] times 44	32 16	the square-side is 44
[44] 30 times 44 [30]	33 : 15	the square-side is 44 [30]
[45] times 45	33 45	the square-side is 45
[45] 30 times 45 30	34 30 15	the square-side is 45 30
46 times 46	35 16	the square-side is 46
46 30 times 46 30	36 02 15	the square-side is 46 30
47 times 47	36 49	the square-side is 47
47 30 times 47 30	37 36 15	the square-side is 47 30
48 times 48	38 24	the square-side is 48
48 30 times 48 30	39 [12] 15	the square-side is 48 30
49 times 49	40 : [01]	the square-side is 49
49 30 times 49 [30]	40 50 15	the square-side is 49 30
50 times 50	41 40	the square-side is 50
50 30 times 50 30	42 30 15	the square-side is 50 30
51 times 51	43 21	the square-side is 51
51 30 times 51 30	44 12 15	the square-side is 51 30
52 times 52	45 04	the square-side is 52
52 30 times 52 30	45 56 15	the square-side is 52 30
53 times 53	46 49	the square-side is 53
53 30 times 53 30	47 42 15	the square-side is 53 30
[54] times 54	48 36	the square-side is 54

[54 30 times] 54 30	49 30 15	the square-side is 54 30
[55 times] 55	50 25	the square-side is 55
[55 30 times] 55 30	51 20 15	the square-side is 55 30
[56 times 56]	52 16	the square-side is 56
[56 30 times 56 30]	53 12 15	the square-side is 56 30
[57 times 57]	54 09	the square-side is 57
[57 30 times 57] 30	55 06 15	the square-side is 57 30
[58 times 58]	56 04	the square-side is 58
[58 30 times 58] 30	57 «1»2 15	the square-side is 58 30
[59 times 59]	58 01	the square-side is 59
[59 30 times 59] 30	59 <:> 15	the square-side is 59 30
[1 times 1]	1	the square-side is 1

[......] iku ...... area ...... 18 times [...]
[......] ......
[............] Tablet of Bel-ban-apli ...... [......]

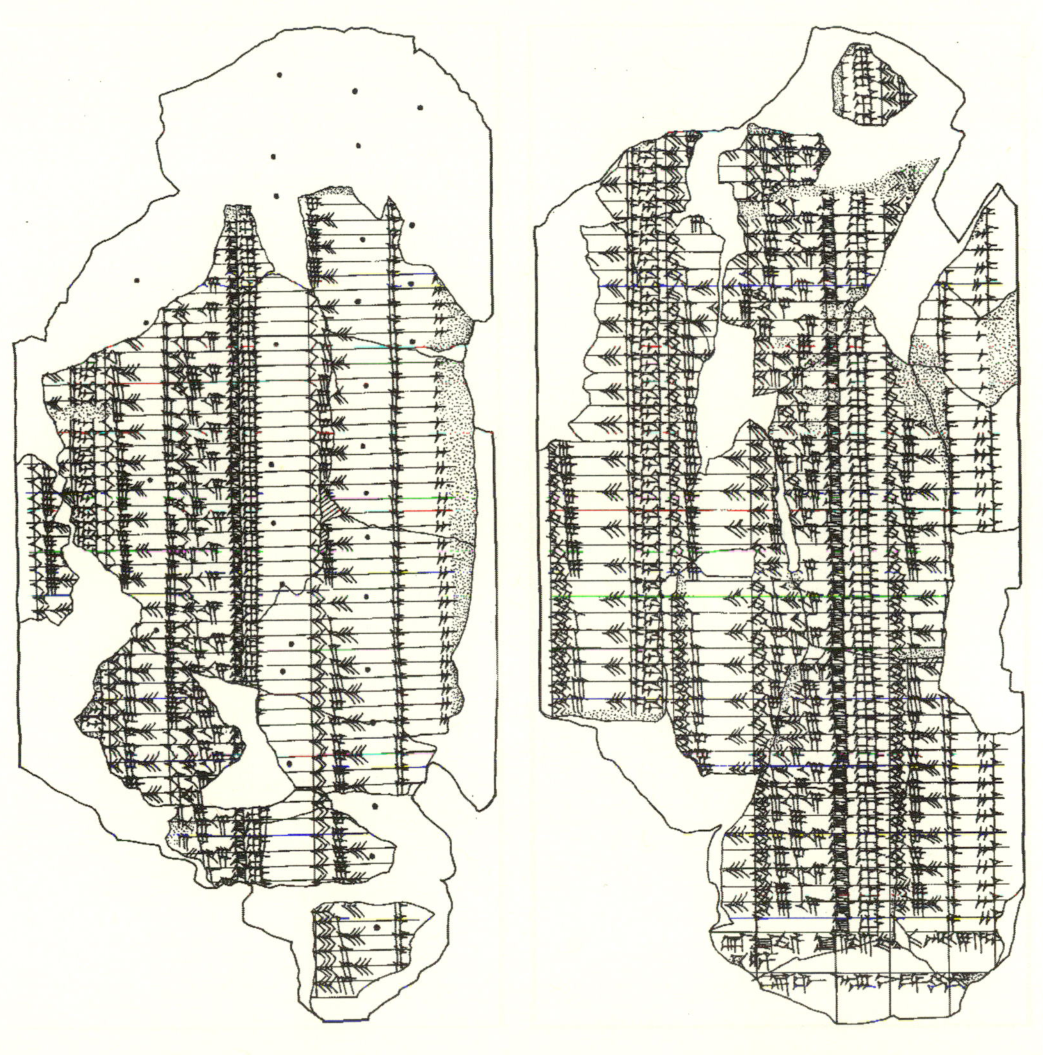

Ash 1924.796 Obverse

Ash 1924.796 Reverse

Mathematical tablets from the house of the Shangû-Ninurta family, late fifth-century Uruk

Between 1969 and 1972 the excavators of Uruk uncovered a house to the east of the city which has become our most important source of information about Late Babylonian mathematical practice. Two different scribal families had successively occupied the house, both maintaining scholarly libraries there. About 180 tablets and fragments found in the house, nine of which are mathematical, can be associated with the Shangû-Ninurta family of incantation-priests. Their datable legal and scholarly tablets span the sixth and fifth centuries BCE, the latest nine of which are from early in the reign of the Persian king Darius II (423–405 BCE), suggesting that the family left the house some time after 412 BCE. The archaeological and archival context of the Shangû-Ninurta family's mathematical tablets is discussed by [Robson forthcoming, chap. 8]; four of them are translated here.

The tablet W 23273 is a compilation of different numerical tables [von Weiher 1993, no. 172; Friberg 1993, text 11]. It starts with a list of integers associated with major deities, then gives various metrological tables of lengths and areas. Between them is a short conversion schema, showing which length unit is equivalent in order of size to which area unit. Each pair of tables first gives the sexagesimal values to the left of the metrological writings, unlike the standard Old Babylonian tables (see Ash 1923.366 on page 88); but when in the second half of each table the format is reversed, the metrological units are written erroneously before the metrological values. Finally, there is a badly understood shadow-clock table before the colophon. This describes the tablet as a copy of an older original, made by Rimut-Anu, the youngest attested mathematical member of the Shangû-Ninurta family.

W 23273

[1]	Anum
[2]	Antum
[3]	Enlil
[4]	Ea
[5]	Sîn
[6]	Enki
[7]	The Seven
8	The Igigi
9	The Anunaki
10	Bel
20	Shamash
30	Sîn
40	Ea
50	Enlil

---

[10]	1 finger
20	2 fingers
30	3 fingers
40	4 fingers
50	5 fingers
1	6 fingers
1 10	7 fingers
1 20	8 fingers
1 30	9 fingers
1 40	1/3 cubit
2 30	1/2 cubit
3 20	2/3 cubit
5	1 cubit
6 40	1 1/3 cubits
7 30	1 1/2 cubits
8 20	1 2/3 cubits
10	2 cubits
15	3 cubits
20	4 cubits
25	5 cubits
30	1/2 rod
35	1/2 rod 1 cubit
40	1/2 rod 2 cubits
45	1/2 rod 3 cubits
50	1/2 rod 4 cubits
55	1/2 rod 5 cubits
1	1 rod
1 30	1 1/2 rods
2	2 rods

2 30	2 1/2 rods	18	18 cables
3	3 rods	19	19 cables
3 30	3 1/2 rods	20	2/3 league
4	4 rods	25	2/3 league 5 cables
4 30	4 1/2 rods	30	1 league
5	5 rods	35	1 league, 5 cables
5 30	5 1/2 rods	40	1 league, [10] cables
6	6 rods	45	1 1/2 leagues
6 30	6 1/2 rods	50	1 2/3 leagues
7	7 rods	55	1 2/3 leagues, 5 cables
7 30	7 1/2 rods	1	2 leagues
8	8 rods	1 30	3 leagues
8 30	8 1/2 rods	2	4 leagues
9	9 rods	2 30	5 leagues
9 30	9 1/2 rods	3	6 leagues
10	[10 rods]	3 30	7 leagues
15	[15 rods]	4	8 leagues
20	[20 rods]	4 30	9 leagues
25	[25 rods]	5	10 leagues
30	[30] rods	5 30	11 leagues
35	[35] rods	6	12 leagues
40	[40] rods	6 30	13 leagues
45	[45] rods	7	14 leagues
50	[50] rods	7 30	15 leagues
55	[55] rods	8	16 leagues
1	1 cable	8 30	17 leagues
1 10	1 10 rods	9	18 leagues
1 20	1 20 rods	9 30	19 leagues
1 30	1 30 rods	10	20 leagues
1 40	1 40 rods	15	30 leagues
1 50	1 50 rods	20	40 leagues
2	2 cables	25	50 leagues
3	3 cables	30	1 sixty leagues
4	4 cables	35	1 10 leagues
5	5 cables	40	1 20 leagues
6	6 cables	45	1 30 leagues
7	7 cables	50	1 hundred leagues
8	8 cables		
9	9 cables	finger 1	10
10	10 cables	fingers 2	20
11	11 cables	[fingers] 3	30
12	12 cables	fingers 4	40
13	13 [cables]	fingers 5	50
14	14 [cables]	fingers 6	1
15	[15 cables]	fingers 7	1 10
16	[16] cables	fingers 8	1 20
17	[17] cables	fingers 9	1 30

cubit 1/3	1 40	rods 1 20	1 20
cubit 1/2	2 30	rods 1 30	1 30
cubit 2/3	3 20	rods 1 40	1 40
cubit 1	5	rods 1 50	1 50
cubits 1 1/3	6 40	cables 2	2
cubits 1 1/2	[7 30]	cables 3	3
cubits [1 2/3	8 20]	cables 4	4
[cubits 2	10]	cables 5	5
[cubits 3	15]	cables 6	6
[cubits 4	20]	cables 7	7
[cubits] 5	25	cables 8	8
rod 1/2	30	cables 9	9
rods 1/2 rod, 1 cubit	35	cables 10	10
rods 1/2 rod, 2 cubits	40	cables 11	11
rods 1/2 rod, 3 cubits	45	cables 12	12
rods 1/2 rod, 4 cubits	50	cables 13	13
rods 1/2 rod, 5 cubits	55	cables 14	14
rod 1	1	cables 15	15
rods 1 1/2	1 [30]	cables 16	16
rods 2	[2]	cables 17	17
rods [2 1/2	2 30]	cables 18	18
[rods 3	3]	[cables 19	19]
[rods 3] 1/2	3 30	[leagues 2/3	20]
[rods] 4	4	[league 1	30]
rods 4 1/2	4 30	[——————	]
rods 5	5	[2	10 fingers]
rods 5 1/2	5 30	4	[20] fingers
rods 6	6	6	30 fingers
rods 6 1/2	6 30	8	40 fingers
rods 7	7	10	50 [fingers]
rods 7 1/2	7 30	12	[1 finger]
rods 8	8	14	[1 10 fingers]
rods 8 1/2	8 30	[16	1 20 fingers]
rods 9	9	18	1 30 fingers
rods 9 1/2	9 30	20	1 40, 1/3 cubit
rods 10	10	30	2 30, 1/2 cubit
rods 15	15	40	3 20, 2/3 cubit
rods 20	20	1	5, 1 cubit
rods 25	25	1 20	6 40, 1 1/3 cubits
rods 30	30	1 30	7 30, 1 1/2 cubits
rods 35	35	1 40	8 20, 1 2/3 cubits
rods 40	40	2	10 cubits
rods 45	45	3	15 cubits
rods 50	50	4	20 cubits
rods 55	55	5	25 cubits
cable 1	1	6	30 cubits
rods 1 10	1 10	7	35 cubits

8	40 cubits
9	45 cubits
10	50 cubits
11	55 cubits
12	1 rod
18	1 1/2 rods
24	2 rods
30	2 1/2 rods
36	3 rods
42	3 1/2 rods
48	4 rods
54	4 1/2 rods
1	5 rods
1 06	5 1/2 rods
1 12	6 rods
1 18	6 1/2 rods
1 24	7 rods
1 30	7 1/2 rods
1 36	8 rods
1 42	8 1/2 rods
1 48	9 rods
1 54	9 1/2 rods
2	10 rods
3	15 rods
4	20 rods
5	25 rods
6	30 rods
7	35 rods
8	40 rods
9	45 rods
10	50 rods
11	55 rods
12	1 [cable]

---

fingers [10	2]
fingers [20	4]
[fingers 30	6]
[fingers 40	8]
[fingers 50	10]
[fingers 1	12]
[fingers 1 10	1]4
[fingers 1 20	1]6
fingers [1 30	1]8
fingers [1 40]	20
cubits [1/2]	30
[cubits 2/3]	40
[cubits 1	1]
[cubits 1 1/3	1] 20
[cubits 1 1/2]	1 30
[cubits 1 2/3]	1 40
cubits 2	2
cubits 3	3
cubits 4	4
cubits 5	5
rods 1/2	6
rods 1/2 1 cubits	7
rods 1/2 2 cubits	8
rods 1/2 3 cubits	9
rods 1/2 4 cubits	10
rods 1/2 5 cubits	11
rods 1	12
rods 1 1/2	18
rods 2	24
rods 2 1/2	30
rods 3	36
rods 3 1/2	42
rods 4	48
rods 4 1/2	54
rods 5	1
rods 5 1/2	1 06
rods 6	1 12
rods 6 1/2	1 18
rods 7	1 24
rods 7 1/2	1 30
rods 8	1 36
rods 8 1/2	1 42
rods 9	1 48
rods 9 1/2	1 54
rods 10	2
rods 15	3
rods 20	4
rods 25	5
rods 30	6
rods 35	7
rods 40	8
rods 45	9
rods 50	10
rods 55	11
rods 1	[12]

Kingship, fate
fingers to [. . .] grains (area)
1 cubit«s» to [. . .] shekels (area)
reeds to [. . .] sar (area)

10 rods to 1 (iku) area
1 cable to 10 (bur) area
1 leagues to 3600 (bur) area
from 6 cables to 3600 (bur) [area]
of [...]-ship
that its factor (?) is not true
that their count is not ......

10	1/2	grain
20	1	grain
30	1 1/2	grains
40	2	grains
50	2 1/2	grains
1	3	grains
1 20	4	grains
1 40	5	grains
2	6	grains
2 20	7	grains
2 30	7 1/2	grains
2 40	8	grains
3	9	grains
3 20	10	grains
3 40	11	grains
4	12	grains
4 20	13	grains
4 40	14	grains
5	15	grains
5 20	16	grains
5 40	17	grains
6	18	grains
6 20	19	grains
6 40	20	grains
7	21	grains
7 20	22	grains
[7] 30	22 1/2	grains, an 8th
[7] 40	23	grains
8	24	grains
8 20	25	grains
8 40	26	grains
9	27	grains
9 20	28	grains
9 40	29	grains
10		a 6th
11 40 shekels		5 grains
13 20 shekels		10 grains
15		a 4th
16 40 shekels		5 grains

18 20	shekels 10 grains
20 1/3	grain
30 1/2	grains
40 2/3	grains
50 5/6	grains
1	1 shekel
1 10	1 shekel, a 6th
1 15	1 shekel, a 4th
1 20	1 shekel, 1/3 grains
1 30	1 shekel, 1/2 grains
1 40	1 shekel, 2/3 grains
1 50	1 shekel, 5/6 grains
2	2 shekels
3	3 shekels
4	4 shekels
5	5 shekels
6	6 shekels
7	7 shekels
8	8 shekels
9	9 shekels
10	10 shekels
11	11 shekels
12	12 shekels
13	13 shekels
14	14 shekels
15	a 4th sar
16	16 shekels
17	17 shekels
18	18 shekels
19	19 shekels
20	1/3 sar
30	1/2 sar
40	2/3 sar
50	5/6 sar
1	1 sar
1 10	1 sar 10 shekels
1 15	1 sar 4th sar
1 20	1 sar 1/3 sar
1 30	1 sar 1/2 sar
1 40	1 sar 2/3 sar
1 50	1 sar 5/6 sar
2	2 sar
3	3 sar
4	4 sar
5	5 sar
6	6 sar
7	7 sar

8	8 sar
9	9 sar
10	10 sar
11	11 sar
12	12 sar
13	13 sar
14	14 sar
15	15 sar
16	16 sar
17	17 sar
18	18 sar
19	19 sar
20	20 sar
30	30 sar
40	40 sar
50	1/2 (iku) area
1	1 sar
1 10	1 10 sar
1 20	1 20 sar
1 30	1 30 sar
1 40	1 (iku) area
2 30	1 1/2 (iku) area
3 20	2 (iku) area
4 10	2 1/2 (iku) area
5	3 (iku) area
5 50	3 1/2 (iku) [area]
6 40	4 (iku) [area]
7 30	4 1/2 (iku) [area]
8 20	5 (iku) [area]
9 10	5 1/2 (iku) [area]
10	1 (eshe) [area]
11 40	1 (eshe) 1 (iku) area
13 20	1 (eshe) 2 (iku) area
15	1 (eshe) 3 (iku) area
16 40	1 (eshe) 4 (iku) area
18 20	1 (eshe) 5 (iku) area
20	2 (eshe) area
21 40	2 (eshe) [1 (iku)] area
23 20	2 (eshe) 2 (iku) area
25	2 (eshe) 3 (iku) area
26 40	2 (eshe) 4 (iku) area
28 20	2 (eshe) 5 (iku) area
30	1 (bur) area
40	1 (bur) 1 (eshe) area
50	1 (bur) 2 (eshe) area
1	2 (bur) area
1 30	3 (bur) area
2	4 (bur) area
2 30	5 (bur) area
3	6 (bur) area
3 30	7 (bur) area
4	8 (bur) area
4 30	9 (bur) area
5	10 (bur) area
5 30	11 (bur) area
6	12 (bur) area
6 30	13 (bur) area
7	14 (bur) area
7 30	15 (bur) area
8	16 (bur) area
8 30	17 (bur) area
9	18 (bur) area
9 30	19 (bur) area
10	20 (bur) area
15	30 (bur) area
20	40 (bur) area
25	50 (bur) area
30	60 (bur) area
35	60 + 10 (bur) area
40	60 + 20 (bur) area
45	60 + 30 (bur) area
50	60 + 40 (bur) area
55	60 + 50 (bur) area
1	2 × 60 (bur) area
1 30	3 × 60 (bur) area
2	4 × 60 (bur) area
2 30	5 × 60 (bur) area
3	6 × 60 (bur) area
3 30	7 × 60 (bur) area
4	8 × 60 (bur) area
4 30	9 × 60 (bur) area
5	600 (bur) area
5 30	600 + 60 (bur) area
6	600 + 2 × 60 (bur) area
6 30	600 + 3 × 60 (bur) area
7	600 + 4 × 60 (bur) area
7 30	600 + 5 × 60 (bur) area
8	600 + 6 × 60 (bur) area
8 30	600 + 7 × 60 (bur) area
9	600 + 8 × 60 (bur) area
9 30	600 + 9 × 60 (bur) area

grains	1/2	10
grains	1	20
grains	1 1/2	30
grains	2	40
grains	2 1/2	50
grains	3	1
grains	4	1 20
grains	5	1 40
grains	6	2
grains	7	2 20
grains	8	2 40
grains	9	3
grains	10	3 20
grains	11	3 40
grains	12	4
grains	13	4 20
grains	14	4 40
grains	15	5
grains	16	5 20
grains	17	5 40
grains	18	6
grains	19	6 20
grains	20	6 40
grains	21	7
grains	22	7 20
grains	[22] 1/2	8th, 7 30
grains	23	7 40
grains	24	8
grains	25	8 20
grains	26	8 40
grains	[27	9]
grains	[28	9 20]
grains	[29	9 40]
grains	[30	a 6th]

grains 4[5	a 4th]
grains 1 [sixty	1/3 shekel]
grains 1 [30	1/2 shekel]
grains 2 [sixties	2/3 shekel]
grains 2 30	5/6 [shekel]
grains 3 sixties	1 [shekel]
grains 3 3600	1 mina
grains 3 sixties	3600 1 talent

---

1 shekels grains 1
square-side 10 [...]

Finished.

[...] I exceeded, Month IV
[...] cubit(s) height
10 Month III ditto
[......] 10 Month II ditto
[......] Month I ditto
[Month I ... shadow falls (?)]
[Month II ... shadow falls (?)]
[Month III ... shadow falls (?)]
[Month IV... shadow falls (?)]
[Month V ...] shadow falls (?)
Month VI 30 shadow falls (?)
Month VII 45 shadow falls (?)
Month VIII 1 shadow falls (?)
Month IX 1 15 shadow falls (?)
Month X 1 30 shadow falls (?)

---

2 12 shadow, 1 40 leagues a day is its square-side.

According to an old tablet, Urukean copy. Rimut-Anu, [son of Shamash]-iddin, descendent of Shangû-Ninurta, [wrote and] checked (it).

Rimut-Anu also owned a table of reciprocals, W 23283+22905, copied for him from an older original by one Nadin, whose relation to Rimut-Anu is unclear (but is unlikely to have been his grandfather of the same name) [von Weiher 1993, no. 174]. After an unusual opening, the table gives the reciprocals of regular numbers from 1 to 4, up to 9 sexagesimal places long, but it is by no means complete (compare Gingerich [1965]). The word 'square-side' is used, erroneously, to mean the reciprocal of an integer; for reasons of space I have left the writing igi 'reciprocal' untranslated. The scribe here uses two vertically aligned wedges (which we represent by a colon) either to represent a blank sexagesimal place or as a word divider between numbers where he has had to break up the tabulation because the numbers are so long.

W 23283+22905

The solution of [1 00] is 1 square-side, 1 00 times 1	«times» < is 1>
50 times 1 is 50	40 times 1 is 40
30 times 1 is 30	20 times 1 is 20
[10] times 1 is 10	10 times 6 is 1 times 1
The scribal [writing (?)] of its 1 tenth (?) is six	
2/3 of 1 is 40	Its half is 30
igi 1	1 square-side
igi 1 : 45	59 15 33 20
igi 1 01 02 06 33 45	58 58 56 38 24
igi 1 01 26 24	58 35 37 30
igi 1 01 30 33 45	58 31 37 35 18 31 06 40
igi 1 [01 4]3 42 13 20	58 19 12
igi [1 02] 12 28 48	57 52 13 20
igi 1 02 30	57 36
igi 1 03 12 35 33 20	56 57 11 15
igi 1 03 16 52 30	56 53 20
igi 1 04	56 15
igi 1 04 48	55 33 20
igi 1 05 [06 15]	55 17 45 36
igi 1 05 [36 36]	54 52 10 51 51 06 40
igi 1 05 [50 37 02 13 20	5]4 40 30
igi 1 05 55 04 41 [15	54] 36 48
igi 1 06 40	54
igi 1 07 30	53 20
igi 1 08 16	52 44 03 45
igi 1 08 20 37 30	52 40 29 37 46 40
igi 1 09 07 12	52 05
igi 1 09 26 40	51 50 24
igi 1 09 07 13	52
igi 1 10 «15» <18> 45	51 12
igi 1 11 06 40	50 37 30
igi 1 11 11 29 03 45	50 34 04 26 40
igi 1 12	50 square-side
igi 1 12 49 04	49 26 18 30 56 15
igi 1 12 54	49 22 57 46 40
igi 1 13 09 «27 49» <34 29> 08 08 53 20 :	49 12 27
igi 1 13 14 31 52 30	49 09 07 12
igi 1 13 43 40 48	48 49 41 15
igi 1 14 04 26 40	48 [36]
igi 1 14 38 58 33 36 :	48 [13 31 06 40]
igi 1 15	4[8]
igi 1 15 51 06 40	47 2[7 39 22 30]
igi 1 15 56 15	47 [24 26 40]
igi 1 16 48	46 [52 30]
igi 1 17 09 37 46 40 :	[46 39 21 36]
igi 1 17 45 36	[46 17 46 40]
igi 1 19 [06 05] 37 30	[45 30 40]
igi 1 19 44 26 40	[45 33 4]5

igi 1 20	[45 square-side]
igi 1 21	[44 26 40]
[igi] 1 21 22 [48 45 :	44 14 12 28] 45
[*About 3 lines missing.*]	
igi [1 ......]	
igi [1 ......]	
igi [1 ......]	
igi [1 ......]	
igi 1 2[......]	
igi 1 2[......]	
igi 1 30	[40 square-side]
igi 1 31 07 30	[39 30 23 13 20]
igi 1 3[2 09] 36	[39 03 45]
igi 1 32 [35] 33 20	[38 52 48]
igi 1 33 15 43 12	[38 24 48 53 20]
igi 1 33 45	[38 24]
igi 1 [34 48] 53 20	3[7 48 07 30]
igi 1 34 55 18 45	3[7 45 33 20]
igi 1 36	3[7 30
igi 1 36 [06 30 1]4 03 45	3[7 27 27 44 11 51 06 40]
igi 1 [......] 20	3[......]
igi 1 [37] 12	3[7 02 13 20]
igi 1 [......] 20	3[......]
igi 1 [......	......]
igi 1 [......	......]
igi 1 [......	......]
igi 1 [......	......]
igi [1 40	36]
igi [1 41 08 08] 53 [20	35 35 44 31 52 30]
igi 1 [41] 15	[35 33 20]
igi 1 [42 2]4	[35 09 22 30]
igi 1 4[3] 40 48	3[4 43 20]
igi 1 4[4] 10	3[4 33 36]
igi 1 45 28 07 30	3[4 08]
igi 1 46 40	3[3 45]
igi 1 48	3[3 20]
igi 1 49 13 36	3[2 57 32 20 37 30]
igi 1 49 21	3[2 45 18 31 06 40]
igi 1 49 44 21 43 4[2 13 20 :	32 48 18]
igi 1 50 35 31 12	[32 32 07 30]
igi 1 51 06 40	[32 24]
igi 1 52 30	[32]
igi 1 53 46 40	[31 38 25 15]
igi 1 53 54 22 30	3[1 36 17 46 40]
igi 1 55 12	31 [15]
igi 1 55 44 26 40	31 [06 14 24]
igi 1 56 38 24	30 [51 51 06 40]
igi 1 57 03 19 10 37 02 1[3 20 :	30 45 16 52 30]
igi 1 57 «57» 53 16 4[8 :	30 31 06 13 52 30]
igi 1 58 31 06 40	[30 32 30]
igi 2	[30 square-side]
igi 2 01 30	[29 37 46 40]

igi 2 02 04 13 [07 30	29 29 28 19 12]
igi 2 [02] 52 4[8	29 17 48 45]
igi 2 03 [01] 07 [30	29 15 49 47 39 15 33 20]
igi 2 [03 27 24 26 40	29 09 36]
igi 2 [04 24 57 32	28 56 06 40]
igi 2 [05	28 48]
igi 2 [06 25 11 06 40	28 28 35 37 30]
igi 2 06 [33] 45	28 [26 40]
igi 2 08	2[8 07 30]
igi 2 09 36	2[7 46 40]
igi 2 10 12 30	[27 38 52 48]
igi 2 11 13 12	[27 26 05 25 55 33 20] «45»
igi 2 11 41 14 04 26 40	27 20 [15]
igi 2 11 50: 09 22 30	27 1[8 24]
igi 2 13 20	27
igi 2 15	26 [40]
igi 2 16 32	26 2[2 21 52 30]
igi 2 16 41 15	26 2[0 14 48 53 20]
igi 2 18 14 24	26 [02 30]
igi 2 18 53 20	25 [55 12]
igi 2 20 37 30	25 [36]
igi 2 22 13 20	25 1[8 45]
igi 2 22 22 48 07 30	25 1[7 02 13 20]
igi 2 24	25
igi 2 25 38 08	24 4[3 09 15 28 07 30]
igi 2 25 48	24 4[1 28 3 20]
igi 2 26 19! «55 28» <08 58> 16 17 46 [40 :	24 36 13 30]
igi 2 26 29 03 45	24 [34 33 36]
igi 2 27 27 21 36	24 [24 50 37 30]
igi 2 28 08 53 20	24 18
igi 2 29 17 57 07 12	[24 06 45 33 20]
igi 2 30	[24 square-side]
igi 2 30	[......]
igi 2 31 42 13 [20	23 43 49 41 15]
igi 2 31 [52 30	23 42 13 20]
igi 2 [33 36	23 26 15]
igi 2 [35 31 12	23 08 53 20]
igi 2 [36 15	23 02 24]
igi [2 38 12 11 15	22 45 20]
igi [2 40	22 30]
[igi 2 42	22 13 20]
[igi 2 43 50 24]	21 58 [21 33 45]
[igi 2 44 01 30]	21 56 [52 20 44 26 40]
[igi 2 46] 40	21 36
[igi 2 48 4]5	21 20
[igi 2 50] 40	21 05 37 [30]
[igi 2 52] 48	20 [50]

[igi 2 53 3]6 40	20 44 09! 3[6]
igi [2 54 5]7 36	20 34 34 <04> 26 [40]
igi [2 55 34] 58 45 55 33 30 :	20 [30 11 15]
igi [2 55 46] 52 20	20 28 [48]
igi [2 57 4]6 40	20 [15]
igi 3	20 square-side
igi 3 [02] 15	19 45! 11 06 [40]
igi 3 [03] 06 19 41 15 :	19! 39 38 [52 48]
igi 3 [04] 19 12	19 31 52 [30]
igi 3 05 11 06 40	19 26 [24]
igi 3 06 37 26 24	[19 17 24 26 40]
igi 3 [07] 30	[19 12]
igi 3 [09 37 46 40	18 59 03 45]
igi 3 [09 50 37 30	18 57 46 40]
[igi 3 12	18 4]5
[igi 3] 12 54 04 [26 40]	18 39 [44 38] 24
[igi 3] 14 [24]	18 3[1] 06 40
[igi 3] 15 18 45	18 25 [55] 12
[igi] 3 16 36 28 48	18 18 37 [58 07] 30
igi 3 16 49 48	18 17 23 [37 17 02] 13 20
igi 3 17 31 51 06 40	18 13 [30]
igi 3 17 45 14 03 45	18 12 [1]6
igi 3 20	18 square-side
igi 3 22 30	17 4[6] 40
igi 3 24 48	17 34 [41 15]
igi 3 27 21 36	17 21 [40]
igi 3 28 20	17 16 4[8]
igi 3 30 <20> «41» 58 31 06 40 :	17 05 09 22 [30]
igi 3 30 56 15	17 [04]
igi 3 33 20	16 52 30
igi 3 36	16 [40]
igi 3 38 27 12	16 28 46 10 18 [45]
igi 3 38 42	16 27 39 15 3[3 20]
igi 3 39 28 43 27 24 26 40 :	16 24 [09]
igi 3 41 11 02 24	16 16 33 45
igi 3 42 13 20	16 12
igi 3 45	16 <square-side>
igi 3 47 33 20	15 49 13 07 30
igi 3 47 <18> «48» 45	15 48 08 53 20
igi 3 50 24	15 37 30
igi 3 51 28 53 20	15 33 07 12
igi 3 53 16 48	15 2!5 55 33 20
igi 3 54 06 3!8 21 14 04 26 40 :	15 22 «39 22 30» <38 26 15>
igi 3 54 22 30	15 21 «21» 36
igi 3 55 55 46 33 36	15 15 <3>1 28 26 15
igi 3 57 02 13 20	15 11 15
igi 4 «24»	15 square-side

According to an old original, Rimut-Anu, son of Shamash-iddin, descendant of Shangû-[Ninurta]. Nadin, his ..., wrote and checked it.

Nadin did indeed check the reciprocal pairs on W 23283+22905, or at least some of them, as he claimed in the colophon. On W 23021, he calculated the pairs between 1 08 20 37 30 and 1 13 14 31 52 30, starting in each case from the second number of the pair and using successive factorization to reduce the number to 1 and then multiply up again to find its reciprocal [von Weiher 1993, no. 176; Friberg 1999]. Compare the standard Old Babylonian method, given in *UET* 6/2 295 (page 148), which is very different. The layout of the calculations on the roughly shaped oval tablet accounts for the fact that the entry 1 09 07 12, 52 05 appears twice in the table of W 23283+22905: once correctly, in the right place; and a second time, two lines later, as 1 09 07 13, 52. The calculation itself is missing from this tablet.

W 23021

52 40 29 37 46 40	1 08 20 37 30	1 04	56 15
2 38 01 28 53 20	22 46 52 30	32	1 52 30
7 54 04 26 40	7 35 37 30	16	3 45
23 42 13 20	2 31 52 30	8	7 30
1 11 06 40	50 37 30	4	15
3 33 20	16 52 30	2	30
10 40	5 37 30	1 reciprocal	1
32	1 52 30		
16	3 45	[50 34 04 26 40	1 11 11 29 03 45]
8	7 30	[2 31 42 13 20	23 43 49 41 15]
4	15	7 [35 06 40	7 54 36 33 45]
2	30	22 [45 20	2 38 12 11 15]
1 reciprocal	1	1 08 16	[52 44 03 45]
		34 08	[1 45 28 07 30]
[52 05	1 09 07 12]	17 04	[3 30 56 15]
[10 25	5 45 36]	[8 32	7 01 52 30]
[2 05	28 48]	[4 16	14 03 45]
[25	2 24]	2 08	[28 07 30]
[5	12]	1 04	[56 15]
[1 reciprocal	1]	32	1 [52 30]
		16	[3 45]
51 50 [24	1 09 26 40]	8	[7 30]
[4] 19 1[2	13 53 20]	4	[15]
[21 36	2 46 40]	2	[30]
1 [48	33 20]	1 reciprocal	[1]
9	[6 40]		
3	[20]	50	1 12
1 [reciprocal	1]	10	6
		2	30
51 12	1 10 18 45	24	2 30
4 16	14 03 [45]	2	30
2 08	28 07 30	1 reciprocal	1

49 22 57 46 40	1 12 54
2 28 08 53	24 18
7 24 26 40	8 06
22 13 20	2 42
1 06 40	54
3 20	18
10	6
1 reciprocal	1

49 09 07 1[2	1 13 14 31 52 30]
4 05 4[5 36	14 38 54 22 30]
20 28 4[5 36	14 38 54 22 30]
1 42 2[4	35 09 22 30]
8 [32	7 01 52 30]
4 1[6	14 03 45]
2 [08	28 07 30]
[1 04	56 15]
[32	1 52 30]
[16	3 45]
[8	7 30]
[4	15]
[2	30]
[1 reciprocal	1]

Rimut-Anu's father Shamash-iddina(m) copied W 23291x, a compilation of mathematical problems with worked solutions and diagrams, most of which explore the relationship between traditional area measure ("reed measure") and Neo-Babylonian seed measure [Friberg et al., 1990]. This is explained further in the Introduction, page 72. Occasionally the problems are introduced with the word "ditto," meaning that the problem-setting scenario is the same as the one before. The original, says the colophon, was a waxed wooden writing board. Such objects rarely survive in the Mesopotamian archaeological record, and never with their surfaces intact. References to writing boards on cuneiform tablets hint at a vast written record now lost to us forever.

W 23291x

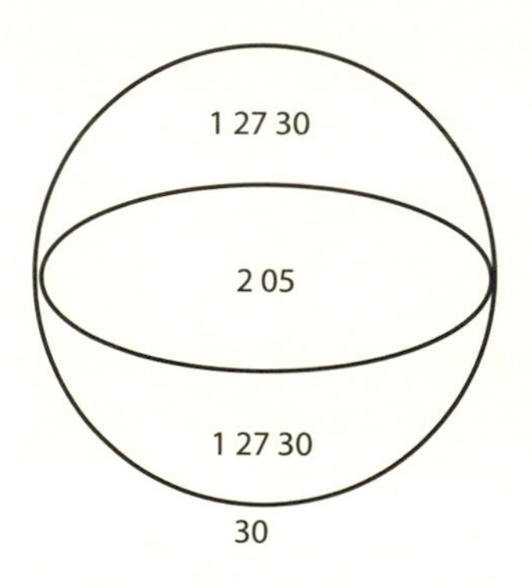

(i) My [dividing line] is 10. What is the extension of the middle?

20 times 10 is 3 20. Because 10 is 1/2 of [20] add 7;30, 1/2 of 15, to 30, so that 37;30. 2;20 times 37;30 is 2 05. The area is 1 iku 25 sar. The crescent moon is 30. What is the area? 30 times 30 is 15 00. Multiply <by> 0;05 50 (so that) 1 27;30. The 1 crescent moon is 1/2 (iku) 37 1/2 sar area. Go times 2. Multiply 1 27;30 by 2 so that 2;55. The 2 crescent moons are 1 1/2 (iku) 25 sar area. Sum them so that they are all 3 (iku) area.

(ii) 1 figure of a circle. 1 curved 1 cable. I took out the digging-out (?) 4 times, 2 rods each. What are each of the areas?

[54 times] 2 is 1 48. The outer digging-out is 1 (iku) 8 sar area. [42] times 2 is 1 24. The second digging-out is 1/2 (iku) 34 sar area. [30] times 2 is 1 00. The third digging-out is 1/2 (iku) 10 sar area. [18] times 2 is 36. The fourth digging-out is 36 sar (area). [12] times 12 is 2 24. Multiply 2 24 by 0;05 so that 12. The fifth, inner digging-out is [12 sar (area)]. Sum them so that they are all 3 (iku) area.

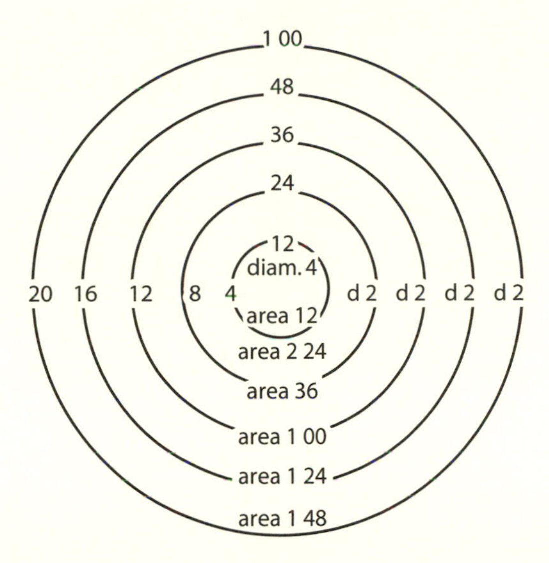

(iii) [A wheel (?).] The dividing lines of what I curved: the *shakkullu* wood is 4 fingers, the fastening band is 1 finger, the rim [3 fingers], the spokes (?) 17 fingers, the hub 1/3 cubit.

[......] The area of the *shakkullu* wood is [1]8 2/3 shekels. The area of the fastening band is 4 shekels and a 4th. The area of the rim is [11] shekels and 3 4ths. The area of 1 spoke (?) is 6 and 3 [8ths] shekels. The area of 6 spokes (?) is 1/2 mina 8 shekels and a 4th. The area of the hub is 2 shekels 15 grains. Sum them so that they are all 1 sar 15 shekels.

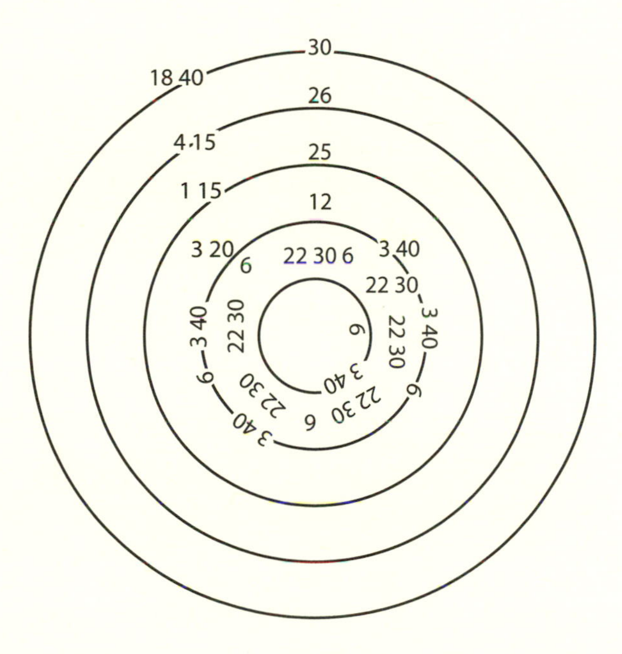

(iv) Reeds, of length 1 rod(-reed), of width 1 rod(-reed), or 1 sar (area). If your cubit is 0;05, <you multiply> the line times ditto and times <1. You take times 1 each. If your cubit is 1> you multiply <times ditto and times> 25. <You take> times 2 24 each.

(v) The length is 1 chain, the width 1 chain. What are the sar (area)? If your cubit is 0;05, there are 5 chains. You multiply 5 by 5 so that 25. The sar are 25. If your cubit is 1, there are 1 00 chains. You multiply 1 times 1 so that <1>. You multiply 1 times 25, so that 25.

(vi) A house of 25 sar (area). What should the square-side be? If your cubit is 0;05, you take 25 each. <5, a chain, is the square-side.> If your cubit is 1, you multiply 25 by 2 24, so that you take 1 00 00 each. <1 00>, a chain, is the square-side.

(vii) [The width is] 4. How long should the length be so that it is 20 sar (area)? If your cubit is [0;05], a 4th is 0;15. You multiply 0;15 times 20 [so that 5]. It is [5], a chain, long. If your cubit is 1, a 4[8th is] 0;01 15. You multiply 0;01 15 by 2 24 so that 3. [You multiply] 3 times [20 so that 1 00].

(viii) Reeds, of length 6 cubits, of width 6 cubits, (or) 1 reed. If your cubit is [0;05, you multiply] the line times ditto and times [2]. Times 0;30 square-side. If your cubit is 1, you multiply the line times [ditto] and <times> 50. Times 1 12 square-side.

(ix) [The length is 2 1/2 rods], the width [2 1/2 rods]. What should the reeds be? If [your cubit is 0;05], you multiply 2;30 times 2;30 so that 6;15. [You multiply 6;15 times 2] so that 12;30. The reeds are 25. [If] your [cubit is 1], 30 each. You multiply 30 times 30 so that 15 00. You multiply [15 00 times 0;00 50] so that 12;30.

(x) [The width is 2. How] long [should the length be] so that it is 20 reeds? Because you do not [understand], raise [......], you multiply times the reciprocal of the constant [of ...] area (?) and times the reeds, so that you will see. If your cubit is [0;05], 2 rods is 2. The reciprocal of 2 is 0;30. You multiply 0;30 times 0;30 [so that 0;15]. You multiply [0;15 times] 10 so that 2 30. You drag it 2 1/2 rods. [If] your [cubit is 1], 2 rods is 24. A 24th is [0;02 30]. You multiply [0;02 30] times 1 12, so that 3. [You multiply] 3 times [10 so that] 30. You drag it 2 1/2 rods.

(xi) [......] so that it is 1 hundred reeds? Because you do not understand, you multiply reeds times the reciprocal of the constant [of] reeds and then [you take] each (square-side). If [your cubit] is 0;05, you multiply 50 times 0;30 so that [25. You take each (square-side)]. The square-side [is a chain]. If [your] cubit is 1, [you multiply 50 times 1 12 so that] 2. ... You take each (square-side). [The square-side is a chain.]

(xii) [*Too fragmentary to translate.*]

(xiii) The square-side of a house is sixty cubits. [I bought] a square-side of 3 cubits for [1/2 shekel]. What did I buy the total house for? Since [you] do not [understand, ...] you multiply the line which is sixty cubits [times ditto, times the reciprocal of] the line [which is] 3 cubits times [ditto and times 30] so that you see. If your cubit <is 0;05>, [the sixty cubits] are 5, the 3 cubits [are 0;15]. [5] times 5 is 25. [You multiply] 0;15 <times> «you multiply» [0;15, so that 0;03 45]. The reciprocal of 0;03 45 is 16.

16 times 25 is 6 40. You multiply 6 40 by 0;30, [so that] 3 20. If your cubit is 1 [the sixty cubits are 1 00], the 3 cubits are 3. 1 00 times <1 00 is 1 00 00> «3». 3 times 3 is 9. [A 9th is 0;06 40.] 0;06 40 times 1 00 00 is 6 40. You multiply 6 40 times 0;30 so that 3 20. You buy the total house for 3 1/3 mina.

(xiv) Ditto. <I bought> a square-side of sixty cubits for 3 <1/3> «2/3» mina. What <did I buy> [a square-side of] 3 cubits for? Because you do not understand, you multiply «times» the line [of 3 cubits] times the reciprocal of «times» the line of [sixty cubits], and you multiply times 3 <1/3> «2/3» mina, so that [you see]. If your [cubit] is 0;05, a 25th is 0;02 24. [0;02 24] times 0;03 45 is 0;00 09. You multiply 0;00 09 times 3 20 so that 0;30. [If] your cubit is 1, a 1 00 00th is 0;00 01. [You multiply] 0;00 01 times 9 [so that 0;00 09]. You multiply [0;00 09] times 3 20 so that 0;30. <You buy> a square-side of 3 cubits for 1/2 [shekel] silver.

(xv) Ditto. For [3] <1/3> «2/3» mina silver I bought a square-side of 1 sixty cubits. What square-side could I buy for 1/2 shekel of silver? Because you do not understand, you multiply the reciprocal of 3 <1/3> «2/3» times 0;30 and then you multiply that which came up for you to see times the line [of sixty] cubits and then you take each (square-side).The reciprocal of 3 20 is 0;00 18. You multiply times 0;30 so that 0;00 09. If your cubit is 0;05, you multiply 0;00 09 by 25 so that 0;03 45. You take 0;15, each (square-side). If your cubit is 1, 0;00 09 times 1 00 00 is 9. <You take> 3, each <square-side>. For 1/2 shekel silver <you buy> a square-side of 3 cubits each.

(xvi) Ditto. For 1/2 shekel of silver I bought a square-side of 3 cubits. What square-side could I buy for 3 <1/3> «5/6» mina of silver? Because you do not understand, you multiply the reciprocal of 1/2 shekel times 3 <1/3> «5/6» mina, and then you take each <square-side>, and then you multiply times 3 cubits so that you see. A 30th is 2. You multiply by 3 20, and then you take 20, each (square-side) of 6 40. If your cubit is <0;05>, you multiply 20 times 0;15 so that the square-side of the house is 5, a chain. If your cubit is 1, you multiply 20 by 3 so that 1 00. The square-side of the house is sixty cubits.

(xvii) The square-side of a courtyard is 20 cubits. The square-side of a baked brick is <2/3> «5/6» cubit. Paving. What is the number of bricks that are in the courtyard? Because you do not know, you multiply «times» the line of the courtyard times the reciprocal of «times» the line of the baked brick so that you see the number of the bricks. If your cubit is 0;05, 1;40 times 1;40 is 2;46 40. You multiply [0;03 20] times 0;03 20, so that 0;00 <1>1 16 40. The reciprocal of 0;00 11 16 40 is 5 24. You multiply 5 24 by 2;46 40 so that 15 00. If your cubit is 1, 20 times [20 is 6 40. 0;40] times 0;40 is 0;26 40. The reciprocal of 0;26 40 is 2;15. [You multiply] 2;15 [times 6 40], so that 15 00, (or) 9 hundred. There are 9 hundred baked bricks.

(xviii) I paved (?) a courtyard with 9 hundred baked bricks of <2/3> «5/6» cubits each (square-side). What is the square-side of the courtyard? You multiply each (square-side) of the baked bricks times 0;40. You take 30, each (square-side) of 15 00. You multiply 30 times 0;40 so that 20. The square-side of the courtyard is 20 cubits.

(xix) [......] total sheepfold. [10 times 10 is 1 40, (or) 1 iku.] 7;30 times 7;30 is 56;15, (or) [56 sar, 15 shekels. 5 times 5 is 25, (or) 25 sar]. 2;30 times 2;30 is

6;15, (or) [6 sar 15 shekels]. [6;15 times 30 is 3] 07;30. The reciprocal of 3 07;30 is 0;00 19 12. 0;00 19 12 [times 1 40 is 0;32 ...]. ...... [0;32 times 1 40] is 53;20. The [...] area is 1/2 iku 3 <1/3> «2/3» sar. You multiply 0;32 times 56;15 so that 30. [The ... area is] 30 sar. You multiply [0;32] times 25 so that 13;20. [The ... area] is 13 1/3 sar. You multiply [0;32] times 6;15 so that 3;20. [The ... area] is 3;20. Sum them all so that the total courtyard is 1 40, (or) 1 iku.

(xx) [Reed measure, of which] the length is 7 cubits, the width 7 cubits, is 1 reed (area), by the small cubit (?). You multiply the line times ditto and 1 12 and then you double that which came up for you to see [...] number of reeds and small cubits. The square-side is 1 10 cubits. What should the reed measure be? 1 10 times 1 10 is 1 21 40. 1 21 40 times 0;01 12 (is 1 38). 1 38 reeds and 1 38 small cubits, or 2 reeds. You double 2 reeds to 1 38, so that the reeds are 1 hundred.

(xxi) 1 reed times 1 reed is 1 reed. 1 reed times 1 cubit is 1 cubit. 1 reed times 1 finger is 1 finger. 1 cubit times 1 reed is 1 cubit. [1] cubit times 1 cubit is 1 small cubit. 1 cubit times 1 finger is 1 grain. [1] finger <times> 1 reed is 1 finger. 1 finger times 1 cubit is 1 [grain]. [1] finger times <1 finger is> 1 small finger. 24 small fingers are [1] grain. 7 grains are 1 finger. 24 fingers are 1 cubit. 7 cubits are 1 reed. 3 fingers and 3 grains are 1 small cubit. 7 small cubits are 1 cubit.

(xxii) 3 sila in seed-measure. What should the reed-measure be? Because you [do not] understand, you multiply the reciprocal of the constant of seed-measure times the constant of reed-measure and times the seed-measure, so that [you see] the reed-measure. [The reciprocal of 48] is 0;01 15. 0;01 15 times 2 is 0;02 30. 0;02 30 times [5 is 0;12 30]. The reed-measure is [25]. Ditto. The reciprocal of 0;20 is 3. You multiply 3 times 50, so that 2 30.

(xxiii) 25 reeds. What should the seed-measure be? Because you do not understand, you multiply the reciprocal of the constant of reed-measure times the constant of seed-measure, and you multiply times the reeds, so that you see the seed-measure. The reciprocal of 2 is 0;30. 0;30 times 0;00 48 is 0;00 24. You multiply 0;00 24 times 12;30 so that 0;05. The seed-measure is 3 sila. Ditto. The reciprocal of 50 is 0;01 12. 0;01 12 times <0;20> is 0;00 24.

Seed-measure and reed-measure, finished. Copy of a wooden writing-board, written and collated according to its original. Tablet of Shamash-iddina, son of Nadin, descendant of Shangû-Ninurta, incantation-priest, Urukean.

Tablets belonging to the Sîn-leqi-unninni family, fl. 200 BCE in Uruk

Two members of the Sîn-leqi-unninni family of lamentation-priests are responsible for the latest datable cuneiform mathematical tablets currently known. Unlike the Shangû-Ninurta family, archaeology has revealed nothing of their domestic lives, but their professional careers with the Resh temple in Uruk, around 200 BCE, can be reconstructed through the carefully recorded dates on the colophons of their tablets [Robson forthcoming, chap. 8].

AO 6484 is a compilation of mathematical problems with worked solutions, whose subjects (with the exception of the series in the first two questions) are recognizably similar to Old Babylonian examples [Thureau-Dangin 1922, 33; MKT, I 96–107; TMB, 76–81]. Its early date of publication gave rise to the interpretation that Late Babylonian mathematics was not substantially different from its predecessors; but we now have a much larger, richer context to put it in. Anu-aba-uter's professional title, "Scribe of *Enuma Anu Ellil*," refers to the great Babylonian series of celestial omens "When the gods Anu and Ellil." He was, in other words, a qualified priestly astronomer. (Here the colon stands for a cuneiform sign which represents either an empty sexagesimal place or basic punctuation.)

AO 6484

By the command of Anu and Antu, may it go well.

(i) Scale (a ladder), which is for (?) 2, from 1 to 10. Sum so that 8 [32 : take out 1 from 8 32, so that] the remainder is 8 31 : Add 8 31 to 8 32, so that 17 03 [......]

(ii) Squares from 1 by 1 : 1, to 10 by 10 : 1 40. Establish the count. Multiply [1 by 0;20 : 1/3], so that 0;20 : multiply 10 by 0;40 : two-thirds, so that 6;40. 6;40 [and 0;20 are 7]. Multiply 7 by 55, so that 6 25 : the count is 6 25.

(iii) The height of a brick wall is 10 cubits. It is open 1 cubit from the width of the brick wall, the height of a log is 1 cubit [......]. How much from the top of the brick wall should I move away in order to see it? The reciprocal of 1 [is 1 ......] Multiply 1 by 10 cubits, so that 10 : you move away 10 cubits in order to see it.

(iv) The height of a log is 1 cubit. 10 [......]. the reciprocal of 1 is 1 : 1 by [......]

(v) 1 2/3 cubits times 1 2/3 cubits. How much is the seed (measure)? [Multiply 2;46 40 by 21 36, so that 1 00 00]. Multiply 1 00 00 by 0;01 48 so that 1 48. The seed is 1 hundred 8.

(vi) Length, width, and diagonal are 40, and the area is 2 00. The length is 15, [the width 8, the diagonal 17].

(vii) A wedge. Each length is 5, the width 6. How [much is the seed (measure)?] Multiply 5 by 5, the lengths, so that 25 : [multiply] 3, half of the width, [by 3 so that 9]. Take out 9 from 25 so that the remainder is 16. How much by how much should you multiply for 16 : [multiply] 4 [by 4 so that 16]. Multiply 4, the diameter, by 3, half of the width, so that 12 : the area is 12. [Multiply] 12 by 21 [36 so that] 4 19 12 : multiply 4 19 12 by 0;01 48 so that 7 4[6;33 36]. The seed is 4 hundred 1 06 and a half.

(viii) The diagonal of a square is 10 cubits. <What is> the length of the square? [Multiply] 10 by 0;42 30 [so that 7;05 is the length]. Multiply 7;05 by 1;25 so that [10;25 is the diagonal].

(ix) The length is 1 cubit, the width 1 cubit, the height 1 cubit [......]. [Multiply] 0;05, cubits, the horizontal, by 0;05, cubits, the horizontal, [so that 0;00 25 :] multiply [0;00 25 by 1, [the vertical so that 0;00 25 : [multiply] 0;00 25 by 6 00 00, the storage, [so that 2 30 ......].

(x) The length is 2 cubits, the width 2 cubits, the height 2 cubits [......]. [multiply] 0;10, cubits, the horizontal, by 0;10, cubits, the horizontal, [so that 0;01 40 :] multiply [0;01 40 by 2, the vertical], so that 0;03 20 : multiply 0;03 20 by 6 00 00, so that [20 00. ......].

(xi) The length is 3 cubits, the width 3 cubits, [the height] 3 [cubits ......]. [Multiply] 0;15, cubits, the horizontal, by 0;15, cubits, the horizontal, [so that 0;03 45] : multiply [0;03 45 by 3], the vertical, so that 0;11 15 : [multiply 0;11 15] by 6 00 00, so that [1 12 30 ......].

(xii) The length is 4 cubits, the width 4 cubits, the height 4 cubits [......]. Multiply [0;20, cubits, the horizontal], by 0;20, cubits, the horizontal, so that 0;06 [40 : multiply [0;06 40 by 4, the vertical, so that 0;26 40] : multiply 0;26 40 by 6 00 00, so that 2 40 00 [......].

(xiii) The length is 5 cubits, the width 5 cubits, the height 5 cubits [......]. Multiply [0;25, cubits, the horizontal], by 0;25, cubits, the horizontal, so that 0;10 25 : [multiply [0;10 25 by 5, the vertical, so that 0;52 05] : multiply 0;52 05 by 6 00 00, so that 5 12 30 [......].

(xiv) A reciprocal and its reciprocal are 2;00 00 33 20. [How much are] the reciprocal and its reciprocal? Multiply [2;00 00 33 20] by 0;30 so that 1;00 00 16 40 : Multiply 1;00 00 16 40 by [1;00 00 16 40 so that 1;00 00 33 20 04 37 46 40]. Take out 1 from its middle so that the remainder is 0;00 00 33 <20> 04 37 46 40. How much [by how much should I multiply so that 0;00 00 33 <20> 04 37 46 40]? Multiply 0;00 44 43 20 by 0;00 44 43 20 so that 0;00 00 33 <20> 04 37 4[6 40 : increase 0;00 44 43 20 to 1;00 00 16 40 so that] 1;00 45 is the reciprocal. Decrease 0;00 44 43 20 from 1;00 00 16 40 [so that 0;59 15 33 20 is its reciprocal].

(xv) A reciprocal and its reciprocal are 2;03. Multiply by 0;30 so that 1;01 30 [: Multiply 1;01 30 by 1;01 30 so that 1;03 02 15]. Take out 1 from its middle so that the remainder is 0;03 02 15. How much by how much [should I multiply so that 0;03 02 15]? Multiply 0;13 30 by 0;13 30 «by» so that 0;03 02 15 : [increase] 0;13 30 [to 1;01 30, so that 1;15 is the reciprocal]. Decrease 0;13 30 from 1;01 30, so that 0;48 is its reciprocal.

(xvi) A reciprocal and its reciprocal are 2;05 26 40. Multiply by 0;30 so that 1;02 43 [20 : multiply [1;02 43 20 by 1;02 43 20], so that 1;05 34 04 37 46 40. Take out 1 from its middle so that [the remainder is 0;05 34 04 37 46 40. How much by how much] should I multiply [so that 0;05 34] 04 37 46 40? [Multiply] 0;18 16 40 [by 0;18 16 40 so that 0;05 34 04 37 46 40.] Increase 0;18 16 40 to 1;02 43 20 so that 1;21 is the reciprocal. [Decrease 0;18 16 40 from 1;02 43 20 so that] the remainder is 0;44 26 40, its reciprocal.

(xvii) A reciprocal and its reciprocal are 2;00 15. Multiply by 0;30 so that 1;00 07 30 : [multiply] 1[;00 07 30 by 1;00 07 30, so that 1;00 15 00 56 15]. Take out 1 from its middle so that the remainder is 0;00 15 00 56 15. How much by how much should I multiply [so that 0;00 15 00 56 15]? Multiply 0;03 53 30 by 0;03 52 30 so that 0;00 15 00 56 15 : [Increase] 0;03 5[2 30 to 1;00 07 30 so that] 1;04 is the reciprocal. Decrease 0;03 52 30 from 1;00 07 30, so that 0;5[6 15 is its reciprocal].

[Tablet of] Anu-aba-uter, scribe of *Enuma Anu Ellil*, son of [Anu-belshunu, descendent of Sîn-leqi-unninni.]

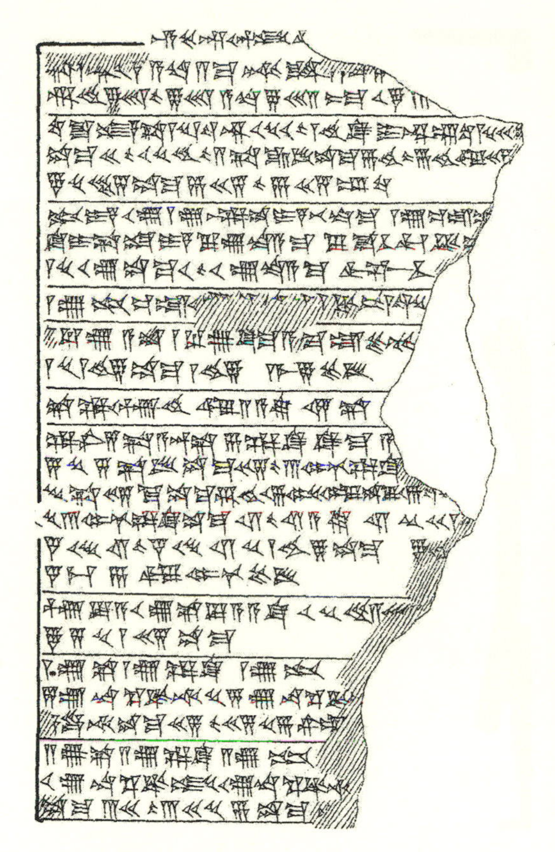

AO 6484 Obverse

AO 6484 Reverse

AO 6555, a compilation of problems about the measurements of Esangila, Marduk's temple in Babylon, and its associated ziggurat Etemenanki, has been studied extensively for the metrological and archaeological information it contains, but was not traditionally considered to be mathematical [Thureau-Dangin 1922, 32; George 1993, no. 13]. In fact, both the terminology and the concern with the relationship between "seed" and "reed" measure place it firmly within the mathematical tradition.

It was copied from a broken tablet by Anu-aba-uter's father, Anu-belshunu, during his apprenticeship under another scholar of the same name in December 226 BCE. He marked damaged passages on the tablet he was copying from with the phrase "NEW BREAK." The compilation is also preserved on a fragmentary tablet from Babylon.

A0 6555

By the command of Anu and Antu, may it go well.

(i) Esangila, the Great Court. Its measurement is 1 (iku) area. The measure of the Court of Ishtar and Zababa is 1/2 (iku) area. Enlarge the Great Court by 2 sar and 1/2 sar from 1 (iku) area, the corner pillar of Ubshu-ukkinna. Reduce the small courtyard by 2 sar and 1/2 sar from 1/2 (iku) area : the recessed doorway which is next to the half-door. If you do not understand length, width, area, the length of the Great Court is 11;23 20, the width of the Great Court is 9. 11;23 20 times 9 is 1 42;30 : 1 42;30 times 0;18 : 30;45. If you do not understand 30;45 it is 1 (iku) 2 sar 1/2 sar area in seed (measure) by the large cubit, the measurement of the Great Court.

(ii) The length of the Court of the deities Ishtar and Zababa is 10;33 20, the width of the Court of Ishtar and Zababa is 4;30. 10;33 20 times 4;30 : 47;30 : 47;30 times 0;18 : 14;15. If you do not understand 14;15 it is 40 sar 7 1/2 sar in seed (measure) by the large cubit, the measurement of the [small] courtyard.

(iii) Total 1 1/2 (iku) area in seed (measure) using the *arû* cubit, the sum of both courtyards: the precinct of the six gates Ka-mah, Ka-Utu-ea, Ka-gal, Ka-Lamma-rabi, Ka-hegal, and Ka-ude-babbara, entrance to the god, which open around the Ubshu-ukkinna courtyard for the rites of the temple. Esangila and the precinct of Ka-sikila are not calculated.

(iv) For [you] to see the measurements of the base of (the ziggurat) Etemenanki, length and width. The length is 3 chains, the width 3 chains, by the cable-cubit. To multiply their calculation, 3 [times 3] is 9 : 9 times 2 is 18. If you do not understand 18 it is 3 (bariga) seed (measure) by the small cubit. The base of Etemenanki. The height is in accordance with the length [and width].

(v) Otherwise, for [you] to see the measurements of the base of Etemenanki, length and width. The length is [10 rods], the width 10 rods by the *arû* cubit. In order to multiply its calculation, [10 times 10 is] 1 40 : 1 40 times 0;18 : 30. If you do not understand 30, it is 30, where 1 (iku) area is 5 (ban) seed (measure) [by the] large [cubit]. The base of Etemenanki. Length, width, and height are 10 rods square [by] the *arû* cubit.

(vi) The measurements of the 6 chapels whose name is the ziggurat. The eastern wing. Chapel of the god: the length is 2, the width 0;40. On either side, the chapels of the deities Nabû and Tashmetu: the length is 0;45 each, the width 0;40 each. Two northern rooms, of the deities Ea and Nuska: Ea's room: the length is 1;25, the width 0;30; Nuska's room: the length is 0;35, the width 0;35. The southern room, room of the deities Anu and Enlil: the length is 1;10, the width 0;30. The western rooms: the twin and the stairwell behind it. The outer front, the bedroom: the length is 2;05, the width 0;30. The inner front: the length is 1;40, the width 0;20. The stairwell, according to them, the length and the width. The courtyard the length is 1;40, the width 1;05, the courtyard is roofed. The bed: the length is 9 cubits, width 4 cubits, bed and chair are correct (?) and surrounded by bolts (?). A second bed is located in the courtyard. Eastern gate, southern gate, western gate, northern gate.

(vii) The measurements, length, width, and height NEW BREAK: its name is Ziggurat of Babylon. The length is 15 rods, the width 15 rods, the height 2 1/2 rods: the lower mud wall. The length is 13 rods, the width 13 rods, the height 3 rods: the second stage. The length is 10 rods, the width 10 rods, the height 1 rod: the third stage. The length is 8 1/2 rods, the width 8 1/2 rods, the height 1 rod: the fourth stage. The length is 7 rods, the width 7 rods, the height 1 rod: the fifth stage. <The length is

5 1/2 rods, the width 5 1/2 rods, the height 1 rod: the sixth stage.> The length is 4 rods, the calculation of the width 3 1/2 rods, the height 2 1/2 rods: the upper shrine, the seventh <stage>, the summit building.

The knowing may show the knowing; the unknowing shall not see it. Written, made good, and checked against an old tablet from Borsippa.

> 18 sar 1 rod 3 sila and 20 of rod : 50 sar = 1 ubu = 1 (ban) 3 sila
> 2 ubu = 1 iku = 3 (ban) : 6 (iku) = an eshe = 3 (bariga)
> 3 eshe = a bur = 1 gur 4 bariga : 60 bur = a shar = 1 hundred 8 gur.

Tablet of Anu-belshunu, son of Anu-balassu-iqbi, descendant of Ahi'utu, Urukean. Hand of Anu-belshunu, son of Nidintu-Anu, descendant of Sîn-leqi-unninni. Uruk, 26th day of Month IX, year 1 23, Seleucus the king.

V. Appendices

V.a. Sources

These are listed in the order in which they appear in the chapter. The reference is to the initial publication, but see the text for further references. All references are to works in the Bibliography below.

W 19408,76	Nissen et al. 1993, fig. 50
VAT 12593	Deimel 1923, no. 82
TSS 50	Jestin 1937
TSS 671	Jestin 1937
IM 58045	Friberg 1990, 541
HS 815	Pohl 1935, no. 65
Ash 1924.689	Gelb 1970, no. 112
AO 11409	Foster and Robson 2004, no. 12
BM 106444	Robson 2003–4
AOT 304	Thureau-Dangin 1903, no. 413
BM 80150	Pinches 1963, no. 42
Ash 1924.447	Robson 2004, 15
Ash 1924.451	Robson 2004, 15
YBC 11924	MCT, 23
Ash 1931.137	Robson 2004, no. 19
Ash 1923.366	Robson 2004, no. 9
BM 22706	Nissen, et al. 1993, fig. 128
IM 73355	Arnaud 1994, no. 55
BM 15285	Gadd 1922
IM 55357	Baqir 1950
Db_2-146	Baqir 1962
YBC 4675	MCT, text B
YBC 6967	MCT, text Ua
BM 13901	Thureau-Dangin 1936
A 24194	MCT, text T
Strasbourg 363	MKT, I 244–45
Strasbourg 364	MKT, I 248–50
Strasbourg 367	MKT, I 259–60
Strasbourg 368	MKT, I 311
YBC 4673	MKT, III 29–34
YBC 4657	MCT, text G
YBC 4663	MCT, text H
YBC 4662	MCT, text J
YBC 7997	MCT, text Pa
Haddad 104	Al-Rawi and Roaf 1984
Strasbourg 362	MKT, I 239–43
AO 8862	MKT, I 108–23
BM 85194	MKT, I 142–93

BM 96957	Robson 1996
VAT 6598	Weidner 1916
YBC 7302	MCT, 44
Ash 1931.91	Robson 2004
YBC 7289	MCT, 42
YBC 7290	MCT, 44
YBC 11126	MCT, 44
Ash 1922.168	Robson 1999, 273–74
UET 6/2 211	Robson 1999, 245–72
UET 6/2 222	Robson 1999, 245–72
UET 6/2 293	Robson 1999, 255
UET 6/2 236	Friberg 2000, 124
UET 6/2 295	Sachs 1947
UET 6/2 274	Friberg 2000, 113–16
YBC 7243	MCT, text Ue
BM 80209	Pinches 1963, no. 39
Plimpton 322	MCT, text A
AO 17264	Thureau-Dangin 1934
HS 245	Horowitz 1998, 179–82
Sm 162	Horowitz 1998, 179–82
Ash 1924.796	MKT, II pl. 34
W 23273	von Weiher 1993, no. 172
W 23283+22905	von Weiher 1993, no. 174
W 23021	von Weiher 1993, no. 176
W 23291x	Friberg et al. 1990
AO 6484	Thureau-Dangin 1922, 33
AO 6555	Thureau-Dangin 1922, 32

V.b. References

ETCSL = Black et al. 1998–

MCT = Neugebauer and Sachs 1945

MKT = Neugebauer 1935–37

TMB = Thureau-Dangin 1938

Al-Rawi, F. N. H. and M. Roaf. 1984. "Ten Old Babylonian Mathematical Problems from Tell Haddad, Himrin." *Sumer* 43: 175–218.

Arnaud, D. 1994. *Texte aus Larsa: Die epigraphischen Funde der 1. Kampagne in Senkereh-Larsa 1933.* Berline Beiträge zum Vorderen Orient. Texte, 3. Berlin: Dietrich Reimer Verlag.

Bahrani, Z. 1998. "Conjuring Mesopotamia: Imaginative Geography and a World Past." In L. Meskell (ed.), *Archaeology under Fire: Nationalism, Politics and Heritage in the Eastern Mediterranean and Middle East.* London: Routledge, 159–74.

Baqir, T. 1950. "An Important Mathematical Problem Text from Tell Harmal (on a Euclidean Theorem." *Sumer* 6: 39–54.

Baqir, T. 1962. "Tell Dhiba'i: New Mathematical texts." *Sumer* 18: 11–14.

Black, J. A., G. Cunningham, J. Ebeling, E. Flückiger-Hawker, E. Robson, J. Taylor, and G. Zólyomi. 1998–. *The Electronic Text Corpus of Sumerian Literature.* http://etcsl.orinst.ox.ac.uk. Oxford: Oriental Institute.

Black, J. A., G. Cunningham, E. Robson, and G. Zólyomi. 2004. *The Literature of Ancient Sumer.* Oxford: Oxford University Press.

Brack-Bernsen, L. and O. Schmidt. 1990. "Bisectable Trapezia in Babylonian Mathematics." *Centaurus* 33: 1–38.

Bruins, E. M., and M. Rutten. *Textes mathématiques de Suse.* Mémoires de la Mission archéologique en Iran, 34. Paris: Geuthner.

Burkert, W. 1972. *Lore and Science in Ancient Pythagoreanism.* Cambridge, MA: Harvard University Press.

Cavigneaux, A. 1999. "A Scholar's Library in Meturan?". In T. Abusch and K. van der Toorn (eds.), *Mesopotamian Magic: Textual, Historical, and Interpretative Perspectives.* Groningen: Styx, 253–73.

Dalley, S. M. 2000. *Myths from Mesopotamia.* 2nd ed. Oxford: Oxford University Press.

Damerow, P. 2001. "Kannten die Babylonier den Satz des Pythagoras? Epistemologische Anmerkungen zur Natur der babylonischen Mathematik." In Høyrup and Damerow (2001), 219–310.

Deimel, A. 1923. *Die Inschriften von Fara, II: Schultexte aus Fara.* Wissenschaftliche Veröffentlichungen der Deutschen Orient Gesellschaft, 43. Leipzig: J. C. Hinrichs.

Fermor, J. and J. M. Steele, 2000. "The Design of Babylonian Waterclocks: Astronomical and Experimental Evidence." *Centaurus* 42: 210–22.

Foster, B. F. 1996. *Before the Muses: An Anthology of Akkadian Literature.* 2nd ed., Potomac, MD: CDL Press.

Foster, B. R. and E. Robson, 2004. "A New Look at the Sargonic Mathematical Corpus." *Zeitschrift für Assyriologie und vorderasiatische Archäologie* 94: 1–15.

Fowler, D. H. 1999. *The Mathematics of Plato's Academy: A New Reconstruction.* 2nd ed. Oxford: Oxford University Press.

Fowler, D. H. and E. Robson. 1998. "Square Root Approximations in Old Babylonian Mathematics: YBC 7289 in Context." *Historia Mathematica* 25: 366–78.

Frank, C. 1928. *Strassburger Keilschrifttexte in sumerischer und babylonischer Sprache.* Berlin: Walter de Gruyter.

Friberg, J. 1981. "Methods and Traditions of Babylonian Mathematics, II: an Old Babylonian Catalogue Text with Equations for Squares and Circles." *Journal of Cuneiform Studies* 33: 57–64.

Friberg, J. 1990. "Mathematik." In D. O. Edzard (ed.), *Reallexikon der Assyriologie und vorderasiatischen Archäologie* 7. Berlin/New York: Walter de Gruyter, 531–85.

Friberg, J. 1993. "On the Structure of Cuneiform Metrological Table Texts from the –1st Millennium." In H. D. Galter (ed.), *Die Rolle der Astronomie in Kulturen Mesopotamiens.* Grazer Morgenländische Studien, 3. Graz: RM Druck- und Verlagsgesellschaft, 383–405.

Friberg, J. 1999. "A Late Babylonian Factorization Algorithm for the Computation of Reciprocals of Many-place Regular Sexagesimal Numbers." *Baghdader Mitteilungen* 30: 139–61.

Friberg, J. 2000. "Mathematics at Ur in the Old Babylonian Period." *Revue d'Assyriologie et d'Archéologie Orientale* 94: 97–188.

Friberg, J. 2001. "Bricks and Mud in Metro-mathematical Cuneiform Texts." In Høyrup and Damerow (2001), 61–154.

Friberg, J., H. Hunger, and F. N. H. al-Rawi. 1990. "'Seed and Reeds": a Metro-mathematical Topic Text from Late Babylonian Uruk." *Baghdader Mitteilungen* 21: 483–557.

Gadd, C. J. 1922. "Forms and Colours." *Revue d'Assyriologie et d'Archéologie Orientale* 19: 149–59.

Gelb, I. J. 1970. *Sargonic Texts in the Ashmolean Museum, Oxford.* Materials for the Assyrian Dictionary, 5. Chicago: University of Chicago Press.

George, A. R. 1993. *Babylonian Topographical Texts.* Orientalia Lovaniensia Analecta, 40. Leuven: Peeters.

Gingerich, O. 1965. "Eleven-digit Regular Sexagesimals and their Reciprocals." *Transactions of the American Philosophical Society* 55, no. 8.

Hilprecht, H. V. 1906. *Mathematical, Metrological and Chronological Tablets from the Temple Library of Nippur.* Philadelphia: University of Pennsylvania.

Horowitz, W. 1998. *Mesopotamian Cosmic Geography.* Mesopotamian Civilizations, 8. Winona Lake: Eisenbrauns.

Høyrup, J. 1982. "Investigations of an Early Sumerian Division Problem, c. 2500 BC." *Historia Mathematica* 9: 19–36.

Høyrup, J. 1996. "Changing Trends in the Historiography of Mesopotamian Mathematics: An Insider's View." *History of Science* 34: 1–32.

Høyrup, J. 2001. "The Old Babylonian Square Texts BM 13901 and YBC 4714: Retranslation and Analysis." in Høyrup and Damerow (2001), 155–218.

Høyrup, J. 2002. *Lengths, Widths, Surfaces: A Portrait of Old Babylonian Algebra and its Kin.* Berlin: Springer Verlag.

Høyrup J. and P. Damerow, eds. 2001. *Changing Views on Ancient Near Eastern Mathematics.* Berliner Beiträge zum Vorderen Orient, 19. Berlin: Dietrich Reimer Verlag.

Huehnergard, J. and C. Woods. 2004. "Akkadian and Eblaite." In R. D. Woodard (ed.), *The Cambridge Encyclopedia of the World's Ancient Languages.* Cambridge: Cambridge University Press, 218–287.

Jestin, R. 1937. *Tablettes sumériennes de Shuruppak conservées au Musée de Stamboul.* Paris: E.de Boccard.

Jones, A. 1996. "Babylonian Astronomy and its Greek Metamorphoses." In F. J. Ragep and S. Ragep (eds.), *Tradition, Transmission, Transformation.* Leiden: Brill, 139–55.

King, L. W. 1912. *Cuneiform Texts from Babylonian Tablets, &c., in the British Museum* 33. London: Trustees of the British Museum.

Kuhrt, A. 1995. *The Ancient Near East, c. 3000–330 BC.* London: Routledge.

Larsen, M. T. 1996. *The Conquest of Assyria: Excavations in an Antique Land, 1840–1860.* London: Routledge.

Lloyd, S. 1980. *Foundations in the Dust: The Story of Mesopotamian Exploration.* 2nd ed., London: Thames and Hudson.

Matthews, R. 2003. *The Archaeology of Mesopotamia: Theories and Approaches.* London: Routledge.

Melville, D. J. 2002. "Computation Rations at Fara: Multiplication or Repeated Addition?' In Steele, J. M. and A. Imhausen (eds.), *Under One Sky: Mathematics and Astronomy in the Ancient Near East.* Alter Orient and Alters Testament, 297. Münster: Ugarit-Verlag, 237–52.

Michalowski, P. 1993. *Letters from Early Mesopotamia.* Atlanta: Scholar's Press.

Michalowski, P. 2004. "Sumerian." in R. D. Woodard (ed.), *The Cambridge Encyclopedia of the World's Ancient Languages.* Cambridge: Cambridge University Press, 19–59.

Nemet-Nejat, K. R. 2002. "Square Tablets in the Yale Babylonian Collection." In Steele, J. M. and A. Imhausen (eds.), *Under One Sky: Mathematics and Astronomy in the Ancient Near East.* Alter Orient and Alters Testament, 297. Münster: Ugarit-Verlag, 253–81.

Neugebauer, O. 1935–37. *Mathematische Keilschrift-Texte,* 1–3. Quellen und Studien zur Geschichte der Mathematik, Astronomie und Physik, A3. Berlin: Verlag von Julius Springer.

Neugebauer, O. 1951. *The Exact Sciences in Antiquity.* Copenhagen: Munksgaard.

Neugebauer, O. and A. J. Sachs. 1945. *Mathematical Cuneiform Texts.* American Oriental Series, 29. New Haven: American Oriental Society.

Nissen, H. J., P. Damerow, and R. K. Englund. 1993. *Archaic Bookkeeping: Early Writing and Techniques of Economic Administration in the Ancient Near East.* Chicago: University of Chicago press.

Pinches, T. G. 1963. *Miscellaneous Texts.* Cuneiform Texts from Babylonian Tablets in the British Museum, 44. London: Trustees of the British Museum.

Pohl, A. 1935. *Vorsargonische und sargonische Wirtschaftstexte.* Texte und Materialen der Frau Professor Hilprecht Collection, 5. Leipzig: J. C. Hinrichs.

Powell, M. A. 1976. "The Antecedents of Old Babylonian Place Notation and the Early History of Babylonian Mathematics." *Historia Mathematica.* 3: 417–39.

Powell, M. A. 1990. "Mathematik." In D. O. Edzard (ed.), *Reallexikon der Assyriologie und vorderasiatischen Archäologie* 7. Berlin/New York: Walter de Gruyter, 457–530.

Roaf, M. 1990. *Cultural Atlas of Mesopotamia and the Near East.* New York and Oxford: Facts on File.

Robson, E. 1996. "Building with Bricks and Mortar: Quantity Surveying in the Ur III and Old Babylonian Periods." In K. R. Veenhof (ed.), *Houses and Households in Ancient Mesopotamia.* Istanbul: Nederlands Historisch-Archaeologisch Instituut te Istanbul, 181–90.

Robson, E. 1997. "Three Old Babylonian Methods for Dealing with 'Pythagorean' Triangles." *Journal of Cuneiform Studies* 49: 51–72.

Robson, E. 1999. *Mesopotamian Mathematics 2100–1600 BC: Technical Constants in Bureaucracy and Education.* Oxford Editions of Cuneiform Texts, 14. Oxford: Clarendon Press.

Robson, E. 2001. "The Tablet House: A Scribal School in Old Babylonian Nippur." *Revue d'Assyriologie* 95: 39–67.

Robson, E. 2001a. "Neither Sherlock Holmes nor Babylon: A Reassessment of Plimpton 322." *Historia Mathematica* 28: 167–206.

Robson, E. 2002. "More than Metrology: Mathematics Education in an Old Babylonian Scribal School." In Steele, J. M. and A. Imhausen (eds.), *Under One Sky: Mathematics and Astronomy in the Ancient Near East.* Alter Orient and Alters Testament, 297. Münster: Ugarit-Verlag, 325–65.

Robson, E. 2002a. "Guaranteed Genuine Babylonian Originals: The Plimpton Collection and the Early History of Mathematical Assyriology." In C. Wunsch (ed.), *Mining the Archives: Festschrift for C.B.F. Walker.* Dresden: ISLET, 245–92.

Robson, E. 2003. "Tables and Tabular Formatting in Sumer, Babylonia, and Assyria, 2500–50 BCE." In M. Campbell-Kelly, M. Croarken, R. G. Flood, and E. Robson (eds.), *The History of Mathematical Tables from Sumer to Spreadsheets.* Oxford: Oxford University Press, 18–47.

Robson, E. 2003–4. "Review of Høyrup and Damerow (2001)." *Archiv für Orientforschung* 50: 365–62.

Robson, E. 2004. "Mathematical Cuneiform Tablets in the Ashmolean Museum, Oxford." *SCIAMVS* 5: 3–65.

Robson, E. forthcoming. *Mathematics in Ancient Iraq: A Social History.* Princeton and Oxford: Princeton University Press.

Sachs, A. J. 1947. "Babylonian Mathematical Texts, 1: Reciprocals of Regular Sexagesimal Numbers." *Journal of Cuneiform Studies* 1: 219–40.

Saggs, H. W. F. 1960. "A Babylonian Geometrical Text." *Revue d'Assyriologie et d'Archéologie Orientale* 54: 131–46.

Said, E. W. 1978. *Orientalism.* London: Routledge & Kegan Paul.

Scheil, V. 1915. "Les tables 1 igi-x-gal bi etc." *Revue d'Assyriologie et d'Archéologie Orientale* 12: 195–98.

Smith, D. E. 1907. "The Mathematical Tablets of Nippur." *Bulletin of the American Mathematical Society* 13: 392–98.

Smith, D. E. 1923–25. *History of Mathematics. Vol. 1, General Survey of the History of Elementary Mathematics. Vol. 2, Special Topics of Elementary Mathematics.* Boston: Ginn & Co.

Thureau-Dangin, F. 1903. *Recueil de tablettes chaldéennes.* Paris: Ernest Leroux.

Thureau-Dangin, F. 1922. *Tablettes d'Uruk à l'usage des prêtres du temple d'Anu au temps des Séleucides.* Textes cuneiforms du Louvre, 6. Paris: Geuthner.

Thureau-Dangin, F. 1934. "Une nouvelle tablette mathématique de Warka." *Revue d'Assyriologie et d'Archéologie Orientale* 31: 61–69.

Thureau-Dangin, F. 1936. "L'equation du deuxième degré dans la mathématique babylonienne d'après une tablette inédite du British Museum." *Revue d'Assyriologie et d'Archéologie Orientale* 33: 27–48.

Thureau-Dangin, F. 1938. *Textes mathématiques babyloniens.* Ex Oriente Lux, 1. Leiden: Brill.

Tripp, C. 2002. *A History of Iraq*. 2nd ed. Cambridge: Cambridge University Press.

Van De Mieroop, M. 2004. *A History of the Ancient Near East, ca. 3000–323 BC*. Oxford: Blackwell.

Van Der Waerden, B. L. 1954. *Science Awakening*. Groningen: Noordhoff.

Van Der Waerden, B. L. 1983. *Geometry and Algebra in Ancient Civilizations*. Berlin: Springer.

von Weiher, E. 1993. *Spätbabylonische Texte aus Uruk*, 4. Ausgrabungen der Deutschen Forschungsgemeinschaft in Uruk-Warka. Endberichte 12. Berlin: Mann.

Walker, C. B. F. 1987. *Cuneiform*. London: British Museum Press.

Weidner, E. F. 1916. "Die Berechnung rechtwinkliger Dreiecke bei den Akkadern um 2000 v. Chr." *Orientalistische Literaturzeitung* 19: 257–63.

Westenholz, A. 1975. *Early Cuneiform Texts in Jena: Pre-Sargonic and Sargonic Documents from Nippur and Fara in the Hilprecht-Sammlung vorderasiatischer Altertümer*. Copenhagen: Munksgaard.

Whiting, R. M. 1984. "More Evidence for Sexagesimal Calculations in the Third Millennium BC." *Zeitschrift für Assyriologie und vorderasiatische Archäologie* 74: 59–66.

3 Chinese Mathematics

Joseph W. Dauben

Preliminary Remarks

The standard view of the history of Chinese mathematics is that it was utilitarian, authoritarian, and basically conservative, placing greater emphasis on traditional methods and classical texts than upon innovation and radically different approaches to established or even newly conceived problems or applications. There can be no doubt that much of Chinese mathematics was devised for utilitarian purposes. The earliest of the mathematical classics, the *Zhou bi suan jing* (*Mathematical Classic of the Zhou Gnomon*), is devoted primarily to astronomical and calendrical matters. Similarly, the *Wu cao suan jing* (*Mathematical Classic of the Five Government Departments*) is a compendium of problems concerning feudal administration.

That Chinese mathematics was authoritarian is reflected in traditional approaches to texts in China. Classical texts in all fields were revered and closely studied, and among those in mathematics, none was greater or more influential than the *Jiu zhang suan shu* (*Nine Chapters on Mathematical Procedures*) with its commentary by the third-century mathematician Liu Hui. The conservative nature of Chinese mathematics is also clear from the way in which classic texts were canonized in the Tang Dynasty, and became the core foundation for the teaching of mathematics, which was also the basis of an examination system that depended more upon memorization and the repetition of established knowledge than upon innovation or novel solutions to traditional problems.

But in studying the material collected for this Sourcebook, in this chapter much of it drawn from classical texts of early Chinese mathematics, it is important to ask to what extent are the above generalizations about Chinese mathematics valid, and to what extent do they deserve to be revised? What this chapter on China intends to provide is an overview of some of the highlights of ancient Chinese mathematics, focusing upon examples of historical interest, as well as upon methods and problems that are either especially typical of Chinese mathematics, or that prove to be unusually innovative in their methods and procedures.

I. China: The Historical and Social Context

Mathematics is closely connected with economic development and commerce, and for China it was the famous “Silk Road” that served to connect China, albeit indirectly and over great distances, with the ancient civilizations of the Mediterranean and Near East. The “Silk Road” was a term first coined by the nineteenth-century German geographer Ferdinand von Richthofen. Similar references may have been made even earlier in Greek by the Byzantines of the sixth to twelfth centuries CE, but Richthofen introduced the plural "Seidenstrassen" that soon found its way into French and English. As Korean and Japanese scholars became interested in the commerce and exchange of goods between East and West, a variety of routes were identified stretching from Constantinople to the Korean peninsula and Japanese islands. The English "Silk Road," however, in the singular, is more misleading than von Richthofen's reference to multiple roads or avenues of trade over which goods and ideas were transmitted. There was the most familiar "silk road" of popular imagination across the deserts, namely, the "oasis route" that ended at Xi'an. But there were also a northern steppe route and a southern route by sea, both of which were equally influential for both mercantile and intellectual commerce. But of these, at least in the period between the Han and Tang dynasties, it was the oasis route that was the most active, although few people ever completed the trip along the entire 5000 miles (8000 km) of this so-called silk route. And as silk and spices were moving West toward the Mediterranean, so much glass passed over the silk route from west to east that Japanese and Russian scholars who study trade along it now routinely state that it was a "silk route" to the west, but a "glass road" to the east.

Sasanian Persian glass bowl, fifth-sixth century BCE.

Among the most exquisite examples of the material goods that made their way to China from the West along these silk routes is a Sasanian Persian glass bowl of the fifth or sixth century CE, excavated in 1983 from the tomb of Li Xian (d. 569) and his wife Wu Hui (d. 547 in Guyuan, Ningxia) [Juliano and Lerner 2001, 97].

The transmission of mathematics along these various routes was recently the subject of a conference held at the Rockefeller Foundation Research and Conference Center in Bellagio, Italy [Dold-Samplonius et al. 2002]. Such transmission over the centuries has taken many forms, of texts, problems, methods, and instruments. But what has always been crucial to the vital progress of the sciences, including mathematics, has been dialogue, whether between rivals arguing the pros and cons of different systems, theories, and techniques, or discussion among colleagues, teachers, students, and those who write the textbooks and commentaries that traditionally have been the main means of preserving and advancing knowledge of mathematics in particular. In this chapter of the Sourcebook, the diverse and ingenious applications of mathematics in a variety of areas, including theoretical and even foundational matters related to the rigor and correctness of mathematical procedures, will be investigated, from the earliest evidence yet found of mathematics in China, to the end of the Ming Dynasty, around the beginning of the seventeenth century CE.

The earliest record of mathematics in ancient China dates to the Shang Dynasty (c. sixteenth–eleventh centuries BCE), in the form of inscriptions on bones and tortoise shells. Prior to that, there is evidence of rudimentary geometric designs on pottery and bone utensils. During the Shang Dynasty, improvements in agriculture fostered a division of labor, and archaeologists have excavated circular granaries, bronze weapons, and ceremonial vessels, all evidence of an exchange economy that led to the first production of coins. Agricultural and ceremonial demands soon led to the first calendars in China, all of which would have required early development of Chinese mathematics.

The first writing in China, including the first written numerals, appeared in the late Shang dynasty but was virtually unknown until, "in 1899, an important discovery led to a complete revision of all that was then known about ancient China" [Martzloff 1997, 180]. What archaeologists discovered were thousands of inscribed bones and tortoise undershells or plastrons excavated at Xiaotun, just to the northwest of Anyang in Henan—on the site of the last capital of the Shang Dynasty.

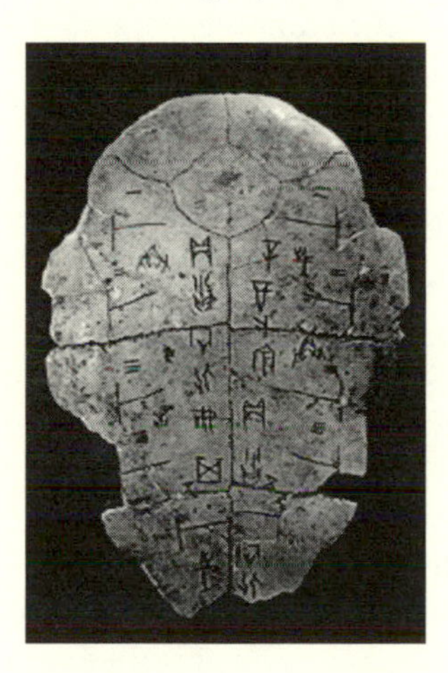

Tortoise plastron with the earliest yet known written Chinese numerals.

New finds over the past century have added tens of thousands more examples to these, and thanks to ongoing and intense study of these materials, their role as divinatory records is now appreciated. Among the nearly 5000 characters represented on the oracle bone inscriptions are the earliest recorded numerals in China. These note, among others, the number of enemy prisoners taken or killed during combat, birds and animals caught while hunting, different animals sacrificed for ritual purposes, and so on.

The numbers found on the oracle bones range from 1 to 30,000, with special characters for 10, 100, 1000, and 10,000. Another ancient script developed for bronze inscriptions mostly dates to the Zhou period (eleventh century–221 BCE). This script too reflects the basically decimal character of Chinese mathematics. By the Han dynasty (206 BCE–220 CE), the characters and notation for writing numbers had become well-established and were virtually the same as the numeration system used in China today. The numbers 1–10 and the powers of 10 up to 10^4 are written as follows:

1	2	3	4	5	6	7	8	9	10
一	二	三	四	五	六	七	八	九	十

10^2	10^3	10^4
百	千	萬

Numbers were written using a decimal system. For example: 5263 was written as 五千二百六十三,, using the following characters for units and powers of ten:

five (五 *wu* "five") thousands (千 *qian* "thousand"),
two (二 *er* "two") hundreds (百 *bai* "hundred"),
six (六 *liu* "six") tens (十 *shi* "ten"), and
three (三 *san* "three"), for which unit or "ones" is understood.

This system made it easy to write numbers where we would today include zeros to indicate that there were no tens or hundreds. For example, the Chinese would write 5006 as follows: 五千六, namely "five thousands and six." This is clearly different from the number 56, which was written: 五十六, "five tens and six."

Given what Jean-Claude Martzloff has called the "dispositional" character of this way of writing numbers [Martzloff 1998, 205], the Chinese had no need of a zero to distinguish positional values, since each value was paired with a specific character denoting its power of ten. This also made it possible to express numbers into the billions, and even larger had there been any need to do so. For example, the largest number found in the early Chinese mathematical classic text, the *Nine Chapters*, appears in Chapter 4: "Short Width," wherein Problem 25 gives the volume of a sphere as 1,644,866,437,500 *chi*^{3}, or in Chinese (with 億 *yi* representing 10^8) as

一萬六千四百四十八億六千六百四十三萬七千五百

(HINT: to read this number correctly, note that there are 16,448 *yi* and 6643 *wan* [10^4], plus 7500).

When computing with such numbers, the various powers of ten represented by 十, 百, 千, 萬, etc., could be easily correlated with a positional system, which is exactly what the counting board provided, namely a flat surface upon which counting rods representing numbers could be placed (more about counting rods below). Individual columns represented the ascending powers of ten from right to left, just as we do today. For example, the numbers above introduced would correspond on the counting board to each of the following configurations:

	10^{12}	10^{11}	10^{10}	10^9	10^8	10^7	10^6	10^5	10^4	10^3	10^2	10	(1)
5263										五	二	六	三,
5006										五			六,
56												五	六,
sphere	一	六	四	四	八	六	六	四	三	七	五		

Chinese numerals and the writing of numbers are intimately linked to the method of computation used in ancient China. Counting rods, representing units, were set out on the surface where position reflected powers of ten. This facilitated in particular computations with fractions and algorithmic methods for extracting roots and solving equations, and solving at its most advanced level systems of linear equations. But more of this below.

Just as the earliest Chinese numbers are associated with divination in the Shang Dynasty oracle bone and turtle shell inscriptions, the divine origins of arithmetic and geometry are repeatedly stressed in the earliest records we have from Chinese history. One common legend is often depicted on Han Dynasty stone reliefs.

女娲 Nü Wa on the left, 伏羲 Fu Xi on the right, from a Han Dynasty tomb at Wu Liang temple.

Here Fu Xi—the first of the "Three Sovereigns"—is shown on the right holding a *ju* or carpenter's square.[1] In some versions of this legend Fu Xi is said to have invented both the carpenter's square and the compass, or *gui*—which is held in the above depiction by his consort Nu Wa (on the left). According to the Chronicles of the famous Chinese historian Sima Qian, the Emperor Yu of Xia (who reigned in the twenty-first century BCE), when attending to floods, carried with him "a plumbline in his left hand and a gnomon and compass in his right" in order to do the surveying required to bring the floods under control [Li and Du 1987, 3].

In addition to the close association of numbers with divination and the early origins of mathematics for practical purposes of land surveying, there were also astronomical uses to which mathematics was put for record-keeping of observations such as comets. Numbers were needed to record the times of their appearance and durations, and mathematics was crucial to the computation of calendars. And while land surveyors were adept at measuring distances over uneven terrains, astronomers were interested in determining the extent of the universe, including the "height of the sun," by means we shall examine shortly.

[1]From a Han burial chamber, illustrated in [Li and Du 1987, 2], although there the illustration as published is upside down. The carpenter's square, also known as the try-square, the framing-square, and by many other names, is an L-shaped tool used in ancient China, as now, to determine right angles and perpendicular lines.

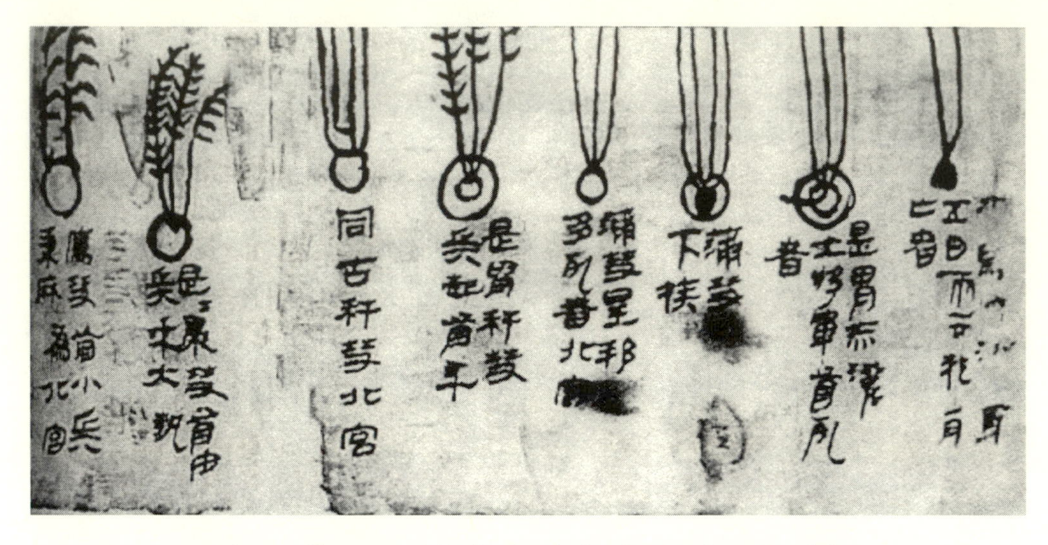

Drawing on silk, 350 BCE. This drawing depicts dozens of comets and shows perceived differences in their heads, cores, and tails.

Not only numbers but geometry was also an integral part of Chinese cosmology and the astronomical models used to chart the position of the heavens, predict astronomical phenomena, and to set the agricultural calendar and various ceremonial rites of the Chinese year depended on geometry and number alike.

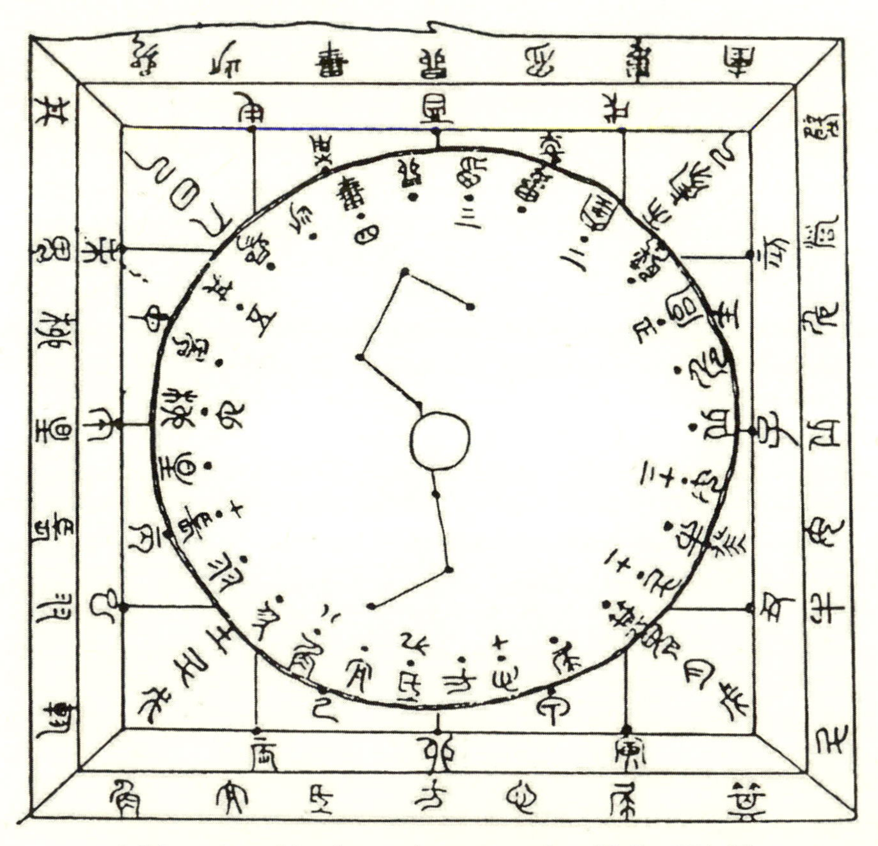

A Shi cosmic model, early second century BCE, from [Cullen 1996, 45].

Mathematics was also crucial in commerce, for the barter economy where exchange of grains by various measures of volume and weight led not only to standardization of weights and measures, but also to strict regulation by the government. Standard measures according to government regulations were manufactured and distributed throughout the country, and any deviation from the standard amounts, especially cheating by local merchants or magistrates, was severely punished. Several such standard measures survive; the one depicted here bears an inscription describing in detail its construction and the measures by volume for which it was the standard.

Standard measure (Bronze), Western Han (Xin) Dynasty (9–23 CE), from *Trésors du Musée national du Palais, Taipei*, a special number of *Connaissance des Arts* 127 (1998), Catalogue of the exhibition at the Grand Palais (Paris), p. 75.

Religion, too, played its part, and the most dramatic foreign influence to affect China was the introduction of Buddhism from India early in the modern era. This had enormous consequences for astronomy and mathematics alike, and in one text in particular it may be reflected in the writing of very large numbers on the order of those addressed in Archimedes' *Sand Reckoner* (see discussion of the *Shu shu ji yi* in section VI).

Who Chinese mathematicians were, and the audiences and purposes for which they wrote the handbooks and treatises that have survived the ravages of war, book burning, and the usual destruction of property over time, are not easy matters to determine. Virtually nothing is known even of major figures like Liu Hui, let alone their students or the circumstances in which they wrote their mathematical treatises or commentaries.

What can be said of the texts translated here is that they served primarily two purposes, as records of mathematical knowledge and improvements made by successive generations of mathematicians, and as didactic handbooks for those who taught mathematics. Indeed, many of the received texts are so terse and laconic that without the aid of a teacher fully conversant with the material in question, the beginning student or nonspecialist would have great difficulty in knowing exactly what the procedures and methods were that the texts prescribe for individual problems and their solutions.

As Zhao Shuang says in his preface to the *Zhou bi suan jing*, in referring to the preservation of various astronomical texts, "Successive ages have preserved them, and the officials have taken them in hand, to enable themselves 'in respectful accord with august Heaven, reverently to grant the seasons to the people.' . . . I feared lest it should be cast aside, [or be thought] hard to penetrate, so that those who discuss the heavens should get nothing from it" [Cullen 1996, 171]. Zhao goes on to describe how he devised diagrams "in accordance with the text" to help make the book clearer. "My sincere hope was to demolish the high walls and reveal the mysteries of the halls and chambers within." His final admonition: "Perhaps in time gentlemen with a taste for wide learning may turn their attention to this work." Thus Zhao indicates his multiple reasons for writing mathematical books: to preserve knowledge, to provide explanations to make the mathematics easier to understand, and to serve as handbooks for anyone interested in availing themselves of their contents on their own.

At various times in Chinese history, a canonical set of mathematical "classic texts" was used as the basis for official teaching, most notably at the Tang dynasty's Imperial Academy, the 国子学 *guozixue* (lit. School for Sons of the State), and as the basis for questions on the official state examination for mathematics. In 656 CE, Li Chunfeng was charged with preparing an official, annotated edition of selected texts, which then became canonized as the 十部算經 *Shi bu suan jing* (Ten Books of Mathematical Classics), most of which as a result thereafter bore as part of their title "suan jing." Although there were originally a dozen in all, some of these were lost in the course of the centuries, and a set of ten classics was first printed in the Northern Song Dynasty in 1084. No copies survive, but this collection was fortunately reprinted in the Southern Song Dynasty in 1213, and a partial version of this collated set of the "Ten Classics" survives in the Shanghai Library. Another great encyclopedic collection of Chinese literature, the early fifteenth-century Ming dynasty compilation of the 永樂大典 *Yongle dadian* (*Great Encyclopedia of the Yongle Reign*, 1403–1407), also printed collations of some of the *Ten Classics*, and the great eighteenth-century imperial collection, under the editorial supervision of Dai Zhen, included editions of each of the *Ten Classics* in the 四庫全書 *Siku quanshu* (Complete Library of the Four Branches of Books). The two most recent and widely used modern collations with commentaries on the *Ten Classics* are those by Qian Baocong, 算经十书 *Suanjing shishu* (The Ten Mathematical Classics, 1963), and Guo Shuchun and Liu Dun, 算经十书 *Suanjing shishu* (1998).

II. Methods and Procedures: Counting Rods, The "Out-In" Principle

Counting Rods

A distinctive feature of Chinese mathematics was the use of counting rods laid out on a flat surface or "counting board" to carry out arithmetic computations. This led to a highly developed set of algorithms that were quick and easy to perform with great accuracy not only for multiplication and division, but also for the computation of square and cube roots. The nature of the counting board and the procedures it naturally suggested also led Chinese mathematicians to introduce the concept of negative numbers and a very flexible means of solving simultaneous linear and higher degree equations in multiple unknowns. The early mathematical classics of the Chinese canon basically assumed a knowledge of how to use the counting board and a general facility with computations using the counting rods. Only two of the ancient classic texts explain in any detail use of the counting rods, namely, the 孫子算經

Sun Zi suan jing (*Mathematical Classic of Master Sun*; also referred to as *Master Sun's Mathematical Manual*), and 夏侯陽算經 *Xiahou Yang suan jing* (*Mathematical Classic of Xiahou Yang*; also referred to as *Xiahou Yang's Mathematical Manual*), from which much of the following account is derived. Examples drawn from the 九章算術 *Jiu zhang suan shu* (*Nine Chapters on Mathematical Procedures*; also referred to as the *Nine Chapters on the Art of Mathematics*) and later classic texts will illustrate how square roots were derived, and how the solution of simultaneous equations were carried out using the counting board.

There were two forms for representing numbers by counting rods, the "vertical form" and the "horizontal" form:

Arabic	1	2	3	4	5	6	7	8	9
Chinese	一	二	三	四	五	六	七	八	九
Vertical	𝍠	𝍡	𝍢	𝍣	𝍤	𝍥	𝍦	𝍧	𝍨
Horizontal	𝍩	𝍪	𝍫	𝍬	𝍭	𝍮	𝍯	𝍰	𝍱

In the case of vertical numbers, the units are simply an alignment of counting rods from 1 to 5; 6 however begins a new series, a single rod placed horizontally stands for the first five, and then successively laying down one vertical rod for 6, another for 7, and so on, up to 9. In Chinese mathematics, there is a standard phrase "five never stands alone," meaning that in the case of the vertical rods, the single horizontal rod 𝍩 that stands for 5 in the number 6 is never used to stand for 5 alone; in the vertical system, the single horizontal rod standing alone always stood for either 10, 1000, etc.; similarly, the single vertical bar 𝍠, which stands for 5 in the horizontal forms for 6, 7, 8, and 9, never stands alone as a horizontal number for 5, but always stands alone as 1, 100, etc.

The ease with which addition could be performed is visually clear from even such a simple example as the sum of 7 and 8: The two horizontal rods across the top of 7 and 8 each represent 5; added together these become 10, represented by the single horizontal bar in the tens column,—, and the remaining 7 units IIIIIII are consolidated with a horizontal rod for 5, leaving—𝍦, which is easily read as 17 after a simple, algorithmic manipulation of the counting rods and without requiring memorization of the fact that 8 + 9 = 17.

On the counting board, numbers would be laid out in columns from right to left, from the lowest to higher powers of ten. The system was positional, alternating vertical and horizontal forms so as not to confuse which power of ten a given numeral represented. For example, the number 529 would be written as 𝍤𝍪𝍨, and 722 as 𝍦𝍪𝍡.

Addition of these two numbers would then have proceeded as follows. Note that unlike our present approach to addition of numbers, ancient Chinese practice began with the numerals in the highest rank of the power of ten on the counting board, and then proceeded from left to right, performing each stage of the addition in a descending order by power of ten:

	529		𝍤	𝍪	𝍨
	<u>+ 722</u>		𝍦	𝍪	𝍡
step 1	12 ··	𝍩	𝍡		
step 2	124 ·	𝍩	𝍡	𝍬	
step 3	1251	𝍩	𝍡	𝍬	𝍠

Subtraction proceeded similarly, beginning with the highest power of ten and proceeding to work toward the right until ending with the final subtraction of the numeral in the unit position. Multiplication used a slightly different layout on the counting board from addition or subtraction, but proceeded in an algorithmic and mechanical way. The two numbers to be multiplied were placed on the counting board with the multiplier on top and the unit of the multiplicand under the highest valued numeral of the multiplier, with an empty line between the two as follows:

529

× 722

Multiplication then began with the largest valued numeral of the multiplier, in this case 5, and the result of multiplying 5 × 722 was then written down in the space between the multiplier and multiplicand, in effect multiplying the multiplicand by 500, and the product now occupies the "hundreds" place on the counting board:

	529
step 1	3610
	× 722

Note that the "zero" is indicated on the counting board by leaving the place holder empty (here ∘ denotes that the place in question is blank, in the above example indicating there are no "hundreds" at this stage of the multiplication). Having multiplied by the 5 in the hundreds place of the multiplier, the 5 was then removed and the multiplicand moved one place to the right, to indicate that the next multiplication would be by the 2 in the tens position. But now, since multiplying by that 2 is effectively multiplying by 20, as each unit of the multiplier is multiplied, beginning with the 7 in the hundreds place, the result is added to the product already obtained from multiplying 5(00) by 722:

	29
step 2a	3610
	× 722

2 × 7 = 14, added to 3610 in the hundreds place above the multiplicand gives:

	29
step 2b	3750
	× 722

2 × 2 = 4, added to 3750 adds 4 in the tens place above the multiplicand:

	29
step 2c	3754
	× 722

	29
step 2d	37544
	× 722

Now finished with the multiplication by 2 in the tens place, the algorithm is again followed to reset the counting board for the last multiplication by the 9 in the units place of the multiplier, moving the multiplicand to the right by one place to indicate multiplication by the unit of the multiplier, 9, which then proceeded as follows:

	9
step 3a	37544
	× 722

9 × 7 = 63, added to 37544 in the hundreds place above the multiplicand gives:

	9
step 3b	38174
	× 722

9 × 2 = 18, added to 38174 in the tens place above the multiplicand:

	9
step 3c	38192
	× 722

9 × 2 = 18, added to 38192 in the units place above the multiplicand:

	9
step 3d	381938
	× 722

Division was handled on the counting board as the inverse operation to multiplication, and could also be used as a check on the correctness of a given multiplication. For example, consider the problem of dividing 381,938 by 529. There was special terminology for the divisor, 法 *fa*, the dividend, 實 *shi*, and the quotient, 商 *shang*. The dividend was placed above the divisor on the counting board. Note that the layout of the counting board is exactly the same as the final position in the solution for the problem of multiplication, as above:

商

實 381938

法 ÷ 529

The first step is to move the divisor to the left, to initiate division of 3819 by 529:

商

step 1 實 381938

法 ÷ 529

Since 5 goes into 38 at most 7 times, place 7 in the hundreds place over the dividend in the row designated 商 *shang* for the quotient, then multiply the 7 (in the hundreds place) by the 5 (in the hundreds place), and subtract 35 (in the 10,000s place) from 381,938, leaving 319(38); then multiply 7 in the hundreds place in the quotient times the 2 in the tens place in the divisor, subtracting 14 in the 1000s place from 319(38), leaving 179(38); finally, multiply the 7 in the hundreds place in the quotient by 9 in the units place in the divisor, subtracting 63 from 179(38) in the hundreds place, leaving 116(38):

商 7

step 1a 實 381938

法 ÷ 529

35

319

14

179

63

116

In preparation for the next step in the division process, move the divisor one place to the right, in order to begin the process of dividing 1163(8) by 529. Since 5 goes into 11 at most 2 times, place 2 in the tens place over the dividend in the row designated 商 *shang* for the quotient. Multiply the 2 (in the tens place) by the 5 (in the hundreds place), and subtract 10 (in the 1000s place) from 11,638, leaving 163(8); then multiply 2 in the tens place in the quotient times the 2 in the tens place in the divisor, subtracting 4 in the hundreds place from 163(8), leaving 123(8); finally, multiply the 2 in the tens place in the quotient by 9 in the units place in the divisor, subtracting 18 from 123(8) in the 10's place, leaving 105(8):

商 72

step 1b 實 11638

法 ÷ 529

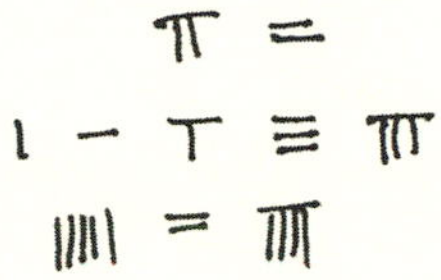

11638

10

163

40

123

18

1058

In preparation for the final step in the division process, move the divisor one place to the right, in order to begin the process of dividing 1058 by 529. Since 5 goes into 10 at most 2 times, place 2 in the unit's place over the dividend in the row designated 商 *shang* for the quotient:

商 722

step 1c 實 1058

法 ÷ 529

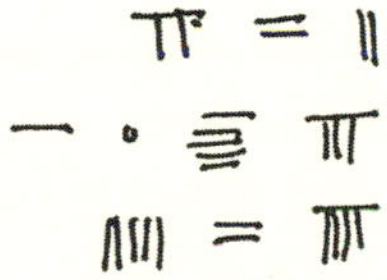

Multiply the 2 (in the units place) by the 5 (in the hundreds place), and subtract 10 (in the hundreds place, i.e., 1000) from 1058, leaving 58; then multiply 2 in the units place in the quotient times the 2 in the tens place in the divisor, subtracting 4 in the tens place from 58, leaving 18; finally, multiply the 2 in the units place in the quotient by 9 in the units place in the divisor, subtracting 18 from 18. Since there is no remainder, the division of 381,938 by 529 leaves a quotient of 722, which confirms the correctness as well of the preceding product of 529 by 722.

Once Chinese mathematicians became adept at the algorithmic nature of the processes of multiplication and division, it was a straightforward matter of carrying out these arithmetic operations quickly, and with great accuracy. The final answers would then have been recorded, using the Chinese written decimal notation.

The "Out-In" Principle

Although Chinese mathematics is often described as being primarily algorithmic, dealing with computations carried out with counting rods in specified sequences of steps according to carefully specified procedures not unlike the steps in a computer program [see Chemla 1987,

1990], one of its most straightforward yet versatile techniques, especially in the context of proof, lies squarely in the domain of geometry. The "out-in" principle has its counterpart in ancient Greek mathematics, where it proved to be an equally powerful if simple means of demonstrating the equivalence of what on sight alone may not seem to be equivalent areas at all. The method and its efficacy are immediately obvious from a very simple diagram. Given a rectangle, under what circumstances may it be said that the areas III and III′ ruled off by intersecting perpendicular lines *AA′* and *BB′* are equal?

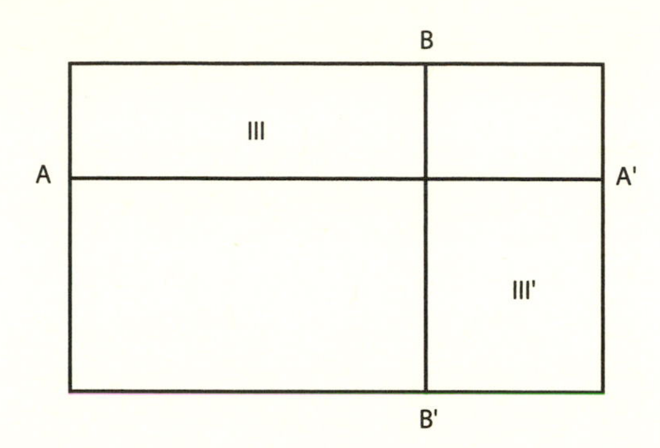

The answer is easy to see once the basic trick of the "out-in" principle is understood, namely, when the two perpendicular lines *AA′* and *BB′* intersect on the diagonal of the rectangle. When this occurs, it follows immediately from one of the most basic assumptions familiar from the axioms of Euclidean geometry (and a self-evident principle just as obvious to Chinese mathematicians as well, even if they did not state it explicitly as an axiom with which they were dealing), namely, if equals are subtracted from equals, then the results are equal.

With reference to the "out-in" rectangle, this may be applied with the following parts of the diagram in mind:

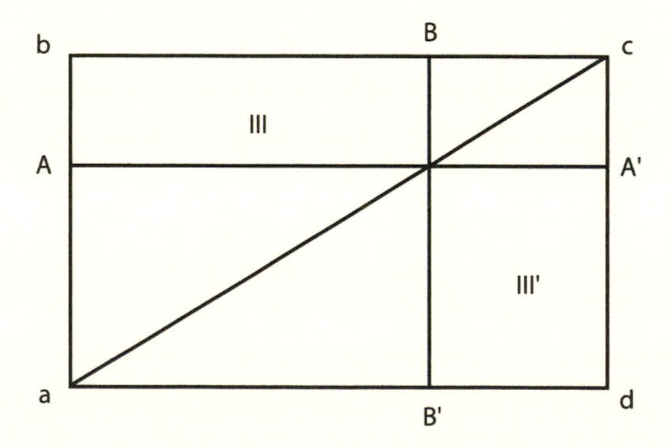

Consider first the diagonal *ac*, which divides the rectangle *abcd* into two equal parts, *abc* and *adc*. Similarly, the two perpendicular lines *AA′* and *BB′* divide the rectangle into two similar rectangles, one in the lower left, the other in the upper right portion of the larger rectangle, both of which are divided by the diagonal into two equal parts, one above, the other below the diagonal:

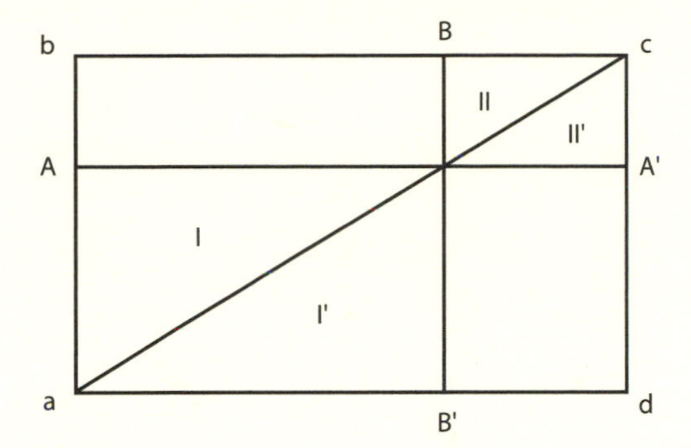

These are labeled I and I′ in the lower rectangle, II and II′ in the upper rectangle; since these two areas I and I′, II and II′, are equal to each other, when subtracted from the two equal areas *abc* and *adc*, the resulting differences are again equal, namely, the two remaining rectangles III and III′ must be equal. This simple yet straightforward "out-in" principle was applied to prove many ingenious results in the Chinese mathematical classics, as will soon become clear from the first appearance of this method in the *Zhou bi suan jing*. The terminology "out-in" comes from regarding the portion of the rectangle below the diagonal as "out," and that above the diagonal as "in." Depending upon the application in mind, the portions of the areas considered "out" have their corresponding areas that are considered "in." For applications of the method, see the discussion below of the *Zhou bi suan jing* and the *Jiu zhang suan shu*.

III. Recent Archaeological Discoveries: The Earliest Yet-Known Bamboo Text

Over a two-month period in December–January 1983–1984, the tomb of a bureaucratic official was unearthed at a Western Han Dynasty site, near Zhangjiashan, in Jiangling county, Hubei Province, China. Found in the tomb were a number of books on bamboo strips, including works on legal statutes, military practice, and medicine. But for historians of mathematics, the real treasure here was a previously unknown book on some 200 bamboo strips, the *Suan shu shu*, or *A Book on Numbers and Computations.*[2] This is the earliest yet discovered work devoted specifically to mathematics from ancient China. It is a significant document not only for the history of mathematics generally, but for the history of Chinese mathematics in

[2]Although when it was first discovered the archaeologists who excavated the *Suan shu shu* referred to it in English as a *Book of Arithmetic*, it is more accurately described as *A Book on Numbers and Computations*. This is clearly the intended meaning of the characters, especially *suan*, which to be more precise refers to calculations using counting rods on a counting board, the traditional means of carrying out arithmetic computations in ancient China. In fact, the *Suan shu shu*, strictly speaking, is not a book on arithmetic; it is certainly not a systematic introduction to the subject of calculation, or even how to carry out calculations, since the methods are assumed known. What the book offers is instead a collection of problems that exemplify various arithmetic applications, operations with fractions, distributions by proportion, calculations of areas of farming fields in a variety of shapes, volumes, and so on.

particular, since it raises naturally important questions of context and its relation to the other mathematical classics that have survived in printed form at least from the Southern Song Dynasty (1127–1279) and in later editions of the earliest mathematical classics. What makes the *Suan shu shu* especially important, however, is that it is not a printed work from a later time, but an ancient work written on bamboo strips, which when tied together originally formed a bamboo roll. The recovered pieces thus constitute an authentic document from the very time at which it was created.

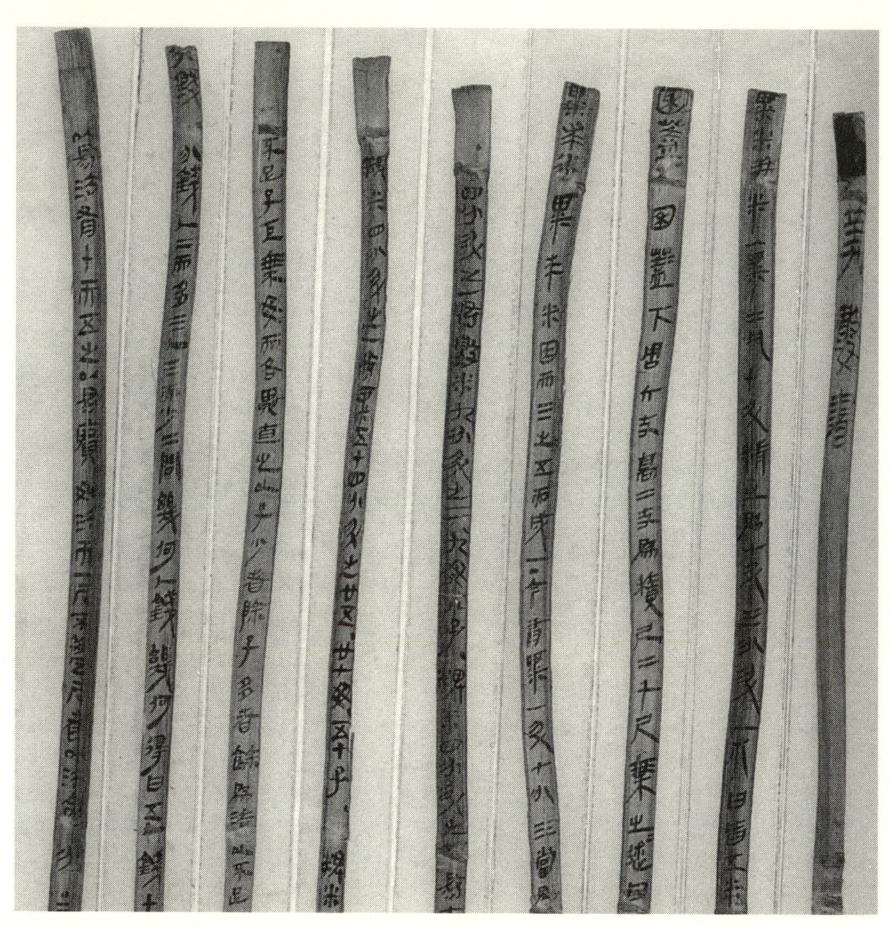

Examples of the bamboo strips on which the *Suan shu shu* are written. The first on the right is the back of slip number 6, on which three characters are written, *Suan shu shu*, taken to be the title of the work itself. Reproduced in [Peng Hao 2001].

Of the 200 bamboo slips forming the *Suan shu shu*, about 180 are in reasonably good or legible condition. The *Suan shu shu* is in the form of a collection of arithmetical problems, 69 of which are computational in nature, along with 92 *suan ti*, or paragraph titles announcing a given problem or method. For the sake of comparison, the famous Chinese mathematical classic, *Jiu zhang suan shu* (*The Nine Chapters on Mathematical Procedures*) in the version collated and annotated by Liu Hui, is more than three centuries later and contains 246 computational problems. One of the questions which naturally arises concerns what may have been the possible connections between the *Suan shu shu* and the *Jiu zhang suan shu*, based upon the two texts now available for comparison.[3] The extent of the differences is sometimes striking. Consider for example, the method described in the *Suan shu shu* (in a section devoted to "Increase and Decrease"):

> To increase or decrease a fraction: to increase a fraction increase its numerator; to decrease a fraction increase its denominator. [Li and Du 1987, 58]

This is of particular interest because, although it is a basic insight into the nature of fractions and how to increase or decrease their value or relative magnitude, nothing like this appears in the *Jiu zhang suan shu*. This suggests at least that whatever the connection between the two works may have been, the *Suan shu shu* did not serve as a direct model for the *Jiu zhang suan shu*. Moreover, there are important matters covered in the *Jiu zhang suan shu* that do not appear at all in the *Suan shu shu*, for example methods for extracting square and cube roots, procedures for solving simultaneous linear equations using matrix methods on the counting board, and virtually all of the famous ninth chapter of the *Nine Chapters* devoted to properties of right triangles and the equally important continuation of the *Nine Chapters* by Liu Hui in what he intended as a tenth section or *juan*. This later became a classic text in its own right, the *Hai dao suan jing* or *Sea Island Mathematical Classic*. However, as we shall see in a moment, there are other problems and methods that do appear in both the *Suan shu shu* and the *Jiu zhang suan shu* that are either the same or so closely similar that they immediately raise natural questions about the possible connections between the two works. But before turning to such matters, it will be helpful to provide a brief description of the general contents of the *Suan shu shu*.

The *Suan shu shu* begins with a section entitled *Xiang cheng* (Self-multiplication), which amounts to a multiplication table for various units and unit fractions (beginning with $1 \times 1 = 1$ and proceeding to $1/5 \times 1/5 = 1/25$, sometimes with units of length, sometimes without), followed by basic rules for multiplying, increasing or decreasing, dividing, simplifying, and adding fractions (addition coming last because it is the most complicated in that it requires the determination of common denominators in the general case before addition of the fractions in question can be performed). This all comprises the first eight sections of the *Suan shu shu*. Then begins a series of problems whose solutions employ the methods just introduced for handling fractions, starting in section nine with a problem, *Jing fen* (dividing fractions), which is described as "the method of finding one person's share"; given a sum of $7 + 1/3 + 1/2$ to be divided among 5 people, what is one person's share? As the problem explains, this requires finding a least common denominator (6), then adding the adjusted numerators $(42 + 2 + 3)$ and dividing the result by 6 and again by 5, so that each person receives 47/30.

Other problems involve subtracting fractional amounts of gold and determining proportional payments or taxes depending upon one's value, status, or wealth. Another series of problems applies the rule of three to determine the value x of an amount b if it is known that an amount a is worth c, that is, $x : b = c : a$, or $x = bc/a$. Another group involves determination of proportional amounts, for example (section twenty), if a given total of grass and straw are mixed in a ratio $3 : 2$, how much of each should be divided between two horses? Other

[3]For English translations of the *Suan shu shu*, see [Cullen 2004], and [Dauben 2006]. The *Jiu zhang suan shu* has been translated into Russian [Berezkina 1957], German [Vogel 1968], English [Shen 1999], and French [Chemla and Guo 2004]. Book Nine on right triangles has also been translated into English [Swetz and Kao 1977].

problems deal with the computation of interest, taxes, and prices under various conditions, and the determination of relative amounts or values of different kinds of grain given specific conversion factors. While most problems in the *Suan shu shu* are solved by direct arithmetic computations, some involve the method of excess and deficiency. For example (section 51), if a certain amount of money is distributed 2 to a person, with a remainder of 3, or if distributed 3 to a person, with a deficit of 2, how much money is distributed to how many people in all?

Beginning with section 53 devoted to *Fang tian* (square farmland), the rest of the *Suan shu shu* is comprised of problems related to calculations of areas and volumes. In fact, the title of this section of the *Suan shu shu*, "Square farmland," is the title of the entire first chapter of the *Nine Chapters.* Section 53 poses the question of how many square *bu* are there in 1 *mu* (equivalent to 240 square *bu*), a problem the *Suan shu shu* resolves by approximating the square root of 240 using the method of excess and deficiency (more on this below). From problems of area the *Suan shu shu* then progresses to determinations of the volumes of various regular three-dimensional figures, including cylinders, haystacks, cones, frustums, and wells. Two complementary problems require determining the largest side of a square that can be cut from a circle of given diameter, and in turn finding the largest circumference of a circle that can be drawn within a square of given side. The *Suan shu shu* ends with four sections devoted to farmland whose area is known; from different widths, the corresponding lengths are to be determined. Also treated is the conversion of equivalent measures of farmland using different units, including *mu*, *qing*, and *li* (as if converting given acreage to equivalent areas measured in square miles or meters).

Suan shu shu *(Book of Numbers and Computations)*

1. Mutual multiplication[4]

1 *cun* times 1 *cun* is 1 (square) *cun*.
[1 *cun*] times 1 *chi* is 1/10 (square) *chi*.
[1 *cun*] times 10 *chi* is 1 (square) *chi*.
[1 *cun*] times 100 *chi* is 10 (square) *chi*.
[1 *cun*] times 1000 *chi* is 100 (square) *chi*.

1/2 *cun* times 1 *chi* is 1/20 (square) *chi*.
1/3 *cun* times 1 *chi* is 1/30 (square) *chi*.
1/4 *cun* times 1 *chi* is 1/40 (square) *chi*.
1/5 *cun* times 1 *chi* is 1/50 (square) *chi*.
1/6 *cun* times 1 *chi* is 1/60 (square) *chi*.
1/7 *cun* times 1 *chi* is 1/70 (square) *chi*.
1/8 *cun* times 1 *chi* is 1/80 (square) *chi*.

1/2 times 1 is 1/2.
[1/2] times 1/2 is 1/4.

[4]The following text, a multiplication table for various units and unit fractions, is reconstructed from the original text in which at least one or more bamboo slips having been misplaced, the multiplications are given in a slightly different order. Throughout this translation, minor scribal errors and mistakes in calculation are corrected without notice; where mistakes represent nontrivial elements of the mathematics or procedures involved, they are duly noted.

1/3 times 1 is 1/3.
[1/3] times 1/2 is 1/6.
[1/3] times 1/3 is 1/9.

1/4 times 1 is 1/4.
[1/4] times 1/2 is 1/8.
[1/4] times 1/3 is 1/12.
[1/4] times 1/4 is 1/16.

1/5 times 1 is 1/5.
[1/5] times 1/2 is 1/10.
[1/5] times 1/3 is 1/15.
[1/5] times 1/4 is 1/20.
[1/5] times 1/5 is 1/25.

3. Further multiplications

1/3 times 1/3 is 1/9.

1/2 *bu* times 1/2 *bu* is 1/4 [square *bu*].
1/2 *bu* times 1/3 *bu* is 1/6 [square *bu*].

1/3 times 2/3 is 2/9.

1/5 times 1/5 is 1/25.
1/4 times 1/4 is 1/16.
1/4 times 1/5 is 1/20.
1/5 times 1/6 is 1/30.

1/7 times 1/7 is 1/49.
1/6 times 1/6 is 1/36.
1/6 times 1/7 is 1/42.
1/7 times 1/8 is 1/56.

1 times 10 is 10.
10 times 10,000 is 100,000.
1000 times 10,000 is 10,000,000.
1 times 100,000 is 100,000.
10 times 100,000 is 1,000,000.
1/2 times 1000 is 500.
1 times 1,000,000 is 1,000,000.
10 times 1,000,000 is 10,000,000.
1/2 times 10,000 is 5,000.
10 times 1000 is 10,000.
100 times 10,000 is 1,000,000.
1/2 times 100 is 50.

5. Fractions to be divided

Any fraction that should be halved, double its denominator; if it should be one-thirded (i.e., divided by 3), multiply its denominator by 3; if it should be divided into

4 parts, multiply its denominator by 4; if it should be divided into 5 parts, multiply the denominator by 5; if it should be divided into 10 or 100 parts, then multiply its denominator by 10 or 100; thus it is possible to divide however one wishes.

6. Dividing fractions

Even if (the fraction should be divided into) an arbitrarily large number of parts, use this (method) presented (above).

7. Simplifying fractions

The rule for simplifying fractions says: Take the numerator and subtract it (successively) from the denominator; also take the denominator and subtract it (successively) from the numerator; (when) the amounts of the numerator and denominator are equal, this will simplify it (the fraction will be simplified). Another rule for simplifying fractions says: if it can be halved, halve it; if it can be (successively) divided by a certain number, divide by it. Yet another rule says: Using the numerator of the fraction, subtract it from the denominator; using the remainder as denominator, subtract it (successively) from the numerator; use what is equal to (both) numerator and denominator as the divisor; then it is possible to divide both the numerator and denominator by this number. If it is not possible (lit. if there is not enough) to subtract but it can be halved, halve the denominator and also halve the numerator. 162/2016, simplified, is 9/112.[5]

This method is equivalent to the well-known Euclidean algorithm, which proceeds by repeatedly subtracting the numerator from the denominator until the amount that remains is too small for any further subtraction; this remainder is then subtracted repeatedly from the numerator. For example, in reducing the fraction 162/2016, we first calculate that $2016 = 12 \times 162 + 72$, then that $162 = 2 \times 72 + 18$. Since 18 can now be subtracted 3 times from 72 leaving a remainder of 18, it is what Chinese mathematicians called the "equal number"; that is, it is the greatest common number that divides both numerator and denominator, and dividing both numerator and denominator by 18 gives the simplified result, $162/2016 = 9/112$.

8. Adding fractions

The rule for adding fractions says: if the denominators are of the same kind (equal), add the numerators together; if the denominators are not of the same kind, but (some) can be doubled (to make the denominators equal), then double (them); if (some) can be tripled (to make the denominators equal), then triple (them); if (some) can be quadrupled (to make the denominators equal), then quadruple (them); if (some) can be quintupled (to make the denominators equal), then quintuple (them); if (some) can be sextupled (to make the denominators equal), then sextuple (them); likewise, the numerator should be doubled, [so] double it; When multiplied by 3, 4, or 5 times like the denominators, and if the denominators are the same amount, (then) add the numerators together. If they (the denominators) are (still) not of the same kind, then mutually multiply (all of) the denominators together as the divisor,

[5]Compare the method here to that of problem 5 of Chapter 1 of the *Nine Chapters* as well as to the commentary there.

and (after) cross-multiplying the numerators with the denominators, add them together as the dividend; and then divide.

Here the *Suan shu shu* introduces a technical expression, *hu cheng*, for cross-multiplication of numerators *zi* by the denominators *mu* of the other fractions to be added together, which served to "homogenize" the numerators, while multiplying the denominators together yielded a common denominator, which was called "equalizing" the fractions. The term for cross-multiplying numerators and denominators, *zi hu cheng mu*, occurs here in the *Suan shu shu*, but the technical terms for equalizing and homogenizing fractions, making it possible to add them together, were later innovations that are reflected in the language and terminology of the *Nine Chapters*.

Now there are 2/5, 3/6, 8/10, 7/12, and 2/3; how much is there (when added together)? (The answer) says: 2 57/60 *qian*; The rule follows (the one) above.

5 people divide 7 and 1/3 and 1/2 *qian*;[6] each one should receive 1 17/30 *qian*. The method says: below (on the counting board) there are three parts (7, 1/3, and 1/2); taking 1 as 6, then multiply 6 times the number of people as the divisor; also take 6 times the amount of *qian* as the dividend. (Divide the dividend by the divisor to give the amount each of the five people should receive.)

This problem would have been solved on the counting board, first putting down the amount of money, 7, 1/3, and 1/2; the least common denominator would have been determined (6) to "equalize" the amounts, multiplied by the number of people among whom the money is to be divided to give the divisor (30); then the numerators of the fractions would have been "homogenized" by multiplying each of the three parts of the *qian* by 6 and adding the results (42 + 2 + 3) as the dividend (47). 47/30 = 1 17/30 *qian* then yields the final answer.

Another rule says: multiply the denominators together as the divisor, cross-multiply the numerators with the denominators as the dividend. Divide the dividend by the divisor.

This one is (another) rule: if (some) need to be multiplied by 10, multiply by 10; if (some) need to be multiplied by 9, multiply by 9; if (some) need to be multiplied by 8, multiply by 8; if (some) need to be multiplied by 7, multiply by 7; if (some) need to be multiplied by 6, multiply by 6; if (some) need to be multiplied by 5, multiply by 5; if (some) need to be multiplied by 4, multiply by 4; if (some) need to be multiplied by 3, multiply by 3; if (some) need to be doubled, double. Stop when the denominators are all the same. When the denominators are all the same, add the numerators together.

11. Buying wood together

3 people (buy) wood together. One of them pays the merchant 5 *qian*, another pays 3 *qian*, and the other pays 2 *qian*. Now there are 4 *qian* left, and (they) want to divide proportionally (*cui fen*) the remaining amount of money (among themselves). The one who paid 5 (*chu wu zhe*) should get 2 *qian*; the one who paid 3 should get 1 1/5 *qian*; and the one who paid 2 should get 4/5 *qian*. The rule says: Add the amounts

[6]The term *qian* refers to the ancient coin of the realm; for a discussion of money and economy in ancient China, see [Loewe 1974].

in *qian* (each) of the 3 people paid as the divisor; then take the 4 *qian* (left over) and multiply by the amount in *qian* each paid; dividing gives the amount of *qian* (each) one should receive.

12. Fox goes through customs

A fox, raccoon, and hound go through customs, and (together) pay tax of 111 *qian*. The hound says to the raccoon, and the raccoon says to the fox: since your fur is worth twice as much as mine, then the tax you pay should be twice as much! How much should each one pay? The result says: the hound pays 15 6/7 *qian*, the raccoon pays 31 5/7 *qian*, and the fox pays 63 and 3/7 *qian*. The method says: let each one double the other; adding them together $(1 + 2 + 4)$, 7 is the divisor; taking the tax, multiplying by each (share) is the dividend; dividing the dividend by the divisor gives each one's (share).

14. Woman weaving

There is a woman in the neighborhood who is displeased with herself (i.e., with her weaving), but happy that every day she doubles her weaving. In five days (she) weaves five *chi*. How much does she weave in the first day, and how much in every day thereafter? The answer: the first day she weaves 1 38/62 *cun*; then 3 and 14/62 *cun*; then 6 and 28/62 *cun*; then 12 56/62 *cun*; then 25 and 50/62 *cun*. The method says: Put down the values 2, 4, 8, 16, 32; add these together as the divisor; Taking the 5 *chi*, multiply this by each of them (2, 4, 8, 16, 32) as the dividend; dividing the dividend by the divisor gives the (amount of) *chi*. If (the amount in) *chi* is not even, multiply by 10 and express the remainder in *cun*. If (the amount in) *cun* is not even, use the remainder to determine (the fractional amount left over in) *fen*.[7] Wang already checked (this).

24. Pieces of silk

A piece of silk 22 *cun* wide and 10 *cun* long is worth 23 *qian*. Now if one wants to buy a piece 3 *cun* wide and 60 *cun* long, the question is how many *cun* (in length from the original bolt of silk 22 *cun* wide) is this and how many *qian* is it all worth? (The answer) says: 8 2/11 *cun*, (which is) worth 18 9/11 *qian*. The method: taking 22 *cun* as the divisor, and multiplying the length and width together (of the amount of silk one wants to buy) as the dividend, the dividend divided by the divisor gives the amount in *cun*. Also taking the amount of *cun* in 1 *chi* (multiplied by the width of the original piece of silk) as the divisor, and taking the product of the amount (of silk purchased) in *cun* times the amount in *qian* that 1 *chi* is worth as the dividend, dividing the dividend by the divisor gives the amount in *qian*.

The method for the second half of the problem appears to be a straightforward application of the rule of three, but it unnecessarily introduces a unit of *chi* that appears nowhere in the original problem. Moreover, this complication leads to a missing element from the method, namely, the width of the original bolt of silk.

---

[7]Compare this problem to problem 4 of Chapter 3 of the *Nine Chapters* (see below, p. 246). In particular, note the identity of the problems but the slight difference in the solution method.

25. Amounts of interest

100 *qian* are lent with interest for 1 (lunar) month at 3 [*qian*]. If 60 *qian* are now lent and repaid before the end of the month in 16 days, how much is the interest? The answer says: 24/25 *qian*. The method says: calculate the product of 100 *qian* and 1 month (30 days), the product of the *qian* and the number (of days) serves as the divisor (100 × 30); put down (on the counting board) the money loaned multiplied by the amount of interest on 100 *qian* for 1 month (60 × 3); and also multiplying by the number of days (16) as the dividend; dividing the dividend by the divisor gives the interest in *qian*.

The Chinese historian of mathematics and expert on the *Nine Chapters*, Guo Shuchun, in comparing this problem to problem 20 of Chapter 3 of the *Nine Chapters*, notes that here, "the instructions are purely general, referring only to 'multiplying the number of coins lent by the number of days'." Guo Shuchun contrasts the similarities and differences between these two problems as follows: "[In the problem from the *Nine Chapters*], concrete data such as 'take one month as 30 days to multiply 1000 coins,' 'the interest 30' and '9 days' respectively correspond to 'the product of one month and 100 coins,' 'the monthly interest of 100 coins' and 'the number of days in the method of the *Suan shu shu*.' Evidently the method given in the *Suan shu shu* is more abstract, and acts as a universal formula for interest problems; while the one in the *Nine Chapters* is a specific algorithm for the given question" [Guo Shuchun 2000, 37].

42. [Given] Millet, find [the equivalent in value of] husked millet

(Given) millet, to find (the equivalent in value of) husked millet, multiply by 3 and divide by 5. Now there is 1 3/7 *sheng* of millet; what should be the equivalent amount (in value) of husked millet? (The answer) says: the equivalent in husked millet is 6/7 *sheng*. The method says: multiply each of the denominators together as the divisor, take 3 times 10 as the dividend.

43. [Given] Husked millet, find [the equivalent in value of] millet

Taking husked millet, to find (the equivalent in value of) millet, multiply by 5 and divide by 3. Now there is 6/7 *sheng* of husked millet, how much is the equivalent amount (in value) of millet? The answer: the equivalent (in value) of millet is 1 3/7 *sheng*. The method: multiply each of the denominators together as the divisor, take 5 times (6 as the dividend).

48. Feathering arrows

The norm: one person in one day makes 30 arrows or feathers 20 arrows. If one now wishes to have one person both make arrows and feather them, in 1 day how many can be made? The answer: 12. The method: combine the arrows and feathering as divisor; taking the arrows, mutually multiply with the feathering as the dividend.

51. Dividing coins

[When] dividing coins, if each person gets 2 coins, then there are three coins too many; if each person gets 3 coins, then there are 2 coins too few. How many people and how many coins are there? The answer says: 5 people and 13 coins. Cross-multiply the excess and deficiency by the denominators (and combine the products) as the dividend, add the numerators together as the divisor; if there is always an excess or likewise a deficiency, cross-multiply the numerators and denominators and put each down (on the counting board) separately; subtract the smaller numerator from the larger numerator; the remainder is the divisor; use the deficiency as the dividend.[8]

53. Square fields

(Given) a field of 1 *mu*, how many (square) *bu* are there? (The answer) says: 15 15/31 (square) *bu*. The method says: a square 15 *bu* (on each side) is deficient by 15 (square) *bu*; a square of 16 *bu* (on each side) is in excess by 16 (square) *bu*. (The method) says: combine the excess and deficiency as the divisor; (taking) the deficiency numerator multiplied by the excess denominator and the excess numerator times the deficiency denominator, combine them as the dividend. Repeat this, as in the "method of finding the width."[9]

This problem is another indication of how much earlier the *Suan shu shu* is in the historical evolution of Chinese mathematics than the *Nine Chapters*. In the former, the square root is approximated by the "excess and deficiency" method; in the *Nine Chapters*, the method of finding square roots has been highly developed into an algorithm with its own technical designation as the *Kai fang shu* or "Method of Finding the Square Root" (lit. "Method of Opening the Square" [Qian Baocong 1963, vol. 1, 150]. This latter method allows determination of the square root by an iterative procedure based upon successive approximations by completing squares. For a discussion of the method, see [Lam Lay Yong 1980, 413–15], and [Lam Lay Yong 1994, 44–46]. The method as given in the *Nine Chapters* is given below, in section V.

56. [Volume of a] hayloft

(The dimensions of a) *chu tong* as well as a *fang que*: the lower width is 1 *zhang* 5 *chi* (15 *chi*), the (lower) length is 3 *zhang* (30 *chi*), the upper width is 2 *zhang* (20 *chi*), the (upper) length is 4 *zhang* (40 *chi*), the height is 1 *zhang* 5 *chi* (15 *chi*), and the volume is 9250 (cubic) *chi*. The method says: multiply together the upper width and length and the lower width and length, and again (take the) upper length added to the lower length, multiplied by the upper width, and (take the) lower length added to the upper length multiplied by the lower width, add them all together, multiplying by the height, divide by 6.

Chu by itself refers to hay; a *chu tong* (hayloft) and *fang que* (square watchtower) are both geometric solids, and *chu tong* is specifically the subject of Problem 19 in Chapter 5 (Construction Consultations) of the *Nine Chapters*. The dimensions there are similar but not identical.

---

[8]An entire chapter of the *Nine Chapters* is devoted to "excess and deficit" problems (Chapter 7). The problem here is solved by the same method given there.

[9]The appropriate method is actually found in problem 65, "Finding the length," rather than in problem 64, "Finding the width." According to the method described here, using the method of excess and deficiency to solve this problem leads to the following computation: $(15 \times 16 + 16 \times 15)/(16 + 15) = 480/31 = 15\ 15/31$ *bu*.

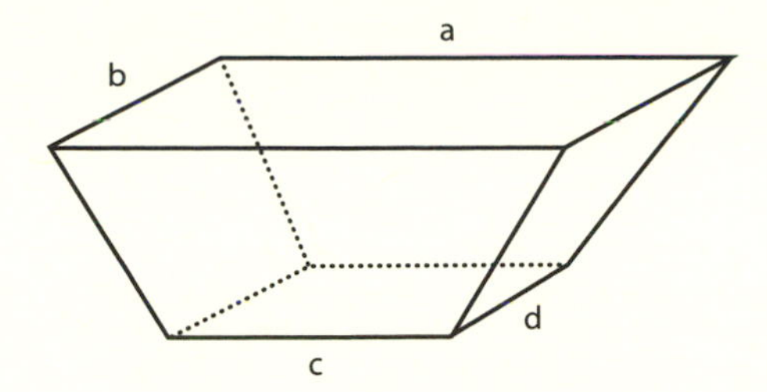

According to the method above, the *Suan shu shu* gives the following formula for the volume of the *chu*: $(ab + cd + (a + c)b + (c + a)d)(h)(1/6) = (800 + 450 + 70 \times 20 + 70 \times 15)(15)/6 = 9250$ cubic *chi*. Note that this is not the same as the method to be found for the same figure in the *Nine Chapters*, which calls for the equivalent computation: $((2a + c)b + (2c + a)d)(h)(1/6)$.

57. [Volume of a] Cone of millet

A cone of millet is 5 *chi* high, the bottom circumference is 3 *zhang*, the volume is 125 (cubic) *chi*. 2 *chi* 7 *cun* is one *dan*, equivalent to 46 8/27 *dan* (of millet). Its method says: multiply the lower circumference by itself, multiply by the height, divide by 36. The volume is 4500 (cubic) *chi*.[10]

64. Finding the width

If the length of a field is 30 *bu*, how wide is it if it is a field 1 *mu* (in area)? (The answer) says: the width is 8 *bu*. The method says: use 30 *bu* as the divisor, and use 240 (square) *bu* as the dividend. Finding the length may also be done this way.

66. Diminishing widths

The method relying upon diminishing widths says: first put (down on the counting board) the width, that is to say: in the denominator's place there is a certain number of *bu*; letting one stand for this number, letting one-half stand for (1/2) this number, letting 1/3 stand for (1/3) this number, etc.; combine the parts taking all of them relying upon the parts combined as the divisor; now combining the field of 240 (square) *bu* put (down on the counting board), and also letting 1 stand for the certain number (of *bu*; i.e. multiply by this number); taking the accumulated *bu*, divide by the accumulated *bu*, dividing gives the length in *bu*. If it is not an exact (whole number of) *bu*, use the divisor to determine its fraction (lit. parts). (The method) goes on to say, to check this (lit. going back), then multiply the width by the length, and check that it is a field of 240 (square) *bu* or 1 *mu*. If its length has no fractional part, the combined divisor no parts, to check (the result) multiply it by what stands for "little ten."

66a. Smaller width: If the width is 1 and one-half *bu*, taking 1 as 2 and one-half as 1, combine them taking 3 as the divisor; then put (down on the counting board) 240 (square) *bu*, and again taking 1 as 2, multiply (both together as the dividend); dividing gives the length in *bu*, which is a length of 160 *bu*. As a result of multiplying (this) by 1 *bu* and one-half *bu*, (there is a field of 1 *mu*).

66b. Next there is 1/3; taking 1 as 6, one-half as 3, and 1/3 as 2, combine them taking 11 (as the divisor; then put down on the counting board the field of 240 square *bu*; again taking 1 as 6 multiply both together as the dividend; dividing) gives a length

---

[10]Compare this problem to problem 13 of Chapter 5 of the *Nine Chapters*. Note that the last line here reflects a lapse in the mathematical calculation, which only follows the computation of the volume of the cone as far as $(30)^2(5) = 4500$, but omits the division by 36 to give the volume of the cone.

of 130 10/11 *bu*. (As a result, taking 1 *bu*, one-half *bu*, and 1/3 *bu*), multiplying (by 130 10/11 *bu*) gives a field of 1 *mu*.

The text continues to consider additions of unit fractions of increasing denominator, finally considering the case for which the given area of 1 *mu* or 240 square *bu* is divided by 1 + 1/2 + 1/3 + 1/4 + 1/5 + 1/6 + 1/7 + 1/8 + 1/9 + 1/10:

66h. Next there is 1/10; taking 1 as 2520, one-half as 1260, 1/3 as 840, and 1/4 as 630, 1/5 as 504, 1/6 as 420, 1/7 as 360, 1/8 as 315, 1/9 as 280, and 1/10 as 252, combine them taking 7381 as the divisor; (then put down on the counting board the field of 240 square *bu*; again taking 1 as 2520, multiply both together as the dividend; dividing) gives a length of 81 6[939]/7381 bu (As a result, taking 1 *bu*, one-half *bu*, 1/3 *bu*, 1/4 *bu*, 1/5 *bu*, 1/6 *bu*, 1/7 *bu*, 1/8 *bu*, 1/9 *bu*, and 1/10 *bu*), multiplying (by 81 6939/7381 *bu*) gives a field of 1 *mu*.[11]

From the preceding excerpts it is apparent that the *Suan shu shu* goes beyond the statements of specific problems and related step-by-step recipes, and offers certain entirely general principles that transcend individual problems and address instead forms and appropriate methods for dealing with broad categories of problems. This is certainly true of the bamboo slip cited earlier from the *Suan shu shu*, which states a very general principle about how to increase or decrease the value of a fraction as well as in problem 25. The real point here is that both the *Suan shu shu* and the *Jiu zhang suan shu* are often quite general in this regard, and indeed, now that we have the entire text (or as much of it as is legible) of the *Suan shu shu*, I believe the level of abstractness and generality achieved in *A Book on Numbers and Computations* can be readily appreciated.

What is also clear is that the relation of the *Suan shu shu* to the *Nine Chapters* is more complex than the latter simply being a linear descendant of the former. At best, the *Suan shu shu* may be one of several earlier examples of mathematical works used by Zhang Cang who originally organized the *Nine Chapters* into a version now lost but which served as the basis for Liu Hui's edition and commentary on the *Nine Chapters* in 263 CE.

Nevertheless, there are some remarkably direct connections between the *Suan shu shu* and the *Nine Chapters*. To begin with, of the 60-odd problems in the *Suan shu shu*, many include technical terms that also appear in the *Nine Chapters*, and in five cases the problem titles in the *Suan shu shu* are exactly the same as the titles for entire chapters of the *Nine Chapters*, namely:

- Field measurement (areas)
- Millet and rice (proportional exchange rates)
- Distribution by proportion
- Short width (division problems)
- Excess and deficiency

What the *Suan shu shu* does not include are problems dealing with square roots, cube roots, *fang cheng* (rectangular array, i.e., matrix methods), or *gou-gu* (right-angled triangles), which represent important advances in mathematics made between the writing of the *Suan shu shu* and the *Nine Chapters*. A full discussion of the similarities and differences of these two

[11]Compare this to problem 11 of Chapter 4 of the *Nine Chapters*.

important works of early Chinese mathematics awaits further study, but for now, one last point should be made about one feature that typifies both works.

Indeed, there is a fundamental methodological similarity that we can now say was not the innovation of Liu Hui or Zhang Cang, but was a characteristic of early Chinese mathematics that links it to Chinese philosophy in an intrinsic way. According to the *Yi jing*, the *Book of Changes*, "all the world falls into its own category." And as Guo Shuchun has recently emphasized, in referring to both the *Suan shu shu* and the *Nine Chapters*,

> To discuss the general properties and definitions of numbers and figures, which is the abstract mathematical theory, is an important aspect of mathematical researches in the times of the Warring States. [Guo Shuchun 2000, 38]

As Chen Zi says to Rong Fang, two interlocutors in the early astronomical Chinese text, the *Zhou bi suan jing*, possibly written sometime midway between compilation of the *Suan shu shu* and the *Nine Chapters* (see Section IV):

> Now among the methods [which are included in] the [mathematical] Way, it is those which are concisely worded but of broad application which are the most illuminating of the categories of understanding. If one asks about one category, and applies [this knowledge] to a myriad affairs, one is said to know the [mathematical] Way.

This is all explained again in a single sentence:

> Therefore, it is the ability to distinguish categories in order to unite categories which is the substance of how the worthy one's scholarly patrimony is pure, and of how he applies himself to the practice of understanding.

In conclusion, then, what the *Suan shu shu* reflects is a very general principle embodied both in Chinese philosophical thought and in mathematics as it was actually applied in practice—namely that classification is indispensable for the sage to learn and master knowledge. The official who either compiled or made use of the *Suan shu shu* was clearly interested in having at the ready a handbook of problem types and basic methods for solving problems of day-to-day mercantile or government administration.

The current version of *A Book on Numbers and Computations* may require considerable reordering of the text and preliminary rearrangement of its various parts before a more satisfactory reading of its elements and overall intent is possible. But as the earliest yet known work on mathematics from ancient China, it reflects the emphasis placed upon the practical significance of mathematics for agriculture, taxation, and commerce. But in addition to being of benefit to the empire, mathematics was also an art, and the *Suan shu shu* also reflects an early stage in the mastery of mathematical knowledge that later works presented here, like the *Nine Chapters*, would go on to surpass, sometimes in remarkable ways.

IV. Mathematics and Astronomy: The *Zhou bi suan jing* (*Mathematical Classic of the Zhou Gnomon*) and Right Triangles (The *Gou-gu* or "Pythagorean" Theorem)

The earliest of the mathematical classics, the *Zhou bi suan jing*, was most probably compiled no later than the first century BCE. It is also known variously (depending upon different transliterations), as the *Chou pi suan ching* (Wade Giles), or *Chou pei suan ching* (Needham).

Among the first to write about this work in English, Mikami Yoshio does not try to translate the title, but refers to it throughout his book, *The Development of Mathematics in China and Japan*, as the *Chou-pei*, having pointed out that *Suan ching* means "a mathematical classic" or "a sacred book of arithmetic" [Mikami 1913, 4]. Joseph Needham, in one of the early volumes of his celebrated series, *Science and Civilisation in China*, translates the title as "The Arithmetical Classic of the Gnomon and the Circular Paths of Heaven" [Needham 1959, 19]. This is based on a liberal understanding of the character *Zhou*, which is usually taken to refer in the title not to the orbits of the planets but to the Zhou dynasty, and the *bi* to the gnomon whose shadow is the basis for determining the distance between earth and sun, one of the first mathematical calculations to be found in the *Zhou bi*. Thus Christopher Cullen suggests rendering the title as "The Gnomon of the Zhou [dynasty]" [Cullen 1996, xi]. In what follows, the text is referred to simply as the *Zhou bi* or *Mathematical Classic of the Zhou Gnomon*.

The author of this work is unknown, and it is likely that the version that survives is a compilation of texts written at various times before the first century BCE. Primarily devoted to calendrical theories, it is a careful record of the *gai tian* theory which took the heavens to be a hemispherical dome suspended over a flat, presumably square earth. Written in two parts, the first part of the book opens with a preface attributed to Zhao Shuang, a third-century CE commentator on the work. This sets the stage for what follows, and very briefly praises the importance of astronomy and mentions two theories of the cosmos, the *hun tian* theory, and the *gai tian* theory. The former is attributed to a work by Zhang Heng (78–139 CE), the *Ling xian*, in which the earth is taken to be spherical and surrounded by concentric spheres of heaven. Needham explains that according to this theory, "Hun Thien (the Celestial Sphere) corresponded to the conception of spherical motions centred on the earth which had been developing slowly among the Greek pre-Socratics, and came to be associated particularly with Eudoxus of Cnidus" [Needham 1959, 216]. The *gai tian* theory is that espoused in the *Zhou bi*. The figures Xi and He referred to in Zhao's preface were legendary figures (as Cullen describes them, two "archetypal official astronomers") whom the equally legendary Emperor Yao of remote antiquity called upon "to bestow the seasons on the people" [Cullen 1996, 3, 43].

The opening preface is followed by a famous dialogue attributed to one Shang Gao, about whom nothing is known, and there is considerable debate about both who may have written this part of the text, and when it was most likely composed. The dialogue is between the Duke of Zhou, a historical figure who dates back to the eleventh-century BCE kingdom of Zhou, and the aforementioned Shang Gao, who is described as a skilled mathematician. As the dialogue opens, the Duke of Zhou mentions Bao Xi, a mythological figure to whom is ascribed the origins of human culture and invention of the first eight of the famous trigrams of the *Yi jing*. We have already met this Bao Xi, alternatively (and better) known as Fu Xi, who is depicted in section I, where he is shown with the carpenter's square, which he is also said to have invented. This of course is especially apt with respect to the *Zhou bi*, in which the gnomon plays an essential role in the geometry of proportionality upon which many of the mathematical calculations in the text depend.

As the dialogue progresses, the Duke of Zhou and Shang Gao speak of the mythological origins of mathematics and emphasize in particular the feats of Fu Xi, who used mathematics adroitly to measure the heavens and regulate the calendar. Part II of the *Zhou bi* is devoted to the latter, whereas Book I is devoted to explaining the mathematics that is needed to make sense of the astronomy in Book II. As such, it is the former that primarily interests us here.

Using the shadow cast by the sun at midday, the *Zhou bi* explains how it is possible to determine the distance from earth to the sun by a straightforward calculation invoking the proportional lengths of gnomons, their shadows, and applications of the "rule of three." Whereas the modern reader will immediately see these as results that follow from the similarity of triangles, it is remarkable that in ancient Chinese there was no word (and apparently no mathematical concept) of triangle per se [see Raphals 2002]. Instead, Chinese geometry speaks of the *gou* (or base/shadow), the *gu* (or height/gnomon), and *xian* (the hypotenuse that joins the extreme edge of the *gou* and *gu*). This dialogue also includes the first statement in Chinese of what in the West is known as the "Pythagorean theorem," along with various additional comments that have been the subject of considerable debate among historians of mathematics over whether these constitute proofs of the *gou-gu* (Pythagorean) theorem or not. We shall have more to say about this below.

Given that the applications of all the mathematics presented in the *Zhou bi* can be used to help measure the heavens, the impressive results recounted in Shang Gao's dialogue are such that the prince, Zhou Gong, is prompted to exclaim: "Ah! Mighty is the art of numbers!" Indeed, as Shang Gao is prompted to say just a few paragraphs later, "The lineal together with the number rules and guides everything in the universe as is desired" [Mikami 1910, 5, 7]. To give some idea of the major differences that can occur between authors in translating the same text, Cullen renders these same passages as follows: "How grandly you have spoken of the numbers!" and "Through its relation to numbers, what the trysquare does is simply to settle and regulate everything there is" [Cullen 1996, 174].

As for the applications of the "Pythagorean theorem" in the *Zhou bi*, these led the historian of mathematics Mikami Yoshio to conjecture that Pythagoras may indeed have learned his famous theorem from the Chinese: "is he not said to have traveled in the east to Babylon and perhaps further on probably to India? Will it not be probable that he had seen the theorem that was destined to be connected with his name on his travel? Might he not have encountered it brought from China in some unknown way? The lapse of time between him and Chou-Kong (Zhou Gong) justly makes the matter an open question" [Mikami 1910, 7]. More blatantly, this idea served to provoke the same possibility in the title of a translation of Chapter 9 of the *Jiu zhang suan shu* published in 1977: *Was Pythagoras Chinese? An Examination of Right Triangle Theory in Ancient China* [Swetz and Kao 1977]. Although the translators insist their title was meant only half-seriously, more to provoke readers' interest than to argue that Pythagoras had actually heard of earlier results in China, the fact remains that from a very early date Chinese mathematicians seem to have appreciated the special properties of right triangles. Additionally, the dialogue goes on to refer to gnomons, the circle and square, and methods of measuring heights and distances that cannot be measured by ordinary means.

A second dialogue then follows between two new figures, Chen Zi and Rong Fang, about whom again nothing historical is known. It is here that, by considering the sun's shadow and estimating its length at different latitudes, the sun's diameter is measured by means of a sighting tube. This section originally included a diagram subsequently lost but later reconstructed as the *Ri gao tu* (Diagram of the Sun's Height). This is basically a base-height (shadow-gnomon) diagram, from which it is a fairly straightforward matter to see how the known lengths of the shadows and distances between gnomons can be used to determine the height of the sun, or its distance (the diagonal "hypotenuse" length) from a given observer. The rest of the *Zhou bi* then turns almost exclusively to astronomical matters.

As for the mathematics of the *Zhou bi*, it is primarily of interest here for its application of the gnomon-shadow relations and the results that follow from the relations that can be derived from the out-in configuration. The *Zhou bi* also makes use of fractions, their multiplication and division, and the finding of common denominators (all of which is covered in the *Suan shu shu*; see section III). Square roots are also taken (although the method is not explained) as a means of applying the *gou-gu* (Pythagorean) relation to solving various forms of the familiar relations $x^2 + y^2 = z^2$ for right-angled triangles (although again, the term "triangle" does not appear in the Chinese texts). An approximation appears for the square root of 5, to the nearest integer "and a bit" [Needham 1959, 22].

In reading the following translated parts of the *Zhou bi*, the reader should keep in mind that here it is assumed that the value π for the ratio of the circumference to the diameter of the circle is 3, and that ancient Chinese astronomers calculated the length of the year to be 365¼ days. Therefore, they divided the circle into 365¼ degrees, so that in one day the sun is taken to move exactly one Chinese degree in the heavens.

Zhou bi suan jing (Mathematical Classic of the Zhou Gnomon)

The Preface of Zhao Shuang

Amongst the high and the great, nothing is greater than heaven, Amongst the deep and the broad, nothing is broader than earth. Their bodies are immense and wide; their shapes extend upwards and outwards through mysterious clarity. One may examine orbital movements back and forth by means of the arcane counterparts (the heavenly bodies) but one cannot grasp that vast extent of space directly. One may check on length and shortness by means of the shadow instrument, but the hugeness of things cannot be measured out in graduations.

Even if one exhausts one's spirit in the effort to understand the transformations [of Heaven and Earth], one cannot trace their mysteries to the ultimate limit. One can seek out the obscure and draw out the hidden but still be unable to gain complete knowledge of their subtleties. Thus strange doctrines have been put forward, and dubious principles have been brought forth. Subsequently there came into being the *hun tian* and *gai tian*. If one takes the two together, one may fill in all the gaps in [one's knowledge of] the way of heaven and earth, and have a means to make visible all the obscurities in them.

The *hun tian* has the text of the *Ling xian* [by Zhang Heng], while the *gai tian* has the methods of the *Zhou bi*. Successive ages have preserved them, and the officials have taken them in hand, to enable themselves [like Xi and He in high antiquity] "in respectful accord with august Heaven, reverently to grant the seasons to the people." With my mediocre talents and shallow scholarly attainments I [can only regard such matters as if] I was looking up at a neighboring high mountain, and taking respectful note of the track of its shadow. But during a few leisure days of convalescence I happened to look at the *Zhou bi*. Its prescriptions are brief but far-reaching; its words are authoritative and accurate. I feared lest it should be cast aside, [or be thought] hard to penetrate, so that those who discuss the heavens should get nothing from it.

I set out at once, therefore, to construct diagrams in accordance with the text. My sincere hope was to demolish the high walls and reveal the mysteries of the halls and chambers within. Perhaps in time gentlemen with a taste for wide learning may turn their attention to this work.

The Book of Shang Gao

Long ago, the Duke of Zhou asked Shang Gao, "I have heard, sir, that you excel in numbers. May I ask how Bao Xi laid out the successive degrees of the circumference of heaven in ancient times? Heaven cannot be scaled like a staircase, and earth cannot be measured out with a footrule. Where do the numbers come from?"

Shang Gao replied, "The patterns for these numbers come from the circle and the square. The circle comes from the square, the square comes from the trysquare, and the trysquare comes from [the fact that] nine nines are eighty-one."[12]

"Therefore fold a trysquare so that the base is three in breadth, the altitude is four in extension, and the diameter is five aslant. Having squared its outside, halve it [to obtain] one trysquare. Placing them round together in a ring, one can form three, four and five. The two trysquares have a combined length of twenty-five. This is called the accumulation of trysquares. Thus we see that what made it possible for Yu to set the realm in order was what numbers engender."

The Duke of Zhou exclaimed, "How grandly you have spoken of the numbers! May I ask how the trysquare is used?"

Shang Gao said, "The level trysquare is used to set lines true. The supine trysquare is used to sight on heights. The inverted trysquare is used to plumb depths. The recumbent trysquare is used to find distances. The rotated trysquare is used to make circles, and joined trysquares are used to make squares."

"The square pertains to Earth, and the circle pertains to Heaven. Heaven is a circle, and Earth is a square. The numbers of the square are basic, and the circle is produced from the square. One may represent Heaven by a rain-hat. Heaven is blue and black; Earth is yellow and red. As for the numbers of Heaven making a rain-hat, the blue and black make the outside, and the cinnabar and yellow make the inside, so as to represent the positions of Heaven and Earth."

"Thus one who knows Earth is wise, but one who knows Heaven is a sage. Wisdom comes from the base [of the right-angled triangle] and the base comes from the trysquare. Through its relations to numbers, what the trysquare does is simply to settle and regulate everything there is."

"Good!" said the Duke of Zhou.

---

[12]The "nine nines" is a reference to the multiplication table, which in ancient China students would have learned from 9 times 9, working backward to 1 times 1. The point is that a knowledge of basic mathematics was essential to all that follows here. Needham translates this portion of the text as follows, which makes the connection with multiplication a bit more transparent: "The art of numbering proceeds from the circle and the square. The circle is derived from the square and the square from the rectangle. The rectangle originates from (the fact that) $9 \times 9 = 81$ (i.e., the multiplication table or the properties of numbers as such.)" Needham then adds in a note: "Chao Chün-Chhing [Zhao Shuang, the third-century CE commentator on the *Zhou bi*] explains that it is necessary to know the properties of numbers before one can work with geometrical figures. Note how radically different this is from the Euclidean method, in which actual numerical values are irrelevant provided the basic axioms and postulates are accepted. Here an arithmetical square is significantly given" [Needham 1959, 22].

The Book of Chen Zi

Long ago, Rong Fang asked Chen Zi, "Master, I have recently heard something about your Way. Is it really true that your Way is able to comprehend the height and size of the sun, the [area] illuminated by its radiance, the amount of its daily motion, the figures for its greatest and least distances, the extent of human vision, the limits of the four poles, the lodges into which the stars are ordered, and the length and breadth of heaven and earth?"

"It is true," said Chen Zi.

Rong Fang asked, "Although I am not intelligent, Master, I would like you to favor me with an explanation. Can someone like me be taught this Way?"

Chen Zi replied, "Yes. All these things can be attained to by mathematics. Your ability in mathematics is sufficient to understand such matters if you sincerely give reiterated thought to them."

At this Rong Fang returned home to think, but after several days he had been unable to understand, and going back to see Chen Zi he asked, "I have thought about it without being able to understand. May I venture to enquire further?"

Chen Zi replied, "You thought about it, but not to [the point of] maturity. This means you have not been able to grasp the method of surveying distances and rising to the heights, and so in mathematics you are unable to extend categories." This is a case of limited knowledge and insufficient spirit. Now amongst the methods [which are included in] the [mathematical] Way, it is those which are concisely worded but of broad application which are the most illuminating of the categories of understanding. If one asks about one category, and applies [this knowledge] to a myriad affairs, one is said to know the Way. Now what you are studying is mathematical methods, and this requires the use of your understanding. Nevertheless you are in difficulty, which shows that your understanding of the categories is [no more than] elementary. What makes it difficult to understand the methods of the Way is that when one has studied them, one [has to] worry about lack of breadth. Having attained breadth, one [has to] worry about lack of practice. Having attained practice, one [has to] worry about lack of ability to understand. Therefore one studies similar methods in comparison with each other, and one examines similar affairs in comparison with each other. This is what makes the difference between stupid and intelligent scholars, between the worthy and the unworthy. Therefore, it is the ability to distinguish categories in order to unite categories which is the substance of how the worthy one's scholarly patrimony is pure, and of how he applies himself to the practice of understanding.[13] When one studies the same patrimony but cannot enter into the spirit of it, this indicates that the unworthy one lacks wisdom and is unable to apply himself to practice of the patrimony. So if you cannot apply yourself to the practice of mathematics, why should I confuse you with the Way? You must just think the matter out again.

---

[13]This a very important concept, and a key to understanding how ancient Chinese thinkers conceived of mathematics and the role of the mathematician, which was not to elaborate theorems and proofs, but to recognize broad categories of problems and the methods appropriate for solving a particular type or category of problem. Thus Christopher Cullen prefers to translate the title of what is commonly referred to as the *Nine Chapters* as the "Mathematical Methods in a Nine-fold Categorisation" [Cullen 2002, 284]. Thus given a problem, if it could be associated with the correct category, it would then be a simple matter of applying the corresponding method to get the correct solution.

Rong Fang went home again and considered the matter, but after several days he had been unable to understand, and going back to Chen Zi he asked, "I have exerted my powers to the utmost, but my understanding does not go far enough, and my spirit is not adequate. I cannot reach understanding, and I implore you to explain to me."

Chen Zi said, "Sit down again, and I will tell you." At this Rong Fang returned to his seat and repeated his request. Chen Zi explained to him as follows:

"16,000 *li* to the south at the summer solstice, and 135,000 *li* to the south at the winter solstice, if one sets up a post at noon it casts no shadow. This single [fact is the basis of] the numbers of the Way of Heaven."

"The *zhou bi* is eight *chi* in length. On the day of the summer solstice its [noon] shadow is one *chi* and six *cun*. The *bi* is the altitude [of the right-angled triangle], and the exact [noon] shadow is the base. 1000 *li* due south the base is one *chi* and five *cun*, and 1000 *li* due north the base is one *chi* and seven *cun*. The further south the sun is, the longer the shadow."

"Wait until the base is six *chi*, then take a bamboo [tube] of diameter one *cun*, and of length eight *chi*. Catch the light [down the tube] and observe it: the bore exactly covers the sun, and the sun fits into the bore. Thus it can be seen that an amount of eighty *cun* gives one *cun* of diameter. So start from the base, and take the *bi* as the altitude. 60,000 *li* from the *bi*, at the subsolar point a *bi* casts no shadow. From this point up to the sun is 80,000 *li*. If we require the oblique distance [from our position] to the sun, take [the distance to] the subsolar point as the base, and take the height of the sun as the altitude. Square both base and altitude, add them and take the square root, which gives the oblique distance to the sun. The oblique distance to the sun from the position of the *bi* is 100,000 *li*. Working things out in proportion, eighty *li* gives one *li* of diameter, thus 100,000 *li* gives 1250 *li* of diameter. So we can state that the diameter of the sun is 1250 *li*."

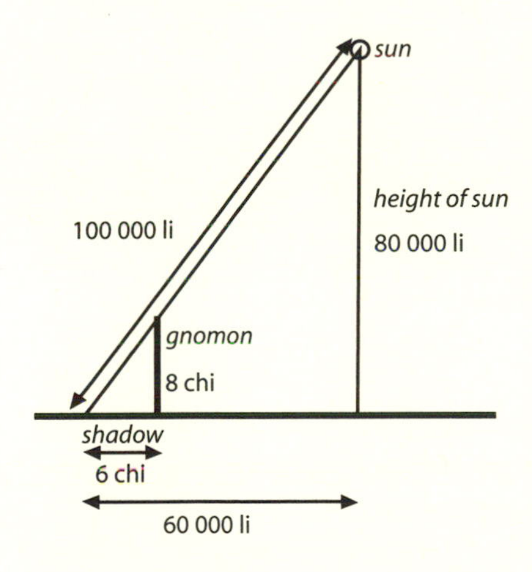

The basic idea here is to set up a system of similar shadow-gnomon/base-height (*gou-gu*) relations. The diagram will help readers work out the various numerical relations discussed above and in what follows, in order to determine both the "height" of the sun and its "diameter." The latter is based upon the proportional numbers drawn from the sighting tube, as illustrated. The reason why Chen Zi instructs Rong Fang to wait until the base (or shadow cast by the gnomon) is 6 *chi* (or 60 *cun*) is so that the ratio of base-height-hypotenuse (*gou-gu-xian*) will be 60-80-100 (what we would term a 3-4-5 right triangle). For additional discussion of the mathematical details under discussion here, see [Cullen 1996].

"Method: the *zhou bi* is eight *chi* long, and the decrease or increase of the base is one *cun* for a thousand *li*."

"Therefore it is said: the [celestial] pole is the length and breadth of heaven."

"Now set up a gnomon eight *chi* tall, and sight on the pole: the base is one *zhang* three *cun*. Thus it can be seen that the subpolar point is 103,000 *li* to the north of Zhou."

Rong Fang asked, "What is meant by [the term] *zhou bi*?" Chen Zi replied, "In ancient times the Son of Heaven ruled from Zhou. This meant that quantities were observed at Zhou, hence the term *zhou bi*. *Bi* means *biao*, gnomon."[14]

16,000 *li* south [of Zhou] on the day of the summer solstice, and 135,000 *li* south on the day of the winter solstice, there is no shadow at noon. From this we can see that from the pole south to noon at the summer solstice is 119,000 *li*, and it is the same distance north to midnight. The diameter overall is 238,000 *li*, and this is the diameter of the solar path at the summer solstice. Its circumference is 714,000 *li*.

From the summer solstice noon to the winter solstice noon is 119,000 *li*, and it is the same distance north to the subpolar point.

Thus from the pole south to the winter solstice noon is 238,000 *li*, and it is the same distance north to midnight. The diameter overall is 476,000 *li*, and this is the diameter of the solar path at the winter solstice. Its circumference is 1,428,000 *li*.

From the equinoctial noon north to the subpolar point is 178,500 *li*, and it is the same distance north to midnight. The diameter overall is 357,000 *li*, [and this is the diameter of the solar path at the equinoxes]. Its circumference is 1,071,000 *li*.

Therefore it is said: the lunar path always follows the lodges, and the solar path is in exact [correspondence] with the lodges.

[If one measures] south to the summer solstice noon and north to the winter solstice midnight, or south to the winter solstice noon and north to the summer solstice midnight, in both cases the diameter is 357,000 *li* and the circumference is 1,071,000 *li*.

From the division of day and night at the spring equinox to the division of day and night at the autumn equinox, there is always sunlight at the subpolar point. From the division of day and night at the autumn equinox to the division of day and night at the spring equinox, there is never sunlight at the subpolar point. Therefore, at the time of the division of day and night at the spring and autumn equinoxes, the area illuminated by the sun extends just up to the pole. This is the equal division of Yin and Yang.

The winter and summer solstices are the greatest expansion and contraction of the [diameter of the] solar path, and the extremes of length and brevity of the day and night. The spring and autumn equinoxes are [the times when] Yin and Yang are equally matched. [As for] the counterparts of day and night, day is Yang and night is Yin. From the spring equinox to the autumn equinox is the counterpart of day. From the autumn equinox to the spring equinox is the counterpart of night.

Therefore the illumination of the noon sun at the spring and autumn equinoxes reaches north to the subpolar point, and the illumination of the midnight sun likewise reaches south to the pole. This is the time of the division between day and night.

Therefore it is said: the illumination of the sun extends 167,000 *li* to all sides. The distance to which human vision extends must be the same as the extent of solar illumination. The extent of vision from Zhou reaches 64,000 *li* north beyond the pole, and 32,000 *li* south beyond the winter solstice noon point.

---

[14]Recently the "Zhou" gnomon is more commonly associated with the Zhou or Western Zhou dynasty, rather than with a specific place named "Zhou." Even Needham, who translates "Zhou" figuratively as a "circular path" of heaven, notes that the "Zhou" of the *Zhou bi* is often taken to be a reference to the dynasty; for details, see [Needham 1959, 19] and [Cullen 1996, xi].

At noon on the summer solstice the solar illumination extends 48,000 *li* south beyond the winter solstice noon. It extends 16,000 *li* south beyond the limit of human vision, 151,000 *li* north beyond Zhou and 48,000 *li* north beyond the pole.

At midnight on the winter solstice the extent of solar illumination southwards falls short of the limit of vision of the human eye by 7000 *li*, and falls 71,000 *li* short of the subpolar point.

At the summer solstice the illumination of the sun at noon and the illumination of the sun at midnight overlap by 96,000 *li* across the pole. At the winter solstice the illumination of the sun at noon falls 142,000 *li* short of meeting the illumination of the sun at midnight, and falls 71,000 *li* short of the subpolar point.

On the day of the summer solstice, if one sights due east and west of Zhou, then from the subsolar points directly due east and west of Zhou it is 59,598½ *li* to Zhou. On the day of the winter solstice the sun is not visible in the regions due east and west, [however] by calculation we find that from the subsolar points it is 214,557½ *li* to Zhou.

All these numbers give the expansion and contraction of the solar path.

At the winter and summer solstices, observe the measurements of the pitch-pipes and listen to the sound of the bells. [Do this by] day at the winter solstice and by night at the summer solstice.

From the extent of the difference of the figures and the limit of solar illumination, the diameter of the four poles is 810,000 *li*, and the circumference is 2,430,000 *li*.

From Zhou southwards to the [furthest] place illuminated by the sun is 302,000 *li*, and northwards to the [furthest] place illuminated is 508,000 *li* from Zhou. The distances east and west [from Zhou to the furthest points illuminated] are each 391,683½ *li*. Zhou is 103,000 *li* south of the center of heaven, and therefore the east-west measurement is shorter than the central diameter by just over 26,632 *li*.

508,000 *li* to the north of Zhou. 135,000 *li* on the day of the summer solstice. The diameter of the solar path at the winter solstice is 476,000 *li*, and its circumference is 1,428,000 *li*. The overall extent of solar illumination measures just over 391,683 *li* east to west through Zhou.

The Square and the circle

These are the methods of square and circle.

The square and circle are of universal application in all activities of the myriad things. The compasses and the try square are deployed in the work of the Great Artificer. A square may be trimmed to make a circle, or a circle may be cut down to make a square. Making a circle within a square is called "circling the square;" making a square within a circle is called "squaring the circle."

Zhao Shuang's essay on Pythagoras' theorem

Diagrams of base and altitude, circle and square.

The base and altitude are each multiplied by themselves. Add to make the hypotenuse area. Divide this to open the square (i.e., take the square root), and this is the hypotenuse.

In accordance with the hypotenuse diagram, you may further multiply the base and altitude together to make two of the red areas. Double this to make four of the red areas. Multiply the difference of the base and the altitude by itself to make the central yellow area. If one [such] difference area is added [to the four red areas], the hypotenuse area is completed.

Figure 3.1 is usually referred to as the "hypotenuse" diagram, or *xian tu*. It is ascribed to Zhen Luan, a sixth-century CE commentator who refers to the diagram in his commentary on Zhao Shuang's discussion of the relations between the base, height, and diagonal (or hypotenuse) connecting the base and height. Further commentary was supplied by Li Chunfeng in the seventh century CE. See [Cullen 1996, 206–17]. Martzloff notes, in discussing this same figure, that it provides a direct proof of the *gou-gu* (Pythagorean) relationship without requiring any shift of areas. The two triangular areas below (or "outside") the square on the *xian*/hypotenuse correspond directly with the two triangular areas at the top and right of the "inside" portion of the *xian* square. This offers yet another example of how useful the "out-in" approach to geometric proofs can be. See [Martzloff 1997, 298–99].

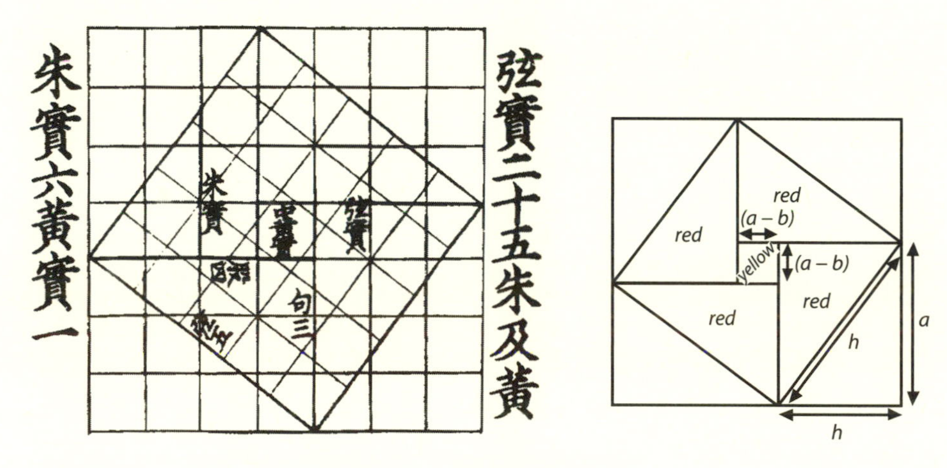

FIGURE 3.1. Hypotenuse diagram.

Subtract the difference area from the hypotenuse area, and halve the excess. Take the difference as the "auxiliary divisor." Divide to open the square, and once more you obtain the base [see fig. 3.2]. If you add the difference to the base, you get the altitude. Whenever you add the base and altitude areas, you form the hypotenuse area. [Either of these] may be a square within, or a trysquare without. The shapes are transformed, but the quantities are balanced; the forms are different, but the numbers are equal.

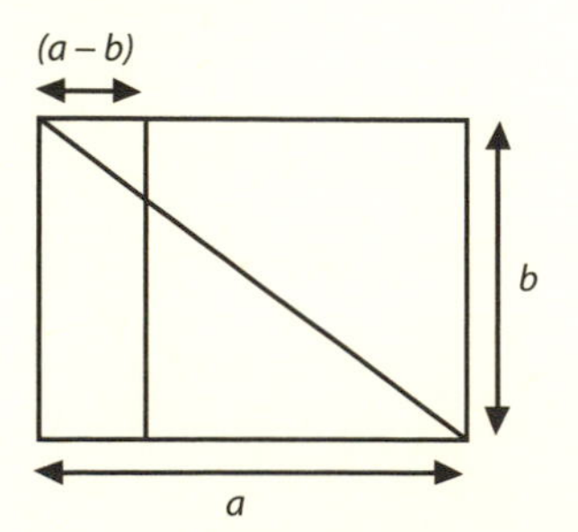

FIGURE 3.2. $(a - b)$ as auxiliary divisor.

The trysquare of the base area has the altitude-hypotenuse difference as its breadth, and the altitude-hypotenuse sum as its length, while the altitude area is the square within [see fig. 3.3]:

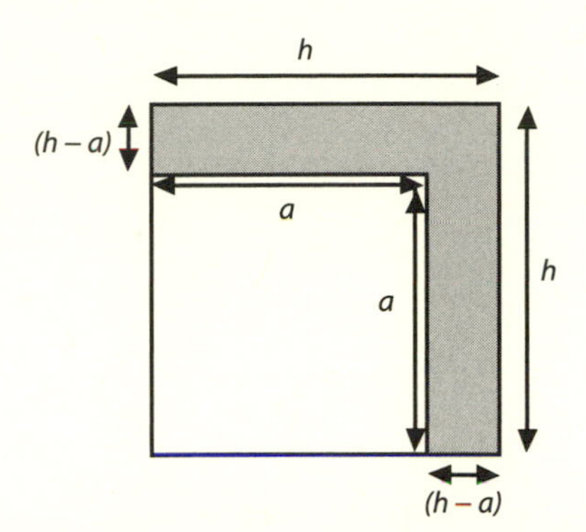

FIGURE 3.3. Base area trysquare.

Subtract the area of the trysquare base from the hypotenuse area; "open" the excess and it is the altitude. [Make] twofold [use of] the altitude on both sides as the "auxiliary divisor"; open the corner of the trysquare base [which now remains] and it is the altitude-hypotenuse difference [see fig. 3.4]:

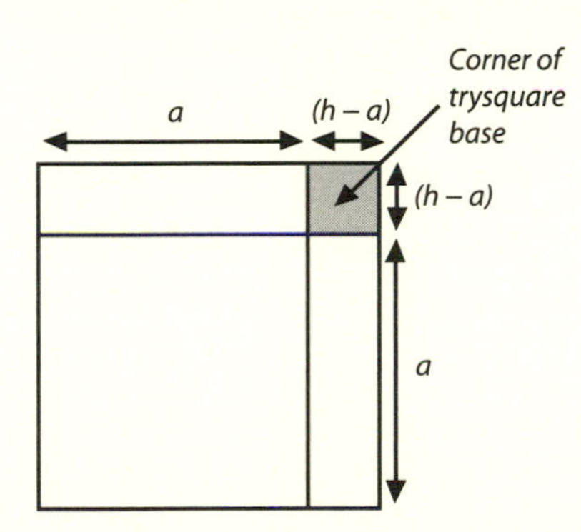

FIGURE 3.4. Corner of trysquare base.

Add [it] to the altitude to make the hypotenuse.

Divide the base area by [this] difference, and you obtain the altitude-hypotenuse sum. Divide the base area by [this] sum, and thus you obtain the altitude-hypotenuse difference.

Let the sum [of altitude and hypotenuse] be multiplied by itself, and add it to the base area to create a [new] area. Take double the sum as the divisor, and what you obtain is then the hypotenuse [see fig. 3.5]:

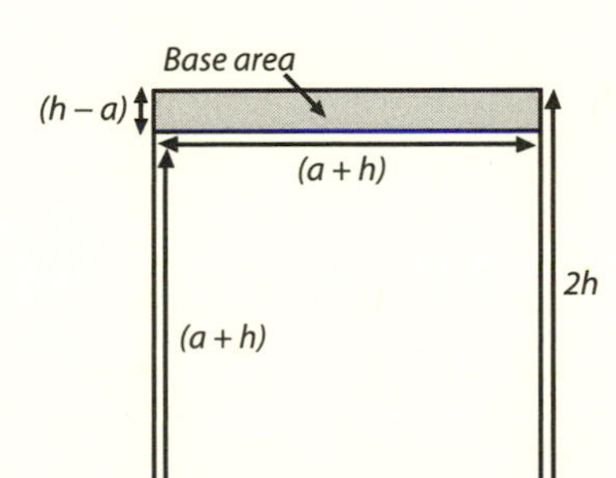

FIGURE 3.5. Hypotenuse from sum and difference of altitude and hypotenuse (not to scale).

Subtract the base area from the sum multiplied by itself, [and] divide [by double the sum as before] to make the altitude [see fig. 3.6]:

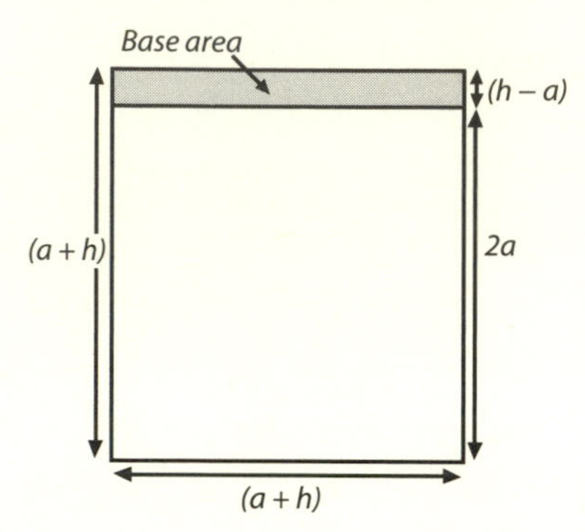

FIGURE 3.6. Altitude from sum and difference of altitude and hypotenuse (not to scale).

The trysquare of the altitude area has the base-hypotenuse difference as its breadth, and the base-hypotenuse sum as its length, while the base area is the square within [see fig. 3.7]:

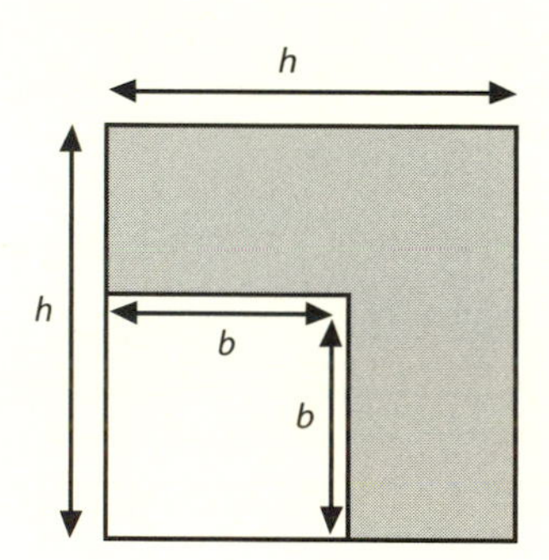

FIGURE 3.7. Altitude area trysquare.

Subtract the area of the trysquare altitude from the hypotenuse area; "open" the excess and it is the base. [Make] twofold [use of] the base on both sides as the "auxiliary divisor"; "open" the corner of the trysquare altitude [which now remains] and it is the base-hypotenuse difference [see fig. 3.8]:

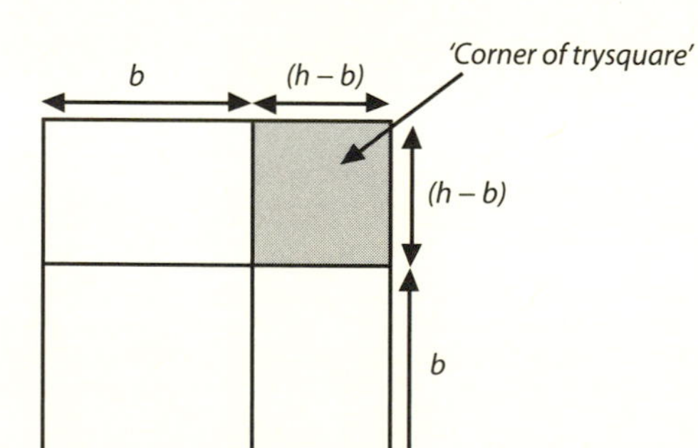

FIGURE 3.8. Corner of trysquare altitude.

Add [it] to the base to make the hypotenuse.

Divide the altitude area by [this] difference, and you obtain the base-hypotenuse sum. Divide the altitude area by [this] sum, and thus you obtain the base-hypotenuse difference.

Let the sum [of base and hypotenuse] be multiplied by itself, and add it to the altitude area to create a [new] area. Take double the sum as the divisor, and what you obtain is then the hypotenuse [see fig. 3.9]:

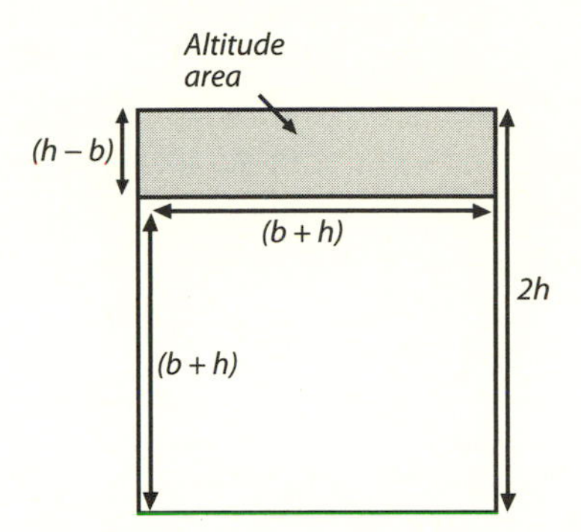

FIGURE 3.9. Hypotenuse from sum and difference of base and hypotenuse (not to scale).

Subtract the altitude area from the sum multiplied by itself, [and] divide [by double the sum as before] to make the base [see fig. 3.10]:

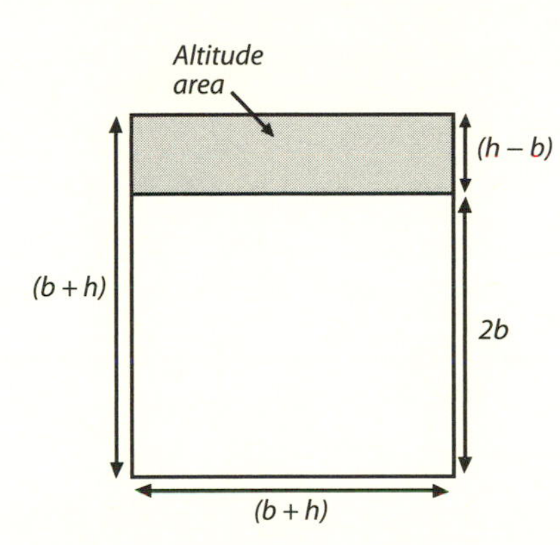

FIGURE 3.10. Base from sum and difference of base and hypotenuse (not to scale).

Multiply the two differences together, double [the resulting area], and "open" it. [As for] what is obtained, [if] you increase it by the altitude-hypotenuse difference, you make the base. [If] you increase it by the base-hypotenuse difference, you make the altitude. [If] you increase it [by] both differences, you make the hypotenuse [see fig. 3.11]:

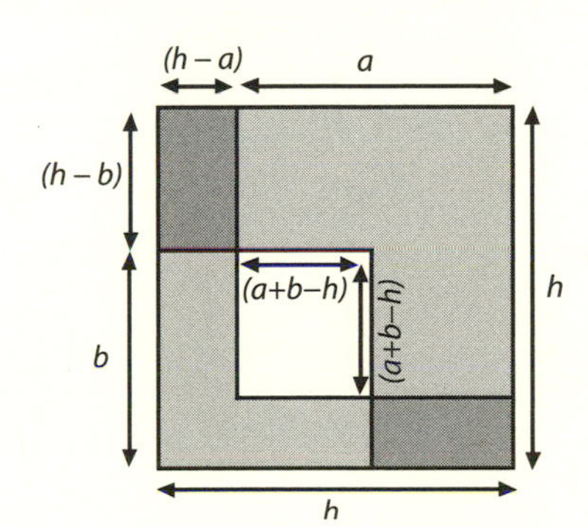

FIGURE 3.11. To obtain the three sides from $(h - a)$ and $(h - b)$, perhaps showing the base and altitude trysquares "plaited round" *pan huan* as described in "The Book of Shang Gao," the paragraph quoted (p. 217) beginning "Therefore fold a trysquare. . . . "

As for the fact that you see the "sum area" (i.e., the square of the sum of base and altitude) when you double the hypotenuse area and set out the base-altitude difference area [in addition to it], examining [the matter] by means of the diagram, doubling the hypotenuse area fills the outer large square, [leaving as] excess the yellow area. Now the excess of this yellow area is just the base-altitude difference area. If you take away the difference area [from double the hypotenuse area], and "open" the excess, you get the outer large square, [and then] the side of the outer large square, which is the base-altitude sum. Let this sum be multiplied by itself, [take] double the hypotenuse area, then subtract [the result] from it, "open" the excess, and you obtain the central yellow square. The side of this central yellow square is the base-altitude difference. Subtract this difference from the [base-altitude] sum and halve, to make the base. Add the difference to the sum and halve, to make the altitude.

Let the hypotenuse be [used] twice to make the joining of length and breadth [of the trysquares referred to earlier]. Let whichever of base or altitude appears [as its "trysquare" in the case considered] be multiplied by itself to give an area. [Take] four [of these] areas and subtract [them] from it (i.e., the square whose side is double the hypotenuse). "Open" the excess, and what you get is the difference [between length and breadth]. Subtract this difference from the joining [of length and breadth], halve the surplus and you have the breadth. Subtract the breadth from the hypotenuse, and you have what you seek (i.e., the so far unknown side) [see fig. 3.12]:

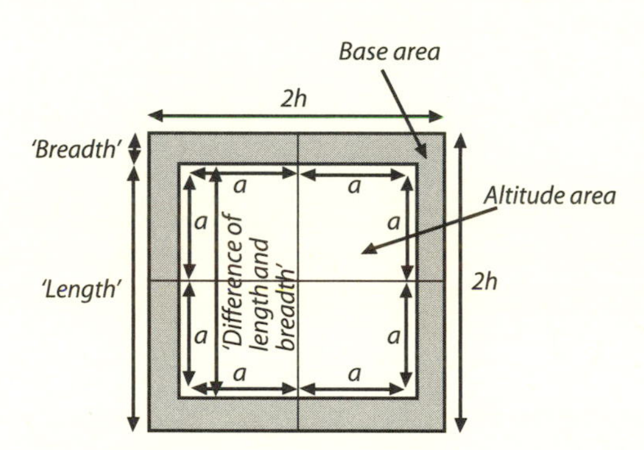

FIGURE 3.12. Using sum and difference of length and breadth of a trysquare (in this case the base trysquare is shown).

If you observe the alternation of compasses and trysquare, how they return and repeat [their operations] together, how they mutually penetrate [but have] an allotted [function], there is something to be gained from each [instance]. Thus they systematize the multitude of rules, and vastly order the manifold principles. threading through the obscure and entering into the subtle, hooking up the profound and reaching afar. So [as the text of the *Zhou bi* says], what they do is simply to settle and regulate everything there is.

V. The Chinese "Euclid," Liu Hui

Liu Hui, about whom little is known except that he lived in the third century CE, is often called the Chinese "Euclid," because he is responsible for the most important ancient Chinese mathematical text, the *Jiu zhang suan shu*. Although he was not the author, it is his edition with numerous commentaries which formed the basis for the text which was central to the Chinese study of mathematics. Liu was also the author of the *Hai dao suan jing*, a work he considered a supplement to the *Jiu zhang*, but which subsequently became an independent Chinese classic.

V.a. The *Nine Chapters*

The oldest of the Chinese classic texts devoted exclusively to mathematics is the *Jiu zhang suan shu* (*Nine Chapters on Mathematical Procedures*). Like the *Zhou bi* (*Mathematical Classics of the Zhou Gnomon*), there is no agreement as to exactly how this title should be translated; Mikami called it the *Arithmetic in Nine Sections* [Mikami 1913, 8], whereas Needham translated the title as *Nine Chapters on the Mathematical Art* [Needham 1959, 24], as do [Li and Du 1987, 33]. Jean-Claude Martzloff prefers *Computational Prescriptions in Nine Chapters* [Martzloff 1997, 127], and Christopher Cullen has opted for *Nine Categories of Mathematical Methods* [Cullen 2002]. The most recent, authoritative translation of this work, by Karine Chemla and Guo Shuchun, calls it *Les Neuf chapitres sur les procedures mathématiques* (*The Nine Chapters on Mathematical Procedures*) [Chemal and Guo 2004]. Hereafter, it will be referred to as simply the *Nine Chapters.*

If the *Nine Chapters* existed as a complete work prior to the infamous burning of the books by the first emperor of a unified China, Qinshi Huang, in 208 BCE, no copies survive; the work which has been handed down was compiled by Zhang Cang and Geng Shouchang in the Former Han dynasty, probably sometime in the first century BCE, and later given a detailed commentary in the third century CE by Liu Hui. Although virtually nothing is known about Liu Hui apart from what he says in his preface to this work (reproduced below), it is his version of the work with commentary that was later chosen to comprise one of the *Shi bu suan jing* (*Ten Books of Mathematics Classics*) to be studied for the official mathematical examinations, and to serve as texts when mathematics was taught at the Imperial College during the Tang dynasty (618–907 CE). For these purposes the *Ten Classics* were collated with a commentary by Li Chunfeng (604–672 CE) and a retinue of other scholars. This edition included the *Nine Chapters*, and it is back to this version of the text with its commentaries that current editions may be traced. The earliest printed versions include one partial copy printed in 1213 CE during the Southern Song Dynasty (now in the Shanghai library), and another version included in the fifteenth-century *Yongle dadian.* Although this work too is only known in fragmentary form, the *Nine Chapters* was again collated and edited by Dai Zhen for the Qing dynasty compilation, the *Siku quanshu* (*Complete Library of the Four Branches of Literature*). The ten classics were again collated and edited with commentary by Kong Jihan in 1773 as the *Suanjing shishu* (*The Ten Mathematical Classics*).

As for the *Nine Chapters*, it is a rich guide to the mathematics of ancient China because it comprises the entire range of then-known mathematics, from basic operations for computing with fractions to difficult problems involving matrix methods and applications of the *gou-gu* (base-height) relation (equivalent to the "Pythagorean" theorem) to which all of Chapter Nine is devoted. As Wang Ling noted in his dissertation (unpublished) on this work, throughout the entire history of Chinese mathematics it holds a "nuclear position":

> Its influence is reflected in all subsequent Chinese mathematical works. Its problems stimulated the creation of a score of new topics of study. It set up a model for mathematical language and a pattern for computation. Calculators and accountants followed its path; calendar experts and astronomers copied and borrowed its technical terms. Written two thousand years ago, the *Chiu chang* opens the first chapter in the history of Chinese mathematics and has remained a kind of "mathematical bible" ever since. [Wang Ling, quoted from Martzloff 1997, 127]

As for the contents of the *Nine Chapters*, based upon some problems very similar to those found in the *Suan shu shu*, it seems clear that parts drew from earlier mathematical works of the pre-Qin period. Other parts of the *Nine Chapters*, however, embody significant innovations of much later mathematics, and reflect the state of the mathematical art in the first century BCE. These include new algorithms for approximating square-roots, matrix methods for solving simultaneous equations, and a variety of problems applying the *gou-gu* relation, all of which are absent from the *Suan shu shu.*

The portions of the text of the *Nine Chapters* reproduced here include the commentary by Liu Hui and Li Chenfeng's group. The comments by Liu Hui (Liu:) and Li Chunfeng and colleagues (Li:) are meant to explain the meaning of the text or details of the methods introduced, and are indented following the passages on which they are commenting. Additional explanatory information is provided in footnotes.

The Nine Chapters of the Mathematical Art

The Preface

The Preface to the *Nine Chapters* is missing from the Southern Song printing of 1213 known from the one incomplete surviving copy in the Shanghai Library, and is printed without attribution or title in the version in the *Yongle dadian.* But its Southern Song editor, Bao Huanzhi, in a work preliminary to the actual printing of the *Ten Classics*, quotes passages from the preface to the *Nine Chapters* and explicitly attributes the text to Liu Hui [Chemla 2004, 747].

In his preface, Liu Hui offers the traditional version of the origin of mathematics, that Fu Xi, one of the "four sages," created the eight trigrams and the nine nines algorithm to "coordinate the variations in the hexagrams." The "nine nines" algorithm is the multiplication table from nine times nine to one times one, upon which facility in both multiplication and division depended. A few lines later, the "power of duality" and the "four diagrams" are also mentioned, both of which have an intimate connection to the eight trigrams and the classic book of ancient Chinese divination and philosophy, commonly known as the *Yi jing* (*Book of Changes*). The parts of the *Yi jing* that concern specifically matters of the lines and hexagrams are now commonly referred to at the *Zhou yi*, and it is this that concerns us here. For all of Chinese mathematics was said to derive from the two primordial opposite principles, the *yin* and *yang*, embodied among other many possible manifestations in the solid — and broken lines – – of the *Zhou yi.* From these the four diagrams (the four possible figures that can be generated from combinations of *yin* and *yang* taken as pairs), the eight trigrams, and in fact the entire set of 64 hexagrams of the *Yi jing* could be generated from combinations of the solid and broken lines [see Needham 1983, vol. 5, 55–58].

———	— —	———	— —
———	———	— —	— —

The Four Diagrams.

qian ☰	*dui* ☱	*li* ☲	*zhen* ☳
(heaven)	(lake)	(fire)	(thunder)
乾	兌	離	震
xun ☴	*kan* ☵	*gen* ☶	*kun* ☷
(wind)	(water)	(hill)	(earth)
巽	坎	艮	坤

The Eight Trigrams.

次序	卦形	卦名	次序	卦形	卦名	次序	卦形	卦名	次序	卦形	卦名
1	䷀	乾	17	䷐	随	33	䷠	遯	49	䷰	革
2	䷁	坤	18	䷑	蛊	34	䷡	大壮	50	䷱	鼎
3	䷂	屯	19	䷒	临	35	䷢	晋	51	䷲	震
4	䷃	蒙	20	䷓	观	36	䷣	明夷	52	䷳	艮
5	䷄	需	21	䷔	噬嗑	37	䷤	家人	53	䷴	渐
6	䷅	讼	22	䷕	贲	38	䷥	睽	54	䷵	归妹
7	䷆	师	23	䷖	剥	39	䷦	蹇	55	䷶	丰
8	䷇	比	24	䷗	復	40	䷧	解	56	䷷	旅
9	䷈	小畜	25	䷘	无妄	41	䷨	损	57	䷸	巽
10	䷉	履	26	䷙	大畜	42	䷩	益	58	䷹	兑
11	䷊	泰	27	䷚	頤	43	䷪	夬	59	䷺	涣
12	䷋	否	28	䷛	大过	44	䷫	姤	60	䷻	节
13	䷌	同人	29	䷜	坎	45	䷬	萃	61	䷼	中孚
14	䷍	大有	30	䷝	离	46	䷭	升	62	䷽	小过
15	䷎	谦	31	䷞	咸	47	䷮	困	63	䷾	既济
16	䷏	豫	32	䷟	恒	48	䷯	井	64	䷿	未济

The sixty-four hexagrams are comprised of 2^6 combinations of the two *yin-yang* dualities, the solid lines — and broken lines – – of the *Zhou yi* [Shen, Crossley, and Lun 1999, 55].

The Preface also refers to the "Nine Arithmetic Arts," described in the *Zhou li* (*Rites of Zhou*), an ancient classic text that in one section outlines prescriptions for the education of the nobility, which was meant to include the six arts, one of which was the *Jiu shu* or "Nine Arithmetical Arts," doubtless a mathematical model of sorts for the *Nine Chapters.* But since the latter includes sections devoted to much later innovations already mentioned, including chapters on matrix methods and the *gou-gu* relation, the "Nine Arithmetical Arts" could not have been precisely those to which each of the *Nine Chapters* is devoted.

Liu Hui recounts the fate of the *Nine Chapters,* how after the Qin emperor called for the destruction of the Confucian classics, which included mathematical texts like the "Nine Arithmetic Arts," these were later reconstructed by Zhang Cang and Geng Shouchang, who brought them up to date with more recent terminology. Liu Hui then recounts his own training as a mathematician. Apparently self-taught, he understood the importance of mathematics to depend upon classification, for knowing the type of problem one had to solve would make clear the method that should be applied for solution of that particular category of problem. To aid in understanding the text, Liu Hui mentions diagrams he has provided which unfortunately have been lost.

Liu Hui brings his preface to a close with an account of his own contribution to mathematics. He describes the procedure already familiar from the *Zhou bi* whereby an understanding of the properties of the *gou-gu* base-altitude figure made it possible to determine the height of the sun and its distance from a given observer. But this method was not mentioned in the *Nine Chapters.* However, knowing the "double difference" method from the "Nine Arithmetic Arts," Liu Hui says he found that it could be applied to "superhuman distances," and that Zhang and Geng, in editing the *Nine Chapters,* had failed to include all the methods that mathematics had to offer. And so Liu Hui wrote a new chapter at first to serve as an appendix of sorts, a tenth section for the *Nine Chapters,* on this very important method of surveying over distances otherwise unmeasurable. This work eventually became a classic text on its own, the *Hai dao suan jing* (*Sea Island Mathematical Classic*), which introduces the double difference method for surveying, as explained below. As Liu Hui says, once the distance of the sun from the observer can be determined, "why not measure high mountains, vast rivers and seas?" Indeed, the power of mathematics is that it may be applied, as he says at the end of his preface, "to the most tricky problems."

Preface of Liu Hui

Fu Xi created the eight trigrams in remote antiquity to communicate the virtues of the gods and parallel the trend of events in earthly matters, [and he] invented the nine-nines algorithm to co-ordinate the variations in the hexagrams. Subsequently, the Yellow Emperor marvelously transformed and extended these so that [he] regulated the calendar [and] harmonized the music scale. [These were] used to investigate the cosmic principles, and thereby the subtle and exquisite power of duality and the four diagrams could be effectively harnessed. It is recorded that Li Shou invented numbers but the details of this are not known. According to [history] the Duke of Zhou drew up the Ethical Codes in which were the Nine Arithmetic Arts. The *Nine Chapters* follows in the tradition of the Nine Arithmetic Arts.

The Qin tyrant (Qin Shihuang) ordered the burning of books. Classics and literature were ruined or lost. Years after that, in the [Former] Han [Dynasty (206 BCE–24 CE)], the Marquis of Beiping, Zhang Cang (c. 250–152 BCE) [and] the Deputy Minister of Agriculture and Finance, Geng Shouchang (c. 1st century BCE) are noteworthy for their proficiency in mathematics. Working on the basis of the remains of incomplete old manuscripts each, as they saw fit, revised and supplemented [the text]. Consequently, examining their titles and comparing them with the ancient ones, [they] differ somewhat, and the contents contain much recent terminology.

I read the *Nine Chapters* as a boy, and studied it in full detail when I was older. [I] observed the division between the dual natures of Yin and Yang [the positive and negative aspects] which sum up the fundamentals of mathematics. Thorough investigation shows the truth therein, which allows me to collect my ideas and take the liberty of commenting on it. Things are known to belong to various classifications. Just as the branches of a tree are to its trunk, so are a multitude of things to an archetype. Therefore I have tried to explain the whole theory as concisely as possible, with spatial forms shown in diagrams, so that the reader should have a reasonably good all-round understanding of it.

As one art of the six gentlemanly arts, the Nine Arithmetical Arts was originally a compulsory course at the Imperial College for the training of aristocratic talent. Though it is called the Nine Arithmetical Arts, they can reach both the infinitesimal and the infinite. The course, nevertheless, is not particularly difficult using the Methods which have been handed down from generation to generation, just like the compass and gnomon in measurement, with which we draw figures. Nowadays enthusiasts for mathematics are few, and many scholars may not be erudite enough to be good at it.

The Administrative System of the Zhou [Dynasty (11th century–256 BCE)] stipulated that one of the duties of the Minister of Home Affairs is to use an eight-*chi*-tall vertical rod to measure the length of shadow at noon on the summer solstice; the place where the shadow stretches for 1 *chi* 5 *cun* is called the earth center. It is said that the sun is then 15,000 *li* south of it, as can be inferred from the rules. In fact, the *Nine Chapters* only mentions a form of distance measurement with four rods, as well as the measurement of the height of a hill by a tree. All the points are within walking distance, visible to one another, without involving such superhuman distances. Thus the rules given by Zhang et al. have not yet exhausted all the numerical problems, but I found the double difference in the Nine Arithmetical Arts, which was just meant for such exotic problems. It can be used for all types of measurement involving extreme heights, depths or distances. It is so called because the differences taken as rates between the corresponding sides of right-angled triangles are applied twice. Set up two poles, 8 *chi* high each, south to north on a plain, as far apart as possible in the city of Luoyang, and measure the shadows at noon simultaneously. Take the difference between the shadows as divisor, the height of the rod multiplied by the distance between the rods as dividend and add the quotient to the height of a rod and [we] get the distance of the sun from the plain. Take the southern shadow multiplied by the rod-to-rod distance as dividend and divide it by the difference of the shadows. The quotient would be the horizontal distance from the southern rod to the sun. Let the horizontal distance and the sun-plain distance be the *gou* and *gu*, respectively. The hypotenuse of the right-angled triangle is the distance of the sun from the surveyor. When we observe the sun southward through a tube one *cun* in diameter, if the sun just fills the tube, then the length of the tube is assumed to be the *gu*, and the tube diameter the *gou*. Take the sun-surveyor distance as the large *gu*, and its *gou* equals the diameter of the sun. Once celestial bodies can be measured, then why not measure high mountains, vast rivers and seas? Existing historical records, I think, should have contained some of the cosmic phenomena with their descriptions and data, so that the outstanding academic achievements could have been preserved. Therefore I have written the Rule of Double Differences, with some comments, so as to facilitate our search for the original meaning. This is appended to the Chapter on Right-angled Triangles. Height measurement involves two poles, and depth measurement two gnomons. If another point is added, we ought to observe thrice; a fourth observation is needed if the additional point is not in the same plane as the other points. By analogy and extension, these rules can be applied to the most tricky problems. Curious gentlemen, read the text thoroughly.

Chapter 1: *Fang tian* (Field Measurement)

If translated explicitly, the title of this chapter in Chinese is "square fields," and it is often referred to as such in works on Chinese mathematics. This section which opens the *Nine Chapters* curiously begins where the *Suan shu shu* ends, namely, with a problem devoted to finding the area of a rectangular field given its width and length. It soon jumps to a series of problems about how to calculate with fractions, including the basic arithmetic operations with fractions and how to simplify them (i.e., applying a method equivalent to the Euclidean algorithm to reduce a fraction to lowest terms). The technique known as "uniformizing and homogenizing" is used to add and subtract fractions, by providing common denominators (uniformizing) and then adjusting numerators accordingly (homogenizing) before adding or subtracting. This is all explained by Liu Hui in his commentary to problem 9. Given two fractions a/b and c/d, the procedures amount to finding the common denominator bd by "uniformizing" the denominators, and then cross-multiplying the numerators and denominators to "homogenize" the fractions, yielding ad/bd and bc/bd.

The chapter ends with twenty problems devoted to the computation of areas. However, despite this chapter's title, the areas are not square fields but rectangles, triangles, trapezoids, circles, rings, and even sectors of circles. The first of these given here is problem 26: "Now given a triangular field ... ." The translation here of "triangles," however, is unfortunate because, as we have seen, ancient Chinese did not have an explicit mathematical term for "triangle." Instead, the shape of the field in question here is referred to as a *gui*, a ceremonial object the top of which was an isosceles triangle. It is important to keep in mind this unusual feature of Chinese geometry, that an object so familiar in the west as a "triangle" did not have a special term of its own in Chinese. For discussion of this in greater detail, see [Raphals 2002].

What also deserves emphasis here is the commentary by Liu Hui, who explains the means of computing the area of a *gui* as follows: "Halving the base means filling the vacancy with the surplus transforms [the triangle] into a rectangle." This can be understood as a "proof" of the computation for such figures by appeal to the "out-in" principle, as in the figure.

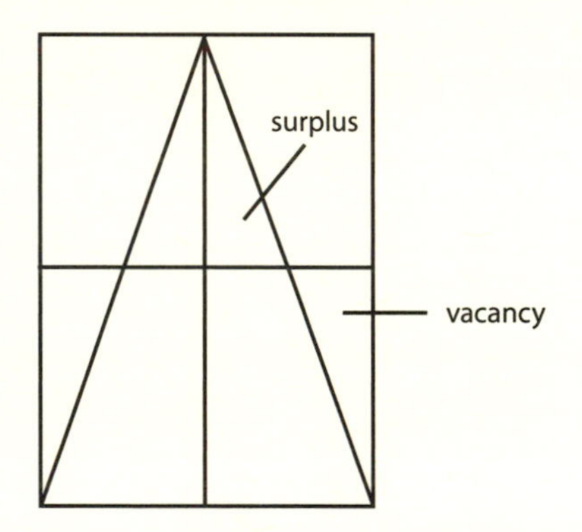

By moving the "surplus" areas at the top of the diagram (inside the *gui*) to the equivalent "vacancy" areas at the bottom of the diagram (outside the *gui* on both the left and right), the *gui* is immediately seen to be equivalent to the rectangular area that comprises the lower half of the diagram, whose area is $(1/2)ab$, where a is the length of the base, and b the height of the *gui*. This application of the "out-in" method is applied in the proofs offered to justify the computations for the areas in problems 26 and 27.

Problem 32 is the most important in the chapter, and includes the rule for computing the area of a circle given its diameter and circumference. From the data in the problem it is clear that the formula assumes the value for π, the ratio of the circumference to the diameter of the circumference, to be 3. Liu Hui knows that this is a very poor approximation, and this prompts one of the longest commentaries in the *Nine Chapters*, which is devoted to determining a more precise ratio. Li Chunfeng and his group in turn comment on Liu Hui's remarks, as well as other parts of the text.

Chapter 1: Field Measurement

Liu: For determining the boundary and area of a field.

1. Now given a field 15 *bu* broad and 16 *bu* long. Tell: how much field? Answer: 1 *mu*.

Rule for Rectangular Fields: Multiply the number of *bu* in breadth by that in length to obtain the *bu* product. Dividing by the conversion factor of 240 [square] *bu* in 1 *mu* gives the number of *mu*. 100 *mu* make 1 *qing*.

> **Liu: The product is the area of the field. Multiplying the breadth by the length is generally called area.**

5. Now given a fraction 12/18. Tell: reducing it [to its lowest terms], what is obtained? Answer: 2/3.

Rule for Reduction of Fractions: If [the denominator and numerator] can be halved, halve them. If not, lay down the denominator and numerator, subtract the smaller number from the greater. Repeat the process to obtain the greatest common divisor. Reduce them by the greatest common divisor.

> **Liu: Reduce a fraction means: a quantity of things cannot always be an integer, but sometimes must be represented by a fraction. A fraction may be tedious to handle. Taking 2/4, 4/8 is tedious, while 1/2 is simple. They are written differently, but equal in value. In practice the divisor and dividend, [while] mutually related, are often mismatched; therefore in arithmetic we have to learn rules for fractions. To reduce a fraction by the greatest common divisor means to divide. Subtract the smaller number from the greater repeatedly, because the remainders are nothing but the overlaps of the greatest common divisor; therefore divide by the greatest common divisor.**

9. Given fractions 1/2, 2/3, 3/4, and 4/5. Tell: combining them, what is obtained? Answer: 2 43/60.

Rule for Addition of Fractions: Each numerator is multiplied by the denominators of the other fractions. Add them as the dividend, multiply the denominators as the divisor. Divide; if there is a remainder, let it be the numerator and the divisor be the denominator. In the case of equal denominators, the numerators are to be added directly.

> **Liu: "Each numerator is multiplied by the denominators of the other fractions" means a simple fraction indicates rough division, while a reducible fraction means fine division. Whether rough or fine, they are all of equal value. Forms of fractions are too complex to share a common denominator without fine division. Multiplication leads to fine division, and only by reducing fractions to a common denominator can they be combined. In general, multiplying a numerator by the denominators [of other fractions] is called homogenizing, and multiplying [all] the denominators is called uniformizing. Here uniformizing implies a common denominator for all the fractions. Homogenizing means retaining the same values for the transformed fractions. As the saying goes, "Things of one kind come together." This is true also of numbers. Fractions with a common denominator cannot be separated, whereas those with different denominators cannot be added. In other words,**

fractions with a common denominator can be added even if the numerators are quite different, whereas fractions with different denominators cannot be added even if the numerators are close to each other. The rules of homogenizing and uniformizing are very important and the most intricate operations on fractions, so they always bring satisfactory results just like untying knots using a [pointed] stick. Multiplying [the denominators] means fine division and reducing means rough division; the rules of homogenizing and uniformizing are used to get a common denominator. Are they not the key rules of arithmetic? An alternate rule is to make numerators homogeneous by dividing the common denominator by the respective denominators. Consider the quotients as rates, and multiply these rates by the corresponding numerator to homogenize.

18. Given 3 1/3 persons share 6 1/3 and 3/4 coins. Tell: how much does each person get? Answer: Each gets 2 1/8 coins.

Rule for the Division of Fractions: Take the number of persons as divisor, the number of coins as dividend. Divide. If either the dividend or the divisor is a [mixed] fraction, convert it to an improper fraction. If both of them are [mixed] fractions, convert them to improper fractions with a common denominator.

Liu: Multiply the numerator and denominator of the dividend by the denominator of the divisor, and multiply those of the divisor by the denominator of the dividend. That is to say, if both the dividend and the divisor are fractions, we first convert them [as mixed fractions] to improper [fractions], and then multiply both the dividend and the divisor by the denominator of the other.

26. Given a triangular field with base 5 1/2 *bu* and altitude 8 2/3 *bu*. Tell: what is the area? Answer: 23 5/6 [square] *bu*.

Rule [for Triangular Fields]: Multiply half the base by the altitude.

Liu: Halving the base means filling the vacancy with the surplus transforms [the triangle] into a rectangle. Alternatively, multiply half the altitude by the base. Multiplication of half the base by the altitude is for getting its mean. Divide the area by the number of [square] *bu* per *mu* giving the number of *mu*.

27. Now given a right-angled trapezoidal field with bases 30 *bu* and 42 *bu* respectively and altitude 64 *bu*. Tell: what is the area? Answer: 9 *mu* 144 [square] *bu*.

Rule [for Right-angled Trapezoidal Fields]: Multiply half the sum of the bases by the altitude, or half the altitude by the sum [of the bases]; divide by the number of [square] *bu* per *mu*.

Liu: Halving the sum means filling the vacancy with the surplus.

32. Given a circular field, the circumference is 181 *bu* and the diameter 60 1/3 *bu*. Tell: what is the area? Answer: 11 *mu* 90 1/12 [square] *bu*.

> Li et al.: Taking the rates of 3 to 1, a diameter of 60 1/3 *bu* corresponds to the 181 *bu* circumference. While taking the precise rate (i.e., $C:d = 22/7$), the diameter is 57 13/22 *bu*.
>
> Liu: Taking Liu's rate (i.e., $C:d = 157/50$), it (the area) should be 10 *mu* 208 113/314 [square] *bu*.
>
> Li et al.: Taking the precise rate, it should be 10 *mu* 205 87/88 [square] *bu*.

Rule [for Circular Fields]: Multiplying half the circumference by the radius yields the area of the circle in [square] *bu*.

> Liu: Taking half the circumference as length and the radius as width; hence multiplying the two gives the area in [square] *bu*. Given a circle of diameter 2 *chi*, consider its inscribed regular hexagon. The rate of the diameter of the circle to its perimeter is 1 to 3.
>
> As in the figure, 3 times the product of the side of the hexagon by the radius yields the area of the inscribed regular dodecagon. Dividing again; then 6 times the product of the side of the dodecagon by the radius yields the area of a 24-gon. The larger the number of sides, the smaller the difference between the area of the circle [and that of its inscribed polygons]. Dividing again and again until it cannot be divided further yields a regular polygon coinciding with the circle, with no portion whatever left out. Outside a regular polygon there are co-apothems. Multiply the sides by the co-apothem giving the area [of the polygon plus these co-apothem rectangles] is larger than that of the circle. As the number of sides increases [without limit] the polygon coincides with the circle and no co-apothem exists, so that [the polygon plus those co-apothem rectangles] is no longer larger than the circle. Take the perimeter [of the inscribed *n*-gon] to multiply the radius; compare that with the area of the inscribed 2*n*-gon; it is twice as large. Hence the Rule "Multiplying half the circumference by the radius yields the area of the circle in [square] *bu*." According to this method, the more precise rate is different from a rate of 3 to 1, which is the perimeter of the [inscribed] hexagon [to the radius]. The difference between a polygon and a circle is just like that between the bow (arc) and its chord, which can never coincide. Yet such a tradition has been passed down from generation to generation and no-one cares to check it. So many scholars followed the tradition that their error has persisted. It is hard to accept without a convincing derivation. Planar forms are either curvilinear or rectilinear. A study of the rates for a square and its inscribed circle may be trivial, but it will lead to far-reaching consequences. By elaborating on this point, one gets wide applications. Let us verify the

facts by diagrams for a more precise rate. Assertions without facts are not sound, so notes and comments in detail are given as follows.

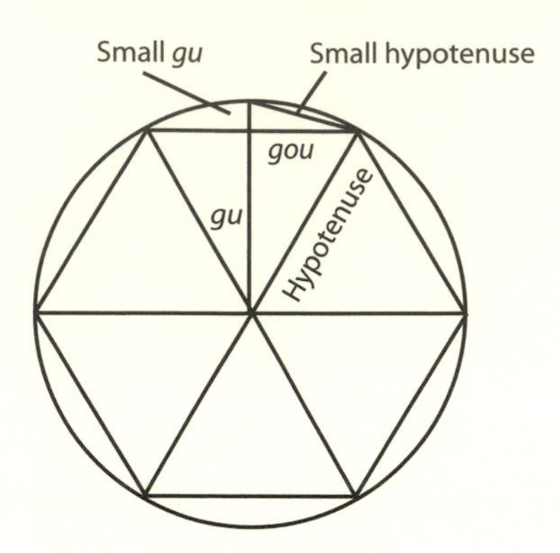

The rule for calculating the dodecagon from hexagon: Let the diameter of a circle be 2 *chi*; and 1 *chi* the radius [one side of the hexagon] be the hypotenuse, and half the side, 5 *cun*, be the *gou* (base) of a right-angled triangle, now find the *gu* (altitude). From the square on the hypotenuse, subtract the square on the *gou*, or 25 [square] *cun*, and the remainder is 75 [square] *cun*. Extract the square root up to *miao*, or *hu* and one digit down once more to the tiny decimal, which is taken as numerator with 10 as its denominator, i.e. 2/5 *hu*. Now the root is the *gu*, 8 *cun* 6 *fen* 6 *li* 2 *miao* and 5 2/5 *hu*, which is 1 *cun* 3 *fen* 3 *li* 9 *hao* 7 *miao* 4 3/5 *hu* less than the radius, the latter number is called the small base; half a side of a hexagon is called a small altitude. From these one gets the [small] hypotenuse. The square on it would be 267,949,193,445 [square] *hu*, where the residue has been neglected. Extracting the square root, get the side of the inscribed dodecagon.

The rule for calculating the 24-gon from the dodecagon: as above, let the radius be the hypotenuse, half a side is the *gou*, now find the *gu*. A quarter of the square on the small hypotenuse is 66,987,298,361 [square] *hu*, where the residue has been neglected. Take it as the square on the *gou*, and subtract it from the square on the hypotenuse. Extract the square root of the remainder, giving a *gu* of 9 *cun* 6 *fen* 5 *li* 9 *hao* 2 *miao* 5 4/5 *hu*. Subtracting it from the radius, the remainder is 3 *fen* 4 *li* 7 *miao* 4 4/5 *hu*, i.e. the small *gou*. Take half a side [of the dodecagon] as the small *gu*, and one gets the small hypotenuse. The square on the small hypotenuse would be 68,148,349,466 [square] *hu*, where the residue has been neglected. Extracting the root, get the side of the [inscribed] 24-gon.

The rule for calculating the 48-gon from the 24-gon: as above, let the radius be the hypotenuse, half a side the *gou*, now find the *gu*. A quarter of the square on the small hypotenuse is 17,037,087,366 [square] *hu*, where the residue has been neglected. Take it as the square on the *gou*, and subtract it from the square on the hypotenuse. Extract the square root of the remainder, giving a *gu* of 9 *cun* 9 *fen* 1 *li* 4 *hao* 4 *miao* 4 4/5 *hu*. Subtracting it from the radius, the remainder is 8 *li* 5 *hao* 5 *miao* 5 1/5 *hu*, i.e. the small *gou*. Take half a side of the 24-gon as the small gu, one gets the small hypotenuse. The square on the small hypotenuse would be 17,110,278,813 [square] *hu*, where the residue has been neglected. Extracting the square root, one gets the small hypotenuse of 1 *cun* 3 *fen* 8 *hao* 6 *hu*, where the residue has been neglected. This is the side of an [inscribed] 48-gon, which, when multiplied by the radius, 1 *chi*, and then by 24, yields an area of 3,139,344 million [square] *hu*. Dividing it by 10 billion, we have 313 584/625 [square] *cun*, i.e., the area of a 96-gon.

The rule for calculating the 96-gon from the 48-gon: as above, let the radius be the hypotenuse, half a side the *gou*, now find the *gu*. A quarter of the square on the small hypotenuse is 4,277,569,703 [square] *hu*, where the residue has been neglected. Take it as the square on the *gou*, and subtract it from the square on the hypotenuse. Extract the square root of the remainder, getting a *gu* of 9 *cun* 9 *fen* 7 *li* 8 *hao* 5 *miao* 8 9/10 *hu*. Subtracting it from the radius, the remainder is 2 *li* 1 *hao* 4 *miao* 1 1/10 *hu*, i.e. the small *gou*. Take half a side of the 48-gon as the small *gu*, one gets the small hypotenuse. The square on the small hypotenuse would be 4,282,154,012 [square] *hu*, where the residue has been neglected. Extracting the square root, one gets a small hypotenuse of 6 *fen* 5 *li* 4 *hao* 3 *miao* 8 *hu*, where the residue has been neglected. This is a side of the [inscribed] 96-gon which, when multiplied by the radius 1 *chi*, and then by 48, yields an area of 3,141,024 million [square] *hu*. Dividing by 10 billion, we get 314 64/625 [square] *cun*, i.e. the area of a 192-gon. When the area of a 96-gon is subtracted from it, the remainder 105/625 [square] *cun* is called the difference of areas. Doubling it, get 210/625 [square] *cun*, which is the total area of the 96 co-apothem rectangles outside the 96-gon of which the width and length are a co-apothem and a side of the 96-gon respectively. Adding the total area to that of the 96-gon, get 314 169/625 [square] *cun*, which is larger than the area of the circle. Therefore, it is convenient to take the integral part, 314 [square] *cun* of the area of a 192-gon, as the definite rate for the area of the circle, and neglect the fractional part.

The area of a circle divided by the radius, 1 *chi*, and doubled yields a circumference of 6 *chi* 2 *cun* 8 *fen*. The diameter squared yields the area of the [circumscribed] square as 400 [square] *cun*. By comparison with the area of the [inner] circle, the rates are 157 [area of circle] and 200 [area of square]. [In other words,] the area of the [circumscribed] square 200 contains its inscribed circle 157, which is still a bit less. In the diagram for a segmental field, the square contains an inscribed circle, which in turn contains an inscribed square; the area of the inner square is half the area of the outer one. So, the area of the circle 157 contains its inscribed square 100. Moreover, reducing the diameter, 2 *chi*, with its circumference 6 *chi* 2 *cun* 8 *fen* gives the reduced rates of circumference 157 and diameter 50. Note that the former is still a bit less [than the true value].

In the Royal Arsenal of the *Jin* Dynasty (265–420 CE) there was stored a bronze *hu* cast under the reign of Wang Mang in the Xin Dynasty (9–24 CE) [see the example depicted on p. 193 in Section I above], on which was inscribed: "2 LEGAL STANDARD MEASURING *HU*. THE INNER SQUARE IS 1 *CHI* SQUARE, OUTSIDE IS A CIRCLE. THERE ARE EXCESSES OF 9 *LI* 5 *HAO*. THE CIRCULAR BASE HAS AN AREA OF 162 [SQUARE] *CUN*. THE DEPTH IS 1 *CHI*. IT HOLDS 1620 [CUBIC] *CUN*. THE VOLUME OF THE *HU* IS DEFINED AS 10 *DOU*." By taking the above-mentioned rate, get the base area of 161 [square] *cun* and more, nearly the same as in the inscription.

The difference of areas 105/625 [between the 96-gon and the 192-gon] is a small amount. By the rate information contained in the area of a dodecagon, take 36/625 from the difference of areas and add it to the area of a 192-gon. Take the sum as the area of the circle, i.e. 314 4/25 [square] *cun*. The diameter squared yields the area of the [circumscribed] square as 400 [square] *cun*. Reducing it by the area of its inscribed circle, get rates of 5000 to 3927, which means that if the area of a square is 5000, that of its inscribed circle will be 3927; and if the area of a circle is 3927, that of its inscribed square will be 2500. Dividing the area of the circle 314 4/25 [square] *cun* by its radius 1 *chi* and doubling it, we obtain 6 *chi* 2 *cun* 8 8/25 *fen* as the circumference. Reducing the diameter 2 *chi* with the circumference we obtain the sought rate of the circumference to its diameter: 3927 to 1250. Such a rate should be precise enough. However, such a method is still too rough to use. So calculate the side of a 1536-gon for the area of a 3072-gon. With the residue omitted, obtain the same result as estimated, which is proved to be true.

Li et al.: In antiquity, the rate was always taken as 3 to 1; with this, however, one would have the circumference too short but the diameter too long, though it is correct for a hexagonal field. Why is this so? Suppose the side of the hexagon is 1 *chi*, its diameter [diagonal] would obviously be 2 *chi*, i.e., with the perimeter rate [being] 6, [or] 1 and 3.

We would like to give another example to elucidate this point. Make six equal equilateral triangles, with 1-*chi* sides and arrange them round a common point forming a regular hexagon with diagonal 2 *chi*. With the common point as center, and a side as radius, we draw a circle. The circumference passes through all the six corners of the hexagon, and encompasses the six sides, for the distance between two opposite sides is shorter than the diagonal. Here, the perimeter is 6 *chi*, the diameter 2 *chi*, and each side 1 *chi*, so that the distance between two opposite sides is less than 2 *chi*, therefore rates of 3 and 1 would mean a shorter circumference, and a longer diameter. This indicates that rates of 3 and 1 are not precise, but a simple shortcut. Liu Hui considered this as too crude, so sought new rates. But it is hard to determine a number exactly coincident with the product of the circumference and diameter, and neither of the rates Liu gave us seems to be exact. Zu Chongzhi considered them as rough enough to be studied further. In compiling and commenting on the book we have collected all the relevant views, examined which is right and which is wrong, and come to the conclusion that Zu's is the most precise.[15] So we add our humble opinion to Liu's comments for the study of future scholars.

Another Rule: One-fourth the product of the circumference and the diameter.

Liu: The circumference is still the perimeter of a polygon. The area of a circle should be the product of half the circumference by its radius, therefore the product of the circumference by its diameter is to be divided by 4. According to Liu's rate, multiplying the given circumference by 50, then dividing it by 157, get the diameter [and conversely] multiplying the given diameter by 157, and dividing it by 50, get the circumference. The new rate should still be slightly smaller—the diameter obtained by means of the circumference is too long, while the circumference obtained by means of the diameter is too short. The areas obtained by means of the diameter are somewhat smaller, while areas obtained by means of the circumference will be slightly larger.

Li et al.: According to the precise rate, multiplying the circumference by 7, and dividing it by 22, gives the diameter. Multiplying the diameter by 22 and dividing it by 7 gives the circumference. Calculate the required area by the Rule.

Another Rule: One-fourth the product of three times the diameter squared.

Liu: The diameter squared gives the area of the circumscribed square. Multiplying the square by 3 and dividing it by 4 implies that the [inner] circle occupies 3/4 of the square. If a side of the [inscribed] hexagon is multiplied by the radius, the product is 1/4 of the external square, which when tripled also equals 3/4 of the external square, which equals the area of the inscribed dodecagon. It is not precise to take it as the area of the circle. According to Liu's rate, the diameter should be squared and then multiplied by 157 and divided by 200.

Li et al.: According to the precise rate, the diameter squared, multiplied by 11 and divided by 14 yields the area of the circle.

---

[15]Zu determined the ratio of circumference to diameter to be between 3.1415926 and 3.1415927. For details, see [Li and Du 1987, 84] and [Martzloff 1997, 278–82]. In what follows, Li Chunfeng et al. understood Liu Hui's approximation of π to be better than using 3, but regarded 22/7 as an even more precise rate for π, whence their distinction of the "precise" rate here.

Another Rule: The circumference squared and divided by 12.

> Liu: The ratio of the perimeter of an inscribed regular hexagon to the diameter of the circle is 3 to 1, so the perimeter squared is 9 times the diameter squared, and equivalent to 12 [inscribed] dodecagons. So one-twelfth of that is the area of a dodecagon. Here the circumference squared is larger than 9 squares on the diameter, so that one-twelfth is not the exact area of a dodecagon. It is too large to be taken as the area of the circle. Just take one-twelfth of the perimeter of a hexagon. According to Liu's rate, the circumference squared, multiplied by 25 and divided by 314 yields the area of the circle. The rate is 25 for the area of the circle and 314 for the circumference squared. Suppose the circumference is 6 *chi* 2 *cun* 8 *fen*, its square is 394,384 [square] *fen*; and suppose the area of the circle is 394,384 [square] *fen*, one gets the rates by dividing both by 1256.

> Li et al.: Squaring the side gives the area of a square. To find the area of a circle by means of its circumference one resorts to the rate for the circumference. Yet this Rule takes rates of 3 and 1. The Rule for Circular Fields calls for the product of half the circumference and the radius. But here it says "the circumference squared and divided by 12." Why? To halve the circumference, one should divide it by 2; to find the radius from the circumference, one should divide it by 6. Therefore, the circumference squared should be divided by 2 times 6, i.e. 12. According to the precise rate, multiply it by 7, and divide it by 88.

Chapter 2: *Su mi* (Millet and Rice)

It may seem odd that the title of this section of the *Nine Chapters* includes rice when this commodity is significantly missing from the list of grains given in the conversion table that is at the head of this chapter. This is because, while *mi* is usually translated today as rice, in ancient Chinese it had a much more general meaning, and was more often a word used for a particular kind of millet, a grain that was much more common and abundant in ancient China than rice. Thus Jean-Claude Martzloff is technically correct in translating these terms, *su* and *mi*, as "decorticated" and "non-decorticated" grains, based upon an early commentator of the Song dynasty, Li Ji, who provided a glossary of terms that appear in the *Nine Chapters* [Martzloff 1997, 132]. A better translation of the Chinese than "millet and rice" would in fact be "grains."

Beginning with a table of conversion rates for the values of different grains, pulses, and their levels of refinement (e.g., hulled, milled, highly milled, coarse, and cooked), this chapter is not just about the computation of grain equivalencies, but includes a wide variety of commodities such as bricks, bamboo, silk, etc. Although most of the problems here are solved by the same method, the "rule of three," the last nine problems (38–46) do not involve proportions. Two specific rules are introduced in this chapter, the *Jinglu* (a rule for finding the unit price given the total cost of a certain commodity), and the *Qilu* (a rule for finding the cost and amounts of two different grades of a given commodity given their total price). These problems do not involve the rule of three, but proceed on the assumption that the difference in price between the superior and inferior items is a single coin. Thus problem 38 asks how many thick versus thin stalks of bamboo are bought for how much each if there are 78 bamboo altogether costing 576 coins. The method calls for a division of 576 by 78 (yielding 7 30/78), where the remainder 30 gives the number of thick bamboo stalks bought for 8 coins each; the remaining 48 thin stalks then cost one coin less, 7 coins per stalk.

Chapter 2: Millet and Rice

Liu: For barter trade and exchange of goods.

Millet and Rice Exchange Rule

millet rate	50	cooked imperial millet	42
hulled millet	30	soya beans	45
milled millet	27	small beans	45
highly milled millet	24	sesame seed	45
imperial millet	21	wheat	45
fine crushed wheat	13 ½	paddy	60
coarse crushed wheat	54	fermented beans	63
cooked hulled millet	75	porridge	90
cooked milled millet	54	cooked soya beans	103 ½
cooked highly milled millet	48	malt	175

Liu: The various exchange rates in the table are in proportion. They can be mutually converted by selecting the corresponding rates. Reduce whenever reducible. The [Exchange] Rule holds in other cases.

Rule of Three: Take the given number to multiply the sought rate. [The product] is the dividend. The given rate is the divisor. Divide.

Liu: This is a general rule. [The concept of] rate can be applied in various problems in the *Nine Chapters*. As the saying goes: "Knowing the past one can predict the future. Shown one corner [of a square], one can infer the other three." This Rule can resolve complicated and tricky problems and overcome the barriers between different quantities [of different items]. Take the rates from the given situation, clearly distinguishing their positions in the array, [then] homogenize and uniformize [the columns]. All of these ultimately depend on this Rule.

A few is the beginning of many; the unit is the foundation of number. Therefore the discussion of rates must be based on the unit. According [to the Exchange Rule] millet 5 and hulled millet 3. That is, millet 5 per unit, hulled millet 3 per unit. To exchange millet for hulled millet, first regard the [given] millet as unit. [One] unit means reducing [the given amount] by 5. That is 5 as unit. Multiply it by 3, that is [one] unit as 3. Hence the rates for [one] unit are equivalent, that is 5 to 3. If dividing before multiplying or there are fractions, so invert [the procedures in]

> the Rule. Looked at as integers, 5 *sheng* of millet is equivalent to 3 *sheng* of hulled millet. Looked at as fractions, 1 *dou* of millet is equivalent to 3/5 *dou* of hulled millet. Use denominator 5, numerator 3. Converting millet to hulled millet: multiply by the numerator and divide by the denominator. Hence the sought rate is always the [numerator and the given rate is the] denominator.

1. Now millet, 1 *dou*, is required as hulled millet. Tell: how much is obtained? Answer: As hulled millet: 6 *sheng*. Method: Taking millet is required as hulled millet, multiply by 3, divide by 5.

> Li et al.: The general rule is "Take the given number to multiply the sought rate. [The product] is the dividend. The given rate is the divisor. Divide." This problem takes the millet required as hulled millet. So the millet is the given number. 3 is the rate for hulled millet, so 3 is the sought rate. 5 is the rate for millet, so 5 is the given rate. The rate for millet is 50 and the rate for hulled millet is 30. Shift positions backwards to find [the answer]. Hence the method says "3" and "5."

9. Now 8 *dou* 6 *sheng* of millet is required as cooked highly milled millet. Tell: How much is obtained? Answer: As cooked highly milled millet 8 *dou* 2 14/25 *sheng*. Method: Taking millet is required as cooked highly milled millet multiply by 24, divide by 25.

> Li et al.: The rate for cooked highly milled millet is 48. Halve both the rates before multiplying and dividing.

14. Now 10 *dou* 8 2/5 *sheng* of millet is required as wheat. Tell: How much wheat is obtained? Answer: As wheat 9 *dou* 7 14/25 *sheng*. Method: As a general rule, taking millet is required as soya beans, small beans, sesame seed [or] wheat, multiply by 9, divide by 10.

> Li et al.: The rates for all the four [cereals] are 45, they are sought for a given amount of millet. In short, reduce both rates by the greatest common divisor, the sought rate is 9 and the given rate is 10. Hence the method "multiply by 9, divide by 10."

32. Now pay 160 coins to purchase 18 bricks. Tell: How much is each [brick]? Answer: 8 8/9 coins a brick.

33. Now pay 13,500 coins to purchase 2350 bamboo. Tell: How much is each [bamboo]? Answer: 5 35/47 coins a bamboo.

The *Jinglu* Rule [A] (Direct Rule [A]): Take the purchasing rate as divisor, the number of coins paid as dividend. Divide.

> Liu: According to this Rule, the number of coins paid is the given number, one [item] is the sought rate, the number of items purchased is the given rate. By applying the Rule [of Three], obtain the sought number. A number multiplied by one remains unchanged, so [we] need not apply

multiplication. Directly divide the number of coins paid by the number of items purchased, obtaining the unit price. If there is a remainder, reduce by the greatest common divisor, and express the number as a fraction.

34. Now pay 5785 coins to purchase 1 *hu* 6 *dou* 7 2/3 *sheng* of lacquer. Tell: How much is 1 *dou*? Answer: 1 *dou* [is] 345 15/503 coins.

37. Now pay 13,670 coins to purchase 1 *dan* 2 *jun* 17 *jin* of silk. Tell: How much is 1 *dan*? Answer: 1 *dan* [is] 8326 178/197 coins.

The *Jinglu* Rule [B] (Direct Rule [B]): Take the sought rate to multiply the number of coins as dividend, take the purchasing rate as divisor. Divide.

Li et al.: According to the Rule of Three, the number of coins paid is the given number, 1 *dou* is the sought rate. So take it to multiply the number of coins, and multiply again by the denominator as dividend. The amount purchased is the given rate. If there are fractions, uniformize [the denominators], the numerator of the improper fraction is the divisor. Divide to obtain the unit price. If there is a remainder, it should be considered as numerator, [and] the divisor as denominator. Express the number of coins less than 1 by a fraction.

38. Now pay 576 coins to purchase 78 bamboos. [They are] classified into thicker and thinner ones. Tell: How much does each of them cost? Answer: 7 coins for 48 [thinner ones], 8 coins for the remaining 30.

43. Now pay 12,970 coins to purchase 1 *dan* 2 *jun* 28 *jin* 3 *liang* 5 *zhu* of silk. It is classified as superior and inferior [in quality]. Tell: How much does each [type] cost per *liang*? Answer: 4 coins per *liang* for 1 *dan* 1 *jun* 17 *jin* 14 *liang* 1 *zhu* of one type, 5 coins per *liang* for the remaining 1 *jun* 10 *jin* 5 *liang* 4 *zhu*.

The *Qilu* Rule: Lay down the amount purchased in *dan*, *jun*, *jin* and *liang* as divisor. Take the given rate (i.e., the number of *zhu* in the unit measure for the unit price) to multiply the number of coins as dividend. Divide. The remainder is [the number of *zhu* in weight of] the superior quality; subtracting the remainder from the divisor gives [the number of *zhu* in weight of] the inferior quality. Reduce the weight in *zhu* to *dan*, *jun*, *jin*, *liang* and *zhu* by continued division by the number of *zhu* in these units. The [final] remainder is the number of *zhu*.

Liu: If no fraction is desired [in the answer], then [for instance] in Problem 38, "pay 576 coins to purchase 78 bamboos" means divide the number of coins by the number of bamboos to obtain [coins a bamboo], with a remainder of 30 [coins]. So increase the price by 1 [coin] for each of the 30 [thicker] ones. The remainder is just the same as the number of the superior quality. Hence the rule says: "The remainder is [the number of the] superior quality." Initially 78 is the divisor, now subtract from it [the number of the] superior quality, then the remainder is [the number of] the inferior quality. Hence the Rule says: "Subtracting the remainder from the divisor gives [the number of] the inferior quality." "Reduce the weight in *zhu* to *dan*, *jun*, *jin*, *liang* and *zhu* by

continued division by the number of *zhu* in these units." This means the number of *zhu* of the superior and inferior quality items are divided by the number of *zhu* in each of the weight units: *dan, jun, jin, liang,* respectively. The [final] remainder is the number of *zhu* which is just what the problem required.

45. Now pay 620 coins to purchase 2100 arrow feathers. They are classified as superior and inferior [in quality]. Tell: How much does each [type] cost? Answer: 3 feathers a coin for 1140 of one type, 4 feathers a coin for the remaining 960 feathers.

The Inverse *Qilu* Rule: Take the number of coins paid as divisor, multiply the given rate by the amount purchased as dividend. Divide. The remainder is the number of coins for the inferior quality items. Subtract the remainder from the divisor to give the [number of coins for the] superior quality items. Multiplying them, respectively, by the corresponding number of items a coin, obtain the number of items.

Li et al.: The *Qilu* Rule means coins are more [and] items are less; the Inverse *Qilu* Rule means coins are less [and] items are more. More and less are opposites, so says the Inverse *Qilu* Rule.

Liu: According to the *Qilu* Rule, pay 620 coins to purchase 2100 arrow feathers. Reverse the order: 240 coins at 4 feathers a coin, for the remaining 380 coins, 3 feathers a coin. This is purchasing items at different prices according to their difference in quality. Thus multiply [the number of] feathers a coin by the number of coins. This is the Inverse *Qilu* Rule.

Li et al.: In the *Qilu* Rule the number of items is the divisor and that of the coins is the dividend. Conversely, in the Inverse *Qilu Rule,* the number of items is the dividend, [and] the number of coins is the divisor. Here the remainder, [which is] less than the divisor, is the total number of coins paid in the problem. Subtract the remainder from the divisor. Obtain the number of coins paid for the superior quality. The larger number of items a coin multiplies the number of coins for the inferior quality, and the smaller number of items a coin multiplies the number of coins for the superior. Hence the Rule reads: "Multiplying them, respectively, by the corresponding number of items a coin, obtain the number of items."

Chapter 3: *Cui fen* (Proportional Distribution)

The 20 problems in this chapter concern direct, inverse, continued, and compound proportion, some of which may be solved by the rule-of-three method presented in Chapter 2. The "Proportional Distribution Rule" describes the procedure that would be followed on the counting board to solve these problems. Problem 2, for example, asks how 5 *dou* of millet should be distributed among three animals, a cow, a horse, and a sheep, knowing the sheep eats half as much as the horse, and the horse eats half as much as the cow. This means the "rates" for the sheep, horse, and cow are 1, 2, and 4, respectively; the sum of the rates yields the divisor, 7; multiplying each of the rates times the 5 *dou* as dividend means that the amount to be ascribed to the sheep in *dou* is 5/7, the horse 10/7, and the cow 20/7. The answer for the problem is given, however, in terms of *sheng*, there being 10 *sheng* to the *dou*. This was done no doubt to avoid "improper" fractions, reducing all of the answers to the same units with only proper fractions in each, so that the sheep owes 50/7 or 7 1/7 *sheng*, the horse 1 *dou* 4 2/7 *sheng*, and the cow 2 *dou* 8 4/7 *sheng*. Note that Problem 4 is virtually the same problem found in the *Suan shu shu*, which attests that it was a popular problem of long standing among Chinese mathematicians.

Chapter 3: Proportional Distribution

Liu: For distribution of grain and taxation.

Proportional Distribution Rule: Lay down the rates for distribution, then add; take the sum as divisor; multiply the amount to be distributed by each rate as dividends. Divide. If there is a remainder, take it as numerator, and the divisor as denominator.

Liu: Proportional Distribution means unequal distribution. The distribution rates laid down are reduced rates. If they have a common factor, reduce them. Add up the rates, then distribute by the rates. Originally there is only one number. Multiply it by each rate and divide it by their sum. Multiplication and division cancel each other, so the given amount has just been distributed. Thus the amount remains unchanged, though the shares differ according to the rates. In the light of the Rule of Three, each rate for distribution is the sought rate, the sum is the given rate, the given amount is the given number. By the Rule of Division, assuming there are three persons in family A, two persons in B, and one person in C, the 6 persons in total share 12 objects. Each gets 2. To know how many objects each family gets, lay down the number of persons [in each family] and multiply by the number of objects for each person. Now by the proportional Distribution Rule, multiply then divide.

2. Now given a cow, a horse and a sheep have eaten up the seedlings of someone's field. The landlord demands 5 *dou* of millet as compensation. The shepherd says, "My sheep eats half as much as the horse." The horse owner says, "My horse eats half as much as a cow." The compensation is to be distributed according to the rates. Tell: how much should each repay? Answer: The cow owner repays 2 *dou* 8 4/7 *sheng*, the horse owner 1 *dou* 4 2/7 *sheng* and the shepherd 7 1/7 *sheng*. Method: Lay down the rates for distribution: cow 4, horse 2 and sheep 1. Take their sum as divisor. Multiply 5 *dou* by each rate as dividend. Divide, giving the number of *dou*.

Li et al.: The problem says that a sheep eats half as much as a horse, a horse eats half as much as a cow, i.e., four sheep eat as much as a cow, and two sheep eat as much as a horse. The Method sets down [the rates] a sheep 1; a horse 2; and a cow, 4. This means that the rates are reduced to a common denominator for distribution.

4. Now given a skillful weaver, who doubles her product every day. In 5 days she produces [a cloth of] 5 *chi*. Tell: how much does she weave in each successive day? Answer: On the first day she weaves 1 19/31 *cun*; on the second day, 3 7/31 *cun*; on the third day, 6 14/31 *cun*; on the fourth day, 12 28/31 *cun*; and on the fifth day, 25 25/31 *cun*. Method: Lay down the rates for distribution: 1, 2, 4, 8, and 16. Take their sum as divisor. Multiply 5 *chi* by each rate as dividend. Divide, giving the number of *chi*.

Inverse Proportional Distribution Rule: Lay down the rates for distribution as the unchanged Rates, the continued product by the other rates as the changed rate. The changed rates replace the unchanged ones as rates for distribution.

8. Now given five officials of different ranks: *Dafu*, *Bugeng*, *Zanniao*, *Shangzao*, and *Gongshi*. They should pay a total of 100 coins. If the payment is to be shared in accordance with ranks, the higher pays the less, and the lower pays the more. Tell: how much should each pay? Answer: *Dafu* pays 8 104/137 coins, *Bugeng* pays 10 130/137 coins, *Zanniao* pays 14 82/137 coins, *Shangzao* pays 21 123/137 coins and *Gongshi* pays 43 109/137 coins. Method: Lay down the reciprocals of the ranks as rates for the shares. Take their sum as divisor. Multiplying the 100 coins by each rate, we have each dividend. Divide, giving the number of coins.

> Liu: The weight of the ranks is 5 for *Dafu* and 4 for *Bugeng*, [etc.]. To make the higher rank get more, a *Dafu* gets 5 shares, a *Bugeng* gets 4 shares, [etc.]. The number of persons is the denominator and the number of shares is the numerator. The denominators are uniformized, then the numerators can be homogenized. The numerators are the rates for distribution. Hence the rates for distribution are 5, 4, ... Here, let the higher rank pay less, and 5 *Dafu* pay 1 portion, 4 *Bugeng* pay 1 portion, [etc.]. Hence [the rule] is called the Inverse Proportional Distribution Rule. The number of officials [of the same rank] to share one portion varies according to their rank. If different denominators, the numerators are not homogeneous, multiply the numerator by the [other] denominators, and then divide [the denominator] by each of the denominators so as to get the inverse rates. Alternatively, uniformize the denominators and then divide by each of the denominators, so as to get the inverse rates for distribution. Take the sum of rates a divisor. Multiply the amount to be distributed by each rate as dividend. Divide, giving the answer.

20. Now given one lends 1000 coins at a monthly interest of 30 coins. Tell: given one lends 750 coins for 9 days, how much interest? Answer: 6 3/4 coins. Method: Take one month as 30 days to multiply 1000 coins as divisor. Multiply the interest 30 by number of coins lent, [and] again by 9 days as dividend. Divide, giving the number of coins.

> Liu: Multiply 30 days by 1000 coins as divisor; it is 30,000, i.e., 30,000 coins, the daily interest is 30 coins. Take 9 days to multiply the coins lent as the coins lent for one day, which according to the Rule of Three are

the given number. The interest, 30 coins, is the sought rate, 30,000 coins are the given rate. Alternatively, divide the monthly interest, 30 coins, by 30 days in a month. The daily interest is 1 coin. Multiply the daily interest by the coins lent for one day as dividend, take 1000 coins as divisor. Pay much attention to the rates; they should be converted into 1. Thus 30 days may be multiplied by the principal or divide the interest. They are equivalent.

Chapter 4: *Shao guang* (Short Width)

All of the 24 problems in this chapter concern division with fractions, or the computation of square and cube roots, or problems involving the area of the square and the volume of a sphere. The first eleven problems begin with a given rectangular area of 1 *mu*, and given widths that increase by $1/n$; the goal is to determine the length of the rectangle. Thus beginning with the first problem with a width of $1 + 1/2$, to the last in the series, $1 + 1/2 + 1/3 + \cdots + 1/12$, to find the length that in each case will yield a rectangle 1 *mu* in area requires first uniformizing and homogenizing the fractions, and then dividing the area to find the length of the side in question. Note that Li Chunfeng, in his commentary on Problem 11, points out that it is "better to solve problems in a simplified way," and offers the lowest common denominator, 27,720, instead of the "uniformized" denominator of 82,160 used in the original statement of the method for this problem, which is too large by a factor of 3.

Problems 12 and 16 are devoted to finding the side s of a square given its area S. The technical term in Chinese for finding square roots is *kai fang*, which literally means "open the square," which the Chinese method proceeds to do as follows. The algorithm for extracting the square root s begins by determining the greatest whole number a whose square is less than the given area S, i.e., $a^2 \leq S$. If $a^2 = S$, then a is the square root of S; assuming $a^2 < S$, taking a as the first approximation of the square root, what remains from the square S is the gnomon as in the red/blue portion of the figure. In Liu Hui's commentary on the square root method, he refers to the given area S as the *shi*, and the first approximation, the a, as the *fa*. This terminology stems directly from the vocabulary for division, where the *shi* and *fa* are the dividend and divisor, respectively, and shows the close affinity between division and the method of root extraction. The first approximation, or the area a^2, is the "yellow" area to which Liu Hui refers. The second approximation of the square root involves the computation of $2a$ (Liu Hui calls this the "determined *fa*") and then finding the largest integer b such that $a^2 + 2ab + b^2 \leq S$. The gnomon represented by the area $2ab + b^2$ is what Liu Hui calls the "red" area in the diagram to which he refers. The rationale behind this computational procedure is apparent from the figure.

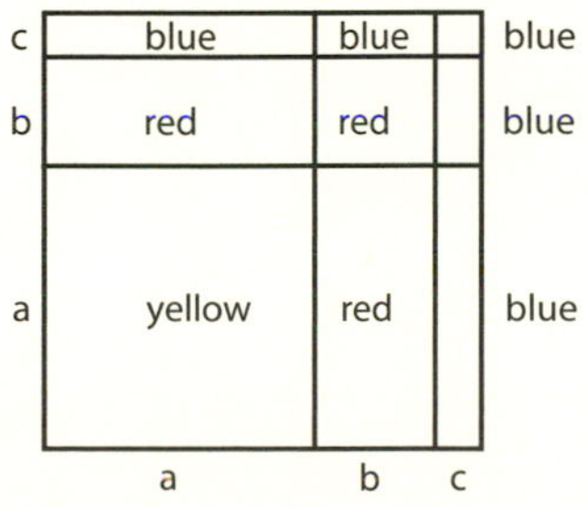

c blue blue blue
b red red blue
a yellow red blue
a b c

If $a^2 + 2ab + b^2 = S$, then the "extraction" is finished and the square root has been determined. But if $a^2 + 2ab + b^2 < S$, then the algorithm proceeds to finding the largest value c such that $a^2 + 2ab + b^2 + 2ac + 2bc + c^2 \leq S$ (the gnomon $2ac + 2bc + c^2$ is what Liu Hui calls the "blue" area in his commentary).

In Problem 12, to find the square root of 55,225, the algorithm begins by finding the largest integer whose square is less then 50,000, i.e., 200. Since this problem is solved algorithmically on the counting board using counting rods, it would begin with the amount 55225 on the board (the *shi*, the amount first put down on the counting board to begin solving the problem), from which the 200 squared would be placed on the board and subtracted from the 55525 leaving as a remainder the "gnomon" of 15225. To find the second approximation of the square root, b, the amount $2a$ would be placed on the board, i.e., 400, and then the largest integer value for b would be determined, in this case 30, so that the sum of 12000($2ab$) and 900(b^2) is now subtracted from the "gnomon" leaving 2325 on the board. The final step is now to determine the largest integer c such that the remaining "gnomon" of 2325 is further reduced by the amount $2ac + 2bc + c^2$. In this case it turns out that $c = 5$, and since 2000($2ac$) + 300($2bc$) + 25(c^2) = 2325, the "gnomon" is exhausted and we have thus computed the exact square root of 55,225.

Although the problems in the *Nine Chapters* all turn out to be perfect squares, the commentators know that in reality not all numbers are so amenable to extraction, and in such cases, as Liu Hui says, as for the square root of 10, an approximation may be obtained such that "the more the digits, the finer the fractions, till the number omitted from the area of the red areas is negligible," meaning that the approximation of the square root can be determined to whatever degree of accuracy one may wish. Some have even taken Liu Hui's qualification—"To express it by the side of [the square on which is the given *shi*] is the only true way"—to mean that he knew the side of the square of area 10 was "irrational," but this is a controversial matter about which there is no general agreement among historians of Chinese mathematics [see Chemla and Keller 2002].

Problem 18 makes use of the square root algorithm to determine the circumference of a circle given its area. All of the problems involving circles as originally given in the *Nine Chapters* proceed on the assumption that the ratio of the circumference C to the diameter d of a circle is 3, an approximation that was common in early Chinese (and western) mathematics. But as Liu Hui and later commentators are quick to point out, this is a very bad approximation. (See commentary in chapter 1.)

Problem 19 extends the "*kai fang*" method for extracting square roots to an analogous method for extracting cube roots, known as "*kai li fang*," or "open the cube." To find the side of the cube whose volume is 1,860,867 [cubic] *chi*, the algorithm begins with the approximation of the largest integer a such that $a^3 \leq S$ (in this case $a = 100$), which upon subtraction when cubed from the volume of the given cube leaves a 3-dimensional gnomon of 860,867. The second approximation of b such that $a^3 + 3a^2b + 3ab^2 + b^3 \leq S$ results in $b = 20$ and a remaining three-dimensional gnomon of 132,867. The final step results in the exact cube root of 1,860,867 equal to 123. In the diagram, x_1 represents the hundreds digit, y_1 the tens digit, and z_1 the units digit.

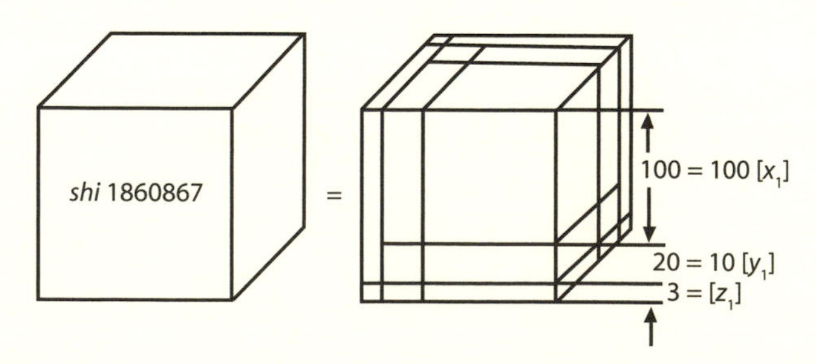

The final problem in Chapter 4 of the *Nine Chapters* is the famous and difficult problem concerning the volume of a sphere. In this case, given a sphere of 4500 (cubic) *chi*, the diameter is determined as 20 *chi.* This assumes that the ratio of the volume of a cube to that of the inscribed sphere is 16 : 9, and that $\pi = 3$. This prompts the longest commentary in the *Nine Chapters*, and Liu Hui has to admit that in fact the complete solution of the problem is beyond his abilities. The *mou he fang gai* (joined umbrellas) and the *yang ma* to which he refers in his commentary on the sphere are illustrated here.

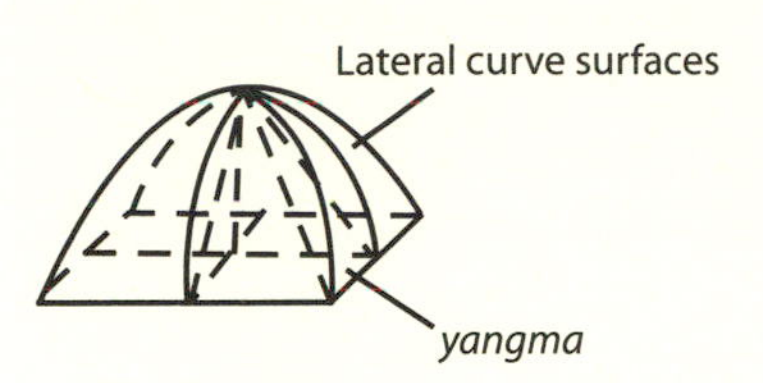

Although Liu attempted a construction to approximate the volume of a sphere in terms of two cylinders circumscribing the sphere at right angles to each other, he was unable to calculate the volumes of the "square umbrella" figures as he described them—and candidly admitted: "Wanting to get the volume of the shape, afraid of losing the truth, not daring to guess, I wait for a capable person to solve it" [based on the translation in Li and Du 1987, 85].

This comment is followed by further remarks by Li Chunfeng and his collaborators, who state that it was Zu Geng who found the solution to the problem. In fact, some 250 years after Liu Hui, the mathematician Zu Chongzhi (429–500 CE) devoted himself to "the error on the sphere (in the *Nine Chapters*), which Zhang Heng mentioned but left uncorrected. It is a scar on mathematics … I have corrected these errors of the past …" [Li and Du 1987, 85]. The commentary in the *Nine Chapters*, however, ascribed this discovery to Zu Chongzhi's son, but in any case, it was Zu senior and junior who deserve the credit in China for having solved the problem of finding the volume of the sphere. Their method is akin to that of Cavalieri by matching cross-sections of planes cut through sections of the sphere and a pyramid of height and side equal to the radius of the sphere. Thus Zu Chongzhi managed to compute the volume of the sphere of diameter d as $V = (1/2)d^3$, or at the "precise rate" for π, $V = (11/21)d^3$.

Chapter 4: Short Width

Liu: For determining areas (volumes) of squares (cubes) and circles (spheres) and their relations.

Short Width Rule: Lay down the integer *bu* and the fractional parts [of the width of a rectangular field]. Multiply the integer and all the numerators by the largest denominator. Divide each numerator by its denominator. Lay [the results] down on the left. Successively multiply the numerators and the reduced integers by the [next larger] denominator. [Continue the process till] all the fractional parts are turned into integers. Add as divisor. Multiply the number of [square] *bu* in a *mu* by the least common multiple of the integers as dividend. Divide giving the number of *bu* in length.

Li et al.: A field of 1 *mu* is 1 *bu* in width and 240 *bu* in length. Now increase its width by shortening the length. Hence [this Rule is] called Short Width. Multiply the integer *bu* by the denominator for the common denominator, and multiply numerators by the denominator to homogenize the numerators. When all the fractions have the same denominators, they can be added together as divisor. The Rule for Addition of Fractions is not suitable here for the continued multiplication of so many terms. It would be too complicated. Thus the new Rule is meant to simplify calculation.

Liu: In the Rule, take the width as divisor, [and] the number of [square] *bu* in a *mu* as dividend. There are fractions in the divisor, uniformize the denominators [and] homogenize the numerators. Take the uniformized denominators to multiply the divisor as dividend. Add the homogenized numerators together and take the sum as divisor. Now multiply the integer and each of the fractions by the denominator, and divide each product by the denominator. The sum is the divisor. So both the divisor and dividend are increased to the same extent, therefore the quotient remains unchanged. Divide giving the number of *bu* in the length of the field.

1. Now given a [rectangular] field whose width is 1 1/2 *bu*. Assume the area is 1 *mu*. Tell: what is its length? Answer: 160 *bu*. Method: The denominator is 2. Take the unit to be 2, and 1/2 to be 1. Adding them get 3, which is the divisor. Lay down 240 [square] *bu*, again take the unit to be 2. Multiply as dividend. Divide giving the number of *bu* in length.

11. Now given a [rectangular] field whose width is 1 1/2 *bu*, plus one-third, one-fourth, one-fifth, one-sixth, one-seventh, one-eighth, one-ninth, one-tenth, one-eleventh, and one-twelfth. Assume the field is 1 *mu*. Tell: what is its length? Answer: 77 29,183/86,021 *bu*. Method: The largest denominator is 12. Take the unit to be 83,160, one-half to be 41,580, one-third to be 27,720, one-fourth to be 20,790, one-fifth to be 16,632, one-sixth to be 13,860, one-seventh to be 11,880, one-eighth to be 10,395, one-ninth to be 9240, one-tenth to be 8316, one-eleventh to be 7560, and one-twelfth to be 6930. Adding them get 258,063, which is the divisor. Lay down 240 [square] *bu*, again take the unit to be 83,160. Multiply as dividend. Divide giving the number of *bu* in length.

Li et al.: It is better to solve problems in a simplified way. Here the largest denominator is 12, it is better to take the unit to be 27,720, one-half to be 13,860, one-third to be 9240, one-fourth to be 6930, one-fifth to be 5544, one-sixth to be 4620, one-seventh to be 3960, one-eighth to be 3465, one-ninth to be 3080, one-tenth to be 2772, one-eleventh to be 2520, one-twelfth to be 2310. Adding them get 86,021, which is the divisor. Lay down 240 [square] *bu*, again take the unit to be 27,720. Multiply as dividend. Divide giving the answer. Here the solution is appropriate, and simpler than the original one.

12. Now given an area 55,525 [square] *bu*. Tell: what is the side of the square? Answer: 235 *bu*.

16. Given again an area 3,972,150,625 [square] *bu*. Tell: what is the side of the square? Answer: 63,025 *bu*.

The Rule for Extracting the Square Root:

Liu: The area of a square is given, find its side.

Lay down the given area as *shi*. Borrow a counting rod to determine the digital place. Set it under the unit place of the *shi*. Advance [to the left] every two digital places as one step.

Liu: This means that if the area is counted by hundreds, the side [is counted] by tens, and if the area [is counted] by ten thousands, the side [is counted] by hundreds.

Estimate the first digit of the root. The estimated number multiplied by the borrowed rod is regarded as *fa*. Then carry out the subtraction.

Liu: First estimate the side of yellow area *A*. To subtract repeatedly the product of the first digit of the root and the borrowed rod means to subtract the self-product of the side of *A* from the *shi*.

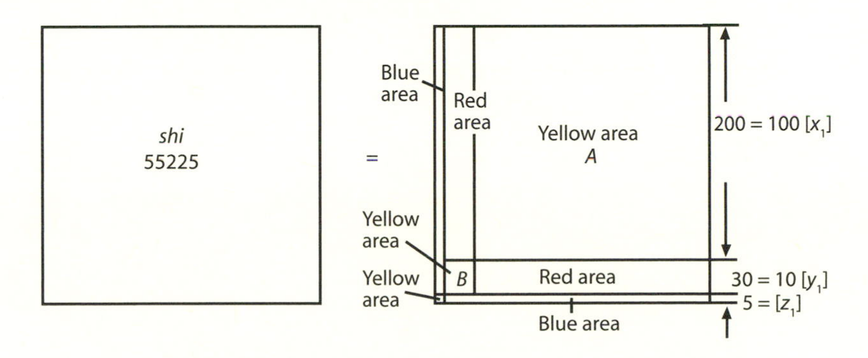

This diagram identifies the colored regions to which Liu Hui refers in his commentary on the process for extracting the square root.

After that, double the *fa* as determined *fa*.

Liu: Double the *fa* so as to join the two red areas as a longer one for determining its length in preparing the second subtraction, hence the determined *fa*.

Prepare the second subtraction. Move the determined *fa* one digit [to the right] and set the borrowed rod as before.

Liu: In order to subtract the red areas, double the square obtained as the determined *fa*. Move one digit, then estimate the second digit of the root. Multiply and subtract. For a correct digital position, move the determined *fa*.

Estimate the second digit of the root. Multiply it by the borrowed rod and subjoin the product to the determined *fa*.

Liu: As has been said before, this is to subtract the yellow area *B* between the two red areas.

Then [carry out] the second subtraction. Subjoin to the determined *fa* a second time.

Liu: Subjoin the determined *fa* to the side of the yellow area *B*, so as to lengthen the side of the two blue areas.

Proceed with the operation in the same manner. If there is a remainder, [the number] is called unextractable, it should be defined as the side on which the square has the area of the *shi*.

Liu: One may also take as the answer: a fraction with the borrowed rod plus the determined *fa* as denominator, but it is only a rough approximation which cannot be used. In general, the square root of an area should, when squared, be restored to itself. [The determined *fa*] without the borrowed rod [the answer] is too small, whereas with the borrowed rod, it is too large. So neither can be used to determine the exact value. To express it by the side of [the square on which is the given *shi*] is the only true way. For example, 10 divided by 3 is equal to 3 with remainder 1. It may be restored to the dividend 10. If one does not express it by the side, one may continue to subjoin the determined *fa* as before, to find the digits, with the digits as numerators, and 10, 100, ... as the denominators. The more the digits, the finer the fractions, till the number omitted from the area of the red areas is negligible.

If the *shi* has a fractional part, reduce it to [an] improper [fraction]. Extract the square root of both the numerator and denominator, which are considered as dividend and divisor respectively, and divide. If the denominator is unextractable, multiply the numerator by the denominator. Extract the root of the product, then divide the root by the denominator.

Li et al.: If the denominator can be extracted, regard it as formed by multiplying two denominators. There remains one denominator after the extraction, thus take one denominator as divisor in division. If the denominator is unextractable, it is multiplied by itself, i.e., there would be two denominators, the square root of which remains one. Thus the square root of the product should be divided by one denominator, and the side to be found is obtained.

Something more about the Rule: "Extracting the square root." Find the side of the square with given area.

"Borrow a counting rod." A borrowed rod is used only for correct determination of the digital place, and the rod itself does not join in the operation. To find the side of the square with given area it is therefore set at the bottom.

"Advance [to the left] every two digital places as one step." If the side of a square is 10, its area is 100. If the side is 100, its area is 10,000. Thus when the borrowed rod is set at the 100s place, the corresponding first digit of the root is at 10, and when it is at the 10,000s, the corresponding first digit of the root is at the 100's.

"Estimate the first digit of the root. The estimated number multiplied by the borrowed rod is regarded as the *fa*. Then carry out the subtraction." First estimate the side of the yellow square A. In fact, the area of a square is the product of two sides. Thus in extracting the square root one should multiply the two sides, i.e. from the *shi* subtract the self-product of the side.

"After that, double the *fa* as the determined *fa*." The number extracted has a remainder, as the subtraction should be carried on. Join the two red areas into a longer one, so as to determine its length for the second subtraction. Hence the determined *fa*.

"Prepare the second subtraction: Move the determined *fa* one digit [to the right]." In order to subtract the red area, take twice the area as the determined *fa*. Move right and estimate the second digit of the root. Multiply and subtract. So move the determined *fa* and then subtract.

"Set the borrowed rod as before. Estimate the second digit of the root. Multiply it by the borrowed rod, and subjoin the product to the determined *fa*. Then [carry out] the second subtraction." This is for subtracting the yellow area B between the two red areas.

"Subjoin to the determined *fa* a second time." It means to add the determined *fa* to the yellow area B so as to lengthen the two blue areas. Continue the operation for the required result.

18. Again given a circular area 300 [square] *bu*. Tell: what is the circumference? Answer: 60 *bu*.

Liu: According to Liu's rate, the circumference should be 61 19/50 *bu*.

Li et al.: According to the precise rate, the circumference should be 61 41/100 *bu*.

The Area-Circumference Rule: Lay down the given number of [square] *bu*. Multiply it by 12. Extract the square root of the product giving the circumference.

Liu: The Rule takes a rate of diameter 1 to circumference 3. It is the reverse operation of the Rule for Circular Fields. According to Liu's rate, the area should be multiplied by 314 and divided by 25. Extract the square root of the result giving the circumference. Its square root is the diameter. The solution finds the circumference thus obtained is a little bit short. If it is multiplied by 200 and divided by 157, then its square root is the diameter, which is too long.

Li et al.: In Liu's comment there were no such words as "And its square root is the diameter." They might be a redundancy in copying. According to the precise rate, it should be multiplied by 88 and divided by 7. According to a rate of circumference 3 to diameter 1, let the circumference be 6 and the diameter 2. Half the circumference multiplied by the radius yields an area of 3, and the square of the circumference 6 is 36. Reducing both of them by the greatest common divisor, we have the rate for the circumference [square] 12. The product, square of circumference should be multiplied by 1, and then divided by 12; the real result would be 3. It remains unchanged with the multiplication by 1, as when it is divided directly by 12, obtaining the area. For restoration, lay down the area 3, have it multiplied by 12, and then restore the square

of the circumference. In general, by squaring a number, and then extracting its square root, one restores the original number. Similarly, extracting the square root, one obtains the circumference.

19. Now given a volume of 1,860,867 [cubic] *chi*. Tell: what is the side of the cube? Answer: 123 *chi*.

Liu: The length, width and height of a cube are equal. If the volume of a cube is known extracting its cube root is to find its side.

The Rule for Extracting the Cube Root: Lay down the given number of cubic *chi* as the *shi*. Borrow a counting rod to determine the digital place. Set it under the unit place of the *shi*. Advance [to the left] every three places as one step.

Liu: This is to say, if the volume is counted by thousands, its side [is counted] by tens, and if the volume by millions, its side by hundreds.

Estimate the first digit of the cube root. The square of the digit estimated, multiplied by the borrowed rod, is regarded as the *fa*. Then carry out the subtraction.

Liu: "The square of the digit estimated, multiplied by the borrowed rod, is regarded as the *fa*. Then carry out the subtraction" means subtract the *fa* from the given volume.

After that, triple the *fa* as the determined *fa*.

Liu: For the second subtraction triple the *fa* so as to put the three cubes together for determining their total top-face area. Hence the determined *fa*.

Prepare the second subtraction. Move the determined *fa* one digit [to the right].

Liu: In carrying out the second subtraction the top-face areas are all square numbers. They should be moved a digit in order to determine their thickness. In extracting the square root, count the area by the hundreds, and the side by the tens, whereas in extracting the cube root, count the volume by the thousands and the side by the tens. However, the determined *fa* has square numbers, so in the second subtraction thousands should be changed to hundreds, i.e. move one digit [to the right].

Triple the number obtained and arrange it in the *zhonghang*.

Liu: This is for determining the total length of the three square bars.

Borrow another rod in the *xiahang*.

Liu: This is for expressing the volume of the cubic block, whose size is indeterminate. The borrowed rod is only to note the digit place of the units.

Move them [to the right]: the former two digits, and the latter three digits.

> Liu: The upper *fa*, the square plates with two sides of the cube, should be drawn one digit back so as to determine their thickness. The *zhonghang*, the square bar with one side of the cube, should be drawn back two digits so as to determine its base area. The *xiahang*, the cubic corner block without the side of the cube, should be drawn back three digits so as to determine its volume.

Estimate the second digit of the cube root. Multiply the digit by the *zhonghang*, then the square of the digit is multiplied by the *xiahang*.

> Liu: [The *zhonghang*] is the total volume of one layer of three square bars. [The *xiahang*] is the total volume of one layer of cubic corner blocks.

Subjoin both to the determined *fa* for the second subtraction.

> Liu: The total volume of one layer of square plates, square bars and cubic corner blocks is known. The second division is to cut off, layer by layer, the volume of the three kinds of solids introduced above.

Double the *xiahang*. Subjoin it to the *zhonghang* once more. Again add the sum to the determined *fa*.

> Liu: In a word, twice the *zhonghang* and thrice the *xiahang* are subjoined to the determined *fa*, because each of the three bars has two lateral faces attached to the two neighboring square plates respectively, and three faces of the cubic corner block are attached to the end faces of the three neighboring bars. They are ready for the third subtraction. It is difficult to convey the whole idea in words, and the relation must involve geometrical models to make it clear to us.

Continue in like manner. If there is a remainder, [the number] is called unextractable.

> Liu: One might as well consider the determined *fa* as denominator, but the answer would only be a rough approximation. It is better to repeatedly extract the cube root of the remainder to more digits.

If the *shi* has a fractional part, reduce it to [an] improper [fraction]. Extract the cube root of both the numerator and denominator, which are considered as dividend and divisor respectively, and divide. If the denominator is unextractable, multiply the numerator by the square of the denominator and extract the cube root of the product, then divide the root by the original denominator.

[We present here the counting rod procedure for problem 19, extracting the cube root of 1,860,867. Here *Y* stands for the cube root, *S* for the *shi*, *F* for the *fa*, *Z* for the *zhonghang*, *X* for the *xiahang*, and *J* for the borrowed rod.]

	(i)	(ii)	(iii)	(iv)	(v)
Y			100	100	100
S	1 860 867	1 860 867	1 860 867	1 860 867	860 867
F				1 000 000	1 000 000
Z					
X					
J	1	1 000 000	1 000 000	1 000 000	1 000 000

	(vi)	(vii)	(viii)	(ix)	(x)
Y	100	100	100	100	100
S	860 867	860 867	860 867	860 867	860 867
F	3 000 000	300 000	300 000	300 000	300 000
Z			3 000 000	3 000 000	30 000
X				1 000 000	1000
J	1 000 000	1 000 000	1 000 000	1 000 000	1000

	(xi)	(xii)	(xiii)	(xiv)	(xv)
Y	120	120	120	120	120
S	860 867	860 867	860 867	132 867	132 867
F	300 000	300 000	364 000	364 000	364 000
Z	30 000	60 000	60 000	60 000	60 000
X	1000	4000	4000	4000	8000
J	1000	1000	1000	1000	1000

	(xvi)	(xvii)	(xviii)	(xix)
Y	120	120	120	123
S	132 867	132 867	132 867	132 867
F	364 000	432 000	43 200	44 289
Z	68 000	36 000	360	1080
X	8000	1000	1	9
J	1000	1000	1	1

23. Now given a volume of 4500 *chi*. Tell: what is the diameter of the sphere? Answer: 20 *chi*.

Li et al.: According to the precise rate, if the diameter is 20 *chi*, the volume should be 4190 [cubic] *chi*, 10/21 [cubic] *cun*.

Diameter of a Sphere Rule: Lay down the given number of [cubic] *chi*. Multiply it by 16, divide it by 9. Extract the cube root of the result, giving the diameter of the sphere.

Liu: The Rule takes the rate of circumference 3 to diameter 1. Suppose the area of a circle is 3/4 that of its circumscribed square, then the volume of a cylinder would also be 3/4 of its circumscribed cube. Furthermore, suppose that the volume of the cylinder means square rate 12 and that of its inscribed sphere means circle rate 9, so that the volume of the sphere is 3/4 that of the circumscribed cylinder. The square of 4 is 16, and the square of 3 is 9, thus the volume of the sphere is 9/16 that of the circumscribed cube. Therefore, we multiply the given volume of the sphere by 16, and divide it by 9, to obtain the volume of the circumscribed cube. The diameter of the sphere is equal to the side of the circumscribed cube, therefore by extracting its cube root we have the diameter required. But the supposition is incorrect. How can this be verified? Take eight cubic blocks with 1-*cun* sides to form a cube with a 2-*cun* side. Cut the cube horizontally by two identical cylindrical surfaces perpendicular to each other, 2 *cun* both in diameter and in height, then their common part looks like the surface of two four-ribbed umbrellas put together. This solid is called a *mouhefanggai* (joined umbrellas). It is composed of eight solid blocks in the form of *yangma*, which have two lateral cylindrical surfaces. Now see that the joined umbrellas have the square rate, and the inscribed sphere has the circle rate. Is there no defect in thus attributing to the cylinder the square rate? With a rate of circumference 3 to diameter 1 the area of a circle would be a little too small. With the cylinder as sphere rate, the volume of the sphere would be somewhat too large. With adjustment, the rate of 9 to 16 would be an approximate value, but is still a little too large. Now we consider the space outside the joined umbrellas and inside the cube. It narrows down gradually, but is difficult to quantify. The solid is formed by a mixture of squares and circles. The sections vary in thickness so irregularly that the solid cannot be compared to any regular block. I am afraid that it would be unreasonable to make conjectures neglecting the [difference in] shape [between the sphere and the joined umbrellas]. Let us leave the problem to whomever can tell the truth.

Li et al.: Zu Geng said that both Liu Hui and Zhang Heng took the cylinder for the square rate and its inscribed sphere for the circle rate. And therefore Zu suggested a new hypothesis himself. His Diameter of the Sphere Rule says: "Double the given volume and extract its cube root, and we have the diameter of the sphere." But why? Take a cubic block, and with its left rear lower edge as axis and the side as radius, draw a cylindrical surface. Now remove the right-upper part and put the parts together. Again draw a horizontal cylindrical surface and remove the front upper part. Now the cubic block is divided into four parts: inside the surface is called the inner block; whereas the other three outside the surface are called outer blocks. Again put the blocks together, and cut them horizontally. Now we explain in terms of right-angled triangles. Let the height be the *gou*, the side of the section of the inner block be the *gu,* and the side of the original cube be the hypotenuse. According to the *Gougu* Rule, subtract the square on the *gou* from the square on the hypotenuse, and we obtain the difference as the square on the *gu*. Therefore, subtract the square on the height from the square on the side of the cube to obtain the difference as the area of the section of the inner block. The area of the horizontal face of the cube is the total area of the sections of the four blocks. So the square on the height is the total area of the sections of the [three outer] blocks. And this is true whatever the height. Indeed, as the saying goes: "All roads lead to one goal."

[The diagram shows how Zu calculated the volume of one eighth of the "joined umbrellas" by comparing the region in the cube outside that solid to the volume of a pyramid, or *yangma*. The shaded areas are equal for each value of the height *h*; thus the total volumes are also equal.]

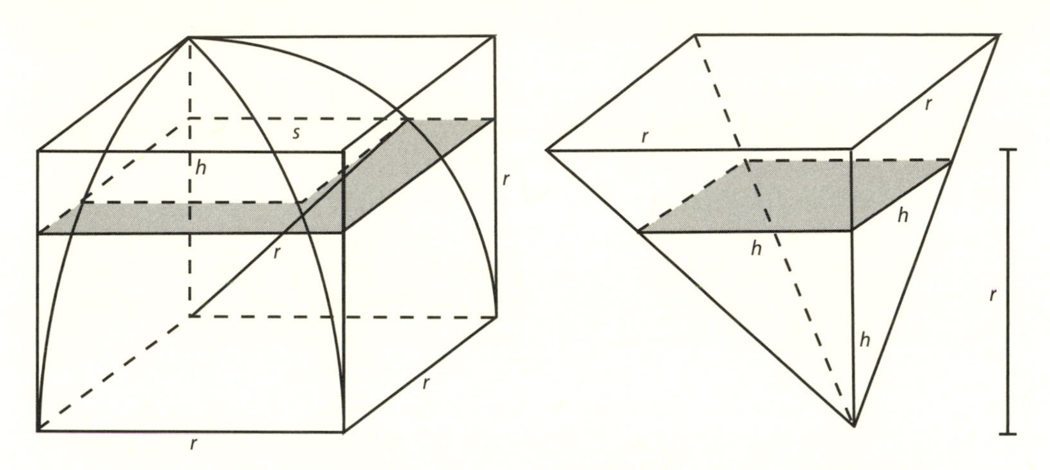

Let us analyze this in further detail by analogy from the general case. Take a *yangma* of equal sides and height, and setting it inverted, cut a horizontal section and take away its upper part. Then the square of the height of the section is equal to the area of the section. Combine the three outer blocks together, and the corresponding section areas of the two solids are equal everywhere, so their volumes cannot be unequal. From this point of view, if the sections of the three outer blocks are combined as one, they form a *yangma*. Now if the volume of a cube is divided into three equal parts, then obviously the *yangma* occupies one part, and the inner block two. Combine eight small cubes into a large one, and eight inner blocks into joined umbrellas. The inner block occupies two-thirds of a small cube. Undoubtedly the joined umbrellas also occupy two-thirds of one large cube. Multiply 2/3 by the circle rate 3, and divide by the square rate 4. Simplifying, get the rate for the sphere. Thus a sphere occupies one half of the circumscribing cube. Since this is a precise result, everything is clear. Zhang Heng followed the ancients blindly, making himself a laughing stock for later generations. Liu Hui stuck to the facts, and failed to revise it. I am afraid either the problem was too difficult to solve or they had not thought about it deeply. According to the precise rate, the volume of the sphere is the cube of its diameter, multiplied by 11 and divided by 21. Here the cube of the diameter is to be found [from

the known volume], i.e., multiplied by 21 and then divided by 11. In general, extract the cube root of the cube of a number to restore it to the original number. Thus extracting the cube root, get the diameter of the sphere.

Chapter 5: *Shang gong* (Construction Consultations)

This chapter is devoted to 28 problems involving volumes of a variety of solids and various computations concerning manpower needs in construction projects such as the building of city walls, ditches, dams, and canals, among others. The first problem depends upon various rates for earth, mud and loam given in the "earthwork conversion" rule, and application of the familiar "rule of three," as Liu Hui points out in his commentary. Problems 2–7 all require computation of the volumes of solids whose cross sections are trapezoids, for which the formula $V = (1/2)(a + b)(h)(l)$ is applied, where a and b are the upper and lower widths of the trapezoid, h the height and l the length of the solid in question.

Problem 9, which asks for the volume of a fort given its circumference C and height h, proceeds on the usual assumption that $\pi = 3$. Thus, the cross-sectional area A is given by $A = C^2/4\pi$, and the volume of the fort is the product of the square of the circumference and the height of the fort, divided by 12. Problem 10 breaks down the computation of the volume for the frustum of a pyramid into several component parts, including the *qiandu* (a right triangular prism) and *yangma* (a corner pyramid). As illustrated in the figure, the frustum is made up of 1 central cuboid, 4 *yangma*, and 4 *qiandu*.

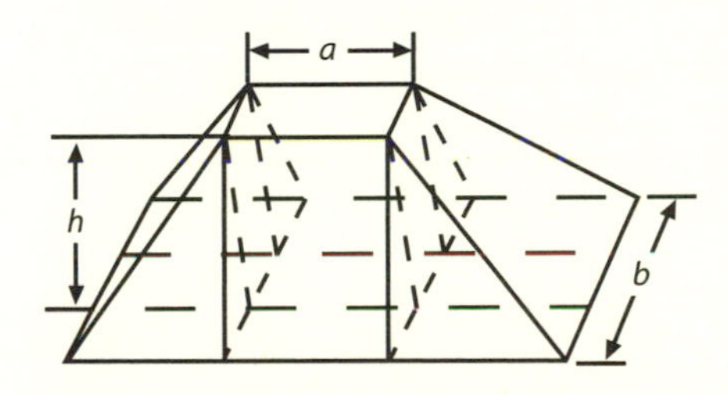

Note that Liu Hui not only provides verification that the formula given in this problem is correct, but he also offers an alternative way of achieving the same result.

Problem 12 involves the volume of a pyramid with square base, whereas Problem 15 concerns a *yangma* pyramid with rectangular base. In this latter problem, another solid, a *bie'nao* is used; this solid is a pyramid with a triangular base. Problem 13 concerns the volume of a circular cone, and Problems 23 and 25 conical piles of grain, the latter piled in a corner and therefore 1/4 the volume of a free-standing conical pile of grain. From the volumes of the piles in question and the ratios of capacity for different grains, Problems 23 and 25 ask what the capacity of the volumes will be in *hu* for millet and rice, respectively.

Chapter 5: Construction Consultations

Liu: For calculating the number of laborers, the volume of earthworks, and the capacity of warehouses.

1. Now given an excavation of 10,000 [cubic] *chi* of mud. Tell: how much rammed earth or loam comes from it? Answer: Rammed earth, 7500 [cubic] *chi*; loam 12,500 [cubic] *chi*.

The [Earthwork Conversion] Rule: An excavation of 4 of mud, makes 5 of loam, 3 of rammed earth, and a cavity of 4. Mud excavated for loam: multiply [it] by 5; for rammed earth: triple it, then divide both by 4. Loam for mud excavated: quadruple it; for rammed earth, triple it, then divide both by 5. Rammed earth for mud excavated: quadruple it; for loam, multiply it by 5, then divide both by 3.

Li et al.: The Method is an application of the Rule of Three. Let 10,000 [cubic] *chi* of excavated mud be the given number; 3, the rate for rammed earth, and 5, the rate for the loam, be the sought rates. The excavation rate of 4 is the given rate. Applying the Rule of Three gives the answer.

8. Now given a square fort, 1 *zhang* 6 *chi* square and 1 *zhang* 5 *chi* high. Tell: what is the volume? Answer: 3840 [cubic] *chi.*

The Rule [for a Square Fort}: Square the side, multiply it by the altitude, giving the number of cubic *chi.*

9. Now given a circular fort with a circumference of 4 *zhang* 8 *chi* and an altitude of 1 *zhang* 1 *chi.* Tell: what is the volume? Answer 2112 [cubic] *chi.*

Liu: According to Liu's rate, the volume should be 2017 131/157 [cubic] *chi.*

Li et al.: According to the precise rate, the volume is 2016 [cubic] *chi.*

The Rule [for a Circular Fort]: Square the circumference, multiply it by the altitude, [and] divide by 12.

Liu: The Rules in this Chapter take a rate of circumference 3 to diameter 1, but this is incorrect. By Liu's rate, it should be: square the circumference, multiply it by the altitude and then by 25, and divide it by 314. The area of the circle here is just the same as the area of a circular field; so multiply the area by the altitude.

Li et al.: Using the precise rate, multiply by 7 and divide by 88.

10. Now given a frustum of a pyramid, with a lower section 5 *zhang* square, an upper section 4 *zhang* square, and an altitude of 5 *zhang.* Tell: what is the volume? Answer: 101,666 2/3 [cubic] *chi.*

The Rule [for a Frustum of a Pyramid]: Multiply the side of the upper square by that of the lower. Square each of them and add [the products]. Multiply [the sum] by the altitude and divide by 3.

Liu: In this Chapter, the *qiandu* [right triangular prism] and *yangma* blocks combine to form a cuboid. So mathematicians have designed three kinds of blocks, called *qi,* to model different solids. Suppose a frustum of a pyramid has an upper section 1 *chi* square, a lower section 3 *chi* square, and an altitude of 1 *chi.* The blocks used are 1 cuboid in the center, 4 *qiandu* at the sides, and 4 *yangma* at the corners. "Multiply the side of the upper square by that of the lower" gives 3 [square] *chi;* "multiply by the altitude" gives a volume of 3 [cubic] *chi.* This gives 1 central cuboid and 4 lateral *qiandu.* The square of the lower side is

9 [square] *chi*; multiply it by the altitude, giving 9 [cubic] *chi*. This gives 1 central cuboid, 2 *qiandu* on four sides, and 3 *yangma* at four corners. The square of the upper side multiplied by the altitude gives 1 [cubic] *chi*, which is another central cuboid. Now the sum of the three kinds of blocks is three times as many as in the actual frustum of a pyramid. Therefore, dividing it by 3 gives the volume in [cubic] *chi*. The total number of blocks used is 27: 3 cuboids, 12 *qiandu*, and 12 *yangma*, with which three frusta of a pyramid can be constructed, i.e., equal to 13 cuboids. This is the verification. Alternatively, square the difference between the sides of the square sections, multiply by the altitude, divide by 3. This gives 4 *yangma*. Multiply [the sides of] the upper and lower squares and then the altitude. This gives the central cuboid and the 4 lateral *qiandu*. Add them, giving the volume of the frustum of a pyramid.

12. Now given a pyramid, with base 2 *zhang* 7 *chi* square, and an altitude of 2 *shang* 9 *chi*. Tell: what is the volume? Answer: 7047 [cubic] *chi*.

The Rule [for a Pyramid]: Square the side, multiply by the altitude, divide by 3.

Liu: In the Rule, let the pyramid have a base 2 square *chi*, an altitude of 1 *chi*. There are 4 *yangma*. According to [the first part of] the Rule, this gives 12 *yangma*, which form 3 pyramids. Therefore divide by 3, giving the volume.

13. Now given a circular cone with a base circumference of 3 *zhang* 5 *chi* and an altitude of 5 *zhang* 1 *chi*. Tell: what is the volume? Answer: 1735 5/12 [cubic] *chi*.

The Rule [for a Circular Cone]: Square the base circumference, multiply by the altitude, divide by 36.

Liu: Using Liu's rate, the volume is 1658 13/314 [cubic] *chi*. In the Rule, the base circumference of the circular cone is taken as the base sides of a pyramid. Square the side of the pyramid, multiply it by the altitude, then divide by 3, giving the volume of a larger pyramid, whose base area is 12 times that of the cone. Now finding [the volume of] a circular cone, divide by 12. Thus divide successively by 3, then by 12, i.e., by 36. Using Liu's rate one should square the base circumference and multiply by the altitude, then multiply by 25 and divide by 942. The ratio of the cone to [its circumscribed] pyramid is 157 to 200. After squaring the diameter, one should multiply by 157 and divide by 600. The reason is the same as in the case of the frustum of a cone.

Li et al.: Using the precise rate the volume is 1656 47/88 [cubic] *chi*. Using the precise rate, multiply by 7 and divide by 264.

15. Now given a *yangma*, with a breadth of 5 *chi*, a length of 7 *chi* and an altitude of 8 *chi*. Tell: what is the volume? Answer: 93 1/3 [cubic] *chi*.

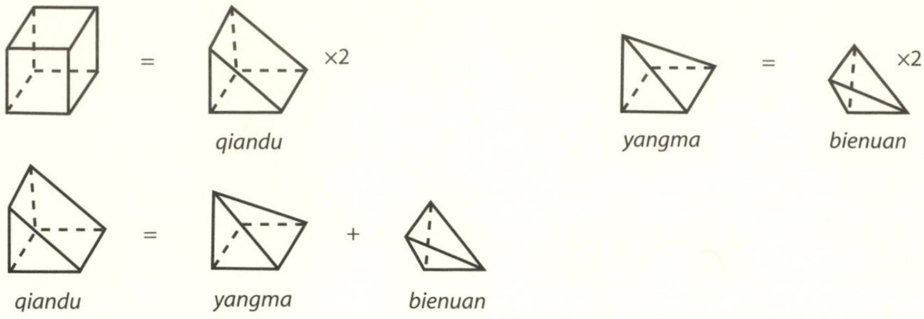

This illustrates how a cube can be cut along its diagonal into two solids called *qiandu*; cutting the *qiandu* on its diagonal produces a *yangma* and a *bianuan* (*bie'nao*, *biannao*). A *yangma* can be cut in half yielding two *bienuan*. From [Martzloff 1997, 282].

The Rule [for a *yangma*]: Multiply the breadth by the length, then multiply by the altitude, divide by 3.

Liu: In the Rule the form of a *yangma* is a corner of a pyramid. Nowadays a corner of a hipped roof is called a *yangma*. Take a cube 1 *chi* in each of breadth, length, and altitude. The product is 1 [cubic] *chi*. Cutting the cube on a diagonal gives two right triangular prisms (*qiandu*). Dissecting a *qiandu* on a diagonal gives a *yangma* and a *bie'nao*. The *yangma* occupies 2 parts and the *bie'nao* 1 part. These are fixed rates. Two *bie'nao* constitute 1 *yangma*, and 3 *yangma* constitute 1 cube. So divide by 3. This is clear when their spatial relationship is verified by standard blocks. Halve all the *yangma* [of a cube] into a total of 6 *bie'nao*. After looking at the parts, it is easy to understand their formal equivalence. Blocks may vary in dimensions; a cuboid can also be divided into 6 *bie'nao*, which are not congruent. But the three sides are equal correspondingly, hence their volumes are equal. There are *bie'nao* different in shape, and *yangma* of different forms. Nevertheless, such *yangma* do not coincide, making it difficult to compare their volume. How so? To cut a cube on a diagonal into *qiandu*, the cube is divided into two equal parts, whereas to cut a *qiandu* into a *yangma* [and a *bie'nao*], the *qiandu* is also divided into two parts, one vertically and the other horizontally. Suppose the *yangma* is the inner block and the *bie'nao* the outer one. Even if the blocks vary in dimensions, there is still this constant rate between them. That is why two solids different in shape can also be equal in volume. Suppose a *bie'nao* 2 *chi* each in breadth, length, and altitude, which is composed of 2 *qiandu* and 2 *bie'nao* all red. Suppose again a *yangma* 2 *chi* in each of breadth, length, and altitude, consisting of a cube, 2 *qiandu* and 2 *yangma*, all black. Combine the red and black blocks into a *qiandu*, 2 *chi* each in breadth, length, and altitude, and then halve its breadth and length, and also its altitude, so that the red and black *qiandu* constitute, in such a case, a cube with an altitude of 1 *chi* and a square base of 1 [square] *chi*. Every two *bie'nao* make one *yangma*. The remaining two kinds of blocks, similar in shape to their corresponding originals, constitute a cube. Cubes from blocks different from the original forms have a rate of 3, while a cube with the original form of the blocks has a rate of 1. Even with *qiandu* of different breadth, length and altitude the same conclusion holds. If the remaining blocks [*bie'nao* and *yangma*] are still in the ratio of one part to two parts, the rates of 1 and 2 are fixed. Therefore the argument is not false. To exhaust the calculation, halve the remaining breadth, length and altitude respectively, and an additional three quarters can thus be determined. The smaller the halves, the finer the remainder. Extreme fineness means infinitesimal, which is formless. In that case, how can one have a remainder?

Exhaustive calculation flows from the situation without involving counting rods. The form of the *bie'nao* is different from a practical object, and that of a *yangma* also shows variations in dimensions. Nevertheless, without the definite volume of the *bie'nao*, there is no way to survey the *yangma*, and without the definite volume of the *yangma*, there is no way to know the volume of solids such as cones and frusta, which are fundamental to the calculations of the tasks of laborers and volumes.

The Rules for *Chutong*, *Quchi*, *Pangchi*, and *Minggu* are all the same: Double the upper length, add [to it] the lower length; also double the lower length, add [to it] the upper length. Multiply the sums by the corresponding breadths. Add; multiply [the sum] by the altitude or depth, divide by 6.[16]

19. Now given a *chutong* with a lower breadth of 2 *zhang*, a length of 3 *zhang*; an upper breadth of 3 *zhang*, a length of 4 *zhang*; an altitude of 3 *zhang*. Tell: what is the volume? Answer: 26,500 [cubic] *chi*.

The Rule [for a Pile of Cereal]: Square the base circumference, multiply by the altitude, and divide by 36.

Liu: It is just like the Rule for Circular Cones. Using Liu's rate, "square the base circumference, multiply by the altitude," then multiply by 25, divide by 942.

If against a wall, divide by 18.

Liu: It occupies half a cone. Using Liu's rate, "square the base circumference, multiply by the altitude," then multiply by 25, divide by 471. The circumference against a wall is half the whole circumference. The square of the former occupies one-fourth of the square of the latter. This requires half the divisor for the whole circumference as divisor.

[16]For a diagram of a *chutong*, see problem 56 of the *Suan shu shu* in section III. Compare the method here to the method there.

If at an inner corner, divide by 9.

Liu: A pile at an inner corner occupies one-fourth of a cone. Using Liu's rate, "square the base circumference, then double," multiply by the altitude, then by 25 and finally divide by 471.

Li et al.: Using the precise rate, multiply by 7, and for the pile on the ground, divide by 264; for the pile against a wall, divide by 132; for the pile at the corner, divide by 66.

1 *hu* of millet measures 2 7/10 [cubic] *chi* in capacity. 1 *hu* of rice, 1 62/100 [cubic] *chi*, and 1 *hu* of soya beans, small beans, sesame or wheat is 2 43/100 [cubic] *chi* each.

23. Now given a pile of millet on the ground, with a base circumference of 12 *zhang*, an altitude of 2 *zhang*. Tell: what is the volume? How many *hu* of millet is it? Answer: The volume is 8000 [cubic] *chi*. There is 2962 26/27 *hu* of millet.

Liu: Using Liu's rate, it would be 7643 49/157 [cubic] *chi*. [Also, there] would be 2830 1210/1413 *hu* of millet.

Li et al.: Using the precise rate, it would be 7636 4/11 [cubic] *chi*. It would be 2828 28/99 *hu* of millet.

25. Now given a pile of rice at an inner wall corner, with a base circumference of 8 *chi*, an altitude of 5 *chi*. Tell: what is the volume? How many *hu* of rice is it? Answer: The volume is 35 5/9 [cubic] *chi*. It is 21 691/729 *hu* of rice.

Liu: Using Liu's rate, it would be 33 457/471 [cubic] *chi*. [There would be] 20 36,980/38,151 *hu* of rice.

Li et al.: Using the precise rate, 33 31/33 [cubic] *chi* [and] 20 2540/2673 *hu* of rice.

Chapter 6: *Jun shu* (Fair Taxes)

This chapter consists of 28 problems concerning the distribution of grain and assessment of conscripted labor according to differing populations and their relative distances from a given location. No general methods are considered in this chapter, and each problem is solved according to methods already introduced in previous chapters.

Problem 2, for example, concerns the fair distribution of military service from five counties with differing populations and distances from the frontier, where 1200 adults are to serve for one month. The method explains how to determine the rates $4:5:4:3:5$ at which each county should contribute adults for service; determining the actual number of individuals each county should contribute is then simply a matter of multiplying the total number of soldiers needed by the appropriate rate and dividing by the total of the individual county rates $(4+5+4+3+5=21)$. Thus for County A with 1200, whose rate is 4, the answer is 4(1200)/21, or 228 4/7. Since a whole number of adults is needed, Liu Hui explains in his commentary that the final answer requires rounding up to 229 adults from County A.

Problems 12 and 14 are pursuit problems solvable by applying the "rule of three." Problem 18 concerns the division of an amount of money among five people in ratios of $1:2:3:4:5$, also solvable by applying the "rule of three." But Liu Hui in his commentary then considers a similar problem of dividing another sum among 7 people, to show that a slightly different approach is required, and that in fact an alternative approach can also be taken with respect to the solution of the problem for proportional 5-way distribution of cash.

Problem 21 concerns a classic "meeting problem," wherein two people moving at different rates begin at different times from different places and the problem is to determine when they will meet. Liu Hui explains how the problem is solved by applying the "uniformizing and homogenizing" rule, which means considering the distance to be covered by the faster (A) at a rate of 1/5 the distance per day compared with the slower (B) at a rate of 1/7 the distance to be covered per day. These two rates can be "uniformized and homogenized" as 7/35 and 5/35 of the distance covered each day; since B starts two days earlier covering 2/7 or 10/35 of the distance, the remaining distance to be covered is 25 of 35 units. Now traveling simultaneously, A and B together can travel 12 units/day, and thus 25/12 yields the answer of 2 1/12 days until the two travelers meet. Compare this problem with one involving variable rates of accelerating and decelerating motions in Chapter 7 of the *Nine Chapters*, Problem 19.

The last problem considered here, Problem 26, is a classic problem in which water flows in 5 channels at 5 different rates into a cistern, the problem being to determine the time it will take to fill the cistern if all 5 channels are open simultaneously. Of particular interest here is that two alternative methods are provided for the solution of this problem, and Liu Hui in his commentary calls attention to parallels with other problems in this chapter. Note that Li Chunfeng and his colleagues seem to have felt it necessary to explain exactly how the procedure Liu Hui describes in the first method was to be carried out on the counting board, whereby the fractions that should be added together would need to be "uniformized and homogenized" before the addition could proceed, with 15 the "uniformized" common denominator and 74 the sum of the "homogenized" numerators. The answer is then 1/(74/15), that is 15/74 of a day to fill the cistern, or as Li Chunfeng and his group put it: "So divide the uniformization by the sum of the homogenizations to obtain the answer."

Chapter 6: Fair Taxes

Liu: To regulate the different expenses for distance and service, etc. in transportation.

2. Now frontier guard-duty is distributed among five counties. County A is near the frontier, with 1200 adults; County B is 1 day away from the frontier, with 1550 adults; County C, 2 days away, with 1280 adults; County D, 3 days away, with 990 adults; County E, 5 days away, with 1750 adults. The total service is 1200 soldiers a month. Assume the guard-duty is distributed in accordance with the number of adults and the distance from the frontier. Tell: how many adults does each county dispatch? Answer: County A, 229 adults; County B, 286; County C, 228; County D, 171; and County E, 286.

Method: Divide the number of adults of each county by the sum of days on guard-duty and in travel for each to get each rate of distribution, i.e., A, 4; B, 5; C, 4; D, 3; and E, 5. Take their sum as divisor. Multiply the [total] number of adults of each county [needed for service in a month] by each rate as dividend. Divide. If the quotients contain fractions, round them.

Liu: Here we also consider the number of days as the basis for fair distribution and imitate the number of adults dispatched as a task of service in transportation. County A is no distance with only 30 days [1 month] in service as rate. For the rate of proportional distribution, County B should send one more adult for every 30 adults so that the service is based on one adult a day, and that brings us the rate of fair distribution. Multiplying the number of adults from each county by the sum of days on guard-duty and in travel, and dividing it by the adults of each county gives the rate of adults, 5 5/7 days of individual service. Round the quotients with fractional parts. Here the fractional part of D is the smallest, so combine it with E, not with B, for D is near E. Combine the surplus [of E] with B. There is a comparatively smaller fractional part in C. Also combine it with B. Combine the surplus [of B] with A, which is now an integer. The fractional parts of A and C are equal, and the two are equidistant from B. Do not combine A with B because one has to combine something lesser with something greater.

7. Now a laborer is hired to carry 2 *hu* of salt. For 100 *li* he gets 40 coins. Assume he carries 1 *hu* 7 *dou* 3 1/3 *sheng* of salt for 80 *li*. Tell: how much will be paid? Answer: 27 11/15 coins.

Method: Multiply the number of sheng in 2 *hu* of salt by 100 *li* as divisor. Multiply 40 coins by the number of sheng for the present task, and then use 80 *li* as dividend. Divide, giving the coins paid.

Liu: In the Method, the number of sheng in 2 *hu* of salt multiplied by 100 *li* is 20,000 *li*, which represents carrying 1 *sheng* of salt 20,000 *li*. This is the given rate in the Rule of Three. Multiply the number of *sheng* for the present task by the number of *li* covered to get the number of *li* for 1 *sheng* of salt, i.e., the given number in the Rule of Three. 40 coins is the sought rate.

12. Now a good walker covers 100 *bu*, while a poor walker 60 *bu*. Assume the latter goes 100 *bu* ahead of the former, who catches up with him. Tell: in how many *bu* will the two come abreast? Answer: 250 *bu*.

Method: From 100 *bu* for the good walker subtract 60 *bu* for the poor walker, the remainder, 40 *bu*, is considered as divisor. Multiply 100 *bu* for the good walker by 100 *bu* for the poor walker who goes ahead as dividend. Divide, giving the required number of *bu*.

Liu: In the Method, "from 100 *bu* subtract 60 *bu*, the remainder 40 *bu*" is the rate for the poor walker going ahead, 100 *bu* for the good walker is the catching-up rate. Reduce them to the going-ahead rate, 2, and the catching-up rate, 5. In terms of the Rule of Three, 100 *bu* for the poor walker going ahead is the given number; 5, the sought rate; and 2, the given rate. Apply the Rule of Three giving the required number of *bu*.

14. Now a hare runs 100 *bu* ahead. A dog pursuing at 250 *bu* is 30 *bu* short. Tell: in how many more *bu* will the dog catch up with the hare? Answer: 107 1/7 *bu*.

Method: Lay down 100 *bu* of the hare running ahead; subtract 30 *bu* of the dog lagging behind. The remainder is the divisor. Multiply 30 *bu* of the dog lagging behind by the number of *bu* of the dog pursuing as dividend. Divide, giving the number of *bu* required.

> **Liu: In the Method, from 100 *bu* of going ahead subtract 30 *bu* of lagging behind. The remainder, 70 *bu*, is the rate of the hare going ahead, and 250 *bu* of the dog pursuing is the rate of pursuit. Reduce them to the rate of going ahead 7, and the rate of pursuit 25. According to the Rule of Three, 30 *bu* of lagging behind is the given number; 25, the sought rate, and 7, the given rate. Apply the Rule, giving the answer.**

18. Now given 5 persons are to share 5 coins. Let the sum of the two greater [shares] equal that of the three lesser. Tell: how much does each get? Answer: A gets 1 2/6 coins; B, 1 1/6 coins; C, 1 coin; D, 5/6 coin; and E, 4/6 coin.

Method: Lay down the cone-shaped rates for distribution. The sum of the two greater is 9, while that of the three lesser is 6. 6 is less than 9 [by] 3. Add 3 to each of the rates for the shares. Take the sum as divisor. Multiply the coins to be shared by each of the shares for each dividend. Divide, giving the coins required.

> **Liu: The cone-shaped rate in the Method means the rate for shares, like a cone with common difference 1, initial term 1; next 2; and then 3, 4, and 5. The terms should be different. Consider 5, 4, 3, 2, 1 as the rates. In the Problem, make the sum of the two greater equal to that of the three lesser. The difference between greater and lesser is 1 person, but that of the coins is 3. Add 4 to each rate of the greater rates, get twice 3, whereas by adding 3 to each lesser rate, get thrice 3 persons. The difference between greater and lesser persons is 1 person, and that between the coins is 3. Add 3 to each of the rates so that their sums are equal. According to the Rule of Three, the sum of the two sums is the given rate; each of the rates [with 3 added] is the sought rate; 5 coins is the sought number. Apply the Rule, giving the required coins, the sums of the greater and lesser of which are equal.**

21. Now A starts from Chang'an to Qi, taking 5 days. B starts from Qi to Chang'an taking 7 days. Assume B starts 2 days earlier than A. Tell: when will they meet? Answer: 2 1/12 days.

Method: Add 5 days and 7 days as divisor. From 7 days subtract the 2 days of B starting earlier. Multiply the remainder by the days of A as dividend. Division gives the number of days to be found.

> **Liu: In the Method, "Add 5 days and 7 days as divisor" means adding the homogenizations as divisor. Applying the Homogenization and Uniformization Rule to 5 days of A's journey and 7 days of B's journey,**

get, in 35 days, 7 journeys for A and 5 journeys for B. Their sum, 12 journeys, takes 35 days. It means the rates of A and B starting in opposite directions. But to turn the [number of] days into journeys, divide the days, so consider the sum as divisor. "From 7 days subtract 2 days of B starting earlier" refers to the same starting time for A and B: when A starts. They cover only the distance remaining from what B covers in 2 days. Consider 7 as the rate for the distance from Qi to Chang'an of B, and 5 as the rate for the [remaining] distance to Chang'an after B starts 2 days earlier. The Problem considers the remaining distance, so take 5 instead of 7, by which multiply the rate of A, 5 days [from Chang'an to Qi] to get 25 days. In these days A covers as many as 7 journeys and B 5 journeys. A covers 1/5 journey a day, and B 1/7. By the Homogenization and Uniformization Rule, A completes 7/35 journey a day, and B 5/35. If the distance from Qi to Chang'an is cut in 35 equal parts, A covers 7 parts a day, and B 5 parts a day. Now B started off 2 days earlier, and has walked 10 parts. The remaining distance to Chang'an is 25 parts, so multiplying the remainder after subtracting 2 days of B gives 25 parts.

26. Now given a cistern which is filled through five canals. Open the first canal and the cistern fills in 1/3 day; with the second, it fills in a day; with the third, in 2 1/2 days; with the fourth, in 3 days; with the fifth in 5 days. Assume all of them are opened. Tell: how many days are required to fill the cistern? Answer: In 15/74 day.

Method: Lay down the amount filled in a day by each canal. Add as divisor, take 1 day as dividend. Divide, giving the answer.

Liu: In the Method, "with the first canal, the cistern fills in 1/3 day" means that it fills 3 times a day; with the second, the cistern fills in a day; "with the third, it fills in 2 1/2 days" means 2/5 of a cistern in a day; "with the fourth, the cistern fills in 3 days" means 1/3 of a cistern in a day; "with the fifth, the cistern fills in 5 days" means 1/5 of a cistern in a day. Add them together and we have 4 14/15 times a full cistern.

Alternative Method: Lay down the number of days and times for a full cistern [in two columns]. Multiply the times for a full cistern by the corresponding number of days. Add as divisor. Take the continued product of the number of days as dividend. Divide, giving the answer.

Li et al.: In the Method "with the first canal, the cistern fills in 1/3 day" means that it fills 3 times a day; with the second, it fills once a day; "with the third, it fills in 2 1/2 days" means twice in 5 days; with the fourth, it fills once in 3 days; with the fifth it fills once in 5 days. So arrange the numbers of days in the right column and the times of the full cistern in the left column. "Multiply the number of days by the corresponding times of a full cistern" means to homogenize the [number of] times a cistern fills and "the continued product of the number of days" means to uniformize the days. Homogenization of the numbers of full cisterns leads to uniformization of the days. So divide the uniformization by the sum of the homogenizations to obtain the answer.

Chapter 7: *Ying bu zu* (Excess and Deficit)

The method of excess and deficiency is used in this chapter to solve a variety of problems in which the numbers involved are placed on the counting board to constitute an array analogous to a matrix, upon which clearly stated operations are then performed to eliminate variables until the equivalent of one unknown is left on the board with its solution. This then yields the solution for the rest of the problem by one or more substitutions, depending upon the problem. In modern terms, the "Rule of Double False Position" is used to solve sets of simultaneous linear equations. For example, Problem 2 in this chapter concerns a number of people who purchase a certain number of chickens. If everyone contributes 9, there is an excess of 11; if everyone contributes 6, the deficit is 16 (thus the double "false" positions, 9 and 6; assuming everyone contributes 9 results in an excess of 11; assuming everyone contributes 6 results in a deficit of 16). The question is, how many people are there, and what is the price of each chicken?

Interpreting the problem algebraically gives the two equations

$$9n - 11 = p,$$
$$6n + 16 = p,$$

for which n is the number of people, and p the price of the chickens:

The excess and deficit rule now states that the contributed rates (9 and 6) are "homogenized" by cross-multiplying the first equation by the deficit of the second, and the second equation by the excess of the first:

$$(16)(9n) - (16)(11) = 16p,$$
$$(11)(6n) + (11)(16) = 11p.$$

This serves to eliminate the "uniformized" excess and deficit, and leaves $210n = 27p$. When reduced to lowest terms, this leaves $70n = 9p$, for which the result $n = 9$ and $p = 70$ provides an immediate solution to the problem.

The reasoning of the Chinese method is virtually the same as the "algebraic" version just given, but it is laid out slightly differently on the counting board, and Liu Hui's solution proceeds to explain exactly what to do: first, lay down the contributed rates and below these, the corresponding excess and deficit, which on the counting board would have displayed the given numbers as follows (using counting rods) from left to right, beginning with the "contributed rates" above, the excess and deficit below:

$$\begin{array}{cc} 6 & 9 \\ 16 & 11 \end{array}$$

Cross multiplying would then leave

$$\begin{array}{cc} 66 & 144 \\ 16 & 11 \end{array}$$

The "homogenized" contributed rates are then added to form the "dividend" 210, and the excess and deficit are combined to serve as the "divisor," which on the counting board would now in fact assume these two positions:

$$\begin{array}{c} 210 \\ 27 \end{array}$$

The rule goes on to say that the difference of the contributed rates (in this case 9 – 6 = 3) serves as a means of reducing the divisor and dividend:

70
9

And thus, as the Excess and Deficit Rule concludes: "The [reduced] dividend is the price of an item. The [reduced] divisor is the number of people," and indeed, the solution to Problem 2 is 9 people, with 70 the price of the chickens.

Problem 6 offers a slight variation on the above by stating the problem in terms of two deficits. In this case, instead of adding the amounts of the excess and deficit, the smaller of the two deficits is subtracted from the larger, and the resulting difference takes into account the fact that two deficits are involved (a similar modification is required for problems involving two excess amounts rather than one excess and one deficit amount).

Problems 17, 18, and 19 then all involve clever ways of interpreting the given conditions of the problem to generate two sets of excess and deficit conditions, to which the "rule of excess and deficit" may then be applied to solve the problem.

Chapter 7: Excess and Deficit

Liu: For the treatment of intricate and implicit [problems].

2. Now chickens are purchased jointly; everyone contributes 9, the excess is 11; everyone contributes 6, the deficit is 16. Tell: the number of people, the chicken price, what is each? Answer: 9 people, chicken price 70.

The Excess and Deficit Rule: Display the contribution rates; lay down the [corresponding] excess and deficit below. Cross-multiply by the contribution rates; combine them as dividend; combine the excess and deficit as divisor. Divide the dividend by the divisor. [If] there are fractions, reduce them. To relate the excess and the deficit for the articles jointly purchased: lay down the contribution rates. Subtract the smaller from the greater, take the remainder to reduce the divisor and the dividend. The [reduced] dividend is the price of an item. The [reduced] divisor is the number of people.

Liu: Let the bottom terms cross-multiply the top; [combine and] then uniformize by the common denominator. [If it is] not reducible, multiply by the common denominator. [Lay down] the contribution rates. Subtract the smaller from the greater, this is called the assumed difference, which is taken to be the lesser assumption. Then combine the excess and deficit to be the determined dividend. Therefore reducing the determined dividend by the lesser assumption then gives the divisor to be the number of people, [and] reducing the dividend gives the item price. The [sum of the] excess and deficit are related to the lesser assumption. If it is not reducible, then multiply it as denominator by the assumed difference which is used to reduce the divisor and the dividend.

The Alternative Rule: Combine the excess and deficit as dividend. Take the contribution rates; subtract the smaller from the greater, the excess is the divisor. Divide the dividend by the divisor obtaining the number of people. Multiply it by the contribution rates; subtract the excess, [or] add the deficit for the item price.

> **Liu: This Rule says that the sum of the excess and deficit is the difference [in the contribution rates] of all the people. Take the rates, subtract the smaller from the greater, the excess is the difference [in rates] for one person. Take the difference for one person to divide the difference for all the people. Thereby obtain the number of people.**

6. Now sheep are purchased jointly; everyone contributes 5, the deficit is 45, everyone contributes 7, the deficit is 3. Tell: the number of people, the sheep price, what is each? Answer: 21 people, sheep price 150.

The Double Excess and Double Deficit Rule: Lay down the contribution rates, with the corresponding excesses [or] deficits below. Cross-multiply by the contribution rates. Subtract the smaller from the greater. The surplus is the dividend. [Take] the two excesses [or] the two deficits, subtract the smaller from the greater. The surplus is the divisor. Divide the dividend by the divisor. If there are fractions, uniformize [the denominators]. In a joint purchase there may appear two excesses and two deficits. Lay down the contribution rates. Subtract the smaller from the greater, [take] the surplus to reduce the divisor and the dividend. The [reduced] dividend is the item price, the [reduced] divisor is the number of people.

> **Liu: For the Rule of Double Deficits, both assumptions yield results less than the answer. The reason for the variation is like that for double excesses. It may have the same numerical values, but the conditions are changed. The dividend is obtained from the difference of products of the deficits cross-multiplied [by the contribution rates]. Then what remains is why it is not sufficient. The surplus is therefore taken as dividend since there is no deficit to subtract from. When the contribution rates are surpluses, which is the double excess [case], both assumptions yield results which are greater than the answer. Assume items are being purchased jointly, everyone contributes 8, the excess is 3; everyone contributes 9, the excess is 10. Homogenize the assumptions, uniformize the double excesses. The double excesses are both 30. Subtract the homogenized terms from each other, the surplus is taken as dividend because there are no excess numbers. Subtract the smaller excess from the greater, the surplus is the divisor. The homogenized 80 is 10 [times] the [first] assumption with the corresponding excess 30. That is 3 homogenized by 10. The homogenized 27 is 3 [times] the [second] assumption with the corresponding excess 30. That is 10 times [the homogenization factor] 3. Now assume the double excesses are 10 and 3, subtract 27 from 80, take the surplus 53 as dividend; then subtract 3 from 10, take the surplus 7 as divisor. Therefore subtract the smaller contribution rate from the greater. The surplus is called the "assumed difference." Since the assumed difference is the lesser assumption, the difference of the double excesses**

can be the dividend. Therefore dividing the divisor by the lesser assumption, one obtains the number of people, dividing the dividend by this number one obtains the item price.

The Alternative Rule: Lay down the contribution rates. Subtract the smaller from the greater. The surplus is the divisor. [Take] the double excess, [or] the double deficit. Subtract the smaller from the greater, the surplus is the dividend. Divide the dividend by the divisor to obtain the number of people. Multiply it by the contribution rates, subtract the excess [or] add the deficits. This is the item price.

Liu: Lay down the contribution rates, subtract the smaller from the greater obtaining the difference [of the contribution rates] for one person. The difference of the double excesses [or] the double deficits is the difference [of the contribution rates] for all people. Therefore dividing it by the difference of one person gives the number of people. Multiplying it by the contribution rate, [then] subtracting the [corresponding] excesses. [or] increasing it by the [corresponding] deficit gives the item price.

17. Now 1 *mu* of good farmland costs 300 coins, 7 *mu* of poor farmland costs 500 coins. Now a total of 1 *qing* farmland is bought, the price is 10,000 coins. Tell: the good and poor farmland, how much of each? Answer: Good farmland 12 1/2 *mu*; poor farmland 87 1/2 *mu*.

Method: Assume 20 *mu* of good farmland, 80 *mu* of poor farmland, excess 1714 2/7 coins; assume 10 *mu* of good farmland, 90 *mu* of poor farmland, deficit 571 3/7 coins.

Liu: Suppose 20 *mu* of good farmland, [this] costs 6000 coins; 80 *mu* of poor farmland costs 5714 2/7 coins. Compared with 10,000 [coins], there is a surplus of 1714 2/7. Assume 10 *mu* of good farmland, [it] costs 3000; 90 *mu* of poor farmland costs 6428 4/7. Compared with 10,000 the deficit is 571 3/7 coins. Use the Rule of Excess and Deficit to solve.

18. Now [there are] 9 pieces of gold, [and] 11 pieces of silver weighing the same. 1 piece is exchanged; the former is 13 *liang* lighter. Tell: what is the weight of one piece each of gold and silver? Answer: Gold weighs 2 *jin* 3 *liang* 18 *zhu*; silver weighs 1 *jin* 13 *liang* 6 *zhu*.

Method: Assume 3 *jin* for gold, 2 5/11 *jin* for silver, the deficit is 49, [lay out] in the right column. Assume 2 *jin* for gold, 1 7/11 *jin* for silver, excess 15. [Lay out] in the left column. Use the denominator to multiply the numbers in each column. Use the excess and deficit to cross-multiply the rates. [Take] the sum as dividend. [Take] the excess plus deficit as the divisor. Divide the dividend by the divisor, to obtain the weight of the gold. Multiply the denominator by the divisor and use [it] to divide, to obtain the weight of the silver. Reduce the fraction.

Liu: According to the Method, assume 9 pieces of gold [and] 11 pieces of silver; both weigh 27 *jin*. Divide by 9 for gold; get 3 *jin*. Divide by 11 for silver; get 2 5/11 *jin*. These are, respectively, the weight of 1 piece of gold or silver. From the 27 *jin* of gold subtract the weight of 1 piece of

gold and add that [weight] to the silver; from the 27 *jin* of silver subtract 1 piece of silver and add that to the gold. Afterwards the gold weighs 26 5/11 *jin* [and] the silver weighs 27 6/11 *jin*. Subtract the smaller from the greater. Then gold is lighter by 17 5/11 *liang*. Compared with 13 *liang*, it is over by 4 5/11 *liang*. After multiplying by the denominator, the numerator is 49 which is the deficit. Now assume the 9 pieces of gold weigh 2 *jin* a piece so that 9 pieces weigh 18 *jin*. 11 pieces of silver weigh a total of 18 *jin* also. Divide by 11, obtain 1 7/11 *jin* a piece. This is the weight of 1 piece of silver. Now from the 18 *jin* of gold subtract 1 piece of gold and add that to the silver. Also subtract 1 piece of silver [from the silver] and add that to the gold. Then gold weighs 17 7/11 *jin*. [and] silver weighs 18 4/11 *jin*. Subtract the smaller from the greater. Then gold is lighter by 8/11 *jin*. Compared with 13 *liang*, it is less by 1 4/11 *liang*. After multiplying by the denominator, the numerator is 15. Solve by the Excess and Deficit Rule. Divide the dividend by the divisor. Obtain the weight of [1 piece of] gold. "Multiply the denominator by the divisor and use [it] to divide" means it is the denominator for silver. Hence it is uniformized. It is then necessary to use the common denominator to divide to obtain the weight of silver. Reduce the remainder [if possible] to simplify the answer.

19. Now a good horse and an inferior horse set out from Chang'an to Qi. Qi is 3000 *li* from Chang'an. The good horse travels 193 *li* on the first day and daily increases by 13 *li*; the inferior horse travels 97 *li* on the first day and daily decreases by 3 *li*. The good horse reaches Qi first [and] turns back to meet the inferior horse. Tell: how many days [till they] meet and how far has each traveled? Answer: 15 135/191 days [till they] meet, the good horse traveled 4534 46/191 *li*, the inferior horse traveled 1465 145/191 *li*.

Method: Assume 15 days, deficit 337 1/2 *li*. Assume 16 days, excess 140 *li*. Cross-multiply the excess and deficit by the assumed numbers [and] add to the dividend. Excess plus deficit as divisor. Divide the dividend by the divisor to obtain the number of days. Simplify the remainder by the *dengshu* [divide by the greatest common divisor] and express as a fraction.

Liu: To find the distance traveled by the good horse, multiply the daily additional increase in distance by 14. Halve it. Add to the distance traveled by the good horse on the first day. Multiply by 15 days to obtain the distance traveled in 15 days. Again multiply the daily additional increase in distance by 15 days. Add the distance traveled by the good horse on the first day. Then multiply by the numerator of the fractional day, and divide it by the denominator. Add the distance already traveled by the good horse to obtain the total distance traveled [by the good horse]. To find the distance traveled by the inferior horse: multiply 1/2 *li* by 14, halve it again. [Then] subtract from the distance traveled by the inferior horse on the first day. Multiply by 15 days to obtain the distance traveled by the inferior horse in 15 days. Next multiply 15 days by 1/2 *li*.

subtract from the distance traveled by the inferior horse on the first day. Multiply by the numerator of the fractional day and divide by the denominator of the fractional day. Add the distance already traveled. Obtain the total distance traveled by the inferior horse.

The half can be calculated by the "rule of 1/2." In the operation of adding half, half *li* is reduced to half denominator, to which add the remainder fraction, then divide the sum by the denominator to obtain the fractional part [of the answer]. "Assume 15 days, deficit 337 1/2 *li*" means that in 15 days the good horse travels 4260 *li*. Subtract the distance to Qi, 3000 *li*, and then the return to meet the inferior horse is 1260 *li*. In 15 days the inferior horse travels 1402 1/2 *li*. Adding the distances traveled by the good and inferior horses, the total distance is 2662 1/2 *li*. Compared with 3000 *li*, it is less by 337 1/2 *li*. Hence it says "deficit."

"Assume 16 days, excess 140 *li*" means that in 16 days the good horse travels 4648 *li*. First subtract the distance to Qi 3000 *li*, and the return to meet the inferior horse is 1648 *li*. In 16 days the inferior horse traveled 1492 *li*. Add the distances traveled by the good and the inferior horses, obtaining 3140 *li*. Compared with 3000 *li*, the surplus is 140 *li*. Hence it says "excess." Cross-multiply the assumptions by the excess and deficit. Combine them as dividend. Combine the excess and deficit as divisor. Divide the dividend by the divisor, obtaining the number of days. This is the answer which has no excess and no deficit.

Multiplying the distances traveled by the good and the inferior horse on the first day by 15 days gives the distance traveled in 15 days at constant speed. To find the initial and final increased or decreased speed, combine 1 and 14, multiply by 14, and then halve. This is the mean product. Multiply by the increased or decreased distance due to increasing or decreasing speed. The product is the distance traveled with increased or decreased speed, respectively. Add to, or subtract from, the respective distance at constant speed, to obtain the distance traveled in 15 days.

To solve for the last day, multiply the distance traveled on the 16th day by the numerator of the fraction of a day and divide by the denominator for the fraction of a day. Obtain the respective distances traveled on the fractional day. Then combine each with the distance traveled in [the previous] 15 days to obtain the answer. To find the distance traveled by the inferior horse, the remaining 1/2 *li* in [the distance traveled by] the inferior horse, the divisor as part of the whole, calculate the half of the *li* by the rule for halving. Add to it the fractional part, this gives the answer.

Chapter 8: *Fang cheng* (Rectangular Arrays)

This chapter represents one of the most innovative and advanced features of ancient Chinese mathematics, the use of matrix methods to solve simultaneous sets of linear equations. These procedures show close affinities with the solutions encountered in the previous chapter where the Excess and Deficit Method, along with the Rule of Double False Position, led to ingenious solutions to complicated problems. The matrix method as described in this chapter led Chinese mathematicians to introduce the concept of negative numbers for the first time, as they naturally arose in the context of turning the matrix into triangular form in the process of eliminating variables.

In all, there are 18 problems in Chapter 8, eight of which involve two equations in two unknowns, another six with three equations in three unknowns, two problems with four equations in four unknowns, and one equation with five equations in five unknowns (Problem 18). Problem 13 has five equations in six unknowns, and therefore only has an indeterminate solution. As Lam Lay Yong says of "Rectangular Arrays,"

> From the solution of a pair of linear equations in two unknowns, the ancient Chinese evolved a general procedure to solve a general system of linear equations. By tabulating numerals in an array, they broke through the shackle of rhetorics and invented a mathematical notation. The *fang cheng* method is surprisingly modern. It is similar to the method of the triangular form described in our textbooks—its name is due to the fact that after the elimination process, the remaining non-zero numerals form a triangle of the matrix. [Lam 1994, 35]

In the first annotated comment Liu Hui makes on this chapter, that it is about "mixed positive and negative [numbers]," he directly acknowledges one of the most powerful features of the new method—its distinction between these two kinds of number as a necessary result of generalizing the calculations performed on the counting board, where sometimes to put a matrix into triangular form it is necessary to subtract larger quantities from smaller ones, leaving a negative remainder on the counting board. The need for negative numbers is first apparent in this chapter in the solution to Problem 4.

Prior to that, however, Problem 1 serves as a paradigmatic example illustrating the "Array Rule." In accordance with the statement of the problem, a modern reader would interpret the given conditions in terms of three linear equations as follows (t standing for top grade paddy, m for medium grade paddy, and l for low grade paddy):

equations	corresponding matrix
(I) $3t + 2m + 1l = 39$	3 2 1 39
(II) $2t + 3m + 1l = 34$	2 3 1 34
(III) $1t + 2m + 3l = 26$	1 2 3 26

The corresponding Chinese array as it was laid down on the counting board simply rotates the above matrix by 90 degrees, and lays down the numbers for each of the three equations vertically, from top to bottom, proceeding from right to left. According to Liu Hui's commentary, the three sets of equations would be laid out as follows (starting from right to left, I, II, and III denote the columns respectively on which the matrix operations will subsequently be performed to put the matrix into triangular form):

III	II	I
1	2	3
2	3	2
3	1	1
26	34	39

To find the measure (in *dou*) of one bundle of each level of grain, it is necessary to eliminate variables, beginning with the second column. This involves "homogenizing and uniformizing" the columns, terminology already familiar from the Rule of Double False Position and the Excess/Deficit Method. The first step is to "homogenize and uniformize" column II with respect to column I, which means multiplying each entry in column II by 3 (the top position in column I) and then subtracting 2 times each entry in column I from the "homogenized" column II, thereby obtaining the 0 needed at the top of column II to begin the process of getting a matrix of triangular form. Thus, following these steps results in the following matrix:

III	II	I
1	0	3
2	5	2
3	1	1
26	24	39

The next target is to eliminate the first value at the top of column III, which can be done by multiplying every entry in column III by 3 (again, because of the 3 in the top position in column I), and then subtracting every value in column I from its corresponding value in column III, which now gives:

III	II	I
0	0	3
4	5	2
8	1	1
39	24	39

The final target in the triangular form process is to eliminate the 4 in the second row of column III, which can be done by "homogenizing" columns II and III by multiplying every entry in column III by 5 and then subtracting 4 times every entry in column II, which then leaves:

III	II	I
0	0	3
0	5	2
36	1	1
99	24	39

Having transformed the matrix into a triangular form with zeros above the diagonal, it is now possible to "read off" the amount in *dou* of the lowest rate of paddy as 99/36 *dou* (in the algebraic version of the matrix, this corresponds to $36l = 99$; one bundle of low grade paddy therefore yields 99/36 *dou*, which upon simplification yields the answer: one bundle of low grade paddy yields 2 27/36 = 2 3/4 *dou*). From this the amount of middle grade paddy can be directly determined from column II as 4 1/4 *dou*, and then the amount of high grade paddy from column I follows as 9 1/4 *dou* per bundle.

Following Problem 3, the "Sign Rule" is introduced in anticipation of the occurrence, in the solution of Problem 4, of negative numbers. In his commentary, Liu Hui explains how red rods represent positive numbers on the counting board, black the negative numbers, and how operations involving the various arithmetic operations with positive and negative numbers should be performed. The Sign Rule is then applied in Problem 4, where the following matrix results from the conditions of the problem:

II	I
7	5
–5	–7
25	11

"Homogenizing" the values in columns I and II to eliminate the 7 at the top of column II is achieved by multiplying every entry in column II by 5, then multiplying every element in column I by 7, and then subtracting corresponding entries in column I from those "homogenized" in column II, leaving:

II	I
0	5
24	–7
48	11

The answer then follows immediately, and as before can be read directly from the counting board, namely that one bundle of lower grade paddy yields 48/24 or 2 *sheng*.

Problem 10, following Liu Hui's instructions, leads to the following matrix, from which the solution then follows easily:

II	I
2/3	1
1	1/2
50	50

The last problem we consider here, Problem 17, involves a matrix that requires negative numbers in order to "homogenize and uniformize" the columns in order to put the matrix into triangular form. It therefore requires, in addition to the basic Array rule, application of the Sign Rule as well.

Chapter 8: Rectangular Arrays

Liu: Treating mixed positive and negative [numbers].

1. Now given 3 bundles of top grade paddy, 2 bundles of medium grade paddy, [and] 1 bundle of low grade paddy. Yield: 39 *dou* of grain. 2 bundles of top grade paddy, 3 bundles of medium grade paddy, [and] 1 bundle of low grade paddy, yield 34 *dou*. 1 bundle of top grade paddy, 2 bundles of medium grade paddy, [and] 3 bundles of low grade paddy, yield 26 *dou*. Tell: how much paddy does one bundle of each grade yield? Answer: Top grade paddy yields 9 1/4 *dou* [per bundle]; medium grade paddy 4 1/4 *dou*; [and] low grade paddy 2 3/4 *dou*.

The Array Rule: [Let Problem 1 serve as example,] lay down in the right column 3 bundles of top grade paddy, 2 bundles of medium grade paddy, [and] 1 bundle of low grade paddy. Yield: 39 *dou* of grain. Similarly for the middle and the left column.

Liu: Given several different kinds of item, display [the number for] each as a number in an array with their sums [at the bottom]. Consider [the entries in] each column as rates, 2 items corresponds to a quantity twice, 3 items corresponds to a quantity 3 times, so the number of items is equal to the corresponding [number]. They are laid out in columns [from right to left, and] therefore called a rectangular array. [Entries in each] column are distinct from one another and [these entries] are based on practical examples. This is the general rule [for arrays]. It is difficult to comprehend in mere words, so we simply use paddy to clarify. Lay down the middle and left column like the right column.

Use [the number of bundles of] top grade paddy in the right column to multiply the middle column then merge.

Liu: The meaning of this rule is: subtract the column with smallest [top entry] repeatedly from the columns with larger [top entries], then the top entry must vanish. With the top entry gone, the column has one item absent. However, if the rates in one column are subtracted [from another column], this does not affect the proportions of the remainders. Eliminating the top entry means omitting one item from the sum (*shi*). In this way, subtract adjacent columns from one another. Determine whether [the sum is] positive or negative. Then one can obtain the answer. First take "top grade paddy in the right column to multiply the middle column." This means homogenizing and uniformizing. To homogenize and uniformize means top grade paddy in the middle column also multiplies the right column. For the sake of simplicity, one omits saying "homogenize and uniformize." From the point of view of homogenizing and uniformizing the reasoning is natural.

Again multiply the next [and] follow by pivoting.

Liu: Again eliminate the first entry in the left column.

Then use the remainder of the medium grade paddy in the middle column to multiply the left column and pivot.

Liu: Again, use the two adjacent columns to eliminate the medium grade paddy.

The remainder of the low grade paddy in the left column is the divisor, the entry below is the dividend. The quotient is the yield of the low grade paddy.

Liu: After eliminating the top grade and medium grade paddy, the remaining [bottom entry] is the yield of not just one bundle of low grade paddy. To reduce the yield on all the bundles one should take the number of bundles of paddy as divisor. Display this. Take the number of bundles of low grade paddy and multiply [the entries in] the second column. Merge, eliminate the entries of the low grade paddy. The *shi* is

> then divided by the number of bundles to give the number of *dou* [of medium grade paddy]. The calculations involved are complicated and inefficient, hence an alternative rule is introduced for simplification. However, if the old rule has to be used, this is a variation.

To solve for the medium grade paddy, use the divisor [of the left column] to multiply the *shi* in the middle column, then subtract the value of the low grade paddy.

> Liu: These are values for the medium [and] low grade paddy. Calculate the yield for one bundle of low grade paddy first. Substitute in the middle column to find [the yield of] the medium paddy. First take the displayed value. Subtract the *shi* [in the left column]. Although the yield of one bundle of low grade paddy is given in the left column by taking the divisor as denominator, from the point of view of rates this is not possible. Hence first take the divisor to multiply the constant [in the middle column] to uniformize and use the divisor as [common] denominator, then subtract the low grade paddy constant. Take the yield of one bundle of low grade paddy. Let it multiply the number of bundles of low grade paddy [in the middle column]. This is the *lieshi* displayed for the low grade paddy [in the middle column]. Subtract it from the *shi* which is the value for the medium grade paddy.

To solve for the top grade paddy also take the divisor to multiply the *shi* of the right column then subtract the values of the low grade and the medium grade paddy.

> Liu: This is the value for the 3 types of paddy in the right column. Now the values for the low grade and medium grade paddy have been found. Let them multiply the number of bundles of paddy in the right column [respectively]. Then, as before, subtract the *lieshi* [in the right column] from the displayed values.

Divide by the number of bundles of top grade paddy. This is the yield of the top grade paddy. The constants are divided by the divisors. Each gives the *dou* of yield [in one bundle].

> Liu: Treat the 3 values similarly. If the remainder is smaller than the divisor, take the latter as denominator. Simplify the denominators and numerators when possible.

The Sign Rule:

> Liu: Now there are two opposite kinds of counting rods for gains and losses, let them be called positive and negative [respectively]. Red counting rods are positive; black counting rods are negative. Alternatively distinguish [positive as] upright and [negative as] slanting. The rule for rectangular arrays [comprises] operations on the red and black entries from left to right. However whether to add or subtract varies, so red and black [counting rods] are used to cancel one another. The operations of subtraction or addition depend on the two types of entries in each

> column. Whether adding or subtracting, the result appears as the bottom entry [in each column]. To comment on these two rules, suppose paddy is used to demonstrate the intended meaning of the rules. Then mixing the red and black [rods] suffices to compare top and low [grade paddy]. For adding and subtracting it is sufficient to operate on adjoining entries, the constants may be sufficiently different to cope with varying rates. If positive without extra, make negative, if negative without extra make positive: the rates do not wander.

Like signs subtract.

> Liu: This is red subtract red, black subtract black. Subtract one column from another to eliminate the first entry. So if the first entries are of like sign, this rule should be applied. If the first entries are of opposite sign the rule below should be applied.

Opposite signs add.

> Liu: Subtracting one column from another depends on appropriate entries with the same signs. Opposite signs [entries] are from different classes. If from different classes, they cannot be merged but are subtracted. So merging red by black is to subtract black, merging black by red is to subtract red. Red and black merge to the original [color], this is "adding"; they are mutually eliminating. This is by eliminating [the top entries] using addition and subtraction to achieve the bottom constant. The prime purpose of the rule is to eliminate the first entry; the magnitudes of entries in other positions are of no concern, either subtract or merge them. The reasoning is the same, not different.

Positive without extra, make negative; negative without extra, make positive.

> Liu: "Without extra" is "without merging." When nothing can be subtracted, put the subtrahend in its place [with color changed], subtracting the resulting constant from the bottom constant. This rule is also applicable to columns whose entries are of mixed signs. In the rule, entries with the same signs subtract their constant terms, entries with opposite signs add their constant terms. This is positive without extra, make it negative; negative without extra, make it positive.

Opposite signs subtract; same signs add; positive without extra, make positive; negative without extra, make negative.

> Liu: This rule uses "opposite signs subtract" as an example to illustrate how, with the above rule, they complement each other. Positive and negative are used to express that they are opposites to permit operating on these two classes. To say negative does not necessarily mean less, to say positive does not necessarily mean more. Thus interchanging the red and black rods in any column is immaterial. So

> one can make the first entries to be of opposite sign. These Rules are in fact but one. So applying these two rules, interchange every operation and sign in carrying out the calculation processes from the top entries to the bottom entries. This is the method. Also subtract to eliminate [the leading entry] of each column. There is no limit on how many [columns] until [there remains] one entry and one constant. Using the signed operations of addition and subtracting as in the "New Rule" is suitable for any column and not restricted by which column and which position.

4. Now there are 5 bundles of top grade paddy. Subtract 1 *dou* 1 *sheng* from the yield. [Then] this is equivalent to [the yield of] 7 bundles of low grade paddy. Now there are 7 bundles of top grade paddy. Subtract 2 *dou* 5 *sheng* of yield. This is equivalent to [the yield of] 5 bundles of low grade paddy. Tell: what is the yield of 1 bundle of top [and] low grade paddy? Answer: One bundle of top grade paddy 5 *sheng*. One bundle of low grade paddy 2 *sheng*.

Method: By rectangular arrays. Lay down 5 bundles of top grade paddy as positive, 7 bundles of low grade paddy as negative. The subtracted yield 1 *dou* 1 *sheng* as positive. Apply the Sign Rule.

> Liu: "Subtract 1 *dou* 1 *sheng* of the yield from the 5 bundles of top grade paddy" means the remainder is equal to that of 7 bundles of low grade paddy. Change its sign and let them subtract. The difference is 1 *dou* 1 *sheng* which is the remaining yield of the top grade paddy. According to the Sign Rule, there is no limit on the number of entries in each column. However, the constant and [other entries] from top to bottom are displayed in turn. Each individual column forms a group of rates. Whether for adding or subtracting, different positions in the same column are for different objects. The merging and subtracting of the upper entries results in the bottom [constant].

10. Now there are 2 persons A [and] B. Each has an unknown amount of coins. A gets 1/2 of B's, then [has] 50 coins. B gets 2/3 of A's, then [has] 50 coins also. Tell: what is the amount of coins A [and] B has each? Answer: A has 37 1/2 coins. B has 25 coins.

Method: Solve by the Array Rule; add and subtract.

> Liu: In the problem, A's coins plus 1/2 B's coins are 50; 2/3 of A's coins plus B's coins are 50 also. Multiply the denominators by the integers and add the product to the corresponding numerators. The numbers in the columns are then determined, i.e., 2 A's and 2 B hold 100 coins, 2 A's and 3 B's hold 150 coins. Solve this by the Array Rule. Follow this example in solving problems with fractions.

17. Now given 5 sheep, 4 dogs, 3 hens, and 2 rabbits cost 1496 coins in total; 4 sheep, 2 dogs, 6 hens, and 3 rabbits cost 1175 coins; 3 sheep, 1 dog, 7 hens, and 5 rabbits cost 958 coins; 2 sheep, 3 dogs, 5 hens, and 1 rabbit cost 861 coins.

Tell: how much is each of them? Answer: A sheep costs 177, a dog 121, a hen 23, and a rabbit 29.

Method: Solve by the Array Rule. Calculate by the Sign Rule.

Chapter 9: *Gou-gu* (Base-Height)

This is perhaps the best-known chapter of the *Nine Chapters*, and has attracted considerable attention in the west because of its application of what is commonly known as the "Pythagorean" theorem for right triangles, namely, given the base a and height b of a right triangle, the square of its hypotenuse c is equal to the sum of the squares of the base and height, that is, $a^2 + b^2 = c^2$. This chapter is the only part of the *Nine Chapters* to have been translated into English on numerous occasions, and readers may find it instructive to compare the different versions of the chapter as translated by [Gillon 1977], [Swetz and Kao 1977], [Lam 1994, 38–41], and the one presented here [Shen, Crossley, and Lun 1999, 439–517].

All the 24 problems presented involve right triangles, though not all the problems require application of the *gou-gu* ("Pythagorean") relation. It is important to keep in mind, as previously mentioned, that there was no specific term for "triangle" in Chinese, and so to translate the title of this chapter as "Right-Angled Triangles" is somewhat misleading, since the text refers consistently to the "base-height" relation in connection with the corresponding *xian* or "hypotenuse." But the apparent similarities between the Chinese *gou-gu* relation and the Greek Pythagorean theorem have proven so striking as to raise the question of possible transmission of the idea from west to east, or from east to west, and Swetz and Kao go so far as to title their book devoted to a translation and commentary on Chapter 9: *Was Pythagoras Chinese? An Examination of Right Triangle Theory in Ancient China.* In fact, the similarities between the treatment of right-triangle properties between east and west go further than the Pythagoren theorem/*gou-gu* relation, because in Chinese the word for hypotenuse, *xian*, means lute string or the string that might be pulled across to form the diagonal of a square by land surveyors. Similarly, the word *hypotenuse* in Greek means "that which is pulled across," again a reference to a string or rope in the tradition of the origins of geometry in land measurement by knotted ropes of the sort associated with the Egyptian *harpedonaptai.*

Despite such tantalizing similarities between mathematics of the right triangle (or base-height relation) east and west, they may stem largely from the fact that when dealing with right triangle figures, or the figures represented by gnomons and the shadows they cast, right triangles present a number of invariant properties when considered on a flat plane that would have been apparent to Chinese and Greek mathematicians alike, requiring no correspondence between the two to reach similar results quite independently of one another.

Among the most fundamental base-height properties exploited in the *gou-gu* chapter of the *Nine Chapters* are those set out directly in the first three problems, which present the paradigmatic relations between the base-height squares and the equivalence of their sum with the square on the hypotenuse or *xian* square. All three problems are limited to the case where the *gou* (a) is 3 *chi* in length, the *gu* (b) is 4 *chi*, and the *xian* (c) is 5 *chi*. In the three problems, two of the values are given and the third is calculated.

In his commentary on the "*Gou-gu* Rule," Liu Hui refers to a diagram that is now lost, but was later reconstructed by Dai Zhen for the edition of the *Nine Chapters* that appears in the *Yongle dadian*. From this it is apparent how Liu Hui proceeds to prove the universal validity of the *gou-gu* relation by a judicious appeal to the "out-in" complementary method, as suggested by consideration of the following figures:

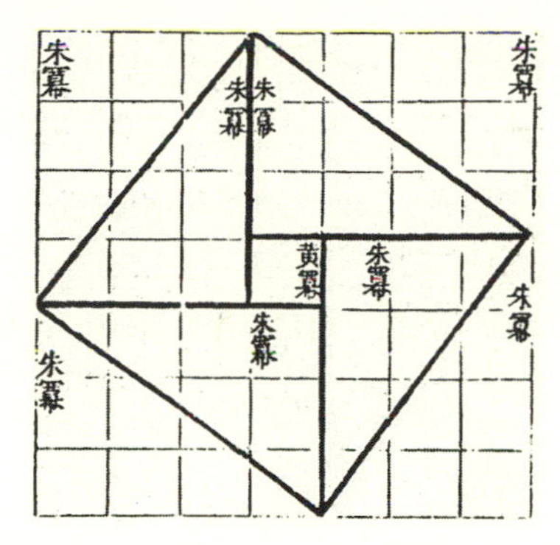

The illustration above was reconstructed by Dai Zhen for the *Siku quanshu* edition of the *Zhou bi suan jing*. Although it does not follow exactly the same color-coding of the diagram as Liu Hui's commentary on the base-height relation in Chapter Nine of the *Nine Chapters*, it does serve to make clear the "out-in" character of the proof that Liu Hui offers for the *gou-gu* relation; at the top, the diagram is labeled "*Xian* Diagram," or "*Hypotenuse* Diagram."

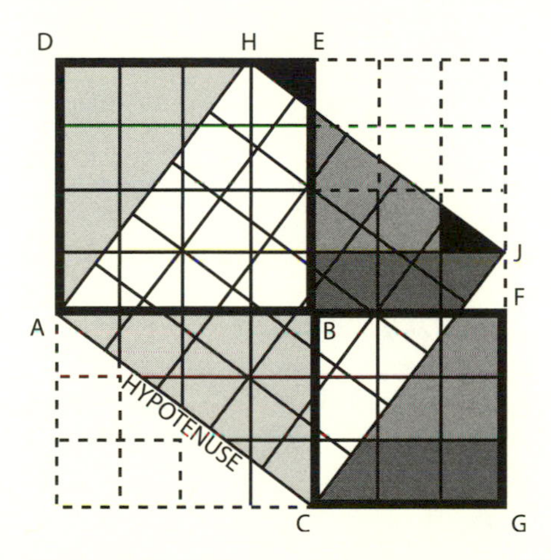

This proof makes use of the so-called "out-in" technique taken as an axiom by ancient Chinese mathematicians. The diagram shows that the sum of the squares on the base (*gou*) *BC* and height (*gu*) *AB* in the diagram, namely the squares *BCGF* and *ABED*, is equal to the square of the hypotenuse (*xian*) *AC*, namely the square *ACJH*. In accordance with the out-in principle, those parts of the two small squares (the shaded portions of *BCGF* and *ABED*) that are on the outside (the "excess") of the large square (*ACJH*) correspond exactly with the "deficit" areas (the shaded portions of *ACJH*) insidc thc *xian* squarc. Since the areas of the two *gou* and *gu* squares may thus be equated with the area of the *xian* square, Liu Hui has thus proved that this equality holds for all such *gou-gu* figures.

Having established the *gou-gu* relation, the *Nine Chapters* proceeds to apply it to a number of classic problems that are familiar from mathematical handbooks wherever the "Pythagorean theorem" was known. Problem 6, for example, considers a reed in the center of a pond that is 10 *chi* in diameter. The reed extends 1 *chi* above the water, but just touches the top of the pond when drawn to its edge. To determine the length of the reed is equivalent to finding the length of the hypotenuse *xian* in the following *gou-gu* figure.

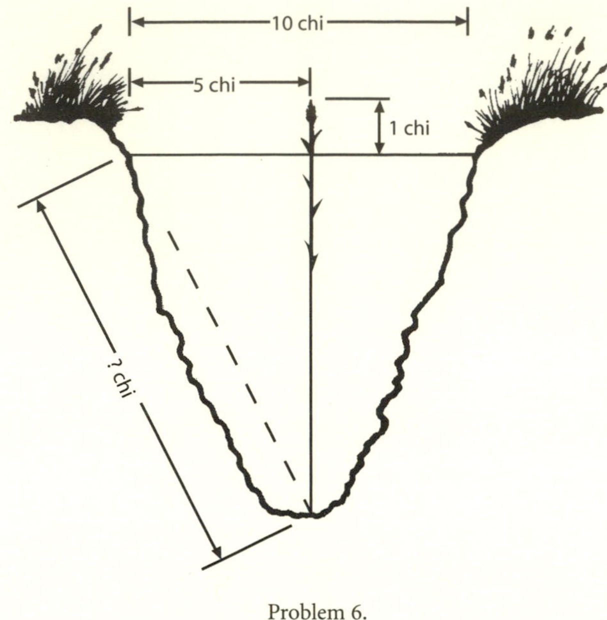

Problem 6.

From the figure, if x denotes the portion of the reed that is underwater, then the length of the reed is $x + 1$ and the *gou-gu* relation may be expressed (algebraically) as follows: $5^2 + x^2 = (x + 1)^2$. Thus $25 + x^2 = x^2 + 2x + 1$, and from $2x = 24$ it follows that $x = 12$ and that the total length of the reed is 13 *chi* or 1 *zhang* 3 *chi*.

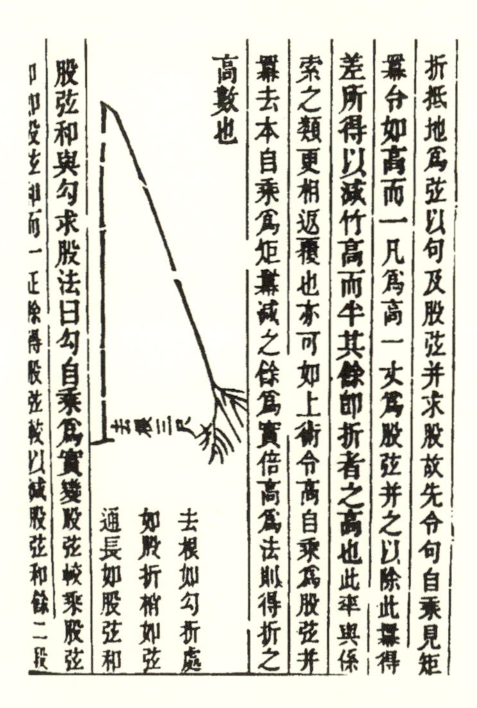
折抵地為弦以句及股弦并求股故先令句自乘見矩
冪令如高而一凡為高一丈為股弦并之以除此冪得
差所得以減竹高而半其餘即折者之高也此率與係
索之類更相返覆也亦可如上術令高自乘為股弦并
冪去本自乘為矩冪減之餘為實倍高為法則得折之
高數也

股弦和與勾求股法曰勾自乘為實變股弦較乘股弦
和即股弦和而一正除得股弦較以減股弦和餘二段

Yang Hui's illustration from *A Detailed Analysis of the Mathematical Methods in the Nine Chapters* of 1261. Problem 13 in the *Nine Chapters* applies the *gou-gu* relation to a bamboo stalk originally 10 *chi* high. When broken as in the illustration, the top falls 3 *chi* away from the base of the stalk. How tall is the stalk that is left standing? Algebraically, if x is the upright length of the broken bamboo (*gu*), then the hypotenuse (*xian*) is $10 - x$ and the *gou-gu* relation may be expressed as $x^2 + 3^2 = (10 - x)^2$. From this is follows that $x^2 + 9 = 100 - 20x + x^2$, and $20x = 91$, or $x = 91/20 = 4\ 11/20$ *chi*. Problem 13: Broken Bamboo.

Problem 16 is an important problem related to the circle inscribed between the *gou*, *gu*, and *xian* lines of the *gou-gu* figure. Although the problem as stated in the *Nine Chapters* specifies that the base *gou* = 8 *bu* and *gu* = 15 *bu*, Liu Hui's commentary offers several solutions to this problem that are entirely general and hold for any *gou-gu* figure. Based upon a color-coded diagram as reconstructed here, Liu Hui explains how the diameter of the inscribed circle may be determined by constructing the following rectangle in such a way that its width is equivalent to the diameter of the circle, and its length is the sum of the *gou*, *gu* and *xian*. Thus if *gou* = a, *gu* = b and *xian* = c, then the diameter D of the inscribed circle is $D = (2ab)/(a + b + c)$.

Liu Hui also stated several equivalent means of expressing the diameter as well:

$$D = a - (c - b)$$

$$D = \sqrt{2(c - a)(c - b)}.$$

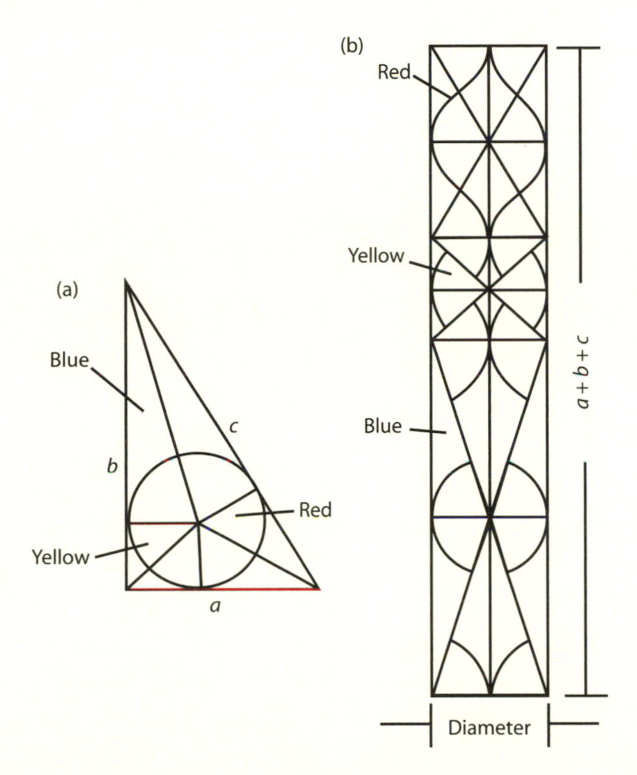

The last of the problems we consider here from the ninth chapter of the *Nine Chapters* is Problem 23, which does not involve the *gou-gu* relation in terms of the squares of *gou*, *gu* and *xian*, but is solved by an elaborate scheme of similar *gou-gu* figures. With reference to the following diagram, the difference DE (*gou*) between the height of the mountain and the height of the tree with respect to the distance EB (*gu*) between the mountain and the tree is the same as the difference BC (*gou*) between the height of the tree and the height of the observer with respect to the distance CA (*gu*) between the tree and the observer, i.e., $DE : EB = BC : CA$, from which it follows that the height of the mountain is the sum of the height of the tree GB and $DE = (EB)(BC)/CA$, or $DH = GB + (EB)(BC)/CA$.

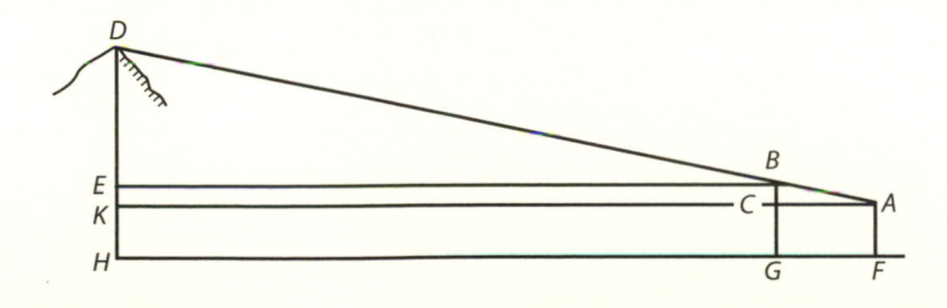

Chapter 9: Right-Angled Triangles

Liu: For the treatment of altitude, depth, length, and width.

1. Now, given a right-angled triangle, the lengths of its *gou* and *gu* are 3 *chi* and 4 *chi* respectively. Tell: what is the length of the hypotenuse? Answer: 5 *chi*.

2. Now given a right-angled triangle, the lengths of its hypotenuse and *gou* are 5 *chi* and 3 *chi* respectively. Tell: what is the length of its *gu*? Answer: 4 *chi*.

3. Now given a right-angled triangle, the lengths of its *gu* and hypotenuse are 4 *chi* and 5 *chi* respectively. Tell: what is the length of its *gou*? Answer: 3 *chi*.

Gou-gu Rule: Add the squares of the *gou* and the *gu*, take the square root [of the sum] giving the hypotenuse. Further, the square of the *gu* is subtracted from the square on the hypotenuse. The square root of the remainder is the *gou*. Further, the square of the *gou* is subtracted from the square on the hypotenuse. The square root of the remainder is the *gu*.

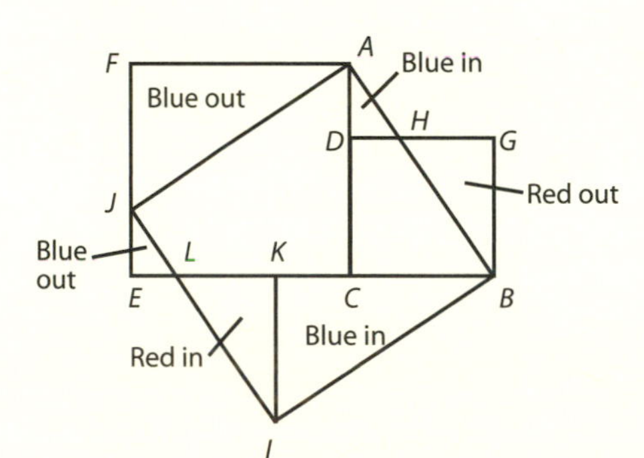

Li Huang's proof of the *gou-gu* Rule (Pythagorean Theorem) in *A Detailed Commentary on the Nine Chapters with Diagrams* (c. 1810). This diagram gives one of many possible interpretations for the color-coded argument provided below by Liu Hui in his commentary on the *gou-gu* Rule.

Liu: The shorter side [of the perpendicular sides] is called the *gou*, and the longer side the *gu*. The side opposite to the right angle is called the hypotenuse [*xian*]. The *gou* is shorter than the *gu*. The *gou* is shorter than the hypotenuse. They apply in various problems in terms of rates of proportion. Hence [I] mention them here so as to show the reader their origin. Let the square on the *gou* be red in color, the square on the *gu* be blue. Let the deficit and excess parts be mutually substituted into corresponding positions, the other parts remain unchanged. They are combined to form the square on the hypotenuse. Extract the square root to obtain the hypotenuse.

6. Given a reed at the center of a pond 1 *zhang* square, which is 1 *chi* high above the water. When it is drawn to the bank, it is just within reach. Tell: what are the depth of the water and the length of the reed? Answer: The water is 1 *zhang* 2 *chi* deep and the reed 1 *zhang* 3 *chi* tall.

Method: Square half the side of the pond. From it we subtract the square of 1 *chi*, the height above the water. Divide the remainder by twice the height above the water to obtain the depth of the water. The sum of the result and the height above the water is the length of the reed.

> Liu: Here take half the side of the pond, 5 *chi*, as *gou*, the depth of the water as the *gu*, and the length of the reed as the hypotenuse. Obtain the *gu* and the hypotenuse from the *gou* and the difference between the *gu* and the hypotenuse. Therefore, square the *gou* for the area of the gnomon. The height above the water is the difference between the *gu* and the hypotenuse. Subtract the square of this difference from that of the area of the gnomon; take the remainder. Let the difference between the width of the gnomon and the depth of the water be the *gu*. Therefore construct [a rectangle] with a width of 2 *chi*, twice the height above the water. Its length is the depth of water to be found.

13. Now given a bamboo 1 *zhang* high, which is broken so that its tip touches the ground 3 *chi* away from the base. Tell: what is the height of the break? Answer: 4 11/20 *chi*.

Method: Square the distance from the base and divide this by its height. Subtract the result from the height and halve the difference to obtain the height of the break.

> Liu: Here we consider 3 *chi*, the distance from the base, as the *gou*, the height left standing as the *gu*, [and] the length of the fallen tip as the hypotenuse. Find the *gu* using the *gou* and the sum of the *gu* and the hypotenuse. Hence, first square the *gou* to obtain the area of a gnomon [which has the area of the *gou* square]. The total height of the bamboo, 1 *zhang*, is the sum of the *gu* and the hypotenuse. Divide the square of the *gou* by that to obtain the difference [between the *gu* and the hypotenuse]. The Method is ... similar to the above Method: square the height, i.e., the sum of the area of the square on the *gu* plus the hypotenuse. From the square [of the total height] subtract the square of the distance from the base. Consider the remainder as dividend, twice the height as divisor. Divide, giving the height of the break.

16. Now given [a right-angled triangle whose] *gou* and *gu* are 8 *bu* and 15 *bu* respectively. Tell: what is the diameter of its inscribed circle? Answer: 6 *bu*.

Method: 8 *bu* is the *gou*, 15 *bu* is the *gu*. Find the hypotenuse. Add these three as divisor. Take the *gou* to multiply the *gu*. Take twice [the product] as dividend. Divide, giving the diameter.

> Liu: The product of the *gou* by the *gu* is the chief subject in the figure, in which there are three pairs of figures, red, blue, and yellow. Doubling them, get four of each. Copy them onto a small piece of paper, and cut them out. Arrange them upright, slantwise or upside down by attaching the equal sides together so as to form a rectangle with the diameter as the width and the sum of the *gou*, the *gu* and the hypotenuse as the length. That is why we add the *gou*, *gu*, and hypotenuse as divisor. From the figure, [the tip of] the blue triangle on the *gu* is equidistant from the *gou*, the *gu*, and the hypotenuse. Measuring this distance on the *gou* and *gu* by a compass horizontally and vertically, one surely gets a small square.

Further, draw a "middle hypotenuse" [parallel to the hypotenuse] through the center to observe the convergence. There appear smaller right-angled triangles on the *gou* and the *gu* respectively. The smaller *gou* on the *gu* [triangle] and the smaller *gu* on the *gou* [triangle] are both sides of the small square, i.e., the radius of the inscribed circle. Therefore the relevant data are all proportional. Let the *gou*, the *gu*, and the hypotenuse be the rates, and take their sum as divisor. Consider the product of the *gou* [of the original triangle] by the corresponding rate as dividend. Divide, giving the smaller *gu* on the *gou* [triangle]. Consider the product of the *gu* [of the original triangle] by the corresponding rate as dividend. Then one gets the smaller *gou* on the *gu* [triangle]. They are differently expressed, but the way is the same to get the divisor and dividend.

Alternatively, subtract the difference between the *gu* and the hypotenuse from the *gou*; subtract the difference between the *gou* and the hypotenuse from the *gu*; or subtract the hypotenuse from the sum of the *gou* and the *gu*, to obtain the required diameter of the inscribed circle. Divide the product of the difference between the *gou* and the hypotenuse by the difference between the *gu* and the hypotenuse. Extract the square root, giving the diameter.

23. Now to the west of a tree there is a hill whose height is unknown. The distance between the tree and the hill is 53 *li* and the tree is 9 *zhang* 5 *chi* high. A person standing 3 *li* to the east of the tree observes that the summit of the hill and the tree-top are aligned. Assume his eyes are at height 7 *chi*. Tell: what is the height of the hill? Answer: 164 *zhang*, 9 *chi*, 6 2/3 *cun*.

Method: Lay down the height of the tree [95 *chi*]; subtract the height of the eyes, 7 *chi*. Take the remainder to multiply 53 *li* as dividend. Take the distance, 3 *li*, from the person to the tree as divisor. Divide, and add the height of the tree to the quotient to obtain the height of the hill.

Liu: By the *Gougu* Rule, from the height of the tree subtract that of the eyes. The remainder 8 *zhang* 8 *chi* is considered as the *gou* rate; 3 *li*, the distance from the surveyor to the tree as the *gu* rate; and 53 *li*, the distance from the tree to the hill, as the given *gu*. Find the *gou*, to which add the height of the tree to get the height of the hill.

V.b. The *Sea Island Mathematical Classic*

In his Preface to the *Nine Chapters*, Liu Hui makes clear that the received text was incomplete, that there was an important category of problems it had failed to include, namely, those solved by the "double difference method." This surveying method relied upon a pair of related "differences," and hence the name—the difference between two poles, and the difference between the shadows the two poles cast (or the difference between the horizontal distances from the base of each pole to the ground point of sightings from the ground to the top of each pole when aligned with a specific object, say the top of a mountain whose height and distance from the poles are to be determined). The fact that Liu Hui considered this the crowning achievement of his own work as a mathematician is evident from the fact that he devoted more than one third of his Preface to describing what the "double difference" method involves. (See preface above.)

Although Liu Hui may have conceived this work as nothing more than an appendix to the *gou-gu* section of the *Nine Chapters*, when the "Ten Classics" were collated and edited in the Tang dynasty, Liu Hui's separate work was elevated to the status of a classic on its own. The treatise itself takes its name from the first problem, which concerns the determination of the height of an island at sea, as calculated from a position quite distant from the island in question. When printed as part of the Northern Song edition of the *Ten Classics* in 1048 CE, it was given the specific title of *Mathematical Classic* (*Suan jing*), although the title is often translated into English as the *Sea Island Mathematical Manual.*

The text of the *Hai dao suan jing* was all but lost by the beginning of the Qing dynasty, and virtually unknown. The text that survives today is based upon the version included in the *Great Encyclopedia of the Yongle Reign*, which in turn was later incorporated into the encyclopedic collection edited by Dai Zhen in the *Complete Library of the Four Branches of Literature.* It was translated into French by Louis van Hee in 1932, and into English by Frank Swetz and Ang Tianse in 1986, by Lam Lay Yong and Shen Kangshen in 1986, and again by Swetz in 1992. The version reproduced here is taken from the most recent English translation by Shen Kangshen, John Crossley, and Anthony Lun in their edition of the *Nine Chapters* [Shen, Crossley, and Lun 1999].

Considering the method that is the focus of the *Sea Island Mathematical Classic*, the basic idea of the "double difference" method may be obtained from a brief consideration of its application in the opening problem. With respect to the following diagram, the first difference is that between the two poles, d; the second difference is that between the two distances KH and GC, the distances from the two poles upon which sightings from H and C, respectively, to the top of the island D depend. Thus the second difference, $GC - KH$, provides the rationale for calling this the "double difference" method.

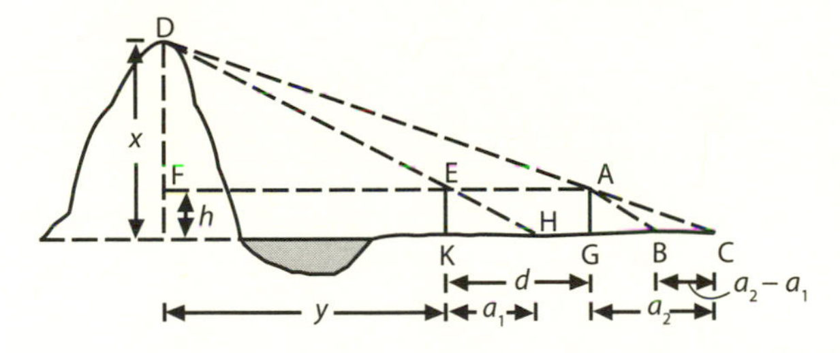

Justification for the double difference method is often explained in terms of the "similar" triangles DEA and ABC (see, e.g., Li and Du 1987, 77), and it is easy to see that the result follows from the fact that the ratio of the height of the mountain minus the height of the poles $(x - h)$ to the height of the poles (h) is the same as the ratio of the distance between the poles (d) to the difference between the sightings $(a_2 - a_1)$, i.e., $(x - h) : h = d : (a_2 - a_1)$. This approach only requires drawing the line AB parallel to EH, giving the difference between the ground points H and C from which the top of the island D is sighted from the two poles KE and GA, respectively. The similarity of the two obtuse triangles DEA and ABC then gives the result directly. But there are no examples of arguments like this involving obtuse triangles in any ancient Chinese mathematical texts.

Indeed, it must be stressed that this approach is not the way in which an ancient Chinese mathematician like Liu Hui would have proved the validity of the double difference method. Although the *Sea Island Mathematical Classic* offers no justifications for the techniques by which its nine problems are solved, based upon the proofs offered for the gnomon-shadow method of determining the height of the sun in the *Zhou bi suan jing*, and Liu Hui's proof of the *gou-gu* ("Pythagorean") theorem in Problem 1 of Chapter 9 of the *Nine Chapters*, a proof of the validity of the double difference method can be given for the sea island problem in a straightforward way, using the out-in complementary principle, as follows.

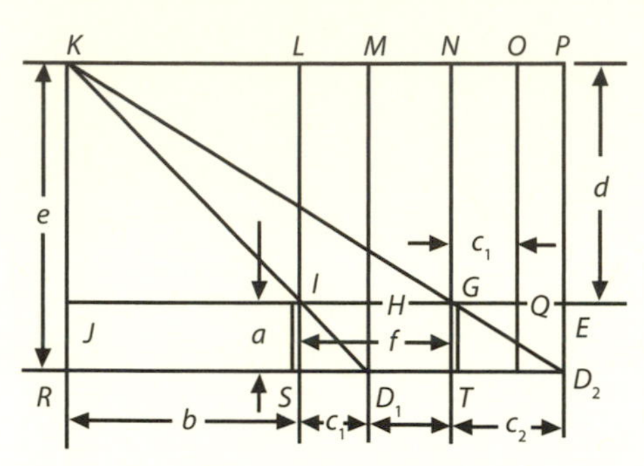

By the out-in complementary principle, and the equal areas on the two sides of the diagonal KD_2 drawn from the top of the sea island K to the ground point D_2 from which K is sighted across the top of the gnomon TG, the rectangles $TGJR$ and $GEPN$ are equal. Similarly, considering the equal areas on both sides of the diagonal KD_1 drawn from the top of the sea island K to the ground point D_1 from which K is sighted across the top of the gnomon S_1, the rectangles $SIJR$ and $IHML$ are equal. Now, since the rectangles $IHML$ and $GQON$ are equal, subtracting $SIJR = GQON$ from $TGJR = GEPN$ gives $TGIS = QEPO$. The "Sea Island" problem thus involves the following known elements:

1. TG, the height of the gnomon is known;
2. GI, the difference between the distances of the two gnomons is known; and
3. QE, the difference between the two "shadows" or distances TD_2 and SD_1 from the bases to the ground points of the two gnomons from which the top of the sea island is sighted along the tops of the two gnomons, is known.

Consequently, it follows that the height d can easily be determined, thanks to the "double differences" as follows: since $TGIS = QEPO$, $(TG \times GI) = (QE \times EP)$, and since $EP = d$, then $d = (TG \times GI)/QE$, and the height of the sea island is consequently $d + TG$.

In the case of finding the height and distance from the poles of the sea island in problem 1, two sightings are required. Other problems in the *Sea Island Mathematical Classic* require 3 or 4 sightings to reach their solutions, although the reasoning in these cases is exactly the same as that in the case of the "double difference" method applied in Problem 1. Of the nine problems, three (Problems 1, 3, and 4) require two observations, four require three (Problems 2, 5, 6, and 8), and two require four (Problems 7 and 9). Problems 1 and 7 are given here. In the latter problem, two sightings of the bank and two of the stone at the bottom of the pool are required. So there is not only the difference between the positions of the two gnomons to consider, but also the differences between the sightings from the two gnomons to both the bank of the pool and to the stone at the bottom of the pool. (However, as Ang and Swetz point out, no allowance is made for the refractive index of the water, "rendering the two observations inaccurate" [Ang and Swetz 1986, 19]. They go on to suggest that this is evidence for the pedagogical nature of the problems in the *Sea Island Mathematical Classic*, rather than its aiming for a realistic analysis of actual applied mathematical problems.) Note that although an "alternative method" is provided in the case of Problem 7, the nineteenth-century mathematician Li Huang pointed out in *A Detailed Commentary on the Nine Chapters with Diagrams*, that "The alternative method is only correct by coincidence, it is not true for the general cases" [Shen, Crossley, and Lun 1999, 553].

Sea Island Mathematical Classic

1. Now survey a sea island. Erect two poles of the same height, 3 *zhang*, so that the front and rear poles are 1000 *bu* apart. They are aligned with the summit of the island. Move backwards 123 *bu* from the front pole, sighting at ground level, and find that the summit of the island coincides with the tip of the pole. Move backwards 127 *bu* from the rear pole, sighting at ground level, and find that the summit of the island also coincides with the tip of the pole. Tell: what are the height of the island and its distance from the [front] pole? Answer: The height of the island is 4 *li* 55 *bu*; it is 102 *li* 150 *bu* from the [front] pole.

Method: Multiply the distance between the poles by the height of a pole as dividend. Take the difference in distance from the points of observation as divisor and divide. Adding the quotient to the height of a pole, get the height of the island.

> Li et al.: Here the summit of the island refers to the top of a hill. Poles are the tips of vertically standing rods. The line of sight passes through the tip of the pole and the summit of the island. The distance 123 *bu* from the pole is the length of shadow of the front pole. The line of sight also passes through the tip of the rear pole and the surveyor. The distance is 127 *bu* from the pole. The difference between the two distances from the pole is called the *xiangduo*, which is considered as divisor. The distance between the poles, 1000 *bu*, is called the *biaojian*. Multiply [it by] the height of a pole for the dividend. Divide and add the height of the pole to obtain the height of the island, 1255 *bu*. There are 300 *bu* in a *li*, so one gets 4 *li* 55 *bu*, the height of the island.

To find the distance of the island from the front pole, multiply the distance between the two poles by the distance moved backwards as divisor. Divide to obtain the distance between the island [summit] and the [front] pole in *li*.

> Li et al.: In the Method, it is better to say: multiply the distance between the poles by the distance from the front pole, giving 123,000 [square] *bu*. Take the difference from the points of observation (*xiangduo*), 4 *bu*, as divisor. Divide to obtain 30,750 *bu*. Divide again by 300 *bu* in a *li* to have 102 *li* 150 *bu*, the distance between the island and the [front] pole.

7. Now survey a clear pool with a white stone at the bottom. On the bank set up a gnomon, whose *gou* is 3 *chi* high. Sight downward from the tip of the *gou* to the

opposite bank. The line of sight cuts the lower *gu* at a length of 4 *chi* 5 *cun*. The line of sight downward to the stone cuts the lower *gu* at a length of 2 *chi* 4 *cun*. Next set up a second gnomon 4 *chi* above the first one. Sight again downward from the tip of the *gou* to the opposite bank, and the line of sight cuts the upper *gu* at a length of 4 *chi*. The line of sight to the stone cuts the upper *gu* at 2 *chi* 2 *cun*. Tell: what is the depth of the pool? Answer: 1 *zhang* 2 *chi*.

Method: Multiply the difference between the lengths of the upper and lower *gu* in sighting the [opposite] bank by the length of the upper *gu* in sighting the stone. The product is taken as the upper rate. Next, multiply the difference between the lengths of the upper and lower *gu* in sighting the stone by the upper *gu* in sighting the [opposite] bank. The product is taken as the lower rate. Multiply the difference between upper and lower rates by the distance between the two gnomons as dividend. Multiply the two differences, i.e., those between the lengths of the upper and lower *gu* in sighting the bank and the stone respectively. Let the product be the divisor. Divide, giving the depth of the pool.

Li et al.: In the Method, multiply the difference, 5 *cun*, between the lengths of the upper and lower *gu* in sighting the bank by the length of the upper *gu*, 22 *cun*, in sighting the stone. The product 110 [square] *cun* is regarded as the upper rate. Then multiply the difference, 2 *cun*, between the lengths of the upper and lower *gu* in sighting the stone by the length of the upper *gu*, 40 *cun*, in sighting the bank. The product 80 [square] *cun* is taken as the lower rate. Multiply the difference between the two rates, 30 [square] *cun*, by the distance between the two gnomons, 40 *cun*. Let the product, 1200 [cubic] *cun*, be the dividend. Multiply the two differences, i.e., 2 by 5, to get 10 as divisor. Divide, giving the depth of the pool, 1 *zhang* 2 *chi*.

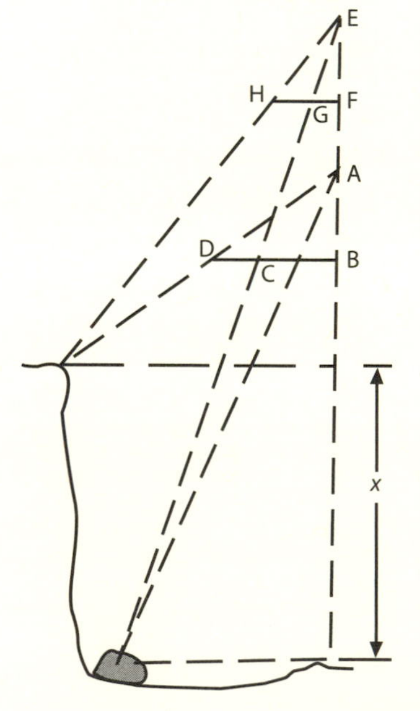

Alternative Method: Consider the sum of the two differences, i.e., those between the lower and upper *gu* in sighting the bank and stone, as divisor, and the product of the difference between the lower *gu* in the respective observation by the distance between the two gnomons as dividend. Divide, giving the depth of the pool.

Li et al.: In the Alternative Method, the difference between the upper and lower *gu* in sighting the bank is 5 *cun*, and that between the upper and lower *gu* in sighting the stone is 2 *cun*. Their sum is 7 *cun*, which is considered as divisor. Next, multiply the difference, 21 *cun*, between the lengths of the lower *gu* in sighting the stone and the bank by the distance, 40 *cun*, between the two gnomons. Let the product 840 [square] *cun* be the dividend. Divide, giving the depth of the pool, 120 *cun*; shift back one digit to obtain 1 *zhang* 2 *chi*.

VI. The "Ten Classics" of Ancient Chinese Mathematics

In the long history of Chinese mathematics, more works have been lost than preserved due to acts of war, vandalism, willful or accidental destruction by fire, and the simple wear and tear of daily use. Only one actual text on bamboo strips has as yet been recovered from archaeological excavations, the *Suan shu shu*, discussed above. And works like the *Zhou bi suan jing* and the *Nine Chapters* were rescued from oblivion by their early collators and commentators like Zhao Shuang and Liu Hui. The first effort to collect the most important ancient works together and establish canonical versions (or replace missing texts with alternative mathematical works) was made in the Tang dynasty, when a number of texts that later would be designated as "classics" were annotated by the Tang scholar Li Chunfeng and a team of colleagues. These works were adopted as the official texts both for teaching at the Imperial University and as the works upon which the country-wide examinations in mathematics would be based. According to the *History of the Old Tang Dynasty*,

> The *jianhou* (Astronomical Observer) named Wang Sibian had presented a memorial to the Throne reporting that ten computational canons [such as the *Wucao* [*suanjing*] or the *Sunzi* [*suanjing*] were riddled with mistakes and contradictions. [Consequently] Li Chunfeng together with Liang Shu, a *suanxue boshi* (Erudite of Mathematics) from the *guozijian* (Directorate of Education), and Wang Zhenru, a *taixue zhujiao* (Instructor from the National University) and others were ordered by imperial decree to annotate ten computational canons [such as the] *Wucao* or the *Sunzi*. Once their task was completed, the Emperor Gaozu [ruled from 618 to 627] ordered that these books be used at the *guoxue* (National University). [Martzloff 1997, 123]

The works that comprised the Tang collection included the following titles:

Ten Books of Mathematical Classics (*Shi bu suan jing*):

Mathematical Classic of the Zhou Gnomon
Nine Chapters on Mathematical Procedures
The Sea Island Mathematical Classic
Mathematical Classic of Master Sun
Mathematical Classic of Zhang Qiujian
Mathematical Classic of Xiahou Yang
Mathematical Classic of the Five Government Departments
Mathematical Procedures in the Five Classics
Zhui shu (exact meaning unknown; sometimes translated as *Method of Interpolation*)
Classic Continuation of Ancient Mathematics

Not all of these were ancient works, and the last on the list was actually written by Wang Xiaotong of the Tang dynasty. Also, an earlier work sometimes counted among the "classics," the *San deng shu* (Three Levels of Numbers), was also lost. Attributed to Dong Quan, this work on writing large numbers was probably similar to the section on notations for large numbers presented in the *Mathematical Classic of Master Sun*. By the time the "Ten Classics" were first printed in the Northern Song dynasty in 1048 CE, the *Zhui shu* (*Method of Interpolation*) by Zu Chongzhi had been lost and was replaced in the list of "Ten Classics" by the *Memoir on Some Traditions of the Mathematical Art*. In fact, it was in the Northern Song edition that these works were renamed as *suanjing*, "mathematical classics," although some retained their traditional titles, like the *Nine Chapters on Mathematical Procedures*, even though it was the most venerated of the "Ten Classics."

In 1126, when the capital of the Northern Song dynasty was besieged, the imperial library and archives were dispersed. During the subsequent Southern Song dynasty the "Ten Classics" were gathered anew with considerable difficulty by Bao Huanzhi, who oversaw their preparation for a version of the texts that was newly printed about 1213 CE. Although the Northern Song edition no longer exists, the Southern Song edition survives in one imperfect copy containing six of the ten "classics." Of these, the version of the *Nine Chapters* only includes the first five chapters, and the introduction is missing. This copy is now preserved in the Shanghai library, and recently has become widely available through a photoreproduction published in 1981.

The later history of the "Ten Classics" was no less precarious than their pre-Tang treatment and their subsequent history down to the Ming dynasty. By the early Qing, they were again all but unknown, despite having been printed twice, once in the Northern Song and again in the Southern Song dynasty. One of China's first efforts to reestablish all of the great works of the past, including mathematical and other scientific works, was the Ming dynasty's *Yongle dadian* (*Great Encyclopedia of the Yongle Reign*), compiled in the early fifteenth century. Between 1403 and 1407, this effort managed to find copies of many by then very rare works, including the *Zhou bi*, the *Nine Chapters*, the *Sea Island Mathematical Classic*, the *Mathematical Classic of Master Sun* (in 2 chapters rather than 3), the *Mathematical Classic of the Five Government Departments*, the *Mathematical Classic of Xiahou Yang*, and the *Mathematical Procedures in the Five Classics* (missing were the *Mathematical Classic of Zhang Qiujian*, the *Classic Continuation of Ancient Mathematics*, and the *Memoir on Some Traditions of the Mathematical Art*).

Thanks to private libraries and the collections of one family in particular, many Southern Song works were either bought or copied by hand for the Jigu library of Mao Jin and his son Mao Yi at Changshu, which included copies of several works missing from the *Yongle dadian*, including the *Mathematical Classic of Zhang Qiujian* and the *Classic Continuation of Ancient Mathematics* (*Ji gu suan jing*). The works of the Mao family's Jigu library are now in the Palace Museum in Beijing (except for the copies of the *Mathematical Classic of Xiahou Yang* and the *Classic Continuation of Ancient Mathematics,* which at some point were also lost). The remaining work missing from the *Yongle dadian*, the *Memoir on Some Traditions of the Mathematical Art* (*Shu shu ji yi*), is known from a copy from a collection of the "Governor of the two Jiangs," now in the Beijing University Library [Li and Du 1987, 228–29].

Like its Ming predecessor, the Qing dynasty also undertook a vast compilation of great works of the past. Under the patronage of the Qian Long emperor, the *Complete Library of the Four Branches of Books* (*Siku quanshu*) was completed between 1773 and 1781. Seven sets were eventually produced, for which the collation and commentary on the mathematical texts were overseen by the mathematician Dai Zhen, along with a team of scholars that included Chen Jixin, Guo Zhangfa, and Ni Tingmei. The basis for the collations of the *Complete Library* drew upon the *Great Encyclopedia of the Yongle Reign*, as well as copies of several texts from the Mao family's Jigu library. Dai Zhen also published the mathematical works separately in several other collections, including the *Imperial Collection of Treasure Accumulating Editions* (*Wuyingdian juzhenban congshu*, 1774). Meanwhile, Kong Jihan also edited a collection that he entitled *The Ten Mathematical Classics* (*Suanjing shishu*), which he also based on texts from the Jigu library and from the *Yongle dadian.* Kong's edition appeared in yet another encyclopedic work known as the *Collected Works of the Ripple Pavilion* (*Weiboxie congshu*, 1773).

Due to all of this editorial activity, by the end of the eighteenth century the "Ten Classics" of ancient Chinese mathematics were widely available for the first time. Not only did this enable many Chinese mathematicians to profit by careful study of these works of the past, but many historical studies and later editions were also produced. The most recent and authoritative collations with contemporary commentaries on the "Ten Classics" are the editions due to Qian Baocong (*Suanjing shishu*, 1963), and Guo Shuchun and Liu Dun (*Suanjing shishu*, 1998).

We have already examined the contents of three of these "Ten Classics." In this section we consider excerpts from two others and discuss some noteworthy problems as well as new methods that arise in more than one of the ancient texts. These subjects not only have generated interest among Chinese mathematicians, but also appear later in Indian, Islamic, and European mathematical treatises.

VI.a. Numbers and arithmetic: The *Mathematical Classic of Master Sun*

Like most of the mathematicians who wrote or commented on what were to become the Chinese classical texts, virtually nothing is known of the author of the *Sunzi suan jing* (*Mathematical Classic of Master Sun*). Some historians have conflated the author of this work with the well-known (and much earlier) author of *Master Sun's Art of War*, another Sunzi (also known as Sunwu). Given that Problem 4 in Chapter 3 of the *Mathematical Classic* deals with a Buddhist sutra, and the fact that Buddhism was not introduced to China until the first century CE, it is clear that the *Mathematical Classic of Master Sun* must be, at least in part, a later work. Moreover, due to the use of certain tax rates that were common in the period 280–473 CE, most historians of mathematics now agree that the *Sunzi suan jing* is most probably a work of the fourth or early fifth century CE [see Lam and Ang 1992, 4–7]. It is also likely that the present version of the *Sunzi suan jing* is a compilation of several previous versions of the text, perhaps written or collated at different times, and hence various chronological inconsistencies occur, although these ultimately do not affect the mathematics. One Problem, 3.36, has even been regarded as spurious, since it depends upon divination and cannot be regarded as a serious mathematical problem (except for the means by which a certain number is computed, a calculation that then determines, based upon whether it is even or odd, the sex of an unborn child).

The most distinctive feature of the *Mathematical Classic of Master Sun*, whoever its author may have been, is the valuable record it provides of exactly how computations were to be performed with counting rods on the counting board. Of the earlier known mathematical works discussed previously, all take for granted a knowledge of arithmetic operations with counting rods, including various basic computations with fractions as well as higher operations such as the extraction of square and cube roots. The author of the *Sunzi suan jing*, however, realized that it was necessary to explain in detail how such computations were to be performed, and thus the text provides step-by-step instructions about how to carry out the calculations for problems that, in most cases, we have already encountered. Explicit procedures are given for the addition, subtraction, multiplication, and division of fractions, and for their simplification. Step-by-step details are also provided showing how the calculations of square and cube roots are to be performed as algorithms on the counting board. The discussion of these arithmetic operations ends with a mathematical table already familiar from the *Suan shu shu*, the "Nine Nines," providing the multiples of integers from nine times nine to one times one. This sort of Nine Nines table, which every school child was expected to memorize, is also known from a manuscript found at Dunhuang, a version that actually uses the counting rod notation to write down the result of each multiplication:

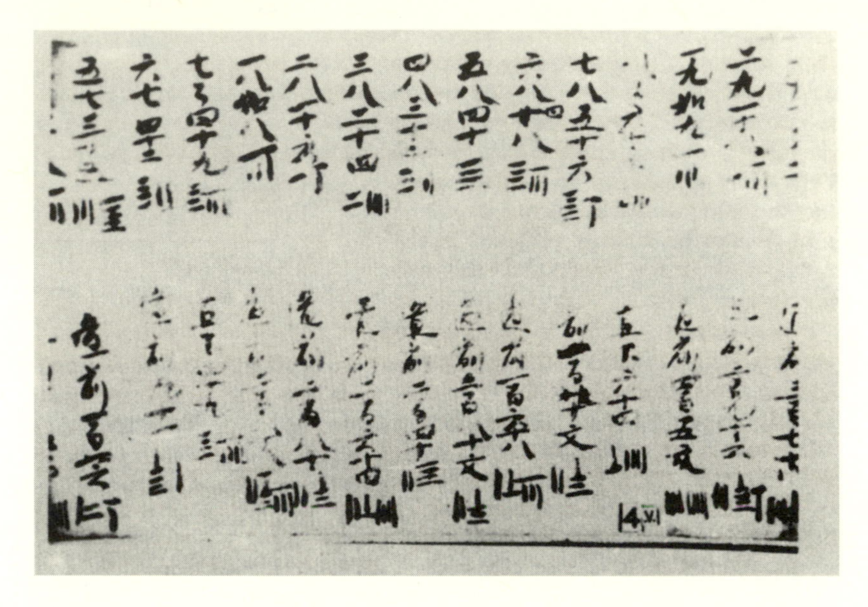

From "A Short Course on Calculation Techniques" from a Dunhuang scroll of the Tang Dynasty. Note the bottom of each column ends with the counting rod configuration for the answer to the multiplication in question [Li and Du 1987, 15].

The *Mathematical Classic of Master Sun* is comprised of three books, introduced by a brief preface. Here Sunzi follows a standard format in which the virtues of mathematics are extolled much as they were by Zhao Shuang in the Preface to the *Zhou bi*, and by Liu Hui in his Preface to the *Nine Chapters.* Likewise, Sunzi notes how remarkable it is that mathematics can be used to measure the heavens, unite the cosmic realms of *yin* and *yang*, relate the three luminous bodies, the four seasons, the five phases and six arts, in addition to facilitating all sorts of practical applications in everyday life. Thus mathematics holds the secret for understanding all things, and when one studies it "readily like a youth with an open mind, one is instantly enlightened." But appreciating the diverse uses of mathematics and understanding how it may be successfully applied to myriad problems requires concentration and discipline, as Sunzi's preface warns.

The first book opens with a discussion of lengths, weights, and measures of volume. Lists are provided of various equivalencies, including exchange rates for grains of various grades of refinement similar to those already encountered at the beginning of Chapter 2 on "grains" in the *Nine Chapters.* The most innovative mathematical feature introduced in the *Sunzi suan jing*, however, is a notation for writing very large numbers, from 10^4 to 10^{80}. This turns out to be very similar to the method Archimedes devised for expressing large numbers in his well-known work, *The Sand Reckoner*, in which he devised a means of writing down a figure larger than the number of grains of sand in the universe.

Chapters 2 and 3 of the *Sunzi suan jing* contain 28 and 36 problems, respectively, of which a representative sampling is provided here. These include various types of problems solved by methods familiar from the *Nine Chapters*, including the method of excess and deficiency, matrix methods, and applications of the *gou-gu* relation. Chapter 3 covers generally more difficult problems in arithmetic, including problems of pursuit (Problem 3.35, cast in terms of the return of three sisters at various intervals), a version of the "washing bowls" problem (3.17) that was later criticized by Zhang Qiujian (see below), and the most famous of the problems in the *Sunzi suan jing*, Problem 3.26, "Master Sun's Problem," the earliest example of what has become known as the "Chinese Remainder Problem."

Mathematical Classic of Master Sun

Preface

Master Sun says: Mathematics [governs] the length and breadth of the heavens and the earth; [affects] the lives of all creatures; [forms] the alpha and omega of the five constant virtues (i.e., benevolence, righteousness, propriety, knowledge, and sincerity); [acts as] the parents for *yin* and *yang*; establishes the symbols for the stars and the constellations; [manifests] the dimensions of the three luminous bodies (i.e., sun, moon, and stars); maintains the balance of the five phases (i.e., metal, wood, water, fire, and earth); [regulates] the beginning and the end of the four seasons; [formulates] the origins of myriad things; and [determines] the principles of the six arts (i.e., propriety, music, archery, charioteership, calligraphy, and mathematics).

[The function of mathematics] is to investigate the assembling and dispersing of the various orders [in nature], to examine the rise and fall of the two *qi* (i.e., *yin* and *yang*), to compute the alternating movements of the seasons, to pace out the distances [of the celestial bodies], to observe the intricate signs of the way of the heavens, to perceive the physical features of the earth, to locate the positions of the celestial and terrestrial spirits, to verify the [causes] of success and failure, to exhaust the principles of morality, and to study the temperament of life. [The field of mathematics covers] the use of the compass and the carpenter's square to regulate squares and circles, the fixing of standard measures to estimate lengths, and the establishment of measures to determine weights. [These measures] are split [to the accuracies of] *hao* and *li* [for lengths], and *shu* and *lei* [for weights].

[Mathematics] has prevailed for thousands of years and has been used extensively without limitations. If one neglects its study, one will not be able to achieve excellence and thoroughness. There is indeed a great deal to master when one views mathematics in perspective. When one becomes interested in mathematics, one will be fully enriched; on the other hand, when one keeps away from [the subject], one finds oneself lacking intellectually. When one studies [mathematics] readily like a youth with an open mind, one is instantly enlightened. However if one approaches [mathematics] like an old man with an obstinate attitude, one will not be skilful in it. Therefore if one wants to learn mathematics [fruitfully], one must discipline oneself and aim for perfect concentration; it is through this way that success in learning is assured.

Chapter 1

In the common model of large numbers [where *wan* represents 10,000 or 10^4],

wan wan (10^8) is called *yi,*
wan wan yi (10^{16}) is called *zhao,*
wan wan zhao (10^{24}) is called *jing,*
wan wan jing (10^{32}) is called *gai,*
wan wan gai (10^{40}) is called *zi,*
wan wan zi (10^{48}) is called *rang,*
wan wan rang (10^{56}) is called *gou,*
wan wan gou (10^{64}) is called *jian,*
wan wan jian (10^{72}) is called *zheng,*
wan wan zheng (10^{80}) is called *zai.*

Chapter 2

9. Now there is the base of a house, 3 *zhang* in the north-south direction and 6 *zhang* in the east-west direction, which is to be laid with bricks. Find the number [of bricks] required if 5 pieces cover an area of 2 *chi.* Answer: 4,500 pieces.

Method: Put down the east-west [length] of 6 *zhang* and multiply it by the north-south [length] of 3 *zhang* to obtain 1,800 *chi.* Multiply by 5 to give 9,000 *chi,* and divide by 2 to get the answer.

14. Now there is a square field with a mulberry tree at the center. The distance of one corner [of the field] to the mulberry tree is 147 *bu.* Find the area of the field. Answer: 1 *qing* 83 *mu* and an odd lot of 180 *bu.*

Method: Put down the distance from the corner to the mulberry tree, 147 *bu,* and double it to obtain 294 *bu.* Multiply by 5 to give 1,470 *bu* and divide by 7 to obtain 210 *bu.* Multiply this by itself to obtain 44,100 *bu* and divide by 240 *bu* to get the answer.

26. Now there are three persons A, B, and C who hold certain sums of money. A says to B and C, "If one half of each of your money is added to mine, the result is 90." B says to A and C, "If one half of each of your money is added to mine, the result is 70." C says to A and B, "If one half of each of your money is added to mine, the result is 56." How much money does each of the three men hold originally? Answer: A 72, B 32, and C 4.

Method: First put down in [three] positions the amounts declared by the three persons, and multiply each by 3 obtaining 270 for A, 210 for B, and 168 for C. Halving each yields 135 for A, 105 for B, and 84 for C. Once again put down 90 for A, 70 for B, and 56 for C, and halve each of them. Subtract [the latter] A and B from [the previous] C, [the latter] A and C from [the previous] B, and [the latter] B and C from [the previous] A, to yield the original amounts of money held by each person.

28. Now there is a gang of robbers who stole an unknown quantity of thin silk from a warehouse. In the distribution of the silk among themselves, it is heard that if each person is given 6 *pi,* there is a surplus of 6 *pi,* and if each person is given

7 *pi*, there is a deficit of 7 *pi*. Find the number of persons and the amount of thin silk. Answer: 13 robbers and 84 *pi* of thin silk.

Method: First put down on the upper right, the amount each person gets, 6 *pi*, and on the lower right, the surplus, 6 *pi*. After this put down on the upper left, the amount each person gets, 7 *pi*, and on the lower left, the deficit, 7 *pi*. Cross-multiply and add the results to obtain the quantity of thin silk. Add the surplus and deficit to obtain the number of persons.

Chapter 3

4. Now there is a book on Buddhism with a total of 29 chapters. Each chapter has 63 characters. Find the total number of characters. Answer: 1,827.

Method: Put down 29 chapters and multiply by 63 characters to get the answer.

17. Now there was a woman washing bowls by the river. An officer asked, "Why are there so many bowls?" The woman replied, "There were guests in the house." The officer asked, "How many guests were there?" The woman said, "I don't know how many guests there were; every 2 persons had [a bowl of] rice, every 3 persons [a bowl of] soup, and every 4 persons [a bowl of] meat; 65 bowls were used altogether." Answer: 60 persons.

Method: Put down 65 bowls, multiply by 12 to obtain 780, and divide by 13 to get the answer.

26. Now there are an unknown number of things. If we count by threes, there is a remainder 2; if we count by fives, there is a remainder 3; if we count by sevens, there is a remainder 2. Find the number of things. Answer: 23.

Method: If we count by threes and there is a remainder 2, put down 140. If we count by fives and there is a remainder 3, put down 63. If we count by sevens and there is a remainder 2, put down 30. Add them to obtain 233 and subtract 210 to get the answer. If we count by threes and there is a remainder 1, put down 70. If we count by fives and there is a remainder 1, put down 21. If we count by sevens and there is a remainder 1, put down 15. When [a number] exceeds 106, the result is obtained by subtracting 105.

27. Now there are six-headed four-legged animals and four-headed two-legged birds [put together]. [A count] above gives 76 heads and [a count] below gives 46 legs. Find the number of animals and birds. Answer: 8 animals, 7 birds.

Method: Double the number of legs and subtract from this the number of heads. Halve the remainder to get the number of animals. Multiply the number of animals by 4 and subtract [the product] from the number of legs. Halve the remainder to get the number of birds.

35. Now there are 3 sisters. The eldest returns once every 5 days, the second returns once every 4 days, and the youngest returns once every 3 days. Find the number of days before the 3 sisters meet together. Answer: 60 days.

Method: Put down on the right 5 days for the eldest sister, 4 days for the second sister, and 3 days for the youngest sister. For each numeral, arrange 1 counting rod on the left. [By performing] cross-multiplication, the number of times each sister returns is obtained. The eldest returns 12 times, the second 15 times, and the youngest 20 times. Next multiply each by the [corresponding] number of days to get the answer.

36. Now there is a pregnant woman whose age is 29. If the gestation period is 9 months, determine the sex of the unborn child. Answer: Male.

Method: Put down 49, add the gestation period, and subtract the age. From the remainder take away 1 representing the heaven, 2 the earth, 3 the man, 4 the four seasons, 5 the five phases, 6 the six pitch-pipes, 7 the seven stars [of the Dipper], 8 the eight winds, and 9 the nine divisions [of China under Yu the Great]. If the remainder is odd, [the sex] is male and if the remainder is even, [the sex] is female.

Methods for Expressing Arbitrarily Large Numbers

The brief section above from Chapter 1 is all that the text says about a means of writing very large numbers; there is no further commentary. There is, however, an alternative account for representing arbitrarily large numbers in another of the "Ten Classics," the *Shu shu ji yi* (*Memoir on Some Traditions of the Mathematical Art*), attributed to the Han dynasty scholar Xu Yue. Some historians of mathematics believe this to be a sixth-century forgery by the work's commentator, Zhen Luan, due in part to anachronistic terminology that would have made it impossible for Xu Yue to have written the work at any time during the Han dynasty. Whatever the date and authorship of this work may be, the problem of arbitrarily large numbers arises in the context of questions about the creation of the world and whether there can be an end to numbering or numbers. If there were, this would imply that the universe must be finite, bounded, and limited, and in turn so must the created world of the "myriad things." Considered in terms of time, this would mean that duration must be finite and that time cannot be infinitely divisible. If, on the other hand, time is infinite or infinitely divisible, then there must be a way to name, to specify, and to number the infinity of duration or the instantaneous moments that in aggregate constitute time. Such instantaneous moments are expressed in the *Shu shu ji yi* with a very specific pair of characters, "cha na," which have no known counterpart elsewhere in Chinese literature. Instead, the word seems to have been borrowed from a Sanskrit term that arises in the context of Buddhist discussions of time. Indeed, very small, infinitesimal amounts of time, "cha na," are instantaneous moments, the snap of a finger, the wink of an eye. Were the number of "cha na" that comprised the duration of time bounded, or infinite in number?

The *Shu shu ji yi*, in answering that there can be no limit to numbers, provides an impressive system for writing down virtually any arbitrarily given number, however large. The system does so using only ten symbols, but each symbol takes on a distinctive magnitude depending on the level of a three-tiered hierarchy of magnitudes to which it may be assigned. The text puts this rather obscurely, "not knowing the 'three,' means that one cannot talk about knowing the 'ten'," but what this really means is that one can only know the value that the character "jiang" assumes, for example, if one knows to which of three levels of magnitude "jian" is assigned: 10^{12}, 10^{64} or 10^{1024}. The latter is a very large number, but not the largest among the third or highest upper rank, which is 10^{4096}, as the table of the ten symbols and the three hierarchies makes clear:

Markers	*Lower degree*	*Medium degree*	*Higher degree*
wan 萬	10^4	10^4	10^4
yi 億	10^5	10^8	10^8
zhao 兆	10^6	10^{16}	10^{16}
jing 京	10^7	10^{24}	10^{32}
gai 該〔垓，陔〕	10^8	10^{32}	10^{64}
zi 梓〔秭〕	10^9	10^{40}	10^{128}
rang 讓〔壤〕	10^{10}	10^{48}	10^{256}
gou 溝	10^{11}	10^{56}	10^{512}
jian 間	10^{12}	10^{64}	10^{1024}
zheng 政	10^{13}	10^{72}	10^{2048}
zai 載	10^{14}	10^{80}	10^{4096}
Modern formulae $n = 1, 2, 3, \ldots$	10^{n+3}	10^4 and 10^{8n}	$(10^4)^{2^{n-1}}$

The *Shu shu ji yi* explains all of this as follows:

> In the method produced by the Yellow Emperor, numbers have ten degrees. In practice [these ten degrees] are used in three ways. The ten degrees are *yi, zhao, jing, gai, zi, rang, gou, jian, zheng,* and *zai*. The "three degrees" [i.e., the three ways according to which the "ten degrees" are manipulated] are the higher, the medium and the lower. According to the lower [degree of] numbers, numbers are transformed progressively by 10; for example, ten myriads are called *yi*, ten *yi* are called *zhao* and ten *zhao* are called *jing*. According to the medium degree, numbers are transformed progressively by a myriad; for example, a myriad myriad is called *yi*, a myriad myriad *yi* is called *zhao* and a myriad myriad *zhao* is called *jing*. According to the higher degree, numbers are modified when available numbers are exhausted; for example [in this system] a myriad myriad is called *yi*, *yi yi* is called *zhao* and *zhao zhao* is called *jing*. [Martzloff 1997, 99]

What makes the account of the *Shu shu ji yi* of interest (as opposed to the more limited scale of large numbers mentioned in the *Sunzi suan jing* without comment) is that it shows how, in addition to purely practical or theoretical concerns of mathematics, philosophical or metaphysical issues could also be inspiration for Chinese mathematicians. Here the question of time, conceived cyclically, led to a unique mathematical solution to the endless generation of numbers by means of a method that was itself appropriately cyclical. Whether originally conceived by Xu Yue or reconceived or even fabricated by Zhen Luan in what seems to be a clearly Buddhist context of concern for great cycles of time and the duration of the universe, the invention of the hierarchies of numbers was a uniquely Chinese contribution. Above all, it provided a remarkable correspondence between the endless cycles of time and the revolutions of the heavens which know no limits, and the endless generation of numbers, which also knows no limit.

In conclusion, to the question originally posed, presumably by Xu Yue—"Is there any end to number?"—the answer in the mathematical microcosm was "no," for the same ten symbols could be transformed within three different levels or cycles to write or name virtually any conceivable number. And the "cha na," those instantaneously small moments of time—the mere snap of a finger—accounted not only for the infinite duration and seamless continuity of time, but showed how the inspiration of Indian culture, through Buddhist concepts, in the hands of Chinese mathematicians, could be transformed into something quite new and equally profound.

The Chinese Remainder Problem

When Sunzi first presented what has subsequently come to be known as "Sunzi's Problem" or the "Chinese Remainder Problem," it was presented in a very straightforward way, with a solution given in terms of a "method" that was specific to the numbers of the original problem, and explained for computation on the counting board. Following the method described by Sunzi, the number required is determined as follows:

$$70 \times 2 + 21 \times 3 + 15 \times 2 - 210 = 23.$$

The "general" instructions given after the basic method Sunzi's rule imply a method of solving the same problem for different remainders r_1, r_2, and r_3:

$$70 \times r_1 + 21 \times r_2 + 15 \times r_3 - 105n.$$

It seems clear from the above, then, that Sunzi had noted the following congruences:

$$70 \equiv 1 \pmod 3 \equiv 0 \pmod 5 \equiv 0 \pmod 7$$
$$21 \equiv 1 \pmod 5 \equiv 0 \pmod 3 \equiv 0 \pmod 7$$
$$15 \equiv 1 \pmod 7 \equiv 0 \pmod 3 \equiv 0 \pmod 5$$

Thus $70 \times 2 + 21 \times 3 + 15 \times 2 = 233$ satisfies the given congruences, and since any multiple of $105 = 3 \times 5 \times 7$ is divisible by 3, 5, and 7, to find the smallest number that satisfies the above congruences is simply a matter of subtracting 105 as many times as it takes to reach the smallest number less than 105, which in turn gives the answer to the remainder problem [for details, see Katz 1998, 198].

In the case of this specific problem, the numbers are simple enough that it would not have been difficult to find the answer by trial and error. But this would not help to account for the method given, and from this alone it seems likely that Sunzi had a more general strategy in mind for approaching simple, integer problems of congruence. Later, in 1275, Yang Hui's *Continuation of Ancient Mathematical Methods* (*Xugu zhaiqi suanfa*) includes five different versions of the problem, each following this example. Meanwhile, the remainder problem also appeared in works by such Arabic mathematicians as Ibn Tāhir and Ibn al-Haytham, and in Fibonacci's *Liber Abaci* of 1202, where the remainder problem is given using the same moduli and rule as those found in the *Mathematical Classic of Master Sun*, after which it was a topic that was often addressed by European mathematicians.

In China, however, it was not until the thirteenth century that a general method was presented for solving such congruence problems. This method, explained by Qin Jiushao, will be considered in section VII.

VI.b. The *Mathematical Classic of Zhang Qiujian*

Like the *Sunzi suan jing*, the *Zhang Qiujian suan jing* (*Mathematical Classic of Zhang Qiujian*) provides valuable information about ancient Chinese mathematics because it also gives explicit, step-by-step instructions for how the counting rods should be set down and manipulated on the counting board in order to carry out basic arithmetic computations, especially for fractions. As the preface warns, "anyone who studies mathematics should not be afraid of the difficulty of multiplication and division, but should be afraid of the mysteries of the 'interconnection of parts'" [Martzloff 199, 194]. Here the "interconnection of parts," or "tong fen," means the "equalization" of fractions to find a common denominator. That this was still regarded as difficult mathematically suggests the challenges to the arithmetic of fractions that persisted, centuries after the early classics had been written.

As for other classics about which very little is known, inferences based upon economic information included in one of the problems (2.13) suggest the *Zhang Qiujian suan jing* was compiled sometime between 466 and 485 CE. The version that survives is based upon the incomplete edition of the Ten Classics from the Southern Song printing now in the Shanghai Library, for which the copy of the *Mathematical Classic of Zhang Qiujian* is also incomplete. Nevertheless, several problems missing from the end of second and the beginning of the third book have been reconstructed in a collation of the text suggested by Ho Peng-Yoke based upon versions of the missing problems mentioned in other texts [Ho 1965].

As in the preface to the *Sunzi suan jing*, Zhang Qiujian emphasizes the difficulties encountered with the arithmetic of fractions, especially the techniques used to "equalize" or determine common denominators. The preface also criticizes the work of Xiahou Yang, so there can be no doubt that the latter predates the *Mathematical Classic of Zhang Qiujian*. Zhang's criticism also makes clear that Chinese mathematicians did not simply praise the works of their predecessors, but were indeed capable of very telling and trenchant commentaries that were as ready to draw attention to shortcomings as they were to add new methods and mathematical innovations of their own.

In the surviving edition of the *Mathematical Classic of Zhang Qiujian*, Chapter 1 consists of 15 problems, Chapter 2 of 22 problems, and Chapter 3 of 38 problems. In Chapter 1, the first six problems concern arithmetic operations with fractions, only the sixth of which, the most difficult, is provided with a step-by-step method explaining its solution with the manipulation of counting rods on the counting board. The solution to Problem 1.8 depends upon knowing that there are 300 *bu* to 1 *li*, and 6 *chi* to 1 *bu*. To "put down the circumference in *bu*" means to express with counting rods on the counting board $(452 \times 300) + 180 = 135{,}780$ *bu*. If each soldier is 10 *bu* from the next, "move backwards by one place" means moving back on the counting board one unit from 135,780 to 13,578, giving the number of soldiers. If they are now 4 *chi* apart, the contracted circumference is $4 \times 13{,}578 = 54{,}312$ *chi*, or 9,052 *bu*, which translates to 30 *li* 52 *bu* for the smaller circle.

Problem 1.17 is of interest for the glimpse it provides of the hierarchy of ancient Chinese nobility, according to which the proportional distribution of gold is divided by rank among five different classes of noblemen. Problem 1.29 is similar to Problem 7.18 in the *Nine Chapters*, where a solution is given using the Rule of False Position, but in the *Zhang Qiujian suan jing* an alternative, direct method is described to solve the problem. The answer depends upon the fact that 1 *liang* is equivalent to 24 *zhu*.

In Chapter 2, Problem 2.13 involves a system of nine levels of households that was in effect only briefly in the Northern Wei dynasty from 466–484 CE, upon which Qian Baocong based his dating of the *Zhang Qiujian suan jing*. Problem 2.17 is similar to Problem 7.2 in the *Nine Chapters*, which there is solved by the method of excess and deficit (as noted at the end of the problem), but again Zhang Qiujian provides a more direct method.

In Chapter 3, Problem 3.10 concerns the determination of the height of the frustum of a square-based pyramid given its capacity and the dimensions of the lower and upper sides of the frustum. This is a much more complicated problem than the straightforward determination of the volume of the frustum of the pyramid that appears in the *Nine Chapters*, but it uses the same formula as given there, namely, for a and b the lengths of the upper and lower sides of the squares of the frustum, and h the altitude, the volume $V = \frac{1}{3}(ab + a^2 + b^2)h$. The problem begins with the conversion of the amount of millet, 938 22/81 *hu*, into the equivalent volume in *chi* (1 *chi* 6 *cun* 2 *fen* is equivalent to 1 *hu*). Problem 3.14 depends directly for its solution of three linear equations in 3 unknowns on the *fangcheng* method to which all of Chapter 8 in the *Nine Chapters* is devoted, but without further explanation in the *Zhang Qiujian suan jing*. Problem 3.24 concerns the perimeter of a city whose shape is not specified, and so the answer is given to cover the several possibilities, the shape being possibly a square, a rectangle, or a circle.

Problem 3.37 of Chapter 3 is the celebrated "washing bowls" problem. Zhang Qiujian includes this, as he mentions in the preface, to show how he can improve upon the solution given for the same problem that had already appeared in the *Sunzi suan jing* (Problem 3.17). Sunzi's solution uses the least common denominator 12 to "equalize" the data of the problem, a simplification that makes the solution less transparent and particular to this given statement of the problem. The solution Zhang offers is by no means limited to the specific numbers given in the problem. By equalizing the terms of the problem by simply multiplying the number of cups together, and then homogenizing the number of people, he provides an algorithm for determining the dividend *shi* and divisor *fa* which results in a general solution for all types of such problems. Finally, the last of the problems in Chapter 3, problem 3.38, is the famous "Hundred Fowls Problem."

Mathematical Classic of Zhang Qiujian

Chapter 1

6. Divide 6587 2/3 and 3/4 by 58 1/2. How much is it? Answer says: 112 437/702.

Method says: Put down 6587 in the upper position. Separately, put down the fractional parts 3 [and 2] in lower right position with 2 on the left. Below them place the fractional parts 4 and 3 with 3 on the left. Cross-multiply to obtain 12 as the denominator and 17 as the numerator. Divide by the denominator to obtain 1 with a remainder of 5. Add 1 to that in the upper position to obtain 6588, multiply by the denominator 12 and add the numerator 5 to obtain 79061. Next, multiply this by the denominator of the divisor which is 2 to obtain 168122. Put down 58 of the divisor in the lower position, multiply by 2 and add 1 from the numerator to obtain 117. Multiply by the denominator 12 of the dividend to obtain 1404 as divisor (*fa*). Divide the *shi* [which is 158122] to obtain 112 and halve both the divisor and the remainder to obtain 437/702.

8. Now there is an encirclement for hunting whose circumference is 452 *li* 180 *bu*; the soldiers are so arranged on the encirclement such that each person is 10 *bu* apart [from the next]. It is now desired to lessen the space between the soldiers to 4 *chi*. Find the contracted circumference. Answer says: 30 *li* 52 *bu*.

Method says: Put down the circumference in *bu*. Move backwards by one place. Multiply by 4 and [the product] is in *chi*. [Convert] to *bu* and divide to obtain the contracted circumference.

17. Now there are 59 *jin* 1 *liang* of gold from the official coffer which are given to 9 princes, 12 dukes, 15 marquis, 18 viscounts and 21 barons. Each prince is given 5 *liang* of gold more than each duke, each duke is given 4 *liang* of gold more than each marquis, each marquis is given 3 *liang* of gold more than each viscount, and each viscount is given 2 *liang* of gold more than each baron. Find the amount of gold given to each prince, duke, marquis, viscount, and baron. Answer says: Prince 1 *jin* 6 *liang*, duke 1 *jin* 1 *liang*, marquis 13 *liang*, viscount 10 *liang*, baron 8 *liang*.

Method says: Put down the numbers for princes, dukes, marquis, viscounts and barons. Multiply the number for princes by 14, that for dukes by 9, that for marquis by 5, and that for viscounts by 2. Add [the products] and subtract the sum from the amount of gold in *liang*. Divide the remainder by the total number of persons. Add to the quotient the differences in the quantities to obtain the amount of gold for each prince, duke, marquis and viscount. The quotient with no addition is the amount of gold for each baron.

29. Now there are 7 cubes of gold whose weight is the same as 9 cubes of silver. When a cube of each is interchanged, the weight of the gold is 7 *liang* lighter. What is the weight of each cube of gold and silver? Answer says: A cube of gold weighs 15 *liang* 18 *zhu*, a cube of silver weighs 12 *liang* 6 *zhu*.

Method says: Multiply the cubes of gold and silver and then multiply by one half of the difference in weight; let the product be the *shi*. Multiply [separately] the cubes of gold and the cubes of silver by the difference in both numbers of cubes and let each [product] be the *fa*. Divide the *shi* by the *fa*.

Chapter 2

13. Now there are certain rates [of contribution among households] where [the average] contribution per household is 3 *pi* of thin silk. The households are divided into nine classes according to their wealth. The difference in contribution between households [belonging to two consecutive classes] is 2 *zhang*. There are 39 households in upper upper class, 24 households in middle upper class, 57 households in lower upper class, 31 households in upper middle class, 78 households in middle middle class, 43 households in lower middle class, 25 households in upper lower class, 76 households in middle lower class and 13 households in lower lower class. Find the amount of silk contributed by each household from the nine classes. Answer says: Each household of upper upper class contributes 5 *pi* of silk. Each household of middle upper class contributes 4 *pi* 2 *zhang* of silk. Each household of lower upper class contributes 4 *pi* of silk. Each household of upper middle class contributes 3 *pi* 2 *zhang* of silk. Each household of middle middle class contributes 3 *pi* of silk. Each household of lower middle class contributes 2 *pi* 2 *zhang* of silk. Each household of upper lower class contributes 2 *pi* of silk. Each household of middle lower class contributes 1 *pi* 2 *zhang* of silk. Each household of lower lower class contributes 1 *pi* of silk.

Method says: Display [on the counting board the areas of] eight classes of households [in order] to find their accumulative differences: upper upper class 16, middle upper class 14, lower upper class 12, upper middle class 10, middle middle class 8, lower middle class 6, upper lower class 4, middle lower class 2; multiply each by its number of households and add [the products]. Multiply the total number of households by the average contribution of silk per household. Subtract this result from the [above] sum and divide the remainder by the total number of households. The result is the contribution of each household from the lower lower class. Next add successively the difference in contribution [between households of two consecutive classes] to obtain the contributions of silk from the households of the other eight classes.

17. Now there is a man who invests money in Luo where the profit is 2/5 [of the amount]. The first time he returns with 16000, the second time he returns with 17000, the third time he returns with 18000, the fourth time he returns with 19000, the fifth time he returns with 20000. After the five trips, he has completely withdrawn his original investment and profit. Find the original investment. Answer says: 35326 5918/16807 *qian*.

Method says: Put down the amount he brought back from the last trip and multiply by 5. Add to this 7 times the amount he brought back from the fourth trip. Multiply [the sum] by 5. Add to this 49 times the amount he brought back from the third trip. Multiply [the sum] by 5. Add to the 343 times the amount he brought back from the second trip. Multiply [the sum] by 5. Add to this 2401 times the amount he brought back from the first trip. Multiply [the sum] by 5. Divide by 16807 to obtain the original investment. This is one method; the Rule of False Position is another method.

Chapter 3

10. Now there is a pit [in the shape of a frustum of a pyramid with a square base]. [A side of] the upper square is 8 *chi* and that of the lower square is 1 *zhang* 2 *chi*. It can hold 938 22/81 *hu* of millet. Find the depth. Answer says: 1 *zhang* 5 *chi*.

Method says: Put down the [amount of] millet and [convert this] into the volume in *chi*. Multiply by 3 and let [the product] be the *shi*. Square [a side] of the upper square and [a side] of the lower square, multiply [the two sides] and let the sum [of the three terms] be the *fa*. Divide the *shi* by the *fa*.

14. Now there are three persons A, B, C and it is not known how much money each has. A Says, "If I have two thirds of B's money and one third of C's money, then my money can reach 100." B says, "If I have two thirds of A's money and one half of C's money, then my money can reach 100." C says, "If I have two thirds of each of A's and B's money, then my money can reach 100." Find how much money each person has. Answers says: A 60, B 45, C 30.

Method says: A 3, B 2, C1, money 300. A 4, B 6, C 3, money 600. A 2, B 2, C 3, money 300. Use the *fang cheng* method to obtain the answer.

24. Now there is a city whose perimeter is 20 *li*. It is desired to set deer antlers [to surround the city] at 3 *chi* apart in 5 rows. Find the total number of antlers used. Answer says: 60100 antlers. If the city is circular, 60060 antlers are used.

Method says: Put down the perimeter of the city in *li* and [convert this to] *chi*. Divide by 3 and multiply by 5. Next, put down 5 and multiply it by 3; square the product and divide by 3 squared; multiply the result by 4. Add this to the number in the upper position to obtain the total number [of antlers]. If the city is circular, put down the perimeter of the city in *li* and [convert this to] *chi*. Divide by 3 and multiply the result by 5. Next, add 1, 2, 3 and 4 to obtain a total of 10 and multiply by 6. Add this [to the number above] to obtain the required sum.

37. Now there was a woman washing cups by the river. An officer asked, "Why are there so many cups?" The woman replied, "There were guests in the house, but I do not know how many there were. However, every 2 persons had [a cup of] thick sauce, every 3 persons had [a cup of] soup and every 4 persons had [a cup of] rice; 65 cups were used altogether." Find the number of persons. Answer says: 60 persons.

Method says: Arrange the numbers representing persons in the right [column] and put down the corresponding numbers representing cups in the left [column]. Mutually multiply the numbers representing cups by the numbers representing persons. Add the [three products] and let [the sum] be the *fa*. Multiply all the numbers representing persons together and multiply this by the given number of cups. Let [the product] be the *shi*. Divide the *shi* by the *fa*.

38. Now one cock is worth 5 *qian*, one hen 3 *qian*, and 3 chicks 1 *qian*. It is required to buy 100 fowls with 100 *qian*. In each case, find the number of cocks, hens and chicks bought. Answer says: 4 cocks worth 20 *qian*, 18 hens worth 54 *qian*, 78 chicks worth 26 *qian*. Another answer: 8 cocks worth 40 *qian*, 11 hens worth 33 *qian*, 81 chicks worth 27 *qian*. Another answer: 12 cocks worth 60 *qian*, 4 hens worth 12 *qian*, 84 chicks worth 28 *qian*.

Method says: Add 4 to the number of cocks, subtract 7 from the number of hens and add 3 to the number of chicks to obtain the answer.

The Hundred Fowls Problem

Although the *Nine Chapters* includes an indeterminate problem among those considered in the chapter devoted to "Rectangular Arrays" (Problem 8.13, which involves five simultaneous linear equations in six unknowns), among indeterminate problems in general one of the best known is that which appeared for the first time in the *Mathematical Classic of Zhang Qiujian* as Problem 3.38, namely the problem of buying 100 fowls for 100 coins. The method offered by Zhang Qiujian, however, is very terse, and offers little indication as to how such problems might be approached in general.

It is possible that this problem was simply solved by trial and error. Or, proceeding more systematically, if $x =$ the number of cocks, $y =$ the number of hens, and $z =$ the number of chicks, then the problem is equivalent to solving the two equations

$$x + y + z = 100,$$

$$5x + 3y + (1/3)z = 100.$$

One way to solve this system would be to ignore the cocks, thus giving a system in only two unknowns. It is then straightforward to find that $x = 0$, $y = 25$, and $z = 75$ is one solution. Then adding 4 cocks and 3 chicks adds 21 to the price, whereas removing 7 hens reduces 21 from the price. Thus from $(0, 25, 75)$ the following three solutions follow directly: $(4, 18, 78)$, $(8, 11, 81)$, and $(12, 4, 84)$.

Perhaps regarding Zhang Qiujian's explanation of the method as insufficient, Zhen Luan (fl. 570 CE) tried to improve on it by explaining that to solve the problem required dividing

1000 by 9 to obtain the number of hens, then subtract the remainder 1 from the divisor 9 to obtain the number of cocks. Unfortunately, the logic of Zhen's method has escaped virtually all modern commentators. As we shall see in section VII, Yang Hui also attacked this problem, and gave a more detailed explanation for a solution method. Outside China, versions of the problem appear in the works of, among others, Alcuin of York in the eighth century, Mahavira in the ninth century, Abu Kamil in the tenth century, Bhaskara in the twelfth century, Leonardo of Pisa in the thirteenth century, and al-Kashi in the fifteenth century.

VII. Outstanding Achievements of the Song and Yuan Dynasties (960–1368 CE)

When they were first printed in the Northern Song dynasty, just as in the Tang dynasty the "Ten Classics" of Chinese mathematics were meant to set the standards for teaching at the Imperial Academy, and for the national examinations. But mathematics was not regularly taught and seems to have fallen in and out of favor, so that there was no continuous history of its being an established part of any curriculum. For example, in 1084 when it was ordered by imperial decree that the subject should be reestablished, the buildings of the Institute of Mathematics, which had fallen into disrepair, were also renovated. But in less than two years it was decided to abolish the subject again, because

> after the introduction [of mathematics], staffing the school and setting up the civil service examinations [in mathematics], there was a general feeling that it was extravagant and did not really help in the running of the country. [Li and Du 1987, 109]

In the Southern Song dynasty mathematics was finally abolished once and for all and was not reestablished. Under such circumstances it is little wonder that there was no significant interest in the subject, and it is easy to understand why even the famous and most important texts of the "Ten Classics" were eventually lost, and except for the *Zhou bi*, were almost entirely unknown. Among the few notable works of this period was one written by Shen Kuo (1031–1095). His *Mengqi bitan* (*Dream Pool Essays*), which included a section on "techniques," addressed the problem of writing very large numbers (on the order of 10^{43}), as well as the summation of series. The "technique of small increments" is concerned with the equivalent of packing problems involving equal difference series, for which Shen Kuo's research laid the foundation in China.[17] Some of his ideas were further developed by Yang Hui (see below). In geometry, Shen Kuo advanced a formula for the approximation of the arc of a circle s given the diameter d, sagita v, and length of the chord c subtending the arc. He called this method the "technique of intersecting circles," and for the arc in question he gave its approximate length as $s = c + 2v^2/d$. Shen Kuo was also interested in simplifying the techniques for counting-rod computations, and devised short cuts to simplify the algorithmic procedures used on the counting board. As he said in the *Dream Pool Essays*, "in arithmetic seeing simplicity use it, seeing complexity change it, don't stick to one method."

Several centuries later Yang Hui likewise was also interested in further simplifying algorithmic procedures whenever possible [Li and Du 1987, 178]. Mention should also be made of an important change underway in this period, the increasing use of the abacus, which

---

[17]See, for example, the thesis by Andrea Eberhard, which surveys the history of the summation of series in the history of Chinese mathematics. For details, consult [Eberhard 1999].

eventually came to replace the counting board as a convenient means of calculating. Unfortunately, it did not facilitate algebraic thought the way the counting board did by virtue of positions serving as physical means of working algebraically, as explained below in the introduction to the "tian yuan" method. But the abacus did serve to accelerate the dull repetition of complex computations, and in this context, rhymes or mnemonic devices were often used to remind the calculator what needed to be done. For example, Shen Kuo described a method of continued addition as "Dividing by 9, increase by 1; dividing by 8, increase by 2, etc.," which, as Li and Du point out, was a forerunner of the verse "9, 1, bottom add 1; 9, 2, bottom add 2," which was also a rhymed scheme for the method of repeated addition [Li and Du 1987, 181].

When the Jin army captured the Northern Song capital in 1127, the wooden block-prints used to publish the mathematical classics, along with many other works, were destroyed and the imperial library dispersed. Soon thereafter, the Southern Song dynasty was established in the south, and the Mongols came into power in the north. But despite the division of the country, there were four mathematicians whose works represent an unexpected rebirth of mathematics in China of a very high order.

In the south, Qin Jiushao wrote the *Shushu jiuzhang* (*Mathematical Treatise in Nine Sections*, 1247), while in the north Li Zhi wrote several influential works, among them the *Ceyuan haijing* (*Sea Mirror of Circle Measurements*, 1248), and the *Yigu yanduan* (*New Steps in Computation*, 1259), the latter being an introduction of sorts to the former. A generation later, Qin Jiushao was followed in the south by Yang Hui, whose first influential work was a careful study of the *Nine Chapters*, the *Xiangjie jiuzhang suanfa* (*A Detailed Analysis of the Mathematical Methods in the Nine Chapters*, 1261). Almost immediately thereafter his *Riyong suanfa* (*Computing Methods for Daily Use*, 1262) was printed, and a decade or so after that there appeared his *Yang Hui suanfa* (*Yang Hui's Methods of Computation*, 1274–1275). And lastly, in the north several decades later Zhu Shijie published his *Suanxue qimeng* (*Introduction to Mathematical Studies*, 1299) and *Siyuan yujian* (*Precious Mirror of the Four Elements*, 1303).

Among the major achievements of these works are impressive results dealing with the solution of higher degree equations and new methods for the solution of simultaneous congruences. These include the "method of the celestial element" and the "method of four unknowns," which use algebraic methods and elimination techniques for the solution of simultaneous higher degree polynomial equations. As might be expected in a society where considerable emphasis was placed upon local business and commerce, numerous works on commercial arithmetic were in circulation as well. Mathematics also continued to be indispensable in the work of astronomers, surveyors, and engineers, but in what follows, emphasis is placed upon more theoretical achievements of mathematics in China during the Song and Yuan dynasties.

VII.a. Qin Jiushao: *Shushu jiuzhang* (*Mathematical Treatise in Nine Sections*, 1247)

Qin Jiushao was born in Sichuan and was raised in the capital of the Southern Song dynasty (today Hangzhou). Later, as a government official, he held various minor posts and for a while was district commander in Sichuan. It was a tumultuous period, during part of which his area was under siege by the Mongol army. It was under such difficult circumstances that Qin Jiushao nevertheless found the time and inspiration to write his *Mathematical Treatise in Nine Sections*. It may be that he found solace and a respite from the distress and worry of difficult times in his mathematics. Given his apparent reverence for number and no doubt numerology of the sort represented in the *Zhou yi* (*Yi jing*), it may not be a coincidence that his *Mathematical Treatise* was divided into nine parts with each part consisting of nine problems. In addition to astronomy, the calendar, meteorology, military matters (arrangement of tents), and commercial affairs (calculation of interest), mathematical subjects include sections on indeterminate analysis, field measurement, surveying (applying the *gou-gu* and double-difference methods), determination of "fair taxes," architectural constructions, and problems related to the capacity of warehouses and the transportation of various commodities.

In his preface Qin emphasizes the part that mathematics plays in virtually every aspect of life, from divination and the place of mathematics in the *Zhou yi* to the role of gnomons and shadows in surveying heaven and earth alike. His references to the Ho-t'u (*He tu*) and Lo-shu (*Luo shu*) diagrams are full of mystical mathematical significance, for these were figures associated with magic squares. The *He tu* was said to have been created by a river and associated with a dragon-horse whose footprints along the river left the figures of the eight trigrams. The *Luo shu* was said to have appeared out of the Yellow River, on the back of a divine tortoise, where the first nine numbers were arranged as a magic square [see Needham 1959, 56–59].

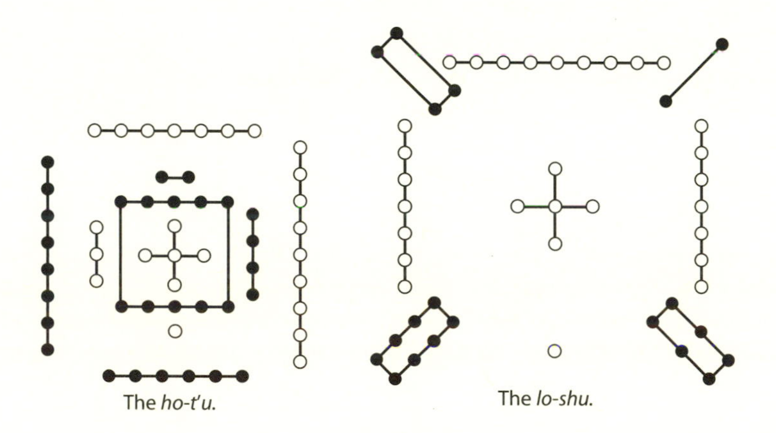

The *ho-t'u*. The *lo-shu*.

But apart from applications determining calendars or carrying out the usual work of government administrators in calculating the needs of corvée labor, Qin Jiushao laments the neglect of mathematics due to the failure of the mathematicians themselves to teach results they preferred to keep to themselves. Consequently, even mathematicians who could manage the basics of arithmetic could no longer handle the extraction of roots or solve indeterminate analysis.

The great innovation of his own work, the *ta-yen* (*da yan*) rule, is explained in the first section of the *Mathematical Treatise in Nine Sections* devoted to indeterminate analysis. This method, which he called the "great extension method of finding one"—held mystical associations for Qin Jiushao as well, who seems to have connected numerology with his method for solving linear congruences. The Preface of the *Mathematical Treatise* links the *Tao* (*Dao*) to numbers, and explains Qin's own progress in mathematics as due to "being in touch with the spiritual powers." That his *da yan* method was special is reflected in the emphasis he places on the fact that it is the one rule of mathematics not to be found in the *Nine Chapters*, and while calendar makers made "considerable" use of it, mathematicians as yet apparently had not. The *da yan* rule is a procedure for solving equations of the form $xP \equiv 1 \pmod{m}$ for x. This is one of the essential steps in solving the Chinese Remainder problem, namely the solution of n congruences $N \equiv r_i \pmod{m_i}$ for which the m_i are assumed to be co-prime. The problem reduces to finding the least integer $N = (\sum r_i x_i (M/m_i)) - \theta M$, where $M = \prod m_i$ and $x_i P_i \equiv 1 \pmod{m_i}$, where each $P_i \equiv M/m_i \pmod{m_i}$. By suitable choice of the integer θ, it is possible to determine the least integer N satisfying the given conditions.

Although the method for solving the Chinese Remainder problem given in the *Sunzi suan jing* suggests that the author apparently understood the general principle at work here, it was not explicitly revealed, and Sunzi's method only proceeds in terms of the numbers specific to "Sunzi's Problem." This was clearly not enough to satisfy the more general mathematical interests of Qin Jiushao, who devised a more detailed procedure. In fact, he carried out each step of this problem in terms of numbers laid down upon the counting board (for details, see [Li and Du 1987, 163–65], where all of the steps as they would have appeared on the counting board are clearly illustrated).

It is the step leading to the determination of the terms x_i such that $x_iP_i \equiv 1 \pmod{m_i}$ that also explains the name Qin Jiushao gave to his solution of the Chinese Remainder Theorem, namely, "the method of finding one by the great extension" (*dayan qiu yi shu*), which goes back to the *Yi jing*. There the method of divination begins with 50 tally sticks, from which one is put aside before dividing the remainder into two groups of *yin* and *yang* (further discussion of the significance of the name of Qin Jiushao's method may be found in [Needham 1959, 119] and [Li and Du 1987, 162–63]).

Qin Jiushao considers nine different problems involving complicated congruences; for example, his Problem 3 involves four moduli: 54, 57, 72, and 75 (explained in considerable detail in [Martzloff 1997, 315–23]). Moreover, mathematical complications immediately arise in congruences related to astronomical applications, since the m_i represent periods of motions of the planets or phases of the moon, for example, in which the congruences are not integral. Qin Jiushao considers more complicated situations as well in which the m_i may be decimal fractions, and explains what to do in cases for which the m_i may not be mutually prime. Although Chinese mathematicians did not use an explicit concept of prime numbers, the necessity in the case of congruences that the m_i contain no common factors was apparent to a mathematician like Qin Jiushao.

The systematic treatment of congruences in Qin Jiushao's *Mathematical Treatise in Nine Sections* is closely related to calendrical applications in which he seems to have been especially interested, and which here take the form of pursuit problems that seek to determine the time elapsed from a given initial position until a second realignment (of planets, for example) will occur. Such pursuit problems arise in Chapters 1 and 2 of the *Mathematical Treatise*, and although the method for determining the "number of years from the initial position" was obviously meant primarily for applications in astronomy, Qin Jiushao used the method for solving congruence problems in a variety of mathematical contexts. Li and Du point out that it was not until Euler and Gauss investigated congruences systematically that results were achieved on a par with those given by Qin, which prompts them to conclude that "In this regard Qin Jiushao's research was more than 500 years earlier than the work in Europe" [Li and Du 1987, 166].

Qin's treatise contains several dozen problems involving the solution of quadratic equations, cubic equations (Problem 6.6), quartic equations (Problems 3.1, 3.8, 4.6), and even one tenth degree equation (Problem 4.5). The coefficients might be positive or negative, integral or decimal—but the method of extracting roots is completely general. In explaining his solutions, Qin Jiushao often provides diagrams showing explicitly the counting rod configurations used to illustrate a step-by-step application of the algorithm for the extraction of roots by iterated multiplication. Problem 3.8, for example, gives 21 diagrams to illustrate the solution of the problem "Finding the area of an annulus." See [Martzloff 1997, 233–44].

Qin Jiushao's Shushu jiuzhang *(Mathematical Treatise in Nine Sections)*

Preface

The Six Arts of the teaching of the Chou were truly made complete by mathematics. It is something that scholars and great officers have always esteemed. Their application was based on the idea that "the Great Void generates the One, and it oscillates without ending."

If we aim at the great, we can be in touch with the spiritual powers and thus live conformably with our destinies; if we aim at the small, we can settle the affairs of this life, and by classification deal with the myriad phenomena. How could this be possible merely by peering at the shallow and familiar? Thus in ancient times they arranged the stalks [for prognostication] in order to calculate the [auspicious days] and fixed the pitch pipes so as to determine the *ch'i*. By means of the plumb and the carpenter's square they cleared their rivers and with the gnomon shadow template they measured [the length of] the shadow.

The greatness of heaven and earth was enclosed by these methods and could not exceed their scope—much less the multiplicity of human life between [heaven and earth]. After [the coming of] the *Ho-t'u* diagram and the *Lo-shu* diagram, they [the ancients] unriddled and explained these secrets. With the eight diagrams and the nine categories, they intricately pieced together an understanding of the essentials and subtleties [of heaven and earth]. They reached their apogee in the application of [these methods to] the *Ta-yen* and *Huang-chi* calendars.

As for changes in the affairs of men, there was nothing that could not be accounted for. As for the dispositions of the spirits, there was nothing that could be hidden. The sages comprehended it wondrously; they talked about it and handed down the crude outlines; the common man found it obscure; and for that reason none of them were aware of it. If you look at essential aspects of their meanings, the numbers and the *Tao* do not derive from two [different] bases.

The Han period was not far remote from antiquity. There were people like Chang Ts'ang, Hsü Shang, Ch'êng-ma Yen-nien, Kêng Shou-chang, Chêng Hsüan, Chang Hêng, and Liu Hung. Some of them investigated the *Tao* of heaven and their methods were handed down to posterity; some of them calculated the allocation of labor and they arrived at effective results in their time. In later generations, scholars were very proud of themselves and, considering [these arts] inferior, did not teach [or discuss] them. Those studies were almost defunct [through neglect]. Only calendar-makers and mathematicians were able to manage multiplication and division, but they could not comprehend square-root extraction or indeterminate analysis. In case there were calculations to be performed in the government offices, one or two of the clerks might participate but the position of the mathematicians was never held in esteem; their superiors left things to them and let them do as they pleased; [but] if those who did computations were only that sort of man, it was merely right that they should be disdained.

In the field of music, there are conductors who can only arrange the sounds of the bells and sounding stones, but is it permissible to say that "to produce complete harmony with heaven and earth" [which the Great Music is said to do] merely consists in this?

As for mathematical books, today there remain the works of more than thirty schools. [This part of mathematics], where heavenly configurations and celestial motions are calculated, is called "the technique of threading." Divination and fate calculation are referred to as *san-shih*; all of these are called "esoteric mathematics," to emphasize their esoteric nature.

What is contained in the *Nine Chapters* are the nine computations of the *Chou-li*. What is related to squares and circles is "geometry." These are all called "exoteric mathematics," in contrast to "esoteric mathematics." Their applications are connected with each other, and we cannot divide them into two [different arts].

Only the *ta-yen* rule is not contained in the *Nine Chapters*, for no one has yet been able to derive it [from other procedures]. Calendar-makers, in working out their methods, have made considerable use of it. Those who consider it as belonging to "equations" are wrong.

Moreover, the things of the world are great in number, and people of ancient times made their calculations before things happened. When their prognostications were calculated, they acted on them. They looked up [to heaven] and looked down [to earth], the plans of men and the plans of spirits, there was nothing they did not draw attention to. By so doing they avoided faults in the results. This we can ascertain in every chapter of the historical records. But later generations, in their undertakings and in their enterprises, were but rarely capable of reflection and contemplation. Gradually laws of heaven and affairs of men have become vague and imperfectly known. Should we not search out the reasons for this?

I, Chiu-shao, am stupid and uneducated and not versed in the arts. But in my youth I was living in the capital, so that I was enabled to study in the Board of Astronomy; subsequently I was instructed in mathematics by a recluse scholar. At the time of the troubles with the barbarians, I spent some years at the distant frontier; without care for my safety among the arrows and stone missiles, I endured danger and unhappiness for ten years. My heart was withered, and my vital power fell away. [But] I knew truly that none of these things was without its "number," and I let loose my imagination among these numbers.

I made inquiries among well-versed and capable [persons] and investigated mysterious and vague matters. If I attained some crude understanding, it was by what is called "being in touch with the spiritual powers and living conformably with destiny." As for the details [of the mathematical problems], I set them out in the form of problems and answers meant for practical use. I collected many of them and, not wanting to cast them aside, I selected eighty-one problems and divided them in nine classes; I drew up their methods and their solutions and elucidated them by means of diagrams.

Perhaps they will serve as material for gentlemen of broad knowledge to peruse in their spare time, for although [mathematics is] a minor art it is worth pursuing. Thus I wish to offer this work to my colleagues. If they say that their skill [in the minor arts] is complete and that this is merely for people like astronomers and provincial clerks, and ask why this should deserve to be used throughout the empire, will that not show them to be benighted!

In the ninth month of the seventh year of *Shun-yu*, Ch'in Chiu-shao of Lu prefecture wrote this.

Chapter 1: Indeterminate Analysis (Ta-yen lei)
The general *ta-yen* computation method is as follows:

A.

1. Set up all the *wên-shu* (problem numbers).
2. First of all, one has to join the numbers ... by twos and find their common factors.
3. a. Reduce the odd numbers; do not reduce the even ones.
   b. Sometimes one reduces and gets 5, and the other number is 10. In this case, one has to reduce the even number and not the odd one.
4. Sometimes all the numbers are even. After all the numbers have been reduced, we may keep only one even number.
5. a. Sometimes after reducing all the numbers there still remain numbers with common factors. Provisionally set them up until you can reduce them with the others [of the same class].
   b. Finally, find the common factors of the ones you have provisionally set up and reduce them [by those common factors] ...
6. If you reduce the odd number, do not reduce the even number, but multiply it. Or if you reduce the even number, do not reduce the odd number, but multiply it.
7. Sometimes one can reduce the one as well as the other. But of the ones that still have common factors, again find their mutual divisor by the method of successive divisions. By this mutual divisor reduce the one and multiply the other.
8. Then you get the *ting-shu* (definite numbers) ...
9. In finding the *ting-shu*, do not allow two numbers to be mutually divisible, or one to become too large. If one [of the numbers] becomes too large, then you have to "compensate."
10. If you do not want to borrow, it is allowed to have the number 1 [as *ting-shu*].

B.

11. You multiply the *ting-shu* with each other, and you get the *yen-mu*.
12. a. You divide the *yen-mu* by all the *ting-shu* and you obtain the *yen-shu*.
    b. Or you set up all the *ting-shu* as factors (*mu*) in the right column, and before all these, you set up the *t'ien-yüan* as factor (*tzŭ*) in the left column. By the *mu* you mutually multiply the *tzŭ* and you get the *yen-shu* too.
13. From all the *yen-shu* you subtract all the [corresponding] *ting-mu* as many times as possible. The part that does not suffice any more is called the remainder (*chi*).

C.

14. On the *chi* and the *ting-shu* one applies the *ta-yen ch'iu-i* (method of searching for one). With this method one will find the *ch'êng-lü* (those of which one gets the remainder 1).
15. The *ta-yen ch'iu-i* method says: Set up the *chi* at the right hand above, the *ting-shu* at the right hand below. Set up the *t'ien-yüan* 1 at the left hand above.

16. First divide the "right below" by the "right-above," and the quotient obtained, multiply it by the 1 of "left-below."
17. Set it up at the left hand below [in the second disposition]. After this, in the "upper" and "lower" of the right column, divide the larger number by the smaller one. Transmit [the numbers to the following diagram] and divide them by each other. Next bring over the quotient obtained and [cross-]multiply with each other. Add the "upper" and the "lower" of the left column.
18. One has to go on until the last remainder (*chi*) of the "upper right" is 1 and then one can stop. Then you examine the result on the "upper left"; take it as the *ch'êng-lü*.
19. Sometimes the *chi* is already 1; this is then the *ch'êng-lü*.

D.

20. Draw up all the *ch'êng-lü* and multiply with the corresponding *yen-shu*, and you will obtain the *fan-yung*.

E.

21. Add up the *fan-yung* and examine the *yen-mu*.
22. If [the sum of the *fan-yung*] surpasses the *yen-mu* by 1, then [the *fan-yung*] are the *chêng- yung*.
23. If [the sum of] the *fan-yung* surpasses a multiple of the *yen-mu* [with 1], then examine the *yüan-shu*.
24. If two numbers [*yüan-shu*] are divisible by the same factor, one subtracts half [the *yen-mu*] [from the *fan-yung*].
25. If there are three numbers of the same class, one divides the *yen-mu* by 3 and diminishes the three numbers with it.
26. All these numbers are the *chêng-yung-shu*.
27. Sometimes you get 1 as *ting-mu*, or if the *yen-shu* and the *yen-mu* are the same, there is no *yung-shu*.
28. We have to examine, among the *yüan-shu*, the ones that are mutually divisible, and borrow the *chêng-yung* from the largest numbers.
29. Among the *yüan-shu* find by twos the common factors. With the common factors reduce the *yen-mu* and you get the *chieh-shu*.
30. With the *chieh-shu* diminish the existing [*yung-shu*] and increase with this [amount] the nonexisting ones. These are the *ch'êng-yung*.
31. Or otherwise, if you wish to place numbers in the vacant places, these will be the *yen-mu*. Reduce the *yen-mu*. Multiply the number obtained with a factor of choice.
32. You borrow proportionally and put them in the places [which are empty].
33. Or, if you wish, after reducing, do not borrow. You are allowed to leave these places empty.

F.

34. After that, multiply the *chêng-yung* with *shêng-yü*. You obtain the *tsung*.
35. Add up the *tsung*.
36. Subtract from it the *yen-mu* as many times as possible.
37. What is not more sufficient [to subtract the *yen-mu* from] is the *lü-shu* looked for.

Problem 4. To compute the amounts of the treasuries: There are seven treasuries of district cities. The daily revenues in full strings [of *kuan*] are the same. The full amount of the annual tax is now being collected. [a *kuan* is a string of 1,000 coins]. Because recently ready money has been scarce, each treasury is allowed to adjust the old rate to the local standard of the place in question. Treasury *A* has a remainder of 10 coins; *D* and *G* have a remainder of 4 coins; *E* has a remainder of 6 coins. In the other treasuries there is no remainder. The local standard of treasury *A* is 12 *wên* [per hundred]. [These standards] proportionally decrease by 1 coin up to *G*. Find the daily draft of all the treasuries in "full strings" as originally collected, with the nominal value and [the daily draft] according to the "old standard" as actually collected; and also the part collected in a great and in a small month. [Respectively a month of 30 days and 29 days.]

Answer: the daily draft of all the treasuries in "full strings" as originally collected amounts to 26 *kuan* 950 *wên*. [The number of coins (*wên*) on 35 strings (*kuan*).] The nominal value is 35 *kuan wên*. [In the daily draft at "old standard"] Treasury *A* collects 224 *kuan* 583 *wên* $\left[\frac{26950 \times 100}{12} = 224{,}583\right]$; Treasury *B*, 245 *kuan*; Treasury *C*, 269 *kuan* 500 *wên*; Treasury *D*, 299 *kuan* 444 *wên*; Treasury *E*, 336 *kuan* 806 *wên*; Treasury *F*, 385 *kuan*; Treasury *G*, 449 *kuan* 166 *wên*.

Method: We solve by the *ta-yen* rule. Draw up the local standard of treasury *A*. Subtract from it the proportional decreasing number. We get all the original standards of all the treasuries. Find their mutual common divisors. Reduce the odd numbers, do not reduce the even numbers, and you get the *ting-shu*. Multiply by each other all the *ting-shu* and this is the *yen-mu*. Reduce the *yen-mu* by the *ting-shu*, and you get the *yen-shu*. Those of which the *yen-shu* is the same as the *yen-mu* are to be omitted; they have no [*yen-shu*]. (These which have no *yen-shu* we take as belonging to the same class.) From [the *yen-shu*] subtract as many times as possible the *ting-mu*; the remainders are the *chi-shu*. Apply the *ta-yen ch'iu-i* to the *chi-shu* and the *ting-shu* to find the *ch'êng-lü*. Multiply the *yen-shu* by these and you get the *yung-shu*. Of those that have no [*yen-shu*], find the greatest common denominator of the *yüan-shu* belonging to the same class (i.e., mutually divisible). Reduce the *yen-mu* and the numbers you get are the *chieh-shu*. Next draw up the remainders of the treasuries that have remaining cash. Multiply by the original *yung-shu*. Add up and you get the *tsung-shu*. Subtract therefrom the *yen-mu* as many times as possible. The remainder is the number of "full strings" of the daily draft of all treasuries. Multiply them by the number of days of a great and of a small month. These are the dividends. Reduce them by the original standard and you get the "old standard" value.

Solution: Draw up the local standard of *A*: 12. Continually decrease by 1. You get 11, being that standard of treasury *B*; 10, begin the standard of treasury *C*; 9, being the standard of treasury *D*; 8, being the standard of treasury *E*; 7, being the standard of treasury *F*; 6, being the standard of treasury *G*. You get the original standards of all the treasuries.

Find the mutually common divisors. After finishing the reduction you get $A = 1$, $B = 11$, $C = 5$, $D = 9$, $E = 8$, $F = 7$, $G = 1$. These are the *ting-shu*. Put 1 before them as "factor." First multiply all the *ting-shu* by each other; you get 27,720; this is the *yen-mu*. Next cross-multiply all the *ting-shu* by the factor 1. You get $A = 27{,}720$, $B = 2{,}520$, $C = 5{,}544$, $D = 3{,}080$, $E = 3{,}465$, $F = 3{,}960$, $G = 27{,}720$. These are the *yen-shu*.

Next investigate the *yen-shu*. Those which are the same as the *yen-mu*, omit all of them. Next, subtract as many times as possible the *ting-shu* from the original *yen-shu*, and you get the *chi-shu*. *A* has no [*chi-shu*], *B* has 1, *C* has 4, *D* has 2, *E* has 1, *F* has 5, and *G* has none. These are the *chi-shu*.

Next investigate these which have the *chi-shu* 1, and consider 1 as *ch'êng-lü*. Some have [as *chi-shu*] the number 2 or more; use them as *chi-shu* in the right column. Put the *ting-shu* in the upper right space and the *t'ien-yüan* 1 in the upper left space. Apply the rule *ta-yen ch'iu-i*. Examine by multiplication and division until the remainder is 1 and then stop [the operation]. Consider the result in the left upper space as the *ch'êng-lü*. *A* has none, *B* has 1, *C* has 4, *D* has 5, *E* has 1, *F* has 3, *G* has none. These are the *ch'êng-lü*. Draw them up in the right column, opposite to the *yen-shu* in the left column.

Multiply the opposite numbers of both columns. These are the *yung-shu*. The number in *A* has no *yung-shu*, *B* has 2,520, *C* has 22,176, *D* has 15,400, *E* has 3,465, *F* has 11,880, *G* has none.

Next, in order to examine those that have no *yung-shu*—only *A* and *G* coincide in the same class—we arrange to borrow them. We draw up the so-called *yüan-shu* and examine them. Now we examine *A* with 12, *G* with 6, *C* with 10, and *E* with 8; all are even; they are of the same class. The *yung-shu* of *E* is 3,465. This number is too small. We cannot borrow from it. Only the *yung-shu* of *C*, being 22,176, is the largest. Therefore we have to borrow from it. Then find the greatest common divisor of 12, 10, and 6; you get 2. By the greatest common divisor 2, reduce the *yen-mu* 27,720; you get 13,860. This is the *chieh-shu*. Then subtract it from the *yung-shu* of *C*, 22,176; the remainder is 8,316. This is the [new] *yung-shu* of *C*. Then the number 13,860, which is lent, is the dividend. We consider the original greatest common divisor 2 as divisor, and divide by it. We get 6,930. This is the [new] *yung-shu* of *A*. We subtract the *yung-shu* of *A* from the lent number 13,860; the remainder is also 6,930. This is the [new] *yung-shu* of *G*. If you do not wish to make the *chieh-shu* of *A* and *G*, it is the same. We can make use of some [other] reductions if we wish to. Take 1/3 from 13,860; you get 4,620, being the *yung-shu* of *A*. Take 2/3, you get 9,240, being the *yung-shu* of *G*. Draw them up in the right column.

Then examine all the treasuries that have no remainder. We get the three treasuries *B*, *C*, and *F* without remainder. First we omit their *yung-shu*. Then draw up the remainders of the four treasuries *A*, *D*, *E*, and *G* in the left column. Multiply them respectively by their own *yung-shu*. Then for *A* we get 46,200; for *D*, 61,600; for *E*, 20,790; and for *G*, 36,960. These are the *tsung-shu*.

Add up these four *tsung-shu*; you get 165,550. As many times as possible subtract the *yen-mu* 27,720. The remainder is 26,950. It is the rate searched for. Reduced to *kuan*, it is 26 *kuan* 950 *wên*. It is the equal number of the daily drafts of all the treasuries.

[Explanation: In modern terms, this problem is to find N such that $N \equiv 10 \pmod{12} \equiv 0 \pmod{11} \equiv 0 \pmod{10} \equiv 4 \pmod{9} \equiv 6 \pmod{8} \equiv 0 \pmod{7} \equiv 4 \pmod{6}$. Qin first reduces the moduli to relatively prime moduli, $m_1 = 1, m_2 = 11, m_3 = 5, m_4 = 9, m_5 = 8, m_6 = 7, m_7 = 1$. The product M of these moduli is 27,720 and the quotients M/m_i are then 27,720, 2520, 5544, 3080, 3465, 3960, and 27,720, respectively. Qin next finds the remainders $P_i \equiv M/m_i$ as 0, 1, 4, 2, 1, 5, and 0, respectively, and then solves the congruences $x_iP_i \equiv 1 \pmod{m_i}$ for each i. The results for the congruences where P_i is not 0 are $x_2 = 1$, $x_3 = 4$, $x_4 = 5$, $x_5 = 1$, and $x_6 = 3$. Qin could then have found the sum $(\sum r_ix_i(M/m_i))$ to be 82,390, from which he would have subtracted $2 \times 27{,}720$ to get the final result of his problem, $N = 26{,}950$. In actual practice, he does something slightly more complicated, but still gets the final result.]

Problem 5: There are three farmers of the highest class. As for the rice they got by cultivating their fields, when making use of full *dou*, [the amounts] are the same. All of them go to different places to sell it. After selling his rice on the official market of his own prefecture, *A* is left with 3 *dou* and 2 *sheng*. After selling his rice to the villagers of Anji, *B* is left with 7 *dou*. After selling his rice to a middleman from Pingjiang, *C* is left with 3 *dou*. How much rice did each farmer have initially and how much did each one sell? Note: The *hu* [a dry measure] of the Crafts Institute [local office for *A*] is worth 83 *sheng*, that of Anji is worth 110 *sheng*, and that of Pingjiang is worth 135 *sheng*. Answer: Total amount of rice: 7380 *dou* to be divided among the three men, or 2460 *dou* each; amount of rice sold by *A*: 296 *hu*; by *B*, 223 *hu*; by *C*, 182 *hu*.

[Explanation: Since 10 *sheng* equal 1 *dou*, this problem can be expressed as a congruence in *sheng* as $N \equiv 32 \pmod{83} \equiv 70 \pmod{110} \equiv 30 \pmod{135}$. Qin solves this using the same method as above, first reducing the moduli to the mutually prime ones, 83, 110, and 27. The three congruences he needs to solve using rules 14–18 [see below for an explicit example of this] are $65x \equiv 1 \pmod{83}$, $41x \equiv 1 \pmod{110}$, and $4x \equiv 1 \pmod{27}$.]

Chapter 2: Heavenly Phenomena

Problem 3: In the *K'ai-his* calendar, the Superior Epoch is 7,848,183 years. We wish to know the method for calculating this period. The computations are to be made for the year 1207; the observations are from 1204, the year of the beginning of the sexagenary cycle.

[Explanation: In the process of solving this problem, Qin must solve the congruence we would write as $377873x \equiv 1 \pmod{499067}$. He does this in eleven steps, using rules 14–18 and the following diagrams. 457999 is the solution of the congruence.

Step 1	Step 2	Step 3	Step 4
		3	
1 377873	1 377873	4 14291	4 14291
0 499067	1 121194	1 121194	33 6866
	1		8

Step 5	Step 6	Step 7	Step 8
2		3	
70 559	70 559	2689 85	2689 85
33 6866	873 158	873 158	3562 73
	12		1

Step 9	Step 10	Step 11
1		11
6251 12	6251 12	457999 1
3562 73	41068 1	41068 1
	6	

In this problem of solving the congruence $377873x \equiv 1 \pmod{499067}$, Qin applies the Euclidean algorithm to the relatively prime integers 377,873 and 499,067, in the process of which he essentially finds that linear combination of these two integers which gives 1. The coefficient of 377,873 is the solution to the congruence.

Chapter 3: Boundaries of Fields

Problem 8: This problem asks for the area of an annulus, where the diameter of the outer circle is $\sqrt{90}$ and the diameter of the inner circle is 8. Although there is certainly a much simpler way to solve this problem, Qin derives the following equation for this area x, by using the formulas $C = \sqrt{10D^2}$ and $A = \sqrt{D^2C^2/16}$ to relate the circumference C, diameter D, and area A of the relevant circles: $-x^4 + 15245x^2 - 6262506.25 = 0$. Note that the first formula implies that Qin is approximating π by $\sqrt{10}$. The counting board diagrams to produce the solution to this fourth degree equation are then given as follows:

Diagram 1		Diagram 2
Q	Quotient (root), called *shang*	Q
0	Position reserved for root	00
a 626250625	Constant term (*shi*)	a 626250625
0 b	Coefficient of x (*fang*)	0 b
15245 c	Positive (*cong*) coefficient of x^2 (upper *lian*)	15245 c
0 d	Coefficient of x^3 (lower *lian*)	0 d
1 e	Negative coefficient of x^4 (*yi yu*)	1 e

The *cong shang lian* is shifted by two ranks.
The *yi yu* is shifted by three ranks.

Then set 20 *bu* as quotient.

Diagram 3	Diagram 4	Diagram 5	Diagram 6
Q	*Q*	*Q*	*Q*
20	20	20	20
a	626250625	626250625	626250625
626250625	*a*	*a*	*a*
0	0	0	0
b	*b*	*b*	*b*
15245	15245 *c*	15245 *c*	14845
c		400 *c′*	*c*
0	20	20	20
d	*d*	*d*	*d*
1	1	1	1
e	*e*	*e*	*e*
Multiply *Q* by *e*. Add to *d*.	The lower *lian* produce the negative *lian*.	The positive and negative *lian* destroying one another, obtain the positive upper *lian*.	Multiply the *fang* using the quotient and the upper *lian*.

Diagram 7	Diagram 8	Diagram 9	Diagram 10
Q	*Q*	*Q*	*Q*
20	20	20	20
626250625	32450625	32450625	32450625
a	*r*	*r*	*r*
296900	296900	296900	296900
a′	*b*	*b*	*b*
14845	14845 *c*	14845 *c*	14845*c*
c			800 *c′*
20	20	40	40
d	*d*	*d*	*d*
1	1	1	1
e	*e*	*e*	*e*
Take the *fang fa* in order to call the quotient to reduce the dividend (*shi*).	Then multiply the quotient by the *yu* and add the result to the lower *lian*.	Multiply the negative upper *lian* with the lower *lian* and the quotient.	The negative upper *lian* and and the positive *lian* destroy one another.

Diagram 11

Q
20

32450625
r

296900
b

14045
c

40
d

1
e

The quotient and the upper *lian* produce the *fang*.

Diagram 12

Q
20

32450625
r

577800
b

14045
c

40
d

1
e

The quotient and the *yu* are multiplied by one another and incorporated into the lower *lian*.

Diagram 13

Q
20

32450625
r

577800
b

14045
c

60
d

1
e

The quotient and the lower *lian* produce the negative *lian*.

Diagram 14

Q
20

32450625
r

577800
b

14045 *c*
1200 *c*′

60
d

1
e

The negative *lian* and the positive *lian* destroy one another.

Diagram 15

Q
20

32450625
r

577800
b

12845
c

60
d

1
e

The quotient and the *yu* are multiplied and incorporated into the lower *lian*.

Diagram 16

Q
20

32450625
r

577800
b

12845
c

80
d

1
e

The *fang* moves back a rank; the upper *lian* two ranks; the lower *lian* three; the *yu* four.

Diagram 17	Diagram 18	Diagram 19
Q	Q	Q
20	20 *bu*	20 *bu*
32450625	32450625	32450625
r	*r*	*r*
577800	590645 *c*, *b*	590564
b		*m*
12845	0	0
c		
80	81 *e*, *d*	0
d		
1	0	0
e		
No quotient (the second figure of the root is zero). Incorporate the upper *lian* into the *fang*, the *yu* into the lower *lian*.	Add the negative *yu* to the negative *lian* and subtract the result from the positive *lian* and a *fang*, giving "mother."	Find the equal and simplify the fraction.

Diagram 20	Diagram 21
Q	
20 *bu*	20
32450625	1298025
r	
590564	2362256
m	
take the equal number 025 as divisor result	The trimultiplicative root has been extracted. 20 transformations were required. Whence the value of the area of the annular field.

The answer is 20 1298025/2362256

[Explanation: In modern terms, Qin first realizes that the solution is a two digit number beginning with "2." He then uses what amounts to synthetic division to continually divide the original polynomial by $x-20$, thus eventually finding an equation for the second digit.]

Chapter 4: Measuring at a Distance

Problem 5: There is a round town of which we do not know the circumference and the diameter. There are four gates [in the wall]. Three *li* outside the northern [gate] there is a high tree. When we go outside the southern gate and turn east, we must cover 9 *li* before we see the tree. Find the circumference and the diameter of the town.

[Explanation: If we generalize the problem so that the distance from the northern gate to the tree is a and the distance from the southern gate to where we can sight the tree is b, and if we set the diameter of the town to be x^2, then Qin says that x is the root of the equation[18]

$$x^{10} + 5ax^8 + 8a^2x^6 - 4a(b^2 - a^2)x^4 - 16a^2b^2x^2 - 16a^3b^2 = 0.$$

In the special case where $a = 3$ and $b = 9$, the equation becomes

$$x^{10} + 15x^8 + 72x^6 - 864x^4 - 11{,}664x^2 + 34992 = 0.$$

Qin gives the root to be $x = 3$; therefore the diameter of the town is $x^2 = 9$ *li.*]

VII.b. Li Zhi (Li Ye)

Li Zhi (also known as Li Ye, to which he changed his name out of deference to the emperor when he learned that "Li Zhi" was also the name of the third Tang ruler) was born in 1192 at Luancheng in Hebei Province under the Jin dynasty. Given that the Jin and Song dynasties were rivals, often at war with each other, as well as the distance that separated Li Zhi and Qin Jiushao, it is generally believed that they worked quite independently of each other, yet they achieved remarkably complementary results. And like Qin, Li also showed how the data of a given problem were to be positioned on the counting board and then manipulated to reach the solutions for problems he presents.

After passing the civil service examination, Li Zhi later served as the governor of Chun Chou, which fell to the Mongols in 1232. It is said that Li managed to escape from the enemy in disguise, and thereafter lived in poverty, devoting himself to study, including mathematics. When the Jin dynasty was conquered two years later, the Mongol prince Kublai Khan reputedly asked for Li's advice about how to rule, which Li gave freely while declining a position in the government itself. Instead, he devoted himself to his studies and to teaching. When Kublai Khan was made emperor thirty years later in 1264, he again sent for Li Zhi, who once more declined to accept a position in the government. This time his reasons may well have had to do with his advanced age, but he was eventually persuaded to accept a position in the Han Lin Academy. However, within months he resigned from the Academy and died shortly thereafter, at the age of 87, most likely in the year 1265 [Mikami 1913, 80].

Li Zhi, like Qin Jiushao, was concerned with the solution of higher equations and the advance of algebraic methods in terms of the "method of the celestial element" and the "method of four unknowns." These methods were designed to eliminate elements from systems of simultaneous equations in one or more unknowns of arbitrary degree. These methods are similar to the elimination methods for matrices already familiar from the *Nine Chapters*, and indeed, whether employing elimination methods or extracting square or cube

[18] The derivation of this equation is left as an exercise for the reader.

roots, the unique ability of the counting board to reserve certain areas or positions for specific elements of a given problem was the secret to success for Chinese "algebra." But in producing the first truly algebraic works in China, Qin and Li were quite different in their approaches.

Li Zhi's Ceyuan haijing (Sea Mirror of Circle Measurements), *1248*

Only a year later than Qin Jiushao, in 1248, Li Zhi in the north published his *Sea Mirror of Circle Measurements.* In the preface to this work Li Zhi explains that although he had studied mathematics from his youth, he was always troubled by the inability of mathematicians to reach agreement about the circle in particular. He was especially bothered by the diverse values given for the ratio of the circumference to the diameter, which Liu Hui had taken to be 157/50, whereas Zu Chongzhi used the more accurate 355/113. Li says that he was also dissatisfied with previous results determining the area of the segment of a circle in terms of arcs and sagittae. But then, at the age of 54, Li Zhi says he learned of what he refers to as the "tong-yuan chiuyong" (*tong yuan qiu yong*) method. Scholars have debated the true meaning of this phrase, but given the results of the *Sea Mirror*, it is likely that it was meant to express the extent to which the method of the celestial element had impressed him, namely the ease and flexibility that algebra contributed to mathematics.

Whatever Li Zhi may have meant by his reference to "yuan," this is the term used to designate the unknown in the "Tian Yuan" or "heavenly element" method. Earlier, Qin Jiushao had used the term *tian yuan* to refer to the "celestial element" or unknown in his treatment of indeterminate problems, but he did not apply the concept in the solution of equations, as did Li Zhi. On the counting board the unknown element was represented by a counting rod, but in his diagrams of the process Li used the Chinese character *yuan*, which was placed at the side of the coefficient for the unknown. The constant term was then placed below the position of the unknown, and was sometimes identified with the character *tai*. Coefficients for the higher powers of the unknown were then put down on the counting board above the position of the unknown in ascending order. The following (which arises in the problem discussed below) shows how the equation $-4x^4 - 600x^3 - 22500x^2 + 11681280x + 788486400$ would have been represented on the counting board. Negative values of the coefficients are indicated by placing a counting rod at a slant through the rightmost nonzero figure. Both Qin Jiushao and Li Zhi used the symbol O to represent zero in their printed works, which may have been introduced to China from India sometime in the Tang dynasty [Mikami 1913, 83]:

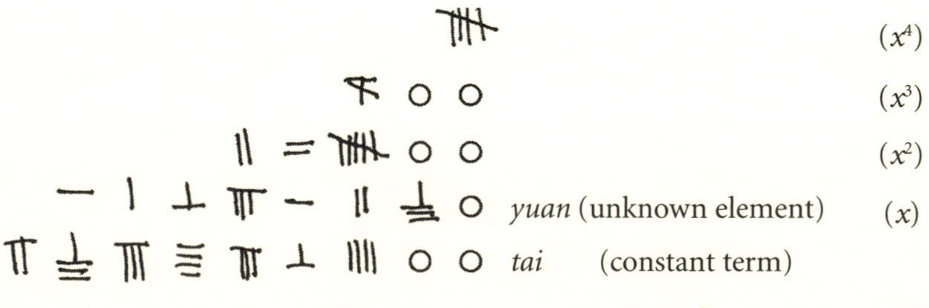

(x^4)

(x^3)

(x^2)

yuan (unknown element) (x)

tai (constant term)

At the very beginning of the *Sea Mirror*, Li Zhi inserts a diagram, a circle inscribed in a right triangle, and then introduces a number of rules from which he proceeds, in the following twelve chapters, to establish various relations between the parts of the figure treated algebraically. This may all be regarded as a continuation of Problem 9.16 from the *Nine Chapters*, which asks for the diameter D of a circle inscribed in a given right triangle of sides a, b and hypotenuse c. The result is given there as $D = 2ab/(a + b + c)$. Liu Hui in his commentary on this problem also gave several alternative formulas for the diameter, including $D = a - (c - b)$ and $D = \sqrt{2(c-a)(c-b)}$.

Li Zhi found additional solutions in the *Sea Mirror*, but unfortunately did not give any of his derivations, so it is only by conjecture that it is possible to reconstruct them (see [Sheng, Crossley, and Lun 1999, 501–504]). We present here one problem from the *Sea Mirror*, a problem similar to one that Qin Jiushao considered in his *Mathematical Treatise in Nine Chapters* (problem 5 of chapter 4).

Chapter 11, Problem 17: There is a tree 135 *bu* from the southern gate [of a circular walled city]. The tree can be seen if one walks 15 *bu* from the northern gate and then 208 *bu* in the eastward direction. Find [the diameter of the circular walled city] as before. The answer says [240 *bu*] as before.

Method: Square the product of the eastern and southern distances [a^2b^2] and let it be the *shi*. Double the product of the southern distance and the square of the eastern distance [$2a^2b$] and let it be the *cong*. Put the square of the eastern distance in the top position. Subtract the sum of the southern and northern distances from the eastern distance. Square the remainder, subtract this from the top position [$a^2 - \{a - (b + b')\}^2$] and place the result in the second position. Next, multiply the sum of the southern and northern distances by the eastern distance and double it. From this, take away the amount in the second position [$2a(b + b') - a^2 + \{a - (b + b')\}^2$] and let this be the first *yilian*. Multiply the eastern distance by 4 and put the product in the top position. Next, add the southern and northern distances and subtract the sum from the eastern distance. After the remainder has been multiplied by 4, subtract the product from the top position [$4a - 4\{a - (b + b')\}$] and call this the second *yilian*. Let 4 be the *xuyu* and solve the quartic equation to obtain the radius.

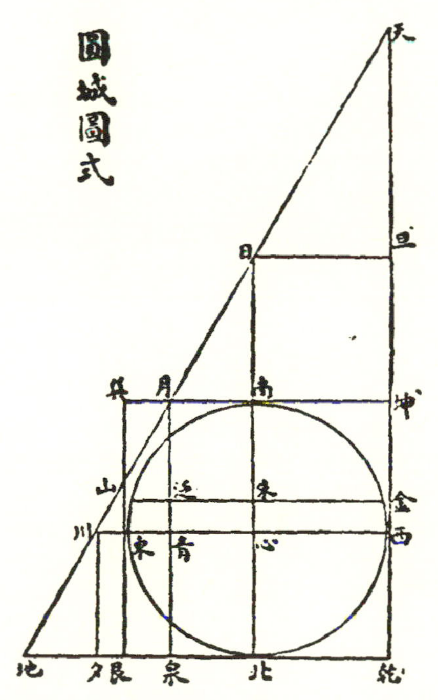

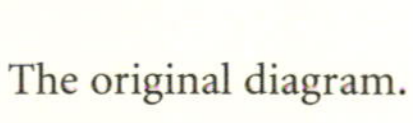
The original diagram.

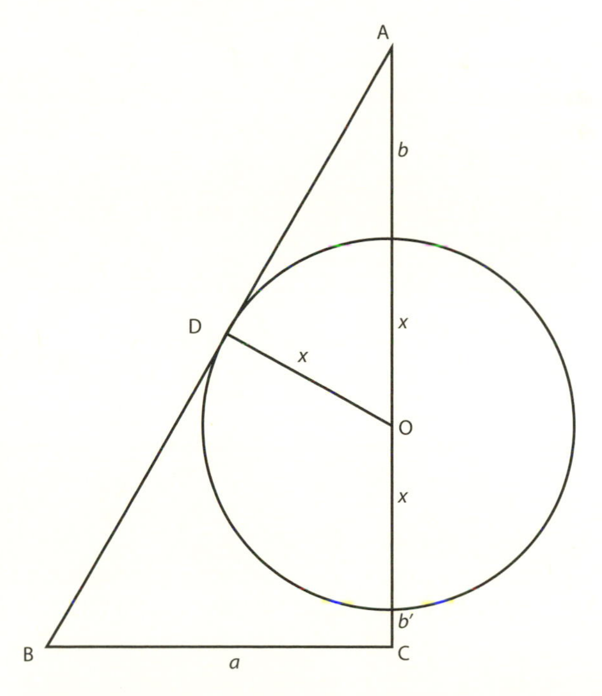

Diagram illustrating the data given in Problem 11.17, i.e., x is the radius of the circle, $a = 208$ *bu*, $b = 135$ *bu* and $b' = 15$ *bu*.

Working: Set up 1 to be the *tianyuan* which represents the radius. Add the *tianyuan* to obtain

1 *yuan* $[x + 135]$
135

which represents the *gaoxian*. Put down the value of the *dagou*, 208, and multiply it by the *gaoxian* to obtain

208 *yuan* $[208x + 28080]$
28080

Divide this product by the *gaogou* to obtain

208 *tai* $[208 + 28080x^{-1}]$
28080

which represents the *daxian*. Square this to obtain

43264 *tai* $[43264 + 11681280x^{-1} + 788486400x^{-2}]$
11681280
788486400

[Put this aside.] Next multiply the *tianyuan* by 2 and add the sum of the southern and northern distances to obtain

2 *yuan* $[2x + 150]$
150

which represents the *dagu*. Subtract from this the value of *dagou*, 208, to obtain the difference

2 *yuan* $[2x - 58]$
–58

Squaring the difference gives

4 $[4x^2 - 232x + 3364]$
–232 yuan
3364

Subtracting this from the quantity set aside gives

–4 $[-4x^2 + 232x + 39900 +$
232 $11681280x^{-1} + 788486400x^{-2}]$
39900 *tai*
11681280
788486400

which is twice the area of the rectangle [formed by *AC* and *BC*]. [Set this aside on the left.] Next, multiply the *dagu* by the *dagou* to obtain

416 *yuan* $[416x + 31200]$
31200

which gives the area of the rectangle. Double this amount to give

832 *yuan* [$832x + 62400$]
62400

which equals the quantity set aside on the left. Equate both quantities to obtain

−4	[$-4x^4 - 600x^3 - 22500x^2$
−600	$+ 11681280x + 788486400 = 0$].
−22500	
11681280	
788486400	

Extract the root of the quartic equation by the *fan* method [i.e. the "*kai fang*" method of the *Nine Chapters* for extracting roots] to give 120 *bu*, which is the radius of the circular city. Hence the answer.

[Explanation: The solution for this problem exploits the similarity of the triangles *ADO* and *ACB*, from which it follows that $AO/AB = x/BC$. Thus $AB = 208(x + 135)/x$. From the *gou-gu* (Pythagorean theorem), $(AB)^2 = (AC)^2 + (BC)^2$, it follows that $(AB)^2 - (AC - BC)^2 = 2(AC \times BC)$, and consequently:

$$-4x^2 + 232x + 39900 + 11681280x^{-1} + 788486400x^{-2} = 832x + 62400.]$$

Li Zhi's Yigu yanduan (New Steps in Computation), *1259*

Some historians of mathematics consider this work an "introduction" to the *Sea Mirror of Circle Measurements*, and while it takes up the same material as the *Sea Mirror*, it does so with a very different way of presenting its algebraic methods. Although the idea is the same, the arrangement of the counting board is inverted, the constant term now appears at the top of the board, the unknown "*yuan*" element comes next, and then the coefficients of the higher powers of the unknown in ascending order. Thus the above equation would appear as follows on the counting board:

𝍦 𝍰 𝍧 𝍬 𝍧 𝍮 𝍣 〇 〇	*tai* (constant term)	
𝍩 𝍠 𝍮 𝍧 𝍩 𝍡 𝍰 〇	*yuan* (unknown element)	(x)
𝍡 𝍪 𝍤 〇 〇		(x^2)
𝍥 〇 〇		(x^3)
𝍣		(x^4)

This format for laying out the coefficients of the equation to be solved was similar to the layout used by Qin Jiushao, who had followed the format used for carrying out the algorithmic steps for extracting roots on the counting board, and it was this format that most mathematicians subsequently chose to adopt. In fact, the *tian yuan* method as well as the use of O for zero had been in use in China prior to the Song dynasty, but any earlier works using such elements have subsequently been lost. Moreover, the use of the "heavenly element method" seems to have been restricted to the areas of what are today Hebei and Shanxi provinces, which were the cultural and commercial centers of the Jin and Yuan dynasties [Li and Du 1987, 139].

Problem 8: There is a circular pond inside a square field and the area outside the pond is 13 and 7½ tenths *mu*. It is only known that the sum of the perimeters of the square and circle is 300 *bu*. Find the perimeters of the square and the circle. Answer: The perimeter of the square is 240 *bu* and the perimeter of the circle is 60 *bu*.

Method: Set up 1 [x] in *tian yuan* [row on the counting board] to stand for the diameter of the circle and multiply this by 3 to obtain the circumference. Subtract this from the sum of the perimeters to obtain [$300 - 3x$]. Square the latter to obtain [$90000 - 1800x + 9x^2$], which is the area of 16 square fields. Put this at the top [of the counting board]. Set up again the *tian yuan* representation for the diameter and multiply it by itself. Multiply this again by 12 to obtain [$12x^2$], which is the area of 16 circular ponds. Subtract this from the representation on top to obtain [$90000 - 1800x - 3x^2$], which is equivalent to 16 portions of the area outside the pond. Place this on the left. Then, display the given area 13 and 7½ tenths *mu* and convert it to obtain 3300 *bu*. Multiply this by 16 to give 52800 *bu* and mutually subtract this amount from that on the left to obtain [$37200 - 1800x - 3x^2 = 0$]. Extract the root from the quadratic equation to obtain 20 *bu* which is the diameter of the circular pond. Next multiply this by 3 to give the circumference.

According to the solution by sections: From the square of the sum of the perimeters, subtract 16 times the given area to form the *shi*. The sum of the perimeters is multiplied by 6 to form the *cong* and 3 *bu* is taken as the *chang fa*.

Interpretation: The areas of 16 circular ponds are equivalent to the areas of 12 squares. From the collection of *cong*, 9 squares are removed and this leaves a remainder of 3 squares. This is the reason why 3 *bu* is the *chang fa*.

The old method: Multiply the sum of the perimeters in *bu* by itself and place it on the top position. Subtract 16 times the given field area from the top position and divide the difference by 6. Let this amount be the *shi*, the sum of the perimeters be the *cong fa*, and 5 tenths be the *lian chang*.

Problem 33: In the middle of a circular field stands a rectangular pond. The area outside the pond is 7300 *bu*. It is only known that the sum of the length and breadth of the pond inside is smaller than the diameter of the field by 55 *bu* and the difference between the length and breadth is 35 *bu*. Find each of the three quantities. Answer: The diameter of the field is 100 *bu*, the length of the pond is 40 *bu*, and the breadth is 5 *bu*.

Method: Set up 1 [x] in the *tian yuan* row to stand for the diameter of the circle, square this, and then multiply by 3 and divide by 4 to obtain [$0.75x^2$], which is the area of the circle. Subtract 7300 *bu* from this to obtain [$0.75x^2 - 7300$], which is the area of the pond. Multiply this by 4 to give [$3x^2 - 29200$], which is the area of 4 ponds. Place this on the left. Set up again the *tian yuan* representation for the diameter and subtract 55 *bu* from this to obtain [$x - 55$], which is the sum of the length and breadth. Square this to obtain [$x^2 - 110x + 3025$], which is the area of 4 ponds plus the square of the difference between the length and breadth. From this take away the square of the difference between the length and breadth, which is 1225 *bu*, to get [$x^2 - 110x + 1800$], which is the area of 4 ponds. This is mutually subtracted with that on the left to obtain [$2x^2 + 110x - 31000 = 0$]. When the root of the quadratic equation is extracted, 100 *bu* is obtained, which is the diameter of the circle. Subtract from this the difference between the diameter and the sum of the length and breadth to obtain the sum of the length and breadth. When the difference between the length and breadth is added to this, two lengths are obtained, and when it is subtracted, two breadths are obtained.

According to the solution by sections: Multiply the given area by 4, minus the square of the difference between the length and breadth, and add the difference of the diameter and the sum of the length and breadth to form the *shi*. Multiply the difference of the diameter and the sum of the length and breadth by 2 to form the *cong* and let 2 *bu* be the *chang fa*.

Interpretation: When the area of 4 ponds is added to the square of the difference of the length and breadth, this gives exactly the square of the sum of the length and breadth.

Old method: Multiply the given area by 4 and place it in the top position. Next, square the difference between the diameter and the sum of the length and breadth and add the result to that in the top position. Subtract from this the square of the difference between the length and breadth. Halve the remainder and let this be the *shi*. Use the difference between the diameter and the sum of the length and breadth as the *cong* and 2 *bu* as the *chang fa*.

VII.c. Yang Hui

Yang Hui was a native of what today is Hangzhou, but little else is known about his life. His contributions to mathematics consist mainly of gathering many lost mathematical works in a search for methods of calculation, including the "method of extracting roots by iterated multiplication." Yang Hui was especially interested in finding simplified methods for algorithms for multiplication and division. Among the best-known features of *A Detailed Analysis of the Mathematical Methods in the Nine Chapters* is the prominence Yang Hui gave to the equivalent of the Pascal triangle in a diagram that accompanied "the source of the method of extracting roots." Yang Hui noted that the diagram was due to Jia Xian and had appeared in *The Key to Mathematics* (*Shi suo suan shu*), which dates the method to the eleventh century.

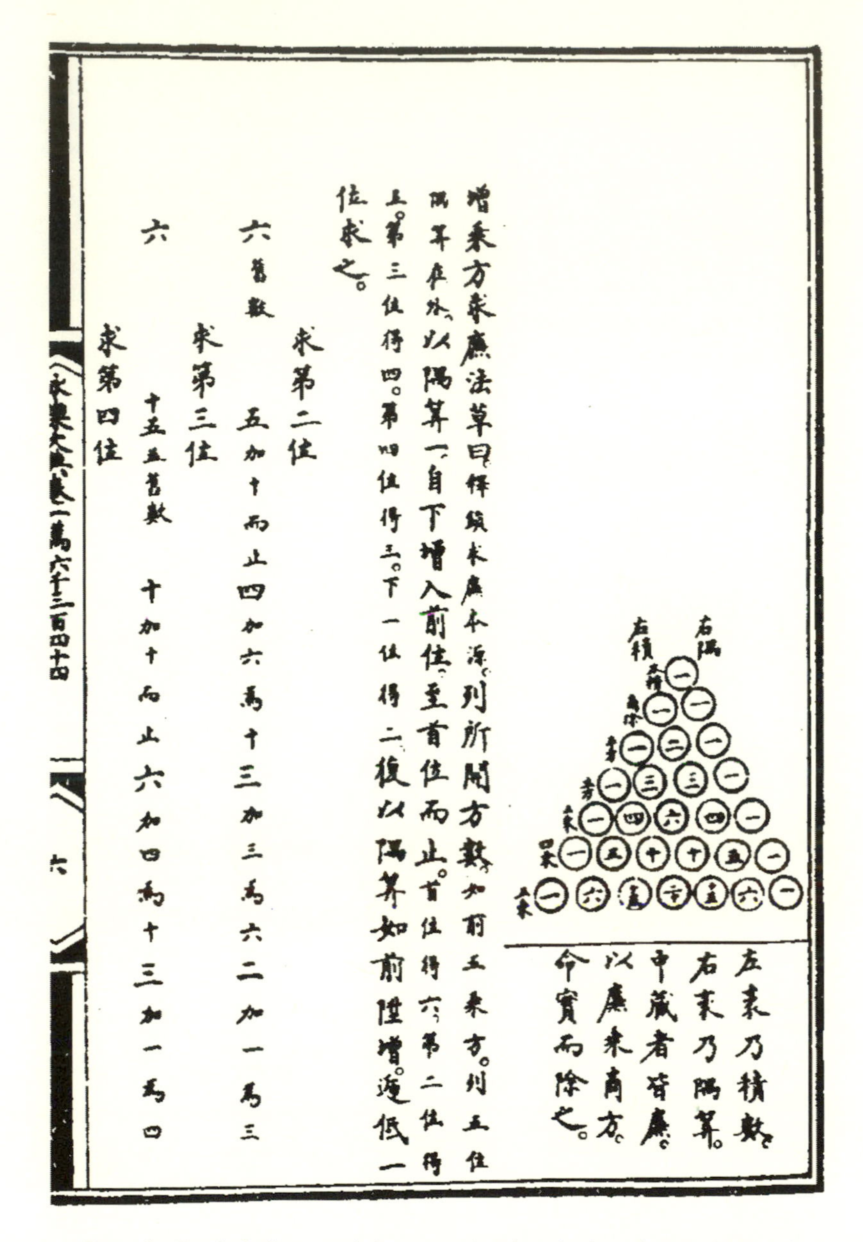

增乘方求廉法草曰釋鎖求廉本源列所開方數如前五乘方列五位
隅算在外以隅算一自下增入前位至首位而止首位得六第二位得
五第三位得四第四位得三下一位得二復以隅算如前陞增遞低一
位求之

求第二位
六舊數 五加十而止 四加六為十 三加三為六 二加一為三

求第三位
六 十五舊數 十加十而止 六加四為十 三加一為四

求第四位

From the *Yongle dadian suanfa* (Computational Methods in the Yongle Grand Encyclopedia, 1407), reproduced in [Guo Shuchun 1993, vol. 1, 1416b].

In the illustration, each line of the diagram represents the coefficients for the expansion of $(a + b)^n$, so that it is very easy to read off the coefficients for $(a + b)^4$, for example, as $a^4 + 4a^3b + 6a^2b^2 + 4ab^3 + b^4$, which appear on the fifth line of the arithmetic triangle as 1 4 6 4 1.

There is also a very similar diagram in Zhu Shijie's *Precious Mirror of the Four Elements,* which extends the arithmetic triangle as far $(a + b)^8$. This diagram is given the title "seven times method" (*qi cheng fang*), since $(a + b)$ multiplied by itself seven times gives the bottom-most line on the diagram. Because the "method of extracting roots by iterated multiplication" easily leads to the coefficients for the expansion of $(a + b)^n$ for any n, Chinese mathematicians knew how to use these coefficients to extract roots of any degree one might wish. In fact, it is possible to use the actual means of generating the terms of the "binomial triangle" to carry out a given root extraction without having to construct the triangle itself. This was an innovation due to Yang Hui, who found a means of streamlining the process by adding the elements of the arithmetic triangle without actually having to lay them out beforehand [see Li and Du 1987, 126].

Among the problems in the *Detailed Analysis of the Mathematical Methods in the Nine Chapters* is the problem of finding the fourth root of 1,336,336. This is understood as finding the solution to the equation $x^4 = 1{,}336{,}336$, for which it turns out that $x = 34$. At first, the methods used by Chinese mathematicians for the solution of equations required that the coefficient of the highest degree of the unknown be 1, and that the remaining coefficients be positive. The first to overcome these restrictions, apparently, was Liu Yi of the late twelfth or early thirteenth century, whose work on this subject, *Discussion on the Old Sources*, is unfortunately lost, but some of Liu Yi's problems and methods were included in Yang Hui's *Practical Rules of Arithmetic for Surveying.*

Yang Hui's Xiangjie jiuzhang suanfa *(A Detailed Analysis of the Mathematical Methods in the* Nine Chapters*)*

The original method for the [binomial] expansion. <This is found in the arithmetic book *Shih So* [Shi suo] (*Unlocking Coefficients*) and this method has been used by Chia Hsien [Jia Xian].>

The unit coefficients of the absolute terms are on the left.

The unit coefficients of the highest powers are on the right.

1

1 1

1 2 1

1 3 3 1

1 4 6 4 1

1 5 10 10 5 1

1 6 15 20 15 6 1

The unit coefficients of the absolute terms lie along the left.
The unit coefficients of the highest powers lie along the right.
The center contains all the other coefficients.
When the coefficients are multiplied by their respective terms, [the sum] forms the expansion which may then be removed.

The working for finding the coefficients of the binomial expansion: <Origin of the method of finding the coefficients by "unlocking">. Arrange the positions for the powers of the expansion. <For instance, in the binomial to the sixth power arrange five positions as shown above and place the unit coefficient of the highest power of the expansion on the outside.> Let the coefficients of the highest powers be 1, and add the numbers in the [two] places above in order to find the number in the place below. [Continue in this manner] till the first position is reached. <The number obtained for the first position is 6 with the number for the second place above it as 5, the number for the third place above as 4, the number for the fourth place above as 3, and the number for the next place above as 2.> Repeat as before with the unit coefficients of the highest powers, adding the [two] numbers above in order to find the number in the position below.

To find the number in the second position:

6 <the previous number> 5 <add 10 and stop> 4 <add 6 to give 10>
3 <add 3 to give 6> 2 <add 1 to give 3>.

To find the number in the third position:

6 15 <the previous number> 10 <add 10 and stop>
6 <add 4 to give 10> 3 <add 1 to give 4>.

To find the number in the fourth position:

6 15 20 <the previous number>
10 <add 5 and stop> 4 <add 1 to give 5>.

To find the number in the fifth position:

6 15 20
first coefficient second coefficient third coefficient

15 <the previous number> 5 <add 1 to give 6>.
fourth coefficient fifth coefficient

Yang Hui suanfa (Yang Hui's Methods of Computation)

This work contains what became a popular outline for studying mathematics, especially the "Outline of the Practice of Calculation" presented in the chapter "Alpha and Omega of Variations on Methods of Computation" in Book One, *Precious Reckoner for Variations of Multiplication and Division.* Among the subjects it treats is a series devoted to packing problems that improved upon the presentation of similar results given earlier by Shen Kuo. These include various problems involving piles of fruit on square, rectangular, and triangular bases. The latter, for example, involved the summation of the series (from top to bottom of the triangular pile of fruit): $1+3+6+\cdots+n(n+1)/2$, for which Yang Hui gives the sum as $n(n+1)(n+2)/6$.

In his Preface to Book One, Yang Hui begins with a reference to the nine nines table for multiplication, and goes on to describe variations on the traditional methods for multiplication and division, which he says reveal the "ingenuity of the mathematicians." In addition to his comments, he provides illustrations and the actual workings of problems as well, admitting that what the book offers will not "unravel all of the subtle techniques," but will at least serve as "stepping stones for the beginner." After a general outline of mathematical studies, which recommends the *Wu cao suan jing* among several other works, now lost, he notes that in addition to seeking shorter methods, after working out a problem in division, the answer should be checked by multiplication.

Yang Hui then explains the terminology for handling fractions, leading up to a discussion of root extraction, which he divides into seven categories. Here, he notes that it is necessary to understand the origins of methods one chooses to apply so that they "will not be forgotten." He also emphasizes that the new methods of root extraction he advances "found no equal in the past." And after criticizing the methods of the *Sea Island Mathematical Classic* for being "obscure and extremely abstruse," he goes on to lament that despite the efforts mathematicians had made, "one still does not understand their purport."

Nevertheless, in Yang Hui's preface to Book Three, *Continuation of Ancient Mathematical Methods*, although he mistakenly associates Liu Hui with the Warring States period (Liu Hui lived during the Eastern Han dynasty), he voices his great admiration for the *Nine Chapters.* He furthermore explains how one day some rare problems came to his attention, which in turn led to his publishing them with amendments and improvements "so that the ancient methods and calculations could be continued." Among these are five problems of indeterminate analysis, the first of which (1.1) is the famous Chinese Remainder Theorem from the *Sunzi suan jing.* In Chapter 2, Problem 2.29 is the "Hundred Fowls Problem" from the *Zhang Qiujian suan jing*, the complexity of which Yang Hui acknowledges by saying that "there is certainly no basic method for this problem." Note that the "original working" is commentary supplied by Zhen Luan and Li Chunfeng.

From the discussion of the "Hundred Fowls Problem", it is clear that with an initial solution to the problem, adding 4 roosters (20 *wen*) and 3 chicks (1 *wen*) will add a total of 7 birds for 21 *wen*; subtracting 7 hens will also subtract 21 *wen*, so that both the number of birds and the total amount they are worth will remain unchanged by these manipulations of the answer to the problem. But beyond this, no further insights into the actual method needed to solve the congruence problem are offered.

Interestingly, in another work, the *Continuation of the Tradition of Strange Computational Methods* (*Xugu zhaiqi suanfa*, 1275 CE), Yang Hui cited a similar problem as follows:

> 100 coins buy Wenzhou oranges, green oranges and golden oranges, 100 in total. If a Wenzhou orange costs 7 coins, a green orange 3 coins and 3 golden oranges cost 1 coin, how many oranges of the three kinds will be bought?"

Yang Hui then offers the following method for solving the problem:

> From 3 times 100 coins subtract 100 coins; from 3 times the cost of a Wenzhou orange, i.e., 21, subtract 1; the remainder is 20. From 3 times the cost of a green oranges, i.e., 9, subtract 1, the remainder, 8. The sum of the remainder is 28. Divide 200 by 28, we have the integer 6. These are the numbers to be found; 6 Wenzhou oranges and 6 green oranges respectively. And then $(200 - 6 \cdot 28)/8 = 4$, this is the difference of the number of Wenzhou oranges and green oranges. Hence the sum of them is 16, whereas the number of golden oranges to be found is 84. [Shen, Crossley, and Lun 1999, 416]

As in other Chinese discussions of this problem, the logic of Yang Hui's solution is not at all clear.

Other problems included by Yang Hui concern a number already encountered in the *Nine Chapters* and the *Sea Island Mathematical Classic*, for example, those involving the relations between squares and circles, as well as the basic *chong cha* method from the *Sea Island*. Yang Hui describes the method in detail, and notes the practical difficulties of actually carrying out the observations called for in such problems. From the language of Liu Hui's text and the accompanying diagram, it is clear that he justified the "double difference method" in terms of the "out-in complementary principle." And as Yang Hui confesses to his readers, his primary reason for writing about all this was because: "Now it is feared that readers of later generations would not know the direct origin of the use of this excellent method belonging to their ancestors."

Book One: Precious Reckoner for Variations of Multiplication and Division

Preface

Arithmetical calculation begins with the rules of the multiplication table from nine times nine. The methods employed in calculation are derived from those of multiplication and division. When the multiplier or divisor begins with the number "one," then the methods of *chia* [jia] (multiplication through addition) and *chien* [jian] (division through subtraction) are used. When unity does not occur in the problem, then it refers to halving (*chê*) [zhe] or doubling (*pei*) [bei]. One adds to the next higher digit in the case of the *chiu kuei* [jiu gui] method (division in nine parts), and subtracts (*sun*) [sun] from the next lower digit in the method known as *hsia ch'êng* [xia cheng] (subtractive multiplication). These supplement the methods of multiplication and division, thus showing the ingenuity of the mathematicians. The learner only knows the existence of the methods of *chia*, *chien*, *kuei*, and *sun* [jia, jian, gui, and sun], but does not realize that their applications may be extended and modified. It is said in the composition *Chin K'ê Fu* [*Jin ke fu*] that the difficulty lies in knowing how to apply the method rather than in understanding the method. This is indeed true.

The author now writes on the varied and numerous methods and adds his comments, illustrations and working to form a book entitled *Ch'êng Ch'u T'ung Pien Suan Pao* [*Cheng chu tong bian suan bao*]. Though it does not unravel all the subtle techniques of the past masters, the author hopes that it will serve as stepping stones for the beginner. The author respectfully puts this in print in order to propagate this art and to hand it down to posterity.

This preface was written on the *hsia-chih* [*T'ung Pien*] (summer solstice day) in the year the cyclical stem-branch characters of which were *chia-hsü* [jia xu] when the reign-title was *hsien-shun* [xian chun] (i.e., 14 June 1274) by Yang Hui of Ch'ien-t'ang [Qian Tang].

Chapter One: Alpha and Omega of Variations on Methods of Calculation

A general outline of mathematical studies

First learn the multiplication tables. <Start with 1 times 1 equals 1 up to 9 times 9 equals 81, from the smaller numbers to the larger ones. Their applications are not shown here.>

Learn the rules of multiplication and how to fix the place-values. <Lesson for one day.> Review the subject on the method of multiplication. <Start with multiplication

to the first place up to the sixth place, and also learn to fix the place-values. This course extends over a period of five days.>

Learn the rules of division and how to fix the place-values. <Lesson for one day.> Review the subject on the method of division. <Start with division to the first place up to the sixth place. Learn to adjust and fix the place-values. This course takes half a month.>

When the rules of multiplication and division are known, then acquire the two books *Wu Ts'ao* [*Wu cao*] and *Ying Yung Suan Fa* [*Ying yong suan fa*] (*Methods of Computation for Practical Use*). Follow the techniques in these books and do two or three problems each day. <The mathematical texts of the various schools are not arranged progressively in their sequence. The two books above are used for the convenience of the beginner.> It is not necessary to comprehend fully the logic of the problems but one must know how a question is asked, which method to apply in order to solve a problem, and when to use multiplication or division. In less than two months one can complete seventy to eighty per cent of the two books *Wu Ts'ao* and *Ying Yung*. In the first chapter of *Hsiang Chieh Suan Fa* [Xiang jie suan fa], there are thirteen problems on multiplication and division, with special reference to their applications. When one goes through carefully the meanings of these examples and also their explanatory notes, the procedure will automatically become clear.

The rules used in the mathematical texts of the various schools do not go beyond the three methods multiplication, division and root-extraction. The tables do not go beyond the two words *ju* [ru] (applied to a unit digit) and *shih* [shi] (applied to a tenth digit), and the manipulation of numbers does not exceed the two positions, horizontal and vertical. However, all these can be stretched and extended beyond bounds. The number of applications of the methods of multiplication and division is unlimited. The *Chih Nan Suan Fa* [Zhi nan suan fa] (*A Guide to Calculation*) seeks shorter methods in multiplication through addition, division through subtraction, division in nine parts, and doubling or halving. How can a student be left without being made to acquire knowledge of all this?

Learn the rules of the multiplication through addition method and also how to fix the place-values. <Lesson for one day.> Review the methods of adding one place, adding two places, and adding to the alternate place. <Three days.>

Learn the rules of the division through subtraction method and also how to fix the place-values. <Lesson for one day.> In the multiplication method the number is increased, while in the division method a certain number is taken away. Whenever there is addition, there is also subtraction. One who learns the division method should test the result by applying the multiplication method to the answer of the problem. This will enable one to understand the method to its origin. Five days are sufficient for review.

In learning the division in nine parts method, one will need at least five to seven days to become familiar with the recitation of the forty-four sentences. However, if one examines carefully the explanatory notes of the art on division in nine parts in the *Hsiang Chieh Suan Fa*, one can then understand the inner meanings of the process and a single day will suffice for committing the tables and their applications to memory. Review the subject on division in nine parts. <One day.>

The method doubling or halving is basically for the application of the multiplication and the division methods. It in fact uses the technique of doubling or halving and hence its name. Actually there is nothing very abstract about it but a knowledge of its applications should be required. The next chapter includes methods and problems pertaining to this. It is only necessary to take one day for review.

[...]

As for the fractional remainder after division, its denominator (*fa*) [fa] is called *fên mu* [fen mu] and its numerator (*shih*) [shi] is called *fên tzŭ* [fen zi]. If the original form of a fraction is to be restored by the process of multiplication, one must employ the method of communicating the numerator and the denominator. If the denominator and the numerator are too cumbersome then the method of reduction must be used. If the fractions are not uniform and it is required to resort to addition, then use the method of addition of fractions. When two fractions need to be compared, make use of the method of finding the least common multiple of the denominators. For averaging non-uniform fractions, use the method of dividing fractions equally. The measurements *chin* [jin] joined with *chu* [zhu] and *liang* [Wu cao] and *p'i* [pi] together with *ch'i* [chi] and *ts'un* [cun] are also similar to fractions. Without the methods of multiplication and division of fractions it is not possible to manage fractions. The study of fractions is an essential and an important part of mathematics. Without learning them, one is not sufficiently armed to understand mathematics. This subject on fractions can be found in the chapter on land surveying of the *Chiu Chang Suan Shu* [Jiu zhang suan shu]. Learning one method a day will provide a general understanding of the subject in less than ten days. Moreover, a further review for two months enables one to explain it to others. The preface of *Chang Ch'iu-chien's Mathematical Manual* [*Zhang Qiujian's Mathematical Manual*] says: "Do not be troubled by the difficulties of multiplication and division, but be concerned over the difficulties of fractions." However, according to the author Yang Hui, fractions in themselves are basically not difficult, but they are cumbersome as from the start they have to be reduced in order that errors may not be made in calculations.

In mathematics, root-extraction constitutes a major item. Under the headings "the right-angled triangle," "besides the essentials," "procedure," and "collection of areas," can be found numerous common examples of root-extraction. The subject may be divided into seven different categories: first, extraction of the square root; second, finding the radius or diameter from the circular area; third, extraction of the cube root; fourth, finding the radius or diameter from the spherical volume; fifth, root-extraction of the fraction; sixth, extraction of the fourth root and roots of higher order; seventh, solving the quadratic equation with positive coefficients of the square and root and negative constant term. These are all contained in the two chapters "diminishing breadth" and "the right-angled triangle." Learn a method a day and work on the subject for two months. It is essential to inquire into the origins of the applications of the methods so that they will not be forgotten for a long time.

Although of the 246 problems in the *Chiu Chang*, none extends beyond the three methods of multiplication, division, and root extraction, the planning of the methods and the arrangements are particularly suitable for going through carefully. For instance, the methods of cross-multiplication and proportion, cross-multiplication and arrangements of ordered numbers, and calculation by tabulation are planned and arranged at the beginning of the chapters.

When the methods of multiplication, division, fractions, and root-extraction in the 246 problems of the *Chiu chang* have been reviewed, then start with the chapters on millet and rice. These will require one day to know thoroughly. The next chapter, distribution by proportional parts, deals with proportional parts. The whole of "short width" is devoted to the addition of fractions and all of "construction consultations" is on substitution of equivalent volumes. The chapter "fair levies" employs the cross-multiplication of ordered numbers. Take three days to peruse each chapter. The applications of the methods in the three remaining chapters, "excess and deficit," "rectangular arrays," and "right-angled triangles," are more complicated so take four days to study each chapter. Make a detailed study of *Chiu Chang* [*Suan Fa*] *Tsuan Lei* [Jiu Zhang Suan Fa Cuan Lei], so that the rules of application are thoroughly known. Then only will the art of the *Chiu chang* be fully understood.

The *Source Book of Matters and Things* says that *kou ku* [gou-gu] and *p'ang yao* [pang yao] were originally two chapters, though now they are generally regarded as one. A detailed study of the meanings of the methods shows there are actually two separate parts. Liu Hui employs the art of *p'ang yao* to transform the methods of "double differences" and "diminishing the area" into nine problems in the *Sea Island Mathematical Manual*. [Liu Yi] applies the art of *kou ku* to the sections *yen tuan* [yan duan] and *so fang* [suo fang] formulating 200 problems in the *I Ku Kên Yüan* [Yi gu gen yuan]. The methods of root extraction have in fact found no equal in the past. The preface of the *Chiu Chang* says, "Some excelled in one or two branches of mathematics and their books were able to form individual schools by themselves." This is indeed believable. However, the setting of the methods and problems in the *Sea Island Mathematical Manual* is obscure and extremely abstruse. None could unravel its mystery. Although Li Shun-fêng [Li Chunfeng] commented on it, he only laid down the methods and did not explain their origins, while the *I Ku Kên Yüan* does not contain detailed explanations of the working. Even after following the techniques of its mathematical exercises, one still does not understand their purport. The above two books resulted from the chapter *kou ku* [gou-gu] in the *Chiu Chang* [Jiu Zhang] and have supplemented the admirable techniques contained in previous books. The preface also says, "The *Chiu Chang* [Jiu Zhang] is to the mathematicians as the six classical books are to the scholars, the *Nan* [*Ching*] [Nan jing] and the *Su* [*Wên*] [Su wen] are to the medical profession and the *Sun Tzŭ* [*Ping Fa*] [Sunzi bing fa] is to the soldier studying military art. The beginner is thus encouraged to understand the *Chiu Chang* [Jiu Zhang].

Book Three: Continuation of Ancient Mathematical Methods for Elucidating the Strange Properties of Numbers

Preface

Mathematics is one of the six established departments of knowledge. Formerly, during the time of Huang-ti [Huang di], this art was originated by the grandee Li Shou

[Li Shou] and was followed by the *Chiu Chang* [*Suan Shu*] written by Chou Kung [Zhou Gong]. Since the period of the Warring States when Liu Hui of the state of Wei wrote the *Hai Tao* [*Suan Ching*] [Hai dao suan jing] up to the time when Chên Luan [Zhen Luan] of Han-chung [Han Zhong] province wrote notes on the *Chou Pi* [*Suan Ching*] [Zhou bi suan jing] and the *Wu Ching* [*Suan Shu*] [Wu jing suan shu] and when Li Shun-fêng [Li Chunfeng] of the T'ang [Tang] dynasty revised and corrected the mathematical texts of the various schools, in each successive generation reputed men of learning have all regarded this as an important department of knowledge.

In this present Sung [Song] dynasty, scholars are selected by means of examinations and the *Chiu Chang* heads the list of mathematical manuals. Therefore the author has great respect for this book and pays special attention to explaining it. Some say that instructions for beginners are lacking and those who learn it for the first time will encounter difficulties. The author uses the methods of multiplication, division, *chia*, and *chien*, and the problems on measures and areas to write a book entitled *Jih Yung Suan Fa* [Ri yong suan fa]. However, the student may have a rough knowledge of the methods of *chia*, *chien*, *kuei* and *pei*, but does not know the applications of their variations. Therefore methods of substitutions in multiplication and division are further added to a new book called *Ch'êng Ch'u T'ung Pien Pên Mo* [Cheng chu tong bian ben mo]. When the author next read the book *I Ku Kên Yüan* written by Mr. Liu I [Liu Yi] of Chung-shan [Zhong Shan] district and the sections on procedures and collections of areas, he discovered that these have excelled during the past in beauty and perfection. How can these works not be expanded for the benefit of later scholars? For this reason the collection *T'ien Mou Suan Fa* [Tian mu suan fa] came into being.

All the four collections of works mentioned above were published, and for once the author thought that his ambition was fulfilled. Unexpectedly, one day Liu Pi-chien [Liu Bijian] and Ch'iu Hsü-ku [Qiu Xugu] brought with them some rare problems from the various mathematical texts and some forgotten literature of old editions, and requested the author to form a collection which they would help financially to publish. As a result, some amendments and improvements were made on the rare problems from the various texts and they were copied so that the ancient methods and calculations could be continued. This collection is entitled *Hsü Ku Chai Ch'i Suan Fa* [Xu gu zhai qi suan fa] (*Continuation of Ancient Mathematical Methods for Elucidating the Strange Properties of Numbers*) and is published with a view to sharing with those who love the subject. It is hoped that the author would not be accused of overstepping his place.

This preface was written on the first day of the winter solstice day, when the cyclical stem-branch designation was *jên-ch'ên* [ren chen], in the year when the reign title was altered to *tê-yu* [de you] (i.e., 14 December 1275) by Yang Hui of Ch'ien-t'ang [Qian Tang].

Chapter One

Five Problems on Indeterminate Analysis

1. There is an unknown number of articles. It is only known that when the number is divided by 3 the remainder is 2; when divided by 5 the remainder is 3; and

when divided by 7 the remainder is 2. Find the original number. <From *Master Sun's Mathematical Manual.*> Answer: 23.

Explanation of the problem: <This is commonly called the Prince of Ch'in's [Qin's] secret method of counting soldiers or the method of repeated shots. If the number exceeds 105, then it should be stated in the problem.> The method of cutting lengths of tubes: Put down 70 which when divided by 3 gives a remainder 1. <The problem states that the remainder is 2, so put down 140.> Put down 21 which when divided by 5 gives a remainder 1. <The problem states that the remainder is 3, so put down 63.> Put down 15 which when divided by 7 gives a remainder of 1. <The problem states that the remainder is 2, so put down 30.> Add the three numbers together <to obtain 233>. Take away the largest multiple of 105 <which is twice 105> to leave a remainder <23> which is the required number.

4. A certain number when divided by 11 leaves a remainder 3, when divided by 12 leaves a remainder 2 and when divided by 13 leaves a remainder 1. Find the number. Answer: 14.

Method: Put down 935 which when divided by 11 gives a remainder 1. <The problem states that the remainder is 3 so put down 2808.> Put down 1573 which when divided by 12 gives a remainder 1. <The problem states that the remainder is 2 so put down 3146.> Put down 924 which when divided by 13 gives a remainder 1. <The problem states that the remainder is 1.> Add the three numbers together <to obtain 6878> and take away the largest multiple of 1716 <which is four times 1716> to leave a remainder <14>. Hence the answer.

Chapter Two

Simultaneous linear equations involving two unknowns

27. Some pheasants and hares are put in the same basket. The top of the basket shows 35 heads <i.e., there are 35 of them> and the bottom shows a total of 94 legs. Find the number of each kind. Answer: 23 pheasants and 12 hares.

The method of simultaneous linear equations: Double the number of heads and subtract this from the number of legs. <Doubling the number of heads does not differentiate the pheasants from the hares, it is just multiplying each one of them by 2 legs. Subtract this quantity from the total number of legs and the remainder gives a surplus of 2 legs for each hare.> Halve the remainder to obtain the number of hares.

Method of solving the number of pheasants first: Multiply the number of pheasants and hares by 4. <A hare has 4 legs.> Subtract from this the total number of legs <94 legs>. The remainder gives the total number of pheasants' legs <46>, hence halve this to obtain the number of pheasants.

Simultaneous linear equations involving three unknowns

29. The price of a cock is 5 *wên* [wen], a hen 3 *wên*, and 3 chicks 1 *wên*. A total sum of 100 *wên* can buy 100 fowls. Find the numbers of cocks, hens, and chicks bought. Answer: 8 cocks costing 40 *wên*; 11 hens costing 33 *wên*; and 81 chicks costing 27 *wên*.

Method in the *Chang Ch'iu-chien Suan Ching* [Zhang Qiujian suan jing]: Add 4 to the number of cocks, subtract 7 from the number of hens and add 3 to the number of chicks. In other words, add and subtract the original solutions [in order to obtain new solutions]. The same treatise says: It is suspected that there may be an omission in the text, though there can never be any proof since the text has been handed down through a very long period. Now on examining the working in detail [it is found that new solutions can be obtained] by adding and subtracting the solutions worked out. There is certainly no basic method for this problem.

Original working: Put down the money, 100 *wên*, as the dividend. Put down also 1 cock and 1 hen, and multiply each by 3, which is the number of chicks, to obtain 3 cocks and 3 hens. Add these numbers to 3 chicks to give a total of 9. This forms the divisor. On dividing the dividend, the quotient 11, which is the number of hens, is obtained. The remainder 1 is subtracted from the divisor 9 to give a remainder 8, which is the number of cocks. Separately, put down the total number of fowls, 100, and from it take away 8 cocks and 11 hens. The remainder is 81, which is the number of chicks. Put down the prices of one cock, one hen and one chick and multiply their respective number by these to obtain the answer.

Working on the assumption of the above method: Put down the said solutions, add 4 to the number of cocks to obtain 12, subtract 7 from the number of hens to obtain 4, and add 3 to the number of chicks to obtain 84. The sum is 100. Hence the answer.

Distribution by proportional parts

41. 100 *wên* are divided among 3 persons. It is known that B gets 2/3 of what A gets and C gets 28 *wên* less than A. Find how much each gets. Answer: A gets 48 *wên*, B gets 32 *wên*, and C gets 20 *wên*.

Method: Since C gets 28 *wên* less than A, add 28 to 100 and let the sum be considered as the total amount of money. The problem says that B gets 2/3 of what A gets, so let the proportional part for A be 3 and that for B be 2. Then the proportional part for C is also 3. Add these together to obtain 8, which forms the divisor. Multiply the total amount of money by each proportional part and divide by the divisor. The respective results are thus obtained though, in the case of C, 28 *wên* has to be subtracted. Hence the answer.

Using a rope to measure a wooden pole

43. There is a wooden pole of unknown length and a rope is used to measure it. The rope is longer than the pole by 4 *ch'ih* [chi] 5 *ts'un* [cun], but if the rope is halved then it is shorter than the pole by 1 *ch'ih*. Find the length of the pole. Answer: 6 *ch'ih* 5 *ts'un*.

Method: Double the difference in length between the pole and half the rope. <Double 1 *ch'ih* to obtain 2 *ch'ih*.> Add this to the excess length of the rope over the pole to obtain the length of the pole. Hence the answer. <The whole rope is longer than one pole by 4 *ch'ih* 5 *ts'un*. If the rope is halved, there is a shortage of 1 *ch'ih*. This means the whole rope measures two poles, but each pole is short of 1 *ch'ih*. Use the method of doubling the shortage and then adding to the excess length. This

means that the lengths of two poles are being made use of and then one of them is concealed.>

Discussion on the square and the circle

The ratio of the circumference of a circle to its diameter taken as 3:1 and the ratio of the side of a square to its diagonal taken as 5:7 have often been discussed by mathematicians but are not easily described in general terms.

In *Chiu Chang* there is a circle whose circumference is 30 *pu* [bu] and its area is 75 *pu*. From Li Shun-fêng's notes, the diameter is 10 *pu*. According to Liu Hui's method, the circumference is multiplied by 50 and then divided by 157. The diameter is 9 87/157 *pu*. Using a more approximate ratio, the circumference is multiplied by 7 and then divided by 22. The diameter is 9 6/11 *pu*. Compare the two ratios and state which is the larger. Use the method of finding the least common multiple of the denominators for finding their common denominator. By Liu Hui's method, the whole diameter is 9 957/1727 *pu*. By the more approximate ratio, the whole diameter is 9 942/1727 *pu*.

If the circumference is known, this technique can be used to find the diameter. Conversely, if the diameter is known, the circumference can also be calculated from this technique.

To find the area from Liu Hui's method: Square the circumference and multiply by 25. Divide by 314 to obtain an area of 71 103/157 *pu*. To find the area using the more approximate ratio: Square the circumference and multiply by 7. Divide by 88 to obtain an area of 71 13/22 *pu*. Compare the two areas by using the method of finding the common denominator. By Liu Hui's method, the area is 71 2266/3454 *pu*. By the more approximate ratio, the area is 71 2041/3454 *pu*.

Both methods agree that the ratio of the circumference of a circle to its diameter is not 3:1. Men in the past have assumed that the ratio of the areas of an inscribed circle and the square is 3:4, hence the adoption of the method that the ratio of the circumference of a circle to its diameter is 3:1.

46. A problem in *Chang Ch'iu-Chien Suan Ching* asks what is the side of a square inscribed in a circle when the diameter of the circle is 2 *ch'ih* 1 *ts'un*. Answer: 1 *ch'ih* 5 *ts'un*. <In Li Shun-fêng's notes, when the square root is extracted, 1 *ch'ih* 4 21/25 *ts'un* is obtained.>

Method: Multiply the diameter in *ts'un* by 5 and divide by 7. Here the ratio of the side of a square to its diagonal is taken as 5:7. <In Li Shun-fêng's notes, there is a fraction 21/25 *ts'un* and the ratio of a square to its diagonal is also taken as 5:7.>

Method based on the "sea-island" problem

53. Across the water there is a pole whose height is not known. Two measuring rods, each 1 *chang* long, stand in front of it at a distance of 15 *ch'ih* apart. A man walks backward 5 *ch'ih* from the front rod and looks at the tip of the measuring rod; the tip of the pole is in the same straight line as his line of vision. Next he walks backwards 8 *ch'ih* from the rod behind and looks at the tip of that measuring rod; the tip of the pole is again in the straight line which is his line of vision. Find the height

of the pole. Answer: The pole is 40 *ch'ih* high. The distance from the front rod to the pole across the water is 25 *ch'ih*.

Method from the *Sea Island Mathematical Manual*: Multiply the distance between the measuring rods by the height of a rod and let the product be the dividend. Find the difference between the distances walked from the measuring rods and let the surplus be the divisor. Hence divide the dividend and add the height of a measuring rod to the quotient to obtain the height of the pole. To find the distance from the front rod to the pole across the water: Multiply the distance between the measuring rods by the distance walked from the front rod and let the product be the dividend. Let the difference of the distances walked from the measuring rods be the divisor and hence divide.

Subtract the smaller remaining perpendicular from the larger remaining perpendicular and let the remainder divide the area formed by the distance between the measuring rods [and the height of a rod measured from the man's view-point]. This gives the height of the pole beyond the tip of the rod. Add to this the height of the rod which includes the height from which the man is looking to give the whole length of the pole. The back measuring rod forms the boundary of the area formed by the larger perpendicular. The difference of the area formed by the smaller perpendicular and the area formed by the larger perpendicular gives the area formed by the distance between the measuring rods and the height of the rods [measured from the man's view-point].

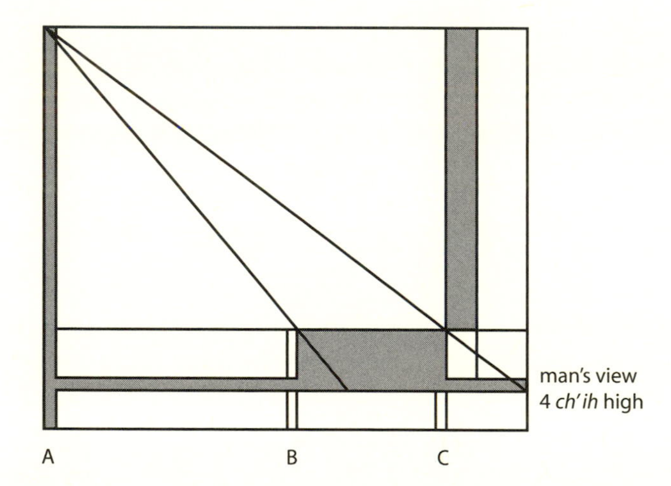

The figure represents both of the diagrams to which Yang Hui refers; the "first diagram" is in reference to the leftmost sighting rod B, from which at five paces to the right the top of A is aligned with the top of B; the "second diagram" is in reference to the rightmost sighting rod C ("behind" B) from which at a distance of 8 paces the top of A is aligned with the top of C. Yang Hui suggests that it would be clearer if the two diagrams were considered separately, although as above, "men of the past have used the two diagrams together."

Explanation of the method: The front measuring rod can be represented in the first diagram, the view being from the tip of the rod to the tip of the wooden pole. The measuring rod behind can be represented in the second diagram, the view being from the tip of this rod to the wooden pole. It is feared that this is not an easy matter to discuss generally, hence the division into two diagrams for the sake of a clear explanation. Men of the past have used the two diagrams together. The problem in the *Sea Island Mathematical Manual* uses the method of *ch'ung ch'a* [zhong cha]. For the first section, the distance of the pole to the front rod gives the length of the area formed by the smaller perpendicular. For the next section, the distance of the pole to the measuring rod behind gives the length of the area formed by the larger perpendicular. Subtract the smaller area from the larger one and the remainder is the area lying between the two measuring rods. This is called the *piao chien chi* [biao jian ji]. Therefore scholars of the past multiply the distance between the measuring rods by the height of a rod [measured from the view-point] and let the product be the dividend. The smaller remaining perpendicular of the front diagram is subtracted from the larger remaining perpendicular of the second diagram. The remainder divides the area lying between the two measuring rods to give the height [of the pole measured from the tip of the rod]. Originally, the dividend is the difference of the area formed by the larger perpendicular and the area formed by the smaller perpendicular, and the divisor is the difference of the smaller remaining perpendicular and the larger remaining perpendicular. Divide the dividend by the divisor to give the height of the pole beyond the boundary. Add to this the height of the measuring rod to include the height from which the man is looking to obtain the whole length of the pole. Now it is feared that readers of later generations would not know the direct origin of the use of this excellent method belonging to their ancestors, therefore two problems on the estimation of the height of a wooden pole across the water are verified by the method of *ch'ung ch'a*. The first problem in the *Sea Island Mathematical Manual* is introduced here so that its good points can be learned and they can be introduced and extended to develop other methods. How can this be a slight amendment?

VII.d. Zhu Shijie: *Siyuan yujian* (*Precious Mirror of the Four Elements,* 1303)

Zhu Shijie sometimes referred to himself as a "resident of Yan," an old name for Beijing, so he was probably born near the later capital of the Mongol emperors, although neither his date of birth nor when he died is known. In fact, virtually nothing is known about the life of this mathematician apart from what may be gleaned from the preface to his most influential work, where it is noted that he was "a famous mathematician and he traveled widely throughout the country for more than 20 years."

In a *Sequel to Biographies of Mathematicians and Astronomers* for the Qing Dynasty, Zhu Shijie is linked with two of his predecessors as follows:

> Together with Qin Jiushao and Li Zhi (or Li Ye), [Zhu Shijie] can be said to form a tripod. Qin contributed positive and negative and the extraction of roots, Li contributed the celestial element and all those contributions stretch back into the past and will survive to thousands of future generations; Zhu includes everything, he has improved everything to such an extent that it is only understood by the gods and has surpassed the two schools of Qin and Li. [quoted from Li and Du 1987, 116–17]

Zhu Shijie, in an *Introduction to Mathematical Studies* (*Suanxue qimeng*, 1299), provides clear rules for multiplication and division using negative numbers. This work also introduced a method for writing large numbers similar to those already encountered in the *Sunzi suan jing* and the *Shu shu ji yi*, using such Buddhist terms as *ji* ("supreme") and *heng he sha* ("sand of the Ganges") for 10^{88} and 10^{96}, and even more expressive terms like *xu yu* ("twinkling of an eye") and *tan zhi* ("snap of the fingers") for very small numbers. But the book seems to have been lost sometime after it was printed in China, although it came to exercise a great influence on the development of mathematics in Japan, and in 1839, when a copy was discovered in a Korean version of 1660, the work again came to the attention of scholars in China.

Meanwhile, Zhu Shijie's greatest contribution to mathematics was his generalization of the "celestial element method" which is the focus of his *Precious Mirror of the Four Elements* (1303). This literal translation of the title's four characters *siyuan yujian* has been widely used since it was first rendered into English as such by Yoshio Mikami in 1913, but Jock Hoe in a recent and very thorough study of this work suggests an alternative title: "Mirror trustworthy as jade relative to the four origins" (quoted with slight variance from [Martzloff 1997, 153]; see also [Hoe 1977, 41–47]). The important qualities that jade was meant to convey to Chinese readers were reflections of its transparency and clarity, and the fact that the results of the "jade mirror" are correct. The image of the mirror in fact is one frequently encountered in titles of Chinese works; it is meant to indicate that they somehow honestly treat a given subject, in this case the mathematics of the work, without distortion. The jade mirror is thus expected to reflect its subject directly in a faithful way. Hoe also argues that *yuan* should not be translated as "element," but as the "source from which all the material universe was born" [Hoe 1977, 45]. But given the prominence of the *tian yuan* method in Zhu's work, it may be that the character *yuan* may have had multiple meanings for him. Thus it may appear in the title as a sort of double-entendre, intended to indicate both the algebraic nature of the method at hand, and the more philosophical connection it was meant to inspire with the "origin of all things," thus indicating its surpassing power to explain all things mathematically.

Thoroughly versed in the basic *tian yuan* method of the celestial element that enabled Chinese mathematicians to deal with equations in one unknown, Zhu followed the results of other mathematicians carefully who had begun to generalize the *tian yuan* method to permit additional unknowns, eventually resulting in Zhu Shijie's more general method known as the *si yuan shu* (method of *four* unknowns). Zu Yi, in his preface to the *Precious Mirror*, expresses the development of the idea as follows:

> Li Dezai of Pingyang region, having published *The Complete Collection for Heroes on Two Principles*, they then had the earth unknown. The critic Liu Runfu, the very talented student of Master Xing Songbu published *Heaven and Earth in a Bag*, which included two problems involving the human unknown. My friend, Master Zhu [Shijie] of Yanshan district explained mathematics for many years. Having explored the mysteries of the three unknowns and sought out the hidden details of the *Nine Chapters*, he set up the four unknowns according to heaven, earth, man, and matter. [Li and Du 1987, 140–41]

What was added over the centuries since the "celestial element" first introduced a single unknown element x to the solution of equations in one unknown was the generalization of this idea to include equations with as many as four unknowns: x, y, z, and u. These were each represented by a different character, *tian* (heaven, x), *di* (earth, y), *ren* (man, z), and *wu* (thing or material, u). This generalization was made possible by an ingenious association of regions of the counting board for each of the possible combinations of variables and their coefficients; the coefficients were placed on the counting board in the position reserved for a particular combination of variables. To make clear the relative positions of all possible combinations of coefficients, the constant term was placed at the very center of the board, associated with the character *tai* (the same character used to designate the constant term in the *tian yuan* method).

Wu (material) element (u)

⋮	⋮	⋮	⋮	⋮	⋮	⋮
....y^3u^2	y^2u^2	yu^2	u^2	zu^2	z^2u^2	z^3u^2....
....y^3u	y^2u	yu	u	zu	z^2u	z^3u....
				yz		
....y^3	y^2	y	太	z	z^2	z^3....
		xu				
....xy^3	xy^2	xy	x	xz	xz^2	xz^3....
....x^2y^3	x^2y^2	x^2y	x^2	x^2z	x^2z^2	x^2z^3....
....x^3y^3	x^3y^2	x^3y	x^3	x^3z	x^3z^2	x^3z^3....
⋮	⋮	⋮	⋮	⋮	⋮	⋮

Di (earth) element (y)

Ren (man) element (z)

Tian (heaven) element (x)

Thus the counting rod configuration for $x + y + z + u$ would have been displayed as follows:

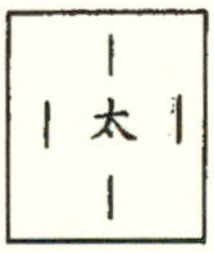

	I	
I	太	I
	I	

Similarly, $x^2 + y^2 + z^2 + u^2 + 2xy + 2xz + 2xu + 2yz + 2yu + 2zu$ would have been displayed as

		I		
	II	O II	II	
I	O	II 太	O	I
	II	O	II	
		I		

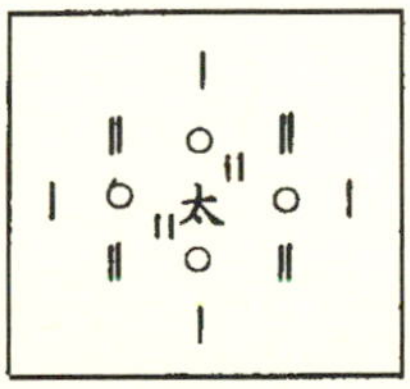

With the above scheme in mind, it is possible to understand the meaning of the description of this method that is given as follows in the preface to Zhu Shijie's work:

> The method of the *Precious Mirror of the Four Elements* is to have the source of all the unknowns [the constant term] in the center and to put the celestial [unknown] element at the bottom, the earth unknown on the left, the human unknown on the right and the material unknown at the

> top, opposites to up and down, proceed and recede to left and right, they interact and change criss-crossing with infinite variety. The hidden parts are revealed of excess and deficit, positive and negative rectangular tables, the method of taking powers and extracting roots; this is concise and deep. This information makes it possible to reveal the central ideas which were not worked out by our predecessors. [Li and Du 1987, 141]

And like the earlier versions of this method as it would have been worked out on the counting board, the procedure is analogous to the elimination methods we have encountered previously. By eliminating one unknown at a time from equations in several unknowns, the technique eventually produces one equation in only one unknown, from which it is then an easy matter working backwards to provide solutions for each of the remaining unknowns.

In the *Precious Mirror*, Zhu Shijie offers four examples, beginning with one unknown, then two, three, and finally a problem involving four unknowns. This is all accomplished in less than five pages in which he describes the method of elimination in very brief terms, and in the rest of the book there is no description of the method at all.

In summarizing the influence and significance of Zhu Shijie's "method of the four unknowns," Li and Du write as follows:

> In his time (13th century CE) the development of the techniques of the celestial element and of the four unknowns, which had started in northern China, reached its zenith. Because of the limitations of recording numbers using counting rods it is obvious that no more than four unknowns can be treated in this way. In Europe the method of elimination for higher degree equations was only described in a systematic way in the eighteenth century in the work of the French mathematician Bézout (1779 CE). [Li and Du 1987, 148]

What is most remarkable is the way in which, once the counting board is conceived as a device for remembering the coefficients of various combinations of unknowns of arbitrary power, the elimination procedure is then essentially a mechanical process, and the mathematician simply becomes a computer, carrying out the prescribed motions mechanically on the counting board. As Jean-Claude Martzloff puts it so evocatively:

> These rod puppets which now chatter together are alive in their own right. They rise, fall, and turn to the left or the right, translating series of polynomial additions, subtractions, multiplications or divisions. When the red camp (positive) meet the black camp (negative) they kill one another, while respecting the sign rules. At the end of this small drama which does not depart from the laws of mathematical etiquette for a moment, the key character in the problem appears: the final resolutory equation. [Martzloff 1997, 261]

In the *Precious Mirror of the Four Elements*, Zhu Shijie made use of virtually every position available on the counting board to represent the coefficients of equations in as many as four unknowns. Of the four paradigmatic examples placed at the beginning of the book to exemplify exactly how each type of problem could be worked out, three are reproduced here, beginning with Problem 2, which involves two equations in two unknowns. Problem 3 involves three equations in three unknowns, and Problem 4 involves four equations in four unknowns. By applying a series of elimination methods, Zhu Shijie was able to reduce these sets of simultaneous equations to a single equation in one unknown, which could then be solved directly. By then working backward to each of the successively higher stages of his elimination procedure, he was able to obtain the full solution for each of the problems presented as models of their kind for students to follow.

Zhu Shijie's Siyuan yujian (Precious Mirror of the Four Elements)

Problem 2

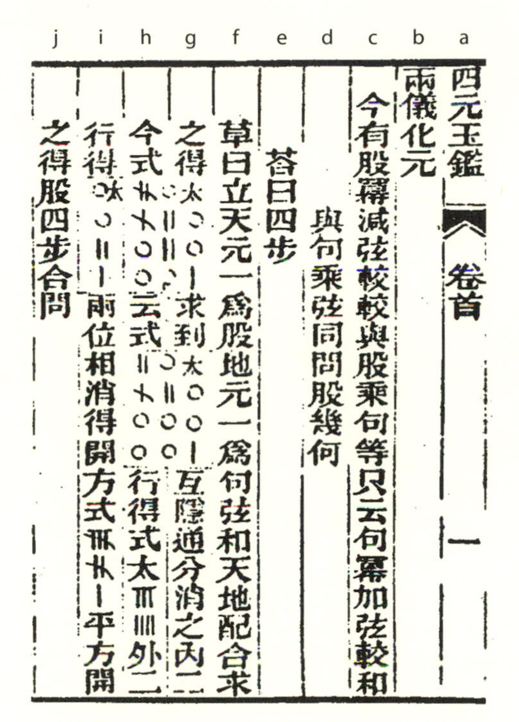

四元玉鑑 卷首 一
兩儀化元
今有股冪減弦較較與股乘句等只云句冪加弦較和
與句乘弦同問股幾何
荅曰四步
草曰立天元一爲股地元一爲句弦和天地配合求
之得……求到……互隱通分消之內二
今式……云式……行得式……外二
行得……兩位相消得開方式……平方開
之得股四步合問

Problem 2 from the edition of Zhu Shijie's *Precious Mirror*, reproduced in [Guo 1993, vol. 1, 1209b]. The title of the work is given at the top of column a: *Siyuan yujian*; the title for Problem 2 is at the top of column b: *Liang yi hua yuan*. The statement of the problem is given in columns c–d. Column e gives the answer, 4 *bu*, and columns e–j are the "working" of the problem which Lam Lay Yong also translates below. From this very limited information, it is virtually impossible to reconstruct the method that Zhu Shijie must have used to solve this problem. Fortunately, a commentary on the *Precious Mirror* by Shen Qinpei of 1829 survives in a manuscript copy in the National Library of China (Beijing): *Siyuan yujian xicao* (*Detailed Solutions for the Precious Mirror of the Four Elements*), photo-reproduced in [Guo 1993, vol. 5, 227–432], and it is upon the procedures given in detail by Shen that the further explanations of the methods used to work out the three problems excerpted here are based.

Subtract from the square of the altitude [of a right-angled triangle] the difference of the hypotenuse and the difference of the altitude and base to equal the product of the altitude and base. It is also given that the square of the base added to the sum of the hypotenuse and the difference of the altitude and base equals the product of the base and hypotenuse. Find the altitude. Answer 4 *bu*.

Working: Put 1 to be the *tianyuan* to represent the altitude. Put 1 to be the *diyuan* to represent the sum of the base and hypotenuse. Combine the *tian* and *di* to obtain the *jinshi* (from the first datum)

$$\begin{matrix} -2 & 0 & \textit{tai} \\ -1 & 2 & 0 \\ 0 & 2 & 0 \\ 0 & 0 & 1 \end{matrix} \qquad \text{[i]}$$

and the *yunshi* (from the second datum)

$$\begin{matrix} 2 & 0 & \textit{tai} \\ -1 & 2 & 0 \\ 0 & 0 & 0 \\ 0 & 0 & 1 \end{matrix} \qquad \text{[ii]}$$

On eliminating these, the inner column is obtained in the form

tai
8
4

and the outer column in the form

tai
0
2
1

On combining both of them, the form for extracting the root is obtained:

–8
–2 [iii]
1

Solve the quadratic equation to give 4 *bu* for the altitude. Hence the answer.

EXPLANATION: Zhu Shijie called this problem *Liang Yi Hua Yuan*, which means "Turning the Two Opposites into One," or "Unifying Yin and Yang," here a reference steeped in Daoist imagery to the fact that Problem 2 deals with an equation in two unknowns. The situation as Zhu Shijie describes it is given in terms of a right triangle. If the base is a, altitude b, and hypotenuse c, then the given data result in two equations involving the base, altitude, and hypotenuse as follows:

[i(data)] $$b^2 - [c - (b - a)] = ba$$

[ii(data)] $$a^2 + c + b - a = ac$$

Setting the *tianyuan* $x = b$ and the *diyuan* $y = a + c$, then the above equations (knowing that $a^2 + b^2 = c^2$), are equivalent to the following two equations:

[i] $$x^3 + 2yx^2 + 2xy - xy^2 - 2y^2 = 0$$

[ii] $$x^3 + 2yx - xy^2 + 2y^2 = 0$$

All Zhu Shijie says about this is "*tiandi peihe quizhi de jinshi*" (coordinate the *tian* and *di* to obtain the desired (sought after) first derived equation). Here "*peihe*" means "combine" or "coordinate," but there is no indication in the text of how this makes it possible to get from the given data to either [i] or [ii]. The steps provided here follow the analysis of the *Precious Mirror* by Kong Guoping [Kong 1999, 376].

To obtain equation [i], subtract $x = b$ from $y = a + c$: $y - x = a + c - b = c + a - b = c - (b - a)$. From $a^2 + b^2 = c^2$ it follows that $b^2 = c^2 - a^2 = (c + a)(c - a)$, and $(c - a) = b^2/(a + c)$. Now write $2a = (c + a) - (c - a) = y - x^2/y$, and $a = (1/2)(y - x^2/y)$. From the given data [i], $b^2 - [c - (b - a)] = ba$; substituting gives $x^2 - (y - x) = x^2 - y + x = x[(1/2)(y - x^2/y)]$, from which it follows (after multiplying by $2y$) that $2x^2y - 2y^2 + 2xy - xy^2 + x^3 = 0$. A suitable rearrangement of terms gives equation [i].

To obtain equation [ii], since $y = a + c$, $c = y - a$, and $c = y - (1/2)(y - x^2/y) = (1/2)(y + x^2/y)$, then $ac = (1/2)(y - x^2/y)(1/2)(y + x^2/y) = (1/4)(y^2 + x^4/y^2)$. Now, from the given data [ii],

$a^2 + c + b - a = a^2 + (c - a) + b = ac$; substituting gives $[(1/2)(y - x^2/y)]^2 + x^2/y + x = (1/4)(y^2 - x^4/y^2)$. Thus, $(1/4)(y^2 - 2x^2 + x^4/y^2) + x^2/y + x = (1/4)(y^2 - x^4/y^2)$; multiplying by $4y^2$ yields $y^4 - 2x^2y^2 + x^4 + 4x^2y + 4xy^2 - y^4 + x^4 = 0$. Dividing by $2x$ and rearranging terms gives a result equivalent to equation [ii]: $x^3 - xy^2 + 2xy + 2y^2 = 0$.

Zhu Shijie called [i] the *jin shi* 今式 or first derived equation, and [ii] the *yun shi* 云式 or second derived equation. These equations [i] and [ii] correspond to the counting board configurations that are also labeled [i] and [ii], respectively, below. The following diagrams will make clear exactly which coefficients correspond to the various terms of the variables x and y for equations [i] and [ii] as given in the *Precious Mirror.*

$-2y^2$	0	*tai*	
$-1xy^2$	$2xy$	0	[i]
0	$2x^2y$	0	
0	0	$1x^3$	

[See the configuration of these coefficients at the top of columns g and h in the figure above]

$2y^2$	0	*tai*		
$-1xy^2$	$2xy$	0		[ii]
0	0	0		
	0	0	$1x^3$	

[See the configuration of these coefficients in the middle of columns g and h in the figure above]

Although the procedure that follows looks forbiddingly complex, as a series of algorithmic transactions on the counting board it would have been a very mechanical process, the basic idea corresponding to a method whereby two equations in the form $A_1y + A_2 = 0$ and $B_1y + B_2 = 0$ are obtained, where A_i and B_i contain only factors of x. From these it is possible to eliminate the y term by multiplying the first equation by B_1 and the second equation by A_1; after subtracting the results, what remains is a polynomial in x alone, $A_2B_1 - A_1B_2 = 0$. This is the elimination method that Zhu Shijie called *hu yin tong fen*, which may be loosely translated as "mutually exchange what is hidden to equalize the parts."

Lam Lay Yong translates this expression as "elimination" or "eliminating," since the idea is to eliminate all the terms containing a specific variable until everything is reduced to a single equation in only one unknown [Lam 1982, 263–269]. In the case of Problem 2, the first step is to eliminate the y^2 terms and then the y terms as well. But all Zhu Shijie says about going from the *jin shi* and *yan shi* equations [i] and [ii] to two equations in x alone is to "mutually eliminate and the inner two columns disappear leaving the equation $4x^2 + 8x = 0$." Cryptic though this may seem, based upon a procedure in Problem 3 wherein Zhu Shijie first sets up two intermediate "left" and "right" equations, it is possible to reconstruct the necessary steps needed to "mutually eliminate" the "inner" and "outer" columns to obtain the solution to the problem.

The first step is to obtain two equations in the form $A_1y + A_2 = 0$ and $B_1y + B_2$; the first equation may be obtained by subtracting the *yun shi* equation [ii] from the *jin shi* equation [i], which gives a third derived equation that Zhu Shijie called the *xiao shi* 消式 or "equation for elimination"[19]:

[a] $$2x^2y - 4y^2 = 0.$$

[19]Note that in the expression *xiao shi* 消式, "xiao" means "to disappear," "vanish," "eliminate," or "remove." Here, in conjunction with "shi" meaning "formula" or "equation," it is translated as "equation for elimination."

This may be simplified as

[a′] $$x^2 - 2y = 0,$$

which serves to eliminate the y^2 term. Zhu Shijie called this equation the *you shi* or "right equation." The second equation, to become the "left equation," may be found by eliminating the y^2 term from the *jin shi* equation [i]. Beginning with the *xiao shi* equation [a′], multiplying by x gives $x^3 = 2xy$; substituting this for x^3 in the *jin shi* equation [i] yields $2yx^2 + 4xy - xy^2 - 2y^2 = 0$; dividing by y now gives

[b] $$2x^2 + 4x - xy - 2y = 0.$$

This is the *zuo shi* or "left equation." On the counting board the coefficients of the two equations would have appeared as follows:

Left Equation		Right Equation	
(B_1)	(B_2)	(A_1)	(A_2)
$2y$	*tai*	$-2y$	*tai*
xy	$-4x$	0	0
0	$-2x^2$	0	$1x^2$

This left equation in the form $A_1y + A_2 = 0$ corresponds to $(x + 2)y - 2x^2 - 4x = 0$; the right equation in the form $B_1y + B_2 = 0$ corresponds to $-2y + x^2 = 0$; and this makes possible the elimination that Zhu Shijie called *hu yin tong fen*. Algebraically, for the right equation, $A_1 = -2$ and $A_2 = x^2$; for the left equation, $B_1 = x + 2$ and $B_2 = -2x^2 - 4x$. To eliminate the y term it suffices to determine $A_2B_1 - A_1B_2$. On the counting board, given the two matrices set-up side by side as above, this entailed the process of multiplying what Zhu Shijie referred to as the two "inside" columns of the matrices for each equation, i.e., $(-2)(-2x^2 - 4x) = 4x^2 + 8x$. On the counting board, A_1B_2 is

tai
8
4

This configuration may be found at the bottom of column h in the figure above. Likewise, the "outside" columns were multiplied together: $(x + 2)(x^2) = x^3 + 2x^2$. On the counting board A_2B_1 is:

tai
0
2
1

This configuration may be found at the top of column *i* in the figure.

Now Zhu Shijie instructs "*liang wei xiang xiao*" (column i), which means to "mutually remove" or subtract one column from the other, i.e., to determine $A_2B_1 - A_1B_2$. This now gives the desired polynomial in x alone, $A_2B_1 - A_1B_2 = x^3 - 2x^2 - 8x = 0$. This is reduced to $x^2 - 2x - 8 = 0$. The corresponding figure on the counting board is

-8
$-2x$
$1x^2$

found near the bottom of column i. Zhu Shijie identifies this as a quadratic equation, whose solution leads to what was asked for, namely $x = b = 4$; i.e. the altitude of the triangle in question is 4 *bu*.

Problem 3

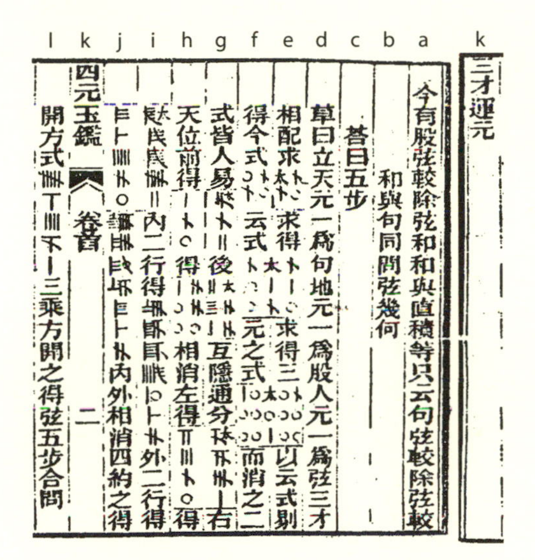
三才運元

今有股弦較除弦和和與直積等只云句弦較除弦較和與句同問弦幾何

荅曰五步

草曰立天元一爲句地元一爲股人元一爲弦三才相配求得[illegible]求得[illegible]求得三[illegible]以云式剔得今式[illegible]云式[illegible]元之式[illegible]而消之二式皆人易[illegible]後[illegible]互隱通分[illegible]右天位前得[illegible]得[illegible]相消左得[illegible]得[illegible]內二行得[illegible]外二行得[illegible]內外相消四約之得[illegible]開方式[illegible]三乘方開之得弦五步合問

四元玉鑑 卷首 二

Problem 3 from the edition of Zhu Shijie's *Precious Mirror* (1303), reproduced in [Guo 1993, vol. 1, 1209b–1210a]. The title for Problem 3 is given at the top of column k (from page 1209b): *San cai yun yun*. The statement of the problem is given in columns a–b (from page 1210a). Column c gives the answer, 5 *bu*, and columns d–l are the "working" of the problem.

The sum of the base, altitude and hypotenuse divided by the difference of the hypotenuse and altitude is equal to the area of the rectangle. It is also given that the sum of the hypotenuse and the difference of the altitude and base divided by the difference of the hypotenuse and base is equal to the base. Find the hypotenuse Answer: 5 *bu*.

Explanation: Zhu Shijie called this problem *San cai yun yun*, which means "The Evolution of the Three Talents" or "Primary Manipulation of Three Unknowns." Li and Du translate this as as "Juggling Three Unknowns," a reference to the fact that solution of this problem involves three equations in three unknowns [Li and Du 1987, 143]. Li and Du also provide a partial explanation for problem 3 [Li and Du 1987, 143–145]. A brief analysis of this problem may also be found in [Wu 2000, 25–26]. See as well [Hoe 1997, 172–211], and [Kong 1999, 377–378].

The problem again concerns a right triangle; if the base is a, altitude b, and hypotenuse c, then the given data result in two equations involving the base, altitude and hypotenuse as follows:

[i data] $$(a + b + c)/(c - b) = ab \quad \text{and}$$

[ii data] $$(c + b - a)/(c - a) = a.$$

Letting the *tianyuan* element x represent the base, the *diyuan* element y represent the altitude, and the *renyuan* element z represent the hypotenuse, i.e., $x = a$, $y = b$, and $z = c$, the cognate equations [i] and [ii] below follow by direct substitution and cross-multiplication from the given data. Zhu's representations of the various equations needed to solve the problem on the counting board are below. Note that while the coefficients of the terms involving powers of x are placed below *tai* in the *tianyuan* portion of the counting board configuration, and those involving powers of y are placed to the left of *tai* in the *diyuan* portion of the counting board, the coefficients involving z are placed to the right of *tai* in the *ren yuan* portion of the board. The coefficients for composite terms like xyz were interpolated as an additional term between the places for y, xy, and x, as we see in the first or *jin shi* 今式 equation:

[i] $$xyz - xy^2 - x - y - z = 0.$$

This would have appeared on the counting board in the following configuration:

0	−1		*tai*	−1		$0y^2$	$-1y$		*tai*	$-1z$
		1			i.e.			$1xyz$		
−1	0		−1	0		$-1xy^2$	$0xy$		$-1x$	$0xz$

The second equation, or *yun shi* 云式, follows from the second set of conditions as stated in the problem:

[ii] $$xz - x^2 + x - y - z = 0.$$

This would have been configured as follows on the counting board:

−1	*tai*	−1		$-1y$	*tai*	$-1z$
0	1	1	i.e.	$0yx$	$1x$	$0xz$
0	−1	0		$0yx^2$	$-1x^2$	$0x^2z$

From the *gou-gu* (Pythagorean) relation, $a^2 + b^2 = c^2$, direct substitution results in the *san yuan shi* 三云式 or "third equation" as follows:

[iii] $$x^2 + y^2 - z^2 = 0.$$

This would have assumed the following configuration on the counting board:

1	0	*tai*	0	−1		$1y^2$	$0y$	*tai*	$0z$	$-1z^2$
0	0	0	0	0	i.e.	$0y^2x$	$0yx$	$0x$	$0xz$	$0xz^2$
0	0	1	0	0		$0y^2x^2$	$0yx^2$	$1x^2$	$0zx^2$	$0x^2z^2$

The first step is to eliminate the y^2 term. Thereafter, solution of this problem will proceed as it did in Problem 2, by then eliminating the y term, leaving an equation for solution in only two variables, whereupon the elimination procedure will be repeated until eventually a single equation in only one unknown remains.

To begin, in order to eliminate the y^2 term from [i], since it appears there as $-xy^2$, multiply (iii) by x, which gives:

$$x^3 + xy^2 - xz^2 = 0.$$

Adding this to the *jin shi* 今式 or first derived equation (i) gives:

[iv] $$xzy - y + x^3 - xz^2 - x - z = 0,$$

Shen Qinpei called this the *xiao shi* 消式 or an "equation for elimination."[20] The *xiao shi* would have appeared on the counting board configuration as:

−1	*tai*	−1	0		$-1y$	*tai*	$-1z$	$0z^2$
	1				$1xzy$			
0	−1	0	−1	i.e.	$0yx$	$-1x$	$0xz$	$-1xz^2$
0	0	0	0		$0yx^2$	$0x^2$	$0zx^2$	$0x^2z^2$
0	1	0	0		$0yx^3$	$1x^3$	$0zx^3$	$0x^3z^2$

To eliminate the y term, Zhu Shijie focused on the coefficients containing the variable y (on the counting board, this amounts to the topmost row of coefficients; algebraically, it means isolating the coefficients with a factor of y, representing the *xiao shi* equation (iv) in the form $A_1y + A_2 = 0$:

$$\text{[iv]} \qquad (xz-1)y + (x^3 - xz^2 - x - z) = 0,$$

for which $A_1 = xz - 1$, and $A_2 = x^3 - xz^2 - x - z$. Similarly, we can rearrange the *yun shi* equation (ii) in the form $B_1y + B_2 = 0$:

$$\text{[ii]} \qquad (-1)y + (-x^2 + xz + x - z) = 0,$$

for which $B_1 = -1$ and $B_2 = -x^2 + xz + x - z$.

With the *xiao shi* matrix of coefficients on the left and the *yun shi* matrix of coefficients on the right, the Chinese method for eliminating the y term calls for multiplying the "outer" columns together and subtracting from this the product of the two "inner" columns. As before, this is equivalent to determining $A_1B_2 - A_2B_1 = 0$. Thus for the two "outer" columns, their product A_1B_2 is

$$(xz-1)(-x^2 + xz + x - z) = -x^3z + x^2z^2 + x^2z - xz^2 + x^2 - xz - x + z.$$

Similarly, the product of the "inner" columns A_2B_1 is

$$(-1)(x^3 - xz^2 - x - z) = -x^3 + xz^2 + x + z.$$

On the counting board the two matrices for these expressions would be:

Left (A_1B_2):			Right (A_2B_1):		
tai	1	0	*tai*	1	0
−1	−1	−1	1	0	1
1	1	1	0	0	0
0	−1	0	−1	0	0

Now, to eliminate y from the equations, calculate $A_1B_2 - A_2B_1 = 0$:

$$x^3 - x^3z + x^2z^2 + x^2z - 2xz^2 + x^2 - xz - 2x = 0.$$

This can be simplified by dividing each term of the equation by x:

$$x^2 - x^2z + xz^2 + xz - 2z^2 + x - z - 2 = 0.$$

[20] [Shen Qinpei 1829, 237a].

The configuration for this equation on the counting board would then have looked as follows:

$$\begin{array}{ccc} -2\,tai & -1 & -2 \\ 1 & 1 & 1 \\ 1 & -1 & 0 \end{array}$$

Zhu Shijie now instructs that z should be replaced by x and x by y. This maneuver serves to rearrange the positions of the coefficients on the counting board in order to facilitate the continuing computation of the problem; this has the effect of rotating the matrix of coefficients clockwise by 90°:

$$\begin{array}{ccc} 1 & 1 & -2\,tai \\ -1 & 1 & -1 \\ 0 & 1 & -2 \end{array}$$

[See the configuration of these coefficients at the top of columns g and h in the figure above.]

This matrix of coefficients corresponds to the following equation:

$$\text{[v]} \qquad y^2 - y^2x + yx^2 + yx - 2x^2 + y - x - 2 = 0.$$

Zhu Shijie called this the *qian shi* 前式 or anterior equation. (Note that Zhu mistakenly gives the wrong coefficient for the x^2 term as +2 [Zhu Shijie, in Guo 1993, vol. 1, 1210]; on the other hand, [Li and Du 1987, 144] mistakenly gives the constant term −2 *tai* as +2 *tai.* [Wu 2000] does not consider this step in the solution of Problem 3.)

The *qianshi* equation (v) is an equation in only two unknowns, x and y. In order to find another equation in the two unknowns x and y, from which it will then be possible to use the same elimination procedure to remove the y terms, Zhu Shijie considered the *san yuan shi* (iii) in the form $A_1y + A_2 = 0$, i.e., $A_1 = y$ and $A_2 = -x^2 + z^2$. Returning again to the *yun shi* equation (ii), in the form $B_1y + B_2 = 0$, we have $B_1 = -1$ and $B_2 = xz - x^2 + x - z$. To eliminate the y^2 term, it is necessary to compute $A_1B_2 - A_2B_1$.

$$\begin{aligned} A_1B_2&: \; y(-x^2 + xz + x - z) = -x^2y + xyz + xy - yz \\ A_2B_1&: \; -1(x^2 - z^2) = -x^2 + z^2. \end{aligned}$$

Consequently, $A_1B_2 - A_2B_1 = 0$ is

$$\text{[vi]} \qquad -x^2y + xyz + xy - yz + x^2 - z^2 = 0.$$

Shen Qinpei also called this *xiao shi* 消式, or an "equation for elimination."[21] The matrix of its coefficients would have appeared on the counting board as follows:

$$\begin{array}{cccc} 0 & tai & 0 & -1 \\ & -1yz & & \\ & 1xyz & & \\ 1 & 0 & 0 & 0 \\ -1 & 1 & 0 & 0 \end{array}$$

[21] [Shen Qinpei 1829, 237a].

In order to eliminate the y term from this *xiao shi* equation (vi) and the *yun shi* equation (ii), consider the *xiao shi* equation [vi] in the form $A_1y + A_2 = 0$, i.e., $A_1 = -x^2 + xz + x - z$ and $A_2 = x^2 - z^2$. As before, considering the *yun shi* equation [ii] in the form $B_1y + B_2 = 0$ with $B_1 = -1$ and $B_2 = xz - x^2 + x - z$, to eliminate the y term it is necessary to compute $A_1B_2 - A_2B_1$, i.e.,

$$A_1B_2: \ (-x^2 + xz + x - z)(-x^2 + xz + x - z)$$
$$= x^4 - 2x^3z - 2x^3 + 4x^2z + x^2z^2 - 2xz^2 + x^2 - 2xz + z^2.$$

$$A_2B_1: \ (-1)(x^2 - z^2) = -x^2 + z^2.$$

On the counting board, these would have appeared as follows, with A_1B_2 on the left and A_2B_1 on the right:

tai	0	1
0	−2	−2
1	4	1
−2	−2	0
1	0	0

tai	0	1
0	0	0
−1	0	0

Consequently, $A_1B_2 - A_2B_1 = 0$ is

$$x^4 - 2x^3z - 2x^3 + 4x^2z + x^2z^2 - 2xz^2 + 2x^2 - 2xz = 0,$$

and after simplifying by dividing each term by x:

$$x^3 - 2x^2z - 2x^2 + 4xz + xz^2 - 2z^2 + 2x - 2z = 0.$$

The following configuration represents this equation on the counting board:

tai	−2	−2
2	4	1
−2	−2	0
1	0	0

In order to facilitate the rest of the computations for this problem on the counting board and in order to bring all of the coefficients into their corresponding places in the *tianyuan* (x) and *diyuan* (y) parts of the counting board, z is replaced by x and x is replaced by y, after which

[vii] $$y^3 - 2y^2x - 2y^2 + 4yx + yx^2 - 2x^2 + 2y - 2x = 0.$$

Zhu Shijie called this the *hou shi* 後式 or posterior equation. The corresponding configuration on the counting board would look as follows:

[vii]

1	−2	2	*tai*
0	−2	4	−2
0	0	1	−2

[See the configuration of these coefficients in the middle of columns g and h in the figure above.]

Like the *qian shi* or anterior equation [v], the *hou shi* or posterior equation [vii] is also now an equation in only 2 variables, x and y. Zhu Shijie proceeded to reapply the "elimination" procedure to reduce the anterior and posterior equations [v] and [vii] to equations containing only terms as high as y^2, and then only y, after which the penultimate step is to remove the y elements as well, leaving only one equation in the single unknown x.

Zhu Shijie, in applying his elimination method, would have proceeded to work with the "inner" and "outer" columns of the two matrices of coefficients on his counting board to produce auxiliary "equations of elimination" from which he could eventually eliminate the various powers of y from his equations. According to Shen Qinpei, Zhu Shijie would most likely have proceeded as follows. Put the *qian shi* or anterior equation

[v] $$y^2 - y^2x + yx^2 + yx - 2x^2 + y - x - 2 = 0$$

into the form $A_1y^2 + A_2 = 0$; thus, $A_1 = 1 - x$ and $A_2 = yx^2 + yx - 2x^2 + y - x - 2$.

Similarly, put the *hou shi* or posterior equation

[vii] $$y^3 - 2y^2x - 2y^2 + 4yx + yx^2 - 2x^2 + 2y - 2x = 0$$

into the form $B_1y^2 + B_2 = 0$; thus $B_1 = y$ and $B_2 = -2y^2x - 2y^2 + 4yx + yx^2 - 2x^2 + 2y - 2x$.

It is now possible to eliminate the y^3 term by the familiar "elimination process," i.e. by determining $A_1B_2 - A_2B_1$ as follows, with the coefficients of the *qian shi* or anterior equation [v] on the left, and the coefficients of the *hou shi* or posterior equation [vii] on the right:

Qian shi (Anterior Equation) *Hou shi* (Posterior Equation)

(A_1)	(A_2)		
1	1	−2	*tai*
−1	1	−1	
0	1	−2	

(B_1)	(B_2)		
1	−2	2	*tai*
0	−2	4	−2
0	0	1	−2

A_1B_2 represents multiplication of the "outer" columns of the two matrices, the outer or left-most column of the anterior equation (v) times the "outer" or right-most columns of the posterior equation (vii):

$$(1 - x)(-2y^2x - 2y^2 + 4yx + yx^2 - 2x^2 + 2y - 2x)$$
$$= -2y^2 + 2y^2x^2 - 3x^2y + 2xy + 2y - 2x - x^3y + 2x^3.$$

A_2B_1 now represents multiplication of the "inner" columns, the two innermost columns on the right of the *qian shi* or anterior equation (v) and the leftmost inner column of the *hou shi* or posterior equation:

$$y(-2x^2 - x - 2 + yx^2 + yx + y) = -2yx^2 - xy - 2y + y^2x^2 + y^2x + y^2.$$

On the counting board, the coefficients of these two equations appear as follows:

Left (A_1B_2)

−2	2	*tai*
0	2	−2
2	−3	0
0	−1	2

Right (A_2B_1)

1	−2	*tai*
1	−1	0
1	−2	0

It now follows that $A_1B_2 - A_2B_1 = 0$ is:

(*) $$x^2y^2 - 3y^2 - x^2y + 3xy + 4y - 2x - x^3y + 2x^3 - xy^2 = 0,$$

which Shen Qinpei called another "*xiao shi*" or equation for elimination. This would have appeared on the counting board as:

−3	4	*tai*
−1	3	−2
1	−1	0
0	−1	2

It is now possible to eliminate the y^2 terms from (∗) by the same sort of elimination technique, this time using (∗) and the previous *qian shi*. Rewriting (∗) as $B_1y^2 + B_2 = 0$, i.e., $B_1 = x^2 - x - 3$ and $B_2 = (-x^3 - x^2 + 3x + 4)y + 2x^3 - 2x$, and recalling that for the *qian shi* in the form $A_1y^2 + A_2 = 0$ we have $A_1 = 1 - x$ and $A_2 = (x^2 + x + 1)y - 2x^2 - x - 2$, then it is possible to eliminate the y^2 term by again computing $A_1B_2 - A_2B_1 = 0$:

$$(1-x)[(-x^3 - x^2 + 3x + 4)y + 2x^3 - 2x] - [(x^2 + x + 1)y - 2x^2 - x - 2](x^2 - x - 3)$$
$$= 7y + 3xy - x^2y - 6 - 7x - 3x^2 + x^3 = 0.$$

Zhu Shijie called this the *zuo shi* 左式 or the "equation on the left."[22] It would have been represented on the counting board as follows:

Zuo (Left) Equation

(B_1)	(B_2)
7	−6 *tai*
3	−7
−1	−3
0	1

(See the configuration of these coefficients at the bottom of columns g and h in the figure above.)

We now use this left equation and again the *qian shi* equation [v], seeking to eliminate the y^2 term from the latter. Rewriting the "left equation" in the form $B_1y + B_2 = 0$, i.e., $B_1 = 7 + 3x - x^2$ and $B_2 = x^3 - 3x^2 - 7x - 6$, and the *qian shi* equation as $A_1y + A_2$ with $A_1 = (1 - x)y$ and $A_2 = (x^2 + x + 1)y - 2x^2 - x - 2$, we determine $A_2B_1 - A_1B_2$ as follows:

$$A_2B_1 = (7 + 3x - x^2)(x^2y + xy + y - 2x^2 - x - 2)$$
$$= -x^4y + 2x^4 + 2x^3y + 9x^2y - 5x^3 - 15x^2 + 10xy - 13x + 7y - 14,$$
$$A_1B_2 = (1-x)y(x^3 - 3x^2 - 7x - 6) = -x^4y + 4x^3y + 4x^2y - xy - 6y.$$

The two expressions would appear on the counting board as follows:

Left (A_2B_1):		Right (A_1B_2):	
7	−14 *tai*	−6	*tai*
10	−13	−1	0
9	−15	4	0
2	−5	4	0
−1	2	−1	0

[22]See Shen Qinpei 1829, 237b. The configuration of this equation in a counting board in the original text by Shen has some mistakes. His calculations actually correspond to the equation $-x^3 + 3x^2 + x^2y + 7x + 3xy + 7y + 6$.

Now $A_2B_1 - A_1B_2 = 0$ may be determined directly as

$$(-x^4y + 2x^4 + 2x^3y + 9x^2y - 5x^3 - 15x^2 + 10xy - 13x + 7y - 14)$$
$$-(-x^4y + 4x^3y + 4x^2y - xy - 6y)$$
$$= 2x^4 - 2x^3y - 5x^3 + 5x^2y - 15x^2 + 11xy - 13x + 13y - 14 = 0.$$

We now put this equation in the form $A_1y + A_2 = 0$, with $A_1 = -2x^3 + 5x^2 + 11x + 13$, and $A_2 = 2x^4 - 5x^3 - 15x^2 - 13x - 14$. Zhen Shijie called this the *you shi* 右式 or "equation on the right."[23] On the counting board, this equation would have been represented, along with the *zuo shi* earlier, as

Zuo (Left) Equation		*You* (Right) Equation	
(B_1)	(B_2)	(A_1)	(A_2)
7	−6 *tai*	13	−14 *tai*
3	−7	11	−13
−1	−3	5	−15
0	1	−2	−5
		0	2

(See the configuration of the *You* Equation coefficients at the top of columns i and j in the figure above.)

It is now possible to eliminate the y terms from the "left" and "right" equations using the same elimination procedure as before. Multiply the two "inner" columns, namely the right column of the *zuo shi* (B_2) and the left column of the *you shi* (A_1):

$$A_1B_2 = (-6 - 7x - 3x^2 + x^3)(13 + 11x + 5x^2 - 2x^3)$$
$$= -2x^6 + 11x^5 + 10x^4 - 43x^3 - 146x^2 - 157x - 78$$

On the counting board:

(A_1B_2)
−78 *tai*
−157
−146
−43
10
11
−2

(See the configuration of these coefficients at the bottom of column i in the figure above.)

Similarly, multiply the two outer columns, namely the left column of the *zuo shi* (B_1) and the right column of the *you shi* (A_2):

$$B_1A_2 = (-x^2 + 3x + 7)(2x^4 - 5x^3 - 15x^2 - 13x - 14)$$
$$= -2x^6 + 11x^5 + 14x^4 - 67x^3 - 130x^2 - 133x - 98$$

---

[23] [Shen Qinpei 1829, 237b].

On the counting board:

$$(B_1A_2)$$

–98 *tai*
–133
–130
–67
14
11
–2

(See the configuration of these coefficients in the middle of column j in the figure.)

(Note that there is an error in the text, doubtless a mistake made in preparing the original manuscript for printing, since the printed text lists the constant term as "–99 *tai*," not "–98." But "–98" is correct, as the final solution to this problem bears out.)

Determining $A_1B_2 - B_1A_2 = 0$ results in a single equation in only one variable,

$$\begin{aligned}&(-2x^6 + 11x^5 + 10x^4 - 43x^3 - 146x^2 - 157x - 78)\\&- 2x^6 + 11x^5 + 14x^4 - 67x^3 - 130x^2 - 133x - 98\\&= -4x^4 + 24x^3 - 16x^2 - 24x + 20 = 0,\end{aligned}$$

which may be simplified as $x^4 - 6x^3 + 4x^2 + 6x - 5 = 0$.

On the counting board, this last equation would assume the following configuration:

–5 *tai*
6
4
–6
1

(See the configuration of these coefficients at the top of column l in the figure.)

Zhu Shijie notes that by solving this fourth-degree equation, the solution for Problem 3 follows directly, namely that $x = 5$, that is, the length of the hypotenuse is 5 *bu*.

Problem 4

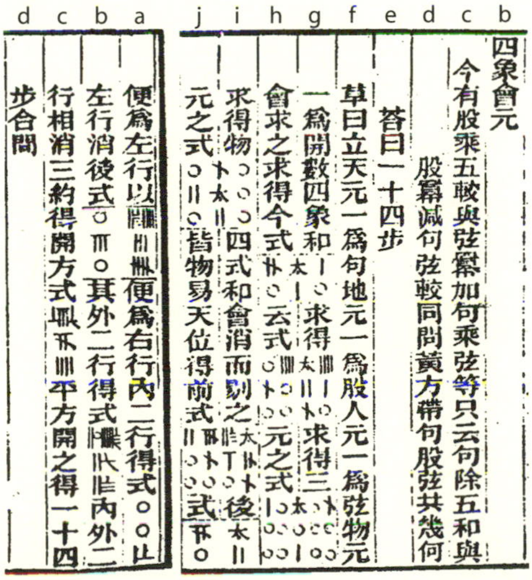
四象會元
今有股乘五較與弦冪加句乘弦等只云句除五和與
股冪減句弦較同問黃方帶句股弦共幾何
荅曰一十四步
草曰立天元一爲句地元一爲股人元一爲弦物元
一爲開數四象和會求得
會求之求得今式
便爲左行以
便爲右行內二行得式○止
左行消後式
其外二行得式
內外二
行相消三約得開方式
平方開之得一十四
步合問

Problem 4 from the edition of Zhu Shijie's *Precious Mirror*, reproduced in [Guo 1993, vol. 1, 1210a–b]. The title for Problem 4 is given at the top of column b (from 1210a): *Si xiang hui yuan*. The statement of the problem is given in columns c–d (from 1210a). Column e gives the answer, 14 *bu*, and columns f–j on page 1210a and columns a–d on page 1210b are the "working" of the problem.

The product of the five differences and the altitude equals the sum of the square of the hypotenuse and the product of the base and hypotenuse. It is also given that the quotient of the five sums and the base equals the square of the altitude minus the difference of the hypotenuse and the base. Find the *huang fang* plus the sum of the base, altitude and hypotenuse. Answer: 14 *bu*.

EXPLANATION: Zhu Shijie called this problem *Si xiang hui yuan*, which may be translated as "Four Elements Gathered Together." Guo Shuchun and and Guo Jinhai call it "Simultaneousness of the Four Phenomena," reflecting the fact that this problem deals with four simultaneous equations in four unknowns [see Guo and Guo 2005, xx]. A detailed analysis of this problem is given in [Martzloff 1997, 265–271]. See as well [Mikami 1913, 92–98], [Hoe 1997, 211–247], and [Kong 1999, 378–379].

The problem once again begins with a set of given data resulting in four equations that are determined in terms of what the text only refers to as the "five differences," the "five sums," and the *huangfang*. In a right triangle with base a, altitude b, and hypotenuse c, the "five differences" are $b-a$, $c-a$, $c-b$, $c-(b-a)$, and $(a+b)-c$; the five sums are $a+b$, $a+c$, $b+c$, $c+(b-a)$, and $(a+b)+c$; while the *huang fang* (yellow square) is $a+b-c$. Jean-Claude Martzloff explains that this is the length of the diameter of the circle inscribed in the right triangle and tangent to its three sides. See [Hoe 1977, 37], [Martzloff 1997, 266], and problem IX-16 of the *Nine Chapters*.

Thus the three given equations are

[i data]: $b(b-a+c-a+c-b+c-(b-a)+(a+b)-c) = c^2 + ac,$

[ii data] $(a+b+b+c+c+a+c+(b-a)+(a+b)+c)/a = b^2 - (c-a),$

[iii data] $a^2 + b^2 = c^2$ (the *gou-gu* or "Pythagorean" relation).

The problem asks to find the sum of the *huang fang* plus the sum of $a+b+c$, i.e.

[iv data] $(a+b-c)+(a+b+c).$

In working out the problem the *tian yuan* element x represents the base, the *di yuan* element y the altitude, the *ren yuan* element z the hypotenuse and the *wu yuan* element u represents what is to be found, i.e., [iv].

As the general configuration of the counting board displayed at the beginning of this section shows, the u terms now occupy the upper half of the board from the central *tai*. Direct substitutions for $x=a$, $y=b$, and $z=c$ result in the following equations reflecting the given data; the corresponding configurations of their coefficients as they would have appeared on the counting board are given to the right of each equation:

[i] $x - 2y + z = 0$ i.e.

−2	*tai*	1
0	1	0

(See the configuration of these coefficients at the top of columns g and h in the figure above.)

Zhu Shijie called [i] the *jin shi* 今式 or "first derived equation."

[ii] $-xy^2 + xz - x^2 + 2x + 4y + 4z = 0$ i.e.

0	4	*tai*	4
−1	0	2	1
0	0	−1	0

(See the configuration of these coefficients in the middle of columns g and h in the figure above.)

Zhu Shijie called [ii] the *yun shi* 云式 or "second derived equation."

[iii]	$x^2 + y^2 - z^2 = 0$	i.e.	1	0	*tai*	0	−1
			0	0	0	0	0
			0	0	1	0	0

(See the configuration of these coefficients at the bottom of columns g and h in the figure above.)

As before, Zhu Shijie called [iii] the *sanyun zhishi* 三元之式 or "third derived equation."

[iv]	$2x + 2y - u = 0$	i.e.	0	−1	0
			2	*tai*	0
			0	2	0

(See the configuration of these coefficients at the top of columns i and j in the figure.)

Zhu Shijie called [iv] the *wuyuan zhishi* 物元之式 or "fourth derived equation."

The first step in a lengthy process of elimination procedures begins the task of reducing the above information to two equations in only the two variables x and y. Upon inspection it is easy to see that the x term can be eliminated from [i] and [iv] by doubling [i] and subtracting [iv] from the result, i.e.,

[TOP]
$$(2x - 4y + 2z) - (2x + 2y - u)$$
$$-6y + 2z + u = 0.$$

This gives what Shen Qinpie called the *shangwei* or "top" equation.[24] The configuration for the coefficients of this equation on the counting board appear as follows:

	[TOP]	
	1	
−6	*tai*	2

Again, it is clear upon inspection that the x^2 term can be removed from equations [ii] and [iii] by simply adding the two together:

$$(-xy^2 + 4y - x^2 + 2x + xz + 4z) + (x^2 + y^2 - z^2) = 0, \quad \text{or}$$

[*]
$$-xy^2 + 2x + zx + 4y + 4z + y^2 - z^2 = 0.$$

On the counting board:

1	4	*tai*	4	−1
−1	0	2	1	0

It is now possible to remove the x terms from this equation and the *wuyuan* equation [iv] by the usual process of elimination. Consider [*] in the form $A_1x + A_2 = 0$, with $A_1 = -y^2 + z + 2$, and $A_2 = 4y + 4z + y^2 - z^2$. Likewise, consider the *wuyuan* equation [iv] in the form

[24] [Shen Qinpei 1829, 238a].

$B_1x + B_2 = 0$, with $B_1 = 2$, and $B_2 = 2y - u$. To eliminate the x term it is necessary to determine $A_1B_2 - B_1A_2 = 0$:

$$A_1B_2\text{:}\ \ (-y^2 + z + 2)(2y - u) = -2y^3 + y^2u + 2zy - zu + 4y - 2u,$$
$$B_1A_2\text{:}\ \ (4y + 4z + y^2 - z^2)(2) = 8y + 8z + 2y^2 - 2z^2$$

So $A_1B_2 - B_1A_2 = 0$ is an equation from which the x terms have been eliminated, what Shen Qinpie called the *zhongwei* or "middle" equation:

$$\text{[MIDDLE]} \qquad -2y^3 - 4y + y^2u + 2yz - 2u - uz - 8z + 2z^2 - 2y^2 = 0.$$

On the counting board:

[MIDDLE]

	1	0	−2		−1	
				2		
−2	−2	−4	*tai*		−8	2

We now seek to eliminate the x terms from the *sanyuan* equation [iii] and the *wuyuan* equation [iv]. In the form $A_1x + A_2 = 0$, for [iii] $A_1 = x$ and $A_2 = y^2 - z^2$, and again, for [iv], $B_1 = 2$, and $B_2 = 2y - u$. Thus to eliminate the first x^2 term, it suffices to determine $A_1B_2 - B_1A_2 = 0$:

$$x(2y - u) - 2(y^2 - z^2) = 2xy - xu - 2y^2 + 2z^2 = 0.$$

Now, in order to eliminate the x terms from this equation, consider it again in the form $A_1x + A_2 = 0$, i.e., $A_1 = 2y - u$, and $A_2 = 2z^2 - 2y^2$. The elimination now proceeds by considering this and the *wuyuan* equation [iv] again as above. $A_1B_2 - B_1A_2 = (2y - u)(2y - u) - (2)(2z^2 - 2y^2)$. This is what Shen Qinpei called the *xiawei*, or "bottom" equation, i.e.,

$$\text{[BOTTOM]} \qquad 8y^2 - 4yu + u^2 - 4z^2 = 0.$$

On the counting board:

[BOTTOM]

		1		
	−4	0		
8	0	*tai*	0	−4

To recap, we now have three equations in three unknowns, from which the x terms have all been eliminated:

$$\text{[TOP]} \qquad -6y + 2z + u = 0,$$
$$\text{[MIDDLE]} \qquad -2y^3 - 4y + y^2u + 2yz - 2u - uz - 8z + 2z^2 - 2y^2 = 0,$$
$$\text{[BOTTOM]} \qquad 8y^2 - 4yu + u^2 - 4z^2 = 0.$$

Beginning with the "top" and "middle" equations, in order to eliminate the terms containing z consider the top equation in the form $A_1z + A_2 = 0$, i.e., $A_1 = 2$ and $A_2 = u - 6y$; for the middle equation in the form $B_1z + B_2 = 0$, we have $B_1 = 2y - u - 8 + 2z$, and $B_2 = -2y^3 - 4y + y^2u - 2u - 2y^2$. We begin the elimination of z by considering $A_1B_2 - B_1A_2 = 0$:

$$A_1B_2\text{:}\ \ 2(-2y^3 - 4y + y^2u - 2u - 2y^2) = -4y^3 - 8y + 2y^2u - 4u - 4y^2,$$
$$B_1A_2\text{:}\ \ (2y - u - 8 + 2z)(-6y + u) = -12y^2 + 8yu - u^2 + 48y - 8u - 12yz + 2zu.$$

Thus $A_1B_2 - B_1A_2 = 0$ is $-4y^3 + 8y^2 - 56y + 2y^2u + 4u - 8yu + u^2 + 12yz - 2uz = 0$. Now, considering this equation in the form $B_1z + B_2 = 0$, we have $B_1 = 12y - 2u$ and $B_2 = -4y^3 + 8y^2 - 56y + 2y^2u + 4u - 8yu + u^2$. It is now possible to eliminate the z terms from this equation by using the top equation and again determining $B_1A_2 - A_1B_2 = 0$, where as above $A_1 = 2$ and $A_2 = u - 6y$.

$$B_1A_2: \quad (12y - 2u)(-6y + u) = -72y^2 + 24yu - 2u^2, \quad \text{and}$$
$$A_1B_2: \quad 2(-4y^3 + 8y^2 - 56y + 2y^2u + 4u - 8yu + u^2)$$
$$= -8y^3 + 16y^2 - 112y + 4y^2u + 8u - 16yu + 2u^2.$$

Therefore, $B_1A_2 - A_1B_2 = 0$ is the equation $8y^3 - 88y^2 + 40yu + 112y - 4u^2 - 4y^2u - 8u = 0$. This would have appeared on the counting board as follows:

			−4
	−4	40	−8
8	−88	112	*tai*

To simplify the continuation of this problem on the counting board, the coefficients in the positions of *wu* on the upper part of the counting board are repositioned to their corresponding positions in the *tian* area on the lower part of the counting board, which means substituting x for u. After simplifying all terms by dividing by the common factor of 4, the following now gives the configuration for this equation on the counting board:[25]

2	−22	28	*tai*
	−1	10	−2
			−1

(See the configuration of these coefficients near the bottom of columns i and j in the figure above.)

Zhu Shijie called this the *qianshi* or the "equation in front." Martzloff refers to this as the "configuration 'before'" [Martzloff 1997, 270], which may now be rewritten as follows:[26]

$$\text{[FRONT]} \qquad 2y^3 - 22y^2 + 10yx + 28y - x^2 - y^2x - 2x = 0.$$

Having successfully eliminated the z terms from the "top" and "middle" equations, consider now the "top" and "bottom" equations.

$$\text{[TOP]} \qquad -6y + 2z + u = 0,$$
$$\text{[BOTTOM]} \qquad 8y^2 - 4yu + u^2 - 4z^2 = 0.$$

As before, for the top equation in the form $A_1z + A_2 = 0$, we have $A_1 = 2$ and $A_2 = u - 6y$; for the bottom equation in the form $B_1z + B_2 = 0$, we get $B_1 = -4z$, and $B_2 = 8y^2 - 4yu + u^2$. To eliminate the z term, the process proceeds by determining $A_1B_2 - B_1A_2 = 0$:

$$A_1B_2: \quad 2(8y^2 - 4yu + u^2),$$
$$B_1A_2: \quad -4z(-6y + u).$$

[25]The configuration of this equation in the original text of the *Siyuan yujian xicao* contains two mistakes. The layout there for the top row is 2 −8 28 *tai* ($2y^3 - 8y^2 + 28y$) and the second row below this is −1 6 −2 ($-1xy^2 + 6xy - 2x$) (as in the figure above; see also [Shen Qinpei 1829, 238a]). The equation is not corrected in [Mikami 1913, 93], [Lam 1982, 268], or in [Hoe 1977, 218, 241]; the correct coefficients, as above, for the y^2 coefficient (−22, NOT −8), and for the xy coefficient (10, NOT 6) are given in [Martzloff 1997, 270]. The counting board configuration as given in the text of the *Siyuan yujian* corresponds to $-x^2 - xy^2 + 6xy - 2x + 2y^3 - 8y^2 + 28y = 0$ [Zhu Shijie 1303, 1210a].

[26][Zhu Shijie 1303, 1210a].

Consequently, $A_1B_2 - B_1A_2 = 0$ is $16y^2 - 8yu + 2u^2 - 24yz + 4uz = 0$.

To eliminate the z term from this equation, write it in the form $B_1z + B_2 = 0$, with $B_1 = 4u - 24y$ and $B_2 = 16y^2 - 8yu + 2u^2$. Again, in consideration with the top equation as before, the z term can be eliminated by determining $A_1B_2 - B_1A_2 = 0$:

$$A_1B_2: \quad 2(16y^2 - 8yu + 2u^2) = 32y^2 - 16yu + 4u^2,$$
$$B_1A_2: \quad (4u - 24y)(-6y + u) = 144y^2 - 48yu + 4u^2.$$

It now follows that $A_1B_2 - B_1A_2 = 0$ is: $-112y^2 + 32yu = 0$. Simplifying this (dividing by $16y$) yields $2u - 7y = 0$. On the counting board:

$$\begin{array}{cc} & 2 \\ -7 & \textit{tai} \end{array}$$

In order to facilitate further computations on the counting board, the coefficient of u in the upper *wu* part of the counting board is moved to its corresponding place in the lower *tian* part of the counting board, (in effect substituting x for u), i.e.:

$$\begin{array}{cc} -7 & \textit{tai} \\ & 2 \end{array}$$

(See the configuration of these coefficients at the bottom of columns i and j in the figure above.)

Zhu Shijie called this the *houshi* or the "equation in back." Martzloff refers to this as the "configuration 'after'" [Martzloff 1997, 270]. The "back" and "front" equations may now be rewritten as

[BACK] $\quad 2x - 7y = 0,$

[FRONT] $\quad 2y^3 - 22y^2 + 10yx + 28y - x^2 - y^2x - 2x = 0.$

With these two equations in only the two variables x and y, it is now possible to remove the y terms through a series of by now familiar eliminations. Begin by considering the "front" equation in the form $A_1y + A_2 = 0$, with $A_1 = 2y^2 - 22y - yx + 10x + 28$, and $A_2 = -x^2 - 2x$, and the "back" equation in the form $B_1y + B_2 = 0$, where $B_1 = -7$ and $B_2 = 2x$. The y^3 term may now be eliminated by determining $A_1B_2 - B_1A_2 = 0$:

$$A_1B_2: \quad (2y^2 - 22y - yx + 10x + 28)(2x) = 4xy^2 - 44xy - 2yx^2 + 20x^2 + 56x,$$
$$B_1A_2: \quad -7(-x^2 - 2x) = 7x^2 + 14x.$$

Consequently, $A_1B_2 - B_1A_2 = 0$ is

$$(**) \qquad 4xy^2 - 44xy - 2yx^2 + 13x^2 + 42x = 0.$$

Continuing the elimination process, this time of the y^2 term, consider (**) in the form $A_1y + A_2 = 0$; thus $A_1 = 4xy - 44x - 2x^2$, and $A_2 = 13x^2 + 42x$. Once again, with the "back" equation in mind as above, the y^2 term may be eliminated by determining $A_1B_2 - B_1A_2 = 0$:

$$A_1B_2: \quad (4xy - 44x - 2x^2)(2x) = 8x^2y - 88x^2 - 4x^3,$$
$$B_1A_2: \quad -7(13x^2 + 42x) = -91x^2 - 294x.$$

Thus $A_1B_2 - B_1A_2 = 0$ is the equation $-4x^3 + 3x^2 + 8x^2y + 294x = 0$. This may be simplified by eliminating a factor of x from each term: $-4x^2 + 3x + 8xy + 294 = 0$. Zhu Shijie called this the *youshi*, or the "equation on the right." The counting board would then have this equation on the right and the "back" equation on the left:

Left			Right	
(B_1)	(B_2)		(A_1)	(A_2)
−7	*tai*		0	294 *tai*
	2		8	3
			0	−4

(See the configuration of these coefficients at the bottom of columns i and j in the figure.)

(See the configuration of these coefficients at the top of columns a and b in the figure.)

Zhu Shijie now sought to eliminate the final y term in the "right" equation as follows. Putting it into the form $A_1y + A_2 = 0$ with $A_1 = 8x$ and $A_2 = -4x^2 + 3x + 294$, the y term may now be eliminated by determining $B_1A_2 - A_1B_2 = 0$. Zhu Shijie proceeds by operating with the "interior" and "exterior" columns of the left and right equations on his counting board. First Zhu Shijie determined the product of the two "inner" columns: $A_1B_2 = (8x)(2x) = 16x^2$, and explicitly gave the configuration of the coefficients on the counting board as

	0 *tai*
A_1B_2	0
	16

(See the configuration of these coefficients at the bottom of column a in the figure.)

Next, Zhu Shijie determined the product of the two "outer" columns, $B_1A_2 = -7(-4x^2 + 3x + 294) = 28x^2 - 21x - 2058$, and gave this configuration of the coefficients explicitly as they would have appeared on the counting board as well:

	−2058 *tai*
B_1A_2	−21
	28

(See the configuration of these coefficients at the bottom of column b in the figure.)

Zhu Shijie then explains how subtracting one from the other results in a second degree equation in the unknown whose solution is the answer to the problem as originally posed; i.e., computing $B_1A_2 - A_1B_2 = 0$ eliminates the y term and leaves $28x^2 - 21x - 2058 - 16x^2 = 0$, or, simplifying, $4x^2 - 7x - 686 = 0$. On the counting board, this is the last configuration to appear at the end of the "working" of the problem:

	−686 *tai*
$B_1A_2 - A_1B_2$	−7
	4

(See the configuration of these coefficients in the middle of column c in the figure.)

Zhu Shijie identified this as a quadratic equation that could easily be solved by known methods, with the result that the desired *huang fang* plus the sum of the base, altitude, and hypotenuse, i.e., $(a + b - c) + a + b + c = 14$ *bu*.

VIII. Matteo Ricci and Xu Guangqi, "Prefaces" to the First Chinese Edition of Euclid's *Elements* (1607)

The Chinese first learned about Western mathematics from Jesuit missionaries who appeared in China during the Ming Dynasty. The most influential for transmitting mathematical knowledge was Matteo Ricci (1552–1610), who arrived in Macao in 1582. Ricci became fluent in Chinese, and as part of the Jesuit program to convert the Chinese to Christianity, sought to impress them with the superiority of Western astronomy, mathematics, and learning in general. A key part of this strategy was the translation of Euclid's *Elements* into Chinese, which he undertook with the help of a Chinese scholar and later convert to Christianity, 徐光启 Xu Guangqi (1562–1633).

Xu was himself an important and influential member of the Ming intelligentsia. As the Grand Secretary of the Wen Yuan Institute (文渊阁 *Wen Yuan Ge*), he was said to have been considered second in China "after the monarch himself." He was certainly well-versed in mathematics, astronomy, agriculture, and even military affairs, and was especially involved in calendar reforms at the end of the Ming Dynasty [Li and Du 1987, 192–93].

Both Matteo Ricci and Xu Guangqi wrote prefaces to their joint translation of the first six books of Euclid's *Elements*. Reaction to the Ricci/Xu translation was generally positive, although most Chinese scholars were interested in the results rather than the method (the axiomatic approach seemed at times repetitive and overly complicated). Many were convinced that there could be nothing new that was not already known to ancient Chinese mathematicians, and this conviction soon encouraged a revival of interest in classical Chinese mathematics which led to a systematic search for lost texts and their serious study. Some mathematicians regretted the fact that only the first six books of Euclid had been translated and even wondered if there were important secrets in the remaining books that Ricci had purposely not wanted to translate. Such concerns were put to rest when the remaining books of Euclid's *Elements* were translated into Chinese in the nineteenth century by Alexander Wylie (1815–1887) and Li Shanlan (1811–1882) [Xu 2005].

Matteo Ricci, who had also made a serious study of the ancient Chinese masters, wrote convincingly in a familiar Chinese style about the universal effectiveness of mathematics to reveal the secrets of nature. Indeed, much of the rhetoric Ricci adopts in his Preface to the *Elements* is reminiscent of prefaces to be found at the beginning of a number of the ancient Chinese mathematical classics, including those that introduce the *Zhou bi* and the *Nine Chapters*. Xu Guangqi was even more emphatic than Ricci in extolling the virtues of Euclid's *Elements*, and of mathematics generally, which he says "is the basic form of the myriad forms, the medium for a hundred schools of learning." If by the "myriad forms" Xu Guangqi meant all things in the created world, then what he says here is nothing less than his belief that the *Elements* holds the key to all of knowledge.

Matteo Ricci's "Preface" to the Chinese translation of Euclid's Elements

In their studies scholars strive to extend their knowledge to the utmost. To extend knowledge to the utmost, one should start from what is evident in order to penetrate into the principles of nature. The principles of nature are subtle and hidden, while human capacities are limited. If one does not base oneself upon what is already understood to deduce what is not yet known, how would one be able to extend one's knowledge?

My country in the Far West, although small as far as its area is concerned, by far surpasses its neighbors by the strict analytical method by which the various schools study the phenomena of nature. Because of that, books abound in which those phenomena are studied carefully. As point of departure in discussions, savants only accept what has been proved by reason, and they do not allow subjective intuitions. They are of the opinion that rigorous investigations with the help of reason are the road to knowledge, and that subjective intuitions only lead to other subjective intuitions. Knowledge can only be called thus if there is no more room for doubt, while, by their very nature, subjective guesses go together with doubt. Even when ideas which are not or only partially based on reason contain elements of truth, doubts remain, and it is always possible to refute them with other principles. They are able to lead people into believing they are right, but they are not capable of bringing people to the point where they let go all doubts. Only fully rational and crystal-clear principles relieve the heart from doubt and convince people that it is impossible to come to false conclusions, and that there are no other undermining principles.

For depth and solidity, nothing surpasses the knowledge that springs forth from the study of mathematics. The mathematician has taken upon him the task of analyzing objects and of determining their boundaries. He dissects them into separate elements in order; if he considers them in their wholeness in order to measure them, he shows how large they are. Number and Measure, in abstraction from the physical particularity of objects, lead to the school of arithmetic and that of geometry, respectively. If we consider those two categories in connection with concrete applications, we get, as far as Number is concerned, the school of Music, which studies how tones should combine to form a harmonious whole; as far as Measure is concerned, we get the school of Astronomy and of the Measurement of Time, which studies how the revolvement of the heavenly bodies can be used to measure time.

Those four major branches split into hundreds of smaller ones, as for example the one which measures the size of the cosmos: the thickness of the various heavenly spheres, the distances of the sun, the moon, and the stars, their size, the diameter of the earth, the height of mountains, hills, and towers, the depth of wells and valleys, the distance between two locations, the size and boundary of fields, palaces, and city walls, and the contents of granaries and warehouses.

Another subdivision of mathematics measures the rays of the sun to gain insight into the sequence of the seasons, the lengths of days and nights, and the hours at which the sun sets and rises. With its help geographical position can be determined, and the exact moment at which the year starts, and within the year the months and days, the equinoxes and solstices, and which years get a leap month and which months an extra day.

Yet another subdivision creates instruments which serve as a model of heaven and earth to observe the positions of the heavenly bodies, to tune the eight kinds of musical instruments, to indicate the time to make daily human life more easy and to regulate worship to the Highest.

There is also a subdivision that gives guidance to the trades and crafts that work with water, earth, stone, or wood; it thus builds city walls, raises towers, designs terraces, and erects palaces from the foundations to the ridge, digs canals, shapes water containers, and builds bridges—and all this not only graceful and beautiful to look at, but also so solid that they will still be intact after thousands of years.

In addition there is a subdivision that produces ingenious methods to move a heavy weight with a small force or to raise it in the air; to transport goods over a long distance; to improve the drainage of flooded fields and to allow the irrigation of arid land; and to let ships move both upstream and downstream. With those mechanical devices sometimes the wind is used, sometimes the stream of water; in other cases pulleys are used or cog-wheels, or use is made of a vacuum.

One subdivision is concerned with sight. Its purpose is to depict reality on the basis of differences in distance, deviation from the vertical and height, in such a way that one can represent on a flat surface the measures of cubes and spheres, enabling one to determine from a distance the measure and true shape of objects. While one draws something on a small scale, one creates the impression of something large; while one operates close at hand, one depicts something that is far away; one draws a circle and creates the illusion that it is a sphere. One is able to paint portraits with all the furrows and convexities, and buildings with shadow and light.

Geography is also a sub-department of mathematics. The purpose of geography is to make miniature representations of all the seas and mountains on earth as well as the individual countries on the continents and the islands in the sea, divided according to administrative regions. Now, if the comprehensive map is a correct representation of reality, and if the regional maps correspond with the overview map, if [the] main structures and details are in the right proportions and there is no room for mistakes and confusion, one is able to calculate the true distances on land and on sea on the basis of the scaled distances on the maps. Thus, with what is near at hand one has knowledge of what is far away, and from what is small one knows what is large. In that way one can determine routes over land and over sea without errors.

All those skills fall directly under the realm of mathematics. [Moreover], as far as all the various professions are concerned, the important principles and the subtle touches depend to a considerable extent on mathematical theory. Thus, a proper foreign policy is unthinkable without a complete knowledge of the shapes of the borders, the distances of the roads to the neighboring countries, and the broad and narrow parts of the territory. Only then is it possible, with the correct diplomatic procedures of negotiation, mutual respect, and hospitality, to avoid unsuspected circumstances, and to avoid that one fears the neighboring countries without ground, or that one underestimates them. If one is not able to calculate how much a country produces, how much money and grain are exported and imported, it is impossible to plan the government. If one knows nothing about astronomy and one has to depend on what others say, in the majority of cases one is delivered to the false promises of quacks. A farmer who does not know beforehand the various seasons of the year has nothing to go by when sowing his crops, and he cannot prepare for droughts and floods in order to safeguard the foundations of the country.

A doctor who does not know how to observe the course of the heavenly bodies, or who, investigating a patient, is not able to decide what is auspicious and what ominous, and blindly applies medicines and needles, is not only useless but might even do great harm. That is why one often sees healthy young people succumbing from insignificant diseases because the magic potion did not work. And that only because the doctor had no insight in the heavenly phenomena! If traders and. merchants are bad at calculating, they will lack insight in the trade of goods, in the calculation of interest over a sum of money, and in the differences between the various categories into which goods are divided. They will either cheat their colleagues or they will be cheated.

These are all examples of how things should not be. But of all the professions who derive their methods from mathematics it is above all the art of warfare—the basis of national security and of the major affairs of state—where the utmost accuracy is indispensable. Therefore, a wise and courageous general will be deeply convinced of the importance of the study of mathematics. To a general who isn't, his knowledge and courage will be of no use. How could a good general trust the fallible senses? In the first place, he has to be able to estimate the amount of food needed for the soldiers and the horses, the distances along the lines of march, the accessibility of the various terrains, where space is sufficient and where the passage is narrow, and the chances of avoiding losses. In the second place, he has to determine in which the battle-order can be arranged: in a circular formation to conceal the size of the army; in horn shape to make it appear larger than it is; in the shape of a new moon to encircle the enemy, or in a point to penetrate the opposing forces and to scatter them. Furthermore, he has to evaluate the efficacy of all offensive and defensive weapons in various circumstances, and he has to investigate the possibilities of improving them and supplying them with the latest novelties. After all, anyone who has read the historical records of the various countries knows that the one that develops the newest and most ingenious weapons, disposes of the means to be victorious in battle, or otherwise to conduct a safe defense.

It is a trifle to gain the victory over a small and feeble army with a strong and large army, but the reverse requires the strength of mind of an intelligent general. I have heard that about 1600 years ago, when Christianity had not yet widely spread, my country in the West was continuously threatened by a united front of neighboring countries. But there were brave warriors who were able, with a small and weakened group of soldiers, to withhold against an army that was ten times as numerous, and to defend isolated cities under siege, and to fend off attack over sea and over land; in a manner similar to that, as it is told, in China Gong Shu and Mozi repulsed each others attacks nine times. Many more examples may be adduced, and each time it was only because they brought mathematical methods into practice and in such a way became familiar with mathematical theory.

From such examples it may be apparent with how many things this art is connected. Its applications are numerous and very important. That is why, in the course of the centuries, the greatest minds in an uninterrupted stream have transmitted the theory to each other and extended it and improved upon it, with the result that it has become a rich and awesome work. That process continued until the more recent antiquity. At my university, I above all got to know one name: Euclid. He brought mathematical theory to great perfection, and he towers high above his predecessors. He has opened new avenues, and has enlightened the path for later generations. His works are numerous and they are of exceptional level. In the books he wrote during his lifetime there is not a single thing that can be doubted. Especially his *Elements* is very exact and can rightly be called a standard work. It makes perfectly clear the way mathematics hangs together and why it does. Everything is contained in his theory and there is nothing that does not follow from it. That is why those who came after him named the book after him. With his other books he already surpassed others, but with this book he surpassed himself. Who now closely takes in this book will not fail to notice how exceptional its order is.

Preceding the propositions and proofs the definitions have been laid down. After that the general principles have been formulated on which the propositions and proofs rest. Next, the propositions follow. They give an explanation of the problem, and a manner of construction or a proof. What comes later, is founded upon results that have been obtained before. The more than five hundred propositions, divided over thirteen books, unroll themselves in a straight line from beginning to end. Nowhere can the order be reversed; it is one unbroken chain. The undoubtable principles at the beginning are extremely simple and clear. Gradually also the hidden and subtle meanings reveal themselves, as for example when one suddenly grasps that the purport of a later proposition was already contained in a previous one. It is difficult to trust upon what is almost beyond one's reach, but with the preceding propositions as a basis, building layer upon layer, step by step, the meanings will be come manifest, like tense eyebrows that relax into a smile. Since times inmemorable it has not happened that an ambitious and intelligent man, with a well-trained mind, was not capable of explaining the whole book. Anyone who devotes himself to the study of mathematics should use this work as a "ladder," however excellent the intelligence with which Heaven has endowed him might be. If someone who has not yet mastered this book still wants to continue his way, this means not only that the student lacks a frame of reference to provide meaning, but also that the teacher has nothing on which to base his explanation. At my university all the books of the various branches of mathematics, of however many parts they consisted, started from this work. Every principle, every theory that is being developed, cites this book as a proof. If one wants to prove something with the help of another book, one should in every case mention its name. But in the case of this book alone, it suffices to cite: Book X, Proposition Y. It is considered as the daily food of the mathematician.

In our times another great scholar has made himself known: Master Ding [Clavius], my teacher at the university. He has further extended the field, and added many details. During my long journey over sea from the West I have visited many well-known countries. All specialists and experts with whom I conversed, told me spontaneously that, although it is not possible to see in the future, Master Ding by far exceeds previous generations. Master Ding, already completely familiar with this book, even produced a commentary upon it. Moreover, he produced two books with additions and further elaborations. Together with the original work, there are altogether fifteen books. He has also provided new proofs for the same subjects in the various books. Thereafter, the work was even more precise and complete. For those who study it after him, it is a ford in the river, a bridge, a shelter in case of danger.

From the very first moment I set foot on Chinese soil, I noticed that those who study mathematics put all trust in their manuals, and that there is no discussion on fundamental issues. Without solid roots and firm fundaments it is difficult to build something up. That is the reason why even the most eminent scholars were not capable of explaining how they had reached their conclusions. Therefore, also those who are right lack the tools to correct those who are wrong. There is no basis to decide for once and for all what is correct. When I was faced with this state of affairs, the plan arose in me to translate this book, and to offer it to the wise and noble of this age to compensate for their trust [in me].

Such were the intents of this humble traveler, but my talents are minute. The grammars of East and West vastly differ, and the meaning of words corresponds in a vague and incomplete manner. As long as one gives oral explanations, it is still possible to do one's best to find solutions, but when wielding the writing brush in order to produce a text, it becomes hard to realize. But from that moment onward I have, at every difficulty, been helped by magnanimous gentlemen. Step by step I have made progress. Oh! How much I have suffered in the laborious process of acculturation! All beginning is difficult! My desire to complete the work had to wait till this day to be fulfilled.

In the year 1601 I went to pay tribute in the capital, and I stayed in the residence for foreigners. In the spring of 1604 the great scholar Xu from Wusong arrived. Because he is a highly cultured man with a long literary training, and, moreover, we spent a lot of time together, I secretly considered that it would be the ideal solution if I could translate the book together with him. Then the occasion occurred that he, having not yet obtained the highest degree in bureaucracy, was going to partake in the Palace Examinations in order to be admitted to the Imperial Academy. On behalf of his study he often consulted me, and we had many personal conversations.

Christianity considers self-cultivation and worship as the most important, and believes that one should not give oneself over to earthly matters. I have, therefore, revealed to him everything I have learned during my training. I have answered him with the most complete explanations of natural phenomena, and I have discussed mathematical theories. In doing so, I also broached the problem of translating Euclid while keeping its essence in tact, as well as the impasse I found myself in. Thereupon Master Xu said: "The only thing that counts is knowing what one is talking about. The worst thing that can happen to a scholar is not knowing something. At present, the continuity in this branch of learning has been lost, and those who want to study it grope in the dark. Now I have been so lucky to come in contact with this book, and, moreover, you appear to be prepared to answer all questions in a magnanimous and generous manner. How could I shrink from the effort! Even if it would cost me all my spare time. If I try to escape the difficulties they only will grow larger. If I face them, they will disappear. I certainly will bring the work to a proper end." Thereafter, he has devoted himself entirely to the task which he took upon himself. I translated orally and he put it into text. We turned the meanings of the original upside down, and investigated them from all angles, in order to find the best equivalent in Chinese. One improved version followed upon another, and altogether there were three revised manuscripts. Xu encouraged me, but I still did not dare to let it rest. Not until the first month of spring, the first six, most fundamental, books had been completed. And even then, only Euclid's original text had been taken up without omissions, and master Ding's commentary had been limited to the first propositions.

The great scholar was very enthusiastic and wanted to complete the whole translation, but I said: "No, let us first circulate this in order that those with an interest make themselves familiar with it. If, indeed, it proves of some value, then we can always translate the rest." Thereupon he said: "Allright. If this book indeed is of use, it does not necessarily have to be completed by us." Thus, we stopped our translation and published it, considering that distributing it and making it public could not wait a single day.

After printing had been completed, I have summarized its general outline. and with it I have preceded the recorded text [i.e. I have written an introduction]. I do not consider myself literate. How would I dare to pose as author, and provide it with my scholarly seal? I only wanted to point out the most essential, and what motivated the translation, to let those who study it have a knowledge of its general intention, and, that they, considering the difficult circumstances, may add to it and render it more perfect to complete the beautiful work.

I hope that enlightened scholars will attentively study its solid principles, and that they jointly will cultivate the various crafts and arts that have been mentioned above. If it will be able to establish merit for, and bring benefit to, the country, then I, who have for so many years dwelt in this country and have received the benefices of a great official, shall have had the opportunity to compensate just one ten thousands of a part for the splendid bounties I have received.

Respectfully written by Li Madou from the Great West in the year dingwei of the Wanli reign period.

Xu Guangqi's "Preface" to the Chinese translation of Euclid's Elements

During the times of Tang and Yu there already were Xi and He who took care of the calendar, as well as the Minister of Works, the Minister of Agriculture, the Forester, and the Director of Ritual Music—if those five Offices would have been deprived of Measures and Numbers, it would have been impossible to fulfill their tasks. Mathematics is one of the Six Arts mentioned in the *Offices of Zhou*, and the (other) five could not have led to any results without the use of Measures and Numbers. Master Xiang and Master Kuang as far as musical sounds are concerned; Lu Ban and Mo Di as far as ingenious devices are concerned: would they have had at their disposal some other kind of trickery? They were experts in bringing [this] method into practice and that is all. Therefore I have said that far back, during the period of the Three Dynasties, those who applied themselves to this calling were immersed in a rock-solid tradition of learning that was passed on from master to pupil—but that has in the end perished in the flames of the Ancestral Dragon. Since the Han dynasty there have been many who haphazardly groped to find their way, like a blind man aiming at a target, vainly shooting in the air without effect; others followed their own judgments, based on outward appearance, like one who lights up an elephant with a candle, and, by the time he gets hold of its head has lost sight of its tail. That in our days that Way has completely disappeared, was that not unavoidable?! The *Jihe yuanben* is the Ancestor of Measures and Numbers; it is that by which one exhausts all the aspects of the square, the round, the plane and the straight, and by which one completely covers the use of compasses, carpenter's square, water-level and measuring rope. Master Li (Matteo Ricci), from his early youth onwards, during the spare time his discussion of the Way [i.e., his religious concerns] left him, has immersed himself in the study of the Arts. Now, this subject constitutes what I called "a rock-solid tradition of learning passed on from master to pupil." Moreover, his teacher, Master Ding [Clavius], is the most famous expert of his age. Therefore, he [Ricci] has gained a thorough knowledge of its [mathematics'] theories. It has already been a long time since he has dwelt with me, stammerer, and every time when the water-clock allowed during our discussions, we were bound to broach this subject: thus I asked him to translate the books on Images and Numbers into Chinese. But he said that as long as this book (Euclid) had not yet been translated, it would be impossible to discuss the other books. Thereupon, we together have translated its most important parts, 6 books in total; and after we completed our task, we went through [the text] again.

Starting from what is clearly perceptible, [this work] penetrates into what is most subtle; from what is doubtful certainty is obtained. Is not the utility of what is useless, the basis of all what is useful?! In truth it can be called "the pleasure-garden of the myriad forms, the Erudite Ocean of the Hundred Schools [of philosophy]." Although it actually has not yet been completed, yet, with it as a reference, it is already possible to discuss the other books. Secretly I have said to myself: "Two thousand years after [our] old learning has become lost, unexpectedly we have the opportunity of repairing the lacunas that have arisen through the incomplete transmission of the texts and the lack of explanations [therein] of [the period] of Tang and Yu and the three Dynasties. The benefit of that for our times is certainly not small! Let me, therefore, together with two or three kindred spirits, print and spread it." The Master [Ricci] said: "This book, as far as its usefulness is concerned, is of no less value than [the works of] the Hundred Schools. Let us hope that there will be people like Xi and He and [Lu] Ban and Mo [Di]. But even those less [gifted] will find great benefit in it: it will train people's inborn intelligence and render it refined and precise." I reflected: "Great usefulness or small usefulness—that, in fact, depends upon the person involved! It is as if entering the [magic forest of] Denglin to fell wood: ridge-poles, beams, rafters and laths are there just to be picked up according to one's liking." However, it should be born in mind, that the Teachings of the Master roughly come in three different kinds: At the highest level the cultivation of one's [moral] self and the serving of Heaven; at the lowest level the "investigation of things and the exhaustion of principles." [The study of] Numbers and Images forms a separate part of the "principles of things." On each [of these levels] [his words are] refined, true, canonical and essential—they really cannot be doubted. Moreover, his methodological explanation and analysis are indeed capable of freeing us from all doubt. I, for my part, earnestly propagate this lowest [part of his teachings], driven by the wish to give priority to what can easily be believed, and to bring people to get the clue of this text; I hope that they, observing the pattern of its design, will know that the Teachings of the Master are trustworthy and cannot be doubted. If that would happen, the benefit of this book would even be much greater! The way in which, as he (Ricci) says, the other mathematical disciples turn this

[book] into use, is explained in general outline in his own preface; I shall not discuss it here.

Xu Guangqi's Various Reflections on the Jihe yanben [Euclid's Elements]

In dedicating oneself to study, there is theory and there is practice. [For both] this book has much to offer: it can help someone who studies the theory to improve his concentration and to refine his intellect; for someone who applies himself to practice it can provide him with fixed methods, and it can bring out his creativity. Therefore, everybody should study it [this book]. I have heard that in a Western country there was an Academy with continuously thousands of students, where one was only admitted to study, after one had been asked if one completely understood this book. Why? One only wanted [to make sure] their mind was finely strung; from its ports many famous scholars sprung. Someone who is able to master this book is able to master anything; the one who loves studying this book, can study anything. In the case of all other things, someone who is capable of doing can talk about it, but also someone who is incapable of doing can discuss it [has an equal right to discuss it]. Only in the use of this book it is the case that one can talk about it if and only if one can do it; if one cannot do it, that immediately entails one cannot talk about it. Why? Because, if when discussing it there is only one small thing one has not yet understood, one immediately is out of words. How could one talk nonsense? Thus, the one who immerses himself in this book will never be without an aid in knowing how to discourse. With all other [fields] of human knowledge, there are things of which one can understand half, or ninety percent, or ten percent; only with the study of geometry is it the case that if one understands it, one understands it completely, if one does not, one is completely in the dark: there is no in-between. If one is extremely gifted but of superficial mind one's talent is of no use. But if someone is of mediocre talent but has his thoughts well-ordered and precise, then the mediocre talent is of use, and by understanding the study of geometry, his exactness will greatly increase! Therefore, to bring the people of this world to return to solid practice, this should be the way to follow.

In sum, its excellence resides in its lucidity and that's all. The usefulness of this book is very broad; it is what is very badly needed in these days. After I had finished the translation, together with a few others who shared my enthusiasm, I had it printed and distributed. Master Li composed a preface, and he was very pleased that the work was spread so promptly. Our intention was that it be made widely public, to let our present generation rapidly become familiar with it. However, those who practice it are rather few; but I have secret hope that after a hundred years everybody will practice it, and that they will then realize that they have started practicing [too] late; they will falsely say I had knowledge of the future, but how could I know in advance [it was so obvious no special gifts were needed]?

There are those who, when they for the first time see this book, are afraid it is obscure and deep and hard to understand, and they insist that I should elucidate the language. I answer them: "The principles of Measures and Numbers, basically have nothing hidden and obscure. As far as the language is concerned, if you today give it a few tries it will become perfectly clear. As long as you have not yet concentrated on it, looking at it, it may seem obscure and deep. It is like when walking through thick mountains you look in all four directions and you do not see a passage; but then you reach a spot and suddenly the passage opens itself. I would like to ask you to spend a day or ten of effort, and once you get the clue, you will realize that in all the books, from beginning to end, the language is crystal clear." The study of geometry is of great benefit in the "extension of knowledge." Firstly, if one understands this, one will realize that what one hitherto has graspingly achieved, and which one makes oneself believe is very ingenious, is all wrong; secondly, if you understand this, you will realize that what we already know is much less—immeasurably less—than what we do not yet know; thirdly, if you understand this, you will realize that your previous patterns of thought were mostly empty and floating, and intangible; [fourthly], if you understand this, you will realize that [the arguments] you previously found valid in discussions have become slippery and unstable. There are five [kinds of] people who cannot study this book: hasty people, coarse people, those satisfied with themselves, jealous people, and the arrogant. Therefore, the study of this book does not only make one's natural talent increase; it is also the foundation of virtue.

IX. Conclusion

From the earliest books written on mathematics in China, scholars were impressed by its ability to measure the immeasurable, and to account for the measuring of everything that could be encountered in the study of the natural world. But having surveyed here if only briefly and episodically the grand sweep of Chinese mathematics from the pre-Qin period down to the Ming and early Qing dynasties, is it possible to characterize Chinese mathematics in any particular way? And is it possible to conceive of Chinese mathematics in terms similar to those Geoffrey Lloyd has applied to comparisons of Greek and Chinese science generally, the former being seen as adversarial, the Chinese as authoritarian? There is no doubt of the dialectical nature of Greek science, and that its constant criticism of theories was adversarial; Chinese science of course could be in its own way authoritarian, especially with the regard for which later generations held the ancient authors and classic texts [Lloyd 1996].

And yet experience in China could certainly be adversarial—especially in a country that was often if not constantly at war. Emperors burned books, feuds were rife between rival factions at court, between competing astronomical theories, and even between mathematicians seeking better methods or solutions to particularly thorny problems, like the value of *pi* or the ratio of the diagonal to the side of the square. But like their Greek counterparts, Chinese mathematicians also appreciated the special character of mathematics and the possibility of proving the correctness of their results. And when mathematicians in China could assert the truth—the veracity of their results by unassailable arguments and methods—then it rose to the level of true authority. No one doubted the established results of the matrix solutions of the *Nine Chapters*, of the double-distance methods of the *Hai dao suan jing*, or the truth of the *gou-gu* relation.

When the "Ten Classics" were finally canonized as a standard set of official works in the Tang dynasty, the fate of mathematics was mixed. Sometimes it was encouraged, sometimes not. At certain times it was taught at the Imperial Academy, the Institute of Astronomy, or the Institute of Records. In the third year of the Xian Qing reign (658 CE), having been reestablished only a few years earlier, it was again criticized and abandoned by official decree for reasons we have already encountered:

> Since mathematics ... leads only to trivial matters and everyone specializes in their own way, it distorts facts and it is therefore decreed that it shall be abolished. ... At the beginning of the Tang a lot of the arguments in the *Mathematical Manual of the Five Government Departments* and others in the *Ten Mathematical Classics* were found to be contradictory. Li Chunfeng and others in the Imperial Academy were ordered to write commentaries, after which the Emperor decreed the new edition of the *Ten Classics* be used throughout the State Institutes. (*Records of Official Books* (656 CE), quoted from [Li and Du 1987, 105])

Apparently, the collations of Li Chunfeng and his colleagues were not so successful that mathematics was valued with respect to the certainty to which the Greeks ascribed it. But later, in the Tang dynasty, mathematics was again rehabilitated. For example, the Tang encyclopedia reports that the Imperial Academy supported thirty students devoted to mathematics. Records of 807 note students enrolled in mathematics at the universities in Chang An (Xi'an) and Luo Yong (Luo Yang), and the Tang encyclopedia also tells us who the students were:

> The scholars in mathematics are in charge of teaching the children of the military and civil officials from class eight down and those of commoners. [Li and Du 1987, 105]

Due to concerns about inconsistencies between the various ancient mathematical texts (and doubtless a conviction that there should be no disagreement about mathematical matters), when the Emperor called for an official compendium of classical works to be approved as the official texts for use at the Imperial Academy (the Guo Zi Jian) and as the basis for civil service examinations in mathematics, it was Li Chunfeng (with a retinue of other experts) who collated and annotated the "Ten Books of Mathematical Classics." These that have been surveyed here serve as a grand summary of the achievements of Chinese mathematics over the 1000 years or so spanning the period from the Han to Tang dynasties.

It should not be forgotten, however, that Chinese mathematics did not develop in isolation, in a cultural vacuum. From the time it was first introduced into China in the first century CE, Buddhism for example greatly influenced China, and along with it, mathematics as well. Not only did Buddhist mathematical and astronomical works come to China, Indian astronomers often held official positions in the state observatory where they employed such concepts as the Greek division of the circle into 360 degrees (although Chinese astronomers continued to favor a circle divided into 365¼ degrees). Tables of sines were also introduced, but again, this does not seem to have attracted the interest of Chinese mathematicians, although the symbol to denote zero—O—may have been the most significant influence of Indian mathematics in China.

Levensita, the State astronomer in the Kai Yuan reign (713–741 CE), translated the Hindu "Catching Nines Calendar" into Chinese. This included methods of computation using Indian numerals, of which the New History of the Tang dynasty said:

> The method is complicated ... and you are lucky if you get it right. (*New History of the Tang Dynasty*, quoted from [Li and Du 1987, 107])

This is testimony to the accuracy and ease with which the Chinese could carry out their own long and complicated computations using the counting board and counting rods, rather than use the methods of Indian mathematics. In addition to the zero, what was influential was the method of recording very large and very small numbers, especially such as *shun xi*, the "twinkling of an eye," and *tan zhi*, the "flick of the finger." To such notational innovations, the Song and Yuan dynasties added the very powerful results of the "tian yuan" method and the means devised not only for solving sets of simultaneous equations, but for finding solutions to equations with as many as four unknowns.

In conclusion, to what extent may Chinese mathematics be regarded as authoritarian as opposed to adversarial, to return to the dichotomy Geoffrey Lloyd considers in his efforts to contrast Greek versus Chinese science? Chinese mathematics was certainly authoritarian in its reverence for classic texts and the sages of the past. But in the best cases such as Liu Hui, this was not an uncritical reverence but a respect that also allowed room for improvement, criticism where warranted, and innovation when necessary, as in the case of the double-distance method Liu Hui proudly added in his new chapter first conceived as an extension of the *Nine Chapters*, but which later became a classic text of its own: the *Hai dao suan jing*.

However, unlike Greek mathematicians who were concerned (some might say obsessed) with certainty, and insisted on proving results in various predetermined ways, by compass and straightedge methods, or the familiar axiomatic Euclidean approach, among others, Chinese mathematicians were not relentlessly adversarial, and when it came to establishing the validity of their mathematics, they did so by examples, by analogy, by looking for categories of problems and showing how established methods suited a particular problem type. In this respect the out-in complementary method was typical, by which the equivalence of apparently dissimilar areas could be established through the obvious equivalences of all other elements in question.

In short, Chinese mathematicians were clearly concerned about justifying their methods and establishing the validity of their results. Their proofs were not axiomatic Euclidean proofs, but they were proofs nevertheless, and clearly they were certainly able to establish the truth or correctness of the solutions they proffered.

In their recent joint venture, Geoffrey Lloyd and Nathan Sivin contrast the way and the word, the *dao* and the *logos*, in a book exploring the similarities and differences between science east and west [Lloyd and Sivin 2002]. Each has its rational unity, its self-consistency. But *logos* in its way was relentless, and led the Greeks in their attempts to come to terms with the irrationality of incommensurable magnitudes, for example. This was something which the Chinese did not formulate, for the way they chose emphasized the harmoniousness of the worlds of nature and number.[27]

While the Greeks may have suffered a certain angst that their mathematics was incomplete, that there were magnitudes in geometry to which no numbers corresponded, whereby they felt forced to restrict much of mathematics to the limits of geometry, the Chinese suffered no such constraints. Not only were their computational abilities with counting rods on the counting board prodigious, but their use of matrix algebra and its corresponding manipulations were equally impressive. And Chinese geometric proofs using the out-in complementary principle were as convincing in their way as the Euclidean axiomatic approach to geometry was in the West.

Chinese mathematicians applied their skills to a diverse body of problems, proving their proficiency in geometry and algebra alike. Recently, in their book *Fleeting Footsteps*, Lam Lay

---

[27]There are historians of Chinese mathematics, however, who do hold that the Chinese knew of incommensurable ratios or quadratic irrationals, as Karine Chemla expresses it. See, among others, [Chemla 1992], [Chemla 1997/98], and [Chemla and Keller 2002].

Yong and Ang Tianse have argued that it was Chinese methods of arithmetic along with its decimal place-valued character that made their way West. When combined with Hindu numerals and Arabic refinements, it was basically a Chinese system that provided the foundations for an effective arithmetic in European hands [see in particular Lam and Ang 1992, 2004]. Likewise, Kurt Vogel has also made a strong case for the transmission from China of the double difference method extolled by Liu Hui in the *Sea Island Mathematical Classic*, which made its way from Cathay to Paris through a series of transmissions and fortuitous transformations [see Vogel 1983].

What is perhaps most compelling about the examples of Chinese mathematics presented here is how similar in tone is the wonder mathematics inspires, not only in the myriad varieties of applications it brings to all aspects of life, but how it also enables the human mind to transcend physical limitations and even measure the immeasurable. As Wang Xiaotong, who criticized Zu Geng's method of interpolation as wrong or incomplete, bragged to the Emperor about his own accomplishments in the treatment of the volumes of irregular solids:

> Your subject has now produced new methods that go further ... He progressed from level surfaces and extended that to narrow and sloping objects ... Your subject thought day and night, studying all the Classics, afraid that any day his eyes would close forever and the future not be seen; he progressed from level surfaces and extended that to narrow and sloping objects [yielding] twenty methods in all in the *Classic Continuation of Ancient Mathematics*. Please request capable mathematicians to examine the worth of the reasoning. Your subject will give a thousand gold coins for each word rejected. [Li and Du 1987, 101–102]

In this sense, the best Chinese mathematicians sought to find ever better applications of what past masters had accomplished. And it was work that could be done with a sense of certainty in its perfection, for Wang Xiaotong must have been confident that he was not about to lose thousands of gold coins for any errors that might be discovered in his results. But the spirit of mathematics was perhaps best summarized by Xu Guangqi in his reflections on the significance of the subject both East and West:

> In truth mathematics can be called the pleasure-garden of the myriad forms, the Erudite Ocean of the Hundred Schools of philosophy.

X. Appendices

X.a. Sources

Suan shu shu (*Book of Numbers and Computations*). Dauben 2006.

Zhou bi suan jing (*Mathematical Classic of the Zhou Gnomon*): Cullen 1995, 171, 174, 176–81, 182, 208–17.

Jiu zhang suan shu (*Nine Chapters on the Mathematical Art*) [Nine Chapters on Mathematical Procedures]: Shen, Crossley, and Lun 1999, selections from pp. 52–514.

Hai dao suan jing (*Sea Island Mathematical Manual*) [Sea Island Mathematical Classic]: Shen, Crossley, and Lun 1999, 539–41, 552.

Sunzi suan jing (*Mathematical Classic of Master Sun*): Lam and Ang 1992, 151–53, 165–67, 171–72, 174, 176, 178–79, 182.

Zhang Qiujian suan jing (*Mathematical Classic of Zhang Qiujian*): Lam 1997, 211–213, 215, 217–218, 220, 222–223, 224, 227, 228, 231, 234–235.

Shu shu jiu zhang (*Mathematical Treatise in Nine Sections*): Libbrecht 1973, 55–63, 328–32, 382–88, 399–400, 409; Martzloff, 1997, 233–244.

Ce yuan hai jing (*Sea Mirror of Circle Measurements*): Lam 1982, 257–59, 264.

Yigu yanduan (*New Steps in Computation*): Lam 1984, 241–142, 252–53.

Xiangjie jiuzhang suanfa (*A Detailed Analysis of the Mathematical Methods in the Nine Chapters*): Lam 1969, 83–84.

Yang Hui suan fa (*Yang Hui's Methods of Computation*): Lam 1977, 5, 11–14, 139–40, 151–53, 164–65, 172–73, 174–76, 183–85.

Si yuan yu jian (*Precious Mirror of the Four Elements*): Lam 1982, 264–69, and new translation by Dauben from the original.

Prefaces to Chinese version of Euclid's *Elements*: Engelfriet 1998, 291–293, 295–296, 454–460.

X.b. Bibliographic guides

Swetz, F. J. and Ang Tainse. 1984. "A Brief Chronological and Bilbiographic Guide to the history of Chinese Mathematics." *Historia Mathematica* 11: 39–56.

Dauben, Joseph W., Karine Chemla, Alexei Volkov, and Xu Yibao. 2000. "Chinese Mathematics." In Albert Lewis, ed., 2nd ed. of J. W. Dauben, ed., *The History of Mathematics from Antiquity to the Present (a Selective Bibliography)*. Providence: American Mathematical Society, on CD-ROM.

Lam Lay Yong. 1985. "Chinese Mathematics." In J. W. Dauben, ed., *The History of Mathematics from Antiquity to the Present (a Selective Bibliography)*. New York: Garland.

Youschkevitch, A. P. 1986. "Chinese Mathematics: Some Bibliographic Comments." *Historia Mathematica* 13: 36–38.

X.c. References

Ang Tianse, and F. J. Swetz. 1986. "A Chinese Mathematical Classic of the Third Century: The Sea Island Mathematical Manual of Liu Hi." *Historia Mathematica* 13: 99–117.

Berezkina, Èl'vira Ivanovna. 1957. "Matematika v devyati knigakh" (Mathematics in Nine Books, an annotated translation of the *Nine Chapters*, in Russian). *Istoriko-matematicheskie issledovaniya* (Studies in the History of Mathematics) 10: 427–584.

Biot, E. 1841–1842. "Traduction et examen d'un ancient ouvrage chinois intitulé: *Tcheou-Pei* littéréralement: 'Style ou signal dans une circonference'." *Journal Asiatique* 11: 593–639. "Note suppléméntaire." *Journal Asiatique* 13: 198–202.

Chemla, Karine. 1987. "Should They Read FORTRAN as if It Were English?" *Bulletin of Chinese Studies* 1(2): 301–316.

———. 1990. "De l'algorithme comme liste d'operations." *Extrême-Orient, Extrême-Occident* 12: 61–87.

———. 1997. "What is at Stake in Mathematical Proofs from Third Century China?" *Science in Context* 10: 227–251.

———. 1997/98. "Fractions and Irrationals between Algorithm and Proof in Ancient China." *Studies in History of Medicine and Science* 15(1–2): 31–54.

Chemla, Karine, and Guo Shuchun. 2004. *Les Neuf chapitres: Le classique mathématique de la Chine ancienne et ses commentaries.* Paris: Dunod.

Chemla, Karine, and Agathe Keller. 2002. "The Sanskrit *karanis*, and the Chinese *mian* (side). Computations with Quadratic Irrationals in Ancient China and India. In Yvonne Dold-Samplonius, Joseph W. Dauben, Menso Folkerts, and Benno van Dalen, eds., *From China to Paris: 2000 Years of Mathematical Transmission.* Stuttgart: Steiner Verlag, 87–132.

Chen Zaixin. 2006. *An English Translation of and Commentaries on the Precious Mirror of the Four Elements.* Ed. Guo Jinhai and Guo Shuchun. Shenyang: Liaoning jiaoyu chubanshe.

Cullen, Christopher. 1994. "How Can We Do the Comparative History of Mathematical Proof in Liu Hui and the *Zhou Bi*?" *Philosophy and the History of Science: A Taiwanese Journal* 3: 59–74.

———. 1995a. "Proof in Liu Hui and the *Zhou bi*." *Philosophy and the History of Science: A Taiwanese Journal* 4: 59–94.

———. 1995b. *Astronomy and Mathematics in Ancient China: the Zhou bi suan jing.* Cambridge: Cambridge University Press.

———. 2004. *The Suan shu shu* 筭數書 *"Writings on Reckoning."* (Needham Research Institute Working Papers 1). Cambridge: Needham Research Institute.

Dauben, Joseph W. 2007. "筭數書 *Suan Shu Shu* (A Book on Numbers and Computations). English Translation with Commentary." *Archive for History of Exact Sciences.*

Dold-Samplonius, Yvonne, Joseph W. Dauben, Menso Folkerts, and Benno van Dalen, eds. 2002. *From China to Paris: 2000 Years of Mathematical Transmission.* Proceedings of a Conference Held at the Rockefeller Foundation Research and Conference Center, Bellagio, Italy, May 2000. Boethius 46.

Eberhard, Andrea. 1999. *Re-Kreationeines mathematischen Konzeptes im chinesischen Diskurs. "Reihen" vom 1.bis zum 19. Jahrhundert.* Boethius 42. Stuttgart: Steiner Verlag.

Engelfriet, P. M. 1998. *Euclid in China: The Genesis of the First Chinese Translation of Euclid's Elements Books I–VI (Jihe Yuanben; Beijing, 1607) and Its Reception up to 1723.* Leiden: Brill.

Fu Dawei. 1991. "Why Did Liu Hui Fail to Derive the Volume of a Sphere?" *Historia Mathematica* 18: 212–38.

Gillon, B. 1977. "Introduction, Translation, and Discussion of Chao Chün-Ch'ing's Notes to the Diagrams of Short Legs and Long Legs of Squares and Circles." *Historia Mathematica* 4: 253–93.

Guo Shuchun 郭書春 ed. 1993. *Shuxue juan* (Mathematical volumes), in the series *Zhongguo kexue jishu dianji tonghuied*, ed. Ren Jiyu. Zhengzhou: Henan jiaoyu chubanshe. In 5 vols.

Guo Shuchun 郭書春, and Liu Dun 劉鈍 (1998), *Jiaodian Suan jing shi shu* 校点《算經十书》 (Collation of the *Ten Mathematical Classics*). Shenyang: Liaoning jiaoyu chubanshe.

van Hee, Louis. 1920. "Le Hai-Tao Souan-King de Lieou." *T'oung Pao* 20: 51–60.

———. 1932. "Le classique de l'île maritime, ouvrage chinois du IIIe siècle." *Quellen und Studien zur Geschichte der Mathematik*, Abteilung B, 2: 255–280.

Ho Peng Yoke. 1965. "The Lost Problem of the *Chang Ch'iu-chien Suan Ching*, a Fifth-Century Chinese Mathematical Manual." *Oriens Extremus* 12: 37–53.

Hoe, John [Jock]. 1977. *Les systèmes d'équations-polynômes dans le Siyuan yujian (1303).* Mémoires de l'Institut des Hautes Edudes Chinoises 6. Paris: Collège de France, Institut des Hautes Etudes Chinoises.

Juliano, Annette L., and Judith A. Lerner, eds. 2001. *Monks and Merchants: Silk Road Treasures from Northwest China.* New York: Asia Society and Harry N. Abrams.

Katz, Victor J. 1998. *A History of Mathematics: An Introduction.* 2nd ed. Reading, MA: Addison-Wesley-Longman.

Kong Guoping. 1999. *Li Ye Zhu Shijie yu jinyuan shuxue* (Li Ye, Zhu Shijie and Mathematics in the Jin and Yuan Dynasties) Heibei: Hebei Science and Technology Publishing House.

Lam Lay Yong. 1969. "On the Existing Fragments of Yang Hui's *Hsiang Chieh Suan Fa.*" *Archive for History of Exact Sciences* 6(1): 82–88.

———. 1977. *A Critical Study of the Yang Hi Suan Fa, a Thirteenth-Century Chinese Mathematical Treatise.* Singapore: Singapore University Press.

———. 1980. "The Chinese Connection Between the Pascal Triangle and the Solution of Numerical Equations of any Degree." *Historia Mathematica* 7: 407–424.

———. 1982. "Chinese Polynomial Equations in the Thirteenth Century." In Li Guihao et al., eds., *Explorations in the History of Science and Technology in China.* Shanghai: Chinese Classsics Publishing House, 231–72.

———. 1987. "The Earliest Negative Numbers. How They Emerged from a Solution of Simultaneous Linear Equations." *Archives internationals d'histoire des sciences* 37: 222–262.

———. 1994. "*Jiuzhang Suanshu*—An Overview." *Archive for History of Exact Sciences* 47: 1–51.

———. 1997. "Zhang Qiujian Suanjing (The Mathematical Classic of Zhang Qiujian): An Overview." *Archive for History of Exact Sciences* 50: 201–240.

Lam Lay Yong, and Ang Tiansi. 1984. "Le Ye and his *Yi Gu Yan Duan.*" *Archive for History of Exact Sciences*, 29(3): 237–265.

———. 1986. "Circle Measurements in Ancient China." *Historia Mathematica* 13: 325–40.

———. 1992, 2004. *Fleeting Footsteps: Tracing the Conception of Arithmetic and Algebra in Ancient China* (including a translation and annotation of *Master Sun's Mathematical Manual*). Singapore: World Scientific.

Lam Lay Yong and Shen Kang Shen. 1984. "Right-Angled Triangles in Ancient China." *Archive for History of Exact Sciences* 30: 87–112.

———. 1986. "Mathematical Problems on Surveying in Ancient China," *Archive for History of Exact Sciences* 36: 1–20.

———. 1989. "Methods of Solving Linear Equations in Traditional Chinese Mathematics." *Historia Mathematica* 16: 107–22.

Li Yan, and Du Shiran. 1987. *Chinese Mathematics: A Concise History*, Trans. John N. Crossley and Anthony W.-C. Lun. Oxford: Clarendon Press.

Libbrecht, Ulrich. 1972. "The Chinese *Ta-yen* Rule: A Comparative Study." *Orientalia Louvaniensa* 3: 179–99.

———. 1973. *Chinese Mathematics in the Thirteenth Century: The Shu-shu chiu-chang of Ch'in Chiu-shao.* Cambridge, MA: MIT Press.

Liu Dun 劉鈍. 1989. "Cong Xu Guangqi dao Li Shanlan" 從徐光啓到李善蘭 (From Xu Guanqi to Li Shanlan). *Ziran Bianzhenfa Tongxun* 自然辯證法通迅 (*Journal of Dialectics of Nature*, Beijing) 11(3): 55–63.

———. 1991. "*Shuli Jingyun* zhong de *Jihe Yuanben* de diben wenti"《數理精蘊》中的《幾何原本》的底本問題 (Original sources of the *Elements* published in the *Essence of Mathematics*). *Ziran Kexueshi Yanjiu* 自然科學史研究 (*Studies in the History of Natural Sciences*, Beijing) 12(3): 88–96.

Lloyd, Geoffrey E. R. 1996. *Adversaries and Authorities: Investigations into Ancient Greek and Chinese Science.* New York: Cambridge University Press.

Lloyd, Geoffrey E. R., and Nathan Sivin. 2002. *The Way and the Word: Science and Medicine in Early China and Greece.* New Haven: Yale University Press.

Loewe, Michael. 1974. *Crisis and Conflict in Han China, 104 BC to AD 9.* London: Allen & Unwin.

Luo Zhufeng 羅竹風, ed. 1991. *Hanyu dacidian* 漢語大詞典 (*Grand Dictionary of Chinese Characters and Vocabularies*), vol. 8. Shanghai, Hanyu dachidian chubanshe.

Martzloff, J.-C. 1993. "Note sur les traductions chinoises et mandchoues des *Eléments d'Euclide* effectuées entre 1690 et 1723." In *Actes du Ve Colloque International de Sinologie de Chantilly, 15–18 septembre 1986*, Ricci Institute, Institut Ricci, Taipei-Paris, 201–212.

———. 1997. *A History of Chinese Mathematics*, Trans. Stephen S. Wilson. New York: Springer-Verlag.

Mikami, Yoshio. 1910. "The Circle-Squaring of the Chinese." *Bibliotheca Mathematica* 10: 193–200.

———. 1913, 1974. *The Development of Mathematics in China and Japan.* Leipzig: Teubner; repr. New York: Chelsea.

Needham, Joseph, ed. 1959. *Science and Civilisation in China*, vol. 3, *Mathematics.* Cambridge: Cambridge University Press.

Nòda, C. 1933. *An Enquiry Concerning the Chou Pei Suan Ching.* Toho Bunka Gakuin Kyoto Kenkyusho Memoirs No. 3. Kyoto: Academy of Oriental Culture, Kyoto Institute.

Qian Baocong 錢寶琮, ed. 1963. *Suanjing shishu* 算經十書 (*Ten Mathematical Classics*). Beijing: Zhonghua shuju.

Qian Baocong 錢寶琮, ed. 1964, 1981. *Zhongguo shuxue shi* 中國數學史 (*A History of Chinese Mathematics*). Beijing, Kexue chubanshe.

Raphals, Lisa. 2002. "When is a Triangle Not a Triangle?" *Ex/Change: Newsletter of Centre for Cross-Cultural Studies* (City University of Hong Kong) 5 (September): 9–11.

Reifler, E. 1965. "The Philological and Mathematical Problems of Wang Man's Standard Grain Measures, the Earliest Chinese Approximation to π." *Jubilee Volume in Honour of Dr. Li Chi,* vol. 1 Taiwan, 387–502.

Shen Kang Shen. 1987. "Parallelism Between Chinese and Indian Mathematics." *Bulletin of Chinese Studies* (Hong Kong University) 2: 171–211.

———. 1988. "Historical Development of the Chinese Remainder Theorem." *Archive for History of Exact Sciences* 39: 285–305.

Shen Kangshen, John N. Crossley and Anthony W.-C. Lun. 1999. *The Nine Chapters on the Mathematical Art: Companion and Commentary.* Oxford: Oxford University Press, and Beijing: Science Press.

Shen Qinpei. 1829. *Siyuan yujian xicao* (Detailed Solutions for the Precious Mirror of the Four Elements). Photo-reproduced in [Guo 1993, vol. 5, 227–432].

Smith, David E. 1931. "Unsettled Questions Concerning the Mathematics of China." *Scientific Monthly* 33: 244–250.

Swetz, Frank J. 1992. *The Sea Island Mathematical Manual; Surveying and Mathematics in Ancient China.* University Park: Pennsylvania State University Press.

Swetz, Frank J., and T. I. Kao. 1977. *Was Pythagoras Chinese? An Examination of Right Triangle Theory in Ancient China.* University Park: Pennsylvania State University Studies No. 40.

Trésors du Musée national du Palais, Taipei. 1998. Special issue *Connaissance des Arts* 127. Catalogue of the exhibition at the Grand Palais, Paris.

Vogel, Kurt. 1968. *Neun Bücher arithmetischer Technik.* Braunschweig: Vieweg.

——— 1983. "Ein Vermessungsproblem reist von China nach Paris." *Historia Mathematica* 10: 360–367. English translation by J. W. Dauben and Benno van Dalen, in [Dold-Samplonius et al., 2002, 1–8].

Volkov, Alexei K. 1994. "Supplementary Data on the Value of π." *Philosophy and the History of Science: A Taiwanese Journal* 3: 95–110.

Wagner, Donald B. 1978a. "Doubts Concerning the Attribution of Liu Hui's Commentary on the *Chiu-Chang Suan-Shu.*" *Acta Orientalia* 39: 199–212.

———. 1978b. "Liu Hui and Tsu Keng-chih on the Volume of a Sphere." *Chinese Science* 3: 59–79.

———. 1979. "An Ancient Chinese Derivation of the Volume of a Pyramid: Liu Hui, Third Century AD." *Historia Mathematica* 6: 164–188.

———. 1985. "A Proof of the Pythagoras Theorem by Liu Hui (third century AD)." *Historia Mathematica* 12: 72–73.

Wang Ling, and Joseph Needham. 1955. "Horner's Method in Chinese Mathematics: Its Origins in the Root-Extraction Procedures of the Han Dynasty." *T'oung Pao* 43: 345–401.

Wang Yusheng 王渝生. 1994. Jihe Yuanben tiyao《幾何原本》提要 (An Introduction to the Chinese *Elements*). In Guo Shuchun 郭書春, ed., *Zhongguo kexue jishu dianji tonghui: Shuxue juan* 中國科學技術典籍通匯: 數學卷 (*A Compendium of Chinese Classics of Science and Technology: Mathematics Section*), vol. 5. Zhenzhou, Henan jiaoyu chubanshe, 1145–1150.

Wu Wenchun [Wu Wenjun]. 1983. "The Out-In Complementary Principle." In *Ancient China's Technology and Science.* Beijing: Institute of the History of Natural Sciences, 66–89.

Wu Wen-Tsun [Wu Wenjun]. 2000. *Mathematics Mechanization: Mechanical Geometry Theorem-Proving, Mechanical Geometry Problem-Solving, and Polynomial Equations-Solving.* Beijing: Science Press, and Dordrecht: Kluwer Academic Publishers.

Wylie, A., and Li Shanlan 李善蘭. 1857. *Jihe yuanben* 幾何原本 (*The Elements*, Books VII–XV). First printed in 1857 in Shanghai under the patronage of Han Yingbi 韩應陛. The translation was reprinted again in 1865 in Nanjing under the patronage of Ceng Guofan 曾國藩, together with the first six Books of the Chinese translation by Matteo Ricci and Xu Guangqi, which was first published in 1607.

Xu Yibao. 2005. "The First Chinese Translation of the Last Nine Books of Euclid's *Elements* and its Source." *Historia Mathematica* 32(1): 4–32.

Yan Dunjie 嚴敦傑. 1943. Oujilide jiheyuanben Yuandai shuru Zhongguo shuo 歐幾里得幾何原本元代輸入中國說 (Euclid's *Elements* had been introduced to China in the Yuan dynasty). *Dongfang zazhi* 東方雜誌 (*The Eastern Miscellany*) 39(13): 35–36.

Zhu Shijie, *Si yuan yu jian* (Precious Mirror of the Four Elements). 1303. In [Guo 1993, vol. 1, 1205–1280].

4 Mathematics in India

Kim Plofker

I. Introduction: Origins of Indian Mathematics

Most of the mathematics we know as "Indian" (originating in the region now called South Asia, comprising modern India, Pakistan, Nepal, Bangladesh, and Sri Lanka) was recorded in Sanskrit, the language of the *Bhagavad Gītā*, the Upaniṣads, and other famous Indian scriptures. The earliest surviving form of Sanskrit, descended from a very ancient Indo-Iranian language, was spoken by people living in the Punjab area sometime around the middle of the second millennium BCE. Other inhabitants of the Indian subcontinent, with different linguistic and cultural heritage, nevertheless blended into the language and lifeways of this "Sanskritic" culture, which eventually spread throughout the region.

Much of the intellectual legacy of this period is impossible to reconstruct with certainty. There are very few absolute chronological markers in the early works, and there is very little trace of them in archaeological evidence—unlike the wealth of original documents preserved in Babylonian clay tablets, for example. There is no definite evidence of writing in Sanskrit, or the vernacular dialects related to it, before about the middle of the first millennium BCE. (The Indus Valley Civilization in northern India had used inscribed symbols many centuries earlier, but their culture had dwindled before the middle of the second millennium. Their sign system has not yet been deciphered, and we do not know whether it influenced literacy in Sanskrit and its relatives.) The writing that survives from this early time is mostly in the form of inscriptions carved in stone: very few other writing materials have been preserved through such long periods in the subtropical Indian climate. Even if there was a common practice of writing on more perishable materials before the engraving of formal inscriptions came into fashion, we are not likely to find much evidence of it.

Moreover, the users of Sanskrit valued it predominantly as a spoken rather than a written language. In fact, writing may have initially been restricted to the vernaculars or non-Sanskrit dialects: the earliest preserved inscriptions and documents are mostly in languages other than Sanskrit. The first known texts in a form of Sanskrit are the "Vedas" (literally "knowledge"), a canon of hymns, invocations, and procedures for religious rituals. The Vedic texts (generally composed in verse or in short prose sentences called *sūtras* to make them easier to memorize) were carefully learned, recited, and handed down orally. The works thus painstakingly preserved were, naturally, the ones most important for the crucial activity of performing the prescribed religious sacrifices. Besides the Vedic hymns and other sacred knowledge, these included later canonical texts on the six subjects called the "limbs of the Vedas":

1. phonetics, by which the archaic Sanskrit of the invocations would be correctly pronounced;
2. grammar, by which its sentences would be understood;
3. metrics, by which the structure of its verses would be preserved;
4. etymology, which explained its vocabulary;
5. astronomy or calendrics, which regulated the daily, monthly, and seasonal timing of the sacrifices; and
6. ritual practice, which ensured the continuity and correctness of the sacrificial tradition.

It is in the ancient Vedas and limbs of the Vedas that we must look for hints of the earliest Indian mathematics. Of course, these texts can reveal only scattered fragments of this era's wider mathematical interests. Remnants of material culture found by archaeologists suggest standards of weights and measures, as well as "ethnomathematical" awareness of, e.g., geometric patterns in ornament. Sanskrit literature, rich though it is—including in later periods whole genres devoted to mathematical topics—can tell us only part of the full story of mathematical thought in India.

II. Mathematical Texts in Ancient India

II.a. The Vedas

The variations in the linguistic style and content of the ancient texts have led scholars to conclude that most of the hymns in the *Ṛg Veda* ("Praise-Knowledge") are the oldest Vedic compositions. After these come the ritual procedures in the *Yajur Veda* ("Sacrifice-Knowledge"), the *Sāma Veda* ("Song-Knowledge") describing the chant melodies used in the rituals, and the *Atharva Veda* ("Knowledge of [the priest] Atharvan") containing prayers, charms, curses, etc. Subsidiary texts such as the Brāhmaṇas and Upaniṣads were composed to expound the ritual details and philosophical principles of the Vedas.

The hymns of the *Ṛg Veda* mention numbers frequently, usually referring to the achievements or lavish gifts of the gods: "sixty thousand horses," "ten thousand cows" (*Ṛg Veda* VIII 46.22); "[The god] Indra destroyed the ninety and nine cities of [the demon] Śambara" (*Ṛg Veda* II 19.6). Some parts of other Vedic texts focus on numbers or the cosmic significance of numbers. For example, one sacrificial ritual in the *Yajur Veda* invokes not only gods and various creatures, but also sets of numbers: "Praise to one, praise to two, praise to three ... praise to nineteen, praise to twenty-nine, praise to thirty-nine ... praise to a hundred, praise to two hundred, praise to all. Praise to two, praise to four, praise to six ..." and so on (*Yajur Veda* VII 2.11–20). One of these sets lists successive powers of ten from a hundred up to a trillion! And ritual was related to the structure of the universe partly by means of number: e.g., there is an eight-day ritual because "there are four [cardinal] directions and four intercardinal directions" (*Yajur Veda* VII 2.3); in the case of a twelve-day ritual, "twelve months are a year and a year is [the creator god] Prajāpati" (*Yajur Veda* VII 2.10).

Possibly some of these numerological ideas may indicate Mesopotamian influence, although there is no definite evidence of transmission. For instance, Prajāpati is said to have successively divided the altar of 720 bricks representing his own body (the days and nights of a 360-day ideal year) into two parts, then three, then four, and so on. But "he did not divide into seven," nor into eleven, thirteen, fourteen, etc. (*Śatapatha Brāhmaṇa* X 4.2)—a pattern rather reminiscent of ancient Babylonian reciprocal tables, where all divisors of 60 that don't produce finite sexagesimal fractions are omitted. (The number 720 itself is the common constant of Babylonian metrology known as the "brick sar.")

(Readers who have heard of the currently popular computation algorithms called "Vedic mathematics" may be wondering where they come into this history. Actually, these algorithms are based on the 1965 book *Vedic Mathematics*, containing arithmetic rules written in the style of Sanskrit *sūtras*. These *sūtras* are often said to represent an ancient mathematical system "rediscovered" or "redeveloped" from the Vedic scriptures, but there is no record of them in traditional study of the Vedas.)

II.b. The *Śulbasūtras*

Direct evidence for Indian mathematical texts as such begins in the early to mid-first millennium BCE, with the first of the compositions known as *Śulbasūtras* ("*Cord-Rules*"). Part of the Vedic "limb" of ritual practice, these works contained rules for laying out and building the brick altars used for fire-sacrifices in this period. Certain shapes and sizes of fire-altars were associated with particular gifts that the sacrificer desired from the gods: "he who desires heaven is to construct a fire-altar in the form of a falcon"; "a fire-altar in the form of a tortoise is to be constructed by one desiring to win the world of Brahman"; "those who wish to destroy existing and future enemies should construct a fire-altar in the form of a rhombus" [Sen and Bag 1983, 86, 98, 111].

The *Śulbasūtra* texts are associated with the names of individual authors, about whom very little is known. Even their dates can only be roughly estimated by comparing their grammar and vocabulary with the more archaic language of earlier Vedic texts and with later works written in so-called "Classical" Sanskrit. The one we shall look at is the oldest according to these criteria, composed by one Baudhāyana probably around 800–600 BCE. It tells the priests officiating at sacrifices how to construct certain shapes using stakes and marked cords. The *sūtras* or rules containing these instructions are extremely concise scraps of Sanskrit prose; translating them usually requires many editorial additions to produce a comprehensible result. They were probably always accompanied by oral expositions or prose commentaries explaining their content in more detail.

Many of the altar constructions involve area-preserving transformations, such as making a square altar into a circular or oblong rectangular one of the same size. We don't know how these geometric procedures originally came to be associated with sacrificial rituals. Various theories of the "ritual origin of geometry" infer that the geometrical figures symbolized religious ideas, and the need to manipulate them ritually inspired the development of the relevant mathematics. It seems at least equally plausible, though, that the beauty and mystery of independently discovered geometric facts were considered spiritually powerful (perhaps like the concepts of number and divisibility mentioned above), and were incorporated into religious ritual on that account.

Baudhāyana-Śulbasūtra

1.1. The various constructions of sacrificial fires are now given.

1.2. We shall explain the methods of measuring areas of their [different] figures [drawn] on the ground. [...]

1.4. Having desired [to construct] a square, one is to take a cord of length equal to the [side of the] given square, make ties at both ends and mark it at its middle. The [east-west] line [equal to the cord] is drawn and a pole is fixed at its middle. The two ties [of the cord] are fixed in it [pole] and a circle is drawn with the mark [in the middle of the cord]. Two poles are fixed at both ends of the diameter [east-west line]. With one tie fastened to the eastern [pole], a circle is drawn with the other. A similar [circle] about the western [pole]. The second diameter is obtained from the points of intersection of these two [circles]; two poles are fixed at two ends of the diameter [thus obtained]. With two ties fastened to the eastern [pole], a circle is drawn with the mark. The same [is to be done] with respect to the southern, the western and the northern [pole]. The end points of intersection of these [four circles] produce the [required] square.

1.5. Now another [method]. Ties are made at both ends of a cord twice the measure and a mark is given at the middle. This [halving of the cord] is for the east-west line [that is, the side of the required square]. In the other half [cord] at a point shorter by one-fourth, a mark is given; this is the *nyañcana* [mark]. [Then] a mark is given at the middle [of the same half cord] for purposes of [fixing] the corners [of the square]. With the two ties fastened to the two ends of the east-west line [*pṛṣṭhyā*], the cord is to be stretched towards the south by the *nyañcana* [mark]; the middle mark [of the half cord] determines the western and the eastern corners [of the square].

1.6. When [the construction of] a rectangle is desired, two poles are fixed on the ground at a distance equal to the desired length. [This makes the east-west line.] Two poles, one on each side of each of the [two above mentioned] poles, are fixed at equal distances [along the east-west line]. A cord equal in length to the breadth [of the rectangle] is taken, its two ends are tied and a mark is given at the middle. With the two ties fastened to the two end poles [on either side of the pole] in the east, the cord is stretched to the south by the mark; at the mark [where it touches the ground] a sign is given. Both the ties are now fastened to the middle [pole at the east end of the *prācī*], the cord is stretched toward the south by the mark over the sign [previously obtained] and a pole is fixed at the mark. This is the south-east corner. In this way are explained the north-east and the two western corners [of the rectangle].

1.7. When the eastern side is desired to be of shorter measure, a mark is given at half [the *tiryaṅmānī*, "breadth"].

1.8. Now another [method]. Ties are made at both ends of a cord of length equal to the measure increased by its half [so that the whole length of the cord is divided into three parts of half the measure each]. In the third [extended] part on the western side a mark is given at a point shorter by one-sixth [of the third part]; this is the *nyañcana*. Another mark is made at the desired point for fixing the corners. With the two ties fastened to the two ends of the east-west line [*pṛṣṭhyā*], the cord is stretched towards the south by the *nyañcana*, and the western and eastern corners [of the square] are fixed by the desired mark.

1.9. The diagonal of a square produces double the area [of the square].

1.10. The breadth [of a rectangle] being the side of a given square [*pramāṇa*, "amount"] and the length the side of a square twice as large [*dvikaraṇī*, "two-maker"], the diagonal equals the side of a square thrice as large [*tṛkaraṇī*, "three-maker"].

1.11. Thereby is explained the side of a square one-third the area of a given square [*tṛtīyakaraṇī*, "third-maker"]. It is the side of a square one-ninth the area of the square [explained in the preceding rule, that is, of the square on the *tṛkaraṇī*].

1.12. The areas [of the squares] produced separately by the length and the breadth of a rectangle together equal the area [of the square] produced by the diagonal.

1.13. This is observed in rectangles having sides 3 and 4, 12 and 5, 15 and 8, 7 and 24, 12 and 35, 15 and 36.

This is the beginning of the text, describing in numbered *sūtras* the construction and transformation of simple geometric shapes. (Later sections explain the traditional placement of altars and the construction of the fancier altar shapes in the form of animals, etc.) The author begins in *sūtra* 1.4 with a rather complicated construction of a square of side *s* (as we would describe it) by means of intersecting circles, shown in figure 4.1. A horizontal east-west line (*pṛṣṭhyā* or *prācī*) of length *s* is drawn, and about its midpoint is made a circle of radius *s*/2 (bolded in the figure). The "second diameter" or perpendicular bisector, connecting the intersections of two circles of radius *s*, fixes the north and south points of this central circle, and a circle with radius *s*/2 is drawn about each of its cardinal points. The intersections of those four circles (marked with solid dots in the figure) are the corners of the desired square.

Alternatively, someone wishing to construct a square can find its right-angled corners by exploiting the proportions of a right triangle as shown in figure 4.2: Mark a cord of length

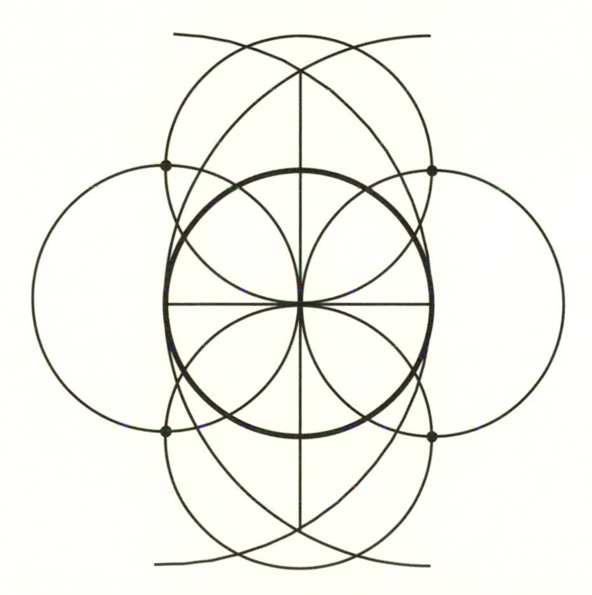

FIGURE 4.1. Śulbasūtra construction of the corners of a square.

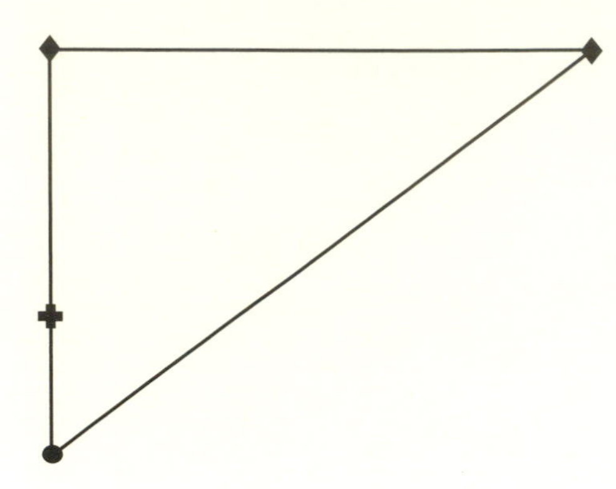

FIGURE 4.2. Śulbasūtra construction of a right angle.

$2s$ in two unequal parts of length $(3/4)s$ and $(5/4)s$. Then attach its ends to two poles (represented by the small solid diamonds) set a horizontal distance s $(= (4/4)s)$ apart along an east-west line. Pull the mark on the cord (the *nyañcana* or "drawing-down" mark, shown as a solid dot) down southward. The taut cord will form the hypotenuse and shorter leg of a 3-4-5 triangle, so the angle between the east-west line (or longer leg) and the shorter leg will be a right angle. The distance $s/2$ can be marked off along the shorter leg (with a small cross in the figure) to fix one corner of the desired square.

The subsequent construction of a rectangle similarly achieves square corners by using an isosceles triangle whose apex, when its base lies on the east-west line, will lie due south or north of the midpoint of the base: hence the southward stretching of a halved cord attached to two poles equidistant from the desired east point. The "eastern side of shorter measure" mentioned in *sūtra* 1.7 refers to the construction of a trapezium, an important plane figure in this ritual geometry because it is the prescribed shape of the so-called "great altar."

Alternatively, in *sūtra* 1.8, for a rectangle of length l one can again use an unevenly divided cord, of length $(3/2)l$, to form a triangle with hypotenuse $(13/12)l$, shorter leg $(5/12)l$, and longer leg l—in other words, a 5-12-13 right triangle, whose shorter leg must be perpendicular to the east-west line on which the longer leg was laid out. The remaining part of chapter 1 states some of these implied facts about triangles and squares more generally.

Chapter 2 discusses various transformations of one geometrical figure into another and ways of modifying their size.

2.1. If it is desired to combine two squares of different measures, a [rectangular] part is cut off from the larger [square] with the side of the smaller; the diagonal of the cut-off [rectangular] part is the side of the combined square. [Alternatively: If it is desired to combine two squares of different measures, a rectangle is formed with the side of the smaller [square] [as breadth] and that of the larger [as length]; the diagonal of the rectangle [thus formed] is the side of the combined square.]

2.2. If it is desired to remove a square from another, a [rectangular] part is cut off from the larger [square] with the side of the smaller one to be removed; the [longer] side of the cut-off [rectangular] part is placed across so as to touch the opposite side; by this contact [the side] is cut off. With the cut-off [part] the difference [of the two squares] is obtained.

2.3. A square intended to be transformed into a rectangle is cut off by its diagonal. One portion is divided into two [equal] parts which are placed on the two sides [of the other portion] so as to fit [them exactly].

2.4. Or else, if a square is to be transformed [into a rectangle], [a segment] of it is to be cut off by the side [of the rectangle]; what is left out [of the square] is added to the other side. [The rule is defective and does not lead to proper geometrical operation.]

2.5. If it is desired to transform a rectangle into a square, its breadth is taken as the side of a square [and this square on the breadth is cut off from the rectangle]. The remainder [of the rectangle] is divided into two equal parts and placed on two sides [one part on each]. The empty space [in the corner] is filled up with a [square] piece. The removal of it [of the square piece from the square thus formed to get the required square] has been stated [in 2.2].

2.6. If it is desired to reduce one side of a square [that is, to make an isosceles trapezium], the reduced side is to be taken as the breadth [of a rectangular portion to be cut off from the square]; the remaining part [of the square] is divided by the diagonal and [one half], after being inverted, is placed on the other side.

2.7. If it is desired to transform a square into [an isosceles] triangle, the square whose area is to be so transformed is doubled and a pole fixed at the middle of its east side; two cords with their ties fastened to it [the pole] are stretched to south-western and north-western corners [of the square]; portions lying outside the cords are cut off.

2.8. If it is desired to transform a square into a double [isosceles] triangle [that is, rhombus], a rectangle twice as large as the square to be so transformed is made; a pole is fixed at the middle of its east side; two cords with their ties fastened to it [the pole] are stretched to the middle points of the southern and northern side [of the rectangle]; portions lying outside the cords are cut off; thereby the [isosceles] triangle on the other side is explained.

2.9. If it is desired to transform a square into a circle, [a cord of length] half the diagonal [of the square] is stretched from the centre to the east [a part of it lying outside the eastern side of the square]; with one-third [of the part lying outside] added to the remainder [of the half diagonal], the [required] circle is drawn.

2.10. To transform a circle into a square, the diameter is divided into eight parts; one [such] part after being divided into twenty-nine parts is reduced by twenty-eight of them and further by the sixth [of the part left] less the eighth [of the sixth part].

2.11. Alternatively, divide [the diameter] into fifteen parts and reduce it by two of them; this gives the approximate side of the square [desired].

2.12. The measure is to be increased by its third and this [third] again by its own fourth less the thirty-fourth part [of that fourth]; this is [the value of] the diagonal of a square [whose side is the measure].

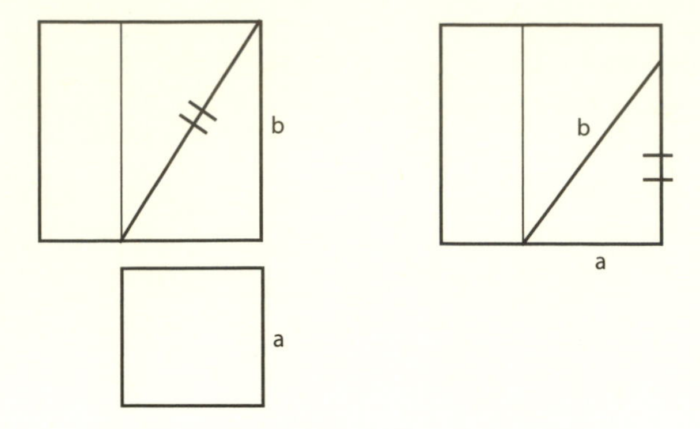

FIGURE 4.3. Adding and subtracting two squares.

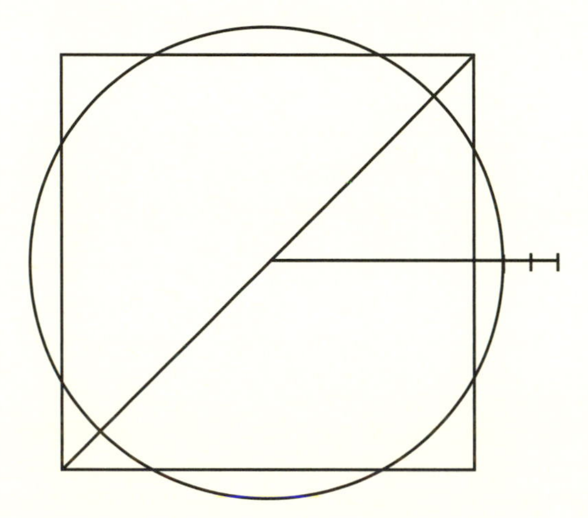

FIGURE 4.4. The "circulature" of a square.

[Fig. 4.3 illustrates the rules in *sūtras* 2.1 and 2.2 for adding the areas of two given squares (of sides a and b, say) into a larger square, and for subtracting the smaller square from the larger. The former rule takes the side of the desired square (marked with a double bar in the diagram on the left side of the figure) as the diagonal of a rectangle with breadth a and length b, or $\sqrt{a^2+b^2}$. In 2.2, the side of a square equal in area to $b^2 - a^2$ is constructed as one leg of a right triangle with hypotenuse b and leg a, formed by rotating one long side of the rectangle ab about a corner till it touches the opposite side. Similar cut-and-paste procedures are used in subsequent *sūtras* to produce other geometrical figures.

These constructions culminate in the transformation of a given square of side s into a circle of the same area with radius r, and vice versa. Here some numerical computations have apparently supplemented simple cord-and-stake manipulations, because the fractions involved are somewhat complicated. The "circulature" and quadrature techniques in 2.9 and 2.10, the first of which is illustrated in figure 4.4, imply what we would call a value of π of 3.088, where either $r = s/2 + (\sqrt{2}s/2 - s/2)/3 = s \cdot (2+\sqrt{2})/6$, or $s = (2r/8) \cdot (7 + 1/29 - 7/(8 \cdot 6 \cdot 29))$. The quadrature in 2.11, on the other hand, suggests that $\pi = 3.004$ (where $s = 2r \cdot 13/15$), which is already considered only "approximate." In 2.12, the ratio of a square's diagonal to its side (our $\sqrt{2}$) is considered to be $1 + 1/3 + 1/(3 \cdot 4) - 1/(3 \cdot 4 \cdot 34) = 1.4142$.]

Evidently, considerable mathematical knowledge had already been amassed by the time these geometrical constructions and numerical computations were recorded. But nothing is said in the texts about how these results were found or verified. Several mathematical commentaries on the *Śulbasūtras* are known, which explain their techniques in more detail, but as far as we can tell the earliest of them is already many centuries later than the base-text. Modern researchers have suggested many ingenious and/or plausible reconstructions of these rules, but we cannot know for sure how the ancient Indian "cord-users" discovered them.

II.c. Mathematics in other ancient texts

Other "limbs" of the Vedas also gave rise to early texts (that is, earlier than the middle of the first millennium CE) that involved some mathematical knowledge. Metrics or prosody, which classified and described the different verse meters, may not sound like a fruitful field for the application of mathematics, but in fact an early metrics text preserves a very interesting computational technique. The work is the *Chandaḥsūtra* (*Rules of Metrics*) of an author named Piṅgala (dated to before 200 BCE), and the technique is a rule for calculating the number of possible meters or syllable patterns for a line of n syllables, where each syllable may be either light or heavy. (Sanskrit heavy and light syllables roughly correspond to the stressed and unstressed syllables of English prosody.) Piṅgala says:

> When halved, [record] two. When unity [is subtracted, record] zero. When zero, [multiply by] two; when halved, [it is] multiplied [by] so much [i.e., squared]. [S. R. Sarma 2003, 130]

The commentaries explain this typically short and incomprehensible *sūtra*. One is to take the desired number of syllables, n, and repeatedly halve it, subtracting 1 from each quotient that happens to be odd. Record the steps by noting a 2 for every halving and a 0 for every subtraction, until the original n is whittled down to 1; diminish it once more by 1 and note down one last 0. For example, finding the number of possible metrical patterns for a line of eleven syllables would proceed as follows:

11	odd, subtract one	and record	0
10	even, halve it	and record	2
5	odd, subtract one	and record	0
4	even, halve it	and record	2
2	even, halve it	and record	2
1	odd, subtract one	and record	0

Now, beginning with the quantity 1, retrace the same steps from finish to start, doubling the current quantity at every recorded 0 and squaring it at every recorded 2. The end result will be 2^n, the required number of possible meters for a line of n light or heavy syllables. In our example above we would work backward through the recorded markers, starting with the final 0 and its corresponding quantity 1, as follows:

0:	double	1	and get	2
2:	square	2	and get	4
2:	square	4	and get	16
0:	double	16	and get	32
2:	square	32	and get	1024
0:	double	1024	and get	2048

And 2048 is indeed the required value of 2^{11}, computed in only six steps rather than by ten successive doublings.

Grammar in the Indian tradition is also gaining increased recognition among historians as a subject with mathematical connections. The classical system of Sanskrit grammar, though it does not deal with mathematical techniques or problems as such, is a formal analysis of natural language structure (indeed, it partly inspired the creation of modern linguistics and phonetics by nineteenth-century scholars). It thus uses many concepts that we now associate with symbolic logic and linguistics. And some scholars argue that grammar, being the paradigmatic analytical science in the Indian intellectual tradition, shaped the methodology and presentation even of mathematical texts.

But the "limb" that ultimately inspired the most Indian mathematical work is undoubtedly that of astronomy/calendrics or *jyotiṣa*. The oldest known Sanskrit text on this subject is the *Jyotiṣavedāṅga* (*Vedic Limb of Astronomy/Calendrics*), dating probably from about the middle of the first millennium BCE. The purpose of this text is to make it possible to calculate the astronomical timing of periodic religious rituals and related timekeeping problems. Simple arithmetic schemes form the basis of these calculations, as in the typical problem of determining the length of daylight. The *Jyotiṣavedāṅga* lays down the relation "six *muhūrtas* [thirtieths of a day] in a half-year [between the solstices]" [Dvivedī 1908, 62]. That is, if one day (from sunrise to sunrise) is 30 *muhūrtas* long, the length of daylight at an equinox is half of that, or 15 *muhūrtas*, and there is a six-*muhūrta* difference between the shortest and longest periods of daylight: 12 and 18 *muhūrtas* at the winter and summer solstices, respectively. The length of daylight for a given day is found as follows: "Whatever has elapsed of the northern half-year [beginning at the winter solstice], or whatever remains of the southern half-year, multiply it by 2, divide [the product] by 61, and add 12 [to the quotient]; that is the length of daylight." [Dvivedī 1908, 66]

That is, on a day d days before or after the winter solstice, the extra time of daylight above the minimum of 12 *muhūrtas* is given as $2d/61$. This is equivalent to $6d/183$, meaning that the six *muhūrtas* of half-yearly variation in daylight are considered evenly distributed over the approximately 183 days of the half-year.

This linear proportion is quite similar to the common Babylonian "linear zigzag functions" for finding the daylight length, and uses the same ratio (3 : 2) for the length of the longest to shortest daylight. Other similarities (such as references to gnomons and water-clocks, and similar time-units) lend plausibility to the hypothesis that some of the methods and parameters of these computations were adapted from Babylonian sources. The transmission presumably occurred via the Persian empire, which in the middle of the first millennium stretched from Mesopotamia to the northwestern parts of India. We see other traces of such contact in ancient inscriptions from northwest India written in an early script derived from Aramaic characters.

Some scholars, however, object that early Indian astronomy must be a great deal older than this period, basing their arguments on astrochronological deductions—astronomical dating of celestial events such as eclipses or sun-star conjunctions mentioned in early texts. The problem with astrochronological arguments, though, is that the dating of the event is only as good as the precision of the source. Since early Sanskrit references to astronomical phenomena don't supply detailed records or precise quantification of measurements, identifying them with datable events in the past necessarily involves a lot of speculative interpretation. This discussion follows instead the more conservative and standard—though still somewhat vague and hypothetical—chronology based on archaeological findings and historical linguistics.

II.d. Number systems and numerals

The scarcity of physical documents from ancient India means that we have only a very incomplete idea of how the writing of numbers emerged and developed. It is common knowledge that the decimal place-value digits we use today were first recorded in India, whence they moved westward via the Islamic world into Europe, undergoing various modifications in form. The Syrian bishop Severus Sebokht famously wrote in the mid-seventh century of the Indian "nine signs" for expressing numbers. But how and when was the decimal place-value system invented?

We have seen that some ancient Vedic texts name successive powers of ten. Except for base-60 time-units and other astronomical measures (which, as noted above, seem to have been derived from Babylonian sources), almost all Indian number words and systems appear to have assumed ten as their base. However, the oldest known physical documents that employ decimal place value are stone inscriptions and inscribed metal deed-plates, none earlier than the second half of the first millennium CE. Earlier inscriptions with numerals use various additive and/or multiplicative number systems, in which the same multiple of different powers of ten is represented by different characters.

Of course, epigraphic writing styles notoriously tend to be conservative (so that a future historian of mathematics drawing inferences only from inscriptions might easily conclude that American mathematicians were working with Roman numerals well into the twentieth century CE!). The content of some earlier texts gives us reason to think that the decimal place-value system was well established, at least among mathematical practitioners, by the early centuries of the first millennium CE. There are occasional hints in nonmathematical texts—for example, the reference to zero in the metrics text of Piṅgala mentioned above, although since in his rule the term "zero" is used merely as an arbitrary marker, it tells us nothing about possible contemporary use of zero as a place-value numeral or as a number. A simile employed by the Buddhist philosopher Vasumitra, perhaps in the first century CE, compares the variable aspects of unchanging realities to counters in merchants' "counting pits": "when [the same] clay counting-piece is in the place of units, it is denoted as one, when in hundreds, one hundred," etc. [McDermott 1974, 196]. Perhaps this "quasi-abacus" structure was already reflected in the place-value writing of numerals.

More solid evidence is provided by a third-century CE Sanskrit version of a Greek astrological text, the *Yavanajātaka* (*Greek Horoscopy*) of Sphujidhvaja, a verse adaptation of a second-century prose translation. We have no ancient manuscripts of this text, so we cannot see directly how its author wrote numerals, but we observe that its verses employ a number-representation scheme sometimes called the "concrete number" or "object-numeral" system. Namely, any number word may be represented by the name of any object or being that is naturally or traditionally found in sets of that number—"moon" and "earth" both represent "one," for example, while "eye" and "twin" mean "two," "limb" is "six" (for the six limbs of the Veda), "tooth" is "thirty-two," and so on. Words for "sky," "void," and "dot" mean "zero," referring to the general concept of emptiness and probably also to the circular shape of its symbol. These synonyms are always arranged in a decimal place-value sense, and thus imply that the reader would assume a decimal place-value representation. However, the order of the digits is the reverse of what we nowadays use, beginning with the least significant. So someone correctly reading the "concrete number" "moon-eye-limb-moon" must automatically think "one-two-six-one" (or one unit, two tens, six hundreds, one thousand) and hence "1621"—not "one plus two and six plus one" or any other additive or multiplicative combination. It is thus reasonable to deduce from the few "concrete numbers" in the *Yavanajātaka* that the decimal place-value system for integers, including a zero symbol, was already familiar to readers of astronomical/astrological texts in the third century. (Severus Sebokht's famous reference to the Indians' using "nine signs" may mean simply that he did not think of writing a zero as expressing a number, not that the Indian decimal digits he encountered lacked a zero-sign.)

Besides the "concrete number" system, there was also an "alphabetic" system, developed probably somewhat later in southern India, which established a one-to-many correspondence between the ten decimal digits and the Sanskrit consonants: each of the thirty-three consonants represented one of the ten digits. Numbers could therefore be expressed, in place-value format, as actual Sanskrit words; some later writers composed entire works that could be read either as literature or as numerical tables.

In addition to obscuring the history of written numerals, the lack of early manuscripts also hampers our efforts to understand the role of mathematical diagrams or figures. Texts occasionally mention seeing or drawing a figure, and the comparatively late manuscripts available to us frequently contain sketches of, e.g., squares and triangles illustrating sample problems in geometry. So we can probably conclude that such visual aids to mathematical understanding always supplemented the verbal rules of the treatises. But the verbal rules generally did not explicitly depend on figures or prescribe specific constructions. In particular, there is little to tell us whether and how most Indian mathematicians used diagrams to arrive at or demonstrate original results. (Except where noted, the figures shown in the original source excerpts are editorial additions, and most have no known counterpart in the manuscript tradition.)

Numerals in Indian texts

The first illustration (fig. 4.5) shows some examples of non-place-value numerals from early Indian inscriptions and coins. (Note that in the third row, the "XX" symbol is mistakenly said to represent 7 instead of 8.) Not every commonly used numeral is attested in any of these sources, and the origins for most of their forms are not definitely known (except for the obvious significance of the one, two, and three strokes representing 1, 2, and 3 respectively). Figure 4.6 shows a page of the famous Bakhshālī Manuscript, a birch-bark manuscript probably copied in northwest India (modern Pakistan) between the eighth and twelfth centuries CE. (Its use of pure place-value numerals is evident, e.g., in the boxed numbers $\begin{matrix}7227\\1200\end{matrix}$ near the top, where the 2s look like our 3s.) Figure 4.7 shows a page of an astronomical table copied in the mid-eighteenth century (courtesy of the Sri Ram Charan Museum of Indology, Jaipur); its numerals are nearly identical to their modern *nāgarī* (Hindi) script forms and can be easily identified with the corresponding glyphs in our own system.

Numerical Symbols between 250 B.C. and A.D. 450

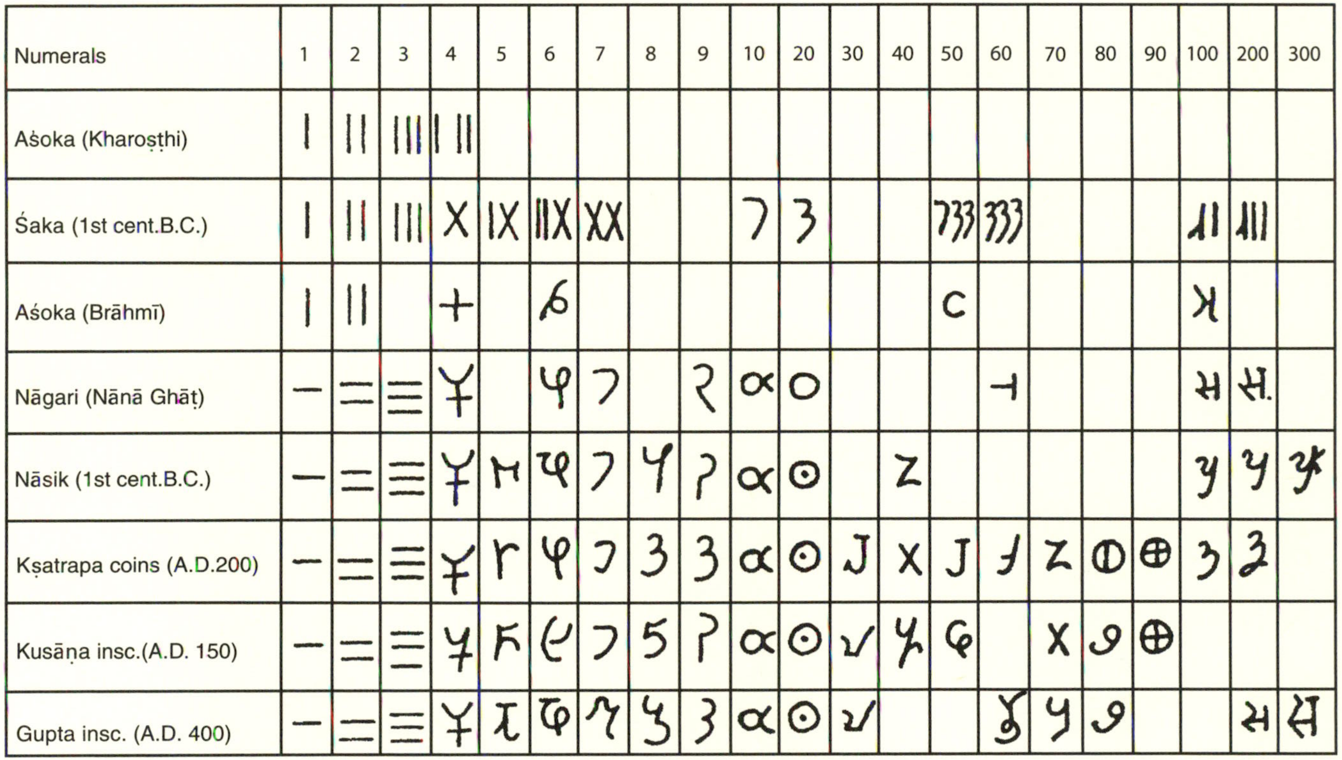

Numerals	1	2	3	4	5	6	7	8	9	10	20	30	40	50	60	70	80	90	100	200	300
Aśoka (Kharoṣṭhi)																					
Śaka (1st cent.B.C.)																					
Aśoka (Brāhmī)																					
Nāgari (Nānā Ghāṭ)																					
Nāsik (1st cent.B.C.)																					
Kṣatrapa coins (A.D.200)																					
Kusāṇa insc.(A.D. 150)																					
Gupta insc. (A.D. 400)																					

Figure 4.5. Early Indian non-place-value numeral forms.

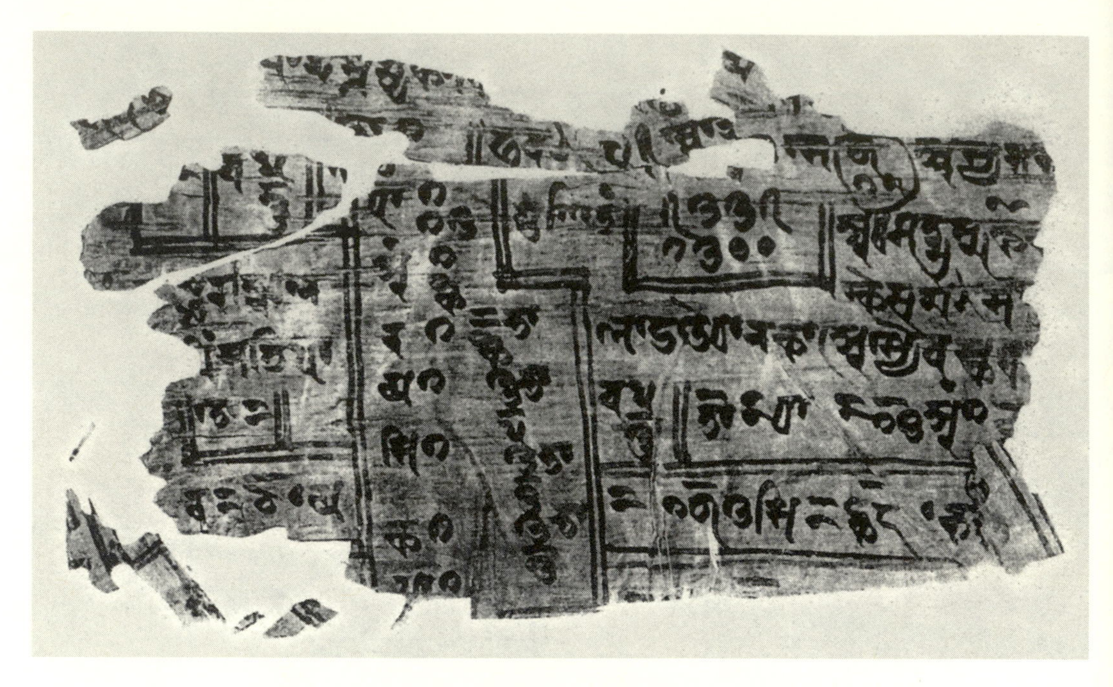

FIGURE 4.6. Place-value numerals in the *Bakhshālī Manuscript*.

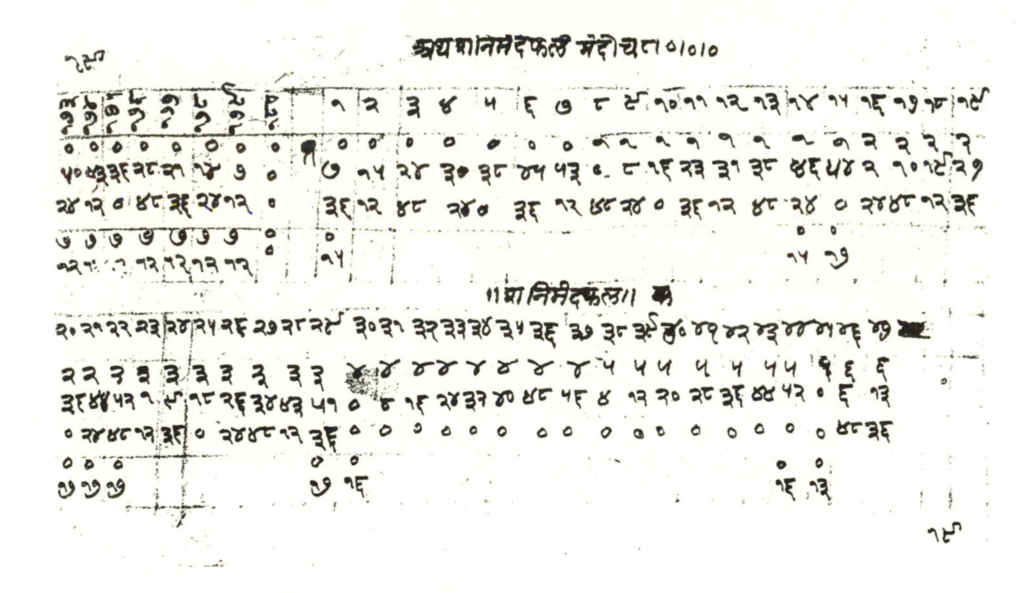

FIGURE 4.7. Numerals in an eighteenth-century manuscript.

III. Evolution of Mathematics in Medieval India

III.a. Mathematics chapters in *Siddhānta* texts

About the middle of the first millennium CE, the textual record for Sanskrit astronomy and mathematics (as well as for many other literary genres) starts to become significantly richer. Not many such works from earlier centuries have survived to the present: besides the

Jyotiṣavedāṅga and the *Yavanajātaka* mentioned above, and one or two other astrological works, we know of only a few fragments and synopses preserved in later texts. Enough remains, however, to tell us that these centuries saw a substantial amount of mathematical activity, including the development of the classical astronomical *siddhānta*, or treatise on mathematical astronomy. Most such texts, like other treatises of the classical period, were composed in verse rather than in abbreviated prose *sūtras*. By the end of the fifth century, writers in this genre had integrated elements from the existing Indian tradition (e.g., various computational techniques, astronomical and calendric concepts, decimal place-value numbers, and geometrical relations) with ones apparently adapted from Greek systems encountered via contact with the Roman Empire (such as plane trigonometry of chords, geocentric cosmological models involving a spherical earth and heaven with their standard reference circles, planetary eccentrics and epicycles, and astrological motivations).

The fundamental aim of the *siddhānta* is both ambitious and simple: namely, to identify the positions of the celestial bodies as seen from any given place on the earth at any given moment in the lifetime of the universe, in order to answer any of a given set of calendric, geographical, and astrological questions. In its basic model, this involves finding the bodies' mean positions at the given time, correcting the mean positions trigonometrically in accordance with their orbital anomalies, and using the resulting true positions to predict the occurrence of sunrise, new and full moons, conjunctions, eclipses, etc. Questions of cosmology, geography, and technical instruments for measuring time and angles were also often addressed in *siddhānta* works.

Interestingly, *siddhāntas* also seem to have been a vehicle for instruction in mathematics more broadly defined. The earliest completely preserved treatise generally classed as a *siddhānta* (though it differs in some ways from the later typical form of the genre) is the well-known *Āryabhaṭīya* of Āryabhaṭa, who was born in 476 CE. Very little is known about Āryabhaṭa's life. He wrote this text probably in Kusumapura, which a later commentator identifies with Pataliputra, which in turn is identified with modern Patna in Bihar; and he evidently taught pupils who preserved and continued his work (probably in Maharashtra in western India). But no reliable source supplies any other significant personal information about him. His *Āryabhaṭīya*, however, remains one of the most important and influential sources for Indian astronomy and mathematics.

The range of mathematical knowledge preserved in the *Āryabhaṭīya*'s second chapter on "Gaṇita" or "calculation," as the following excerpts indicate, is unexpectedly varied. The initial verses discuss basic arithmetic and geometric procedures such as finding square and cube roots and computing areas. The following rules involving sines, shadows, and circles are obviously relevant to astronomical models. But where is the astronomical application of such techniques as finding the sum of terms in an arithmetic progression or computing the interest on a given principal? Certainly they are applied nowhere else in the *Āryabhaṭīya*. It seems clear that Āryabhaṭa considered *gaṇita* to embrace computational techniques in general, not just calculations required for astronomical practice. But he saw nothing incongruous or inappropriate in discussing such techniques in an astronomy text. Perhaps there also existed fuller treatments of non-astronomical *gaṇita* subjects in separate works, as we see in later centuries, but none survive from the time of the *Āryabhaṭīya*.

The "Gaṇita" chapter consists of 33 numbered verses (in the meter called "āryā"), which is a very small space into which to cram such a wide selection of mathematical knowledge. Not surprisingly, such verse treatises were almost always supplemented by prose commentaries to elucidate their meanings. The earliest such commentary that we possess is the *Bhāṣya* ("Commentary") on the *Āryabhaṭīya* composed by one Bhāskara in 629. Like most other Indian mathematicians of this period, Bhāskara as a historical personage is almost completely unknown to us; he probably came from what is now part of Maharashtra or Gujarat, and he wrote two other works on astronomy following the school of Āryabhaṭa, but his life is otherwise a mystery. He is frequently called "Bhāskara I" to distinguish him from the twelfth-century mathematician of the same name (see below).

What his commentary on the *Āryabhaṭīya* does abundantly reveal is the depth and breadth of mathematical thought in Sanskrit at this time. Though Bhāskara's individual virtuosity is undeniable, his frequent references to other authorities make it clear that he was working within a well-established tradition of reflection on mathematical knowledge and its foundations. However, the most important aspects of that knowledge within this tradition are in many ways rather different from what we see in modern mathematics.

The most crucial goal is obviously to ensure that the reader understands the proper interpretation of the (often very cryptic) rule or rules contained in each concise verse. To that end, a commentator usually provides a detailed grammatical analysis of the rule, and frequently one or more worked examples as well. Various arguments to demonstrate the truth of the rule are often given, but the sort of formal rigorous proof that modern mathematicians consider the backbone of their subject is rare. (Occasionally this leads to a blurring of the distinction between accurate and approximate formulas, as in Āryabhaṭa's statements—unchallenged by Bhāskara—that the volume of a regular tetrahedron is half its height times the area of the base, and the volume of a sphere is the area of one of its great circles times the square-root of that area. These rules are probably based on analogy with the accompanying area rules for the corresponding plane figures.) In fact, detailed explanations and demonstrative diagrams are occasionally referred to as "for the dull-witted"! The intelligent student is evidently expected to ponder ambiguous or riddling phrases, into which the poet has packed multiple meanings to convey the greatest possible amount of mathematics in the smallest possible number of words, and infer from them what the linguistically and mathematically consistent interpretation must be.

Neither the *Āryabhaṭīya* nor Bhāskara's *Bhāṣya* is designed for the raw beginner in arithmetic: familiarity with the basic arithmetic operations is assumed, and even the square- and cube-root extraction procedures are treated rather sketchily. More complicated techniques are illustrated in greater detail, and numerous word problems appear in the examples. Bhāskara frequently digresses upon fundamental issues, such as the classification of mathematical topics as arithmetic or geometric, the fact that surds ("*karaṇīs*") have no "statable size," the advantage of place-value numerals, and the equality of a unit arc with its chord. (His musings on the possibility of the place-value principle being exploited by unscrupulous traders foreshadow in some ways the concerns of Italian merchants many centuries later!) Bhāskara's comments also reveal the close intellectual ties between Sanskrit mathematics and grammar, which is in turn linked to logic and philosophy. He carefully scrutinizes Āryabhaṭa's statements not just for their mathematical meaning but also for desiderata such as semantic validity, lack of redundancy, correct terminology, and multiplicity of meanings. All these characteristics, in his perspective, are components of mathematical truth.

Āryabhaṭīya *of Āryabhaṭa with commentary of Bhāskara I*

Chapter 2: Calculation

Mathematics is of two kinds: mathematics of fields and mathematics of quantities. Proportions, pulverisers, and so on, which are specific [subjects] of mathematics, are mentioned in the mathematics of quantities; series, shadows, and so on, [are mentioned] in the mathematics of fields. Therefore, in this way, mathematics as a whole rests upon the mathematics of quantities or the mathematics of fields. [...]

Here as master Āryabhaṭa is starting the treatise, a salutation to a favorite god, in his mind, is urged by devotion indeed:

2.1. Having paid homage to Brahmā, Earth, Moon, Mercury, Venus, Sun, Mars, Jupiter, Saturn, and the group of stars, here Āryabhaṭa tells the knowledge honored in Kusumapura.

"Brahmā" is his favorite. Because a salutation to a favorite god urged by devotion destroys the obstacles that interrupt the work one craves and desires to finish, he starts by paying homage to him.

Or else, a composition briefly stating the topics of the *Svāyaṃbhuva Siddhānta* has been undertaken by the master, and the father of the *Svāyaṃbhuva Siddhānta* is the honored Creator [Brahmā]. Therefore it is proper for him [Āryabhaṭa] to first make a salutation to him [Brahmā]. [...]

As for "Āryabhaṭa," by mentioning his own name, he indicates that there are other works made according to the *Svāyaṃbhuva Siddhānta*; therefore, since there are a great number of works made according to the *Svāyaṃbhuva Siddhānta*, it would not be known by whom this work was composed. This is why he mentions his name. [...]

As for "the knowledge honored in Kusumapura," Kusumapura is Pāṭaliputra; he states the knowledge honored there. [...]

In order to assign places to numbers, he states:

2.2. One and ten and a hundred and one thousand, now ten thousand and a hundred thousand, in the same way a million, ten million, a hundred million, and a thousand million. A place should be ten times the previous place.

One sets forth the places of numbers for the sake of easiness. For, if not, mathematical operations would be difficult because of the lack of any assignment of places to numbers. Why? When placing the abundance of units, many units have to be placed. On the other hand, truly, when the places are settled, that operation to be accomplished with many units, can be performed with a single one only. [...]

Here this may be asked: What is the power of the places, [that power with] which one unit becomes ten, a hundred, and a thousand? And truly if this power of places existed, purchasers would have shares in especially desired commodities. And according to [their] wish what is purchased would be abundant or scarce. And if this was so, there would be the unexpected possibility for things to be different in worldly affairs. [...]

2.3ab. A square is an equi-quadrilateral, and the area/result is the product of two identicals. [...]

[Question:]

In which other cases is there a possibility for the name "square" to be [given to] an undesirable specific equi-quadrilateral field?

It is replied: This kind of equi-quadrilateral with unequal diagonals would have [that name], and this [field made of] two equi-trilateral fields placed as if upright, would have [that name].

[Question:]

What is wrong with the possibility for the name "square" [to be given to these fields]?

It is stated: "And [its] area is the product of two identical [sides]." Therefore, the product of two identical [sides] should give the area, and such is not as wished in the above cases. [...]

[Question:]

In which case [is it right for the name "square" to be given to a specific equi-quadrilateral]?

One should mention the diagonals. Thus, a square is a specific equi-diagonal-equi-quadrilateral field. [...]

When one has sketched an equi-quadrilateral field and divided [it] in eight, one should form four rectangles whose breadth and length are three and four and whose diagonals are five. There, in this way, stands in the middle an equi-quadrilateral field whose sides are the diagonals of the [four] rectangles which were the selected quadrilaterals. And the square of the diagonal of a rectangular field there, is the area in the interior equi-quadrilateral field. [...] And a field is sketched in order to convince the dull-minded. Therefore, every single square is a specific equi-quadrilateral. [...]

[Fig. 4.8 shows "these fields" described in Bhāskara's comment. The sketch "to convince the dull-minded" is shown in figure 4.9. It is not known exactly what Bhāskara means by "dividing it in eight"—eight triangles?]

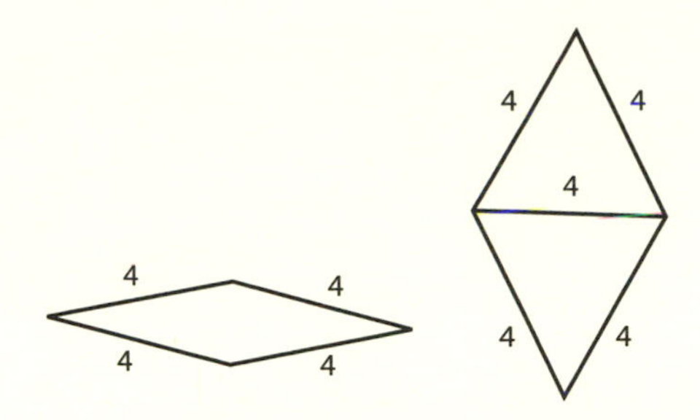

FIGURE 4.8. Bhāskara's non-square equi-quadrilaterals.

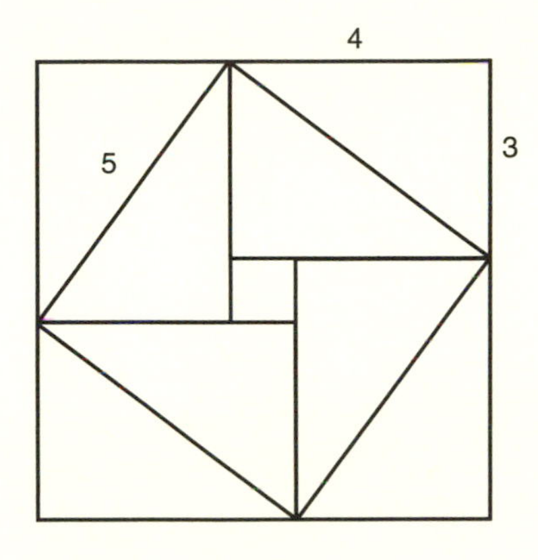

FIGURE 4.9. Bhāskara's division of an equi-quadrilateral field.

He states the latter half of an *āryā* to expose operations on cubes:

2.3cd. A cube is the product of a triple of identicals as well as a twelve-edged [solid]. [...]

In order to compute square-roots, he says:

2.4. One should divide, constantly, the non-square [place] by twice the square-root. When the square has been subtracted from the square [place], the quotient is the root in a different place. [...]

In order to compute cube roots, he says:

2.5. One should divide the second non-cube [place] by three times the square of the root of the cube. The square [of the quotient] multiplied by three and the former [quantity] should be subtracted from the first [non-cube place] and the cube from the cube [place]. [...]

[The "square" and "non-square" places refer to the even and odd powers of 10, respectively. The largest possible perfect square is subtracted from the digit(s) in the highest square place, which are replaced by the remainder; and the root of that perfect square is the highest digit of the desired square-root. The digit(s) in the next non-square place are then divided by twice the root that was just found; the remainder from the division is substituted into the current non-square place, and the process begins again from the next square place.

So, to find the square-root of 1444, we consider that the highest square place is the hundreds place, containing the digits 14. The largest perfect square less than 14 is 9, so the first digit of the desired square-root is $\sqrt{9} = 3$. The digit in the highest square place is now $14 - 9 = 5$, so the next non-square place contains the digits 54. Dividing 54 by $2 \cdot 3 = 6$ (requiring a non-zero remainder) yields 8; then the 54 is replaced by $54 - 6 \cdot 8 = 6$. Now the next square place contains the digits 64, and we must repeat the process, beginning by finding the largest perfect square less than 64. In our example, we have come to the last place, so we take the required perfect square to be $64 = 8^2$; the complete desired square-root is thus 38 and the computation is over.

Similarly, to take the cube-root of a number in decimal place-value notation, we must consider its triples of one "cube" and two "non-cube" places, beginning from the units place. In the number 13824, for example, the highest cube place is the thousands place, containing the digits 13. We subtract from those digits the largest possible perfect cube, or $2^3 = 8$, and save the cubed quantity 2 as the first digit of the desired cube-root. The remainder of the subtraction is $13 - 8 = 5$, so the next place—the so-called "second non-cube place"—now contains the digits 58. "Three times the square of the root of the cube" is $2^2 \times 3$ or 12, with which we divide the 58 in the second non-cube place, yielding a quotient of 4 with a remainder of 10. The 4 becomes the next digit of the desired cube-root. The first non-cube place then contains the digits 102. We must subtract from that 102 "the square of the quotient multiplied by three and the former quantity," or $4^2 \times 3 \times 2 = 96$. The remainder is 6, so the digits in the following (and final) cube place are 64. When we subtract the cube of the quotient, 4^3, from that, the remainder is zero and the procedure ends, leaving 24 as the desired cube-root.]

Now in order to compute the area of a trilateral field, he says:

2.6ab. The bulk of the area of a trilateral is the product of half the base and the perpendicular. [...]

In order to compute the volume of just that equi-trilateral field, he says the latter half of the *āryā*:

2.6cd. Half the product of that and the upward side, that is [the volume of] a solid called "six-edged."

The upward side is a height in the middle of the field. "That" refers to the area. [...] [As for:] "that is the solid," it amounts to: the volume, and that [solid] is six-edged. [...]

Now, when the size of the upward side is known, one can state that the volume is half the product of that and the upward side, but not when [it is] unknown. [...]

Now, in order to compute the area of a circular field, he states:

2.7ab. Half of the even circumference multiplied by the semi-diameter, only, is the area of a circle. [...]

In order to expose the volume, he states:

2.7cd. That multiplied by its own root is the volume of the circular solid without remainder. [...]

[Here Āryabhaṭa discusses some analogous area and volume rules—as noted previously, without indicating that the analogy is only approximate. Half the base times the height gives the area of a triangle, suggesting that the volume of a tetrahedron would likewise be the product of half the area of the base triangle and the "upward side" or height of the solid. Similarly, the rule for the area A of a circle depends on its circumference—A is half the circumference times the radius—so, presumably, it is deduced that the volume V of a sphere of the same radius is determined by the area: $V = A\sqrt{A}$.]

In order to know the area and the size of [the lines whose top is] the intersection—for [isosceles and scalene] quadrilaterals and so on, and for interior ears, he states here an *āryā*:

2.8. The two sides, multiplied by the height [and] divided by their sum, are the "two lines on their own fallings." When the height is multiplied by half the sum of both widths, one will know the area.

An example:

1. Let the earth be fourteen, and the face four units. The two chief ears [should measure] thirteen, tell [the lines] whose top is the intersection and the area.

Setting down:

Procedure: The base [of the inner right-angled trilaterals] is half the difference of the earth and the face, [5]. A perpendicular is established with that base, precisely by means of the computation told separately [2.17ab], and that [perpendicular] is 12. This very perpendicular is the height. Separately, the two sides are multiplied by it, what results is 48, 168. The sum of the sides is 18. The two quotients of the division by this last quantity are the "two lines on their own falling": $2\frac{2}{3}$, $9\frac{1}{3}$, half the sum of the widths is 9; the height multiplied by this is the area of the field, 108. [...]

[The quadrilateral figure discussed here, and shown in figure 4.10, is the traditional trapezium: its base is often called the "earth" and its top the "face"; what Bhāskara calls the "base" is half their difference. By "ears" are meant the diagonals (the same word is also used for the hypotenuse of a right triangle). Their intersection produces the two unequal segments called the "lines on their own fallings."]

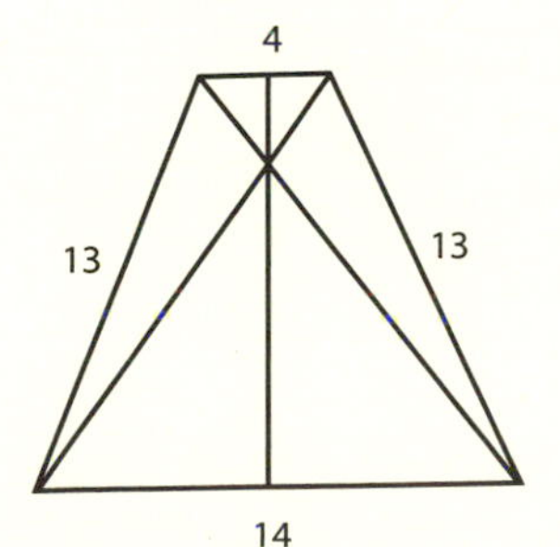

FIGURE 4.10. The trapezium of Bhāskara's example.

2.9ab. For all fields, when one has acquired the two sides, the area is their product. [...]

> Now, because the word "all" expresses "everything without exception," every field, indeed, without exception is referred to [with this rule]. Therefore, because the area of all fields has been established with this very rule, the mentioning of the previously stated rules [2.6ab, 2.7ab, 2.8] was useless.
>
> It is not useless. The verification and the [computation of the] area are told with this [rule].
>
> Verifications of the areas of the [previously] stated fields [are made] since the experts in mathematics Maskari, Pūraṇa, Pūtana, etc., verify the area of all fields in rectangular fields. And it has been said:
>
> Always, when one has sought the area by means of told procedures, then, one should know [its] verification in a rectangular field, since in rectangles the area is obvious.
>
> The computation of the areas of fields unstated [in known rules is possible] just by transforming the desired field into a rectangle.
>
> [Question:]
>
> But how are the computation of the area and the verification acquired with just one effort? Now, if this [rule] was originally made in order [to carry out] a verification, how can that [rule] be for the computation of the area? And, [conversely, if this rule was originally made] in order to compute the area, how can [it be used] for a verification?
>
> There is no drawback. It has been observed that what is originally made for one purpose is the instrument of an other purpose. That is as in the following [case]:
>
> Canals are constructed for the sake of rice paddies. And from these [canals] water is drunk and bathed in. [*Aṣṭādhyāyī*, 1.1.22, *Pātañjala-bhāṣyam*].
>
> It is just like this in this case also. [...]

[This discussion illustrates the careful logical scrutiny given to the meaning and arrangement of mathematical rules. The product of two adjacent sides is said to give the area for all figures (either approximately, or if the figure has been dissected into a rectangle of equal area, exactly). At first glance, this seems to make the previously-stated area formulas for specific figures redundant (provoking, in the commentarial dialogue, the timeless resentment of the student who has been forced to learn information that turns out to be unnecessary). No details are known about the work of the three "experts in mathematics" mentioned by Bhāskara.]

In order to exhibit a chord equal to the semi-diameter of an evenly circular [field], he states:

2.9cd. The chord of a sixth part of the circumference, that is equal to the semi-diameter. [...]

In order to compute an evenly-circular [field] with a Rule of Three, he states:

2.10. A hundred increased by four, multiplied by eight, and also sixty-two thousand is an approximate circumference of a circle whose diameter is two *ayutas* [ten thousands]. [...]

[Question:] Now why is the approximate circumference told, and not indeed the true circumference itself?

They believe the following: There is no such method by which the accurate circumference is computed.

[Objection:] But here it is:

The *karaṇī* which is ten times the square of the diameter is the circumference of the circle.

In this case also, the [rule] "the circumference of a unity-diameter [circle] is ten *karaṇīs*" is merely a tradition and not a proof.

[Objection:] Now some think that the circumference of a field with a unity-diameter when measured directly is ten *karaṇīs*.

This is not so because *karaṇīs* do not have a statable size. [...]

[The ancient term *karaṇī* was translated more or less literally in the *Śulbasūtra* excerpts above as the "maker," or side, of a square of a given size—in other words, the square-root. Here it implies a technical meaning somewhat like that of our "surd": a number whose square-root cannot be stated, an irrational number. The circumference equal to the "*karaṇī* which is ten times the square of the diameter" thus means $\sqrt{10d^2}$. This value $\sqrt{10}$ for the circumference/diameter ratio is an early approximation often seen, for example, in Jain mathematical texts. Bhāskara is well aware that this ratio is not exactly represented either by the quantity $\sqrt{10}$ or by Āryabhaṭa's suggested $\frac{62832}{20000} = 3.1416$. However, it is not known when, how, or to what precise extent the "unstatability" of *karaṇīs* became familiar to Indian mathematicians; they are apparently not discussed here in terms of ratios of integers as in Greek number theory.]

Now, in order to compute chords, he states:

2.11. One should divide the quarter of the circumference of an evenly-circular [field]. And, from trilaterals and quadrilaterals as many half-chords of an even [number of] unit arcs as one desires [are produced], on the semi-diameter. [...]

Moreover: It is proper to say that a unit arc can be equal to its chord; even someone ignorant of treatises knows this; that a unit arc can be equal to its chord has been criticized by precisely this [master].

But we say: An arc equal to a chord exists. If an arc could not be equal to a chord then there would never be steadiness at all for an iron ball on level ground. Therefore, we infer that there is some spot by means of which that iron ball rests on level ground. And that spot is the ninety-sixth part of the circumference. [...]

Procedure: Having drawn a circle with a pair of compasses whose [opening] is equal to the semi-diameter determined by a size as large as [desired], one should divide that [circle] into twelve [equal parts]. And these twelfth parts should be regarded as "*rāśis.*" Now, in the circle which is divided into twelve [equal parts], in the east one should make a line which has the form of a chord, and which penetrates [the circle at] the tips of two *rāśis* from south to north. Likewise in the western part also. In the same way, again, in the southern and northern parts also, one should make chords extending from east to west. And furthermore in the eastern, western, southern and northern directions, in exactly the same way, one should make lines which penetrate [the circle at] the tips of four *rāśis*. Then they should be made into trilaterals [by drawing the diagonals of the rectangles obtained].

And thus a field produced by a circumference is drawn with a pair of compasses with a stick fastened to the opening. In the field drawn in this way all is to be shown.

In this drawing the [whole] chord of four unit arcs is equal to the semi-diameter. Half of that is the [half-]chord of two unit arcs. And that is 1719.

This is the base; the semi-diameter is the hypotenuse; therefore the perpendicular is the root of the difference of the squares of the base and the hypotenuse. That exactly is the [half-]chord of four unit arcs. And that is 2978. When one has subtracted this from the semi-diameter, the remainder is the arrow of [the half-chord of] two unit arcs. The hypotenuse is the root of the sum of the squares of the arrow and the [half-]chord of two unit arcs. And that precisely is the [whole] chord of two unit arcs, which is 1780. Half of that is the [half-]chord of one unit arc, 890. [...]

[Here Āryabhaṭa gives a general description of the structure and determination of the quantities we now call sines, but the actual values of his sine-table are supplied in the previous chapter dealing with astronomical constants. Like most Indian mathematicians, Āryabhaṭa divides the quadrant into 24 arcs of 225 arcminutes. His trigonometric radius is 3438 (approximately equal to the 21600 arcminutes of the circumference divided by 2π), which enables him to use the same length units for arcs and their sines. (Note that in the Indian tradition the sine is not a ratio but the length of a line-segment within the circle, and

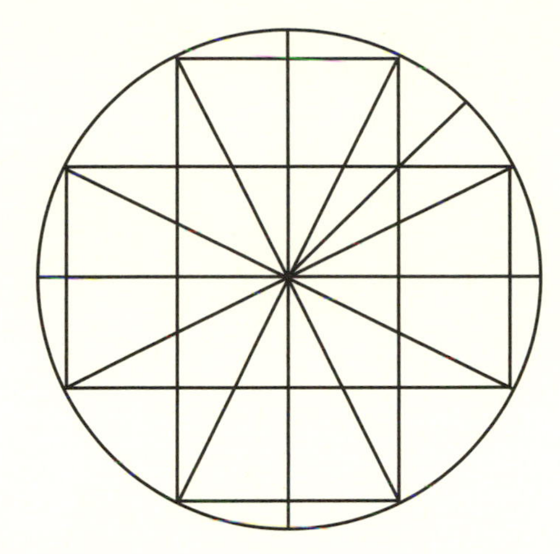

FIGURE 4.11. The computation of sines.

it is not normalized to a radius of unity. Such trigonometric quantities using a non-unity radius R are frequently represented as "Sin" or "Cos" to distinguish them from the modern forms of these functions. The value $R = 3438$ probably did not originate with Āryabhaṭa, but his is the earliest surviving text to record it.)

Āryabhaṭa's 24 sines are as follows: 225, 449, 671, 890, 1105, 1315, 1520, 1719, 1910, 2093, 2267, 2431, 2585, 2728, 2859, 2978, 3084, 3177, 3256, 3321, 3372, 3409, 3431, 3438. (The first of these, a sine-value of 225 equal to its arc, is defended by Bhāskara via a physical argument about the flatness of 1/96 of an iron ball's circumference.) Presumably, all were computed by consideration of the various triangles and rectangles inscribed in a circle, as the above excerpt from Bhāskara's explanation illustrates (see figure 4.11). He evidently assumes the bisection of each *rāśi* (literally "heap" or "quantity," or more specifically "zodiacal sign" or thirty degrees) to give "unit arcs" of fifteen degrees each. Bhāskara will subdivide them further as he goes on to compute all 24 of Āryabhaṭa's sine-values from right triangles. The "arrow" of which he speaks is the "versed sine" or "sagitta" of the arc, i.e., the difference between the radius and the cosine, a quantity obsolete in modern trigonometry but standard in Indian texts.]

He says the computation of the unknown distance and height of a light with the shadows of two gnomons:

2.16. The upright side is the distance between the tips of the [two] shadows multiplied by a shadow divided by the decrease. That upright side multiplied by the gnomon, divided by [its] shadow, becomes the base. [...]

It is as follows: A gnomon has been fixed at some distance from a known light on a pillar of unknown height. Its shadow's [length] is known indeed. At a [certain] distance measured from the tip of its shadow there is a second gnomon. The distance between the tips of the shadows is the distance defined as the tip of the shadow of the previous gnomon [as measured] from the tip of the shadow of that [second gnomon]. It is multiplied by the desired [shadow], the first shadow or the second shadow. [...]

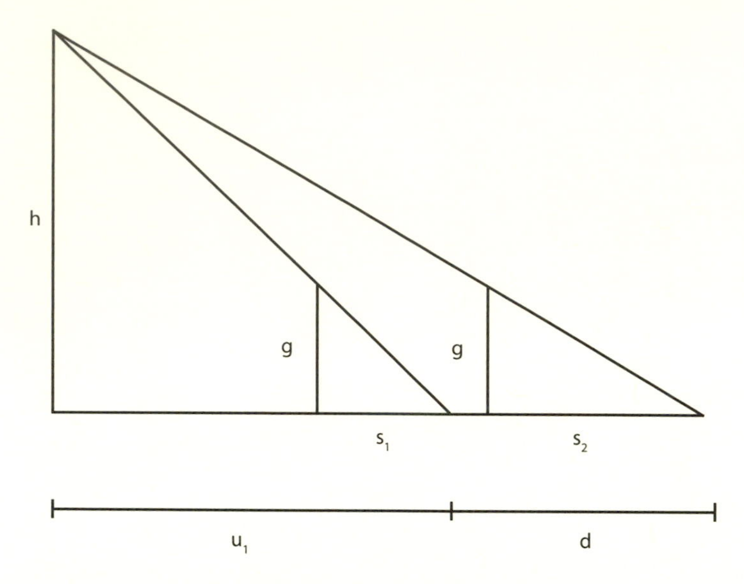

FIGURE 4.12. The double-shadow problem.

An example:

1. The shadows of two equal gnomons are observed to be respectively ten and sixteen *aṅgulas* and the distance between the tips [of the shadows] is seen as thirty. Both the upright side and the base should be told.

Setting down:

Procedure: The distance between the tips of the shadows is 30; it is multiplied by the first shadow, 300; the difference of the [lengths of the] shadows is 6; what has been obtained with this is the upright side, 50. Precisely this upright side is multiplied by [the height of] the gnomon [i.e., 12 *aṅgulas*]; what has been obtained is 600, which when divided by the [first gnomon's] shadow is the base, 60. From the second shadow, too, the upright side is 80; the base is the same, 60. [...]

[Āryabhaṭa introduces the properties of right triangles with some formulas involving shadows cast by gnomons, measured in *aṅgulas* or "fingers." Verse 2.16 is the most complicated of these, allowing the distance and height of a light on a pillar to be found from the known height g and shadows s_1, s_2 of two equal gnomons (see figure 4.12). Each gnomon and its shadow comprise the legs of a right triangle similar to one formed by the height h of the pillar and the distance u from its foot to the shadow-tip. (Somewhat counterintuitively, the word for the leg translated here as "base" actually refers to the vertical height h, while "upright side" means the horizontal distance u_1.) The upright side (say, u_1) is found by proportion from either of the given shadows (say, s_1) and the distance d between the two shadow-tips:

$$\frac{u_1}{s_1} = \frac{d}{s_2 - s_1}.$$

Then the height h is derived from the similar right triangles mentioned above:

$$\frac{h}{g} = \frac{u_1}{s_1},$$

and the solution is complete.]

In order to compute the hypotenuse, he has told:

2.17ab. That which precisely is the square of the base and the square of the upright side is the square of the hypotenuse. [...]

2.17cd. In a circle, the product of both arrows, that is the square of the half-chord, certainly, for two bow [fields]. [...]

Just in this case they relate hawk and rat examples. It is as follows: The half-chord is the base. The upright side is in the space between the center of the circle and the half-chord. The hypotenuse which is the root of the sum of their squares, is the semi-diameter of the circle. And that is explained.

Setting down:

This half-chord is the height of the hawk's position. The space between the circumference and the half-chord is the rat's roaming ground. The half-diameter, which is the hypotenuse, is the hawk's path. The center of the circle is the spot of the rat's slaughter. [...]

An example:

A hawk was resting upon a wall whose height was twelve *hastas*. The departed rat was seen by that hawk at a distance of twenty-four *hastas* from the foot of the wall; and the hawk was seen by the rat. There, because of his fear, the rat ran with increasing speed towards his own residence which was in the wall. On the way he was killed by the hawk moving along the hypotenuse. In this case I wish to know what is the distance not attained by the rat, and what is the distance crossed by the hawk. [...]

Procedure: The square of the height of the hawk is 144; when that is divided by the size of the rat's roaming ground, 24, the quotient is 6. The rat's roaming ground, when increased by this difference, is 30, and when decreased is 18. Their halves in due order, the path of the hawk and the distance to the rat's residence: 15, 9. [...]

["Both arrows" are the two segments s_1, s_2 of the diameter of a circle intersected at right angles by a chord $2h$, dividing the circle into two "bows" (see fig. 4.13). The "hawk and rat" is a classic word problem utilizing the formula $h^2 = s_1 \cdot s_2$: set up as per Bhāskara's explanation above, it gives the radius of the circle (the hawk's dive) and the difference between the radius and the shorter arrow (the remaining distance to the unfortunate rat's hole).]

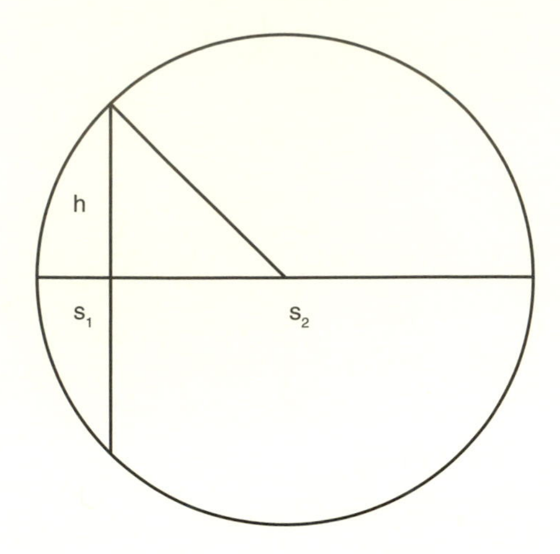

FIGURE 4.13. The perpendicular chord and diameter.

Now, in order to compute the value of series, he has told:

2.19. The desired [number of terms], decreased by one, halved, increased by the previous [number of terms], having the common difference for multiplier, increased by the first term, is the mean [value]. [The result], multiplied by the desired, is the value of the desired [number of terms]. Or else, the first and last [added together] multiplied by half the number of terms [is the value]. [...]

Here, [in this verse], there are many rules which stand in separate stanzas. Their union [is made] according to a [suitable] connection [of words].

"The desired [number of terms] decreased by one, halved, having the common difference for multiplier, and increased by the first term" is the rule in order to compute the mean value.

"The mean [value] multiplied by the desired [number of terms] is the value of the desired [number of terms]" is [the rule] in order to compute the value of the [desired] number of terms.

"The desired [number of terms] is decreased by one, halved, increased by the [number of] previous [terms], having the common difference for multiplier, and increased by the first term" is [the rule] in order to compute the value of the last, the penultimate, etc. [terms].

"The desired [number of terms] decreased by one, halved, increased by the previous [number of terms], having the common difference for multiplier, increased by the first, and multiplied by the desired [number of terms] is the value of the desired [number of terms]" is a [rule] in order to compute the number of as many terms as desired.

In this way these [rules] are united within an *āryā*, without a quarter. We will explain these [rules] in due order in nothing but examples.

An example:

1. The first term of a series is seen as two, and the common difference is told to be three. The number of terms is expressed as five. Tell the values of the mean and of the whole [number of terms].

Setting down: The first term is 2, the common difference is 3, the number of terms is 5. Procedure: The desired number of terms is 5, decreased by one, 4, halved, 2, having the common difference for multiplier, 6, increased by the first term, 8; this is the mean value. Just this multiplied by the desired number of terms produces the whole value, 40. [...]

[See section IV.c. for a discussion of another *Āryabhaṭīya* verse concerning series.]

In order to compute the interest on the capital, he states:

2.25. The interest on the capital, together with the interest [on the interest], with the time and capital for multiplier, increased by the square of half the capital. The square root of that, decreased by half the capital and divided by the time, is the interest on one's own capital. [...]

An example:

1. I do not know the [monthly] interest on a hundred. However, the [monthly] interest on a hundred increased by the interest [on the interest] obtained in four months is six. State the interest of a hundred produced within a month.

	100	0	
Setting down:	1	4	months, 4; interest, 6.
	0	6	

Procedure: The interest on the capital increased by the interest [on the interest], 6; multiplied by the time and the capital, 2400; the square of half the capital, 2500, is increased by that, 4900. Its square root, 70, decreased by half the capital, 20, is divided by the time; the interest on [one's own] capital is what has been produced, 5.

Verification with a Rule of Five: "If the monthly interest on a hundred is five, then what is the interest of the interest [of value five] on a hundred, in four months?"

	1	4
Setting down:	100	5
	5	0

What has been obtained is one. This increased by the [monthly] interest on the capital is six *rūpas*, 6. [...]

In order to explain the Rule of Three, he says an āryā and a half:

2.26. Now, when one has multiplied that fruit quantity of the Rule of Three by the desire quantity, what has been obtained from that divided by the measure should be this fruit of the desire.

2.27ab. The denominators are respectively multiplied to the multipliers and the divisor. [...]

[Question:]

Only the Rule of Three is mentioned here by master Āryabhaṭa. How should different proportions such as the Rule of Five, etc., be understood?

It is replied: Only the very seed of proportions has been indicated by the master; by means of that seed of proportions, the Rule of Five, etc., all indeed, has been established. [...]

An example:

1. I have bought five *palas* of sandalwood for nine *rūpakas*. How much sandalwood, then, should be obtained for one *rūpaka*? [...]

Setting down: 9 5 1

Procedure: Since five *palas* of sandalwood [have been obtained] with nine *rūpakas*, nine is the measure quantity, five is the fruit quantity. Since "how much [has been obtained] with one *rūpaka*?" [is the question], one is the desire quantity. The fruit quantity multiplied by that desire quantity which is one, 5, should be divided by the measure quantity which is nine, $\frac{5}{9}$. In this case, since parts in *palas* are not desired, [one should use] "a *pala* is four *karṣas*"; [the previous result] multiplied by four is $\frac{20}{9}$. What has been obtained is two *karṣas* and two parts [of nine] *karṣas*. 2 *karṣas* and $\frac{2}{9}$. parts of *karṣas*. [...]

[The *trairāśika* or "rule of three quantities" is a pivotal concept in Indian mathematics: if a certain result or "fruit" is obtained with a given "measure," how much is obtained with the "desired" amount? The three quantities are conventionally laid out in the sequence "measure–fruit–desire," as in Bhāskara's example. As Bhāskara notes, the subsequent proportion rules such as the Rule of Five, Rule of Seven, etc., are all just extensions of the *trairāśika*.]

He says, in order to teach the reversed procedure:

2.28. In a reversed [operation], multipliers become divisors and divisors, multipliers, and an additive [quantity] is a subtractive [quantity], a subtractive [quantity] an additive [quantity]. [...]

Here is an example in another case:

1. Two times [a given quantity] is increased by one, divided by five, multiplied by three and again decreased by two, divided by seven; what has been obtained is one. How much was there before? [...]

This procedure is: what has been obtained is one, 1; multiplied by seven, what results is 7; increased by two, 9; divided by three, 3, with five for multiplier, 15; decreased by one, 14; halved, what has been obtained is 7. [...]

In order to show an example of equations, he says:

2.30. One should divide the difference of coins [belonging] to two men by the difference of beads. What has been obtained is the price of a bead, if what is made into money [for each man] is equal. [...]

An example:

1. The first tradesman has seven horses with perpetual strength and auspicious marks, and a hundred *dravyas* are seen by me in his hand. Nine horses and the amount of eighty *dravyas* [belonging to] the second [tradesman] are seen. The price of one horse, and the equal wealth [of both tradesmen] should be told by [assuming] the same price [for all the horses].

Setting down:

7	100
9	80

Procedure: The difference of beads, 2; the difference of coins, 20. This divided by the difference of beads is the price in *dravyas* of one horse, ten, 10. With that price, the price of the horses of the first [tradesman] is 70, of the second is 90. With what exists in the hand of each [tradesman] and with this [price of each one's horses], equal wealth exists [in the hands of] both as well, 170. [...]

Now, the pulverizer computation is stated. In this case there are two *āryā* rules:

2.32. One should divide the divisor of the greater remainder by the divisor of the smaller remainder. The mutual division [of the previous divisor] by the remainder [is made continuously. The last remainder], having a clever [quantity] for multiplier, is added to the difference of the [initial] remainders [and divided by the last divisor].

2.33. The one above is multiplied by the one below, and increased by the last. When [the result of this procedure] is divided by the divisor of the smaller remainder, the remainder, having the divisor of the greater remainder for multiplier, and increased by the greater remainder, is the [quantity that has such] remainders for two divisors. [...]

[As for] "having a clever [quantity] for multiplier," the meaning is: having a multiplier according to one's own intelligence.

[Question:]

But how is the multiplier according to one's own intelligence?

[It should answer this question:] Will this quantity [the remainder], multiplied by what [is sought], give an exact division, when one has added or subtracted this difference of remainders [to the product]? [...]

An example:

2. [A quantity when divided] by twelve has a remainder which is five, and furthermore, it is seen by me [having] a remainder which is seven, when divided by thirty-one. What should one such quantity be?

Setting down: $\begin{matrix} 5 & 7 \\ 12 & 31 \end{matrix}$

Procedure: "One should divide the divisor of the greater remainder by the divisor of the smaller remainder," the remainder is seven above, twelve below. When the mutual division [of these two is made] the quotient is one, and again one, the remainder is two above, five below. Here the clever [quantity is computed]. There is an even number of terms; therefore, will this quantity, multiplied by what [is sought], when one has added the difference of the remainders which is two unities, give an exact division by five? What has been obtained is four unities, which is the clever [quantity]. One should place it below the previously obtained [quantity]. And the quotient of the division is two; what has been obtained should be placed below. With that rule: "The one above multiplied by the one below is increased by the last remainder," what has been obtained is 10. "When divided by the divisor of the smaller remainder"—the remainder is just this, which has "the divisor of the greater remainder for multiplier"—what is produced is 310. "Increased by the greater remainder, [this] is the [quantity that has such] remainders for two divisors," and that is this, 317.

[In Bhāskara's example, integers N, x, and y are sought such that $N = 12y + 5 = 31x + 7$. The "mutual division of divisors" may be understood as a continued division recasting the equation between two unknown quantities with smaller and smaller coefficients, until it is reduced or "pulverized" into a form that can be solved by inspection. In algebraic notation, we may gloss Bhāskara's comment as follows:

$$y = \frac{31x + 2}{12} = 2x + w,$$
$$x = \frac{12w - 2}{7} = 1w + v,$$
$$w = \frac{7v + 2}{5} = 1v + u,$$
$$v = \frac{5u - 2}{2}.$$

At this point an integer solution may be found "cleverly," i.e., by inspection: the equation is satisfied by $u = 2$, $v = 4$. Then we work our way back up the chain of substitutions, multiplying each successive quotient "above" by the term "below" it and adding the subsequent term. Hence $x = 10$ and $N = 317$.

The remaining excerpts from the last two chapters of the *Āryabhaṭīya* illustrate a few of Āryabhaṭa's astronomical ideas and techniques.]

Chapter 3: Time-determination

3.5. The revolutions of the sun are solar years, the conjunctions of sun and moon lunar months, the conjunctions of earth and sun [civil] days, and the turnings of the stars sidereal [days].

> [...] But what are those turnings of the stars? [The same] as turnings of the earth: "Earth 1582237500 [rotations in a *yuga*]," as said in the [first] chapter [on astronomical constants]. How are these turnings of the earth said [to be the same] as turnings of the stars? The constellations are attached to the orb of the stars. They move toward the western direction because of the pushing of the cosmic wind upon it [the orb]. The constellations see the earth as if [it is] turning with its own motion towards the east: thus the direction of the host of stars is [explained] by means of this motion of the earth. Thus it is said: "and the turnings of the stars sidereal [days]." But others quote [it as] "and the turnings of the earth sidereal [days]." [...]

[Here Āryabhaṭa explains the astronomical basis of some fundamental time-units, and causes some difficulty for his commentator. A *yuga* is a canonical time interval in which a celestial body is considered to complete a certain number of revolutions about the earth, starting from the zero-point of celestial longitude. In his chapter on constants, Āryabhaṭa previously assigned the diurnal cycles of a *yuga* to rotations of the earth instead of revolutions of the stars, as Bhāskara notes (see also his verse 4.9 below). Apparently Bhāskara knew of "others" who read "earth" for "stars" in this verse, probably following what Āryabhaṭa actually said. But Bhāskara, like the vast majority of Indian astronomers, rejected the idea of the earth's motion for physical reasons, and so preferred the reading that seemed to make more sense.]

Chapter 4: The Sphere

4.9. In the same way that someone in a boat going forward sees an unmoving [object] going backward, so [someone] on the equator sees the unmoving stars going uniformly westward. [...]

4.39. Divide the distance between [the centers of] the earth and the sun, multiplied by [the diameter of] the earth, by the difference between [the diameters of] the sun and the earth. The quotient is the length of the earth's shadow [measured] from the [perpendicular] diameter of the earth.

[Bhāskara's commentary on the last part of Āryabhaṭa's text does not survive. In verse 4.39, Āryabhaṭa draws on the properties of similar right triangles, as in his shadow-problems in chapter 2, to determine the length of the shadow cast by the earth in a lunar eclipse. If the

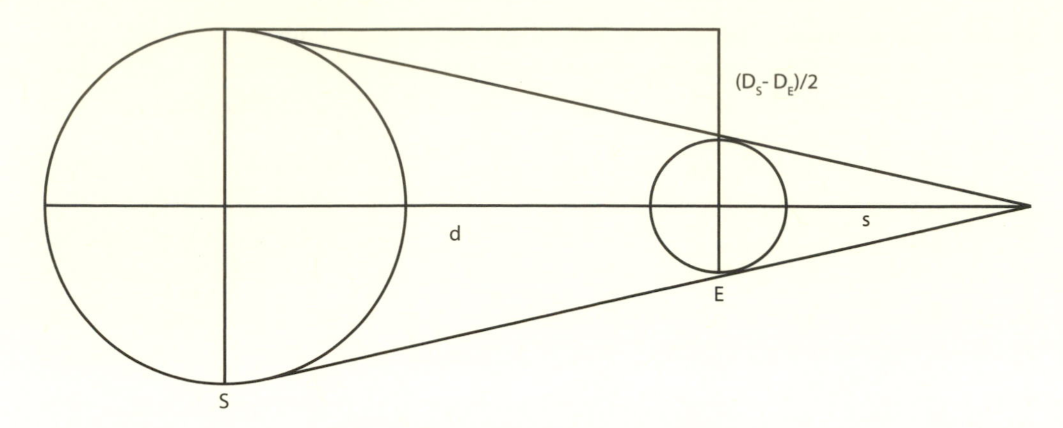

FIGURE 4.14. Computing the earth's shadow.

distance between the centers of the sun and the earth is d, the distance from the center of the earth to the tip of its shadow s, and the difference between the two bodies' diameters $D_S - D_E$, then the expression

$$\frac{s}{D_E} = \frac{d}{D_S - D_E}$$

gives the value of s, as illustrated in figure 4.14.]

[...]

4.50. The [work] named *Āryabhaṭīya* is [equivalent to] the ancient *Svāyaṃbhuva*, always [and] ever. Whoever makes [any] criticism [of it shall have] his good actions and longevity destroyed.

The *Svāyaṃbhuva Siddhānta* that Bhāskara mentions in his commentary on Āryabhaṭa's first verse is alleged to have been spoken by the god Brahmā (one of whose names is Svayaṃbhū) and to have inspired the *Āryabhaṭīya*. This seems to be a reference to an earlier astronomy treatise, the *Paitāmahasiddhānta* (from "Pitāmaha," another name of Brahmā), thought to date from the early fifth century but now surviving only in a corrupt and abridged form. It is structured as a dialogue between Brahmā and a devotee who seeks to learn the science of calculation, and many of its teachings are indeed mirrored in the *Āryabhaṭīya*. The *Paitāmahasiddhānta* also directly inspired another major *siddhānta*, written by a contemporary of Bhāskara: the *Brāhmasphuṭasiddhānta* (*Corrected Treatise of Brahmā*) completed by Brahmagupta in 628. This astronomer was born in 598 and apparently worked in Bhillamāla (identified with modern Bhinmal in Rajasthan), during the reign (and possibly under the patronage) of King Vyāghramukha.

Although we do not know whether Brahmagupta encountered the work of his contemporary Bhāskara, he was certainly aware of the writings of other members of the tradition of the *Āryabhaṭīya*, about which he has nothing good to say. This is almost the first trace we possess of the division of Indian astronomer-mathematicians into rival, sometimes antagonistic "schools." This term should not be taken to imply well-defined academic institutions or lineages; in fact, we know little about their structure beyond the occasional criticisms in their texts. As in the modern quantitative disciplines, mathematical results *per se* apparently provoked little debate; it was in the application of mathematical models to the physical world—in this case, the choices of astronomical parameters and theories—that disagreements arose. For example, the size of the divisions of the *yuga* or astronomical era was a chief point of controversy. The most usual accusation was that of textual heterodoxy, or failure to adhere to the cosmological parameters of a divinely revealed text—which might be about scriptural or, as we have seen in the case of the *Paitāmahasiddhānta*, scientific subjects. Since mathematical and scriptural truth were not necessarily treated as separable, agreement with a sacred authority was considered to be one aspect of scientific validity (heretical as that may seem from the perspective of a modern scientist). As we will see, though, this idea did not always prevent mathematicians from contradicting scriptural models, even as they complained about their rivals doing the same.

Such critiques of rival works appear occasionally throughout the first ten astronomical chapters of the *Brāhmasphuṭasiddhānta*, and its eleventh chapter is entirely devoted to them. But they do not enter into the mathematical chapters that Brahmagupta devotes respectively to *gaṇita* (chapter 12) and the pulverizer (chapter 18). This division of mathematical subjects reflects a different twofold classification from Bhāskara's "mathematics of fields" and "mathematics of quantities." Instead, the first is concerned with arithmetic operations beginning with addition, proportion, interest, series, formulas for finding lengths, areas, and volumes in geometrical figures, and various procedures with fractions—in short, diverse rules for computing with known quantities. The second, on the other hand, deals with what Brahmagupta calls "the pulverizer, zero, negatives, positives, unknowns, elimination of the middle term, reduction to one [variable], *bhāvita* [the product of two unknowns], and the nature of squares [second-degree indeterminate equations]"—that is, techniques for operating with unknown quantities. This distinction is more explicitly presented in later works as mathematics of the "manifest" and "unmanifest," respectively: i.e., what we will henceforth call "arithmetic" manipulations of known quantities and "algebraic" manipulations of so-called "seeds" or unknown quantities. The former, of course, may include geometric problems and other topics not covered by the modern definition of "arithmetic." (Like Āryabhaṭa, Brahmagupta relegates his sine-table to an astronomical chapter where the computations require it, instead of lumping it in with the other "mathematical" topics.)

Brāhmasphuṭasiddhānta *of Brahmagupta*

Chapter 1: Planetary Mean Longitudes

1.32. Nobody at all computes the civil day or the mean [longitudes] using treatises other than [the one] of Brahmā, [such as the ones] by Āryabhaṭa etc., without contradicting scripture.

Chapter 2: Planetary True Longitudes

2.2–5. The sines: the Progenitors, twins; Ursa Major, twins, the Vedas; the gods, fires, six; flavors, dice, the gods; the moon, five, the sky, the moon; the moon, arrows, suns. [...]

[Brahmagupta's sine-table, like much other numerical data in Sanskrit treatises, is encoded mostly in the concrete-number notation that uses names of objects to represent the digits of place-value numerals, starting with the least significant. This is the only time we will translate concrete numbers in these source excerpts literally, to give the flavor of the original sentence, rather than loosely as conventional number words.

There are fourteen Progenitors ("Manu") in Indian cosmology; "twins" of course stands for 2; the seven stars of Ursa Major (the "Sages") for 7; the four Vedas, and the four sides of the traditional dice used in gambling, for 4, and so on. Thus Brahmagupta enumerates his first six sine-values as 214, 427, 638, 846, 1051, 1251. (His remaining eighteen sines are 1446, 1635, 1817, 1991, 2156, 2312, 2459, 2594, 2719, 2832, 2933, 3021, 3096, 3159, 3207, 3242, 3263, 3270.) The *Paitāmahasiddhānta*, however, specifies an initial sine-value of 225 (although the rest of its sine-table is lost), implying a trigonometric radius of $R = 3438 \approx C(')/2\pi$: a tradition followed, as we have seen, by Āryabhaṭa. Nobody knows why Brahmagupta chose instead to normalize these values to $R = 3270$.]

Chapter 7: Lunar Crescent

7.1. If the moon were above the sun, how would the power of waxing and waning, etc., be produced from calculation of the [longitude of the] moon? The near half [would be] always bright.

7.2. In the same way that the half seen by the sun of a pot standing in sunlight is bright, and the unseen half dark, so is [the illumination] of the moon [if it is] beneath the sun.

7.3. The brightness is increased in the direction of the sun. At the end of a bright [i.e., waxing] half-month, the near half is bright and the far half dark. Hence, the elevation of the horns [of the crescent can be derived] from calculation. [...]

7.11–12. The degrees of half the separation of the moon from the sun, multiplied by the moon's diameter, divided by ninety, is the illumination. When [the separation] is [greater than or] equal to ninety degrees, the versine [i.e., versed sine] of twice the degrees (the versine [here] is the [trigonometric] radius increased by the sine of the degrees in excess of ninety) is multiplied by the moon's diameter and divided by twice the radius.

[Brahmagupta discusses the illumination of the moon by the sun, rebutting an idea maintained in scriptures: namely, that the moon is farther from the earth than the sun is. In fact, as he explains, because the moon is closer the extent of the illuminated portion of the moon depends on the relative positions of the moon and the sun, and can be computed from the size of the angular separation α between them. The width L at the widest point of the waxing or waning lunar crescent (measured in "fingers" or digits, like the diameter D_M of the lunar disk) is approximated by the linear proportion $L/D_M = (\alpha/2)/90$, so it will vary from 0 to one-half the width of the disk between new moon and quadrature, as we would expect. When the sun and moon are more than a quadrant apart and the moon is gibbous, L is instead calculated trigonometrically, which we can represent as projecting the illuminated half of its circumference onto a plane perpendicular to the line from the moon to the earth (see fig. 4.15):

$$\frac{L}{D_M} = \frac{\text{Vers}\,\alpha}{2R}$$

where R, as usual, is the trigonometric radius. For those who are puzzled about taking the versed sine of an angle greater than ninety degrees, Brahmagupta helpfully explains that it

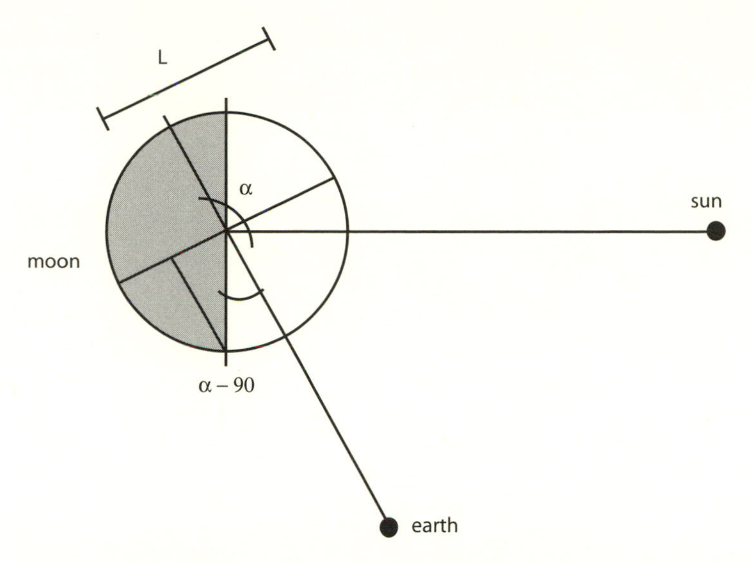

FIGURE 4.15. Projection for the moon's illumination.

is equal to $R + \mathrm{Sin}(\alpha - 90)$. It's not clear why he specifies different rules for the crescent and gibbous lunar phases; either rule could be applied in either case, though the latter is more geometrically exact.]

Chapter 10: Star/Planet Conjunctions

10.62. One is not a master through [knowing] the treatises of Āryabhaṭa, Viṣṇucandra, etc., even when [they are] known [by heart]. [But] one who knows the calculations of Brahmā [attains] mastery.

10.63. Having understood the auspicious, wonderful triple motion of the sun, moon and [lunar] node as spoken by Brahmā, one obtains the worlds of the sun, moon and node, and here [in this world], renown. [...]

10.65–66. Mean motions, true motions, the Three Questions [i.e., direction, place, and time], lunar and solar eclipses, rising and setting [times], the lunar crescent and shadows at any time, the conjunction of planets, [and] the conjunction of stars with planets, are three hundred and seventy-eight *āryā* [verses], in ten chapters. The calculation in the Brahmā [treatise] is without fault.

Chapter 12: Calculation

12.1. Whoever knows separately the twenty operations beginning with addition, and the eight practices ending with shadows—he is a calculator.

12.2. Of two [fractional] quantities the divisors and the numerators when multiplied by the opposite divisors have the same divisor. In addition the numerators are to be added together; in subtraction the difference of the numerators is to be computed.

12.3. Integers are multiplied by the divisor and added to the numerator. The product of the numerators of two or many [fractions] divided by the product of the divisors is [their] multiplication.

12.4. Inverting the divisor and the numerator in division, the divisor [of the fraction] to be divided is multiplied by the divisor [of the inverted fraction] and the numerator is multiplied by the numerator. [This is] the division of two [fractions] reduced to the same denominator.

12.5. The square of the numerator [of a fraction] reduced to the same denominator divided by the square of the divisor is the square [of the fraction]. The square-root of the numerator reduced to the same denominator divided by the square-root of the divisor is the square-root.

12.6. The cube of the last [digit] is to be set down, and, at the first [place] from it, the square of the last [digit] multiplied by three and by the preceding [quantity]. The square of the preceding [quantity] multiplied by three and by the last [digit is set down]. And the cube of the preceding [quantity]. [The result is] the cube.

12.7. The divisor [of the digit] at the second non-cubic [place] is the square of the cube-root multiplied by three; the square of the quotient multiplied by three and by the previous [number] is to be subtracted from the first [non-cubic place] and the cube [of the quotient] from the cubic [place]. [The result is] the root.

12.8. The sum of the numerators [of fractions] having similar divisors, divided by the divisor, is the result in the first reduction of fractions; the numerators are multiplied by the numerators and the divisors by the divisors in the second.

12.9. The upper numerators are multiplied by the divisors in the third reduction of fractions. In the first and second of the two [next cases] the divisors are multiplied by the divisors and the upper numerators by [the divisors] increased or diminished by their own numerators.

[The reader is apparently expected to be familiar with basic arithmetic operations as far as the square-root; Brahmagupta merely notes some points about applying them to fractions. The procedures for finding the cube and cube-root of an integer, however, are described (compare the latter to Āryabhaṭa's very similar formulation). They are followed by rules for five types of combination of fractions:]

$$1.\ \frac{a}{c}+\frac{b}{c} \qquad 2.\ \frac{a}{c}\cdot\frac{b}{d} \qquad 3.\ \frac{a}{1}+\frac{b}{d} \qquad 4{,}5.\ \frac{a}{c}\pm\frac{b}{d}\cdot\frac{a}{c}=\frac{a(d\pm b)}{cd}$$

12.10. In a Rule of Three there are an argument, a result, and a desired quantity; the measures of the first and the last are similar. The desired quantity multiplied by the result and divided by the argument is the result [of the desired quantity].

12.11. The result [of the desired quantity] in an inverse Rule of Three is the product of the argument and the result divided by the desired quantity. In the cases of odd-numbered [proportions] beginning with the Rule of Three and ending with the Rule of Eleven,

12.12. there is a transition of the result on both sides. The result is to be known as the product of many quantities divided by the product of a few. In all the divided [numbers, i.e., fractions], there is transition of the divisors on both sides.

12.13. In bartering goods, there is an exchange of prices first; the rest is the same as what was said [previously].

The operations of the eight practices have been explained.

[...]

12.20. The sum of the squares is that [sum] multiplied by twice the [number of] step[s] increased by one [and] divided by three. The sum of the cubes is the square of that [sum]. Piles of these with identical balls [can also be computed].

[Here the sums of the squares and cubes of the first n integers are defined in terms of the sum of the n integers itself; again, see section IV.c. for more on series.]

12.21. The approximate area is the product of the halves of the sums of the sides and opposite sides of a triangle and a quadrilateral. The accurate [area] is the square-root from the product of the halves of the sums of the sides diminished by [each] side of the quadrilateral.

12.22. The base is decreased and increased by the difference between the squares of the sides divided by the base; when divided by two they are the true segments. The perpendicular [altitude] is the square-root from the square of a side diminished by the square of its segment.

12.23. The square-root of the sum of the two products of the sides and opposite sides of a non-unequal quadrilateral is the diagonal. The square of the diagonal is diminished by the square of half the sum of the base and the top; the square-root is the perpendicular [altitude].

12.24. Subtracting the square of the upright from the square of the hypotenuse, the square-root [of the remainder] is the arm. Subtracting the square of the arm [from the square of the hypotenuse], the square-root [of the remainder] is the upright. The square-root of the sum of the squares of the upright and the arm is the hypotenuse.

12.25. At the junction of the diagonals or at the junction of a diagonal and a perpendicular [altitude] there are an upper and a lower segment. The two segments [of the base, put down] in two places are divided by their sum and separately multiplied by the hypotenuse [diagonal] and by the perpendicular [altitude].

[In a quadrilateral with two pairs of opposite sides a and c, b and d, the approximate area is said to be $(a + c)/2 \cdot (b + d)/2$. If $s = (a + b + c + d)/2$ is the semiperimeter, then the accurate area is $\sqrt{(s-a)(s-b)(s-c)(s-d)}$. This "accurate" area rule is actually exact only for quadrilaterals that are cyclic, discussed further in the following verses. In a triangle, the same area rules apply when one side, say d, is set equal to zero. Verse 22 implies that a triangle's altitude divides its base b into two "segments," which are equal to half of $(b \pm (c^2 - a^2)/b)$ where c is the longer of the other two sides.

According to verse 23, a (cyclic) quadrilateral that is "non-unequal" (specifically, an isosceles trapezium) has each diagonal equal to $\sqrt{ac + bd}$. Verse 25 can be interpreted in a number of ways to give formulas for the various upper and lower segments forming the right triangles created by the intersection of its diagonals and altitudes (see figure 4.16). For example, if the

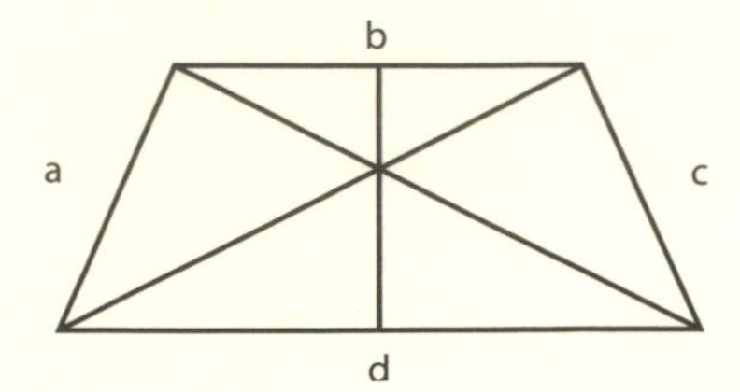

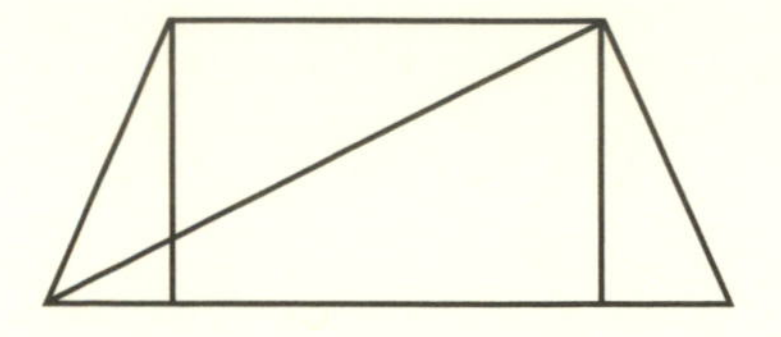

FIGURE 4.16. The segments in a trapezium.

"two segments" mean the top and bottom parallel sides, the rule computes what Āryabhaṭa called the "two lines on their own fallings" (*Āryabhaṭīya* 2.8) and the segments of the diagonals where they cross. Or, consider two altitudes dropped from the top corners of the trapezium, dividing the base into a central segment (equal to the top side) and two equal outer segments. Taking the "two segments of the base" to be the central portion and an outer one, we can use the same rule to find the lengths cut by the intersections of these altitudes and the diagonals.]

12.26. The diagonal multiplied by the side [of a cyclic quadrilateral] with non-unequal flanks and divided by the perpendicular multiplied by two is the heart [i.e., the radius of the circle]. [The radius of a cyclic quadrilateral] with unequal [sides] is half the square-root of the sum of the squares of the side and the counter-side.

12.27. The product of two sides of a [circumscribed] triangle divided by the perpendicular multiplied by two is the heart-string [i.e., the radius]. That [radius] multiplied by two is the diameter of the circle which touches the angles of a trilateral or a quadrilateral.

[Brahmagupta does not explicitly state that he is discussing only figures inscribed in circles, but it is implied by these rules for computing their circumradius. For a "non-unequal" trapezium with oblique side a, altitude h, and diagonal x, the circumradius is $xa/2h$. For an "unequal" or scalene quadrilateral with opposite sides a and c, it is $\sqrt{a^2 + c^2}/2$ (if the diagonals are perpendicular), and for a triangle with base b it is $ac/2h$.]

12.28. One should multiply the sum of the products of the arms adjacent to the diagonals, after it has been mutually divided on either side, by the sum of the two products of the arms and the counter-arms. In an unequal [cyclic quadrilateral] the two square-roots are the two diagonals.

12.29. Imagining a pair of triangles with unequal [sides] separately in the middle of a [cyclic] quadrilateral with unequal [sides], the two segments [are found] by means of the pair of diagonals as before, as are the two perpendiculars separately.

12.30–31. Imagining two triangles within [a cyclic quadrilateral] with unequal sides, the two diagonals are the two bases. Their two segments are separately the upper and lower segments [formed] at the intersection of the diagonals. The two lower segments] of the two diagonals are two sides in a triangle; the base [of the quadrilateral is the base of the triangle]. Its perpendicular is the lower portion of the [central] perpendicular; the upper portion of the [central] perpendicular is half of

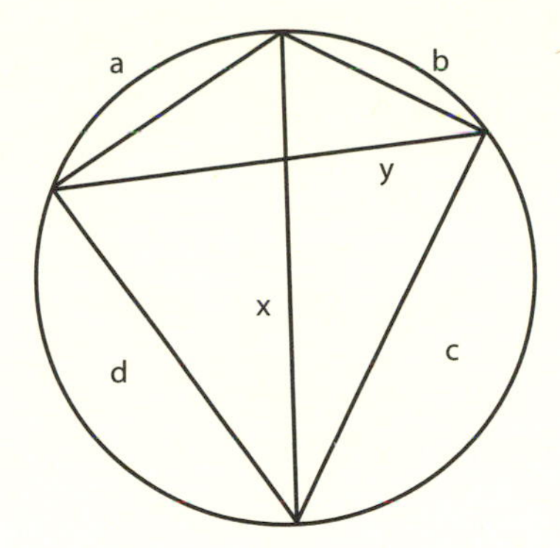

FIGURE 4.17. Brahmagupta's "unequal" cyclic quadrilateral.

the sum of the [side] perpendiculars diminished by the lower [portion of the central perpendicular].

12.32. At a juncture of a diagonal and a perpendicular the two lower portions of the diagonal and the perpendicular [are found] by proportion. [The diagonal and the perpendicular] diminished by these [lower portions respectively] are the upper [segments. The same is the case] in a needle with an intersection.

[For a scalene cyclic quadrilateral as shown in the following figure 4.17, the diagonals x and y are given in verse 28 by

$$x = \sqrt{\frac{(ab+cd)(ac+bd)}{ad+bc}} \quad \text{and} \quad y = \sqrt{\frac{(ad+bc)(ac+bd)}{ab+cd}}.$$

Next is the well-known "Brahmagupta's theorem" about the altitude of a (cyclic) quadrilateral. Brahmagupta describes the various segments of such a quadrilateral, referring also to those of a "needle" figure—i.e., a triangle formed by producing two opposite oblique sides of a quadrilateral until they intersect.]

12.33. The sum of the squares of two unequal quantities is the side [of an isosceles triangle]; twice their product is the perpendicular; and the difference of the squares of the two unequal [quantities] multiplied by two is the base of a triangle with two equal sides.

12.34. The square of [a first] given [quantity is set down] twice [and] divided by a pair of [two other] given [quantities]. The halves of the sums of the [two] results and the given [quantities] are the two sides of a triangle with unequal [sides]; the sum of the halves of the [two] results diminished by the [two] given [quantities] is the base.

12.35. The square of a given arm is divided and diminished by a given [quantity]; half of that is the upright. [When the quotient] is increased by the given [quantity] it is the diagonal of an elongated quadrilateral figure.

12.36. The two diagonals of an elongated [quadrilateral] are the two arms in a quadrilateral with two equal [sides]; the square of the arm [of the oblong] divided by a given [quantity and] diminished by the given [quantity and] divided by two, when increased by the upright, is the base; when diminished [by the upright], the top.

12.37. The square of the diagonal [of an oblong] is three of the sides [of a quadrilateral] with three equal [sides]. The fourth [side is found thus]: subtracting the square of the upright from three times the square of the arm, if it is greater [than the square of the diagonal], it is the base; if less, the top.

12.38. The uprights and arms of two excellent [right triangles] multiplied by each other's hypotenuses are the four sides in [a cyclic quadrilateral with] unequal [sides]. The greatest [side] is the base, the least the top; the other two arms are the pair of sides.

12.39. The height of a mountain multiplied by a given multiplier is the distance to a city; it is not erased. When it is divided by the multiplier increased by two it is the leap of one of the two who make the same journey.

[Here Brahmagupta explains how to construct various figures with rational sides, by manipulating arbitrarily chosen quantities. In essence, he is juxtaposing right triangles to produce, successively, an isosceles triangle, a scalene triangle, an "elongated quadrilateral" or rectangle, an isosceles trapezium, an isosceles trapezium with three equal sides, and a scalene cyclic quadrilateral. Verse 39 tacks on a formula for finding Pythagorean triples: given a length m and an arbitrary multiplier x, construct the sides $a = mx$ and $b = m + mx/(x + 2)$. Apparently phrased as a traditional word problem, the rule represents m as the height of a mountain and mx as the distance from the base of a mountain to a city. Two travelers wish to go from the top of the mountain to the city; one descends the distance m to the base and walks along mx, while his companion (a magician) leaps vertically to the height b and flies along the hypotenuse. See the related "two-monkeys" problem in *Līlāvatī* 156, in section III.c.]

12.40. The diameter and the square of the radius [each] multiplied by 3 are [respectively] the practical circumference and the area [of a circle]. The accurate [values] are the square-roots from the squares of those two multiplied by ten.

12.41. In a circle the square-root from four times the diameter, diminished and multiplied by the versine, is the chord. The square of the chord divided by four times the versine and increased by the versine is the diameter.

12.42. Half of the difference between the diameter and the square-root from the difference between the squares of the chord and the diameter is the small[er] versine. The two diameters diminished and multiplied by the "erosion" and divided by the sum of [the two of them each] diminished by the "erosion" are the two versines.

12.43. The squares of the two half-chords [of two intersecting circles] are divided by a pair of given versines. The two quotients [of which each is] increased [by its own] versine are the two diameters. The sum of the two versines is the "erosion"; the sum of the two quotients is that [sum of the diameters] diminished by the "erosion."

[Brahmagupta accepts $\sqrt{10}$ as an "accurate" value of π as compared to the "practical" value of 3. His geometry of chords focuses on the versed sine and the "erosion," or the width of the double segment shared by two intersecting circles.]

12.44. The result [i.e., volume] of an excavation with equal [sides] is the area [of the base] of the figure multiplied by the height. [The volume] of a needle [i.e., a pyramid] is [that] divided by three. The sums of the sides [in an excavation] with equal tops and bottoms, when divided by the single tops [*ekāgras*], are the "equal rope" [*samarajju*].

12.45–46. The pragmatic computation is the computation of half the sum of the tops and bottoms multiplied by the depth; the superficial computation is half the sum of the computations of the top and bottom multiplied by the depth. Subtracting the pragmatic result from the superficial computation, one should divide the remainder by three; adding the quotient to the pragmatic result, it becomes the accurate result.

12.47. The area of the shape [of a cross-section of a pile] is multiplied by the height. [Or,] half the sum of the top and bottom [breadths] is multiplied by the height [and] the length. [Such] is the computation of the cubic [measure]. The computation of the number of bricks [in a pile] is [the volume] divided by the volume-result of [each] brick.

[After the geometry of plane figures, Brahmagupta discusses the computation of volumes and surface areas of solids (or empty spaces dug out of solids). His straight-forward rules for the volumes of a rectangular prism and pyramid are followed by a more ambiguous one, which may refer to finding the average depth of a sequence of pits with different depths. The next formula apparently deals with the volume of a frustrum of a square pyramid, where the "pragmatic" volume is the depth times the square of the mean of the edges of the top and bottom faces, while the "superficial" volume is the depth times their mean area.]

[...]

12.62. The integer [part] multiplied by the sexagesimal parts of its fractional [part and] divided by thirty is the square of the fractional [part]; it is to be added to the square of the integer [part]. A square and a cube are the products of two or of three equal [quantities].

12.63. Twice the smaller [part] of a quantity increased and multiplied by the greater [part and] added to the square of the smaller [part] is the square. The product of a quantity which has been increased and diminished by an arbitrary [quantity] and added to the square of the arbitrary [quantity] is the square.

12.64. Two squares of an arbitrary smaller quantity are [separately] increased or decreased by two squares of the fractional [part] of the second [quantity and] divided by twice the second quantity, [and] increased or decreased by that,

12.65. in the second place, with which the quotient is equal. The divisor increased or decreased by the result is halved; it is the square-root of the sum or of the difference of the squares, or the second [number] increased and decreased by the result.

12.66. This is just an indication; I will tell the rest in [the chapters on] the origin of sines and on the pulverizer. [This] twelfth chapter, on addition and so on, has 66 *āryā* [verses].

[These are rules for facilitating the computation of squares and square-roots. First, Brahmagupta gives an approximation for the square of a sexagesimal number $a + b/60$, simply omitting the last term in the exact value of the square $a^2 + 2ab/60 + (b/60)^2$. The next verse explains that for some quantity q, $q^2 = (2p + (q - p))(q - p) + p^2$, or else $q^2 = (q + x)(q - x) + x^2$. Finally, the "square-root of the sum or of the difference of the squares" of a sexagesimal number $a + b/60$ and a "smaller quantity" c is apparently approximated by a increased or diminished by a "quotient," $(c^2 \pm (b/60)^2)/2a$.]

Chapter 18: Pulverizer

18.1. Since generally [the answers to] questions can not be known without the pulverizer, therefore I shall explain the pulverizer together with questions.

18.2. [One is] a master of the knowers of treatises by means of knowing the pulverizer, zero, negatives, positives, unknowns, elimination of the middle [term], [reduction to] one *varṇa* ["color" or unknown], *bhāvita*, and the nature [*prakṛti*] of squares.

18.3. Divide the divisor having the greatest remainder [*agra*] by the divisor having the least remainder; whatever is the remainder is mutually divided; the quotient[s] are to be placed separately one below the other.

18.4–5. Multiply the remainder by an arbitrary number such that, when increased by the difference between the two remainders [*agras*], it is eliminated [i.e., there is no remainder]. The multiplier is to be set down [as is] also the quotient. [Beginning] from the last, multiply the next to last by the one above it; [the product], increased by the last, is the end of the remainders [*agrānta*]. Divide [it] by the divisor having the least remainder [*agra*]; multiply the remainder by the divisor having the greatest remainder [*agra*]. Increase [the product] by the greatest remainder [*agra*]; [the result] is the remainder—

18.6. —of the product of the divisors. A *yuga* of two [planets] is the product of divisors; what has lapsed of the *yuga* is the remainder [*agra*] of the two. Thus what has lapsed of a *yuga* of three or more planets is calculated by means of the pulverizer. [...]

18.16. Whoever, when a given remainder of the Sun in revolutions and so on is on a Monday or a Thursday or a Wednesday, tells the zodiacal sign [and so on of the Sun's longitude], he knows the pulverizer.

[Brahmagupta's treatment of the pulverizer is similar to Āryabhaṭa's, and his comments on its use show the subject's astronomical context. In a *yuga*, a celestial body not only completes a canonical number of revolutions but also traverses given numbers of zodiacal signs, degrees, arcminutes, and so on. Therefore, if one knows the number of days d currently elapsed from the start of the *yuga*, one can compute by proportion, say, the number m of accumulated arcminutes traversed by the Sun up to now: $d = pm$, where p is the ratio of days to arcminutes in a *yuga*.

But suppose only part of the required information is available: for example, if one knows the current weekday $w = d \bmod 7$ but not d itself, and the number m_0 of arcminutes in the Sun's current longitude but not the accumulated revolutions, zodiacal signs, and degrees that make up the rest of m. From this "given remainder of the Sun" and the weekday one can set up a first-degree indeterminate equation that satisfies the known conditions. For example, the desired number of days d will be equivalent to $7x + w = (p \cdot 60)y + pm_0$, where x is some integer number of weeks and y is some integer number of degrees (including all the accumulated revolutions) and all the other quantities are given. Then the equation is solved by the method of the pulverizer, and the resulting lump sum of arcminutes can be broken down into accumulated revolutions plus the sign, degree, etc., of the Sun's present longitude. After several additional verses on techniques and examples for this astronomical pulverizer, Brahmagupta turns to some more general algebraic topics, as follows.]

[...]

18.30. [The sum] of two positives is positive, of two negatives negative; of a positive and a negative [the sum] is their difference; if they are equal it is zero. The sum of a negative and zero is negative, [that] of a positive and zero positive, [and that] of two zeros zero.

18.31. [If] a smaller [positive] is to be subtracted from a larger positive, [the result] is positive; [if] a smaller negative from a larger negative, [the result] is negative; [if] a larger [negative or positive is to be subtracted] from a smaller [positive or negative, the algebraic sign of] their difference is reversed—negative [becomes] positive and positive negative.

18.32. A negative minus zero is negative, a positive [minus zero] positive; zero [minus zero] is zero. When a positive is to be subtracted from a negative or a negative from a positive, then it is to be added.

18.33. The product of a negative and a positive is negative, of two negatives positive, and of positives positive; the product of zero and a negative, of zero and a positive, or of two zeros is zero.

18.34. A positive divided by a positive or a negative divided by a negative is positive; a zero divided by a zero is zero; a positive divided by a negative is negative; a negative divided by a positive is [also] negative.

18.35. A negative or a positive divided by zero has that [zero] as its divisor, or zero divided by a negative or a positive [has that negative or positive as its divisor]. The square of a negative or of a positive is positive; [the square] of zero is zero. That of which [the square] is the square is [its] square-root.

18.36. The sum [of two quantities] increased or diminished by [their] difference [and] divided by two is [their] mixture. The difference of [two] squares [of the quantities] divided by the difference [of the quantities themselves] is increased and diminished by the difference [and] divided by two; [this] is the operation of unlikes.

18.37. The perpendicular is the side of a square [*karaṇī*]. Its square divided by an arbitrary [number] is diminished and increased by the arbitrary [number]. The smaller [quantity] is the base; the larger divided by two the arm. Those [numbers] whose product is a square are to be combined;

18.38. the square of the sum of the square-roots of the *karaṇīs* divided by an [arbitrary] number is multiplied by the arbitrary [number] or the square of [their] difference [is so treated]. [The number] to be multiplied is [placed] as many [times as the number of terms in] the multiplier, obliquely one under the other; and their products are added together.

18.39. [The number] to be divided and the divisor are separately multiplied by the divisor with an arbitrary [term] of it negative, and added; [this is done] repeatedly. [The number] to be divided is divided by the divisor [that has] become a single [term]. A square is the product of two identical [numbers].

18.40. The two halves of a *rūpa* increased and decreased by the square-root of the square of the *rūpa* diminished by a given *karaṇī*—the first [the sum] is the *rūpas*, the [one] other than that, the second, is the *karaṇī*—repeatedly.

18.41. The sum and the difference of unknown [numbers], squares, cubes, squares of squares, [and numbers] gone to the fifth or sixth [power, are taken] if they are equal; if [the unknowns] are unequal [they are taken] separately.

18.42. The product of two similar [numbers] is a square; the product of three or more is [a number] gone to that [power]; the product [of numbers] of different types, [that is] the product of colors [*varṇas*] by each other, is the *bhāvita*. The rest is as previously [explained].

[After explaining algebraic sign rules, Brahmagupta prescribes a rule for finding two unknown quantities if their sum and difference are known, or if their difference and that of their squares are known. The subsequent discussion of the *karaṇī*, here apparently meaning a non-square number or its square-root, explicitly recalls its geometric origin. If the given perpendicular $\sqrt{B}$ of an isosceles triangle is a *karaṇī*, values for its side c and base a can be found by manipulating B rather than the unstatable $\sqrt{B}$ itself: choose an arbitrary m and define $a = B/m - m$ and $c = (B/m + m)/2$.

Brahmagupta then lists other techniques for manipulating *karaṇīs* in arithmetic operations. In multiplication of a surd $\sqrt{K}$ by more than one factor we "combine" the factors: $a\sqrt{K} \cdot b\sqrt{K} = ab \cdot K$. To compute $\sqrt{K_1} \pm \sqrt{K_2}$, we seek an arbitrary m such that K_1/m and K_2/m are both squares, so

$$\sqrt{K_1} \pm \sqrt{K_2} = \sqrt{m\left(\sqrt{\frac{K_1}{m}} \pm \sqrt{\frac{K_2}{m}}\right)^2}.$$

The product of a *karaṇī* $\sqrt{K_1}$ with a "multiplier" $\sqrt{K_2} + \sqrt{K_3}$ containing more than one *karaṇī* term is $\sqrt{K_1K_2} + \sqrt{K_1K_3}$. Division of multiple-*karaṇī* terms is accomplished by clearing the *karaṇīs* out of the denominator by means of factors with changed sign, e.g.:

$$\frac{\sqrt{K_1} + \sqrt{K_2}}{\sqrt{K_3} + \sqrt{K_4}} = \frac{(\sqrt{K_1} + \sqrt{K_2})(\sqrt{K_3} - \sqrt{K_4})}{(\sqrt{K_3} + \sqrt{K_4})(\sqrt{K_3} - \sqrt{K_4})} = \frac{(\sqrt{K_1} + \sqrt{K_2})(\sqrt{K_3} - \sqrt{K_4})}{K_3 - K_4}.$$

The *rūpa*, "form" or "unity," is the element of known quantities or constants. Unknowns are represented by the names of different colors, a practice that was apparently already standard when Brahmagupta wrote.]

The seed of equalizing [quantities] of one color.

18.43. The difference between *rūpas*, when inverted and divided by the difference of the unknowns, is the unknown in the equation. The *rūpas* are [subtracted on the side] below that from which the square and the unknown are to be subtracted.

18.44. Diminish by the middle [number] the square-root of the *rūpas* multiplied by four times the square and increased by the square of the middle [number]; divide [the remainder] by twice the square. [The result is] the middle [number].

18.45. Whatever is the square-root of the *rūpas* multiplied by the square [and] increased by the square of half the unknown, diminish that by half the unknown [and] divide [the remainder] by its square. [The result is] the unknown.

[These are Brahmagupta's rules for solving equations in one unknown or "color." The first treats a linear equation where b unknowns plus c *rūpas* are equal to d unknowns plus e *rūpas*; then the unknown $x = (e - c)/(b - d)$. In a quadratic equation where a squares plus b unknowns equal c *rūpas*, x is given by

$$\frac{\sqrt{4ac + b^2} - b}{2a} \quad \text{or} \quad \frac{\sqrt{ac + b^2/4} - b/2}{a},$$

using the so-called "elimination of the middle term."]

18.51. Subtract the colors different from the first color. [The remainder] divided by the first [color's coefficient] is the measure of the first. [Terms] two by two [are] considered [when reduced to] similar divisors, [and so on] repeatedly. If there are many [colors], the pulverizer [is to be used].

[...]

The seed of the *bhāvita* [the product of colors]:

18.60. The product of the *bhāvita* and the *rūpa* increased by the product of the unknowns is divided by an arbitrary [number]; the greater of the arbitrary [number] and the quotient is to be added to the smaller [coefficient]; and the smaller to the larger. [The two sums] divided by the *bhāvita* are [taken] in reverse.

[...]

18.62–63. When one has made arbitrary amounts [for] those [colors] whose product is the *bhāvita*, leaving out [one desired] color, the sum of the [assumed] colors times their [coefficients] is *rūpas*. The product of the [assumed] color-amounts with [the coefficient of] the *bhāvita* thus becomes the [new] coefficient of the desired color [when combined with its original coefficient]. It is solved even without the equation of the *bhāvita*; so why is that done?

[Solving for one variable in terms of another requires isolating it and dividing the remaining terms by its coefficient; if there are more unknowns than equations, one must resort to the pulverizer. An equation containing two unknowns and their product or *bhāvita*, such as $axy = bx + cy + d$, may be solved in two ways. The first technique prescribes dividing $ad + bc$ by some arbitrarily chosen m. Since $ad + bc = (ax - c)(ay - b)$, considering one of the two factors on the right side equal to the arbitrary m allows us to solve for the variable in the other factor, as follows. The resulting quotient n, if larger than m, is added to the smaller of the two coefficients b and c, and m is added to the larger (or if $n < m$, vice versa). The sum involving the x-coefficient b, divided by the coefficient a of the *bhāvita*, is taken "in reverse" to be the value of y, and similarly the other sum becomes the value of x.

Or, we could just assume arbitrary values for all but one of the unknowns and then solve for the remaining "color" directly. As Brahmagupta comments, why bother with the more complicated rule?]

The nature of squares:

18.64. [Put down] twice the square-root of a given square multiplied by a multiplier and increased or diminished by an arbitrary [number]. The product of the first [pair], multiplied by the multiplier, with the product of the last [pair], is the last computed.

18.65. The sum of the thunderbolt-products is the first. The additive is equal to the product of the additives. The two square-roots, divided by the additive or the subtractive, are the additive *rūpas*.

[The so-called "square-nature" methods are ways of solving second-degree indeterminate equations. Here, Brahmagupta explains how to find a solution for what is now commonly known as "Pell's Equation"; we will illustrate the procedure using one of his examples below, namely $83x^2 + 1 = y^2$. The key is to find, for the given "multiplier" N, a solution (a, b) to an auxiliary equation $Na^2 \pm k = b^2$ where $k \neq 1$, and then manipulate a and b to provide a solution to the original equation. If we take our "given square" to be 1 and multiply it by the "multiplier" 83, we want to increase or diminish the result by some quantity to give a perfect square. E.g., $83 \cdot 1^2 - 2 = 9^2$. After we "put down twice" the chosen roots, $\begin{smallmatrix} 1 & 9 \\ 1 & 9 \end{smallmatrix}$, we take the "sum of the thunderbolt-products" (apparently a technical term for cross-multiplication): $1 \cdot 9 + 9 \cdot 1 = 18$. That is the "first" quantity, and the "last" is $83 \cdot 1^2 + 9^2 = 164$. The new "additive" is the square of the previous one: $2 \cdot 2 = 4$, giving $83 \cdot 18^2 + 4 = 164^2$. Then the desired x and y are found by dividing the "first" and "last" by the previous "subtractive": $x = 18/2 = 9$, $y = 164/2 = 82$. The same technique can also be used to form a new solution from two distinct previous solutions, instead of from one solution "put down twice."]

18.66. Computing the square-roots of the additive *rūpas* separately from the square-roots of the given additives or subtractives, whatever are the last and the first square-roots are the given additives or subtractives.

18.67. When four is added, the square of the last square-root decreased by three is halved [and] multiplied by the last square-root; [the result is] a last square-root. The square of the last square-root diminished by one is divided by two [and] multiplied by the first square-root; [the result is] a first square-root.

18.68. If four is subtracted, two squares of the last square-root are [respectively] increased by three and one; [put down] half of [their] product separately [and the same] diminished by one. Multiply by the first [square-root] diminished by one; [the result is] the last [square-root]. Multiply [the other] by the product of the square-roots; [the result is] the first [square-root] pertaining to the last square-root.

18.69. If the multiplier is a square, the additive [quantity] divided and both increased and decreased by some [number] is halved. The first is the last square-root; the second divided by the square-root of the multiplier is the first [square-root].

18.70. If the multiplier is divided by a square [without remainder], the first [square-root] is divided by the square-root of that [square]; if the additive [quantity] is divided by a square [without remainder], then both roots are multiplied by the square-root of that [square].

18.71. The sum of the multipliers multiplied by eight [and] divided by the square of the difference between the multipliers is the quantity; the two multipliers multiplied by three, increased inversely, [and] divided by [their] difference are the two square-roots.

[Other second-degree indeterminate equations can be solved by the techniques described here. For example, if we have a solved auxiliary equation $Na^2 + 4 = b^2$ ("where four is added"), a solution to $Nx^2 + 1 = y^2$ will be $y = b(b^2 - 3)/2$, $x = a(b^2 - 1)/2$.]

18.72. A square is increased and decreased by another square; the sum is divided by the square of half their difference; [the two] are multiplied by that. The sum and the difference are two squares, and also the product increased by one.

18.73. By what *rūpas* a square is diminished and by what increased, the sum of these is divided by an arbitrary [number and] diminished by the arbitrary [number]; the square of half of that, increased by the subtractive [number], is the [desired] quantity.

18.74. By which two [numbers a quantity] is increased and decreased [to form] a square, the difference of these is divided and increased [or] diminished by an arbitrary [number]; the square of half of these, diminished by the additive, is the [desired] quantity of two additives [or] of an additive and a subtractive.

[Systems of simultaneous indeterminate equations involving squares also fall into the "square-nature" category. In verse 72, the sum and difference $(a^2 + b^2)$ and $(a^2 - b^2)$ of two arbitrary squares a^2 and b^2 are computed, and a factor M is defined as "the sum [of that sum and difference] divided by the square of half their difference." Then $x = M(a^2 + b^2)$, $y = M(a^2 - b^2)$ are such that $x + y$, $x - y$, and $xy + 1$ are all squares. If, for given *rūpas* a and b, we want to find some x such that $x + a$ and $x - b$ are both squares, we choose some arbitrary m and define $x = (\frac{1}{2}((a+b)/m - m))^2 + b$. Tinkering with the signs produces expressions for an x that makes $x \pm a$ and $x \pm b$ perfect squares.]

18.75. [He who] computes within a year the square of the remainder on a Wednesday of the minutes of a zodiacal sign, multiplied by ninety-two or by eighty-three [and] increased by one, [that is] a square—[he] is a calculator.

18.76. [Whoever] computes within a year the remainder of the seconds of the [longitude of the] Sun on a Thursday, decreased and multiplied by five [or] by ten, [that is] a square, he is a calculator.

[This astronomical application of "square-nature" methods is artificially modeled on that of the pulverizer. There is no practical astronomical reason to manipulate a known number of arcseconds or arcminutes in an otherwise undetermined longitude so that it produces a perfect square.]

[...]

18.99. With these questions in one's heart one may compose other questions by the thousands; by means of this pronouncement one may thus solve the questions posed by other treatises.

III.b. Transmission of mathematical ideas to the Islamic world

Indian mathematics, of course, did not develop in entire isolation, but we know little about its early interactions with other mathematical traditions. We have seen that some of its topics were apparently inspired by contact with Babylonian and Greek astronomy and astrology (such as the trigonometry of chords which Indian astronomers recast as sines), and foreign scholars in their turn doubtless encountered and absorbed some Indian discoveries. For example, Buddhist travelers brought knowledge of Indian astronomy to China and Central Asia as early as the first centuries CE. And the few surviving fragments of sources on astronomy and astrology in pre-Islamic Iran indicate that they owed much to Sanskrit works. It was perhaps through such intermediaries, as well as Indian commercial contacts, that scholars in West Asia learned about what Severus Sebokht in 662 called the Indians' "valuable methods of calculation" involving decimal place-value numerals, and their "subtle discoveries in astronomy."

More direct transmissions from India to West Asia followed the rapid expansion of Islam. Parts of northwest India were under Muslim political control by the early eighth century, when Muslim merchants were already present on the Malabar coast in the southwest and in other areas. In the second half of the eighth century, a delegation from India arrived at the new caliphate in Baghdad with a *siddhānta* belonging to the school of Brahmagupta, which scholars of the caliph's court translated or adapted into Arabic. The early translations from Sanskrit inspired several other astronomical/astrological works in Arabic; some even imitated the Sanskrit practice of composing technical treatises in verse. Unfortunately, the earliest texts in this genre have now mostly been lost, and are known only from scattered fragments and allusions in later works. They reveal that the emergent Arabic astronomy adopted many Indian parameters, cosmological models, and computational techniques, including the use of sines.

These Indian influences were soon overwhelmed—although it is not completely clear why—by those of the Greek mathematical and astronomical texts that were translated into Arabic under the Abbasid caliphs. Perhaps the greater availability of Greek works in the region, and of practitioners who understood them, favored the adoption of the Greek tradition. Perhaps its prosaic and deductive expositions seemed easier for foreign readers to grasp than elliptic Sanskrit verse. Whatever the reasons, Sanskrit-inspired astronomy was soon mostly eclipsed by or merged with the "Graeco-Islamic" science founded on Hellenistic treatises.

Decimal place-value arithmetic, however, retained its status as a crucial mathematical tool, as well as its name "Indian computation." Arabic algebra too may have been influenced by Indian techniques for calculation with unknowns, although the links are far from clear: for example, we would expect to see in an Indian-inspired algebraic tradition an explicit discussion (such as Brahmagupta provided) of the relations between positive and negative quantities, but Islamic algebra generally does not admit negative numbers. Many other computational techniques in Arabic treatises (such as the "rule of three" and iteration algorithms) owe something to Indian sources, but so far the connections have not been traced distinctly.

In any case, Muslim mathematicians soon came to identify their subject strongly with the methodology and philosophy of Greek mathematics, and presumably grew less receptive to different approaches. Certainly the illustrious Muslim scientist al-Bīrūnī, when he studied Sanskrit exact sciences in India around the turn of the millennium, was frankly chauvinistic about the mathematics he knew, in comparison to its Indian counterpart:

> I began to show [Indian scholars] the elements on which this science rests, to point out to them some rules of logical deduction and the scientific methods of all mathematics, and then they flocked together round me from all parts, wondering, and most eager to learn from me [¼] I can only compare their mathematical and astronomical literature, as far as I know it, to a mixture of pearl shells and sour dates, or of pearls and dung, or of costly crystals and common pebbles. Both kinds of things are equal in their eyes, since they cannot raise themselves to the methods of a strictly scientific deduction. [Sachau 1992, vol. 1, 23, 25]

III.c. Textbooks on mathematics as a separate subject

So far, almost all the Sanskrit mathematics we have seen is incorporated into works dedicated to astronomy. We do not know what other written materials for mathematics instruction were produced along with the early *siddhāntas*, but starting in about the last quarter of the first millennium we encounter several texts devoted to *gaṇita* or calculation in general as a subject in its own right. Probably the first of these is also by far the oldest known physical exemplar of an Indian mathematical text, the so-called Bakhshālī Manuscript (BM), a page of which was shown in fig. 4.6. Discovered in 1881 by a farmer in a field near modern Peshawar in Pakistan, this birch-bark manuscript had apparently been buried for safekeeping and never retrieved. Thus it was protected from the environment and also from being discarded after replacement by a newer copy. Even so, the birch-bark pages had decayed somewhat by the time of its discovery, and their condition has deteriorated further since then. Nonetheless, scholars have been able to recover much of the book's content and to determine that it contained rules and examples for a variety of arithmetic techniques. The manuscript contains no definite information about its date, but its language style and script suggest that it was probably written as far back as the eighth century—almost certainly no later than the twelfth. Much of the content is doubtless even older than the manuscript itself.

The somewhat fragmentary excerpts shown here follow the rule-and-example style of exposition, with the examples first stated as word problems and then laid out numerically, sometimes followed by a verification procedure. Rules (*sūtras*) and examples are presented in verse and the solution is explained in prose. The prose explanations do not always represent the solution procedures quite accurately, suggesting that at least some of the verses are older than the commentary on them. The translation mimics the original notation and layout fairly closely (although *sūtra* labels such as Q2 and N18 are editorial additions). Numbers in the manuscript are set off from the text in boxes, and fractions are expressed numerator over denominator, though without a horizontal bar between them. Negative quantities are denoted by a sign in the form of a "+" following the quantity—most likely derived from an abbreviation of the Sanskrit word for "negative"—rather than the modern "–" preceding it. The notation uses word abbreviations and a standard ordering of the terms, rather than symbolic operators.

The techniques demonstrated in the BM deal largely with solving different types of proportion problems; the three illustrated below are variants of the types that used to be known in English arithmetic as "fellowship" (determining the shares of partners with different amounts of capital), "alligation" or "mixture" (analyzing mixtures of different types of quantities), and "practice" (computation with amounts measured in mixed units). The fundamental law of proportion or *trairāśika*, introduced above in our excerpts from the *Āryabhaṭīya*, is frequently used to verify solutions by other procedures.

The Bakhshālī Manuscript

Sūtra [11]: Having subtracted the numerators [separately] from [their own] denominators, one should make their reciprocals, [and then] reduce [the fractions] to the same denominator. Having then discarded [the denominators], one should point out [the numerators] as the properties.

Example [1 for Sūtra 11]: A jewel is said to be sold to a group of five merchants. What would be the price of the jewel that is said to be sold in this case? ... a half or a third part or a quarter part or a fifth part or a sixth part.

> [...] "[And then] reduce [the fractions] to the same denominator. Having then discarded [the denominators]" [Sūtra 11]. The fractions are reduced to the same denominator. [...] The result is $\frac{120}{60}$, $\frac{90}{60}$, $\frac{80}{60}$, $\frac{75}{60}$, and $\frac{72}{60}$. "Having then discarded [the denominators]" [Sūtra 11]. The result is: 120, 90, 80, 75, and 72. The sum of these is taken; the result is 437. Sixty having been subtracted from this, the remainder is 377. This is the price of the jewel.
>
> [The price of the jewel is also calculated from] the total sum of the properties of four [merchants and from the given part of the property of the remaining one]. Half of the property of the first [merchant] is 60. [The properties of the other four merchants are:] 90, 80, 75, and 72. The sum of the four [merchants], 317, is increased by half of the first's, i.e., by sixty: 377. This is] the property of the first. The total sum of the property of the first and those of the third, fourth, and fifth is 347, [which is] increased by a third part of the second's, 30. The result is 377. This is the property of the second. Again, the total property of the first, second, fourth, and fifth is 357, and a quarter of the third's is 20. [The former], increased by this, is 377. This is the property of the third. Again, [the total sum of the properties of] the first, second, third, and fifth is 362 [...] This [...] Again [...] is the property of the [...]
>
> Now, [the Statement] of the first [case] is made:
>
> | 120
90
80
75
72 | Added up; the result is 437. Sixty should be subtracted from this; the remainder is 377. |
> |---|---|

Now, [the Statement] of the second [case]:

120
30
80
75
72

[where] one third [of the property of the second merchant] is taken. [The sum is] 377, [which is] the property of the second.

Now, [the Statement] of the third [case] is made:

120
90
20
75
72

Thus, [the sum is] 377, [which is] the property of the third.

[The Statement] of the fourth [case] is made.

120
90
80
15
72

Thus, [the sum is] 377, [which is] the property of the fourth.

The Statement of the fifth [case] is made.

120
90
80
75
12

Thus, [the property] of the fifth is 377. This is the price of the jewel, [which is the same] in every case. [...]

[The above problem asks us to determine the price of a jewel purchased by five merchants, said to be equal to the wealth of any four of them plus a certain portion of the wealth of the remaining one. The portion is one-half in the case of the first merchant, one-third for the second, one-fourth for the third, and so on. This is actually an indeterminate system of linear equations.

According to the *sūtra*, if we choose fractions equal to unity minus the reciprocal of each given portion and reduce them to the same denominator (say, d), the numerators $n_1, \ldots, n_5$ represent the respective wealth of each merchant. The example illustrates the computation of the price as equal to $(n_1 + n_2 + \cdots + n_5) - d$.]

[...]

I will next expound the impurity of gold, for which this is a *sūtra*.

[Sūtra 27:] Having multiplied [the weights of] the gold pieces by [their own] impurities, one should then divide their sum by [the weights of] the gold pieces added together. This [result] is indeed the loss [of gold] per unit [weight] (i.e., impurity) [of the alloy].

Example [1 for Sūtra 27]: [Four] gold pieces, the quantities of which are one, two, three, and four *suvarṇas* [respectively], are inferior by one, two, three, and four negative *māṣas* [per *suvarṇa* in order]. [They are melted into] one having a single luster. [What is the impurity of the alloy?]

Statement of those [given numbers] is made.

−1	−2	−3	−4	[impurity in *māṣas*]
1	2	3	4	[weight in *suvarṇas*]

Computation. "Having multiplied [the weights of] the gold pieces by [their own] impurities," says [Sūtra 27]. When one has multiplied [each weight] by [its own] impurity, the result is: 1, 4, 9, and 16 [...] Their sum is 30. The sum of [the weights of] the gold pieces is 10. When one has divided [the 30] by this, the quotient is 3. [This is the impurity of the alloy.]

Verification by means of the *trairāśika*:

$\frac{10}{1}$	$\frac{30}{1}$	$\frac{1}{1}$	result: $\frac{3}{1}$ *māṣas*
$\frac{10}{1}$	$\frac{30}{1}$	$\frac{2}{1}$	result: $\frac{6}{1}$ *māṣas*
$\frac{10}{1}$	$\frac{30}{1}$	$\frac{3}{1}$	result: $\frac{9}{1}$ *māṣas*
$\frac{10}{1}$	$\frac{30}{1}$	$\frac{4}{1}$	result: $\frac{12}{1}$ *māṣas*

Example [2 for Sūtra 27]: There are [four] gold pieces, the quantities of which are one, two, three, and four *suvarṇas* [respectively]. They are inferior by one half, one third, one fourth, and one fifth of a *māṣa* [per *suvarṇa* in order]. What is the impurity [of the alloy made from them]?

1	2	3	4	[weight in *suvarṇas*]
[−]$\frac{1}{2}$	[−]$\frac{1}{3}$	[−]$\frac{1}{4}$	[−]$\frac{1}{5}$	[impurity in *māṣas*]

Computation. "Having multiplied [the weights of] the gold pieces by [their own] impurities" [Sūtra 27]. [The result of] this [computation] is set up:

$\frac{1}{2}$	$\frac{2}{3}$	$\frac{3}{4}$	$\frac{4}{5}$

"One should then divide their sum" [Sūtra 27]. Reduction to the same denominator having been made, [the fractions are] added together: $\frac{163}{60}$. When one has divided [this] 'by [the weights of] the gold pieces added together'—[the sum of the weights of] the gold [pieces] in the present case is 10—when one has divided [the $\frac{163}{60}$] by this, the result is $\frac{163}{600}$. This is the loss [of gold] per *suvarṇa*.

A verification should be made by means of the *trairāśika*.

$\frac{10}{1}$	$\frac{163}{60}$	$\frac{1}{1}$	result: $\frac{163}{600}$ *māṣas*
$\frac{10}{1}$	$\frac{163}{60}$	$\frac{2}{1}$	result: $\frac{163}{300}$ *māṣas*
$\frac{10}{1}$	$\frac{163}{60}$	$\frac{3}{1}$	result: $\frac{163}{200}$ *māṣas*
$\frac{10}{1}$	$\frac{163}{60}$	$\frac{4}{1}$	result: $\frac{163}{150}$ *māṣas*

[The last terms are] added together; the result is $\frac{1630}{600}$ *māṣas*, to be reduced to $\frac{163}{60}$ [*māṣas*].

[The *māṣa* and *suvarṇa* (generally equal to sixteen *māṣas*) are units of weight. If n pieces of gold each have weight w_i *suvarṇas*, and impurity k_i *māṣas* of non-gold material per *suvarṇa* (for $i = 1$ to n), then the combined impurity of all of them when melted together must be

$$\frac{w_1k_1 + w_2k_2 + \cdots + w_nk_n}{w_1 + w_2 + \cdots + w_n}$$

The calculation is verified by computing, by means of the *trairāśika*, the impurity of each of the n original weights as part of the combined total.]

[...]

[2nd solution of Example 1 for Sūtra N18.]

[...] "And squared. When one has added [the square] to it" [Sūtra N18]: 481. The square root [obtained before] is $21\frac{20}{21}$. This is inaccurate. Therefore, [we recall Sūtra Q2 again].

[Sūtra Q2:] The divisor for the remainder from the diminution of the non-square number by the square of the first approximation, is twice [the first approximation]. Division of half the square of that (i.e., the quotient just obtained) by the second approximation [is made]. Subtraction [of the result from the second approximation gives the third approximation]. Less the square.

[The first half of this rule], "The divisor for the remainder [...] is twice," has been already applied, [the result being $21\frac{20}{21}$]. "The square of that,"

21				"the 2nd approx."	21	: divisor
	–					
$\frac{20}{21}$	$\frac{400}{441}$	"half"	$\frac{1}{2}$		$\frac{20}{21}$	

The remainder, [400], should be removed [and preserved separately for verification]. When one has applied [the rule], "a factor into a factor" [Sūtra Q1], [the result is] divided: $\frac{21}{461}$. "The lower should be multiplied by the lower, and the upper by the upper" [Sūtra Q3]. [The result is $\frac{400}{19362}$.] One should [then] subtract the square. There, reduction to the same [denominator] is made:

$\frac{425042}{19362}$	$-\frac{400}{19362}$	Remainder:	$\frac{424642}{19362}$

[This is part of a problem involving the solution of a quadratic equation, and employs an interesting approximation formula for non-integer square roots. If a non-square integer C is equal to $A^2 + r$ where A and r are integers, then

$$\sqrt{C} \approx A + \frac{r}{2A} - \frac{(r/2A)^2}{2(A + r/2A)}.$$

The above equation can be thought of as the second iteration of the well-known root-approximation rule

$$\sqrt{C} \approx A_{i+1} = A_i + \frac{r}{2A_i}.$$

In this example, $C = 481$ and $A = 21$, giving $r/2A = 40/42 = 20/21$. The commentary prescribes, not very clearly, finding "half the square of that" and, after some manipulation of fractions, deriving $\sqrt{481} \approx 424642/19362$.]

[...]

[Sūtra 58:] When a [sales] tax is imposed on garments, calculate correctly the tax [to be paid in cash], by means of the rule of *trairāśika*, from [the fractional portion of] a garment remaining after that [tax in kind] has been taken away.

Example [1 for Sūtra 58]: A [sales] tax on garments is one twentieth [of the value]. A certain man buys forty-two garments by *paṇas*. Two garments are taken away as [a part of] the tax. [...] Also, ten *paṇas* [are paid as the remaining portion of the tax]. What is the price [of a piece of garment]? Tell [me], O learned one!

[When n garments costing a units apiece are purchased, m of them plus a cash value b units are charged as a sales tax, which is $1/p$ the total value:

$$na/p = ma + b.$$

In this example, the price per garment is therefore 100 of the copper coins called *paṇas*.]

[...]

[Mā]rtikāvatī [...] This [book] has been written by the king of the mathematicians, a *brāhmaṇa*, the son of Chajaka, for the sake of Hasika, son of Vasiṣṭha, in order that it may [also] be used by his descendants.

Mathematics stands at the top of all sciences. Mathematics, produced in the mundane world that has a beginning and end, is great. It was after the Creation that [mathematics was] to be generated by Śiva, the Supreme Spirit. Mathematics that was produced, and that begins with (?) [...], is based on numbers. [...]

[This is the colophon or end of the manuscript; Sanskrit colophons typically contain information about the title and author of the work, the place and date of its copying, and the identity of its scribe and of the patron who commissioned the copy, sometimes followed by an auspicious verse. The available portion of the BM colophon has no date, but it appears to mention the locality Mārtikāvatī, probably referring to an ancient city near Peshawar. The "son of Chajaka" may have been the author of the commentary or some part of it, or perhaps just a scribe.]

In the ninth century, a Jain scholar named Mahāvīra (apparently under the patronage of a ninth-century ruler of Maharashtra and Karnataka) composed what is now the earliest Sanskrit textbook on mathematics per se to survive in a complete form. The classification of operations and rules in his *Gaṇitasārasaṅgraha* or "Epitome of Calculation" somewhat resembles that in the work of Brahmagupta, but is much more detailed. The excerpts shown here represent, to borrow Mahāvīra's elaborate eulogistic metaphor, glimpses of the "water," the "bank," the "crocodile," and the "gems" pertaining to his mathematical ocean.

The technical part of the text begins with definitions of units from the "infinitely minute" particle upwards reflecting Jain metaphysical interests in varieties of the infinitely large and infinitely small. Mahāvīra moves on from metrology to the eight basic operations (but unusually, he omits from the list addition and subtraction, presumably as being too elementary, and substitutes instead summation and subtraction of series). The remaining excerpts are taken from his fourth chapter on "miscellaneous problems," dealing with various kinds of formulas for solving quadratics or problems involving square roots, and from the sixth chapter on "mixed problems." At the beginning of the *Gaṇitasārasaṅgraha*, as at the end of the BM, the author offers a lyrical tribute to the importance and ubiquity of mathematical computation; first, though, he salutes his namesake, the founder of Jainism.

Gaṇitasārasaṅgraha *of Mahāvīra*

Chapter 1: On Terminology

Salutation and benediction

1.1. Salutation to Mahāvīra, the Lord of the Jinas, the protector [of the faithful], whose four infinite attributes, worthy to be esteemed in [all] the three worlds, are unsurpassable [in excellence].

1.2. I bow to that highly glorious Lord of the Jinas, by whom, as forming the shining lamp of the knowledge of numbers, the whole of the universe has been made to shine. [...]

1.9. In all these transactions which relate to worldly, Vedic or [other] similarly religious affairs, calculation is of use.

1.10. In the science of love, in the science of wealth, in music and in the drama, in the art of cooking, and similarly in medicine and in things like the knowledge of architecture;

1.11. In prosody, in poetics and poetry, in logic and grammar and such other things, and in relation to all that constitutes the peculiar value of [all] the [various] arts, the science of computation is held in high esteem.

1.12. In relation to the movements of the sun and other heavenly bodies, in connection with eclipses and the conjunctions of planets, and in connection with the *tripraśna* and the course of the moon—indeed in all these [connections] it is utilized.

1.13–14. The number, the diameter and the perimeter of islands, oceans and mountains; the extensive dimensions of the rows of habitations and halls belonging to the inhabitants of the [earthly] world, of the interspace [between the worlds], of the world of light, and of the world of the gods; [as also the dimensions of those belonging] to the dwellers in hell; and [other] miscellaneous measurements of all sorts—all these are made out by means of computation.

1.15. The configuration of living beings therein, the length of their lives, their eight attributes and other similar things, their progress and other such things, their staying together and such other things—all these are dependent upon computation [for their due measurement and comprehension].

1.16. What is the good of saying much in vain? Whatever there is in all the three worlds, which are possessed of moving and non-moving beings—all that indeed cannot exist as apart from measurement.

1.17–19. With the help of the accomplished holy sages, who are worthy to be worshipped by the lords of the world, and of their disciples and disciples' disciples, who constitute the well-known jointed series of preceptors, I glean from the great ocean of the knowledge of numbers a little of its essence, in the manner in which gems are [picked up] from the sea; gold is from the stony rock and the pearl from the oyster shell; and give out, according to the power of my intelligence, the *Sārasaṅgraha*, a small work on arithmetic, which is [however] not small in value.

1.20–23. Accordingly, from this ocean of *Sārasaṅgraha*, which is filled with the water of terminology and has the [eight] arithmetical operations for its bank; which [again] is full of the bold rolling fish represented by the operations relating to fractions, and is characterized by the great crocodile represented by the chapter of miscellaneous examples; which [again] is possessed of the waves represented by the chapter on the rule-of-three, and is variegated in splendor through the luster of the gems represented by the excellent language relating to the chapter on mixed problems; and which [again] possesses the extensive bottom represented by the chapter on area-problems, and has the sands represented by the chapter on the cubic contents of excavations; and wherein [finally] shines forth the advancing tide represented by the chapter on shadows, which is related to the department of practical calculation in astronomy—[from this ocean] arithmeticians possessing the necessary qualifications in abundance will, through the instrumentality of calculation, obtain such pure gems as they desire.

1.24. For the reason that it is not possible to know without [proper] terminology the import of anything, at the [very] commencement of this science the required terminology is mentioned.

Terminology relating to [the measurement of] space

1.25–27. That infinitely minute [quantity of] matter, which is not destroyed by water, by fire and by other such things, is called a *paramāṇu*. An endless number of them makes an *aṇu*, which is the first [measure] here. The *trasarēṇu* which is derived therefrom, the *ratharēṇu*, thence [derived], the hair-measure, the louse-measure, the sesamum-measure, which [last] is the same as the mustard-measure, then the barley-measure and [then] the *aṅgula* are [all]—in the case of [all] those who are born in the worlds of enjoyment and worlds of work, which are [all] differentiated as superior, middling and inferior—eight-fold [as measured in relation to what immediately precedes each of them], in the order [in which they are mentioned]. This *aṅgula* is known as *vyavahārāṅgula*.

1.28. Those, who are acquainted with the processes of measurement, say that five hundred of this [*vyavahārāṅgula*] constitutes [another *aṅgula* known as] *pramāṇa*. The finger measure of men now existing forms their own *aṅgula*. [...]

Names of the operations in arithmetic

1.46. The first among these [operations] is *guṇakāra* [multiplication], and it is also [called] *pratyutpanna*; the second is what is known as *bhāgahāra* [division]; and *kṛti* [squaring] is said to be the third.

1.47. The fourth, as a matter of course, is *varga-mūla* [square root], and the fifth is said to be *ghana* [cubing]; then *ghanamūla* [cube root] is the sixth, and the seventh is known as *citi* [summation].

1.48. This is also spoken of as *saṅkalita*. Then the eighth is *vyutkalita* [the subtraction of a part of a series, taken from the beginning, from the whole series], and this is also spoken of as *śeṣa*. All these eight [operations] appertain to fractions also.

General rules in regard to zero and positive and negative quantities

1.49. A number multiplied by zero is zero, and that [number] remains unchanged when it is divided by, combined with [or] diminished by zero. Multiplication and other operations in relation to zero [give rise to] zero; and in the operation of addition, the zero becomes the same as what is added to it.

1.50. In multiplying as well as dividing two negative [or] two positive [quantities, one by the other], the result is a positive [quantity]. But it is a negative quantity in relation to two [quantities], one [of which is] positive and other negative. In adding a positive and a negative [quantity, the result] is [their] difference.

1.51. The addition of two negative [quantities or] of two positive [quantities gives rise to] a negative or positive [quantity] in order. A positive [quantity] which has to be subtracted from a [given] number becomes negative, and a negative [quantity] which has to be [so] subtracted becomes positive.

1.52. The square of a positive as well as of a negative [quantity] is positive; and the square roots of those [square quantities] are positive and negative in order. As in the nature of things, a negative [quantity] is not a square [quantity]; it has therefore no square root. [...]

Chapter 2: Arithmetic Operations

Summation

2.61. The number of terms in the series is [first] diminished by one and [is then] halved and multiplied by the common difference; this when combined with the first term in the series and [then] multiplied by the number of terms [therein] becomes the sum of all [the terms in the series in arithmetical progression]. [...]

2.65. [Each of] ten merchants gives away money [in an arithmetically progressive series] as a religious offering, the first terms of the [ten] series being from 1 to 10, the common difference [in each of these series] being of the same value [as the first terms thereof], and the number of terms being 10 [in every one of the series]. Calculate the sums of those [series]. [...]

2.69. When, to the square root of the quantity obtained by the addition of the square of the difference between twice the first term and the common difference to 8 times the common difference multiplied by the sum of the series, the common difference is added, and the resulting quantity is halved; and when [again] this is diminished by the first term and then divided by the common difference, we get the number of terms in the series. [...]

2.72. The first term is 5, the common difference 8, and the sum of the series 333. What is the number of terms? [...]

2.93. The first term [of a series in geometrical progression], when multiplied by that self-multiplied product of the common ratio, in which [product the frequency of the occurrence of the common ratio is] measured by the number of terms [in the series], gives rise to the *guṇadhana*. And it has to be understood that this *guṇadhana*, when diminished by the first term, and [then] divided by the common ratio lessened by one, becomes the sum of the series in geometrical progression. [...]

2.99. A certain man [in going from city to city] earned money [in a geometrically progressive series] having 5 *dināras* for the first term [thereof] and 2 for the common ratio. He [thus] entered 8 cities. How many are the *dināras* [in] his [possession]? [...]

Chapter 4: Miscellaneous Problems [on Fractions]

4.4. In the operation relating to the *bhāga* variety, the [required] result is obtained by dividing the given quantity by one as diminished by the [known] fractions. In the operation relating to the *śeṣa* variety, [the required result] is the given quantity divided by the product of [the quantities obtained respectively by] subtracting the [known] fractions from one. [...]

4.23–27. A collection of bees characterized by the blue color of the shining *indranīla* gem was seen in a flowering pleasure-garden. One-eighth of that [collection] became hidden in *aśoka* trees, $\frac{1}{6}$ in *kuṭaja* trees. The difference between those that hid themselves in the *kuṭaja* trees and the *aśoka* trees, respectively, multiplied by 6, became hidden in a crowd of big *pāṭalī* trees. The difference between those that hid themselves in the *pāṭalī* trees and the *aśoka* trees, diminished by $\frac{1}{9}$ of itself became hidden in an extensive forest of *sāla* trees. The same difference, together with $\frac{1}{7}$ of itself, became hidden in a forest of *madhuka* trees; $\frac{1}{5}$ of that whole collection of bees was seen hidden in the *vakula* trees with well-blossomed flower-buds; and that same $\frac{1}{5}$ part was found hidden in *tilaka, kuravaka, sarala* and mango trees, and on collections of lotuses, and at the base of the temples of forest elephants; and 33 [remaining] bees were seen in a crowd of lotuses, that were variegated in color on account of the large quantity of [their] filaments. Give out, O you arithmetician, the [numerical] measure of that collection of bees. [...]

4.29–30. Of a collection of mango fruits, the king [took] $\frac{1}{6}$; the queen [took] $\frac{1}{5}$ of the remainder, and three chief princes took $\frac{1}{4}, \frac{1}{3}, \frac{1}{2}$ [of that same remainder]; and the youngest child took the remaining three mangoes. O you, who are clever in [working] miscellaneous problems on fractions, give out the measure of that [collection of mangoes]. [...]

The rule relating to the *śeṣamūla* variety [of miscellaneous problems on fractions].

4.40. [Take] the square of half [the coefficient] of the square root [of the remaining part of the unknown collective quantity], and combine it with the known number remaining, and [then extract] the square root [of this sum, and make that square root become] combined with half of the previously mentioned [coefficient of the] square root [of the remaining part of the unknown collective quantity]. The square of this [last sum] will here be the required result, when the remaining part [of the unknown collective quantity] is taken as the original [collective quantity itself]. But when that remaining part [of the unknown collective quantity] is treated merely as a part, the rule relating to the *bhāga* variety [of miscellaneous problems on fractions] is to be applied.

4.41. One-third of a herd of elephants and three times the square root of the remaining part [of the herd] were seen on a mountain slope; and in a lake was seen a male elephant along with three female elephants [constituting the ultimate remainder]. How many were the elephants here?

4.42–45. In a garden beautified by groves of various kinds of trees, in a place free from all living animals, many ascetics were seated. Of them the number equivalent to the square root of the whole collection were practicing *yoga* at the foot of the trees. One-tenth of the remainder, the square root [of the remainder after deducting this], $\frac{1}{9}$ [of the remainder after deducting this], then the square root [of the remainder after deducting this], $\frac{1}{8}$ [of the remainder after deducting this], the square root [of the remainder after deducting this], $\frac{1}{7}$ [of the remainder after deducting this], the square root [of the remainder after deducting this], $\frac{1}{6}$ [of the remainder after deducting this], the square root [of the remainder after deducting this], $\frac{1}{5}$ [of the remainder after deducting this], the square root [of the remainder after deducting this]—these parts consisted of those who were learned in the teaching of literature, in religious law, in logic, and in politics, as also of those who were versed in controversy, prosody, astronomy, magic, rhetoric and grammar and of those who possessed the power derived from the 12 kinds of austerities, as well as of those who possessed an intelligent knowledge of the twelve varieties of the *aṅga-śāstra*; and at last 12 ascetics were seen [to remain without being included among those mentioned before]. O [you] excellent ascetic, of what numerical value was [this] collection of ascetics? [...]

[These rather daunting word problems are just specific types of what we would call equations in one unknown. The *bhāga* or "fraction" problems set a desired total x equal to a known quantity c plus various fractional parts $a_1x, a_2x, \ldots, a_nx$; the value of x is therefore given by $c/(1-(a_1+a_2+\cdots+a_n))$. A *śeṣa* or "remainder" problem sets x equal to c plus n successive fractions of the successively diminished remainder of x, so

$$x = \frac{c}{\left(\frac{n}{n+1}\right)\left(\frac{n-1}{n}\right)\cdots\left(\frac{1}{2}\right)}.$$

In a *mūla* or "square-root" problem, one solves a quadratic equation in $\sqrt{px}$ where p is some factor; thus the number of elephants sought in verse 41 is given by $(2/3)x - 3\sqrt{(2/3)x} - 4 = 0$.]

Chapter 6: Mixed Problems

6.216. [The number of] men, times [the number of the ones who are] beloved plus one, diminished by twice [the number] beloved, is [the number of] false [statements]. The square of [the number of] men minus those is [the number of] true statements.

6.217. Five men are enamored of a prostitute and three [of them] are dear [to her]. Then she says to each one [separately], "You are [my only] beloved." How many [of her statements] are true?

The rule for the varieties of combination.

6.218. When [one has] set down a sequence of [numbers] beginning with and increasing by one, above and below, in order and in reverse order [respectively], a reverse-product [in the numerator] divided by a reverse-product [in the denominator] is [the size of] the extension [of the different varieties of combination].

6.219. Oh mathematician, tell [me] now the characters and varieties of the combinations of the flavors: astringent, bitter, sour, pungent, and salty, together with the sweet flavor.

6.220. You tell [me] right away, friend: how many varieties [are there], because of varieties of combination, of a necklace made with diamonds, sapphires, emeralds, corals, and pearls?

6.221. Oh friend knowledgeable in the principles of calculation, tell [me]: how many varieties [are there], because of varieties of combination, of a garland made of the flowers of the pandanus, the *aśoka* tree, the *campaka* tree, and the blue lotus?

[The first of these two rules concerns the truth values of a set of n explicit statements and the $(n^2 - n)$ implicit statements they entail: "A loves (only) B" implies also "A does not love C" and "A does not love D" and so forth. So the prostitute's assurances to her n lovers actually constitute n^2 statements. If she genuinely likes m men, then the same statement made to each of them implies $(m - 1)$ lies about not liking the others. The same statement made to each of the remaining $(n - m)$ disliked men, however, contains m implicit lies and one explicit lie. So the total number of true statements, explicit and implicit, is $n^2 - (m(m-1) + (n-m)(m+1))$.

The rule for determining the number of possible combinations when choosing some number r of n things considers first a fraction made of products of the first n integers, in order and reversed: $(1 \cdot 2 \cdot 3 \cdots n)/(n(n-1)(n-2)\cdots 1)$. Then the "reverse" product of the $(n - r)$ rightmost factors is selected in both numerator and denominator. Their quotient, $((r+1)(r+2)\cdots n)/((n-r)(n-r-1)\cdots 1)$, is the number of possible combinations.]

Other authors of mathematical treatises or mathematical chapters in astronomical treatises in this period include Śrīdhara in (probably) the eighth century, of whose writings three summaries on arithmetic still survive in incomplete form (his work on algebra, and probably a full-length work on arithmetic, are lost); Āryabhaṭa (II) in probably the mid- to late tenth century; Śrīpati and (probably) Jayadeva in the mid-eleventh century. More popular than any of their works were the twelfth-century arithmetic and algebra textbooks by the renowned astronomer Bhāskara (II—no known relation to the seventh-century commentator on the *Āryabhaṭīya*): the *Līlāvatī* ("Beautiful") and *Bījagaṇita* ("Seed-Computation") respectively. Bhāskara, who lived in the Sahyadri region in western Maharashtra, was born in 1114 into a family whose members may have filled hereditary posts as court scholars (at least, it is recorded that his great-great-great-grandfather held such a position under a noble patron, as did Bhāskara's son and some other descendants). Hardly anything is known about the other events of Bhāskara's life; it is speculated that he may have had a daughter named Līlāvatī because of his allusions to a girl so addressed in his book on arithmetic, and his son's son helped to set up a school in 1207 for the study of Bhāskara's writings.

The impact and eventual fate of that school are unknown, but the many thousands of surviving manuscripts of Bhāskara's works leave no doubt that they were very widely read. In fact, the *Līlāvatī* and the *Bījagaṇita* are generally regarded as the best-known—indeed, the standard—mathematics textbooks of the Sanskrit tradition. Comparisons of the content of the *Līlāvatī* with the mathematical chapters by Āryabhaṭa and Brahmagupta suggest the ways in which the subject of *gaṇita* had matured in the course of several centuries:

- The organization of the text systematically presents units, place value, the eight arithmetic operations (now standardized as $+, -, \times, \div, x^2, \sqrt[2]{x}, x^3, \sqrt[3]{x}$), operations with fractions, and operations with zero. These are followed by various algorithms for solving what we would consider linear and quadratic algebra problems; the *trairāśika* and other proportion rules; basic combinatorics formulas; geometry of right triangles, of other rectilinear figures, of circles, of various solid figures, and of shadows; and further results in combinatorics.
- All rules are illustrated by examples, and Bhāskara's own accompanying prose commentary is interspersed among the verses containing the examples and rules. (The verses are represented in this translation by numbered paragraphs.) This "auto-commentary" supplies the numerical statements and solutions for all the examples. Word problems, as we have already seen in the work of Mahāvīra, often display elaborate imagery and rhetorical flourishes. (Sometimes they also inspired artistic representations, as seen in an unknown scribe's careful depiction in

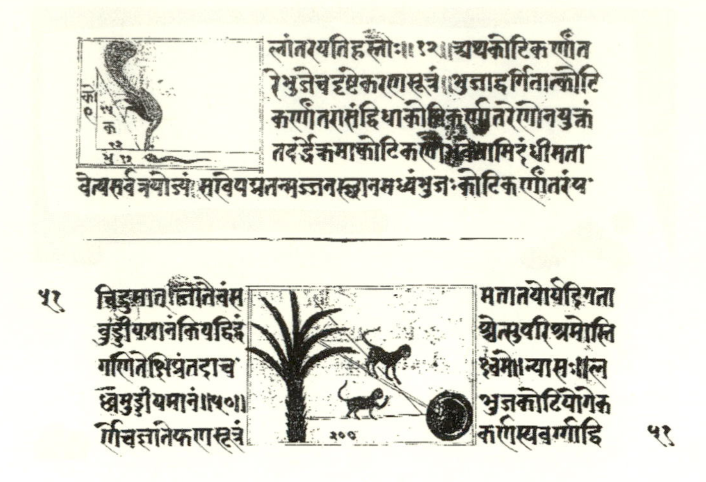

FIGURE 4.18. Pages from a manuscript of the *Līlāvatī*.

a manuscript of the *Līlāvatī* (fig. 4.18, courtesy of the Sri Ram Charan Museum of Indology, Jaipur) of the "snake-and-peacock" and "two-monkeys" problems from verses 151 and 156.)

- Bhāskara explicitly identifies what he considers the foundational issues of his subject. Namely, arithmetic or computation with known quantities are just specific applications of algebra or computation with unknowns (see verse 64); and the fundamental principle of most of these applications is the *trairāśika* or ratio (verses 240–241).
- The coverage of topics is significantly extended, requiring more than 270 verses for the *Līlāvatī*'s arithmetic (and about 200 for the *Bījagaṇita*'s algebra), as opposed to Āryabhaṭa's 30+ or Brahmagupta's 160+ verses for their sketches of the whole of *gaṇita*.

Līlāvatī *of Bhāskara II*

1. Having bowed to [Gaṇeśa] who causes the joy of those who worship him, who, when thought of, removes obstacles, the elephant-headed one whose feet are honored by multitudes of gods, I state the arithmetical rules of true computation, the beautiful *Līlāvatī*, clear and providing enjoyment to the wise by its concise, charming and pure quarter-verses.

2. Two times ten *varāṭakas* [cowrie] are a *kākiṇī* [shell], and four of those are a *paṇa* [copper coin]. Sixteen of those are considered here [to be] a *dramma* [coin, "drachma"], and so sixteen *drammas* are a *niṣka* [gold coin]. [...]

Now, the explanation of the places of numbers.

9. Homage to Gaṇeśa, delighting in the writhing black snake playfully twining about his beautiful neck, bright as a blue and shining lotus.

10. In succession, one, ten, hundred, thousand, *ayuta* [10^4], *lakṣa* [10^5], *prayuta* [10^6], *koṭa* [10^7], *arbuda* [10^8], *abja* [10^9], *kharva* [10^{10}], *nikharva* [10^{11}], *mahāpadma* [10^{12}], *śaṅku* [10^{13}]; after that,

11. *jaladhi* [10^{14}], *antya* [10^{15}], *madhya* [10^{16}], *parārdha* [10^{17}]: these, increasing by multiples of 10, are the designations of the places of the numbers for practical use, produced by the early [authorities].

That is the explanation of the places of numbers.

Now, addition and subtraction: now, the rule of operation for addition and subtraction, in half a verse.

12. The sum or difference of the numerals according to their places is to be made, in order or in reverse order.

Now, an example:

13. Oh Līlāvatī, intelligent girl, if you understand addition and subtraction, tell me the sum of the amounts two, five, thirty-two, one hundred ninety-three, eighteen, ten, and a hundred, as well as [the remainder of] those when subtracted from ten thousand.

Statement: 2, 5, 32, 193, 18, 10, 100. Result from addition: 360. Result when subtracted from *ayuta* (10000): 9640.

That is addition and subtraction.

Now, the method for multiplication: the rule of operation for multiplication, in two and a half verses.

14. One should multiply the last [i.e., most significant] digit in the multiplicand by the multiplier, [and then the other digits], beginning with the next to last, by [the same multiplier] moved [to the next place]. Or, the multiplicand is [set] down repeatedly, corresponding to the [separate] parts of the multiplier, [and] multiplied by those parts; [and the results are] added up.

15. Or, the multiplier is divided by some [number], and the multiplicand is multiplied by that [number] and by the quotient: [that is] the result. Those are the two ways of dividing up the number in this manner. Or [when the multiplicand is] multiplied by [the multiplier's] separate digits, [the product] is added up [from those].

16. Or, [if the multiplicand is] multiplied by the multiplier decreased or increased by a given [number], [the product should be] increased or decreased by the multiplicand times the given [number].

Now, an example:

17. Fawn-eyed child Līlāvatī, let it be said, how much is the number [resulting from] one-three-five multiplied by twelve, if you understand multiplication by separate parts and by separate digits. And tell [me], beautiful one, how much is that product divided by the same multiplier?

Statement: Multiplicand 135, multiplier 12.

Multiply the last digit of the multiplicand by the multiplier; when it is done in the same way [for all the digits], the result is 1620. Or, when the multiplier is divided into two separate parts (4,8), when the multiplicand is multiplied separately by both [of those] and [the products] combined, the result is the same, 1620. Or the multiplier is divided by three, [and] the quotient is 4. When the multiplicand is multiplied by that and by three, the result is the same, 1620. Or when [the multiplier] is divided into its digits (1,2), when the multiplicand is multiplied separately by those and [the products] added up according to their place value, the result is the same, 1620. Or when the multiplicand is multiplied separately by the multiplier minus two (10) and by two, and [those products] are added, the result is the same, 1620. Or when the multiplicand is multiplied by the multiplier plus eight (20), and diminished by the multiplicand times eight, the result is the same, 1620.

That is the method for multiplication.

Now division: the rule of operation for division, in one verse.

18. In division, the divisor [multiplied by some number] is subtracted from the last [digit(s) of the] dividend; that multiplier is the result. But when possible, one should divide after having reduced the divisor and dividend by a common [factor].

Now for the case of division, the statement of the digits of the product and the divisor ([formerly] its multiplier) in the previous example: Dividend 1620. Divisor 12. The quotient from the division is the multiplicand, 135.

Or the dividend and divisor are reduced by three $\left(\frac{540}{4}\right)$, or by four $\left(\frac{405}{3}\right)$. When [those dividends are] divided by their respective divisors, the result is the same, 135.

That is division.

Now squaring: the rule of operation for the square, in two verses.

19. The product of two equal [quantities] is called the "square." Now the square of the last [digit] is set down, and so are the subsequent digits multiplied by the last [digit] times two, each [product] above [the place] of its own [digit]. [Then] when one has moved [to a fresh location] the quantity [to be squared], disregarding the last [digit], [the procedure is performed] again.

[I.e., the square of the n-digit number $a_n a_{n-1} \ldots a_2 a_1$ is found as follows: after writing an a_n^2 above a_n, $2a_n a_{n-1}$ above a_{n-1}, …, $2a_n a_1$ above a_1, write down again the first $n-1$ digits $a_{n-1} \ldots a_2 a_1$ and repeat the process. The square of the entire number is the sum of all the resulting products, according to their place values.]

20. Or, the square of two [separate] parts [of a given number] is their product multiplied by two, added to the sum of the squares of those parts. Or, the square is the product of [two equal] numbers [separately] increased and decreased by a given [quantity], added to the square of that given [number].

Here is an example:

21. Friend, tell [me] the square of nine, of fourteen, of three hundred minus three, and of *ayuta* [10000] plus five, if you know the way to produce squares.

Statement: 9, 14, 297, 10005. The squares of those, produced according to the stated method, are 81, 196, 88209, 100100025.

Or, [there are] two parts of nine—4, 5. The product of those (20), times two (40), is added to the sum of the squares of those parts (41); the result is the same square, 81.

Or, [there are] two parts of fourteen—6, 8. The product of those (48) times two [is] 96. The squares of those parts [are] 36, 64. [96] is added to the sum of those (100); the result is the same square, 196.

Or, the two parts [are] 4, 10; and in the same way the same square is 196.

Or, the number [is] 297. This is decreased and separately increased by three: 294, 300. The product of those (88200) is added to the square of three (9); the result is that same square, 88209. It is always [to be done] in this way. That is the square.

Now, the square-root: the rule of operation for the square-root, in one verse.

22. Having subtracted the [largest possible] square from the last odd [decimal] place [i.e., from the one- or two-digit multiple of the highest even power of 10 contained in the given number], multiply the square-root [of the subtracted quantity] by two. When the [next] even place is divided by that [here, to operate upon the "next place" means to use the remainder from the previous operation with an additional least significant digit brought down from the subsequent decimal place], after subtracting the square of the [integer] quotient from the next odd place after that, set down two times the quotient in the "row" [of successive digits of the answer]. When the [next] even place is divided by the "row," after subtracting the square of the quotient from the next odd place, set down that result multiplied by two in the "row"; [do] this repeatedly. Half of the "row" is the [desired] square-root.

Here is an example:

23. Friend, [you should] know the square-root of four, and similarly of nine, and of the four squares previously given, respectively, if your understanding of this [topic] has been increased.

Statement: 4, 9, 81, 196, 88209, 100100025. [Their] square-roots obtained in order [are] 2, 3, 9, 14, 297, 10005.

That is the square-root.

Now, the cube: the rule of operation for cubing, in three verses.

24. And the product of three equal [quantities] is defined [as] the cube. The cube of the last [digit] is set down, and then the square of the last multiplied by the first and by three, and then the square of the first multiplied by three and by the last, and also the cube of the first. All [those]

25. are added up according to their different place-values; [that] is the cube [of a two-digit number]. [Or if one] considers that [quantity] as [split into] two parts, the last [part may be divided] in the same way repeatedly. Or, in finding squares and cubes, the procedure may be performed [starting with] the first digit.

26. Or the quantity is multiplied by [each of its] two parts, multiplied by three, and added to the sum of the cubes of the parts. The cube of the square-root, multiplied by itself, is the cube of the square number.

Here is an example:

27. Friend, tell me the cube of nine, and the cube of the cube of three, and also the cube of the cube of five, and then the cube-root from the cube, if you [have] a solid understanding of cubes.

Statement: 9, 27, 125. The resulting cubes, in order, are 729, 19683, 1953125.

Or, the quantity is 9, [and] its two parts 4, 5. The quantity is multiplied by both of those (180), multiplied by three (540), and added to the sum of the cubes of the parts (189); the resulting cube is 729.

Or, the quantity is 27, [and] its two parts 20, 7. [It] is multiplied by both of those and multiplied by three (11340), [and] added to the sum of the cubes of the parts (8343); the resulting cube is 19683.

Or, the quantity is 4; its square-root is 2; the cube of that is 8. That, multiplied by itself, is the cube of four, 64.

Or, the quantity is 9; its square-root is 3; the cube of that is 27. The square of that is the cube of 9, 729. The cube of a square number is just the square of the cube of the square-root.

That is the cube.

Now, the rule of operation for the cube-root, in two verses.

28. The first [decimal place] is a cube place; then there are two non-cube [places], and so forth. When one has subtracted from the highest cube [place] the [greatest possible] cube, the [cube] root is put down separately. Divide the next [place] by the square of that [cube-root] multiplied by three;

29. set the result in the "row." Subtract the square of that [quotient] times three, multiplied by the last [digit of the root], from the next [place]; subtract the cube of the quotient from the next [place] after that. [Proceeding] in the same way repeatedly, the "row" becomes the cube-root.

Now the statement of the cubes [given] previously, in order [to find] the [cube-]roots: 729, 19683, 1953125. The roots obtained in order: 9, 27, 125.

That is the cube-root. These are the eight operations.

Now, the eight operations for fractions. Now, the rule of operation for reducing fractions to the same denominator, in one verse.

30. The numerator and denominator [of each fraction] are multiplied by the other's denominator: in this way [they are] reduced to the same denominator. Or, both numerator and denominator may be multiplied by [each other's] reduced denominators, by the intelligent [calculator].

Here is an example:

31. Three, one-fifth, one-third: tell [me], friend, [the values of] those [reduced to] a common denominator, in order to add [them]; and also one sixty-third and one-fourteenth, in order to subtract [them].

Statement: $\frac{3}{1}, \frac{1}{5}, \frac{1}{3}$. Reduced to a common denominator: $\frac{45}{15}, \frac{3}{15}, \frac{5}{15}$. The sum is $\frac{53}{15}$.

Now, the statement in the second example: $\frac{1}{63}, \frac{1}{14}$. [The numerators] are multiplied by [each other's] denominators reduced by seven: 9, 2. Reduced to a common denominator: $\frac{2}{126}, \frac{9}{126}$. The result after subtraction is $\frac{7}{126}$, and when reduced by seven, $\frac{1}{18}$.

That is reduction to a common denominator.

Now, the rule of operation for fractions of fractions, in half a verse.

32. Numerators are multiplied by numerators, denominators by denominators: [that] is the procedure for simplifying fractions of fractions.

Here is an example:

33. One-fourth of one-sixteenth of one-fifth of three-fourths of two-thirds of one-half of a *dramma*, wise one, was given to a beggar by the one [he] begged from. Tell me, dear child, how many *varāṭakas* were offered by that stingy [one], if you know the reduction procedure in arithmetic for fractions of fractions.

Statement: $\frac{1}{1}, \frac{1}{2}, \frac{2}{3}, \frac{3}{4}, \frac{1}{5}, \frac{1}{16}, \frac{1}{4}$. When simplified, the result is $\frac{6}{7680}$; reduced by six, the result is $\frac{1}{1280}$. So one *varāṭaka* was given.

That is the reduction of fractions of fractions. [...]

Now, the rule of operation for operations with zero, in two verses.

45. In addition, zero [produces a result] equal to the added [quantity], in squaring and so forth [it produces] zero. A quantity divided by zero has zero as a denominator; [a quantity] multiplied by zero is zero, and [that] latter [result] is [considered] "[that] times zero" in subsequent operations.

46. A [finite] quantity is understood to be unchanged when zero is [its] multiplier if zero is subsequently [its] divisor, and similarly [if it is] diminished or increased by zero.

Here is an example:

47. Tell [me], what is zero plus five, [and] zero's square, square-root, cube, and cube-root, and five multiplied by zero, and ten divided by zero? And what [number], multiplied by zero, added to its own one-half, multiplied by three, and divided by zero, [gives] sixty-three?

Statement: 0. That, added to five, [gives] the result 5. The square of zero is 0; the square-root, 0, the cube, 0; the cube-root, 0.

Statement: 5. That, multiplied by zero, [gives] the result 0.

Statement: 10. That, divided by zero, is $\frac{10}{0}$.

[There is] an unknown number whose multiplier is 0. Its own half is added: $\frac{1}{2}$. [Its] multiplier is 3, [its] divisor 0. The given [number] is 63. Then, by means of the method of inversion or assumption [of some arbitrary quantity], [which] will be explained [later], the [desired] number is obtained: 14. This calculation is very useful in astronomy.

Those are the eight operations involving zero.

Now, the rule of operation for the procedure of inversion, in two verses:

48. In finding a quantity [i.e., an operand] when [the result] is given, make a divisor [into] a multiplier, a multiplier a divisor, a square a square-root, a square-root a square, a negative [quantity] a positive [one], [and] a positive [quantity] a negative [one].

49. But if it was increased or decreased by its own part, the denominator, increased or decreased by its [own] numerator, is the [corrected] denominator, and the numerator is unchanged. Then the rest [of the procedure] in inversion is as stated [above].

Here is an example:

50. Tell [me], quick-eyed girl, if you know the correct procedure for inversion, the number which, multiplied by three, added to three-fourths of the result, divided by seven, diminished by one-third of the result, multiplied by itself, decreased by fifty-two, having its square-root taken, increased by eight, and divided by ten, produces two.

Statement: The multiplier is 3; [the quantity] added, $\frac{3}{4}$; the divisor, 7; [the quantity] subtracted, $\frac{1}{3}$; the square; [the quantity] subtracted, 52; the square-root; [the quantity] added, 8; the divisor, 10; the given [result], 2. By means of the stated procedure, the resulting quantity is 28.

That is the rule for inversion.

Now, the rule of operation for methods of assumption [of some arbitrary quantity], with reduction of given [quantities] and remainders and simplification of fractional differences, in one verse:

51. As in the statement of the example, any desired number is multiplied, divided, or decreased and increased by fractions. The given quantity, multiplied by the desired [number] and divided by that [result], is the [required] quantity; [that] is called the operation with an assumed [quantity].

Example:

52. What quantity, multiplied by five, diminished by its own one-third, divided by ten, increased by one-third, one-half, and one-fourth of the [original] quantity, is seventy minus two?

Statement: The multiplier is 5; its own part is negative, $\frac{1}{3}$; [the proportion] subtracted, $\frac{1}{3}$; the divisor, 10; parts of the quantity are added, $\frac{1}{3}, \frac{1}{2}, \frac{1}{4}$; the given [answer] is 68.

The quantity assumed here is 3. [It is] multiplied by five (15), decreased by its own third part (10), [and] divided by ten (1). [It is] added to the third, half, and quarter $\left(\frac{3}{3}, \frac{3}{2}, \frac{3}{4}\right)$ of the quantity assumed here (3); the result is $\frac{17}{4}$. The given [answer], 68, multiplied by the assumed [number], is divided by that. The resulting quantity is 48.

So however the [unknown] quantity in an example [when] multiplied or divided by something, or decreased or increased by a part of the quantity, [becomes] the given [answer], that is the way that [some] imagined given quantity [is transformed] via the operation in the [above-]mentioned explanation. Then divide the given [answer], multiplied by the assumed [quantity], by whatever [result] is obtained. The result [of that] is the [desired] quantity. [...]

Now a certain operation [for producing] squares is explained:

60. The square of an assumed number, multiplied by eight and decreased by one, then halved and divided by the assumed number, is one quantity; its square, halved and added to one, is the other.

61. Or else, one divided by double an assumed number and added to that is the first quantity, and one is the other. These give pairs of quantities, the sum and difference of whose squares, when decreased by one, are squares.

Example:

62. Tell [me], my friend, the numbers whose squares, subtracted and [separately] added [to each other] and [then] diminished by one, produce square-roots [i.e., are perfect squares]; [a problem] with which those skilled in algebra, [who] have gone beyond the algebra [techniques] called "six-fold," torment the dull-witted.

In the first rule, the assumed quantity is considered [to be] $\frac{1}{2}$. Its square $\frac{1}{4}$, multiplied by eight, is 2. This, decreased by one (1) and halved, is $\frac{1}{2}$. It is divided by the assumed quantity $\frac{1}{2}$; the resulting first quantity is 1.

Its square (1) is halved $\left(\frac{1}{2}\right)$ and increased by 1 $\left(\frac{3}{2}\right)$. That is the other quantity. So the two quantities are $\frac{1}{1}, \frac{3}{2}$. In the same way, with one as the assumed quantity, the two numbers are $\frac{7}{2}, \frac{57}{8}$; with two, $\frac{31}{4}, \frac{993}{32}$.

Now in the second rule, the assumed quantity is 1. Unity (1) is divided by that multiplied by 2: $\frac{1}{2}$. Added to the assumed quantity, the resulting first quantity is $\frac{3}{2}$; the second is unity, 1. Thus the two quantities are $\frac{3}{2}, \frac{1}{1}$.

In the same way, with two as the assumed quantity, [they are] $\frac{9}{4}, \frac{1}{1}$; with three, $\frac{19}{6}, \frac{1}{1}$; with one-third, $\frac{11}{6}, \frac{1}{1}$.

Or, [the following] rule:

63. The square of the square of an assumed number, and the cube of that number, each multiplied by eight and the first product increased by one, are such quantities, in the manifest [arithmetic] just as in the unmanifest [algebra].

The assumed quantity is $\frac{1}{2}$. The square of its square, $\frac{1}{16}$, multiplied by eight $\left(\frac{1}{2}\right)$, is added to 1; the resulting first quantity is $\frac{3}{2}$. Again, the assumed quantity is $\frac{1}{2}$; its cube, $\frac{1}{8}$, is multiplied by eight. The resulting second quantity is $\frac{1}{1}$. Thus the two quantities are $\frac{3}{2}, \frac{1}{1}$.

Now with one as the assumed quantity, [they are] 9, 8; with two, 129, 64; with three, 649, 216. And [computation can be done] to an unlimited extent in this way in all procedures, by means of assumed quantities.

64. Algebra, [which is] equivalent to the rules of arithmetic, appears obscure, but it is not obscure to the intelligent; and it is [done] not in six ways, but in many. Concerning the [rule of] three quantities, [with all of] arithmetic and algebra, the wise [have] a clear idea [even] about the unknown. Therefore, it is explained for the sake of the slow[-witted].

[The "square-nature" or second-degree indeterminate equation problems, as we have seen in the *Brāhmasphuṭasiddhānta*, are considered a topic in algebra (sometimes called the "six-fold" subject—see the discussion of the *Bījagaṇita* below). Here Bhāskara describes an arithmetic technique to allow the novice (or the "dull-witted"!) unfamiliar with algebra to tackle a somewhat similar problem: finding quantities x and y such that $x^2 \pm y^2 - 1 = z^2$ (though x, y, and z are generally not integers). Bhāskara's first formula starts with an arbitrary or "assumed" quantity a and computes $x = (8a^2 - 1)/(2a)$ and $y = x^2/2 + 1$.]

[...]

Now, the rule of operation for the [rule] of three quantities, in one verse:

73. The [given] amount and the desired [amount], [being of] the same type, are [written] in the first and last [positions, respectively]. The result of that [given amount], [being of] a different type, is [put] in the middle. That, multiplied by the desired [amount] and divided by [the given amount in] the first [position], is the result of the desired [amount]. In the inverse [rule of three quantities], the procedure is reversed.

Example:

74. If two and a half *palas* of saffron are obtained for three-sevenths of a *niṣka*, tell me at once, best of merchants, how much of that [can be bought] with nine *niṣkas*.
Statement: $\frac{3}{7}$ $\frac{5}{2}$ $\frac{9}{1}$. 52 *palas* of saffron and 2 *karṣas* are obtained. [...]

Now the rule of operation in the inverse [rule of] three quantities:

77. [Sometimes] decrease in the result occurs when there is increase in the desired [quantity], or increase when decrease. So the inverse [rule of] three quantities should be known by those who understand calculation.
When there is decrease in the result, when there is increase in the desired [quantity], or increase in the result in the case of decrease [in the desired quantity], then the inverse [rule of] three quantities [is used]. That is as follows:

78. In the case of the cost of living beings [according to their] age, and the weight and alloy of gold, and the subdivision of amounts, the [rule of] three quantities should be inverted.

An example involving the price with respect to the age of a living being:

79. If a woman [slave] sixteen years old is bought for [a price of] 32, what is [the price of one] twenty years old? [If] an ox after two years of labor is bought for four *niṣkas*, then what is [the price of one] after six years of labor?
Statement: 16 32 20. The result is $25\frac{3}{5}$ *niṣkas*.
Statement of the second [problem]: 2 4 6. The result is $1\frac{1}{3}$ *niṣkas*. [...]

Now, the rule of operation in the [rule of] five or more quantities, in one verse:

82. In the case of five, seven, nine, etc., quantities, reverse the result and the divisors. When the product of the larger [set] of terms is divided by the product of the smaller [set] of terms, [that quotient] is the result.

Here is an example:

83. If the interest on one hundred for a month is five, tell [me] what is [the interest] on sixteen when a year has passed? Also, tell [me], mathematician, the time from the principal and the interest, and the amount of the principal when the time and the result are known.
Statement: $\begin{array}{cc} 1 & 12 \\ 100 & 16 \\ 5 & \end{array}$. The resulting interest is $9\frac{3}{5}$.

Now, in order to know the time, the statement is $\begin{array}{cc} 1 & \\ 100 & 16 \\ 5 & \frac{48}{5} \end{array}$. The resulting months are 12.

In order [to find] the amount of the principal, the statement is $\begin{array}{cc} 1 & 12 \\ 100 & \\ 5 & \frac{48}{5} \end{array}$. The resulting principal is 16. [...]

The rule of operation for understanding [variations of light and heavy syllables in] verses, beginning with [those having] equal [quarter-verses], in one and a half *āryā* [verses]:

132. When the accumulation is multiplied by itself, up to the number of terms measuring the syllables in a quarter-verse, the result

133. is the number [of variations of light and heavy syllables] of verses having equal [quarter-verses]. The square and the fourth power of that, each diminished by its own square-root, are [the numbers of the variations for verses] having half-equal and unequal [quarter-verses, respectively].

Example:

134. Tell me quickly the number [of combinations of syllables] of verses in the *anuṣṭubh* meter having equal, half-equal, and unequal [quarter-verses], respectively.
Statement: [Each] subsequent [product] is multiplied by two: 2. The number of terms is 8. The numbers obtained are: for equal meters 256, and then for half-equal ones 65280, and for unequal ones 42949017602.

[Here is a prosody calculation similar to the ancient rule we saw in section II.c. The *anuṣṭubh* meter has 8 syllables in each of its four identical quarter-verses, so the number of distinct syllable patterns that can be made with light and heavy syllables is $2^8 = 256$. "Half-unequal" verses have identical patterns not in all four quarterverses, but only in their two half-verses. So the number of distinct syllable patterns in such a 16-syllable half-verse will be 2^{16}, less the 2^8 excluded cases where all four quarter-verses are identical. Similarly, in a 32-syllable "unequal" verse where all the quarter-verses are different, the number of permissible patterns is $2^{32} - 2^{16}$.]

Now, geometry:

Then the rule of operation for finding any one of the arm, upright, and hypotenuse from the other two, in two verses:

135. Whichever arm is chosen, the other arm of a triangle or rectangle, in the direction diverging from that, is called the "upright" by those who know this [subject].

136. The square-root of the sum of the squares of those two is the hypotenuse. The square-root of the difference of the squares of the arm and the hypotenuse is the upright. The square-root of the difference of the squares of the hypotenuse and the upright is the arm.

Example:

137. When the upright is four and the arm three, what is the hypotenuse? And [using] the hypotenuse [derived from that] upright and arm, tell [me] the arm by means of the upright and the hypotenuse.

Statement: The upright is 4, the arm 3. The square of the arm is 9, the square of the upright 16. The square-root of the sum of the squares of those two (25) is 5. The result is the hypotenuse.

Now the calculation of the upright from the hypotenuse and arm. Statement: The hypotenuse is 5, the arm 3. The difference of the squares of these two is 16. The square-root of that is the upright, 4.

Now the calculation of the arm from the upright and hypotenuse. Statement: The upright is 4, the hypotenuse 5. The difference of the squares of these two is 9. The square-root of that is the arm, 3.

Now, the rule of operation for knowing those [quantities] by another method, in one and a half verses.

138. When the product of the two quantities, multiplied by two, is increased by the square of their difference, [the result] is the sum of their squares. In the same way, the product of their sum and difference is the difference of their squares. Thus it is always to be understood by the intelligent [calculator].

In the previously-stated example beginning "The upright is four"—statement: The upright is 4, the arm 3. When the product of those two (12) multiplied by two (24) is increased by the square of [their] difference (1), The sum of the squares is 25. The square-root of that is the hypotenuse, 5.

Now, the calculation of the upright from the hypotenuse and arm. Statement: The hypotenuse is 5, the arm 3. The sum of those two (8), multiplied by the difference of those same two [quantities] (2), is the difference of the squares, 16. Its square-root is the upright, 4.

Now, deriving the arm. Statement: The upright is 4, the hypotenuse 5. In this way the resulting arm is 3.

Example:

139. If the arm is equal to three plus one-fourth and the upright is the same [amount], then tell me quickly, mathematician, what is the amount of the hypotenuse?

Statement: The arm is $\frac{13}{4}$, the upright $\frac{13}{4}$. The sum of the squares of those two is $\frac{169}{8}$. This hypotenuse is in fact a surd, from its not [having] an [integer] square-root.

A technique for the purpose of knowing its approximate square-root:

140. The square-root from the product of the denominator and numerator, multiplied by [any] desired large square, divided by the denominator multiplied by the square-root of [that] multiplier, is close [to an exact answer].

This is the surd hypotenuse: $\frac{169}{8}$. The product of its denominator and numerator (1352), multiplied by the square of ten thousand, is 13520000. Its square-root is 3677. This is divided by the square-root of the multiplier (100) multiplied by the denominator (800). The quotient is the approximate square-root: $4\frac{477}{800}$. That is the hypotenuse. [Such problems are] always [to be solved] in this way.

Now, the rule of operation in the [production] of right triangles, in two verses.

141. The arm is [some] assumed [quantity]. The quotient of that [arm] multiplied by twice an assumed [number], [divided] by the square of [that] assumed [number] minus one, is the upright. That, separately multiplied by the assumed [number] and decreased by the arm, is the hypotenuse; and this triangle is right-angled.

142. Or else, the arm is [some] assumed [quantity], [and] the square of that divided by an assumed [number] is put down in two places, [separately] diminished and increased by the assumed [number], [and] halved. Those two [results] are the upright and hypotenuse; [they], or the arm and hypotenuse similarly [derived] from an [assumed] upright, are not surds.

Example:

143. If the arm is twelve, quickly state two [of the] various non-surd uprights and hypotenuses [derived] by means of [these] two methods.

Statement: The assumed arm is 12, the assumed [number] 2. The arm multiplied by twice that (4) is 48. [It] is divided by the square (4) of the assumed [number] (2), minus one (3): the quotient is the upright, 16. This is multiplied by the assumed [number] (32) [and] diminished by the arm (12): the result is the hypotenuse, 20. Or, [using] three [as] the assumed [number], the upright is 9, the hypotenuse 15; or [using] five, the upright is 5, the hypotenuse 13, and so forth.

Now [using] the second method. Statement: The assumed arm is 12. Its square (144) is divided by the assumed [number] 2; the quotient is 72. [This, in two separate places], is diminished (70) and increased (74) by the assumed [number] 2, and halved. The two results are the upright and the hypotenuse, 35, 37. Or, with four [as the assumed number], the upright is 16, the arm 20; with five, the upright is 9, the hypotenuse 15.

[These are handy methods for producing Pythagorean triples: e.g., given one side a and an assumed number x, derive the other side $b = 2xa/(x^2 - 1)$ and the hypotenuse $c = bx - a$. A little algebraic manipulation verifies that $a^2 + b^2 = c^2$.]

[...]

The rule of operation when the sum of the arm and hypotenuse is known, and the upright is known separately, in one verse:

151. The square of the pillar is divided by the distance between the snake and [its] hole; the result is subtracted from the distance between the snake and [its] hole. The [place of] meeting of the snake and the peacock is separated from the hole by a number of *karas* equal to half that [difference].

Example:

152. There is a hole at the foot of a pillar nine *hastas* high, and a pet peacock standing on top of it. Seeing a snake returning to the hole at a distance from the pillar equal to three times its height, [the peacock] descends upon it slantwise. Say quickly, at how many *karas* from the hole [does] the meeting of their two paths [occur]?

Statement: The pillar is 9, the distance between the snake and the hole 27. The resulting [number of] *hastas* [from] the hole at the point of their meeting is 12.

[This is simply a variant of the "hawk and rat" problem seen in Bhāskara I's commentary on *Āryabhaṭīya* 2.17.]

[...]

The rule of operation for knowing the upright and hypotenuse when the hypotenuse added in one place to the upright [is known], and the arm is known [separately]:

156. When the height of the palm tree is divided by the distance [from the tree] to the pond plus twice the height of the palm tree, and multiplied by the distance between the palm tree and the pond, the quotient is just the height of the leap. [Compare *Brāhmasphuṭasiddhānta* 12.39 above.]

Example:

157. One monkey came down from a tree of height one hundred and went to a pond [at a distance of] two hundred. Another [monkey], leaping some [distance] above the tree, went diagonally to the same place. If their [total] distances [traveled] are equal, then tell me quickly, learned one, if [you have] a thorough understanding of calculation, how much is the height of the leap?

Statement: The distance between the tree and the pond is 200, the height of the tree 100. The resulting height of the leap is 50. The upright is 150, the hypotenuse 250, [and] the arm 200.

[...]

Now, the rule in the case of a "non-figure":

163. A rectilinear figure given by some [over-]confident [person] where the sum of [all but one of] the sides is smaller than or equal to the other side is to be known as a "non-figure."

Example:

164. When the sides are given by an [over-]confident [person] [as] three, six, two, and twelve in a quadrilateral, or [as] three, six, and nine in a triangle, one must consider that a "non-figure."

These two figures are impossible to demonstrate. [Put] in the places of the sides straight sticks equal to the [proposed] sides; the impossibility of laying out [such an example] will be obvious.

The rule of operation for knowing the base-segments and so forth, in two *āryā* verses:

165. The sum of the two arms in a triangle is multiplied by their difference and divided by the base. The quotient, standing in two [places], is [separately] diminished and increased by the base, [and] halved. [The results] are the segments of the base [adjacent] to those two [arms].

166. The square-root of the difference between the squares of an arm and its own base-segment becomes the altitude. Half the base multiplied by the altitude is the accurate area of a triangle.

Example:

167. In a triangular figure, when the base is equal to fourteen and its arms thirteen and fifteen, then tell [me] quickly the altitude and the two base-segments, and also the amount of equal units known as the area.

Statement: The base is 14, the two arms 13, 15. The resulting base-segments are 5, 9, and the altitude 12, and the area of the figure 84. [...]

The rule of operation for calculating the approximate area of a quadrilateral or triangle, in one verse:

169. Half of the sum of all the sides, [set down] in four places, is [separately] diminished by [each of] the sides. The square-root of the product of those [results] is the approximate area in a quadrilateral; [computed] the same way in the case of a triangle, it is called exact.

Example:

170. The base is equal to fourteen, the opposite [side is] the number nine, and the two arms are equal to thirteen and twelve *karas*. If the altitude is exactly the number twelve, then tell [me] the area in that figure as it is stated by earlier [authorities].

Statement: The base is 14, the opposite side 9, and the two arms 13 and 12. The altitude is 12. By the stated rule, the area of the figure is the square-root of [the non-square number] 19800. The square-root of that is somewhat less than one hundred and forty-one, 141. But that is not the true area in this figure; the true area, by means of the rule to be stated [later, i.e., in verse 173 below], beginning "the [half] part of the sum of the base and [the side] opposite it multiplied by the altitude," is 138.

Here is the statement of the previously given triangle: The base is 14, the two arms 13, 15. By this method too the true area of the triangle is 84. Here [the area] is called inaccurate [only] for a quadrilateral.

Now, a rule for investigating the inaccuracy, in one and a half verses:

171. Since the two diagonals of the quadrilateral are not determined, then how can the area in this [figure] be determined? Its two diagonals computed according to [the method of] the earlier [authorities] are not appropriate in other cases; and [since] with the same sides [there may be] different diagonals, therefore the area is not unique.

In a [given] quadrilateral, when one and another [opposite] corner have approached [each other] [...] they compress the hypotenuse attached to them. But the other two, stretching outward, expand their own hypotenuse. Hence it is said "with the same sides [there may be] different diagonals."

[Verse:] Without specifying either one or the other of the altitudes or diagonals, how can one ask for the definite area of that [quadrilateral] even though it is indefinite? He who asks for or states [that area] is a demon, or completely and permanently ignorant of the determination of quadrilateral figures.

The rule of operation in computing the area of regular quadrilaterals and rectangles, in two and a half verses:

172. Let one diagonal of a quadrilateral with four equal sides be considered as given. Now the square of the side, times four, is diminished by the square of that [diagonal]. The square-root of that is the amount of the second diagonal.

173. In an equal quadrilateral, the product of the unequal diagonals, divided by two, is the accurate area. And in an equal quadrilateral with equal diagonals, as in a rectangle, [the area] is the product of its arm and upright. Otherwise, in a quadrilateral with equal altitudes, the [half] part of the sum of the base and [the side] opposite it, multiplied by the altitude, [is the area].

[It is interesting to note here that Bhāskara is familiar with "earlier" rules for the area and diagonals of quadrilaterals (see *Brāhmasphuṭasiddhānta* 12.21 and 12.28 above), but apparently not with the condition that the quadrilaterals must be cyclic. (Other mathematicians after Brahmagupta's time were evidently aware of this constraint, however.) If cyclicity is not specified then these quantities, as Bhāskara forcefully remarks, cannot be exactly determined for an arbitrary quadrilateral.]

[...]

The rule of operation for calculating the area of an irregular quadrilateral, in half a verse:

183. The sum of the areas of the two triangles standing on each side of the diagonal is exactly the area in this case.

The rule for knowing the base-segments and so on of a [quadrilateral] with equal altitudes, in two verses:

184. In a quadrilateral with equal altitudes, if the base is considered [to be] the base diminished by the [side] opposite, and the two arms are two [adjacent] sides, then its base-segments are computed just as in a triangle, and likewise its altitude too.

185. The base of the quadrilateral is diminished by [one] base-segment; the square-root of the sum of the squares of that and the altitude is the diagonal. When the altitudes are equal, [then] in comparison to the sum of the smallest side and the base, the sum of the [side] opposite the base and the other side is smaller.

Example:

186. The two sides are equal to fifty-two and forty minus one, the [side] opposite [the base] to twenty-five, and the base to sixty.

187. This figure with unequal altitudes was indicated in earlier [verses] and the amounts of its diagonals were determined [to be] fifty-six and sixty-three. State two other diagonals for such [a figure with the specified sides], and also the diagonals of [a quadrilateral] with equal altitudes [and the specified sides].

Statement: Here, considering the greater diagonal equal to sixty-three, the other hypotenuse is separately known [to be] 56. Now, considering the diagonal to be equal to thirty-two (32) instead of fifty-six, the statement in the case of the diagonal to be computed separately [is as follows]: The resulting two non-square segments are 621, 2700. The sum of their square-roots $24\frac{23}{25}$, $51\frac{24}{25}$, is the second hypotenuse, $76\frac{22}{25}$.

Now, if the same figure has equal altitudes: "The base is considered [to be] the base diminished by the [side] opposite": in this way a triangle is assumed for the sake of knowing [the quadrilateral].

Statement: The resulting base-segments are $\frac{3}{5}$, $\frac{172}{5}$, and the resulting altitude is [the square-root] of a non-square, $\frac{38016}{25}$. The result by means of the method for approximating a root is $38\frac{622}{625}$. This, then, is the constant altitude in the quadrilateral. The sum of the squares of the base diminished by the smaller base-segment and of the constant altitude is 5049; this is the square of the diagonal. The square of the second diagonal [derived] in the same way from the larger base-segment is 2176. The two diagonals produced by the method for approximating the square-roots of those [squares] are $71\frac{1}{20}$, $46\frac{13}{20}$. Thus in a quadrilateral with these [given] sides there are many different [pairs of] diagonals.

When [some condition is] determined in such a way, the two diagonals are [uniquely] determined and [may be] calculated. The computation of them according to Brahmagupta and others is as follows:

188. Multiply the sum of the products of the two [sets of] sides adjacent to the diagonal, in both cases [i.e., computed for each diagonal] divided by the other [sum], by the sum of the products of the two [sets of] opposite sides. The square-roots [of those two results] are the diagonals in an irregular [quadrilateral].

Statement: "The products of the two [sets of] sides adjacent to the diagonal": in one case the product of the two [sides] 25 [and] 39 is 975, and similarly the product of the [other] two, 52 [and] 60, is 3120. The sum of the two products is 4095. Similarly, in the second case, when the product of the two [sides] 25 [and] 52 [is made], the result is 1300; and likewise in the second case, when the product of the two [sides] 39 [and] 60 [is made], the result is 2340. The sum of the two products is 3640. The product of two opposite sides 52 [and] 39 is 2028, and subsequently the product of the two [sides] 25 [and] 60 is 1500. Their sum, 3528, is multiplied by that sum 3640; the result is the "first sum," 12841920. It is divided by the sum of the products of the two [sets of] sides adjacent to the first diagonal, 4095; the quotient is 3136. Its square-root, 56, is one diagonal. Similarly for the second diagonal, the sum of the products of the two [sets of] sides adjacent to the first diagonal is multiplied by the sum of the products of the opposite sides; the result is 14447160. [That] is divided by the sum of the products [of the sides] adjacent to the second diagonal, 3640; the quotient is 3969. Its square-root, 63, is the second diagonal.

That is the computation of the diagonals in this irregular figure. He [that is, Bhāskara] shows the difficulty of this method of diagonal-calculation by demonstrating an easier method:

189. The arms and uprights of two assumed right triangles, multiplied by each other's hypotenuses, [become] the sides; in this way some assumed irregular

quadrilateral [is constructed]. Then [the calculation of] the diagonals from the two triangles is as follows:

190. The product of the arms increased by the product of the uprights is one diagonal; the sum of the products of the upright [of each triangle] and the arm [of the other] is the other [diagonal] in this easy computation. The cumbersome [procedure] performed by the earlier [authorities] is not [what] we know.

Statement of two right triangles: [sides 3, 4, 5 and 5, 12, 13]. The arms of those two [triangles] are multiplied by each other's hypotenuses; the uprights are multiplied by each other's hypotenuses; with "the arms and uprights" in this way the result is 25, 60, 52, 39. The largest of those is the base, the opposite [side] is small[est], the other two are the sides; considering [them] in this way, the demonstration of the figure [may be made]. The two diagonals are computed with great ease: 63, 56. The resulting products of the arms of this pair of right [triangles] with each other's uprights are 36, 20. Their sum is one diagonal, 56. The products of the two arms (3, 5) and of the two uprights (4, 12) are 15, 48. Their sum is the other diagonal, 63. Thus the diagonals are [found], and thus it is known easily. Now, when one has exchanged the smaller side and the [side] opposite [the base], a new figure is stated. Statement: [base 60, opposite side 39, sides 25, 52].

[Again, Bhāskara shows himself more familiar with Brahmagupta's rules than with the specific conditions applying to them. He suggests that the diagonals rule known to us from Brahmagupta's text is too "cumbersome," and proposes instead a technique for constructing from two arbitrary right triangles a more restricted type of scalene quadrilateral, whose diagonals can be more easily computed.]

[...]

Now, the rule of operation in the case of a circular figure, in one verse:

199. When the diameter is multiplied by twenty-seven–nine–three (3927), and divided by zero–five–twelve (1250), that is the accurate circumference. Or, when it is multiplied by twenty-two (22) and divided by seven (7), [it] is the approximate [circumference] for practical purposes.

Example:

200. Friend, when the amount of the diameter is seven (7), then tell the amount of the circumference; and considering the amount of the circumference [as] twenty-two (22), [tell] the measure of its diameter.

Statement: The amount of the diameter is 7; the resulting amount of the circumference is $21\frac{1239}{1250}$. Or, the resulting approximate circumference is 22.

Or else, the statement for computing the diameter from the circumference: by reversing the multiplier and divisor, the accurate amount of the diameter is $7\frac{11}{3927}$.

The rule of operation in finding the amount of a circle and a sphere, in one verse:

201. In the case of a circle, one-fourth of the diameter multiplied by the circumference is the area, which, multiplied by four, is the outer [surface] of a sphere, like the net [covering] a ball. And the surface area [found] in this way, multiplied by

the diameter and divided by six, is determined [to be what] is called the volume inside the sphere.

Example:

202. Tell [me], intelligent one, what is the area of a uniform [equidistant?] figure whose diameter is equal to seven? and what is the area like the net on a ball upon the surface of a sphere whose diameter is seven? and what is the cubic amount inside that sphere? if you know the clear *Līlāvatī*.

Statement for demonstrating the area of the circular figure: The diameter is seven, the circumference $21\frac{1239}{1250}$. The area of the figure is $38\frac{2423}{5000}$.

Statement for demonstrating the surface area of the sphere: The diameter is seven, the surface area of the sphere $153\frac{1173}{1250}$.

Statement for demonstrating the volume inside the sphere: The diameter is seven, the volume inside the sphere $179\frac{1487}{2500}$.

Now, the rule of operation for computing those results by another method, in one and a half verses:

203. When the square of the diameter is multiplied by three thousand nine hundred and twenty-seven and divided by five thousand, the accurate area [results]; or when it is multiplied by eleven and divided by fourteen, it is the approximate area for practical purposes. Half the cube of the diameter plus the twenty-first part of itself, is the volume of a sphere.

The rule of operation for finding the chord and arrow, in one and a half verses:

204. The square-root of the product of the sum and the difference of the chord and the diameter [is found]; the diameter, diminished by that and halved, is the arrow. And the square-root of the diameter diminished by the arrow and multiplied by the arrow, multiplied by two, is the chord. When the square of half the chord is divided by and [then] added to the arrow, [that] is [what] they call the amount of the diameter in a circle. [...]

Now, the rule of operation for finding the amount of the sides of figures from a triangle to a nonagon [inscribed] in a circle, in three verses:

206–207. If the diameter of a circle is multiplied by three–two–nine–three–zero–one (103923), by three–five–eight–four–eight (84853), by four–three–five–zero–seven (70534), by zero–zero–zero–zero–six (60000), by five–five–twenty–five (52055), by two–two–nine–five–four (45922), and by one–three–ten–four (41031), respectively,

208. and divided by zero–zero–zero–zero–twelve (120000), the sides of [figures] within a circle, beginning with the triangle and ending with the nonagon, are separately obtained in order. [...]

The rule of operation for determining the distance between [the tip of] the shadow and the lamp, and the height of the lamp, in one and a half verses:

239. [One] shadow, multiplied by the distance between the tips of the two shadows and divided by the difference between [the lengths of] the shadows, is

the [total] ground [between the tip of the shadow and the lamp]. The product of the ground and the gnomon, divided by the shadow, produces the height of the flame of the lamp. What is discussed here is pervaded by the [rule of] three quantities in its different [forms], just as everything is [pervaded] by Hari [Śiva].

Example:

240. Intelligent one, the shadow of a gnomon of twelve *aṅgula*s is seen [to be] eight *aṅgula*s. When one has set down [that gnomon] again in a place two *kara*s beyond the tip of [that] shadow, if its [shadow] is equal to twelve *aṅgula*s, then if you know shadow-calculations, tell [me] the distance from [the tip of] the shadow to the lamp, and the height of the lamp.

Statement: Here, the distance between the tips of the two shadows, in *aṅgula*s, is 52, and the two shadows are 8, 12. The first of these, 8, multiplied by that [distance] 52 (416), is divided by the difference in the amounts of the shadows, 4. The resulting amount of ground is 104. This is the distance between the tip of the first shadow and the base of the lamp: that is the meaning. In the same way, the amount of ground [to] the tip of the second shadow is 156. "The product of the ground and the gnomon, divided by the shadow": thus in both cases the resulting height of the lamp is the same, $6\frac{1}{2}$ *hasta*s. In the same way [everywhere] in shadow-calculations, the determination is performed by considering a [problem in the rule of] three quantities, as follows: The second shadow (12) is greater than the first (8) by so much; if the ground equal to the distance between the tips of the shadows is obtained by that part of the shadow, then what [is obtained] by the first shadow [proportionally]? In this way the amount of the distance between [each] shadow and the base of the lamp is separately obtained. Then, a second [rule of] three quantities: When the gnomon is the upright if the arm is [considered to be] equal to the shadow, then what [is the upright] when the arm is equal to the ground? And exactly the same height of the lamp is thus obtained in both cases.

Thus the [rule of] five or more quantities is completely explained by considering the [rule of] three quantities. Just as this universe is pervaded by Lord Nārāyaṇa [Viṣṇu] (who removes the sufferings of those who worship him and is the sole generator of this universe), with his many forms—worlds and heavens and mountains and rivers and gods and men and demons and so on—in the same way, this whole type of computation is pervaded by the [rule of] three quantities.

If so, then what is [the purpose of] the many [different rules]? He responds:

241. Anything that is calculated in algebra or here [in arithmetic] by means of a multiplier and a divisor is understood by those of clear intelligence as just a [rule of] three quantities. But for increasing the intelligence of dull-witted ones like us, it has been explained by the wise in many different and easy rules. [...]

Now, in the case of the "net of calculation" [or "net of numbers"], the rule of operation for [determining] the different numbers [formed] with specified digits, in one verse:

261. The product of the numbers in the sequence starting with one and ending with the [number of] places is the [number of] different numbers [produced] with

digits limited [to that number of places]. [That product] divided by the number of digits, multiplied by the sum of the [specified] digits, [and] added up in [each of the specified] places, is the sum of the amounts [produced with the specified digits].

Here is an example:

262. Say quickly how many different numbers are produced with two and eight, or with three, nine, and eight, and then with [the digits] starting with two and ending with nine, respectively.

Statement: 2, 8. Here there are two places, 2. There are two numbers in the sequence starting with one and ending with the [number of] places, 1, 2. Product: 2. There are two different numbers thus produced, 2. Now that product is multiplied by the sum of the digits (10): 20. Divided by the number of digits (2): 10. Added in [each of] the two places, it results in the sum of the numbers, 110 [which is indeed the sum of all 2-digit numbers that can be made with 2 and 8]. [...]

271. Although [neither] the multiplier nor divisor nor square nor cube is asked [for], bad [students, although] conceited [about their abilities as] mathematicians, [will] certainly [make] a mistake in [calculating about] this "net of numbers."

That is the "net of numbers" in the *Līlāvatī*.

272. Those who hold at their throats the accurate *Līlāvatī* illustrating elegant sentences, [whose] parts are adorned with excellent [rules for] reduction and multiplication and squaring [etc.], attain ever-increasing happiness and success.

The *Bījagaṇita* takes its title from one of the standard terms for algebra: "computation with seeds," i.e., with undetermined quantities that eventually bear fruit as quantified results. Its chapters cover the following topics: algebraic operations for positive and negative quantities, zero, unknowns, and *karaṇīs* or surds; the "pulverizer" or linear indeterminate equations; the "square-nature" or quadratic indeterminate equations, including the so-called "cyclic method" for solving "square-nature" problems; linear and quadratic equations; equations with more than one unknown; equations involving products of unknowns.

As in the *Līlāvatī*, Bhāskara supplements the rules and sample problems of the *Bījagaṇita*'s numbered verses with an intermittent prose commentary. Most of this autocommentary is silently omitted from the excerpts below (as it is from some Sanskrit editions), but the initial selections include portions of a later and more detailed commentary by the polymath Sūryadāsa. Sūryadāsa (1507–1588 or thereabouts) was descended from a scholarly family of Pārthapura (probably modern Pathri in Maharashtra); he also wrote poems, treatises on astronomy and philosophy, and commentaries on the Vedas. (Compare his use of grammatical and philosophical principles to expound Bhāskara's mathematical rules with the commentary of the first Bhāskara on Āryabhaṭa, some nine centuries earlier.) He was skilled as well in the Sanskrit poetic tradition of elaborately complex and ambiguous sentences, playing on the multiple meanings of words such as "*bhāskara*," literally "sun." Especially interesting is Sūryadāsa's interpretation of infinity as the result of division by zero; note that the treatment of $n/0$ in the more advanced *Bījagaṇita* is different from that in the *Līlāvatī* (verse 45), where $n/0$ is just assigned the status of "having zero as a denominator."

Bījagaṇita *of Bhāskara II with commentary of Sūryadāsa*

Obeisance to Gaṇeśa. Obeisance to Sarasvatī. Obeisance to the elders.

[...]That Bhāskara [or Sun], whose body was revealed in order to lift this world up when it had been destroyed by the power of the time of the Kali [*yuga*], in order to help it when it had been struck down by the darkness of ignorance, having written the manifest mathematics [arithmetic] in accordance with his expressed plan, desiring to expound this exceedingly difficult mathematics of the unknown which is algebra [or the origin], at first, with the wish to accomplish what had been begun, effected by himself the auspiciousness which is in the form of a reverential salutation to a deity [Gaṇapati = Gaṇeśa] in accordance with the standards of behavior of the cultured, the necessity of making which [i.e., calculation or salutation] was made known to him by his hearing what is inferred from the behavior of the learned that is distinguished by its being the special cause of the removal of the obstacles which impede it. Having joined it by means of words having several meanings with usefulness to students, he ties it together with the Upendravajrā meter in a verse [beginning]: "The generator."

1. The generator of the intellect I praise, which the wise men [or Sāṅkhya philosophers] declare to have been imbued by the existing Puruṣa, the unique source of all that is manifest, the unmanifest lord, and the numbered.

"I praise the lord of the intellect," this is the [grammatical] connection of the verbal action and the instrument of action. The meaning is: "I praise, [i.e.] I salute, the lord of the intellect, [i.e.] Gaṇādhipati." Here his [Gaṇapati's] lordship of success and intellect is really established from the evidence of the traditional doctrines and the teachings. Surely in this paying homage to [him] just as the lord of the intellect, if [one asked] "what is the cause?" [the answer is that] it is not [only as such]. Since, because this science of the unmanifest is uniquely feasible through the intellect, such a god [as the lord of the intellect] is to be asked by us [for assistance], having this in view from the beginning, the teacher shall speak. [...]

Having established the auspiciousness characterized by paying obeisance to his chosen deity with the first verse, now, beginning with the book, the teacher, praising *bīja* with the dodge of telling the usefulness of beginning it, with one *Śālinī*-verse tells [the verse beginning]: "Previously mentioned."

2. Previously mentioned [in the *Līlāvatī*] was the manifest whose source is the unmanifest. Since generally questions cannot be very well understood by the dull-witted without the application of the unmanifest [algebra], therefore I tell also the operation of the *bīja* [algebra].

> The sequence is: "Previously mentioned was the manifest. Therefore I tell the operation of the *bīja*." Thinking: "Why therefore?" he says [the phrase beginning]: "Since." From which cause generally questions cannot be very well understood by the dull-witted, that is, by those of little intelligence, without the application of the unmanifest [*avyakta*, i.e., algebra]. The meaning is that [the questions] are excessively difficult to understand. The manifest of what sort? *Avyaktabīja*. That is *avyaktabīja* whose source is the unmanifest. The meaning is that the calculation of the unmanifest has become the cause of the manifest.

The Six-Fold Operation of Positive and Negative Quantities

> Now in connection with describing what is to be explained in the treatise, with respect to all [operations] such as multiplication and division, because of its priority he speaks of the addition and subtraction of positive and negative [quantities] by means of half an *Upendravajrā* verse [beginning]: "In addition, the sum occurs."

3a–b. In the addition of two negative [quantities] or of two positive [quantities], [their] sum occurs; the addition of a positive and a negative [quantity] is [their] difference. [...]

3c–4b. A triad of ones and a quartet of ones are together [both] negative or [both] positive or separately positive and negative or separately negative and positive. Tell me quickly [if] you know the addition of the two positive and negative [quantities]. [...]

4c–d. A positive [quantity] that is going to be subtracted attains negativity; a negative [quantity] positivity. The addition of those [two] is as has been described [previously]. [...]

5a–b. When one subtracts a pair [of ones] from a triad [of ones], [either] a positive from a positive or a negative from a negative, and the reverse, [in each case] say the remainder quickly. [...]

6a–b. A positive pair is multiplied by a positive triad, [or] a negative by a negative, [or] a positive by a negative. What is [the result]? [...]

6c–7b. A positive octet of ones is divided by a positive quartet of ones, [or] a negative by a negative [or] a negative by a positive, [or] a positive by a negative. Say quickly what this [quotient] is [in each case] if you understand [computation] thoroughly. [...]

7c–d. The square of a positive and of a negative [quantity] is positive. The two square roots of a positive [quantity] are positive and negative. The square root of a negative [quantity] does not exist because it is not a square. [...]

8. Oh friend! Tell me quickly the square of a positive triad of ones and [that] of a negative. And quickly tell [me] separately the square root of nine having a positive nature and having a negative nature.

It is clear. Thus the six-fold [operation] of positive and negative [quantities].

The Six-Fold Operation of Zero

Having described in this way the six-fold [operation] of positive and negative [quantities], now he investigates the six-fold [operation] of zero [with the verse beginning]: "In the addition of zero."

9a–b. In the addition of zero or in the subtraction [of zero] a positive or negative [quantity remains] as it was. [But] when it is subtracted from zero it goes to its opposite.

In the addition and in the subtraction of zero a positive or a negative [quantity remains] as it was. The meaning is that, when addition and subtraction are being accomplished by means of "*kha*"—[that is] zero—a positive or negative [quantity] remains "*tathaiva*"—[that is,] as it was determined because in the addition and subtraction of any number whatsoever by zero, the zero does not change the form [of the number]. So when it is subtracted from zero, it goes to its opposite. The meaning is that a positive or negative [quantity] when it is subtracted from zero attains reversal; because it is said that: "A positive [quantity] that is going to be subtracted attains negativity."

Here he proclaims an example [with the verse beginning]: "A positive triad of ones."

9c–d. There are a positive and a negative triad of ones, and there is a zero. Tell [me] what [each] will be when it is added to zero and when it is subtracted from zero.

It has a clear meaning.

Now he describes multiplication by zero [with the verse beginning]: "In the multiplication and so on."

10a–b. In the multiplication and so on of zero [by a quantity the result is] zero. In the multiplication [of a quantity] by zero [the result is] zero. And a quantity divided by zero becomes [a quantity] having zero as its divisor.

In the multiplication and so on of *kha*—[that is,] of zero [by a quantity], a *kha*—[that is,] a zero—is [i.e., results]. The meaning is that [it is a fact] that, when zero is multiplied by any number whatsoever, zero is [the product] because a number multiplied by zero is zero because of the non-existence of its being in the sphere of counting by reason of its independence. Here by the word "*ādi*" ["and so on"] it is to be known that division, square, and square roots are the same. In this way Nārāyaṇa also has defined this incidentally by means of a poetic utterance in his algebra as follows: "On account of multiplication by zero a quantity goes to the state of being zero. But, when it is divided by zero, it does not return to its previous condition [non-zero finite quantity] because it is absorbed in that [infinite] just as a serious yogi who has attained the unique bliss-giving place of Brahma which consists of pure thought because he is pervaded by the *ātman* does not [return] to the path of *saṃsāra* [finite world]."

So a quantity divided by zero becomes one having zero as its divisor.

Here he proclaims an example [with the verse beginning]: "Multiplied by two."

10c–d. Tell me [the results when] zero is multiplied by two [and] divided by three, [when] three is divided by zero, and the square and square root of zero.

It has a clear meaning.

Now he shows that in the science of computation there is another name, infinite, for the number which has a zero as its divisor. Then he skillfully defines the infinity of this [with the verse beginning]: "In this."

11. In this quantity also, which has zero as its divisor, there is no change even when many [quantities] have entered into it or come out [of it] just as at the time of destruction and of creation, when throngs of creatures enter into and come out of [him, there is no change] in the infinite and unchanging one [i.e., Viṣṇu].

In this quantity which has zero as its divisor, even when many numbers have entered into or come out of [it], there is no change. The meaning is that of whatever [quantity] the divisor is zero, when a fractional number is being combined with that [*khahara*, zero-divided] which has the same denominator, there is zeroness in the denominator and the numerator. If [it is asked]: "Surely, since one sees change in the quantity having zero as its divisor at the beginning of its combination with a number divided by one, two, three, and so on, how is it said that no change occurs?," it is true that since it follows from the meaning of the words, there is no change of its state of being [a quantity] which has zero as its divisor in the quantity which has zero as its divisor. Or else the word "*aṅkeṣu*" [i.e., in numbers] here is to be understood "in non-fractional [numbers]."

Now he shows how wonderful his poetry is by confirming the infinity of [the quantity] which has zero as its divisor by the example of Viṣṇu because of the sameness of [his] infinity [with the lines beginning]: "Just as." "Just as at the time of destruction and creation when many throngs of beings enter into and come out of [him], there is no change in the infinite and unchanging [Viṣṇu], so [there is no change in the *khahara*]." The idea is that, when at the time of destruction, beings enter into Viṣṇu and at the time of creation come out of Viṣṇu, there is no change [in him] since he is infinite. This has been stated in the [*Mahā*]*Bhārata* in the *Śāntiparvan* in a conversation between Bhīṣma and Yudhiṣṭhira: "From whom all beings are born at the coming of the first *yuga* and in whom they go to destruction again at the end of the *yuga*."

Thus the six-fold [operation] of zero. [...]

The Six-Fold Operation of the Surd

> Thus having examined the six-fold [operation] of unknowns, now wishing to speak of the six-fold [operation] of surds at the beginning he speaks of their addition and subtraction [with the verse that begins]: "The sum of two [given] surds."

23c–24b. Assuming the sum of two [given] surds to be the great[er surd] and the square root of the product multiplied by two to be the small[er surd], the sum and the difference of these two are [treated] like *rūpas*, [but] one should multiply and divide a square by a square. [...]

> Now he enunciates the addition and subtraction of surds by another method [with the verse that begins]: "Of it divided by the small[er surd]."

24c–25b. The square root of the great[er surd] divided by the small[er surd], increased by one [or] diminished by one, [each] multiplied by itself [and then] multiplied by the small[er surd]—[these] are their sum and difference in order. Or, if the square root [of the above quotient] does not exist, it is put down separately.

> [...] A demonstration is enunciated. In this case, the sum and the difference of the two surds which are to be spoken of, which are measured by two and eight, in accordance with what was said [previously] are 18 [and] 2. Here "two and eight" are assumed to be squares. As much as is the square of the sum of the square roots of these two [numbers], the sum measured by that must exist. And so, when the sum is being effected in accordance with what was said, there is produced a number that is measured by the square of the sum of the square-roots. So, for instance, the surd 2. Its square-root is 1; 25. And the surd 8. Its square-root is 2; 51. The sum of these two is 4; 16. Its square is 18; 12. This is the sum of the two surds. So the difference of the two roots is 1; 26. The square of this is 2;3. It has been demonstrated that this is the difference. [...]

25c–26b. Tell [me] separately the sum and the difference of two surds measured by two and eight, and of two [others] numbered three and twenty-seven, and, oh friend, thinking for a while, of [another] two measured by three and seven, if you know the six-fold [operation] of the surd.

> [...] And so the setting out is: *ka* 2 [and] *ka* 8. Here by [the sūtra] that begins: "Assuming the sum of the two [given] surds to be the great[er surd]," the great[er surd] is 10. Then the product of the two surds is 16. Its square-root is 4. [This] multiplied by two is the small[er surd], 8. As in the case of *rūpa*s, the sum and the difference of these two are 18 [and] 2. These are the surds of sum and difference, *ka* 18 [and] *ka* 2. [...]

[These are a few of Bhāskara's rules for simplifying expressions with *karaṇīs.* The square of the sum or difference of two *karaṇīs,* $(\sqrt{K_1} \pm \sqrt{K_2})^2$, is the sum or difference of the "greater," $K_1 + K_2$, and the "smaller," $2\sqrt{K_1 K_2}$. We can also express the same quantity as follows:

$$(\sqrt{K_1} \pm \sqrt{K_2})^2 = \left(\sqrt{\frac{K_1}{K_2}} \pm 1\right)^2 \cdot K_2.$$

In explaining Bhāskara's example for the square of the sum or difference of $\sqrt{2}$ and $\sqrt{8}$, Sūryadāsa provides for comparison the rough results directly computed from approximate square roots of 2 and 8 as sexagesimal fractions. The prescribed method, on the other hand, will give the values $(\sqrt{2} \pm \sqrt{8})^2 = (2+8) \pm 2\sqrt{2 \cdot 8}$.]

[...]

70. The small[er root] is arbitrary. [That number] by which its [the smaller root's] square multiplied by the "nature" is increased or decreased gives a square-root [that] they say is a positive or negative additive, and that square-root [they say] is the largest root.

71. Having set down the small[er root], the largest [root], and the additive [and] entering these [quantities] or others below them in order, many square-roots are to be computed from them by means of suppositions; therefore supposition is proclaimed. There are thunderbolt multiplications of the greatest and the small[er roots]; their sum [is considered to be] the small[er root]. The product of the two small[er roots] is multiplied by the "nature"; [this] joined to the product of the greatest [roots] [is considered to be] the greatest [root]; [and] the product of the two additives is the [new] additive. Or the difference of the two thunderbolt multiplications is the small[er root]. The product of the two small[er roots] multiplied by the "nature" and the product of the two greatest [roots]—the difference of these [two] is the greatest [root]. Here also the product of the additives is the [new] additive.

72. The additive divided by the square of an arbitrary [number] is the [new] additive. Or else the two [roots] divided by the arbitrary [number] are the two square-roots. The additive is multiplied at the time [when] the two roots are multiplied.

73. One should divide double the arbitrary [number] by the difference of the square of the arbitrary [number] and the "nature"; that is the least root when the additive is one. From that [is found] the greatest [root]; here [the process] is infinite because of the suppositions and from the arbitrary [number].

74. [Example:] What square, multiplied by eight [and] increased by one, is a square? Oh calculator, let it be told. [Example:] Or what square, multiplied by eleven [and] increased by one, is a square, oh friend?

75. Making the small[er] and greatest roots and the additive into the dividend, the additive, and the divisor, the multiplier is to be imagined. When the square of the multiplier is subtracted from the "nature" or is diminished by the "nature" so that the remainder is small, that divided by the additive is the [new] additive. It is reversed [if the square of the multiplier] is subtracted from the "nature." The quotient of the multiplier is the small[er] square-root; from that [is found] the greatest

[root]. Then [it is done] repeatedly, leaving aside the previous square-roots and additives. They call this the circle. Thus there are two integer square-roots increased by four, two, or one. The supposition for the sake of an additive one is from the roots with four and two as additives.

76. [Example:] What square, multiplied by sixty-seven and increased by one, and what, oh friend, multiplied by sixty-one and increased by one, gives a square-root? Tell it, friend, if the square [and] the "nature" are to a high degree spread in your mind like a creeper.

[Skipping to a later section of the *Bījagaṇita*, upon which Sūryadāsa's commentary has yet to be published, we now explore some of Bhāskara's verses on second-degree indeterminate equations. The procedures described in verses 70–74 are similar to the "square-nature" techniques of Brahmagupta, whose "multiplier" N is here called the "nature." Then Bhāskara goes on to specify an alternative, the so-called "circle" or "cyclic" method, which was mentioned (and perhaps first discovered) by the mathematician Jayadeva a century or so earlier.

Again, for a given "nature" N we devise some auxiliary equation $Na^2 + k = b^2$, and we wish to find a solution to $Nx^2 + 1 = y^2$. We start by applying the pulverizer to the roots and additive a, b, and k to find some "multiplier" m_1 such that $am_1 + b$ is divisible by k, and the difference between $m_1{}^2$ and N, also divisible by k, is as small as possible. Then we use this m_1 to calculate new values k_1 and a_1 for the additive and smaller root as Bhāskara prescribes, from which we can compute the new greater root $b_1 = \sqrt{Na_1{}^2 + k_1}$:

$$k_1 = \frac{m_1{}^2 - N}{k}, \quad a_1 = \frac{am_1 + b}{k}.$$

Because $N \cdot 1^2 + (m_1{}^2 - N) = m_1{}^2$ is also a second-degree indeterminate equation (although not a very interesting one), we can "compose" its coefficients in the standard way with those of our original equation $Na^2 + k = b^2$ to get a new equation. Taking "the sum of the thunderbolt-products" $(a \cdot m_1 + b \cdot 1)$ as the smaller root, the "product of the additives" $k(m_1{}^2 - N)$ as the additive, and the "product of the first pair times the multiplier plus the product of the last pair" $(a \cdot 1 \cdot N + b \cdot m_1)$ as the greater root, we get

$$N(am_1 + b)^2 + k(m_1{}^2 - N) = (Na + bm_1)^2.$$

Dividing through all terms by k^2 leaves us with the very expressions we found above for k_1 and a_1. Therefore b_1 must have the form $(bm_1 + Na)/k = (m_1(am_1 + b) - a(m_1{}^2 - N))/k$, and hence is an integer. So $Na_1{}^2 + k_1 = b_1{}^2$ is another auxiliary equation.

Now we can apply the pulverizer again, this time to a_1, b_1, and k_1, to get a new multiplier m_2, from which we can compute k_2, a_2, and b_2 as above. And we continue through the cycles in this way until we find an equation with the additive ± 4, ± 2, or ± 1, which can be manipulated by one of the previously mentioned square-nature methods to provide a solution to $Nx^2 + 1 = y^2$, as desired. It is not known how, or whether, Indian mathematicians showed that this method would always provide a solution after a finite number of cycles.]

[...]

89. The *yāvattāvat* is to be considered the measure of the unknown quantity; in this [matter] from the effort of one doing what is prescribed the two sides are to be made equal by subtracting or adding, multiplying [or] dividing. One should subtract

the unknown on one [side] from the other side; removing the *rūpa*s from the other side, one should divide the remainder of the *rūpa*s by the remaining unknown. The manifest measure of the unknown quantity is produced. Here for two and more unknowns one may imagine by his own intellect a *yāvattāvat* multiplied by two and more or divided or increased or diminished, or sometimes, considering in this way, a known [number].

90. [Example:] One [person] has three hundred coins [and] six horses, another has ten horses having an equal price and a debt of a hundred coins. The two have equal wealth. What is the price of a horse?

91. When half the wealth of the first [person] is increased by two, if the second [person] has wealth equal to that, or if the first is three times more wealthy than the other, tell me in each case the price of a horse.

92. [Example:] The quantity of rubies, flawless sapphires, and pearls of one [person] is five, eight, [and] seven respectively; the number of these gems of another man is seven, nine, [and] six, oh friend; ninety coins [belong to the first], sixty-two [to the second]. They have equal wealth. Oh intelligent knower of algebra, tell [me] quickly the prices for each type of gem.

93. [Example:] One [person] says: "Give me a hundred; I'll be twice as wealthy as you, oh friend." Then the other says: "If you give me ten, I'll be six times [as rich] as you." Tell me how the wealth of these is measured.

94. [Example:] The eight rubies, ten emeralds, [and] hundred pearls that are in your ear-ornament were purchased by me for you at an equal price. The sum of the prices of this triad of gems was half a hundred diminished by three. Tell [me], my dear, the price [of each] separately if, oh auspicious one, you are clever in mathematics. [...]

113. [Example:] One monkey came down from a tree of height one hundred and went to a pond [at a distance of] two hundred. Another [monkey], leaping some [distance] above the tree, went diagonally to the same place. If their [total] distances [traveled] are equal, then tell me quickly, learned one, if [you have] a thorough understanding of calculation, how much is the height of the leap? [...]

120. [Example:] What quantity divided by zero, increased by the first [quantity], diminished by nine, squared, increased by its own square-root, [and] multiplied by zero is ninety? [...]

125. [Example:] A fifth part of a troop of monkeys, minus three, squared, has gone to a cave; one is seen [having] climbed to the branch of a tree. Tell [me], how many are they? [...]

Here the amount of the troop is *yāvattāvat* 1. The fifth part minus three, $\frac{1, -15}{5}$, is squared, $\frac{1, -30, 225}{25}$. That added to the [one] seen [monkey], $\frac{1, -30, 250}{25}$, is equal to the troop. Making the two sides with the same divisor, after removal of the divisor and subtraction, they become 1, –15, 0 [=] 0, 0, –250. Having multiplied both by four and added the square of fifty-five, 3025, the two roots are 2, –55 [=] 0, 45. And here as before the result is a twofold amount, 50, 5. In this case the second is not to be taken due to its inapplicability. People have no confidence in [or "comprehension of"] a manifest [quantity] becoming negative [as in a group of 5/5 – 3 monkeys].

[These verses and the two below are from the *Bījagaṇita*'s sections on linear and quadratic equations in one unknown. Bhaskara designates unknowns not only by color names but by the standard term *yāvattāvat*, "as much as so much." Among his many versified examples, Bhāskara thriftily reuses several from the *Līlāvatī*, such as the above "two-monkeys" problem [*Līlāvatī* 157]. Note also that despite his previous poetic simile about the "infinite and unchanging" Viṣṇu, he still expects students to find finite values for quantities multiplied and divided by zero, as in the *Līlāvatī*.

The standard layout of algebra equations (as far as we can tell from the extant manuscripts) uses sets of terms in the fixed sequence "square-unknown-*rūpas*." The numerical coefficients are often also identified by syllabic abbreviations (not shown here), but there are no explicit operator symbols or equals sign. Negative quantities are indicated by a dot over them rather than by the small cross seen in the BM, but apparently they were not generally accepted as solutions to problems about concrete objects.]

[...]

128. [Example:] In a figure, the arm and upright are equal to fifteen and twenty [respectively]. Then what is the hypotenuse? The demonstration of this well-known calculation should be explained too.

Here the hypotenuse is *yā* 1. When [one has] turned that triangle around, the *yāvattāvat*-hypotenuse is considered the base; and the arm and upright are [its] two sides. The altitude there has triangles on both sides, in both of which the arm and upright are in their previous form [i.e., proportion]. Hence, a [rule of] three quantities: If, when the hypotenuse is *yāvattāvat*, the arm is this, 15, then when the hypotenuse is equal to the arm, what [is the result]? The quotient should be the arm [in the smaller triangle]. That base-segment joined to the [original] arm is $\frac{225}{y\bar{a}\,1}$. Again: If, when the hypotenuse is *yāvattāvat*, the upright is this, 20, then when the hypotenuse is equal to the upright, what [is the result]? The resulting base-segment joined to the upright is $\frac{400}{y\bar{a}\,1}$. The sum of the base-segments is made equal to the *yāvattāvat*-hypotenuse; [since] that is the square-root of the sum of the squares of the arm and upright, the amount of the hypotenuse is obtained. The resulting two base-segments produced with that are 9, 16. Hence the altitude is 12. Observation of the figure: [see figure 4.19].

Now it is explained otherwise: the hypotenuse is *yā* 1. Half the product of the arm and upright is the area of a triangular figure, 150. Another, four-sided, figure [constructed] with a quadruple of this unequal triangle, [with side] equal to the hypotenuse, is considered in order to know the hypotenuse. Thus a square is produced here in the middle [of the large square]; the amount of [its] side, equal to the difference between the arm and upright, is 5, its area 25. Twice the product of the arm and upright is the area of the four triangles, 600. The sum of those all together is the area of the large triangle, 625. When [one has] made that equal to the square of *yāvattāvat*, the result is the amount of the hypotenuse, 25. If [there is] no square-root of the known [quantity] then the hypotenuse is in [the form of] a *karaṇī*.

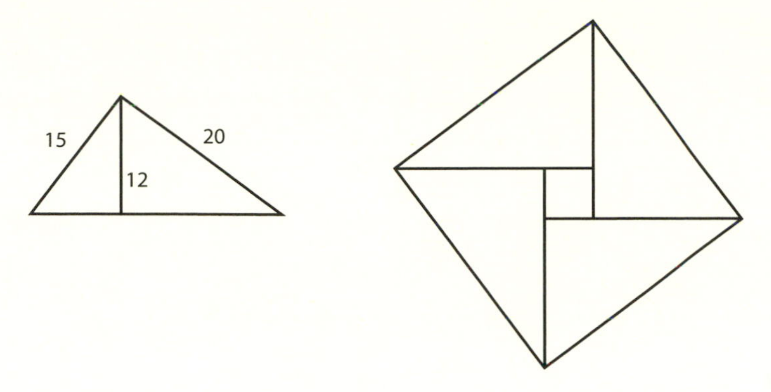

FIGURE 4.19. The right-triangle diagrams described by Bhāskara.

The rule for that operation, in a verse:

129. The product [of the arm and upright], multiplied by two, increased by the square of the difference of the arm and upright, should be equal to the sum of [their] squares, just as [in the case] of two unknowns.

Hence, for the sake of brevity, the square-root of the sum of the squares of the arm and upright is the hypotenuse: thus it is demonstrated. And otherwise, when one has set down those parts of the figure there, [merely] seeing [it is sufficient].

[These verses are presumably the ultimate source of the widespread legend that Bhāskara gave a proof of the Pythagorean theorem containing only the square figure shown in figure 4.19 and the word "Behold!"]

III.d. The audience for mathematics education

Who were the readers for whom copies of these mathematical texts were produced in their thousands? The answer is largely obscured by the lack of records about education and training in the mathematical sciences, but some basic patterns can be deduced. In the first place, livelihoods were to a great extent determined by the four basic social divisions of Hinduism known as the *varṇas* (not to be confused with the word's technical meaning "color" in algebra): namely, the Brāhmaṇas or priests, who dominated learned professions as well as officiating at religious ceremonies; Kṣatriyas or warriors, generally identified as rulers and nobility; Vaiṣyas or merchants, including some skilled craftsmen; and Śūdras or laborers, menial workers and artisans. (The various subgroups of "untouchables" and so forth who were excluded from these divisions were generally too low-status to be acceptable candidates for any kind of education or non-menial work.) Within each *varṇa* were various kin and occupational groups (*jāti* or "caste") within which the members of the group were expected to live, work, and marry. Although this system was in theory a fixed hereditary hierarchy, in practice it was somewhat more fluid: a relatively low-status caste within the Brāhmaṇa *varṇa* might be poor compared to a group of wealthy Vaiṣyas, and an ambitious family or group might improve its position by moving to a new location among strangers and claiming a higher status.

The vast majority of Hindu authors of mathematical texts were Brāhmaṇas, and most belonged to some caste of astronomical/astrological/mathematical practitioners called *jyotiṣīs*. (Non-Hindus who rejected the *varṇa* system, such as Jains and Buddhists, pursued scholarly activities often as religious ascetics or monks.) Most *jyotiṣīs* earned their livings by making calendars and astrological predictions, either for local customers or in the service of noble patrons. Kṣatriyas, Vaiṣyas, and occasionally even Śūdras might have a professional need or an individual liking for mathematical or astronomical knowledge, and a few of them even wrote Sanskrit books on those topics. Presumably there are many more such works by non-Brāhmaṇas in vernacular languages.

The hereditary nature of the mathematical professions meant that training of students was usually a family affair. Numerous authors (including Bhāskara II) mention in the brief introductory verses of their treatises the fact that they studied under their fathers; sometimes an uncle or brother is identified as a teacher. However, there are also references to non-relatives acting as teachers, who presumably received pay for their lessons (either from the students' families or from charitable benefactors) or offered their services as a religious obligation. The latter would be especially true of scholars in monastic or temple institutions. The "school" as an institutional lineage—a voluntary association of teachers and pupils transmitting a specific intellectual heritage based on the works of a founder—is also attested here and there in Indian mathematics (e.g., for the disciples of Āryabhaṭa and for the "Kerala school" discussed below).

Direct addresses to students in some verses of mathematics textbooks like the *Līlāvatī*, using epithets like "best of merchants," suggest that they were intended for the use of non-*jyotiṣīs* too. Evidently some level of mathematical knowledge was quite widespread among the commercial classes and educated people in general. The Brāhmaṇa hereditary specialists in astronomy/astrology nonetheless dominated formal mathematical instruction (at least insofar as it is reflected in Sanskrit textbooks), perhaps much as the mathematics specialists of modern universities dominate the teaching of calculus or statistics to biologists and engineers.

Moreover, not all of the educated people who learned some mathematics were male, judging from some other verses in the *Līlāvatī* and elsewhere addressed to hearers called, e.g., "beautiful lady" or "intelligent girl." Although medieval Indian society restricted women mostly to the domestic sphere, among the educated classes they were evidently expected to know something of calculation, as a tool in household management and as a civilized refinement. One fifteenth-century text recounts a few word problems posed to ladies of a court by mathematicians, indicating that recreational arithmetic was a polite amusement for men and women alike; and lists of the "civilized arts" or accomplishments which ladies were expected to master often included games of chance, prosody, and other activities where calculation was useful. There is no evidence of women's involvement in advanced mathematical studies or astronomy, although there are a few astrology-related texts on omens ascribed to female authors.

III.e. Specialized mathematics: Astronomical and cosmological problems

Notwithstanding the existence of a broader audience for general mathematics education, mathematics remained closely linked to technical astronomy. Astronomers also frequently exploited some specialized techniques seldom or never discussed in general math textbooks. We have already seen that trigonometry, for example, was discussed mostly in an astronomical context. Moreover, astronomers developed many ingenious uses of iterative approximations to solve problems that the closed-form methods available to them could not handle. Interpolation was another subject that attracted interest: e.g., Brahmagupta described a second-order interpolation rule for use in sine tables, very similar to the later Newton-Stirling formula. And it was a common practice among astronomers to approximate complicated trigonometric models by various sinusoidal interpolation rules whose results coincided at a few key points with the exact values. (Ingenious approximation rules of all types were employed especially in astronomical "karaṇas" or handbooks, a sort of abbreviated and simplified *siddhānta* for practical use.) While iteration and interpolation were not classified as mathematical subjects in their own right (although iterative formulas had the special name of "asakṛt" or "not-(just)-once" rules), they were evidently recognized as versatile tools for astronomical calculations.

The example shown below is taken from the *Śiṣyadhīvṛddhidatantra* (*Treatise for Increasing the Intelligence of Students*) composed by one Lalla, probably in the middle of the eighth century. The verses in question are part of his chapter on lunar eclipses (regarded as highly ominous and astrologically significant events), whose goal is to predict the exact time, duration, and appearance of an eclipse. As Lalla explains, this depends on calculating the longitudes of the sun and moon as well as the moon's latitude, which depends on how far away it is from its node, or the intersection of the lunar orbit with the ecliptic.

Śiṣyadhīvṛddhidatantra *of Lalla*

Chapter 5: Lunar Eclipse

5.1. If one wants to ascertain [the time of] a lunar eclipse, one must find the true longitudes of the Sun, Moon, and its ascending node, on the fifteenth *tithi* (i.e., Full Moon day) in the light half of the lunar month, at sunset. [...]

Half-duration of the eclipse

5.14–15. Take the sum or difference in the semi-diameters of the obscuring and the obscured bodies. Square it and subtract from it the square of the Moon's latitude. Find the square root of the remainder. Multiply it by 60 and divide by the difference of the true motions in minutes of the two bodies. The results are, respectively, the approximate half durations of the eclipse and the total eclipse in *ghaṭikās*. When these times are severally multiplied by the true motions of the Sun, Moon and its node and each product divided by 60, the results are, respectively, the motions of the Sun, Moon and its node during these times.

5.16. The Sun's and Moon's motions in minutes should, respectively, be subtracted from their longitudes and the node's motion added to its longitude, if the first half of duration of the eclipse or of the total eclipse is required; the reverse process is to be followed if the second half is required. From these longitudes, again calculate the half duration of the eclipse and of the total eclipse. Repeat the process till the times are fixed.

[Lalla's computational model approximates the ecliptic and the lunar orbit as two intersecting straight lines, with the earth's shadow and the moon represented by circles, the former centered at a point on the ecliptic and the latter on its orbit (see figure 4.20). The lunar latitude β is the perpendicular distance of the center of the moon from the ecliptic. Just at the start of the eclipse, when the two circles are touching, the longitudinal or ecliptic distance between their centers is represented by the longer leg of the right triangle whose shorter leg is the latitude and whose hypotenuse is the sum of their radii. (Lalla's alternative calculation using the difference

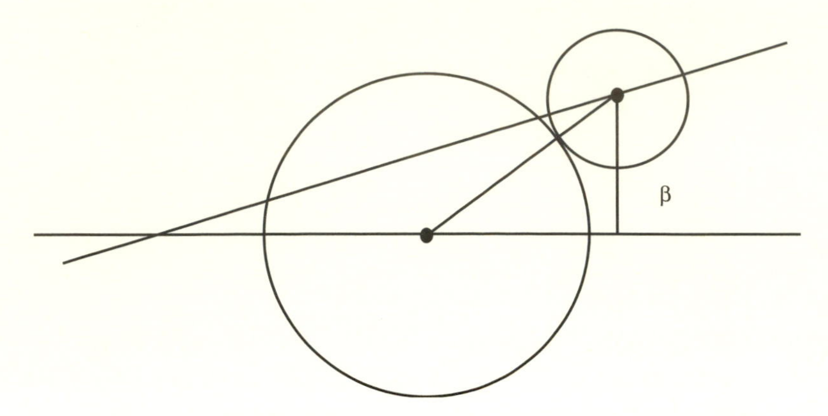

FIGURE 4.20. The earth's shadow and the moon at the start of an eclipse.

of the radii instead of their sum gives the separation at the beginning of the period of totality, when the moon's disk is internally tangent to that of the shadow rather than externally.)

The time interval it will take to reach mid-eclipse, where the longitudinal separation is zero, is found by dividing that separation (expressed in arcminutes) by the speed of the eclipse, that is, the difference in the daily longitudinal motions of the two luminaries. To express this time interval in the standard units of *ghaṭikās* or sixtieths of a day, the quotient is multiplied by 60. In other words, the time interval Δt is given in terms of the radii r_{shadow} and r_{moon}, the daily motions v_{moon} and v_{sun}, and the lunar latitude β by

$$\Delta t = \sqrt{(r_{\text{shadow}} + r_{\text{moon}})^2 - \beta^2} \cdot \frac{60}{v_{\text{moon}} - v_{\text{sun}}}.$$

The longitudinal increments traversed in that interval Δt by the moon and shadow are added to their original longitudes at the start of the eclipse, to give their positions at mid-eclipse. However, due to the quickness of the lunar motion, during that time the moon's latitude will have changed slightly; so if we compute the new latitude corresponding to its new longitude and then recalculate Δt, we will get a different value. Hence Lalla prescribes iterating the computation of the longitudes, latitude, and time until the value of Δt becomes fixed. This is a very common tactic, particularly for calculating time intervals dependent on quantities that change during the course of the desired time.]

IV. The Kerala School

IV.a. Mādhava, his work, and his school

The reverence frequently accorded to earlier teachers, and the longevity of classical Sanskrit as a learned language, meant that the works of mathematicians from preceding centuries continued to be studied and taught. The seven hundred years of layered innovations, repetitions, rediscoveries, confusions, comparisons, and critiques extending from the writings of the first Āryabhaṭa to those of the second Bhāskara served as the springboard for some very remarkable developments in fourteenth- to sixteenth-century Kerala, a region on the southwest coast of the subcontinent. This was the home of the school of Mādhava, a "school" in the more familiar sense of a sequence of direct transmissions from teachers to pupils in the same locality. (We have also used "school" to mean, loosely, the adherents of one of a few characteristic sets of astronomical parameters and the works that used them; in that sense, the "school" of Mādhava adhered to the "school" of Āryabhaṭa.) Their mathematical work was conducted in traditional Keralese joint-family compounds/estates known as "illams," which often supported academic and other activities for family members.

Little is known about Mādhava's personal history and education. He was born probably in the second half of the fourteenth century, and worked for some decades before and after 1400 in an illam at Iriñjālakkuḍa near modern Kochi. The only writings of Mādhava currently known to survive are some astronomical treatises, some of which are dated in the first few years of the fifteenth century. But he is now most renowned for his discoveries in trigonometric power series, preserved only in a few isolated verses. These verses, along with other parts of Mādhava's work, were studied and expounded upon in an illam not far from Mādhava's, home to Mādhava's own pupil Parameśvara and Parameśvara's son Dāmodara. There Dāmodara taught Nīlakaṇṭha and Jyeṣṭhadeva, students from other nearby illams, both of whom in their turn gave instruction to another scholar named Śaṅkara Vāriyar, who worked near the middle of the sixteenth century.

IV.b. Infinite series and the role of demonstrations

The power series that Mādhava's followers so carefully elucidated were equivalent to what we know as Maclaurin series expansions for the sine, cosine, and arctangent. In particular, Mādhava found what is essentially Leibniz' infinite series for the ratio of the circumference of a circle to its diameter, and also derived a numerical value equivalent to $\pi = 3.14159265359$.

In a commentary he wrote on the *Līlāvatī* of Bhāskara (the *Kriyākramakarī*; specifically, in his comment on verse 199 at the start of the section on circles), Śaṅkara discussed at length Mādhava's π-series, drawing on the work of his own teachers Jyeṣṭhadeva and Nīlakaṇṭha. Part of this discussion is reproduced in the excerpt below. The exposition, presumably originating in Mādhava's own rationale for the validity of his series as he explained it to his students, is remarkable for its painstaking verbal demonstrations of the steps in the reasoning behind the formula. While many Sanskrit commentaries suggest some sort of rationale or verification for certain mathematical results, the type of elaborate detailed proof constructed by Śaṅkara, using manipulations of an involved geometric construction to correspond to his operations on quantities, is practically unknown outside the treatises of Mādhava's school.

Kriyākramakarī *of Śaṅkara*

Now, [Bhāskara] begins to mention the method for deriving the circumference in the figure of a circle. Now a verse which is a rule for calculation in the case of a circle.

> When a diameter is multiplied by 3927 and divided by 1250 [the result is] a very accurate circumference. Or when [a diameter] is multiplied by 22 and divided by 7, [the result is] crude and for practical use. [*Līlāvatī* 199]

Here, assuming any desired [quantity] as the diameter, multiply it by 3927 and divide by 1250. The result is the very accurate circumference. Or multiply the desired diameter by 22 and divide by 7. The result is the crude circumference. It is for practical use. In this the three-number rule is thus:

> If a circumference of 3927 belongs to a diameter of 1250, then how great is the circumference of a given diameter?

[If someone asks] how it is known that 3927 is the circumference corresponding to a diameter of 1250, we reply that it is from the words of the teacher [Āryabhaṭa]. The teacher said:

> 8 times 100 increased by 4 and 62 of 1000s is an approximate circumference of a circle with a diameter of 20000. [*Āryabhaṭīya* 2.10]

When these two, the diameter and the circumference have been reduced by 16, [the results are] mentioned here (in verse 199). Here, because, if [one of] these, the diameter and the circumference, measured by a certain measure has no fraction, the other which is measured by the same measure has fractions even after one has gone a long way, just an "approximate circumference" is taught [by the teacher, Āryabhaṭa].

Here, however, indicating a circumference closer [to the true value], some employ another reading:

> When a diameter is multiplied by 355 and divided by 113, [the result is] a very accurate circumference.

This is the preferred reading of those who know reasoning. The teacher Mādhava also mentioned a value of the circumference closer [to the true value] than that:

> Thirty-three, two, eight, eight, three, three, three, four, twenty-seven, eight, two (2827433388233)—the wise said that this is the measure of the circumference when the diameter of a circle is nine *nikharva* [10^{11}].

[...]

An easier way to get the circumference is mentioned by him (Mādhava). That is to say:

1. Add or subtract alternately the diameter multiplied by four and divided in order by the odd numbers like three, five, etc., to or from the diameter multiplied by four and divided by one.
2. Assuming that division is completed by dividing by an odd number, whatever is the even number above [next to] that [odd number], half of that is the multiplier of the last [term].
3. The square of that [even number] increased by 1 is the divisor of the diameter multiplied by 4 as before. The result from these two (the multiplier and the divisor)
   is added when [the previous term is] negative, when positive subtracted.
4. The result is an accurate circumference. If division is repeated many times, it will become very accurate.

Here, though the word "divided by one" has [practically] no meaning, it is intended to demonstrate that the division by odd numbers begins not with three but with one. Whatever diameter is assumed here, multiply that by four and divide it by the odd numbers one, three, five, separately. Then separately sum up the odd-numbered among the results and the even-numbered [terms]. Then subtract the sum of the even-numbered [terms] divided by three, seven, etc. from the sum of the odd-numbered [terms] divided by one, five, etc. The result is the circumference of [a circle] having the desired diameter.

[After discussing various accepted ratios of the circumference C to the diameter d and one of Mādhava's methods for computing C (not discussed here), Śaṅkara undertakes to describe an "easier way." Thereby the circumference is "very accurately" (although as Śaṅkara notes, not exactly) expressed by a finite sum of terms of a series plus a correction term, as follows:

$$C \approx \frac{4d}{1} - \frac{4d}{3} + \frac{4d}{5} - \cdots + (-1)^{n-1}\frac{4d}{2n-1} + (-1)^n \frac{4dn}{(2n)^2+1}.$$

The following demonstration relies on the diagram shown in figure 4.21, reconstructed from Śaṅkara's description below. This excerpt does not include Śaṅkara's subsequent discussion of the correction term $4dn/((2n)^2 + 1)$, which has been investigated elsewhere (Hayashi et al. 1990).]

[Construction:] For demonstrating this, having placed a square whose four sides are equal to a desired diameter, draw a circle in it by means of a compass [whose opening] is equal to half of the diameter and draw the east-west and north-south line(s). Then the circumference touches the middle of each of the four sides of the square in all four directions. Having placed as many dots as you wish with equal intervals at that end of the eastern side which is south from the end of the east-west line, draw lines from the center of the circle to the dots.

[Approximation of arcs by sines:] Then add the square of the distance between the end of that line and the end of the east-west line to the square of the half-diameter, and take the square root of each. The measure of the length of the hypotenuse-line is produced. Then, having multiplied all the segments of the eastern side between

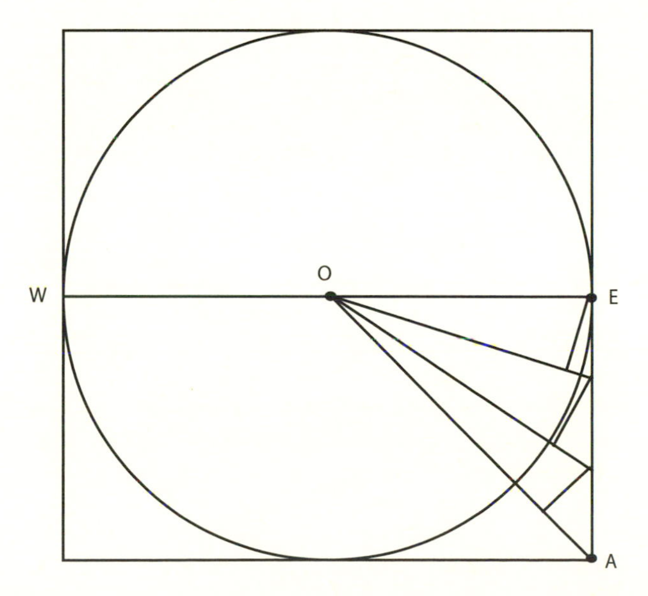

FIGURE 4.21. The diagram described in the circumference series demonstration.

the [hypotenuse-]lines by the half-diameter, divide by the line which passes from the center [of the circle] within [it] to the southern end of each [segment]. The result is the distance from the end of the left [line] of two adjacent lines to the other [line]. That (the distance), passing from the end of the left line to the hypotenuse [which is] the southern line, is placed with its [perpendicular] direction opposite to its (the hypotenuse's). This is the edge of the segment of the arm which is between the lines, which [segment] is the hypotenuse. The end of the southern line is the arm. This figure is similar in shape to the figure whose hypotenuse is the southern line and edge is the half-diameter. The similarity is derived from the fact that the hypotenuse of the desired figure is perpendicular to the edge of the standard figure and the edge of the desired figure is perpendicular to the hypotenuse of the standard [figure].

[The half-side EA, which is equal to the radius $r = d/2$, is divided into n equal segments of length l. Then the bottom endpoint of each equal segment l_i is a vertex of a right triangle with vertical leg $i \cdot l$, horizontal leg $OE = r$, and hypotenuse $h_i = \sqrt{r^2 + (il)^2}$. (Consider the point E to be the zeroth such vertex and the radius OE the zeroth hypotenuse.) These n hypotenuses h_i divide the arc of the circle between E and h_n into n small unequal arcs $c_1, \ldots, c_n$.

From E and from each of the subsequent $n - 1$ vertices on EA, we drop a perpendicular or "edge" e_i onto the hypotenuse h_i, forming a small right triangle with hypotenuse l_i and leg e_i, in the bottom corner of the large right triangle with legs r and il. Owing to the similarity of the large and small triangles, the length of e_i will be given by $e_i = l \cdot r/h_i$.]

Having multiplied the distance derived in this way by the half-diameter, divide by [the length of] the left one of two adjacent lines. The result is the Sine of the part (arc) of the circumference between the lines. [That] is also a part of the circumference when the segment is small. Therefore, multiply each segment by the square of the half-diameter and divide by the mutual product of the line advanced from the end of the southern (line) of each and the line advanced from the end of its northern (line). The results are the parts of the circumference between (the pair of) lines.

There, though it should be divided by the product of the two adjacent lines, even if it is divided by the square of one of those two, the error will not be large in the calculation of a part of the circumference. When it is divided by the square of the left of the two adjacent lines, [the quotient will be] greater than the true value. [When it is divided] by the square of the other (i.e., the right line), [the quotient will be] less. In this case the first, the second, etc. parts of the circumference obtained by [division by] the square of the right line are exactly the same as the second, the third, etc., obtained by [division by] the square of the left line. Therefore whatever is the difference between the first of (the parts) which are obtained (by dividing) by the square of the left line and the last of (those) which are obtained (by dividing) by the square of the right line, that is indeed the difference between the two sums of the parts of the circumference. Half of it is its difference from the true value. Because the [divided] part is small, the difference is very close to zero. Therefore [this] is mentioned assuming the division of the parts multiplied by the square of half of the diameter by the squares of the southern lines.

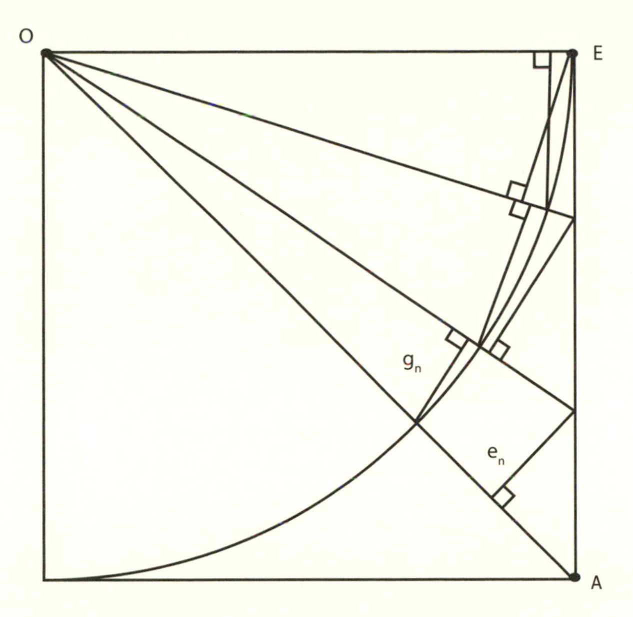

FIGURE 4.22. Detail of the previous diagram.

[Now we raise another small perpendicular g_i from each hypotenuse h_i, at the point where h_i intersects the circle, onto the hypotenuse h_{i-1} above it (see fig. 4.22). These perpendiculars g_i are the Sines of the small arcs c_i, and because they are proportional (again, by the similarity of right triangles) to the corresponding segments e_i, they can be expressed as follows:

$$g_i = \frac{e_i \cdot r}{h_{i-1}} = \frac{lr^2}{h_i h_{i-1}}.$$

The segments g_i are approximately equal to the corresponding c_i, if l is small. Moreover, if we replace the product $h_i h_{i-1}$ in the denominator with the square of the smaller (or what Śaṅkara calls the "left") hypotenuse h_{i-1}, the resulting value of g_i will be slightly greater than its true value. The square of the "right" or "southern" hypotenuse h_i, on the other hand, will produce a g_i that is too small, but not by much. Therefore, the sum of all the arcs c_i corresponding to the n segments of EA is

$$\frac{C}{8} \approx \frac{lr^2}{{h_1}^2} + \frac{lr^2}{{h_2}^2} + \cdots + \frac{lr^2}{{h_n}^2}.$$

Śaṅkara now proceeds to rewrite this summation as follows.]

[Fixation of the Divisor:] All of the parts have the same measure. Therefore, in the calculation of a part of the circumference the multiplicands and the multipliers are [each] of only one kind. The divisors are, however, of various kinds. If the multiplicands and the divisors were of one kind, then, after multiplying [the multiplicands] by the sum of the multipliers and dividing by the divisor once, the sum of the quotients would result. Therefore a method for deriving divisors of one kind should be sought. The calculation of results by dividing by a multiplier without a difference is the very method.

How are results obtained by dividing by a multiplier? It is said that

1. The multiplicand and each result (quotient) is multiplied by the difference between the multiplier and the divisor and divided by the multiplier; whatever the sum of these is to be added to the multiplicand when the multiplier is greater than the divisor.
2. When the multiplier is less [than the divisor], the multiplicand is increased by the even results while the odd [results] are subtracted. The true result is obtained.

The circumference should be calculated by this method.

There in the calculation of the first results multiply [the multiplicand] by the difference between the square of each line and the square of the half-diameter and divide by the square of the half-diameter. [Then] make the sum of the quotients. The difference between the square of the line and that of the half-diameter is the square of the segments of the east side between those two [lines]. These [segments] [make an arithmetic progression which] begins with a portion as [its] first term and has that portion as [its] common difference. Therefore multiply the portion as a multiplicand by the sum of the squares of these and divide by the square of the half-diameter.

Then in the calculation of the second result the first result was mentioned as the multiplicand previously. But it does not exist separately here because only the sum of the [first] results was calculated. Then the second result is also calculated from the multiplicand of the first result.

How is the second result obtained from the multiplicand of the first result? From division by the product of two divisors after multiplying [the multiplicand] by the two multipliers. In both cases the square of the segment of the side between the line and the half-diameter is the multiplier and the square of the half-diameter is the divisor. Therefore, by multiplying the portion by the square of the square of the segment of the side and dividing by the square of the square of the half-diameter the second result should be calculated.

In the calculation of the third result the sixth power of the segment of the side and the half-diameter are the multiplier and divisor [respectively]. In the same way also in the calculation of the following fourth results, etc., powers like the eighth power, etc., are the multipliers and the divisors. The multiplicand is, however, always the portion of the side. Here in the calculation of the sum of each result the sum of the multipliers is assumed to be the multiplier. There in the calculation of the first result it was mentioned previously that the multiplier is the sum of the squares of the sides of the hypotenuse of each line (in an arithmetical progression) beginning with a given portion, having [that] portion as [its] common difference, and ending with the radius. In the same way also in the calculation of the second result, etc., the multiplier is the sum of the fourth power, etc., of those sides. Therefore the method of calculating the sum of the squares, the sum of the squares of squares, etc., should be investigated here. Even if only the sum of the even powers like squares etc., are used here, yet because the sum of the odd powers like the root, the cube, and the fifth power also occur, the sum of both is demonstrated.

[First, Śaṅkara describes a general expression (explained below at the end of the excerpt) for a quantity a multiplied by a multiplier p and divided by a divisor q, both positive:

$$\frac{ap}{q} = a + \frac{a(p-q)}{p} + \frac{a(p-q)^2}{p^2} + \frac{a(p-q)^3}{p^3} + \cdots.$$

Of course, when $p < q$, every second term in this series will be negative. So each of our arcs c_i summing up to $C/8$ can be rewritten as

$$c_i \approx \frac{lr^2}{h_i^2} = l - \frac{l(h_i^2 - r^2)}{r^2} + \frac{l(h_i^2 - r^2)^2}{r^4} - \cdots.$$

Geometrically, $h_i^2 - r^2$ is just the square of the vertical leg il. As a result, we could think of our successive c_i as

$$c_1 \approx \frac{lr^2}{h_1^2} = l - \frac{l(1l)^2}{r^2} + \frac{l(1l)^4}{r^4} - \frac{l(1l)^6}{r^6} + \cdots,$$

$$c_2 \approx \frac{lr^2}{h_2^2} = l - \frac{l(2l)^2}{r^2} + \frac{l(2l)^4}{r^4} - \frac{l(2l)^6}{r^6} + \cdots,$$

$$\vdots$$

$$c_n \approx \frac{lr^2}{h_n^2} = l - \frac{l(nl)^2}{r^2} + \frac{l(nl)^4}{r^4} - \frac{l(nl)^6}{r^6} + \cdots.$$

But Śaṅkara wants to take "the sum of the multipliers" as the multiplier in each "result." In other words, we are to combine the terms with like powers in all of our n series expressions for the arcs $c_1, \ldots, c_n$. Remembering that $nl = r$, this gives us

$$\frac{C}{8} = \sum_{i=1}^{n} c_i \approx r - \frac{l}{r^2}\sum_{i=1}^{n} i^2 l^2 + \frac{l}{r^4}\sum_{i=1}^{n} i^4 l^4 - \frac{l}{r^6}\sum_{i=1}^{n} i^6 l^6 + \cdots.$$

So Śaṅkara next discusses how to express sums of integral powers.]

[Sums of progressions of powers:] There first of all, in the calculation of the sum [of the sides] the last side is equal to the half-diameter. If all sides were equal to the half-diameter, then the half-diameter could be multiplied by the number of the sides. [But] each preceding side in order is shorter. The condition of being shortened begins with the portion as the first term, has the portion as [its] common difference, and ends with the half-diameter decreased by the portion. Therefore the sum of the parts is less than the sum of the sides by just the half-diameter. Then the half-diameter multiplied by the number of the sides is less than the sum [of the sides] multiplied by 2 by just the half-diameter. Therefore, half of (the product of) the half-diameter multiplied by the number of the sides increased by 1 is the sum of the sides. The portion should also be assumed to be the multiplicand here though it is not mentioned. The product of the portion and the number of the sides increased by 1 is greater than the half-diameter by a portion. Only when the portion is small will the result be accurate. In such a case half of (the product of) the half-diameter multiplied by the half-diameter is the product of the portion and the sum of the sides.

Now the calculation of the sum of the squares [of the sides] should be investigated. The side multiplied by itself is indeed the square of the side. There, if all the sides as the multipliers were equal to the half-diameter, then the sum of the squares would be the sum [of the sides] multiplied by the half-diameter. Here, however, only one multiplier is equal to the half-diameter, and the others are somewhat less. These shortened parts also begin with the portion and have the portion as common difference. The sum of the sides multiplied by these in order should be subtracted from the sum [of the sides] multiplied by the half-diameter in order to arrive at the sum of the squares.

Here whatever are the products of the shortened parts and the successively previous sides equal to the radius diminished by its [portions] taken in reversed order beginning with the last term, the sum of these is the sum of the sums multiplied by the portion. In fact, the sum of the sums (i.e., the partial sums of a progression) is called the "sum of the sums." Among them the last sum is the sum of all sides. The second sum from the last is the sum of the sums other than the last side. Then the sum previous to the second sum from the last is the sum of the sides up to that. Thus the previous sums are less than the sum following them by one side each. Therefore there is known a connection of the last side with just one sum, but of the next to last [side] with both the last and the next to last [sums]. In the same way there is the inclusion of the sides preceding the next to the last in the sums [numbered] three, four, and so on. Therefore the sum of the previous sides beginning with the last one multiplied by one, two, three, etc. is arrived at as the sum of the sums. Moreover the previous [sides] are multiplied by the portion multiplied by one, two, etc. Thence [it has] the condition of being the sum of the sums multiplied by the portion.

It was explained previously that the sum of the sides multiplied by the portion is half of the square of the last side. Then this is also half of the sum of the squares. Therefore [this] is called the sum of half of the squares. The sum multiplied by the half-diameter is the sum of the squares and the sum of half of the squares. Then when its own third is subtracted from this, the sum of the squares is obtained. It is [like] grinding flour that the sum of the sides multiplied by the portion is half of the square of the half-diameter. The sum of the squares should also be multiplied by the portion. Then, having multiplied half of the square of the half-diameter by the half-diameter, subtract its own third part. When a third is subtracted from the half, a third of the whole is produced. Therefore, it is arrived at that a third of the cube of the half-diameter is the sum of the squares multiplied by the portion.

[By the basic rule for the sum of an arithmetic progression, the sum of the successive sides $l, 2l, 3l, \ldots, nl = r$ is just $l \cdot \sum_{i=1}^{n} i = r(n+1)/2$. This sum times the "portion" l is $lr(n+1)/2$ or approximately $r^2/2$ when l is small.

Each successive side il is equal to $r - l(n-i)$. The sum of all their squares is therefore given by the sum of the products $(il)(r - l(n-i))$, which can be rewritten as "the sum of the sides multiplied by the half-diameter," minus the sum of the product of each side il with its own "complement" $r - il = l(n-i)$, as follows:

$$\sum_{i=1}^{n} (il)^2 = l(r - l(n-1)) + 2l(r - l(n-2)) + 3l(r - l(n-3)) + \cdots + (n-1)l(r-l) + nl(r)$$

$$= r\sum_{i=1}^{n} il - (l \cdot l(n-1) + 2l \cdot l(n-2) + 3l \cdot l(n-3) + \cdots + (n-1)l \cdot l).$$

This is set up in order to introduce the "sum of the sums multiplied by the portion," or the double sum $l \cdot \sum_{j=1}^{n-1} \sum_{i=1}^{j} il$. That is, a sequence of successive sums of sides can be represented as

$$\sum_{i=1}^{n-1} il = l + 2l + 3l + \cdots + (n-2)l + (n-1)l,$$

$$\sum_{i=1}^{n-2} il = l + 2l + 3l + \cdots + (n-2)l,$$

$$\vdots$$

$$\sum_{i=1}^{2} il = l + 2l,$$

$$\sum_{i=1}^{1} il = l.$$

And if all the corresponding terms in this sequence are multiplied by the portion l and added up vertically, the first result becomes $l \cdot l(n-1)$, the second $l \cdot 2l(n-2)$, and so forth. In other words, each is equal to the product of some side il and its "complement"—that is, equal to some term in the sum of the squares $(il)^2$, above. Consequently,

$$\sum_{i=1}^{n} (il)^2 = r \cdot \sum_{i=1}^{n} il - l \cdot \sum_{j=1}^{n-1} \sum_{i=1}^{j} il.$$

Now, since we said above that $l^2 \sum_{i=1}^{n} i \approx (nl)^2/2$, if n is large,

$$l \cdot \sum_{j=1}^{n-1} \sum_{i=1}^{j} il \approx \sum_{j=1}^{n-1} \frac{(jl)^2}{2} = \frac{1}{2} \sum_{j=1}^{n-1} (jl)^2 \approx \frac{1}{2} \sum_{i=1}^{n} (il)^2.$$

Combining this result with our previous equation, we see that "the sum multiplied by the half-diameter is the sum of the squares and the sum of half of the squares":

$$r \cdot \sum_{i=1}^{n} il \approx \sum_{i=1}^{n} (il)^2 + \frac{1}{2} \sum_{i=1}^{n} (il)^2 = \frac{3}{2} \sum_{i=1}^{n} (il)^2,$$

and therefore "the sum of the squares multiplied by the portion" must be

$$l \cdot \sum_{i=1}^{n} (il)^2 \approx l \cdot \frac{2}{3} \cdot r \cdot \sum_{i=1}^{n} il \approx \frac{2}{3} \cdot r \cdot \frac{r^2}{2} \approx \frac{r^3}{3}.$$

The same reasoning is then followed to get corresponding expressions for the sum of the cubes or of higher powers.]

In the same way, when the sum of the cubes multiplied by the portion has been computed by someone assuming that whatever are the sides which are multipliers of the squares of the sides multiplied by the portion are all equal to the half-diameter as before, whatever is the excess of what is obtained over the real value is investigated here. There because the excess of the assumed multipliers over the real multipliers has the portion as [its] first term and the portion as [its] common difference, according to the mentioned principle the sum of the sums of the squares multiplied twice by the portion is the excess part here. It is also mentioned that the sum of the squares multiplied by the portion is a third of the cube [of the half-diameter]. Therefore it is also arrived at that the sum of the sums of the squares multiplied by the square of the portion is a third of the sum of the cubes multiplied by the portion. From that, the sum of the squares multiplied by the portion when multiplied by the half-diameter is the sum of the sum of the cubes multiplied by the portion and its third. Then when a fourth is subtracted from this, the sum of the cubes multiplied by the portion is obtained.

In this way the successive sums multiplied by the half-diameter are the sums of the sum multiplied by the portion and the next sum of powers. Then for the calculation of the successive sums it is established that each sum should be multiplied by the half-diameter and diminished by its own part divided by the number [of the power] increased by 1.

[A Formula for the subtraction of a part:] Here, add to or subtract from a dividend the result from a numerator and denominator. [The result] has parts. Or in that situation, what is obtained from the denominator increased or decreased by the numerator is to be applied to the divisor in reverse order.

According to this rule [not the dividend but] the divisor can be rectified by its own part. It was previously mentioned that [what is obtained] from the square of the half-diameter [divided] by 2 is the sum [of partial sides] multiplied by the portion. Moreover when it is multiplied by the half-diameter and diminished by [its own] third the sum of the squares multiplied by the portion is derived. There according to the rule of "from a numerator and denominator etc.," the number 2 which was previously a divisor increased by [its] one half becomes the number 3, which should be assumed to be the [new] divisor. In that case the square of the half-diameter multiplied by the half-diameter is the dividend. Therefore [what is obtained] from the cube of the half-diameter divided by the number 3 is the sum of the squares multiplied by the portion. In like manner, each successive [result] from the successive powers involving the half-diameter divided by each of the successive divisors increased by [its own] third, fourth, etc., is the sum of the cubes, etc., multiplied by the portion. Therefore divisors are increased by one at a time.

On the other hand, it was already mentioned that the even sums of powers like the sum of the squares and the sum of the squares of squares, etc., should be multiplied by the portion, but not all [the sums of powers]. Therefore only the odd sums of powers like the third and the fifth power [of the half-diameter] divided by three, five, etc., should be taken. The division of those by the even powers like the square of the half-diameter, etc., is also mentioned. When the successive powers are divided [each] by the power under it, they are all half-diameters. Therefore just the half-diameter should be divided separately by odd numbers like three, five, etc. The results should be subtracted from or added to the half-diameter in order.

[Just as the sum of the squares of the successive sides *il* was previously expressed as "the sum [of the sides] multiplied by the half-diameter" minus "the sum of the sums [of the sides] multiplied by the portion," Śaṅkara similarly recasts the sum of the cubes using the sum of the squares and the sum of the sums of the squares. This allows him to approximate $l \cdot \sum_{i=1}^{n} (il)^3$ as $r^4/4$, and correspondingly for higher powers:

$$l \cdot \sum_{i=1}^{n} (il)^k \approx \frac{r^{k+1}}{k+1}.$$

Śaṅkara points out that the manipulation of such fractional parts is fundamentally the same as for ordinary fractions, referring to a verse (by an unknown author) explaining how to compute the sum or difference of a given fraction and some fractional part of that fraction. (Compare the "combination of fractions" rule given by Brahmagupta in *Brāhmasphuṭasiddhānta* 12.9.)

Finally, we can substitute such a term for each sum of even powers in our previous expression for the circumference:

$$\frac{C}{8} = \sum_{i=1}^{n} c_i \approx r - \frac{l}{r^2}\sum_{i=1}^{n} i^2 l^2 + \frac{l}{r^4}\sum_{i=1}^{n} i^4 l^4 - \frac{l}{r^6}\sum_{i=1}^{n} i^6 l^6 + \cdots$$
$$\approx r - \frac{1}{r^2}\frac{r^3}{3} + \frac{1}{r^4}\frac{r^5}{5} - \frac{1}{r^6}\frac{r^7}{7} + \cdots,$$

which is the series that Mādhava prescribed. Śaṅkara ends by tidying up the remaining loose end of his general rule for a quantity multiplied by a multiplier and divided by a divisor, ap/q.]

[Demonstration of the Expansion Used for the Fixation of the Divisor:]

> If the multiplier is less than [the divisor], add the even results to the multiplicand and subtract the odd [results]. The true result is obtained.

What was said was mentioned just as [a method for] calculation. Its demonstration is indicated [here].

If the multiplier is greater than the divisor, multiply the multiplicand by the difference between the multiplier and the divisor, divide by the divisor, and add what is obtained to the multiplicand; when the multiplier is less [than the divisor], subtract what is obtained. It is easy to understand that the result is to be obtained. In this situation, when the multiplier is less [than the divisor], if the multiplicand is multiplied by the difference between the multiplier and the divisor and divided by the multiplier, then what is obtained is somewhat greater than the true [result]. In order to avoid that, it should be multiplied again by the multiplier and divided by the divisor. In this case, when one has multiplied by the difference between the multiplier and the divisor and divided by the divisor, what is obtained (the second result) is to be subtracted from the first result. If in this situation one should divide by the multiplier, in order to avoid [the slight error] this second result also is to be multiplied by the difference between the multiplier and the divisor and divided by the divisor. The [third] result should be subtracted from the second result. If the third [result] is also divided by the multiplier, for the same [reason] as before the fourth result should be subtracted from the third. When the results are divided in this way by the multiplier repeatedly, the successive [results] are to be each subtracted from its predecessor. Therefore it should be known that the odd-numbered [results] are to be subtracted and the even-numbered to be added with respect to the multiplicand.

In quite the same manner when the multiplier is greater than the divisor, multiply each result by the difference between the multiplier and the divisor and divide by the multiplier; the successive results should each be added to the result preceding itself. Therefore it is to be understood that all [the results] should be added to the multiplicand. Thus even though the results have [been] calculated repeatedly, there is never a completion of applying [the rule]. So, the calculating of results should be stopped without regard for what follows when it is as accurate as you wish.

Here the successive results are less when the difference between the multiplier and the divisor is less than the multiplier; it is inevitable when the multiplier is greater than the divisor. Even when [the multiplier) is less than [the divisor], if the multiplier is greater than half of the divisor, the results one after the other are less. Therefore, when the multiplier is less than half of the divisor, each successive result is greater than its predecessor. It is never known in this situation that the multiplier is less than half of the divisor because the multiplier is the square of the half-diameter and the divisor is the square of the half-diameter increased by the square of each side. What is mentioned [in the *Yuktidīpikā* or "Lamp of Rationales," another work by Śaṅkara]

> Because no side greater than the half-diameter is produced, what is divided by the odd numbers 3, 5, etc., should be subtracted or added separately in order.

is correct.

[Conclusion:] Therefore the half-diameter corrected in this way is part of the circumference between the first and the last lines drawn previously. There the first line extends from the center of the circle toward the East and the last goes toward the South-East. Thus the part of the circumference between those two is exactly one-eighth of the whole circumference. When it is multiplied by eight, it becomes the whole circumference because the ratio is the same in every eighth part of the circumference.

Or else, one should correct [the result] from the dividend multiplied by eight, which is equal to the diameter multiplied by four, divided by three, five, etc., when that (the diameter multiplied by four) is the dividend. And so the circumference is to be obtained.

[We begin by trivially transforming ap/q as follows:

$$\frac{ap}{q} = a + \frac{a(p-q)}{q} = a + \frac{a(p-q)}{p} \cdot \frac{p}{q}.$$

Then we can use the same transformation on the second term of the result:

$$\frac{a(p-q)}{p} \cdot \frac{p}{q} = \frac{a(p-q)}{p} + \frac{a(p-q)}{p} \cdot \frac{(p-q)}{q} = \frac{a(p-q)}{p} + \frac{a(p-q)^2}{p^2} \cdot \frac{p}{q},$$

and continue the "calculating of results" in the same way as long as we wish:

$$\frac{ap}{q} = a + \frac{a(p-q)}{p} + \frac{a(p-q)^2}{p^2} + \frac{a(p-q)^3}{p^3} + \cdots.$$

Again, "when the multiplier is less than the divisor," every term with an odd power of $p-q$ will of course be negative. This justifies our original infinite series expression for the small arcs c_i, and so the demonstration is complete.]

Many features of this exposition—arc-segments so small that they are equal to their Sines, similar right triangles, adjustments to infinite series—strongly recall elements in the seventeenth-century development of the infinitesimal calculus. In consequence, some scholars identify the ingenious work of the Kerala school as a form of calculus, and Mādhava as the true originator of mathematical analysis. More daringly, it is sometimes suggested that the European invention of calculus techniques was inspired by the Keralese discoveries, through a hypothesized chain of transmission involving sixteenth-century Jesuit missionaries in Cochin and a quest for improved trigonometric methods for navigation. While there are indeed many deep and striking similarities between European and Keralese infinitesimal methods, the hypothesis of transmission is probably unsustainable. The trigonometric power series of, e.g., Gregory and Leibniz seem to have grown out of earlier calculus topics such as stereometry, finding tangents and normals, maximum-minimum problems, and quadratures and rectifications in general, which apparently do not figure in these Keralese explorations of the relationships between straight lines and arcs of a circle.

This is not to suggest that the work of Mādhava's school was less brilliant or less deserving of circulation than the developments in seventeenth-century Europe. Rather, what appears to have limited its influence was its comparative isolation, physically and linguistically: some of the Kerala school's works were composed in Malayalam, a Keralese vernacular, and some of the Sanskrit ones were written in Malayalam script, both generally unintelligible to readers in other regions. There is no definite evidence that any of their remarkable results on infinite series were known anywhere outside of Kerala before the nineteenth century.

IV.c. Other mathematical interests in the Kerala school

The abundant output of the aforementioned members of the Kerala school (except Dāmodara, who is not known to have left any extant compositions) is still being studied and published, and much remains unexplored in their writings and those of their more obscure colleagues and pupils. It is already quite clear, however, that Mādhava's infinite series were far from the only outstanding and original accomplishments of these scholars. In addition to astronomical treatises of their own composition, they produced commentaries on several texts by Āryabhaṭa, Bhāskara I, and Bhāskara II, among others. Throughout these works appear various innovations in topics including astronomical computation and observation, iterative approximations, solution of equations, interpolation techniques, and spherical trigonometry.

The following excerpt from a commentary on the *Āryabhaṭīya* by Mādhava's student's son's student Nīlakaṇṭha, while far less complicated than Śaṅkara's exposition of the π-series, illustrates the delight that these mathematicians took in ingenious detailed demonstrations of computational rules. The first formula that Nīlakaṇṭha validates here is Āryabhaṭa's expression for the sum of the squares of the first n natural numbers, which Nīlakaṇṭha conceives as part of a solid with dimensions n, $n + 1$, and $2n + 1$.

Āryabhaṭīyabhāṣya *of Nīlakaṇṭha*

> A sixth part of the triple product of [the term-count, *n*] plus one, [that sum] plus the term-count, and the term-count, in order, is the total of the series of squares. And the square [of the total] of the series [of natural numbers] is the total of the series of cubes. [*Āryabhaṭīya* 2.22]

[From the discussion of *Āryabhaṭīya* 2.19 in section III.a., we recall that "the desired [number of terms] decreased by one, halved, having the common difference for multiplier, and increased by the first term" gives the mean value, and "the mean [value] multiplied by the desired [number of terms] is the value of the desired [number of terms]." In other words, for the arithmetic series with first term and common difference equal to 1, the mean value of the first n terms will be $1 + (n - 1)/2 = (n + 1)/2$, and their sum will be $n(n + 1)/2$. Verse 2.22 now asserts that the sum of the squares and cubes of the first n natural numbers will be

$$\sum_{i=1}^{n} i^2 = \frac{(n+1)(2n+1)(n)}{6} \quad \text{and} \quad \sum_{i=1}^{n} i^3 = \left(\sum_{i=1}^{n} i\right)^2 = \left(\frac{n(n+1)}{2}\right)^2$$

respectively. Nīlakaṇṭha embarks on his demonstration as follows:]

First, then, the explanation of the total of the series of squares is to be shown. Here it is said, "A sixth part of the product of a triple of quantities beginning with the term-count plus one is the total of the series of squares." And, being that this is demonstrated if there is equality of the total of the series of squares multiplied by six and the product of the three quantities, their equality is to be shown.

A figure with height equal to the term-count, width equal to the term-count plus one, [and] length equal to the term-count plus one plus the term-count is [equal to] the product of the three quantities [Fig. 4.23]. But that figure can be made to construct the total of the series of squares multiplied by six. In this way: When the square of the term-count is multiplied by six, there are six figures having the form of the square of the term-count. Combining the figures among them two by two, one produces three figures. Hence the length of them is equal to the term-count multiplied by two; the width is equal to the term-count. Again, one produces in just the same way, by means of squares [times six] of the term-count successively diminished by one, [successive sets of] figures three by three. Hence their width is successively less by one than [that of] the previous ones, [and their] length less by two. And the height of all of them is to be considered as unity.

Again, with these figures one constructs the figure having the form of the product of those three quantities. That is in the following way: First, having taken the figures made with [squares of area equal to] the square of the term-count, and set

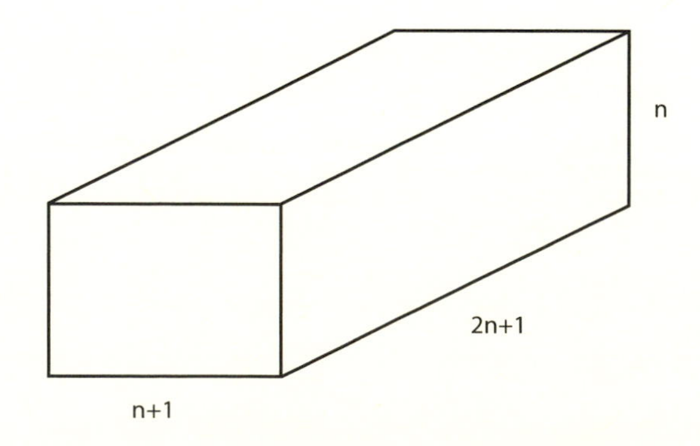

FIGURE 4.23. The triple product in the sum of n squares.

down one of them extending north and south, one should set down the second to the west of it as before, with the form of a [vertical] wall. Then break the third in a part equal to the term-count plus one. When it is thus, one of those two pieces [will have] length equal to the term-count plus one [and] width equal to the term-count. But the length of the other with respect to the former is equal to the term-count minus one. But now that [length] is to be considered a width; and the former width is now to be considered a length, from the suitability of the smaller part to "wideness" and of the larger to "longness." So that [broken piece] has length equal to the term-count and width equal to the term-count minus one. Of those two, set down the piece with length equal to the term-count plus one to the north of the two northern sides, next to [them] and extending east and west.

When it is thus, there are the west and north sides of the figure having the form of the product of the three quantities. There, the north side has length equal to the term-count plus one, because of the figure's being set down in that place in that way. The west side has length equal to the term-count plus one plus the term-count. That side is just the sum of the length of the west figure and the width of the west side of the north figure. It was said there that the length of the west figure is equal to the term-count multiplied by two, and the other measure is one *aṅgula* [digit]. So the length of that side is equal to the term-count plus one plus the term-count. And the height of the two sides is equal to the term-count, because of the two figures' width being as it is.

Again, one should set down the other [piece of the third] figure upon and to the south of the one first set down, extending east and west. The setting down is to be done in such a way that the south side stands upon the south side of [the figure] first set down; [and its] west side is to be made contiguous to the east side of the west side [i.e., the western "wall" of the whole construction]. When it is thus, it is the south side of the figure.

And that has length equal to the term-count plus one. How? It was said that the figure set down to the south is of length equal to the term-count. But that added to the south side, measuring one *aṅgula*, of the west [side] is equal to the term-count plus one. And the height of the south side is equal to the term-count, because the set-down figure of width equal to the term-count minus one was set down upon the [figure] first set down, of height one *aṅgula*.

Again, break one of the figures constructed with the square of the term-count minus one, in a part equal to the term-count. Then one of the two pieces has length equal to the term-count and width equal to the term-count minus one. When there is considered to be an interchange, as before, of the length and width of the other [piece], the length is equal to the term-count minus one, the width is equal to the term-count diminished by two. [...]

[Nīlakaṇṭha begins by considering the rectangular prism formed by the triple product $(n + 1)(2n + 1)(n)$, as shown in figure 4.23. This prism can then be built up from sets of three "domino"-slabs composed of pairs of squares, as shown on the left side of figure 4.24. The first set of "dominoes" has breadth n, length $2n$, and thickness 1. The right side of the figure illustrates the beginning of the process of building up the prism with these three identical slabs. One "domino" (shown with shaded surface) is laid down flat for the "floor," and a second

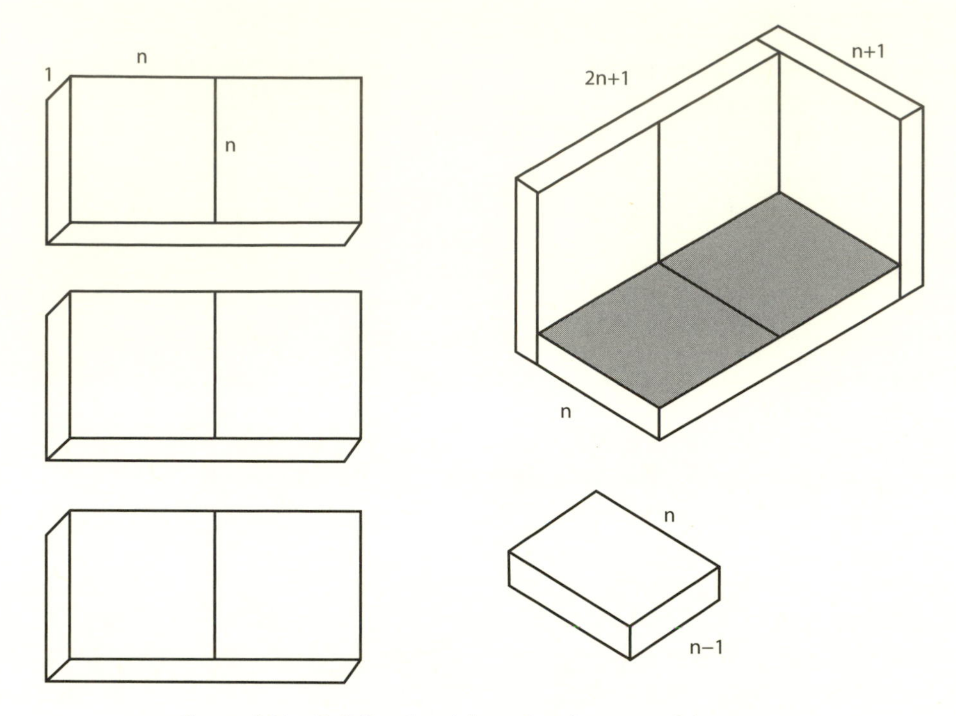

FIGURE 4.24. Building the triple product from sets of six squares.

one is set lengthwise on edge next to it, forming a vertical "wall." The third in the set is broken nearly in half, so that one piece has length $n + 1$ and the other length $n - 1$. The larger broken piece is set on edge at one end. The leftover piece will be placed on edge at the other end, with its edge of length n resting on the "floor" and its shorter edge resting against the long vertical "wall."

Now the next smaller set of three "dominoes," composed of pairs of squares of side $n - 1$, will be laid in exactly the same way on the exposed portion of the "floor," and so on until the n sets of "dominoes" have been nested into the solid prism pictured in figure 4.23. Since each set contained six equal squares, the whole prism must contain six times the sum of the n successive squares. So one-sixth of its total volume must be the sum of those n squares, just as Āryabhaṭa asserted.]

Now, the explanation of the total of the series of cubes. Here it is said, "The square [of the total] of the series [of natural numbers] is the total of the series of cubes." And this is demonstrated if there is equality of the square of the series and the total of the series of cubes. And their equality is demonstrated in the feasibility of making the total of the series of cubes construct the square [of the total] of the series [of first powers]. And the method of that construction is thus:

First, then, construct a figure having the form of the square of [the total of] the series [of natural numbers], with side and counter-side equal to [the total of] the series, and height one *aṅgula*. Beginning from a place [at a distance] equal to the term-count north and west from the corner [in the direction] of Agni [i.e., south-east], break that in the north-south and east-west directions. When it is thus, there are those four figures.

In this case, [the one] in the south-east corner is [a square] with side and counter-side equal to the term-count, because of [its] being broken in a place [at a distance] from the corner equal to that in both directions. [There are two rectangles] north and west of that, with width equal to the term-count. But their length is equal to the series [of natural numbers] diminished by the last term, because of the diminution by a part equal to the term-count in the series having one for [its] common difference and first term, and [thus] because of the equality of the term-count and the last term. The other [figure] is [a square] with side and counter-side equal to [the total of] the series minus the last term, because of the diminution to the south and east by a part equal to the term-count.

Again, break two [pieces] on both sides of [the square] in the corner, in the manner to be stated. There two [rectangles] bordering that [square], with width equal to the term-count minus one, are made; after that, two with width equal to the term-count minus two, and again two with width equal to the term-count minus three. The breaking is to be done so that [all] these [pairs of rectangles] have width successively less by one *aṅgula* than each previous [pair].

When [it is] broken in this way, on both sides there are [identical] figures [equal in] number to the term-count minus one, and equal in length to the term-count. [...]

So the number of those is equal to the term-count minus one. Again, place them one above the other on [the square] in the corner. Then that should be a cubical figure of length and width and height equal to the term-count, because of the length and width of all [of them] being equal to the term-count. Inasmuch as the length and width are equal to the term-count, from the setting-down of [those squares] of height one *aṅgula*, [equal in] number to the term-count, upon the corner [square] of height one *aṅgula*, the height too is in fact equal to the term-count. Thus this figure with the form of the cube of the term-count is produced.

Again, one should construct, by means of the fourth figure with side and counter-side equal to [the total of] the series minus the last term, cubical figures of [volumes equal to the cubes] of the term-count minus one, etc. In that case, the first breaking is to be done in the place [at a distance] from the corner equal to the term-count minus one in both directions; and then in the place [at a distance] equal to the term-count minus two. In the same way, break at the place [at a distance] successively less by one *aṅgula* from each previous [corner]. When this is done in the same way as before, there are those cubical figures of [volumes equal to the cubes] of the term-count minus one, etc.

[Nīlakaṇṭha's proof for the sum of the cubes uses the same sort of geometrical manipulation. Consider a square slab of thickness 1 and side $n(n+1)/2$, as shown in figure 4.25. If a rectangle of width n is cut off from each of two adjacent sides, the resulting four pieces will be a square of side n, a square of side $n(n+1)/2 - n$, and two rectangles of width n and length $n(n+1)/2 - n$.

Each of these two rectangular slabs therefore contains $(n-1)/2$ square slabs of side n. Stack all those $n-1$ square slabs on top of the corner square of side n, and the resulting solid is a cube of side n.

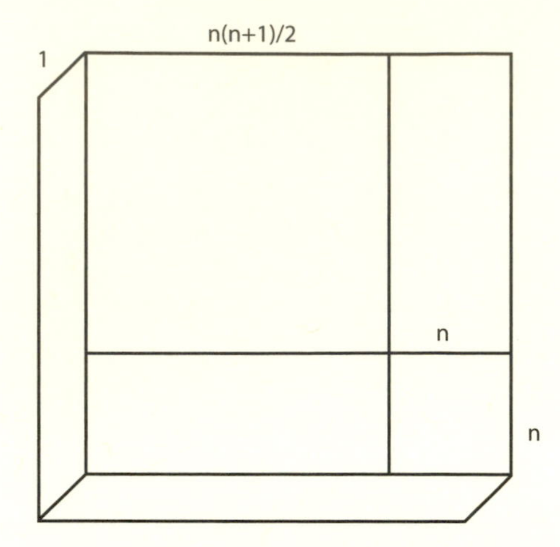

FIGURE 4.25. Building the sum of cubes.

Now take the remaining square of side $n(n+1)/2 - n$ and divide it similarly, by cutting off overlapping rectangles of width $n - 1$. Manipulate the resulting small corner square and the adjacent rectangles in the same way to stack up a cube of side $n - 1$. Repeat the procedure with the remaining corner square, and so on until the original square slab of side $n(n+1)/2$ has been completely dissected into n successively smaller cubes. Thus the sum of those n cubes indeed equals $(n(n+1)/2)^2$.]

V. Continuity and Transition in the Second Millennium

V.a. The ongoing development of Sanskrit mathematics

The lasting predominance of Bhāskara II's mathematics texts, and many historians' ignorance of later work such as that of the Kerala school, helped foster the notion that Indian mathematics after the twelfth century stagnated or decayed. In fact, besides the Keralese work, there was a persistent flow of Sanskrit writing on mathematical subjects in the rest of the second millennium. The classic textbooks continued to be copied, studied, and taught, and new commentaries were written on them; we have already seen a sample of such efforts in the sixteenth-century commentary of Sūryadāsa on the *Bījagaṇita.*

In addition, some new treatises on mathematics, as well as new *siddhāntas*, were composed between the twelfth and sixteenth centuries, though most of them still remain entirely or partially unpublished. One of the best known is the *Gaṇitakaumudī* (*Moonlight of Computation*) completed in 1356 by Nārāyaṇa Paṇḍita, about whose life few other details are available. We do know that he also wrote a partly extant algebra work, the *Bījagaṇitāvataṃsa* (*Garland of Algebra*), mentioned above by Sūryadāsa in his commentary on *Bījagaṇita* 10. The *Gaṇitakaumudī*'s structure is largely based on that of the *Līlāvatī*, but it also contains some more extensive investigations of various topics including permutations and combinations (the "net of numbers"), accompanied by a lengthy section on magic squares. The first of the following excerpts show a few of Nārāyaṇa's combinatorics rules, in particular the diagram called the "partial Meru" (named for Mount Meru in Hindu mythology) for classifying permutations. The latter part is Nārāyaṇa's introduction to the mathematics of magic squares. The *Gaṇitakaumudī*'s numbered verses are supplemented by a prose commentary which may or may not be by Nārāyaṇa himself.

Gaṇitakaumudī *of Nārāyaṇa Paṇḍita*

Chapter 13: Net of Numbers

Now the rules concerning the net of numbers:

13.1. Now I will briefly describe the net of numbers which causes enjoyment for mathematicians, in which those who are jealous, depraved, and poor mathematicians fall down.

13.2. It is applied to dance and music, metrics, medicine, garland-making, and mathematics as well as architecture. Knowledge of these [subjects] is [indeed acquired] by means of the net of numbers. [...]

Rule for the partial Meru:

13.25–27. The horizontal rows of cells measured by the [number of] places and ending with one are to be made as many as there are [cells] underneath each other and above. In the first cell in the first horizontal row one should write one, in the other [cells] zero. In the rows below this, one should write the sequences called increasing placed above. One should multiply these numbers by the product of their own rows. This is called the partial Meru by the virtuous persons. The state of being the enumeration is always produced from the sum of the numbers in the cells on the hypotenuse.

Example: Of what sort is the partial Meru having six places and cells with numbers?

O friend! show that quickly [if] you know the method of the net of numbers.

Here there are 6 places. The partial Meru of six places is produced by means of the operation as described:

1	0	0	0	0	0
	1	2	6	24	120
		4	12	48	240
			18	72	360
				96	480
					600

[...]

The rule for the [method] of extension:

13.49–51. The setting of the indicated numbers with the smaller ones first is called the [base] order. After one has put the smaller [number] under the first larger [number] the rest is as above; [and] the base order in the absence of those; [continue] until the [base] order becomes the reverse order. The method of the extension of numbers in this way [pertains to] the varieties of images and the notes of a lute, *sa, ri, ga, ma, pa, dha,* and *ni.*

Example: ... Of what sort is the extension of the varieties of Viṣṇu's images? O friend! [tell this] if you are able to cross the ocean named the net of numbers.

Here, when one puts a row [of numbers] above Viṣṇu's weapons, that is the base order. The first syllables of *padma* (lotus), *gada* (club), *cakra* (discus) and *śaṅkha* (conch) are the extension:

1	2	3	4
pa	ga	ca	śa

In this way the extension of the images of Viṣṇu [is also constructed]:

1	pa ga ca śa	7	pa ga śa ca	13	pa ca śa ga	19	ga ca śa pa
2	ga pa ca śa	8	ga pa śa ca	14	ca pa śa ga	20	ca ga śa pa
3	pa ca ga śa	9	pa śa ga ca	15	pa śa ca ga	21	ga śa ca pa
4	ca pa ga śa	10	śa pa ga ca	16	śa pa ca ga	22	śa ga ca pa
5	ga ca pa śa	11	ga śa pa ca	17	ca śa pa ga	23	ca śa ga pa
6	ca ga pa śa	12	śa ga pa ca	18	śa ca pa ga	24	śa ca ga pa

And [similarly] for musical notes, namely, *sa, ri, ga, ma, pa, dha,* and *ni.*

Rule for [the number of] an indicated [variation]:

13.52–54. Markers are to be placed in [all] the cells except one of the partial Meru measured by the [number of] places. Whatever is the number at the end of the indicated [variation], as far as that is from the final [number] of the base [order], just as far in the cell below of the partial Meru one should place a marker. There is omission of this [last number] in both the base order and the indicated [variation]. [Do this] again as long as there is such a number. The sum of the numbers which fall in the cells occupied by the markers is the measure of the variations at the indicated [variation] combined with the first number.

Example: O mathematician! tell me quickly! [...] The base order is lotus and club, and discus and conch. How much is the variety of conch, club, discus, and lotus? [...]

The base order in the second example is

1	2	3	4
pa	ga	ca	śa

The indicated variation is 4231. By means of the operation as described the number of the indicated [variation] is produced: 22.

[This "partial Meru" technique is a method for identifying one of a given set of permutations or "variations," such as the various melodies formed with the seven notes of the musical scale, or the images of the god Viṣṇu with his four iconic attributes, the lotus, the club, the discus, and the conch. The "partial Meru" figure is constructed as shown with the cells in its second row containing successive factorials, which are multiplied in the lower cells by successive integers. The method depends upon assigning to the variations a fixed order or "extension," as follows: Choose a particular permutation as the "base order," e.g., 1-2-3. Then write beneath it the next permutation as Nārāyaṇa directs: "put the smaller number under the first larger number," with "the rest as above," or 2-1-3. Continue gradually working from the base order to the reverse base order in this way until you have produced the complete extension:

1 2 3
2 1 3
1 3 2
3 1 2
2 3 1
3 2 1

One could then identify the ordinal number of any variation, or vice versa, by laboriously searching through the whole extension, but the "partial Meru" method allows us to calculate it directly instead. To figure out the ordinal number of the variation given in the example, conch-club-discus-lotus or 4-2-3-1, we compare it to the base order 1-2-3-4, using the four-place partial Meru shown below.

Nārāyaṇa instructs us to start with the "end" or rightmost number of the indicated variation. That rightmost number, 1, is fourth from the right in the base order, so in the rightmost column of the partial Meru we mark the entry in its fourth row, 18. Now "omission" of the 1 from both sequences produces the new sequences 4-2-3 and 2-3-4. We see that the new rightmost number 3 is second from the right in the base order, so in the next column we mark the entry in the second row, 2. And so on through the remaining columns, after which we add up all the marked entries: 18 + 2 + 1 + 1 = 22, the desired number of the indicated variation.

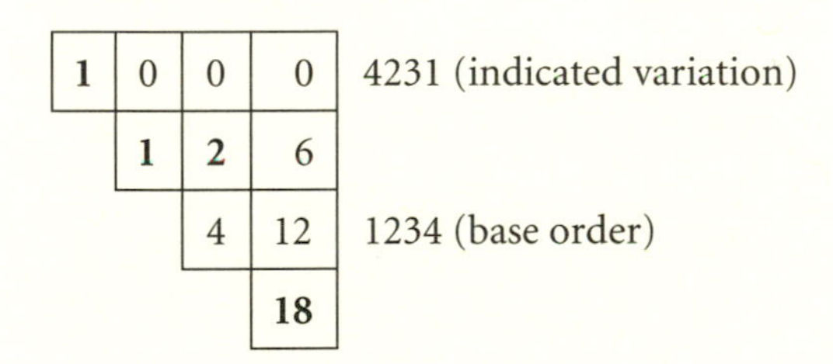

Nārāyaṇa remarks in his second verse that the "net of numbers" methods are applicable to music, and indeed he seems to have adopted some of them directly from musical texts. For example, the thirteenth-century *Saṅgītaratnākara* (*Jewel-Mine/Ocean of Music*) of Sārṅgadeva uses concepts such as the "extension" and "partial Meru" in just this way to classify permutations of notes (melody or *tāna*) and permutations of beats (rhythm or *tāla*).]

[...]

Chapter 14: Magic Squares

The mathematics of magic squares:

14.1. Then [a class of] good mathematics connected with series was taught to the spirit Māṇibhadra, who was interested, by Śiva, the teacher of the three worlds.

14.2. I will describe its essence, called *bhadragaṇita*, in order to surprise good mathematicians, in order to please those who know magical diagrams, and in order to dispel the pride of poor mathematicians.

Terminology:

14.3. A magic square may be of three sorts: having an even *garbha*, having an odd *garbha*, and the odd. Those [magic squares] which are mixed and circular are called *upabhadras*.

14.4. When the number of a magic square is divided by four and there is no remainder, that has an even *garbha*. However, if the remainder is two, that has an odd *garbha*. If the remainder is three or one, that is simply the odd.

14.5. The sum for all magic squares in [computed] by the method of series. The first [term] and the increase for those [magic squares] whose sum is given are to be determined.

14.6. The sum is divided by the number of the magic square; the enumeration of the equal [elements] is the result. As many as there are cells [so great] is the number of terms in the series.

14.7. When a magic square has cells in a square, the root of that is the foot. Here the technical terminology in the mathematics of magic squares has been established by Nārāyaṇa.

Rule for the sum:

14.8. The square of the number of terms is added to the number of terms; a half is the sum [of the terms in a sequence] with one as the first [term] and as the increase. That, divided by the square root of the number of terms, is indeed the result in a given magic square.

Example: In [a magic square] having sixteen cells, in one having cells [numbered by] the square of six, or in one having nine [cells], with one as the first [term] and one as the increase, tell quickly what the result is in each case.

In a magic square of [order] four, *a* 1, *d* 1, *n* 16. In a magic square of [order] six, *a* 1, *d* 1, *n* 36. In a magic square of [order] three, *a* 1, *d* 1, *n* 9. The sums are 136, 666, 45. The results are 34, 111, 15.

Rule for computing the first [term] and the increase:

14.9. The gain of the number of terms minus one is the negative dividend, the number of terms is the divisor, and the property is the addendum. The quotient and the multiplier, produced from the pulverizer, together with the addendum, are the first [term] and the increase [respectively].

Example: In the cells previously mentioned, the properties are four hundred, one thousand two hundred and ninety-six, and one hundred and eighty in order. O learned man! Tell [me] the integer first [term] and the increase if you are proud of your skill in mathematics.

> The setting of the first [example]: *a* 0, *d* 0, *n* 16. Here [by the rule] "the gain of the number of terms minus one" the sum of the number of terms minus one is 120. If one supposes this to be the negative dividend, the number of terms the divisor, and the gain the addendum, the setting for the pulverizer is: dividend: –120, addendum: 400, divisor: 16. Thence reduced: dividend: –15, addendum: 50, divisor: 2. The quotient and the multiplier together with the addendum are: addendum: –15, quantity: 25; addendum: 2, quantity: 0. These two are the first [term] and the increase. When one multiplies the addenda of the quotient and of the multiplier by zero, and adds [the products] to the quantities [respectively], the integer first [term] and increase are 25, 0; [multiplied] by one, they are 10, 2, and by two, –5, 4.
>
> Thus the first [term] and the increase together with the addendum of the second [example] are: addendum: –35, quantity: 1; addendum: 2, quantity: 2. [When they are multiplied] by zero, 1, 2 are produced; by one, –34, 4.
>
> The first [term] and the increase together with the addendum of the third [example] are: addendum: –4, quantity: 20; addendum: 1, quantity: 0. [When they are multiplied] by zero, 20, 0; by one, 16, 1; by two, 12, 2; by three, 8, 3; by four, 4, 4; and by five, 0, 5. Thus there is an infinity [of answers] according to [the number] chosen. Wherever there is a calculation of the first [term] and the increase, the two [of them] are to be known from the pulverizer.

[The characteristics of magic squares composed of successive integers are found by means of the familiar mathematics of arithmetic series. The syllabic abbreviations Nārāyaṇa uses for those quantities are translated here as *a* for the first term, *d* for the common difference, and *n* for the number of terms. The term *garbha* refers to the concentric "shells" of cells in a magic square: a square of even order has an odd *garbha* unless the order is divisible by 4, when it has an even *garbha*. When the sum and order of a magic square are specified, a sequence of numbers satisfying those conditions can be found by a first-degree indeterminate equation, which Nārāyaṇa interprets in terms of the "pulverizer."]

Meanwhile, in mathematical astronomy, certain cosmological questions attracted increasing interest from about the sixteenth century onward. These had to do with the problem of reconciling geometric models of the universe with the traditional cosmology found in sacred texts like the Hindu Purāṇas. We have already seen that scriptural authority was one of the factors that *siddhānta* authors in different schools could use to justify their choice of astronomical parameters. But the earlier astronomers, as we saw in Brahmagupta's remark on the relative distances of the sun and moon in chapter 7 of the *Brāhmasphuṭasiddhānta*, could also argue directly against scriptural models that contradicted their theories. Entire chapters and treatises were now written on the topic of "non-contradiction": that is, reconciling the *siddhānta* models of terrestrial and celestial spheres, and the trigonometric techniques that relied upon them, with sacred cosmology. Since the latter, for example, described the earth as a flat disk many millions of kilometers in diameter, supporting the immense central Mount Meru behind which the celestial bodies appeared to set and rise, making the two systems mutually consistent was not an easy task. Although several astronomers and philosophers tried their hands at it (among them Bhāskara's commentator Sūryadāsa and other members of his family), it is still not clear exactly why "non-contradiction" came to be considered so crucial. Perhaps exposure to foreign cosmologies from the Muslim and Christian worlds reinforced the importance of conforming to the universe as described by scripture.

V.b. Scientific exchanges at the courts of Delhi and Jaipur

Foreign sciences became more and more widespread in India during this period, and had a sporadic but significant impact on mathematical sciences in Sanskrit. From the start of the millennium onward, there were frequent invasions and conquests of various parts of the subcontinent by Muslim rulers from the northwest. Although the resulting endemic warfare disrupted many institutions, the conquests also brought together scientists from the Indian and Islamic traditions, especially at the new courts of Muslim emperors and sultans and some of their Hindu subordinates. Persian was the official administrative language of Muslim imperial courts, but in northern India the natives and the newcomers came to communicate mostly in a mixed vernacular derived both from Persian and from Modern Indo-Aryan dialects related to Sanskrit. This vernacular, the ancestor of modern Hindi (written in *nāgarī* characters) and Urdu (in Persian script), helped mixed teams of scholars to translate texts from Arabic or Persian into Sanskrit, and vice versa.

The courts of the Tughluq and Mughal emperors at Delhi, and of Maharaja Sawai Jai Singh at Jaipur, furnish notable examples of such mathematical cross-fertilization. In the late fourteenth century, under the patronage of Fīrūz Shāh Tughluq, a Jain astronomer named Mahendra Sūri adapted Islamic sources into a Sanskrit treatise on the astrolabe. Around the middle of the seventeenth century, one Nityānanda composed Sanskrit *siddhāntas* and tables based on Islamic models. And in the 1720s and 1730s Jai Singh (Jayasiṃha), the Maharaja of Jaipur, launched an ambitious project to test and compare Indian, Islamic, and European astronomical systems, which involved the translation into Sanskrit of Islamic astronomical treatises and Arabic or Persian versions of Hellenistic works on astronomy, geometry, and spherics.

At the same time, Sanskrit treatises inspired several Muslim scientists to produce translations or new works based upon them. The *Līlāvatī* and *Bījagaṇita* in particular became familiar to Muslims upon their translation into Persian in the sixteenth and seventeenth centuries, under the Mughal emperors Akbar and Shāh Jahān. Moreover, some scientific dictionaries were compiled listing technical terms in Persian and Sanskrit, to make study and translation of foreign texts easier.

The two pages from Sanskrit manuscripts shown in figures 4.26 and 4.27, courtesy of the Sri Ram Charan Museum of Indology, Jaipur, illustrate the influence of non-Indian mathematics on the work of the court astronomers of Jai Singh. The former is an astronomical text showing the influence of Western celestial models in its eccentric orbital circles. (The labeling of points in the figure with letters is another feature inspired by Graeco-Islamic texts.) The latter is an exposition of Euclidean geometry; note the classic "Bride's Chair" diagram for the Pythagorean theorem.

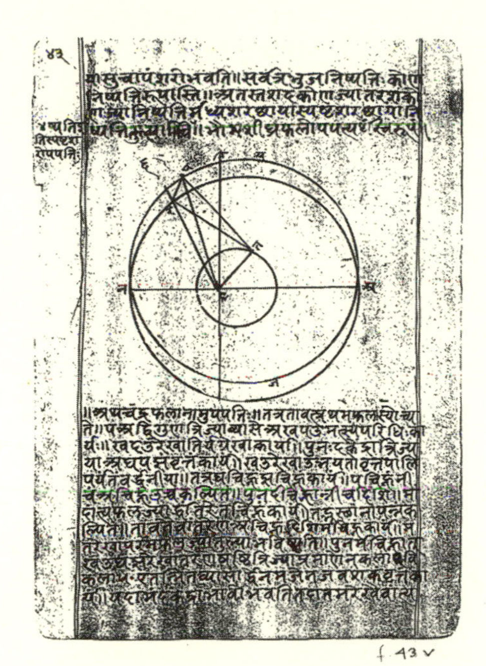

FIGURE 4.26. A Sanskrit astronomical manuscript based on non-Indian works.

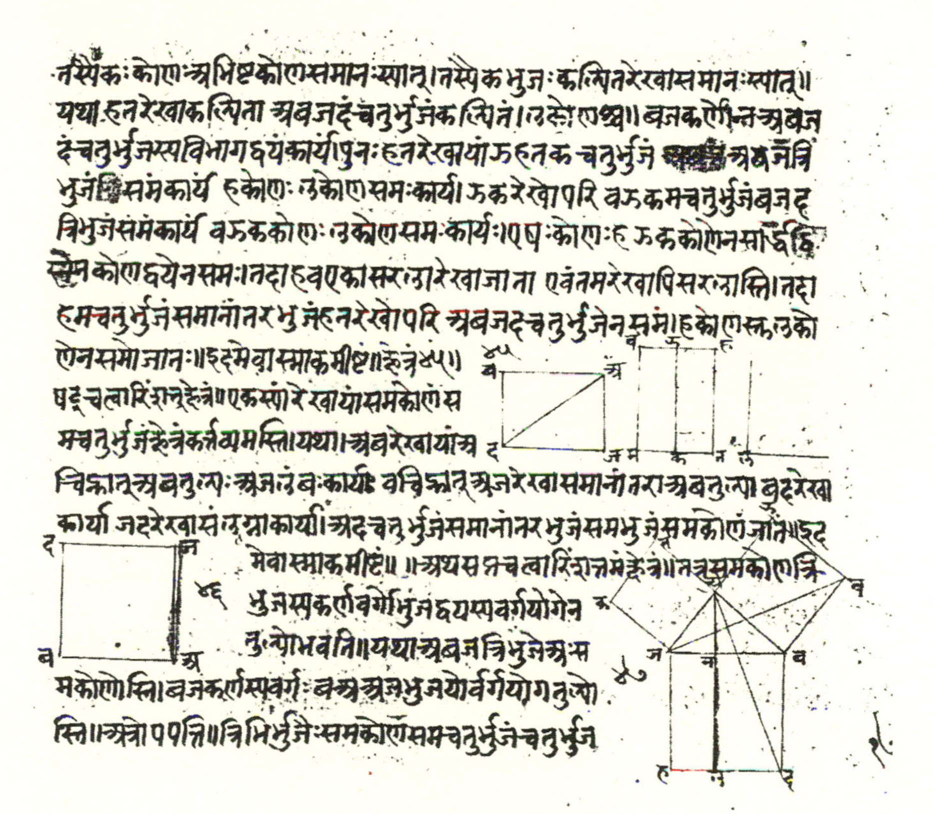

FIGURE 4.27. A Sanskrit manuscript on Euclidean geometry.

V.c. Assimilation of ideas from Islam; mathematical table texts

Outside of royal courts, however, the majority of Indian astronomers and mathematicians probably did not read any Islamic texts or Sanskrit translations of them. Graeco-Islamic cosmological and physical models generally received a cold welcome, although some astronomers in eighteenth-century Benares debated at length whether to incorporate them into the traditional *siddhānta* structure. (And as we have seen, the philosophical crisis concerning "non-contradiction" may have been partly due to encounters with such models.) But the indirect influence of Islamic mathematical astronomy, especially via some of its practical computational tools, was much greater.

The Islamic plane astrolabe in particular gained great popularity, owing to its handiness as a sort of analog computer: when its rotating disks were adjusted for a particular time and place, the positions of celestial bodies could be read directly from its scales, with no calculation required. Its advantages earned the astrolabe the Sanskrit name *yantrarāja* ("king of instruments"), beginning at least in the late fourteenth century when Mahendra Sūri wrote his well-known manual on the astrolabe under that title. Other Sanskrit treatises on the *yantrarāja* were composed in the succeeding centuries, and many astrolabes were constructed with inscriptions in *nāgarī* script for Sanskrit-reading users.

The Islamic *zīj* or astronomical table apparently had an even greater impact on the practice of Indian mathematical astronomy. Sine tables like Brahmagupta's, and other versified lists of numbers in alphabetic or concrete-number notation, had of course formed part of Sanskrit treatises for many centuries, and had been written down in manuscripts in tabular number-boxes. But detailed "table texts" as we know them, graphically ordered rows and columns of numerical values of various functions, seem to have caught on in India after the introduction of *zījes.*

The genre proved immensely popular, doubtless because the use of tables was so much more efficient than working out celestial positions from scratch using the formulas of a *siddhānta* or similar text. Indian astronomers constructed tables of bewildering variety and often brilliant ingenuity; much of the corpus is still unknown, and many of the known tables are still not well understood mathematically. Most tables were designed for use over a span of several decades or even centuries; they usually enumerated planetary positions and/or values of functions for synchronizing the time-units in calendars. Since the Indian calendar required true lunar months to be synchronized with true solar years, and also reckoned several other time-units for astrological purposes, its preparation was a very complicated task that was greatly simplified by the devising of such tables.

Probably the best known of all table texts was the *Tithicintāmaṇi* (*Thought-Jewel of Lunar Days*) composed around 1525 by Gaṇeśa Daivajña, the author of a renowned commentary on the *Līlāvatī.* A manual for calendar-makers, the *Tithicintāmaṇi* provides in a mere eighteen verses all the necessary information for using its tables to identify the true time with respect to any one of the standard sets of luni-solar time-units at any moment in any year (and therefore to construct a complete calendar showing the sequences of these time-units). The comparative simplicity of the instructions, however, is more than compensated for by the complexity of the tables: Gaṇeśa's algorithms for constructing them are still not fully understood. Unusually for an Indian astronomer, he appears to have selected his computational parameters from sources in different schools, according to observational testing of their accuracy:

> [The parameters of] the sun and lunar apogee are [from the] Saura [school]. [That of] the moon diminished by 9 minutes is [the Saura value]. The node [of the moon's orbit] is, however, [from

> the] Ārya [school]. From those, the agreement of [the calculation of] eclipses with observation occurs. What is stated by me is such a *tithi* [a thirtieth of a synodic month, sometimes known as a lunar day]. This [*tithi*] should be accepted in *maṅgala*, *dharma*, *nirṇaya*, and *vidhi* [the performance of various rites] because of [its] agreement with observation. [Ikeyama and Plofker 2001, 275]

VI. Encounters with Modern Western Mathematics

VI.a. Early exchanges with European mathematics

European mathematical sciences began to reach Indian audiences later in the second millennium. As early as the sixteenth and seventeenth centuries some Jesuit missionaries undertook astronomical and geographical observations in India, some of which attracted the attention of monarchs and scholars at Indian courts, where several European telescopes and other instruments were acquired. At Jaipur in the early eighteenth century, Jai Singh obtained copies of Latin works on heliocentric astronomy and the mathematics used for its computations, such as tables of logarithms. For his comparative project on the merits of Indian, Islamic, and European astronomy, scholars at his court used and translated much of this Latin material, with the assistance of European Jesuits. (The frequent involvement of missionaries in scientific activities was partly strategic. Science and technology were viewed by many evangelists as an ally of Christian proselytizing: it was hoped that they would foster belief in the superior rationality of European thought in general and European religion in particular.) There were also translations of European works into Persian and Urdu.

Conversely, from the late seventeenth century onward, Sanskrit mathematical astronomy made its way to Europe. The earliest sources were a few translated texts and summaries provided by missionaries instructed (often sketchily) by Indian *jyotiṣīs*. European scholars found it no easy task to figure out the structure and context of Indian astronomy from these incomplete texts. The material was quite different technically and stylistically from the geocentric and heliocentric world-systems they were familiar with, and even the basic chronology of Indian history was at that time very poorly understood. One of the first salient facts to emerge from this study was that Sanskrit astronomy, like Indian sacred cosmologies, reckoned with very long eras (such as the *yuga*) involving many thousands and sometimes millions or billions of years. While this was troubling to many theologians who feared it might undermine Biblical chronology, it was exciting news to some eighteenth-century astronomers engaged in refining Newtonian celestial mechanics. Indian sources seemed to promise the possibility of testing contemporary dynamical models against a hitherto unknown corpus of astronomical observations from many centuries, if not millennia, in the past.

These hopes were eventually dashed by the discovery that the Sanskrit texts did not actually contain ancient observational records of any use for such purposes. The pendulum of European scholarly opinion seems to have then swung in the other direction, and Indian mathematical astronomy was disparaged as inaccurate, muddled, decayed, merely imitative of Greek sources, and even (according to one very imaginative critic) an outright fraud crafted by second-millennium Brāhmaṇas to discredit Christianity. Serious scholarly attention continued to reveal important technical and historical aspects of Sanskrit treatises (for instance, in H. T. Colebrooke's seminal early nineteenth-century translations of the mathematical work of Brahmagupta and Bhāskara II), but they were no longer considered to have any possible relevance for the development of modern mathematical sciences.

VI.b. European versus "native" mathematics education in British India

European assessments had profound consequences for the role of Indian sciences in Indian education. During most of the early British period under the British East India Company, the chief practical advantage of "native education" from the Company's viewpoint was the resulting supply of Indian employees competent in the Hindu and Muslim juridical systems, who could assist colonial officials in deciding legal questions. To this end, near the close of the eighteenth century, the Company undertook to found some Hindu and Muslim schools more or less on the model of indigenous ones. Such schools generally taught some arithmetic and commercial mathematics, but not the higher-level topics known to the authors of *siddhāntas* and other advanced texts.

Public opinion in the early nineteenth century generally supported the view of many missionaries and many Company officials that the British should more actively foster educational enterprises in India. The motives were varied: education was viewed as an adjunct of missionary work, a check on Indian "superstition and ignorance," a means to revive the neglected study of traditional Indian disciplines, the generator of a useful labor pool for colonial administration, a propaganda tool for the spread of Empire, and so forth. The diversity of the goals inspired hot debate on the content and medium of colonial education. Should Indians be taught their classical disciplines in their classical languages of Sanskrit, Persian, and Arabic? Should they study the equivalent of a modern European curriculum in their native vernaculars? Or should they learn modern subjects via instruction in English?

The second alternative was espoused by many educated Indians convinced that embracing modern science would not only benefit their society but also put Indians and British on a more equal footing. Among these was the mathematician Ramchandra, one of the first Indians to publish mathematical works in English that attracted academic attention in the West. Ramchandra's research specialty, on which he published books in 1850 and 1863, was an innovative approach to the methods of differential calculus. But it took second place to his pedagogical mission of promoting vernacular education, particularly on scientific subjects in support of "modern rational thought." He published (primarily in Urdu) dozens of popular articles and books on astronomical, mathematical, and technological topics, as well as biographical sketches of scientists. Other Indians with similar views produced similar publications; several founded clubs and publications that became the prototypes of modern scientific societies and journals in India.

Education via the revival of "Oriental learning," on the other hand, was often defended by European scholars and Company officials who had learned Asian classical languages and gained some idea of the richness of their literatures. A British official at Bhopal in the 1830's, Lancelot Wilkinson, advocated in particular the teaching of Sanskrit exact sciences. This, he claimed, would not only win the confidence of the Hindu educated castes by showing respect and appreciation for their learning, but would pave the way for a synthesis of traditional and modern sciences in which mathematical consistency and accuracy would be honored over the "fables" of sacred cosmology:

> With the aid of the authority of the Siddhantas, the work of a general and extensive enlightenment may be commenced upon at once, and will be most readily effected, the truths taught by them being received with avidity. To explain and correct their errors will at the same time be easy [...] [T]o give a command and powerful influence over the native mind, we have only to revive that knowledge of the system therein [in *siddhāntas*] taught, which notwithstanding its being by

far the most rational, and formerly the best cultivated branch of science amongst the Hindus [...] has, from the superior address of the followers of the Purans [Purāṇas], and the almost universal practice amongst the jyotishis, of making all their calculations from tables and short formulae, couched in enigmatical verses, been allowed to fall into a state of utter oblivion [...] With what exultation will every man of ingenious mind amongst them receive explanations making plain and clear what is now all unintelligible and dark! They will not stop in simply admitting what is taught in the Siddhantas. Grateful to their European Instructors for bringing them back to a knowledge of the works of their own neglected, but still revered, masters, they will in the fulness of their gratitude [...] also readily receive the additions made during the last few hundred years in the science [i.e., in Europe]. [Wilkinson 1834, pp. 506–509]

Wilkinson's pessimism about the "degeneration" of Indian mathematical astronomy into mythological models and cookbook number-crunching was far from fully justified. His pedagogical scheme, however, had an impact on the history of Indian mathematics out of all proportion to the fewness of its adherents. For he actually trained several students in his synthesized curriculum, and some (especially one Bāpu Deva Śāstrī and his student, Sudhākara Dvivedī) went on to edit, translate, and publish a remarkable number of Sanskrit astronomical and mathematical texts, as well as composing new ones on modern mathematical topics in Sanskrit and Hindi. These scholars, along with Wilkinson, essentially founded the present-day field of textual study of Sanskrit mathematical sciences, although their pedagogical efforts to Sanskritize modern mathematics were less fruitful.

For in 1854 the Company had published its famous "Educational Despatch" rejecting indigenous educational systems in favor of a modern curriculum:

[W]e must emphatically declare that the education which we desire to see extended in India is that which has for its object the diffusion of the improved arts, science, philosophy and literature of Europe: in short of European knowledge. The systems of science and philosophy which form the learning of the East abound with grave errors, and eastern literature is at best very deficient as regards all modern discovery and improvements; Asiatic learning, therefore, however widely diffused, would but little advance our object. [...] We do not wish to diminish the opportunities which are now afforded in special institutions for the study of Sanskrit, Arabic and Persian literature, or for the cultivation of those languages which may be called the classical languages of India. [...] But such attempts, although they may usefully co-operate, can only be considered as auxiliaries and would be a very inadequate foundation for any general schemes of Indian education. [Richey 1965, online excerpt]

Moreover, the preferred linguistic medium of secondary and higher education was to be English, although Indian vernaculars would be used in common schooling, especially for the lower classes:

[A] knowledge of English will always be essential to those natives of India who aspire to a high order of education. [...] [W]hile the English language continues to be made use of as by far the most perfect medium for the education of those persons who have acquired a sufficient knowledge of it to receive general instruction through it, the vernacular languages must be employed to teach the far larger classes who are ignorant of, or imperfectly acquainted with English. [...] We look, therefore, to the English language and to the vernacular languages of India together as the media for the diffusion of European knowledge [...]. [Richey 1965, online excerpt]

VI.c. Assimilation into modern global mathematics

The British government continued these educational policies when it took over the East India Company's administration of India in 1857, founding new English-medium universities at Bombay, Madras and Calcutta. Modeled on European institutions, these presidency universities and the numerous smaller colleges of course included mathematics among the subjects they offered. (Astronomy, on the other hand, was pursued at the research level, mostly by European scientists, at several newly established Indian observatories, but it was not featured in native education.) In many cases, mathematics teaching materials were not only modeled on but directly borrowed from English originals. The early training of the famous number theorist Srinivasa Ramanujan is a case in point: he taught himself higher mathematics from English textbooks, and his matriculation examination at Madras University in 1903 included typical Euclidean problems such as "Prove that the opposite angles of a quadrilateral inscribed in a circle are together equal to two right angles" [Berndt and Reddi 2004, 334].

This Westernized mathematical training naturally influenced the practice of mathematics in India. A mathematics BA from an Indian university would probably become a mathematics teacher rather than a court astronomer, and he would know of European research journals, professional societies, and popular works in his field. Many educated Indians felt, like Ramchandra, that they had a duty to make technical knowledge available to others: they published thousands of textbooks and popular articles on mathematical subjects in local vernaculars. A few Indians were admitted to membership in European mathematical societies, but to accommodate the number of the mathematically inclined, it proved necessary to found similar societies in India. In fact, elementary Western mathematics was so familiar to educated Indians that an Indian periodical around 1880 could satirize the colonial government with a parody of Euclid called "Political Geometry": "Definitions. 1. A political point is that which is visible to the Government but invisible to the people. 2. A line of policy is length without breadth of views" [Nurullah and Naik 1951, 314].

As the twentieth century advanced, and especially after Independence in 1947, nationalistic feeling encouraged a certain amount of "re-indigenization" in educational policy. More textbooks and courses were offered in Indian languages; some universities taught traditional sciences such as *jyotiṣa*, which by this time had become almost exclusively astrological in content. Such changes did not reverse the fundamental trend toward Westernized sciences in education and research. The brilliant achievements of Ramanujan, Harish-Chandra, and their contemporaries and successors were no longer part of "Indian mathematics" as we have explored it here; rather, they were Indian contributions to the "global mathematics" of today.

VII. Appendices

VII.a. Sources

Śulbasūtras. Sen, S. N., and A. K. Bag. 1983. *The Śulbasūtras.* New Delhi: Indian National Science Academy.

Numerals. Figure 4.5: A. K. Bag. 2003. "Need for Zero in the Mathematical System of India." In A. K. Bag and S. R. Sarma eds., *The Concept of Śūnya.* Delhi: Indian National Science Academy, 159–69, p. 161.

Figure 4.6: Hayashi, Takao. 1995: *The Bakhshālī Manuscript: An Ancient Indian Mathematical Treatise.* Groningen: Egbert Forsten, p. 584 (f. 55r).

Figure 4.7: Sri Ram Charan Museum of Indology, Jaipur, ms. 1.20 i 115. F. 19v.

***Āryabhaṭīya* of Āryabhaṭa with commentary of Bhāskara (I).** Chapter 2: Keller, Agathe. 2006. *Expounding the Mathematical Seed: A Translation of Bhāskara I on the Mathematical Chapter of the Āryabhaṭīya,* 2 vols. Basel: Birkhäuser.

Chapters 3–4: K. S. Shukla. 2005. *Āryabhaṭīya of Āryabhaṭa with the Commentary of Bhāskara I and Someśvara.* New Delhi: Indian National Science Academy, 1976. Translated K. Plofker.

***Brāhmasphuṭasiddhānta* of Brahmagupta.** Dvivedī, Sudhākara. 1901–1902. *Brāhmasphuṭa-siddhānta* (*The Pandit* NS 23–24). Benares.

Chapters 2, 7, 10: Translated K. Plofker, 2005.

Chapters 12, 18: Translated D. Pingree, 2000. Revised K. Plofker, 2005.

Bakhshālī Manuscript. Hayashi, Takao. 1995. *The Bakhshālī Manuscript: An Ancient Indian Mathematical Treatise.* Groningen: Egbert Forsten.

***Gaṇitasārasaṅgraha* of Mahāvīra.** Rangacarya, M. 1912. *Ganita-Sara-Sangraha.* Madras: Government Oriental Manuscripts Library.

Chapter 6: Translated K. Plofker, 2005.

***Līlāvatī* of Bhāskara (II).** Āpaṭe, V. G. 1937. *Līlāvatī* 2 vols. (Ānandāśrama Sanskrit Series 107). Puṇe. Translated K. Plofker, 2000, 2005.

Figure 4.18: Sri Ram Charan Museum of Indology, Jaipur, ms. 1.13 i 1. Ff. 48r, 51v.

***Bījagaṇita* of Bhāskara (II) with commentary of Sūryadāsa.** Verses 1–26: Jain, Pushpa Kumari. 2001. *The Sūryaprakāśa of Sūryadāsa,* vol. I. Vadodara: Oriental Institute.

Verses 70–129: Āpaṭe, V. G. 1930. *Bhāskarīya Bījagaṇita* (Ānandāśrama Sanskrit Series 99). Puṇe. Translated D. Pingree, 2000.

Bhāskara's prose commentary: Jhā, Acyutānanda. 2005. *The Bījagaṇita (Elements of Algebra) of Srī Bhāskarāchārya* (Kashi Sanskrit Series 148). Banaras: Jaya Krishna Das Gupta, 1949. Translated K. Plofker.

***Śiṣyadhīvṛddhidatantra* of Lalla.** Chatterjee, Bina. 1981. *Śiṣyadhīvṛddhidatantra of Lalla,* parts I and II. New Delhi: Indian National Science Academy.

***Kriyākramakarī* of Śaṅkara.** Hayashi, T., Kusuba, T., and Yano, M. *Studies in Indian Mathematics: Series, Pi and Trigonometry* (Tokyo, 1997). Translated by Setsuro Ikeyama. Forthcoming.

***Āryabhaṭīyabhāṣya* of Nīlakaṇṭha.** Śāstrī, K. S. 1930. *Āryabhaṭīyabhāṣya of Nīlakaṇṭha* (Trivandrum Sanskrit Series 101). Trivandrum. Translated K. Plofker and H. White, 2003.

***Gaṇitakaumudī* of Nārāyaṇa.** Kusuba, Takanori. 1993. "Combinatorics and Magic Squares in India: A Study of Nārāyaṇa Paṇḍita's "Gaṇitakaumudī," chapters 13–14." Ph.D. dissertation, Brown University.

Assimilation of foreign mathematics. Figure 4.26: Sri Ram Charan Museum of Indology, Jaipur, ms. 1.13 i 16. F. 43v.

Figure 4.27: Sri Ram Charan Museum of Indology, Jaipur, ms. 1.13 i 17. F. 17v.

VII.b. References

Āpaṭe, V. G. 1930. *Bhāskarīya Bījagaṇita* (Ānandāśrama Sanskrit Series 99). Puṇe.

———. 1937. *Līlāvatī* (Ānandāśrama Sanskrit Series 107), 2 vols. Puṇe.

Bag, A. K. "Need for Zero in the Mathematical System of India." In Bag and Sarma (2003), pp. 159–169.

Bag, A. K., and S. R. Sarma, eds. 2003. *The Concept of Śūnya.* Delhi: Indian National Science Academy.

Berndt, Bruce C., and C. A. Reddi. 2004. "Two Exams Taken by Ramanujan in India." *American Mathematical Monthly* 111, 4: 330–339.

Bronkhorst, Johannes. 2001. "Pāṇini and Euclid: Reflections on Indian Geometry." *Journal of Indian Philosophy* 29: 43–80.

Burnett, Charles, et al., eds. 2004. *Studies in the History of the Exact Sciences in Honour of David Pingree.* Leiden: Brill.

Chatterjee, Bina. 1981. *Śiṣyadhīvṛddhidatantra of Lalla*, parts I and II. New Delhi: Indian National Science Academy.

Chemla, Karine, and Agathe Keller. 2002. "The Sanskrit *karaṇīs* and the Chinese *mian*." In Dold-Samplonius et al. (2002), pp. 87–132.

Colebrooke, H. T. 1817. *Algebra, with Arithmetic and Mensuration, from the Sanscrit of Brahmegupta and Bhascara.* London: John Murray.

Dani, Ahmad Hasan. 1986. *Indian Palaeography.* 2nd ed. New Delhi: Munshiram Manoharlal.

Datta, B., and A. N. Singh. 1962. *History of Hindu Mathematics: A Source Book*, parts I and II. Repr. Bombay: Asia Publishing House.

Dhupakara, A. Y. *Taittirīyasaṃhitā.* 1957. 2nd ed. Pāraḍīnagara: Bhāratamudraṇālaya.

Dold-Samplonius, Yvonne, et al., eds. 2002. *From China to Paris: 2000 Years Transmission of Mathematical Ideas.* Stuttgart: Steiner Verlag.

Dvivedī, Sudhākara. 1901–1902. *Brāhmasphuṭasiddhānta* (*The Pandit* NS 23–24). Benares.

———. 1908. *Yājuṣa-Jyautiṣa and Ārca-Jyautiṣa.* Benares: Medical Hall Press.

———. 1912. *Līlāvatī* (Benares Sanskrit Series 153). Benares: Braj Bhushan Das & Co.

Farmer, Steve, et al. 2004. "The Collapse of the Indus-Script Thesis: The Myth of a Literate Harappan Civilization." *Electronic Journal of Vedic Studies* 11 (2).

Gauḍa, V. S., and C. S. Caudhuri. 1994–1997. *Śatapathabrāhmaṇa*, 3 vols. Kāśī: Acyutagranthamālākāryālaya, Saṃ.

Gupta, R. C. 1969. "Second Order Interpolation in Indian Mathematics up to the Fifteenth Century." *Indian Journal of History of Science* 4: 86–98.

Hayashi, Takao. 1995. *The Bakhshālī Manuscript: An Ancient Indian Mathematical Treatise.* Groningen: Egbert Forsten.

Hayashi, T., T. Kusuba, and M. Yano. 1990. "The Correction of the Mādhava Series for the Circumference of a Circle." *Centaurus* 33 (2–3): 149–174.

Ikeyama, Setsuro. 1997. *Studies in Indian Mathematics: Series, Pi and Trigonometry* (Tokyo, 1997). Translated, T. Hayashi, T. Kusuba, and M. Yano Forthcoming.

Ikeyama, Setsuro, and Kim Plofker. 2001. "*The Tithicintāmaṇi* of Gaṇeśa, a Medieval Indian Treatise on Astronomical Tables." *SCIAMVS* 2: 251–289.

Jain, Pushpa Kumari. 2001. *The Sūryaprakāśa of Sūryadāsa*, vol. I. Vadodara: Oriental Institute.

Jhā, Acyutānanda. 1949. *The Bījagaṇita (Elements of Algebra) of Srī Bhāskarāchārya* (Kashi Sanskrit Series 148). Banaras: Jaya Krishna Das Gupta.

Keller, Agathe. 2005. "Making Diagrams Speak, in Bhāskara I's Commentary on the *Āryabhaṭīya*." *Historia Mathematica* 32 (3): 275–302.

———. 2006. *Expounding the Mathematical Seed: A Translation of Bhāskara I on the Mathematical Chapter of the Āryabhaṭīya*, 2 vols. Basel: Birkhäuser.

Kumar, Deepak. 1995. *Science and the Raj, 1857–1905.* Delhi: Oxford University Press.

Kusuba, Takanori. 1981. "Brahmagupta's Sutras on Tri- and Quadrilaterals." *Historia Scientiarum* 21: 43–55.

———. 1993. "Combinatorics and Magic Squares in India: A Study of Nārāyaṇa Paṇḍita's "Gaṇitakaumudī," chapters 13–14." Ph.D. dissertation, Brown University.

Max Müller, F. 1873. *Ṛgvedasaṃhitā.* London: Trübner & Co.

McDermott, Charlene. 1974. "The Sautrantika Arguments Against the Traikalyavada in the Light of the Contemporary Tense Revolution." *Philosophy East and West*, 24 (2): 193–200.

Minkowski, Christopher Z. 2004. "Competing Cosmologies in Early Modern Indian Astronomy." In Burnett et al. (2004), pp. 349–385.

Nurullah, Syed, and J. P. Naik. 1951. *A History of Education in India (During British Period).* Bombay: Macmillan.

Pingree, David. 1970–1994. *Census of the Exact Sciences in Sanskrit*, series A, vols. 1–5. Philadelphia: American Philosophical Society.

———. 1973. "The Mesopotamian Origin of Early Indian Mathematical Astronomy." *Journal for the History of Astronomy* 4: 1–12.

———. 1978. *The Yavanajātaka of Sphujidhvaja*, 2 vols. Cambridge, MA: Harvard University Press.

———. 1981. *Jyotiḥśāstra*. Wiesbaden: Harrassowitz.

———. 1999. "An Astronomer's Progress." *Proceedings of the American Philosophical Society* 143: 73–85.

———. 2002. "Philippe de La Hire's Planetary Theories in Sanskrit." In Dold-Samplonius et al. (2002), pp. 428–53.

Plofker, Kim. 2005. "Derivation and Revelation: the Legitimacy of Mathematical Models in Indian Cosmology." In T. Koetsier and L. Bergmans, eds., *Mathematics and the Divine*, Amsterdam: Elsevier, pp. 61–76.

Rangacarya, M. 1912. *Ganita-Sara-Sangraha*. Madras: Government Oriental Manuscripts Library.

Richey, J. A., ed. 1965. *Selections from Educational Records, Part II (1840–1859)*. Repr. Delhi: National Archives of India. (1854 Educational Despatch reproduced at http://www.mssu.edu/projectsouthasia/history/primarydocs/education/Educational_Despatch_of_1854.htm.)

Sachau, Edward. 1992. *Alberuni's India*, 2 vols. Repr. New Delhi, Munshiram Manoharlal.

Salomon, Richard. 1998. *Indian Epigraphy*. Oxford: Oxford University Press.

Sarasvati Amma, T. A. 1979. *Geometry in Ancient and Medieval India*. Delhi: Motilal Banarsidass.

Sarma, K. V. 1975. *Līlāvatī of Bhāskarācārya with Kriyākramakarī*. Hoshiarpur: VVBIS& IS, Panjab University.

———. 1977. *Tantrasaṅgraha of Nīlakaṇṭha Somayāji*. Hoshiarpur: VVBIS& IS, Panjab University.

Sarma, Sreeramula Rajeswara. 2003. "Śūnya in Piṅgala's Chandaḥsūtra." In Bag and Sarma (2003), pp. 126–141.

Śāstrī, K. S. 1930. *Āryabhaṭīyabhāṣya of Nīlakaṇṭha* (Trivandrum Sanskrit Series 101). Trivandrum.

Seidenberg, A. 1978. "The Origin of Mathematics." *Archive for History of Exact Sciences* 18: 301–342.

Sen, S. N., and A. K. Bag. 1983. *The Śulbasūtras*. New Delhi: Indian National Science Academy.

Shukla, K. S. 1976. *Āryabhaṭīya of Āryabhaṭa with the Commentary of Bhāskara I and Someśvara*. New Delhi: Indian National Science Academy.

———. 1954. "Ācārya Jayadeva, the Mathematician." *Gaṇita* 5: 1–20.

Shukla, K. S., and K. V. Sarma. 1976. *Āryabhaṭīya of Āryabhaṭa*. New Delhi: Indian National Science Academy.

Srinivas, M. D. 2004. "The Methodology of Indian Mathematics." Preprint.

Wilkinson, Lancelot. 1834. "On the Use of the Siddhántas in the Work of Native Education." *Journal of the Asiatic Society of Bengal* 3: 504–519.

Young, Richard F. 1997. "Receding from Antiquity: Indian Responses to Science and Christianity on the Margins of Empire, 1834–1844." *Kokusaigaku-kenkyū* 16: 241–274.

Zastoupil, Lynn, and Martin Moir, eds. 1999. *The Great Indian Education Debate: Documents Relating to the Orientalist-Anglicist Controversy, 1781–1843*. Richmond, VA: Curzon.

5 Mathematics in Medieval Islam

J. Lennart Berggren

I. Introduction

What is the mathematics of medieval Islam?

The mathematics of medieval Islam includes the mathematical theories and practices that grew, and often flourished, in that part of the world where the dominant religious and cultural influence was the religion of Islam. The historical period involved is roughly the 700 years from 750 to 1450 CE, although the earliest works included here were written around 825. Geographically, medieval Islam extended from the Iberian Peninsula through North Africa and the Middle East, to the central Asian republics of the former Soviet Union, Afghanistan, Iran, and even parts of India. Although the mathematical treatises were written in a number of languages, including Persian and Turkish, the principal language used was Arabic and for that reason it is occasionally called Arabic mathematics. This name, however, too easily becomes "Arab mathematics," with the implication that most of its practitioners were Arabs, even though many were (for example) Iranians, Egyptians, Moroccans, and so on. So, the designation "Islamic mathematics" is preferable.[1]

Sources for our knowledge of Medieval Islamic mathematics

We know about medieval Islamic mathematics mainly through documents written in the Arabic language by pen and ink on paper. Chinese taken prisoner in the Battle of Atlakh (ca. 750) taught the Muslims how to make paper, and this knowledge soon spread over the whole of the Islamic world. The mathematical treatises of medieval Islam are found in libraries and private collections

---

The author warmly thanks Jan P. Hogendijk for considerable help and advice in choosing the selections as well as in obtaining the texts of some of them. He also thanks the editor of this volume, Victor Katz, for his advice on what should be included. Needless to say, neither is responsible for any deficiencies in the final list, for which the author is solely responsible.

[1]Islam is a religion, but it created an Islamic culture that included significant, and often thriving, communities of other religious groups—such as Christians and Jews. These groups played an important role in the history of Islamic mathematics, especially during the early period. Yet, Islamic mathematics was mathematics created by a people who were mainly Muslims in a culture whose dominant element was the religion of Islam.

all over the world, but the largest collections are those in the countries that once constituted the medieval Islamic world. Significant, but secondary, collections are found in countries such as England, France, Germany, and Russia that were once colonial powers in the Islamic world.

These treatises are often prose compositions, as one finds in mathematical books and papers today. However, there are also countless tables of numbers, some with hundreds of thousands of entries, which were computed (mainly for astronomical purposes) according to mathematical principles and methods. These almost never have explanations of how the numbers were computed, but patient work by scholars (often with modern computers and statistical techniques) can make these tables provide information on the history of mathematics.

However, not all our sources are of a literary nature. Physical artifacts may also provide important sources, which, with patient detective work and knowledge of the literary sources, may be made to tell their own stories. Such artifacts are often mathematical or astronomical instruments, examples being three world maps in the form of circular disks. These are designed to allow users to find the direction of the holy city of Mecca from a given locality simply by rotating a ruler around the center of the disk. (This mathematical problem is a good example of the many problems for which we have solutions in all three kinds of sources mentioned here: mathematical treatises, tables, and instruments!)

Historical outline

"Islam" is an Arabic word that means "submission." In a religious context it means "submission" to the will of the one god, Allah, as He revealed it to His Prophet Muḥammad and as it was later recorded in Islam's holy book, the Arabic Qur'ān. (The system we have followed in transcribing Arabic letters is explained in [Berggren 1986, 25].) Muḥammad first received these revelations in the early seventh century, and by 645 his message, directed first to his fellow Arabs, had inspired armies and missionaries to spread Islam to neighboring lands and, finally, from Spain to China. Islam in its expansion conquered lands that had been parts of Byzantine or Persian civilization, and bordered lands such as India and China with their own high civilizations. Thus it is not surprising that for a variety of reasons—political, cultural, and religious—emerging Islamic elites began to take an interest in some of this learning. What ensued, from roughly 750 to 900, was Muslim acquisition of whole areas of ancient learning, including mathematics (and its companion, astronomy). The word "acquisition" signifies here the active participation of Muslims in the process, not only as patrons of scholars of other faiths but as interested learners and, increasingly, scholars themselves.

The mathematics, to speak only of the subject of interest here, came principally from three traditions. The first was Greek mathematics, from the great geometrical classics of Euclid, Apollonius, and Archimedes, through the numerical solutions of indeterminate problems in Diophantus's *Arithmetica*, to the practical manuals of Heron. But, as Bishop Severus Sebokht pointed out in the mid-seventh century, "there are also others who know something." Sebokht was referring to the Hindus, with their ingenious arithmetic system based on only nine signs and a dot for an empty place. But they also contributed algebraic methods, a nascent trigonometry, and methods from solid geometry to solve problems in astronomy. The third tradition was what one may call the mathematics of practitioners. Their numbers included surveyors, builders, artisans in geometric design, tax and treasury officials, and some merchants. Part of an oral tradition, this mathematics transcended ethnic divisions and was a common heritage of many of the lands incorporated into the Islamic world.

The efforts of Muslim scholars to master the three traditions just mentioned is hinted at in the Banū Mūsā's ("Sons of Moses" [ibn Shākir]) preface to their translation of Apollonius's *Conics* and Ibn al-Haytham's introduction to his restoration of the Book 8 of that work. From these three traditions, medieval Islam created a mathematics whose contents reflected not only its sources but, as in al-Khwārizmī's application of algebra to inheritance law, the Muslim society that created and sustained it. In fact, al-Khwārizmī, from whom we give a number of selections, drew on Indian mathematics both for his *Short Treatise on Hindu Reckoning*, which introduced Islamic mathematicians to the base-10 positional system of the Hindus, and for the sine function in his astronomical tables. At the same time, in his *Algebra*, he drew on widespread methods of the practitioners for solving equations of the first and second degree as well as equally widespread "cut-and-paste" methods (as they have been called[2]) to demonstrate the validity of the solutions.

Later in the ninth century the three brothers Muḥammad,[3] Aḥmad and al-Ḥasan ibn Mūsā (known collectively as the Banū Mūsā) collaborated in Baghdad with the talented linguist and mathematician Thābit ibn Qurra to make available, in Arabic, works by Apollonius of Perga and Archimedes. By the end of the ninth century not only the basic *Elements* of Euclid, but such books as Archimedes' *On the Measurement of the Circle*, were widely circulated in Arabic. Available also, in good Arabic translations with informed commentary, were such advanced works as Archimedes' *On the Sphere and Cylinder*, Apollonius's *Conics*, and Ptolemy's classic of mathematical astronomy, the *Almagest*. And Muslim mastery of these works led, for example, to Ibn al-Haytham's theorem on the volume of a paraboloid, Abū Sahl al-Kūhī's use of a hyperbola to inscribe an equilateral pentagon in a square, and Aḥmad al-Ṣāghānī describing new projections of the surface of a sphere onto a plane.

Since conics arose both in theoretical problems and in the design of sundials and special forms of astrolabes, it is no surprise that Muslim mathematicians turned their attention to the design of an instrument, known as "the perfect (or "complete") compass," that allowed one to draw conic sections as one would circles with the traditional compass. However, quite apart from these advanced topics, the *Elements* themselves posed a number of challenges to Islamic mathematicians, for example, that of extending their results. And here Abū al-Wafā' al-Būzjānī's construction of a regular pentagon with a compass of fixed opening is but one example of many possible. Also going beyond Euclidean methods in their solution, but still dealing with problems of the type found in the *Elements*, are Abū Sahl al-Kūhī's inscribing an equilateral pentagon in a square and an anonymous treatise on constructing a regular nine-gon.

The challenge of solving geometrical problems stimulated a number of mathematicians to write general treatises on what is now known as heuristics, or (as George Pólya put it in his provocative title) "How to solve it." Among the Muslims who wrote such treatises were al-Sijzī (on strategies for problem solving) and Ibrāhīm Ibn Sinān (on the method of geometrical analysis).

Another challenge that the *Elements* posed concerned the status of the parallel postulate within the deductive structure of the work. This question was first raised by Greek writers who suspected that Euclid had made a postulate out of what was really a theorem. And attempts to

---

[2]The term, due to Jens Høyrup, denotes methods in which two plane figures are shown to be equal by dissecting each of them into collections of pairwise-congruent figures.

[3]The oldest of the three, Muhammad, died in 873.

prove the postulate continued through medieval Islam, attracting the efforts of such notable mathematicians as Omar Khayyām in the eleventh and twelfth centuries and Naṣīr al-Dīn al-Ṭūsī in the thirteenth century.

A third challenge posed by the *Elements* was that of understanding Euclid's classification of irrational magnitudes found in Book X of his *Elements.* For the Greeks these were geometrical magnitudes not measurable by a unit magnitude of the same kind. But in al-Karajī's treatment, written in the late tenth century, we may see the beginnings of an extended concept of real numbers within a deductive framework.

Algebra, as a branch of the general area of computation, also developed remarkably during the ninth and tenth centuries. In his *Algebra* Abū Kāmil shows his expertise in the area with the solution of a computationally complicated problem of solving three non-linear equations in three unknowns. Late in the tenth century al-Karajī knew the binomial theorem and its accompanying table of coefficients, now known as Pascal's Triangle. This theorem led Omar Khayyām to the discovery of how to extract arbitrary whole number roots of numbers. Although Omar's treatise is lost, the knowledge it contained led to a real tour de force in a work by the fifteenth-century Samarkand mathematician, al-Kāshī, in which he shows how to extract the fifth root of a number on the order of ten trillion! Al-Kāshī had not only consummate computational skills but also a command of decimal fractions, which he claimed to have discovered. However, al-Kāshī was mistaken in this, for such fractions had been used five centuries earlier by the Baghdadi mathematician al-Uqlidisi.

Both Euclidean and advanced geometry united fruitfully with algebra in medieval Islam. In the late ninth century Thābit ibn Qurra gave a rigorous geometrical demonstration of the validity of the algebraists' methods of solving quadratic equations. At the beginning of the tenth century Abū Kāmil showed how to use algebra to construct the side of an equilateral pentagon inscribed in a square. As a final example one has, somewhere in the late eleventh or early twelfth century, Omar Khayyām writing on constructing roots of cubic equations by means of intersecting conics. But geometrical and computational traditions of Islamic mathematics also came together in trigonometry, which advanced remarkably in the late tenth century with the introduction of all six trigonometric functions, the computation of tables accurate to the equivalent of 8 decimal places, and the discovery of theorems such as the sine theorem for plane and spherical triangles and the tangent theorem for spherical triangles. This work culminated in Naṣīr al-Dīn's thirteenth-century work, *On the Sector Figure,* in which he gives a number of proofs of these and other theorems as well as a systematic discussion of how to solve plane and spherical triangles. Prior to this work, trigonometry had been discussed as ancillary to astronomy, and one finds applications of that subject to—among many other areas—timekeeping by the sun or stars, by Abū al-Wafā', and to geodesy, by al-Bīrūnī, but Naṣīr al-Dīn's treatise is the first treatment of the subject in its own right, independent of astronomy.

Parts of Naṣīr al-Dīn's *Sector Figure* also had connections with other areas of mathematics, such as its enumeration of configurations of certain types arising from four great circles on a sphere, no three of which pass through the same point. Here one sees counting arguments found both in many areas of combinatorics today and centuries before Naṣīr al-Dīn's time in combinatorial problems that arose in Islam from questions involving prosody and lexicography. And, in the western part of the Muslim world, the North African writer, Ibn Mun'im (d. 1228) had, by Naṣīr al-Dīn's time, gone beyond enumeration to use mathematics and the rules of Arabic word formation to count the number of possible Arabic words having at most ten letters. At present such problems are part of discrete mathematics, and other parts of this broad area were considerably developed in medieval Islam. For example, a tenth-century Baghdadi mathematician generalized the Greek results on sums of divisors of whole numbers[4] to the notion of balanced numbers, i.e., pairs of numbers each having the same sum for their proper divisors. Even more impressive, however, were the achievements of Islamic mathematicians, such as Ibn al-Haytham and al-Būzjānī, in creating methods for forming what we know as "magic squares."

A striking feature of Islamic mathematics, in contrast to Greek mathematics, is the close relation of theory and practice. For example, Islamic architecture is admired worldwide for its imaginative decorative designs. And in a work by al-Būzjānī's student we find a record of a meeting of this great mathematician with artisans and architects to discuss solutions to problems arising when creating modules for use in Islamic tessellations. It is fascinating to see how the mathematicians had to take into account the artisans' objections to what were theoretically nice methods and how the artisans had to understand the difference between exact and approximate methods.

The mathematical instrument known as the astrolabe used the circle-preserving property of stereographic projection to create an analog computer to solve problems of spherical astronomy and trigonometry. Although the astrolabe was a Greek invention, the Muslims added circles indicating azimuths on the horizon. On the one hand such a step was surely useful in a society where one direction, that of Mecca, was sacred, but, on the other, the construction of such circles stimulated geometrical investigations. Also the design of new types of astrolabes stimulated the geometrical imaginations of a number of Muslim scientists, such as Ḥabash al-Ḥāsib, al-Ṣāghānī, and al-Zarqālī. Finally, we have included a number of selections that touch on broader topics of the acquisition of foreign mathematics, the study of mathematics, and mathematicians' attitudes towards their subject.

What the reader needs to know

We do not claim canonicity for all the authors and selections represented here. Granted, it would be hard to claim that the space allotted to Islamic mathematics in this volume had been well used if there were no selections from al-Khwārizmī, Ibn al-Haytham, al-Bīrūnī, Omar Khayyām, and al-Kāshī. But, even given that, the choice of which particular work to excerpt is to a considerable extent arbitrary. The most we claim for our selections is that they are representative both of the range of topics found in medieval Islamic mathematics and of the quality of its better work. Moreover, many of them are items that one is likely to find discussed in modern scholarly treatments of the area.

We have tried, as much as possible, to make each selection comprehensible without reference to other selections from the same work. However, the reader will soon discover that medieval Islamic mathematics was quite sophisticated and much of it was embedded in well-developed theories, that do not admit of simple summaries to provide easy access to readers coming from a modern tradition. To deal with this problem we have tried to use short introductory paragraphs, and occasional footnotes, in our selections, to provide brief explanations of concepts that are unfamiliar or, perhaps worse, "false friends"—concepts (such as that of

[4]Greek work in this area centered around the concepts of perfect and amicable numbers, a perfect number being one, such as 6, equal to the sum of its proper divisors. Amicable numbers are pairs of numbers each of whose proper divisors sum to the other.

"sine" or "ratio") whose modern counterpart of the same name is not the same as the medieval concept. Inevitably, of course, we shall have failed to meet the full extent of every reader's need on this score, and for additional background on most of the concepts arising in the readings we refer the reader to [Berggren 1986, 2003].[5]

Arabic names

We have tried to follow the practice of giving the full name of a mathematician the first time we mention him, and then a short form thereafter. Arabic names will be unfamiliar to many readers, so we give a short explanation here of how to read (and remember!) an Arabic name. Thus, a child of a Muslim family will receive a name (called in Arabic *'ism*) like Muḥammad, Ḥusain, Thābit, etc. After this comes the phrase "son of so-and-so," and the child will be known as Thābit ibn Qurra (son of Qurra) or Muḥammad ibn Ḥusain (son of Ḥusain). The genealogy can be compounded. For example, Ibrāhīm ibn Sinān ibn Thābit ibn Qurra carries it back to the great-grandfather. Later in life one might have a child and then gain a paternal name (*kunya* in Arabic) such as Abū 'Abdullāh (the father of 'Abdullāh). Next comes a name indicating the tribe or place of origin (in Arabic *nisba*), such as al-Ḥarrānī, "the man from Ḥarrān." At the end of the name might come a tag (*laqab* in Arabic), it being a nickname such as "the goggle-eyed" (al-Jāḥiẓ) or "the tent-maker" (al-Khayyāmī) or a title such as "the orthodox" (al-Rashīd) or "the blood-shedder" (al-Saffāḥ). Putting all this together, we find that one of the most famous Muslim writers on mechanical devices (see excerpt below) had the full name Badī' al-Zamān Abū al-'Izz Isma'īl ibn al-Razzāz al-Jazarī. Here the *laqab* "Badī'al-Zamān" means "prodigy of the Age," certainly a title a scientist might wish to earn, and the *nisba* al-Jazarī signifies a person coming from al-Jazīra, the country between the upper reaches of the Tigris and Euphrates Rivers.

Organization

This chapter is organized thematically. Within each section, the excerpts are organized chronologically.

II. Appropriation of the Ancient Heritage

Islamic mathematicians were well aware of their debt to the Greeks. This is illustrated in these two selections. The first selection is an excerpt from the commentary by the Banū Mūsā to their revision of an Arabic translation of part of Apollonius's *Conics* in the late ninth century. The second selection is from ibn al-Haytham's preface to his eleventh century reconstruction of the lost eighth book of that work.

The Banū Mūsā were three sons of Mūsā ibn Shakir, himself a reformed highway robber who ended his career as an astrologer in the service of the Caliph al-Ma'mūn (ruled 813–833). The eldest brother, Muḥammad, was expert in both geometry and astronomy.

---

[5]There are, fortunately, a number of other excellent works that the reader could also consult, a fairly recent source for a list of such works being the section on mathematics in medieval Islam by Jan P. Hogendijk in [Lewis 2000].

The second, Aḥmad, was interested in mechanics, and the youngest, al-Ḥasan, was particularly interested in geometry. The fortune they accumulated in their lifetime allowed them both to purchase ancient books to study as well as to support such distinguished scholars as Thābit ibn Qurra in their translations of ancient mathematical works. (One result of this was an Arabic translation of Apollonius's *Conics*, whose preface we reprint here.) Their most important work is *On the Measurement of Plane and Solid Figures*, to date available in English only in a translation of the twelfth-century Latin translation by Gerard of Cremona [Clagett 1964].

Banū Mūsā on Conics

In the name of God, the merciful, the forgiving. I have no success except through God.

The first book of the treatise of Apollonius on Conics in the revision of the Banū Mūsā and the version of Hilāl b. Abī Hilāl al-Ḥimṣī.

Truly the position of the science of the sections occurring in cones and of the figures and lines occurring in them is in the highest rank in the science of geometry. The ancients used to call the propositions on conic sections "the amazing propositions," and they were of the opinion that whoever has reached the point in the science of geometry where he has mastered the understanding of this science [Conics] has attained the highest rank in the science of geometry. The ancient students of the science of geometry never ceased to be interested in discovering this science, and to labor in the study of it, and to write down what they understood of it, little by little, in their books, until this process reached Apollonius. This man was an Alexandrian; he was interested in this science, and was a man outstanding in the science of geometry and a master of it. So he composed on that subject a treatise in eight books in which he collected the advances in this science made by his predecessors and added what he himself was responsible for discovering. But then this treatise was corrupted and the mistakes in it multiplied in the course of time through the succession of people copying it, one from another. There were two causes for its corruption: one was the cause common to all books which go through a succession of hands in copying, due to the negligence of those who copy them in correcting the copies and comparing them [with the original], and to the differences between books and the obliteration of what is in them before they are renewed by copying. The other cause is peculiar to this treatise and treatises like it, but to no others; for this treatise is an obscure one, the understanding of which is difficult, and only a few people have control of it. But ease of understanding a book is helpful in emending it when the need for that arises. Furthermore, it is a work the copying of which is long and difficult, and the correction of which is a labor.

So for the reasons we have described, corruption took place in this work after Apollonius until there appeared in Ascalon a man, one of the geometers, called Eutocius. He was outstanding in the science of geometry, and there are books which he has composed which bear witness to his powers. So this man, when he realized to what extent corruption had overcome this treatise, assembled for it what he could of the copies found in his time. Hence it was possible for him, through the copies he had assembled and his prowess in the science of geometry, to restore the first four books of this treatise. But in doing so he followed the method of one who does not merely seek to report his restorations on the basis of what Apollonius wrote; rather he collected [manuscripts], preferred [readings] and employed his intelligence in what he could not correct in reporting the exact words of Apollonius on the topic, until he discovered the proof for it.

Thus the investigators of the science of conics confined themselves, after Eutocius, to reading the four books which he had corrected; this is in accord with what Galen says in his censure of the geometers of his time in the work *Water, Air and Places*, which is evidence for the small number of those geometers whose mind aspired to investigate the science of conics at that time, to say nothing of those who lived after Eutocius.

But as for the people of our time, there are few of the geometers among them who possess understanding of the treatise of Euclid on geometry, not to mention what is beyond that. Indeed some of them, in the feebleness of their understanding, have failed to comprehend the beginning of Euclid's work, to say nothing of what is after it, and have replaced Euclid's words there with words of the utmost stupidity and incorrectness. There were some amongst them who went so far as to compose geometrical propositions which they proved—in their own opinion—with proofs contradicting the proofs of Euclid, to the extent that some of them declared, in their proofs, that a cone is half the cylinder. Some of this class of people whom we have described recognized their error after an interval of time had elapsed and renounced it. But some of them continued in their error and persisted in it; their books are [still] to be found in our time, and hence we have refrained from reporting their error.

Now we had got hold of seven books of the eight books which Apollonius composed on conics in the form in which he had composed them. So we wanted to translate and understand them, but that task proved impossible for us because of the excessive number of errors which had accrued in that treatise for the reasons we have described. But we persisted in that for a long time. Then al-Ḥasan b. Mūsā, through his prowess and superiority in the science of geometry, succeeded in the theory of the science of the section of the cylinder when it is cut by a plane not parallel to its base; the circumference of [that] section is a closed curve. He discovered the theory of it and the theory of the basic *symptomata* of the diameters, axes and chords which occur in it. Thus he discovered how to measure it, and was able to make that an introduction to and a way of studying the science of conic sections, for he thought that that would be easier for him in his investigation and more like the [correct] way of proceeding in order in this science. Then he investigated the science of the conic section of a cylinder when the line bounding it is a closed curve [an ellipse], and found that the shape of the section of the cylinder, the theory of which he had discovered, is identical with the shape of a section of the cone. And he discovered the proof of the fact that to every section falling within the cylinder in the way we described, there corresponds some cone in which falls an identical section, and that to every elliptical section of this type there corresponds some cylinder which contains the equivalent section. So al-Ḥasan at that point composed a treatise on what he had discovered of that science and died—may God have mercy on him.

Then Aḥmad b. Mūsā managed to travel to Syria as the one in charge of its post; for he intended to search for manuscripts of that work [Apollonius's *Conics*] in the hope that he would collect thence material for it which would enable him to correct it. But that proved impossible for him. But he got hold of one manuscript of the four books of Apollonius's treatise which Eutocius had restored, although it too had accumulated errors after Eutocius, for the reasons we described. So when Aḥmad got this manuscript, he began to comment on the treatise, beginning with the first four books which Eutocius had restored, since he found that the errors in these were fewer than those in the original of Apollonius's treatise. So he expended toil and hardship on understanding them, until he was done with them. Then his departure from Syria to Iraq took place, and when he returned to Iraq, he went back to commenting on the rest of the seven books which had come down to us from the original treatise of Apollonius. We have already described the state of corruption into which this treatise had fallen due to the multiplicity of errors. However Aḥmad, due to his understanding of the four books which Eutocius restored, had acquired control over the understanding of the rest of the treatise, experience with it, and understanding of the methods employed by Apollonius and the basic principles he set down. Thus, by these means, he was enabled to understand the three remaining books of the seven, so that he completely comprehended them. And he did something for this treatise which is of great use in facilitating the comprehension of it for anyone who wants to read it—something which Apollonius did not do when he composed it, nor Eutocius when he restored it. Namely he examined every premise which one needs for the proof of each of the propositions, mentioned it explicitly in the place where one needs it, and described its position in the treatise.

The man entrusted with the translation of the first four books—under the supervision of Aḥmad b. Mūsā—was Hilāl b. Abī Hilāl al-Ḥimṣī, and the one entrusted with the translation of the three remaining books was Thābit b. Qurra al-Ḥarrānī, the geometer.

After the above, we begin with geometrical theorems which we consider necessary for facilitating the understanding of this treatise, then we follow that with the preface with which Apollonius prefaced his treatise; then, one by one, with the seven books which we succeeded in getting translated and commented. We have already mentioned that the first four came out according to the restoration of Eutocius, and the three following according to the composition of Apollonius.

Ibn al-Haytham on the Completion of the Conics

In the name of God, the Merciful, the Compassionate.

Treatise by Al-Ḥasan ibn al-Ḥasan ibn al-Haytham on the Completion of the *Conics*.

Apollonius mentioned in the preface to the *Conics* that he divided his work into eight books, and he explained the notions which he had discovered and which were contained in each one. He mentioned that the eighth book dealt with problems occurring in conics. But only seven books of this work were translated into Arabic, and the eighth book was not found.

When we studied this work, investigated the notions in it, and went through the seven books many times, we found that it lacked notions, which this work should not leave untreated. Thus we were convinced that the notions which were passed by in the seven books are the notions which were in the eighth book. But he

[Apollonius] postponed them, because it was not necessary for him to use them in the notions contained in the [first] seven books. These notions [in Book VIII] to which we have referred are notions which are made necessary by notions contained in the seven books.

Among these, he explained the ratio in which the tangent divides the axis of the [conic] section, and he explained how we draw a tangent to the [conic] section which makes with the axis an angle equal to a known [angle]. These two notions make it necessary for us to explain how we draw a tangent to the [conic] section such that the ratio of it to the part of the axis which is cut off by it is a known ratio, and to draw a tangent to the [conic] section such that the part of it which falls between the [conic] section and the axis is equal to a known line. Moreover, these motions are among those to the knowledge of which the [human] mind aspires.

Also he explained how we draw a tangent to the [conic] section which makes with the diameter drawn from the place of contact an acute angle equal to an assumed angle. This notion also makes it necessary for us to draw a tangent to the [conic] section which ends at the axis, such that the ratio of it to the diameter drawn from the place of contact is a known ratio.

Another example is that he spoke in the preface to the seventh book of the diameters of the [conic] sections, their classification and their distinction, and alluded to the fact that they have special properties which occur in them in connection with their *latera recta*. Moreover, he says in the preface to this book that the notions which follow in this book are very necessary in the problems which occur among those that will be mentioned in the eighth book. [So the eighth book] contained problems connected with the diameters and their special properties.

Also he explained how we draw from an assumed point a line which is tangent to the [conic] section and meets it in one point. This notion makes it necessary for us to explain how we draw from an assumed point a line which meets the [conic] section in two points, such that the part of it which falls inside the [conic] section is equal to an assumed line, and that we draw a line which intersects the [conic] section such that the ratio of the part of it outside [the conic] to the part of it inside [the conic] is equal to an assumed ratio.

It is inconceivable that the work did not deal with these notions we mentioned and referred to, because they are beautiful notions, the beauty of which is not less than the beauty of what the seven books contain. On the contrary, among them are [some] which exceed in beauty and understanding the propositions that were transmitted. Hence it is most likely that these notions are [the notions] which the eighth book contained. However, he [Apollonius] did not mention them before the eighth book, since he could dispense with using them in the books which were transmitted.

Since in our opinion this state of things is impossible, and since in our mind our good opinion of the author of the work [Apollonius] was strong, the good opinion became predominant in us, and we decided that these notions and similar ones were the notions contained in the eighth book. When our judgment about that had been established, we started to derive these notions, to explain them and to collect them in a book containing them, to replace the eighth book and to be the completion of the *Conics*. We make our derivation of these notions by analysis, synthesis and *diorismos* in order that it become the clearest of the [eight] books.

This is the time we begin the treatise. We ask the help of God.

III. Arithmetic

The Hindu base-ten place-value system reached Baghdad by the ninth century. Many Islamic authors wrote descriptions of this system and the new algorithms it required in order to do calculations within it. The earliest known Islamic treatise on this system was *The Short Treatise on Hindu Reckoning*, written by Muḥammad ibn Mūsā al-Khwārizmī (c. 780–850) around 825. This text is no longer extant in Arabic, but there are several Latin versions which were made in the twelfth century and later. The following selections are translated from a thirteenth century Latin manuscript now in the Cambridge University library. The first excerpt is a general description of the Hindu system, while the second excerpt deals with the multiplication of fractions and the third with division.

Al-Khwārizmī's Treatise on Hindu Reckoning

Algorizmi said: Let us speak praises to God our guide and defender, worthy both to render Him His due and multiply His praise by increasing it, and let us entreat Him to guide us in the path of righteousness and lead us into the way of truth, and to help us in addition with goodwill in these things which we have decided to set out and reveal: concerning the numbering of the Indians by means of IX symbols, by which they set out their universal system of numbering, for the sake of its ease and brevity, so that this work, to be sure, might be made easier for the seeker after arithmetic, i.e., the greatest number as much as the smallest, and whatever there is in it as a result of multiplication and division, also addition and subtraction, etc.

Algorizmi said: since I had seen that the Indians had set up IX symbols in their universal system of numbering, on account of the arrangement which they established, I wished to reveal, concerning the work that is done by means of them, something which might be easier for learners if God so willed. If, moreover, the Indians had this desire and their intention with these IX symbols was the reason which was apparent to me, God directed me to this. If, on the other hand, for some reason other than that which I have expounded, they did this by means of this which I have expounded, the same reason will most certainly and without any doubt be able to be found. And this will easily be clear to those who examine and learn.

So they made IX symbols, whose forms are these: [9 8 7 6 5 4 3 2 1]. There is also a variation among men in regard to their forms: this variation occurs in the form of the fifth symbol and the sixth, as well as the seventh and the eighth. But there is no impediment here. For these are marks indicating a number and the following are the forms in which there is that variation: [5 4 3 2]. And already I have revealed in the book of *al-jabr* and *almuqabalah*, that every number is composite and that every number is put together above one. Therefore one is found in every number and this is what is said in another book of arithmetic. Because one is the root of all number and is outside number. It is the root of number because every number is found by it. But it is outside number because it is found by itself, i.e., without any other number. But the rest of number cannot be found without one. For when you say one, because it is found from itself, it does not need another number. But the rest of number needs one, because you cannot say two or three unless one comes first. Number is therefore nothing else but a collection of ones, and as we said, you cannot say two or three unless one precedes; we have not spoken about a word, so to speak, but about an object. For two or three cannot exist, if one is removed. But one can exist without second or third. Therefore two is nothing but the doubling or repetition of one; and likewise three is nothing but the tripling of this same unity; in this way understand about the rest of number. But now let us return to the book.

I have found, said Algorizmi, that everything that can be expressed in terms of number is also whatever is greater than one up to IX, i.e., what is between IX and one, i.e., one is doubled and two results, and likewise one is tripled and three results; and so on for the rest up to IX. The X is put in the place of one and X is doubled and tripled, just as was done in the case of one; from its doubling results XX, from its tripling XXX, and likewise up to XC. After this C [a hundred] comes back in the place of one and is doubled there and tripled, just as was done in the case of one and X; and there will be produced from it CC and CCC, etc. up to DCCCC [nine hundred]. Again, a thousand is put in the place of one, and by doubling and tripling, as we have said, there result from it II thousand and III thousand, etc. up to infinity according to this method. And I have found that the Indians worked according to these places. Of these, the first is the place of the units, in which is doubled and tripled whatever is between one and IX. The second is the place of the tens, in which is doubled or tripled whatever is from X to ninety. The third is the place of the hundreds, in which is doubled and tripled whatever is from C to DCCCC. Furthermore, the fourth is the place of the thousands, in which is doubled and tripled whatever is from a thousand to IX M. The fifth place is X [ten thousand] in the following way: every time the number rises, places are added. The arrangement of a number will be as follows: everything that will have been one in the higher place will be X in the lower, which is before it, and what will have been X in the lower will be one in the higher, which precedes it; and the beginning of the places will be on the right of the writer, and this will be the first of them and is itself placed there for the units. But when X was put in the place of one and was made in the second place, and its form was the form of one, they needed a form for the tens because of the fact that it was similar to the form of one, so that they might know by means of it that it was X. So they put one space in front of it and put in it a little circle like the letter o, so that by means of this they might know that the place of the units was empty and that no number was in it except the little circle, which we have said occupied it, and [thus] it is shown that the number that is in the following place was a ten and that this was the second space, which is the place of the tens. And they put after the circle in the aforesaid second place whatever they wished from the number of tens from what is between X and XC and these are the forms of the tens; the form of X is thus [10], the form of XX [20]. And likewise the form of XXX is thus [30], and so on up to IX [tens] there will be, clearly, a circle in the first place and a character pertaining to the number itself in the second place. Moreover, one must know this, that the character that signifies one in the first place, in the second signifies X, in the third C, and in the fourth $\overline{\text{I}}$. And likewise the character that in the first place signifies two, in the second signifies XX and in the third CC and in the fourth $\overline{\text{II}}$ and understand likewise about the rest. But let us return to the book.

After the place of the tens follows the place of the hundreds in which is doubled and tripled whatever is from C to DCCCC and its form is just as the form of one put in the third place, thus 100, and the form of two hundred is just as the form of two placed likewise in the third place, thus 200; also the form of three hundred is the form of three placed in the third place, thus 300, and so on up to nine hundred. This place also is followed by the place of the thousands, in which likewise is doubled and tripled whatever is from a thousand to $\overline{\text{IX}}$ [nine thousand]. The form of this is just as the form of one put in the fourth place, thus 1000; the form of two thousand is just as the form of two placed in the fourth place, thus 2000, and so on up to IX thousand; moreover, there are placed before [i.e., to the right of] the character in the fourth place three circles, so that it may be shown what is in the fourth place, just as there were placed [before this character] in the second place one circle and [before the character] in the third place two circles, so that it might be shown what were the places of the tens and hundreds, and this happens when there is not before the number itself another number in the same place. But if, along with the number that is put in these places, there is another number below it, it must be put in that place which is due to it. E.g., if there is along with X some number from those that are below it [i.e. to the right of it], say as in XI or XII, they are placed thus: 11 [or 12]; i.e., in the first place, where the circle was placed, a one is to be placed and in the second place a one also is to be placed which signifies X. Likewise, if there is along with C another number from those that are below it, it is to be placed in the place which is due to it. Let us show this by a particular example and let us say that the number was CCCXXV. When we wanted to put it in its places, we put it as follows: we began from the right of the writer and placed V in the first place and XX in the second going toward the left of the writer, and CCC in the third place, each number in its own place, i.e., the units in the place of the units, which is the first, and the tens in the place of the tens, which is the second, and indeed the hundreds in the place of the hundreds, which is the third, and this is the form 325; and it will be likewise in the other places according to this order; i.e., as often as a number is made larger and the places increase, each kind of number is to be put in its own place that is due to it. But when X or more is gathered in any of the places, it is to be raised to a higher place and from each X a one is to be produced in the higher place. Again, if there is another number in the same place, at which a number arrives by increasing, it is to be added on and they are to be added together and if there is in it X or more, from each X a one is to be made and to be raised to a higher place; i.e., if ten is gathered in the first place, one is to be made from it and placed in the second place, and if in the same place there is likewise a number, it is to be added to it; and if there is X there, one is to be made from it and raised also to the third place. E.g., if in the first place, which is the place of units, you have X, make a one from it and place it in the second place. Moreover, in the first place put a circle just as we have said, so that it may be shown that there are two places. But if there is XI, make a one from the X and put it in the second place as above and send down one into the first. But if you find some number in the second place, where you have placed the very number that you made from X, add it to that. And if there is X, or more, make from X a one and again place it in the third place; and what remains below X, let it remain in its own place. Moreover, what we say of more than ten holds for any large number. E.g., if there is in the second or third place a large number, such as if you find IX in the third place, which is the place of the hundreds, and if there is a X in the second place, make from the X a one and change it to the third place, and there add it to IX and there results X; make a one from the X itself and change it to the fourth place and there it will be a thousand. If, on the other hand, you found XX in the second place you would also make two from it, and adding two to IX in the third place, XI would also result; you would again make a one from the X and change it to the fourth place where it would be a thousand; and there remains a one in the third place and therefore indicates X or more. And this must be known that, because you have changed your number and put it in the following place, you must put it by means of its own characters; i.e., if it is X, instead of it place the character that signifies one in the first place, and if it is XX, instead of it place the character that signifies two in the first place. And understand likewise for the rest. But if there remains in the same place, from which you have changed a number, something from the number, move it down likewise by means of its own characters; i.e., if there remains a one or two, move it down there by the character that signifies the same number; i.e., if there remains one, copy there the character of one, and there remains two, copy there the character of two, etc. But each form will have significance according to the place; i.e., in the first place it will signify units, in the second tens, in the third, hundreds, etc., just as has been said above.

[...]

Now we shall begin to treat the multiplication of fractions and their division and the extraction of roots, if God so wills.

Know that fractions are called by many names innumerable and infinite such as half, third, quarter, ninth and tenth and XIIIth and XVIIIth, etc. But the Indians constructed their fractions out of sixty; for they divided one into LX parts which they called minutes. Once more, [they divided] each minute into LX parts, which they called seconds; and one minute will be out of LX and one second out of three thousand and six hundred and each second is once more divided by LX, and a third will be out of two hundred and XVI thousand and each third is divided by LX fourths, and so on to infinity will be the places. Thus the first place of the degrees is the place in which stands a whole number, and in the second place will be the minutes. In the third also are the seconds and in the IVth the thirds and so on in the IXth and Xth place. And know that every whole number that is multiplied by a whole number yields a whole number, and every whole number multiplied by some fractions yields [a product] according to the nature of that fraction; and two degrees multiplied by two minutes will be IIII minutes and three degrees by six thirds will by XVIII thirds. Minutes also multiplied by minutes will be seconds and seconds by seconds will be thirds[6] and thirds by thirds will be fourths and fourths by fourths will be fifths, because you join both places that you multiply in turn; and what is aggregated from the number of fractions is like that which results from a whole number multiplied in turn. For example, six minutes multiplied by VII minutes will by XLII seconds, because minutes are parts out of LX parts of one integer and when you multiply

[6]This and the next two are clearly incorrect, since the orders of the respective products should be fourths, sixths and eighths. Indeed, the author states the correct rule immediately after this when he says "you join [i.e. "add"] both places."

parts out of LX by [parts out of] LX there will be what results from the multiplication of LX by LX which is three thousand six hundred; and likewise VII seconds multiplied by IX minutes will be LX-three thirds; and all LX of these will also be one second and there will remain three thirds, because minutes are parts out of LX and seconds are parts out of three thousand and six hundred. Thus multiply them in turn and there will result parts of two hundred and XVI thousand which are thirds and are LX [each] out of three thousand six hundred.

And when you want to multiply one and a half by one and a half, make one and a half into minutes and there will XC. Once more make the one and a half by which you wish to multiply in the same minutes and there will likewise be XC; multiply one of them by the other and there will be VIII thousand C seconds; divide the seconds by LX and there will be minutes, because every LX [seconds] make one minute. And there will result for you CXXXV minutes; and divide them by LX and there will be degrees, because every LX minutes make one degree. And this will be one integer from the number; and there will result for you two [degrees] and XV minutes, which are one quarter of one.

[...]

Know that when you wish to divide a number with a fraction by some number with a fraction or a number with a fraction by a whole number or a whole number by a number with a fraction, you must make each number of the same nature, i.e., turn both numbers into the lower place. E.g., if the lower place is of seconds, put each number into seconds; but if there are thirds in one of them and seconds with the other, turn both into thirds, and if there is something with one of them from the fourth or sixth or something else lower than these places, while the other number is an integer, turn both into that place which is lower in both; then divide what you wish by what you wish, after you have made each number of one kind, and what results will be degrees, i.e., a whole number, because in the case of any two numbers that are of one kind, if one of them is divided by the other, what results will be a whole number. E.g., if XV thirds are divided by six thirds, there will result from the uniformity of the division two-and-a-half; because XV thirds make V wholes and when you divide them by VI thirds, which is two wholes, there will result two and a half. And likewise halves are divided by halves and fourths by fourths, minutes also by minutes and seconds by seconds and thirds by thirds. And when you wish to divide X seconds by V minutes, make the minutes seconds, so that they are of one kind of one place; and there will be three hundred seconds; and as long as you wish to divide X seconds by them, X cannot be divided by three hundred. Know therefore that a whole did not result. So place a circle in the place of one and multiply X by LX and there will be six hundred, and when you divide this by three hundred, there will result two, which are two minutes. And know that as regards every number that is divided by another number, if what is extracted from that which is divided is multiplied by that by which it is being divided, the first number will return, i.e., the number being divided. Of this an example is: that when you divide L by X, there will result what is owed to one, i.e., five. And when you multiply that which has resulted to you from the division, i.e., five, by that by which you are dividing, which is X, there will return the first number, i.e., L. When therefore we divided X seconds by V minutes, there resulted what is owed to one, i.e., two minutes. And when we multiplied two minutes, i.e., what resulted to us from the division, by that by which we divided, which is V minutes, X seconds were produced and this is a proof of the division. Likewise, when you wish to divide X minutes by V thirds, change the minutes in thirds, and there will be XXXVI thousand thirds; and divide by V thirds and there will be VII thousand two hundred degrees and this is what is owed to one. And when you wish to check this, multiply VII thousand two hundred degrees by V thirds and there will result XXXVI thousand, and when you divide this by LX there will result VI hundred seconds, and when once more you divide VI hundred seconds, there will be ten minutes.

When you wish to set out a whole number and fractions, put the whole number in a higher place; then put whatever is from the first place, which are minutes, beneath the whole number and the seconds under the minutes and likewise the thirds under the seconds and the rest as you wish according to the places. Of this an example is: that when you wanted to set out XII degrees and XXX minutes, together with XLV seconds and L fourths, we set down XII. After this we put beneath it XXX in the place of the minutes and beneath XXX, XLV in the place of the seconds. In the place of the thirds we put circles, because there was a lack of thirds, and so that we might know that there were still fourths remaining. Then we put under the circles fifty in the place of the fourths and this is their form:

12
30
45
00
50

Decimal fractions first appear in *The Book of Chapters on Hindu Arithmetic*, written in Damascus about 952 by Abū'l-Ḥasan al-Uqlīdisī, about whom we know little else. We give here several excerpts from his book. The first is on multiplying mixed numbers; the second gives the results of halving 19 five times; the third shows how to increase 135 by its tenth, the result by its tenth, etc. up to five times. We also give an excerpt in which al-Uqlīdisī gives reasons for abandoning the dust board for calculation in favor of pen and paper. Finally, we show the Arabic text of part of the second excerpt, which displays the earliest use of a symbol for dividing the integral part of a number from its decimal fraction part.

Al-Uqlīdisī's Chapters on Hindu Arithmetic

We say that the way [to multiply two whole numbers mixed with fractions] is to draw them, make the fractions of one number each, combine numbers to fractions, [and then] multiply and divide by the product of the two numbers of which the parts are derived.

For example, we want to multiply 7 and a half by 5 and a third. We assume them like this:

7	5
1	1
2	3

.

We multiply 7 by two and add the one, which is the one-half; that becomes 15 with 2 below. We multiply 5 by 3 and add the third, which is one; it becomes 16 with 3 below. We multiply 16 by 15 and divide by 6. The outcome is 40.

Again, we want to multiply

$$\left|\begin{array}{cccccc} 19 & & & \text{by } 13 & & \\ & 1 & 1 & & 1 & 1 \\ & 3 & 4 & & 2 & 5 \end{array}\right|.$$

We add their fractions [i.e. $\frac{1}{3}$ and $\frac{1}{4}$, $\frac{1}{2}$ and $\frac{1}{5}$]; one becomes 7 of 12, and the other 7 of 10. The form of that is

$$\begin{array}{ccc} 19 & \therefore & 13 \\ 7 & & 7 \\ 12 & & 10 \end{array}$$

We have shown how that addition is performed. We [now] multiply 19 by 12 and add 7 to that; it becomes 235. We multiply 13 by 10 and add the seven; it becomes 137. We multiply one by the other; we get 32,195. We divide that by the product of 12 by 10, which is 120. The outcome is 268 and 35 of 120.

For combining them there is another way: we draw them like this:

$$\begin{array}{cccccc} 19 & & & \therefore & 13 & \\ & 1 & 1 & & 1 & 1 \\ & 3 & 4 & & 2 & 5 \end{array}$$

We start by multiplying 19 by 3 and the product by 4: to that we add the three and the four. It becomes like the former. Below it we draw the product of 3 by 4. We do the same with the other; we multiply; we multiply 13 by 2 and by 5, and add the two and the five to that. It becomes like the former. We go on as before and get the same result.

[...]

In what is drawn on the principle of numbers, the half of one in any place is 5 before it. Accordingly, if we halve an odd number we set the half as 5 before it, the units place being marked by a sign ' above it, to denote the place. The units place becomes tens to what is before it. Next, we halve the five as it is the custom in halving whole numbers. The units place becomes hundreds in the second time of halving. So it goes always.

For example, we want to halve 19 five times. We say: one-half of 9 is four and a half; we set the half as 5 before the four; next, we halve the 10. We mark the units place. That becomes 95. [The scribe often omitted the sign'.] Now we halve the five and the nine; we get 475. We halve that and get 2'375, the units place being thousands to what is before it, for if we want to state what we have got, we say that halving has led to two and 375 of one thousand. We halve that and get 11875. We halve a fifth time and get 059375, which is 59375 of a hundred thousand.

We relate that saying: one-half, and half of one-eighth, and one-fourth of one-eighth. If we want to retrace that, we start as usual in such cases. It comes back to what it was, but whenever a zero comes at the beginning we drop it and ignore it, because we do not need it. The 19 is regained.

[...]

For example, we want to add to a number one-tenth of it five times. We set that number as usual. We repeat it, brought one place down; we thus know its tenth; we add that to it; we have thus added its tenth once. We draw the outcoming fraction before it and relate it to the units place, after having marked the units place. We add one-tenth of that again, and so on five times.

For example, we want to increase 135 by one-tenth of it five times. We repeat it below, brought one place down, marking the units place. It becomes

135
 135

We add them and get 1485. We add to this one-tenth of it, for the second time, by finding out its tenth. That is

1485
 1485

We add them and get 16335. This is one hundred sixty-three and thirty-five of one hundred; which is one-fourth and one-tenth.

We increase it by one-tenth of it, by finding out its tenth first, and then adding them; they become

163′35
 16 335

If we add them, we get 179′685. What is before the units place is 685 of a thousand, because the units place is fourth with respect to it.

If we increase it by one-tenth of it a fourth time, it becomes 1976535. If we add to it one-tenth of it, it becomes 21741885. We relate that which is before the units place: it is 41885 of hundred thousand.

Thus we have increased 135 by one-tenth five times.

[...]

In this book we state all that is done by Hindi (schemes), not with *takht* or erasure, but with inkpot and paper. This is because many a man hates to expose the *takht* between his hands when he finds the need to use this art of calculation, for fear of the misinterpretation of the attendants or whoever may see it. It belittles him, for it is seen between the hands of the misbehaved who earn their living by astrology in the streets. Moreover, he who calculates on it finds it so difficult to reconsider what he has calculated to the extent that in most cases he repeats it, (not to mention) the exposure of the content to the blowing wind which changes the figures, apart from making the fingers dirty, over and above other things which distort orderliness.

In addition to all that we have said, it (that is, what we here suggest) is simpler and quicker than the arithmetic of the *takht*. Of it we shall show what will be appreciated and considered a novelty by all who see it. It is one of the most curious things done in this arithmetic and the best that is being discussed. I have seen no one of the people of Baghdad discuss it or do anything in it. This is why I mention it and give the working of all its schemes. I shall leave nothing that is done by the *takht* without doing it in this way; with the direction of God.

[...]

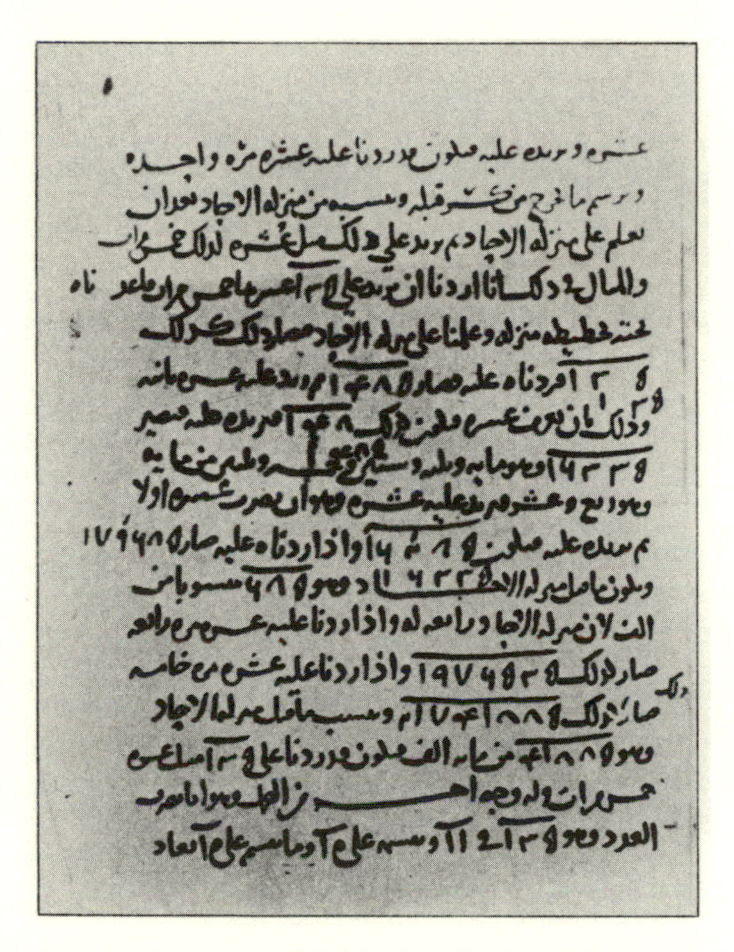

FIGURE 5.1. On this manuscript page from Al-Uqlīdisī, the earliest decimal sign appears, twice in line 10.

Another early work on Hindu numeration whose Arabic text still exists was written by Kūshyār ibn Labbān, born in the region south of the Caspian Sea around 975. He was an accomplished astronomer, whose works exerted considerable influence. In particular, his treatise on arithmetic, *Principles of Hindu Reckoning*, became one of the chief arithmetic textbooks in the Islamic world. We include here excerpts from that text on subtraction, halving, multiplication, and finding square roots.

Kūshyār ibn Labbān's Principles of Hindu Reckoning

On Subtraction

We wish to subtract 839 from 5,625. We set it down according to the first figure.

5625
839 [Fig. 1]

The smaller one is under the larger one. All of the categories correspond, units under units, tens under tens. Then we subtract the 8 from the 6 which is above it. It is not possible to subtract so we subtract it from the 56 which is above it. There remains 48. We put the 4 in place of the 5 because it is of the order of the tens, and the 8 in place of the 6 because it is of the order of the units. It remains as in the second figure.

4825
839 [Fig. 2]

Then we subtract the 3 from the 2 above it. There remains 79. We put the 70 in place of the 80, and the 9 in place of the 2. It remains according to the third figure.

4795
839 [Fig. 3]

Then we subtract the 9 from the 5 which is above it. It is impossible to subtract. We subtract it from 95 which is above it. There remains 86. We put the 80 in place of the 90 and the 6 in place of the 5. It remains according to the fourth figure.

4786
839 [Fig. 4]

That is what we wished to do.

Another kind of Subtraction, Which is Halving

We wish to halve 5,625. We set it down according to the figure.

5625 [Fig. 1]

We halve the first 5 of the number to get 2½. We put the 2 in place of the 5 and put the ½ under it, 30.

5622
30 [Fig. 2]

If it is considered a *dirham* they [the 30] are the *falus*. If it is used as a degree, they are minutes. We halve the 2 in the tens; 1 remains in its place. We halve the 6 after it; a 3 remains in its place according to the third figure.

5312
30 [Fig. 3]

Then we halve the last 5 which is tens of the 3 preceding it. Its half is 25. We put the 20 in place of the 5 because it is the order of the tens in relation to the 3. We add the 5 to the 3 which is from its units. There remains what is according to the fourth figure.

2812
30 [Fig. 4]

On Multiplication

We wish to multiply 325 by 243. The two are placed on the dust board as in the first figure.

325
243 [Fig. 1]

The first order of the multiplier is under the last place position of the multiplicand. Multiply the 3 of the multiplicand by the 2 of the multiplier to give 6. We put it above the 2 of the multiplier beside the 3 of the multiplicand according to the second figure.

6 325
243 [Fig. 2]

If the product were other than 6 and contained tens and units, we would have put the units above the 2 and the tens to the left of the units. We multiply the upper 3 also by the lower 4. We add the 10 to the tens so that 6 becomes 7. It results in what is in the third figure.

72325
243 [Fig. 3]

We multiply the upper 3 by the lower 3 to give 9. We put it above the lower 3 in place of the upper 3. We shift the lower orders one place (to the right). It results in what is shown in the fourth figure.

72925
243 [Fig. 4]

Then we multiply the 2 which is above the lower 3 by the lower 2 to get 4. We add it to the 2 that is above the lower 2 to get 6. then we multiply the upper 2 by the lower 4 to get 8. We add it to the 9 which is above the 4. Then we multiply the upper 2 also by the lower 3 to get 6. We put it above the 3 in place of the upper 2. Then we shift the lower orders one place (to the right). It results in what is shown in the fifth figure.

77765
243 [Fig. 5]

Then we multiply the upper 5 by the lower 2 to give 10. We add it to the tens order (of) that (which) is above the 4. Again, we multiply the 5 by the lower 4 to give 20. We add the 2 to the tens order (of that which is above it) to give 9. Again, we multiply the 5 by the lower 3 to give 15; the 5 is left in its place and the 10 is added to its tens. It results in what is shown in the sixth figure.

78975
243 [Fig. 6]

That is what we wished to do.

Multiplication of Degrees and Fractions

If we wish to multiply degrees with fractions (by degrees with fractions), we convert the degrees with fractions, of the two numbers, to the category of the lowest fraction in each of them. Thus, we multiply the degrees by 60 and add the minutes to it. We also multiply the result by 60 and add it to the seconds, and so on with what follows. Then we multiply the derived fraction of one by the derived fraction of the other.

On Square Root

We wish to extract the root of a square whose number is 65,342. We set it down on the dust board and differentiate the placed positions by means of marking off the digits in twos [from the right] until the last mark is reached. A number is set down. It is such that if we multiply it by itself and subtract it from what is above it of the square, it comes to zero or there remains what is less than the found squared number. We find it to be 2. We put it under the 6 and above it also as in the first figure.

2
65342 [Fig. 1]
2

Then we square it and subtract it from what is above it of the square. We double the 2 in its place and shift it and the upper 2 one place to that shown in the second figure.

2
25342 [Fig. 2]
4

Then we seek a number [the largest possible] which we put under the 3 such that if we multiply it by the lower 4, then by itself, and we subtract it from what is above it of the square, it comes to 0 or the remainder is less than the lower orders. It is 5 which we put under the 3 and also above it as in the third figure.

25
25342 [Fig. 3]
45

Then we multiply it (the 45) by 5 and subtract it from what is above it of the square. Then we double the 5 in its place and shift the lower and upper orders one place as in the fourth figure.

25
2842 [Fig. 4]
50

Then we seek a number which if we multiply it by the lower orders, then by itself, and we subtract it from what is above it of the square, either nothing is left or the remainder is less than the lower orders. It is a 5 which we put under the 2 and above it. We multiply it by the 5 and subtract it from what is above the 5 of the square and we multiply it by itself because that [digit] next to the 5 is a 0. We subtract it from what is above it of the square to give the result shown in the fifth figure.

255
317 [Fig. 5]
511

The uppermost orders give the root of the square. The remainder of the square is parts of the lower orders of 1 by those near it after the last lowest 5 is doubled plus 1 which is always added to it. Then the resulting root is 255 and 317 parts of 511 parts of 1. If we multiply the remainder by 60 and we divide it by 511, it is in the *falus* of *dirhams* or minutes of degrees. That is what we wished to do.

Root of an Integer plus Fractions

If we wish the root of an integer plus fractions, we convert the integer plus fractions to the category of the last [the least] fraction it has. Then when we see that the mark of the fraction is even, we extract its root. When the [mark of the] fraction is odd, we multiply it by 60 once again so that it is converted to a fraction with an even mark. Then its root is extracted, and if what remains of the square is zeros preceded by no number, take half of those zeros and put them before the resulting root.

Section on the result if it is the root of degrees or the root of fractions with an even mark. Then the mark is half of the mark of that fraction. The root of seconds is minutes, the root of fourths is seconds, and so on by analogy.

In this selection from Book 1, Chapter 5, of *The Reckoners' Key*, the Persian mathematician Ghiyāth al-Dīn Jamshīd al-Kāshī (d. 1429) illustrates a procedure for extracting roots of numbers by extracting the fifth root of a number on the order of tens of trillions.[7] As always when one is describing a procedure that uses many numbers several times each there is a risk of losing track of which number the pronoun "it" refers to. The task in translating Arabic in this regard is lightened because single digit numbers in Arabic are referred to as "her" and larger numbers, products, etc. are referred to as "him/it."

This selection is an example of a general procedure which al-Kāshī has just explained. We shall not give this explanation, but we shall explain some of his terminology. Al-Kāshī's context for his discussion is a sequence whose first term is some whole number n and whose successive terms are its integral powers. In relation to the following terms of the sequence, al-Kāshī names the initial term the "first side,"[8] although with reference specifically to the second and third terms he calls it "root" and "cube (!)." The terms of the sequence, from the second onwards, he calls by the general name of "things with sides," although the second and third terms also have the specific names of *māl* and "cube." He then explains how one gives specific names to the following terms, using combinations of *māl* and "cube," by starting "*māl*, cube, *māl māl*, *māl* cube," etc. In general, to go from one term to the next, as he says, one replaces the final occurrence of *māl* in the description by "cube," but if no *māl* appears then one replaces the initial "cube" by *māl māl*. One computes the exponent[9] of any term by taking 2 for each *māl* that occurs in its description, and 3 for each "cube" and adding up these numbers. Al-Kāshī says that although his discussion addressed only "the ascending side" of the sequence, it applies, with obvious changes, to the descending side (or, as we should say, to the powers of $1/n$). Finally, he states that "Every [number] with sides, for which there is found a side that generates it exactly, is called 'rational'

[7]For a discussion of the historical background to extraction of higher roots in medieval Islam, as well as an explanation of the mathematics behind this procedure, and to see how the procedure generates the numbers of Pascal's triangle, consult [Berggren 1986, 53–63]. (The procedure is equivalent to the Ruffini-Horner procedure for the equation $x^5 = a$.)

[8]Influenced by the idea of logarithms, unknown at the time, one might call it the "base."

[9]Al-Kāshī calls this "the number of its rank."

and [a number] for which no [such] side is found is called 'surd.'"[10] (However, we render these terms here by "perfect" and "imperfect," following the choice of vocabulary in Dakhel's translation of a similar section in Book III of the *Key*, where al-Kāshī discusses the root extraction procedure for sexagesimal numbers [Dakhel 1960].)

Al-Kāshī on Extracting a Fifth Root

Example: We want to deduce the first side of this number: 44,240,899,506,197 assuming that it is a *māl* cube, which is in the fifth degree. So we draw the table as described, and we enter the number mentioned (forty-four trillion,[11] two hundred forty billion, eight hundred ninety-nine million, five hundred six thousand, one hundred ninety-seven). And we divide [it] into cycles—the number of places of each cycle being the degree of *māl* cube, which is five—by double lines.[12]

Then we seek the greatest single [digit] whose fifth power we can subtract from the number mentioned. And we found it to be five, which we place in the row of the result above the column of the last perfect [number][13] and below it, at the bottom of the row of the side. And we place its powers[14] in the bottoms of the other [rows]: its square, which is 25, in the row of *māl*; and its cube, 125, in the row of the cube; and its *māl māl*, 625, in the row of *māl māl*;[15] and its *māl* cube, 3125, in the row of the number, below the number, so that the units of each one of them is in the last column of the rational.

Then we subtract from it [i.e., the leftmost cycle] what we placed under the number, and we place the difference under it [the 3125] after drawing between them a row to indicate the erasure of what is above it. Then we add the upper five to the lower five and place the sum, ten, above it [the lower 5] in the row of the side, after drawing a line above it [the lower 5] to indicate the erasure of what is below it. And we multiply the mentioned five [in the row of the result] by the sum [10] and place the result [50] above what was placed in the row of the *māl* [25], so that its units are in the column of the last rational.

And we add it [50] to it [25] and we place the sum above it [50] (after we draw a line between the two of them) and we multiply the five by it [the sum]. And we add the result [375] to what is in the row of the cube, and we multiply it [5] by the result [500]. And we add it [the product] to what is in the row of *māl māl* [and obtain 3125].

Then we add the upper five to the lowest [number, i.e., 10] a second time to the row of the side[16] and we multiply it [5] by it [the sum] and we add the result to what is in the row of the *māl*. And we multiply it [5] by it [what is now in the row of *māl*] and we add the result to what is in the row of the cube [to get 1250]. Then we add the five mentioned, the upper one, to the lower a third time to the row of the *māl*. And we multiply it [5] by it [20 (=15 + 5)] and we add the result to what is in the row

---

[10] This number is, of course, relative to the particular base chosen. Hence 25 is "rational" relative to the base 5 but "irrational" relative to the base 3.

[11] In the Arabic, powers of tens greater than a thousand are described in terms of thousands. So a trillion is "a thousand of a thousand of a thousand of a thousand."

[12] We have used only single lines in the tables below.

[13] This is the column in which the "1" of the largest power of 10^5 less than the number would occur. In the case of this 14-place number it is the eleventh column.

[14] Literally, "the [number] with sides" [that it gives rise to], which we explained in the introduction to this selection.

[15] This is labeled "Row of the square-square" in the table.

[16] The printed Arabic text that we have consulted has, mistakenly, "row of the cube."

of the *māl* [150]. Then we add the upper one [the 5] to the lower [20] a fourth time to the row of the side. So now there results in the rows, above the dividing lines,[17] the following: in the row of the side 25, in the row of the *māl* 250, in the row of the cube 1250, and in the row of *māl māl* 3125.

And now the time has come to copy [i.e., to transpose some numbers a number of places to the right]. And so we copy what is in the row of *māl māl*,[18] which is the row of the second of the number, by one place, and what is in the row of the cube by two places, and what is in [the row of] *māl* by three places, and what is in the row of the side by four places. And so the units place of what is in the row of the side falls in the column following the first column of the next-to-last cycle.[19]

Row of the result	5		
Row of the number	4424 3125 1299	08995	06197
Row of the square-square Row of the second of the number	312 3125 2500 625	5	
Row of the cube Row of the third of the number	12 1250 750 500 375 125	50	
Row of the square Row of the fourth of the number	250 100 150 75 75 50 25	250	
Row of the root Row of the fifth of the number	25 20 15 10 5	25	

[17]These are the lines dividing the table into horizontal strips.

[18]That the only element shifted in this row in the figure is the top number is good evidence that the whole procedure is done on a dustboard in which figures [in the row of the side, *māl*, etc.] are erased when they are used for the last time. So there is only one entry in each of the lower rows when one is working on a dust board.

[19]Arabic is written from right to left, since the last cycle of the number in this example is "4424," the one preceding it is "08995," and its "first" column is the one with the "5" in it.

Then we seek the largest single [digit] in the row mentioned in the deliberations [above], which we find to be three.[20] We place it above the perfect [number][21] preceding the last perfect [number] and under it in the row of the side to the right of the five [in "25"]. And 253 results in the row of the side, and we multiply it [the 3] by that and we add the result to what is in the row of *māl.* And so on until we end up at the row of *māl māl.* And we multiply it [3] in what resulted in it [the line of *māl māl*] and we place the result below the number and [*] we subtract it from the number. Then we add the upper three to what is in the row of the side one time to *māl māl,* and we multiply it [3] in the sum, and we add the result to what is above it [the sum] according to the mentioned pattern, until we reach the row of *māl māl.*

Then we add it [3] to what is in the row of the side a second time to the row of the cube, and so on until we add it to what is in the row of the side a fourth time to the row of the side.[22] And so now there results in this way, in the lines above the dividing line in the row of the side 265, and in the row of *māl* 28,090, and in the row of the cube 1,488,770, and in the row of *māl māl* 39,452,405.

Row of the result	5	3	
Row of the number	1299 1056 243	08995 95493 13502	06197
Row of the square-square Row of the second of the number	39 394 42 352 39 312	45240 52405 20574 31831 81831 5	5
Row of the cube Row of the third of the number	 14 14 13 12	14887 88770 81912 06858 79581 27277 77277 50	70
Row of the square Row of the fourth of the number		28 28090 786 27304 777 26527 768 25759 759 250	090
Row of the root Row of the fifth of the number		 265 262 9 6 253	265

[20]The natural question "Largest with respect to what?" is answered by the following descriptions, up to [*] below, of how one operates with it to obtain the largest possible number less than that in the row of the number.

[21]For the terminology see Note 5 above.

[22]I.e. we stop there, obtaining 265 in the row of the side.

And now it is [again] time to copy. We copy according to the pattern mentioned [earlier]. Then we seek the largest number in the row mentioned and we find it to be six, which we put above the first perfect [number],[23] and under it [6] in the row of the side to the right of the five. And we multiply it by the sum, and we add the result to what is above it. And so on until we end up at the row of *māl māl*. And so we multiply it [6] by what is in it [the row of *māl māl*] and we subtract the result from the number. And so 21 remains in the row of the number below the dividing line.

And if that did not remain in it[24] then the number that we posited would be a perfect *māl* cube. And its first side would be 536, and this is what would result in the row of the result, and the operation would be finished. And so, when 21 remains we know that it is not perfect. And so we require something between the *māl* cube of 536 and the *māl* cube of 537 so that it will be the result belonging to what remains from the number, namely 21.

Row of the result	5	3	6
Row of the number	242 242	13502 13502	06197 06176 21
Row of the square-square Row of the second of the number	41 40 39	26949 91365 35583 90343 45240	58080 90384 67696 17696 5
Row of the cube Row of the third of the number		15399 171 15227 170 15057 169 14887	06560 41496 65064 45448 19616 49616 70
Row of the square Row of the fourth of the number		28 28 28 28 28	72960 16044 56916 16008 40908 15972 24936 15936 090
Row of the root Row of the fifth of the number			2680 2674 2668 2662 2656

And so we add[25] the upper six to what is in the row of the side one time to the row of *māl māl* and we operate with it, according to the procedure mentioned, a second time to the row of the cube. And we operate with it according to that procedure until

[23]This is the number in the units place in the "Row of the number."

[24]That is to say, if the remainder were zero.

[25]Al-Kāshī begins the calculation of the $537^5 – 536^5$.

we add it to it, that far and no further.[26] And so the operation finishes thus. And what results in the four rows we put in another table [see below]. And we add them [the numbers in the table] and we add one to the sum. And it becomes what is between the two successive rational sides, i.e., *māl* cube of 536 and *māl* cube of 537. And it is the conventional result.

Row of *māl māl*	412694958080
Row of the cube	1539906560
Row of *māl*	2872960
Row of the side	2680
Sum of the above plus one	414237740281

IV. Algebra

The earliest extant textbook on algebra is the *Compendium on Calculation by Completion and Reduction*, by Muḥammad ibn-Mūsā al-Khwārizmī. This work was dedicated to the Caliph al-Ma'mūn, who ruled from 813 to 833 and established in Baghdad a research center called the House of Wisdom. We include several excerpts from this text, including parts of the preface, the introduction, the methods of solution of two types of mixed quadratic equations, the geometric justification for one of the solution methods, a discussion of multiplication of algebraic expressions of degree one, and some problems on legacies from the last section of the work.

Al-Khwārizmī's Compendium on Calculation by Completion and Reduction

From the preface

That fondness for science, by which God has distinguished the Imam al-Mamun, the Commander of the Faithful (besides the caliphate which He has vouchsafed unto him by lawful succession, in the robe of which He has invested him, and with the honors of which He has adorned him), that affability and condescension which he shows to the learned, that promptitude with which he protects and supports them in the elucidation of obscurities and in the removal of difficulties—has encouraged me to compose a short work on Calculating by (the rules of) Completion and Reduction, confining it to what is easiest and most useful in arithmetic, such as men constantly require in cases of inheritance, legacies, partition, lawsuits, and trade, and in all their dealing with one another, or where the measuring of lands, the digging of canals, geometrical computation, and other objects of various sorts and kinds are concerned—relying on the goodness of my intention therein, and hoping that the learned will reward it, by obtaining (for me) through their prayers the excellence of the Divine mercy; in requital of which, may the choicest blessings and the abundant bounty of God be theirs! My confidence rests with God, in this as in every thing, and in Him I put my trust. He is the Lord of the Sublime Throne. May His blessing descend upon all the prophets and heavenly messengers!

---

[26] The Arabic is "add it to it to it."

Methods for solving equations

When I considered what people generally want in calculating, I found that it always is a number. I also observed that every number is composed of units, and that any number may be divided into units. Moreover, I found that every number, which may be expressed from one to ten, surpasses the preceding by one unit; afterwards the ten is doubled or tripled, just as before the units were; thus arise twenty, thirty, etc., until a hundred; then the hundred is doubled and tripled in the same manner as the units and the tens, up to a thousand; then the thousand can be thus repeated at any complex number; and so forth to the utmost limit of numeration.

I observed that the numbers which are required in calculating by Completion and Reduction are of three kinds, namely, roots, squares, and simple numbers relative to neither root nor square.

A root is any quantity which is to be multiplied by itself, consisting of units, or numbers ascending, or fractions descending. A square is the whole amount of the root multiplied by itself. A simple number is any number which may be pronounced without reference to root or square.

A number belonging to one of these three classes may be equal to a number of another class; you may say, for instance, "squares are equal to roots," or "squares are equal to numbers," or "roots are equal to numbers."

Of the case in which *squares are equal to roots*, this is an example. "Square is equal to five roots of the same"; the root of the square is five, and the square is twenty-five, which is equal to five times its root.

So you say, "one third of the square is equal to four roots"; then the whole square is equal to twelve roots; that is a hundred and forty-four; and its root is twelve.

Or you say, "five squares are equal to ten roots"; then one square is equal to two roots; the root of the square is two, and its square is four.

In this manner, whether the squares be many or few (i.e., multiplied or divided by any number), they are reduced to a single square; and the same is done with the roots, which are their equivalents; that is to say, they are reduced in the same proportion as the squares.

[...]

I found that these three kinds, namely, roots, squares, and numbers, may be combined together, and thus three compound species arise; that is, "squares and roots equal to numbers"; "squares and numbers equal to roots"; "roots and numbers equal to squares."

Roots and Squares are equal to Numbers; for instance, "one square, and ten roots of the same, amount to thirty-nine dirhams"; that is to say, what must be the square which, when increased by ten of its own roots, amounts to thirty-nine? The solution is this: you halve the number of the roots, which in the present instance yields five. This you multiply by itself; the product is twenty-five. Add this to thirty-nine; the sum is sixty-four. Now take the root of this, which is eight, and subtract from it half the number of the roots, which is five; the remainder is three. This is the root of the square which you sought for; the square itself is nine.

The solution is the same when two squares or three, or more or less are specified; you reduce them to one single square, and in the same proportion you reduce also the roots and simple numbers which are connected therewith.

For instance, "two squares and ten roots are equal to forty-eight dirhams"; that is to say, what must be the amount of two squares which, when summed up and added to ten times the root of one of them, make up a sum of forty-eight dirhams? You must at first reduce the two squares to one; and you know that one square of the two is the half of both. Then reduce everything mentioned in the statement to its half, and it will be the same as if the question had been, a square and five roots of the same are equal to twenty-four dirhams; or, what must be the amount of a square which, when added to five times its root, is equal to twenty-four dirhams? Now halve the number of the roots; the half is two and a half. Multiply that by itself; the product is six and a quarter. Add this to twenty-four; the sum is thirty dirhams and a quarter. Take the root of this; it is five and a half. Subtract from this the half of the number of the roots, that is two and a half; the remainder is three. This is the root of the square, and the square itself is nine.

The proceeding will be the same if the instance be, "half of a square and five roots are equal to twenty-eight dirhams"; that is to say, what must be the amount of a square, the half of which, when added to the equivalent of five of its roots, is equal to twenty-eight dirhams? Your first business must be to complete your square, so that it amounts to one whole square. This you effect by doubling it. Therefore, double it, and double also that which is added to it, as well as what is equal to it. Then you have a square and ten roots, equal to fifty-six dirhams. Now halve the roots; the half is five. Multiply this by itself; the product is twenty-five. Add this to fifty-six; the sum is eighty-one. Extract the root of this; it is nine. Subtract from this the half of the number of roots, which is five; the remainder is four. This is the root of the square which you sought for; the square is sixteen, and half the square eight.

Squares and Numbers are equal to Roots; for instance, "a square and twenty-one in numbers are equal to ten roots of the same square." That is to say, what must be the amount of a square, which, when twenty-one dirhams are added to it, becomes equal to the equivalent of ten roots of that square? Solution: Halve the number of the roots; the half is five. Multiply this by itself; the product is twenty-five. Subtract from this the twenty-one which are connected with the square; the remainder is four. Extract its root; it is two. Subtract this from the half of the roots, which is five; the remainder is three. This is the root of the square which you required, and the square is nine. Or you may add the root to the half of the roots; the sum is seven; this is the root of the square which you sought for, and the square itself is forty-nine.

When you meet with an instance which refers you to this case, try its solution by addition, and if that does not serve, then subtraction certainly will. For in this case both addition and subtraction may be employed, which will not answer in any other of the three cases in which the number of the roots must be halved. And know, that, when in a question belonging to this case you have halved the number of the roots and multiplied the half by itself, if the product be less than the number of dirhams connected with the square, then the instance is impossible; but if the product be equal to the dirhams by themselves, then the root of the square is equal to the half of the roots alone, without either addition or subtraction.

In every instance where you have two squares, or more or less, reduce them to one entire square, as I have explained under the first case.

[...]

Geometrical justification

When a square plus twenty-one dirhams are equal to ten roots, we depict the square as a square surface *AD* of unknown sides. Then we join it to a parallelogram, *HB*, whose width, *HN*, is equal to one of the sides of *AD* [see fig. 5.2]. The length of the two surfaces together is equal to the side *HC*. We know its length to be ten numbers since every square has equal sides and angles, and if one of its sides is multiplied by one, this gives the root of the surface, and if by two, two of its roots. When it is declared that the square plus twenty-one equals ten of its roots, we know that the length of the side *HC* equals ten numbers because the side *CD* is a root of the square figure. We divide the line *CH* into two halves by the point *G*. Then you know that line *HG* equals line *GC*, and that line *GT* equals line *CD*. Then we extend line *GT* a distance equal to the difference between line *CG* and line *GT* to make the quadrilateral. The line *TK* equals line *KM* making a quadrilateral *MT* of equal sides and angles. We know that the line *TK* and the other sides equal five. Its surface is twenty-five obtained by the multiplication of one-half the roots by itself, or five by five equals twenty-five. We know that the surface *HB* is the twenty-one that is added to the square. From the surface *HB*, we cut off a piece by line *TK*, one of the sides of the surface *MT*, leaving the surface *TA*. We take from the line *KM* line *KL* which is equal to line *GK*. We know that line *TG* equals line *ML* and that line *LK* cut from line *MK* equals line *KG*. Then the surface *MR* equals surface *TA*. We know that surface *HT* plus surface *MR* equals surface *HB*, or twenty-one. But surface *MT* is twenty-five. And so, we subtract from surface *MT*, surface *HT* and surface *MR*, both [together] equal to twenty-one. We have remaining a small surface *RK*, or twenty-five less twenty-one or 4. Its root, line *RG*, is equal to line *GA*, or two. If we subtract it from line *CG*, which is one-half the roots, there remains line *AC* or three. This is the root of the first square. If it is added to line *GC* which is one-half the roots, it comes to seven, or line *RC*, the root of a larger square. If twenty-one is added to it, the result is ten of its roots. This is the figure:

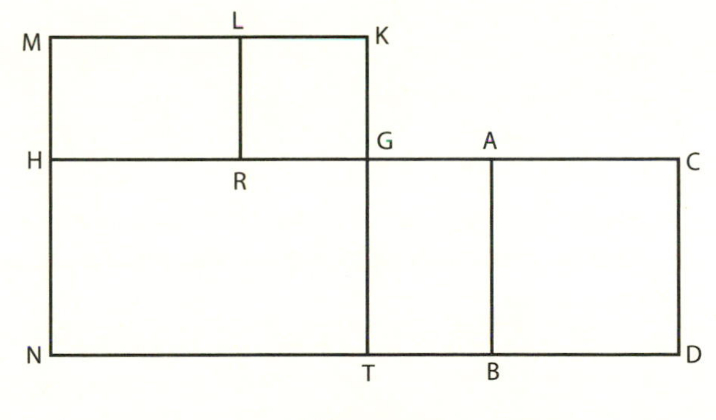

FIGURE 5.2

On multiplication

I shall now teach you how to multiply the unknown numbers, that is to say, the roots, one by the other, if they stand alone, or if numbers are added to them, or if numbers are subtracted from them, or if they are subtracted from numbers; also how to add them one to the other, or how to subtract one from the other.

Whenever one number is to be multiplied by another, the one must be repeated as many times as the other contains units.

If there are greater numbers combined with units to be added to or subtracted from them, then four multiplications are necessary; namely, the greater numbers by the greater numbers, the greater numbers by the units, the units by the greater numbers, and the units by the units.

If the units, combined with the greater numbers, are positive, then the last multiplication is positive; if they are both negative, then the fourth multiplication is likewise positive. But if one of them is positive, and one negative, then the fourth multiplication is negative.

For instance, "ten and one to be multiplied by ten and two." Ten times ten is a hundred; once ten is ten positive; twice ten is twenty positive, and once two is two positive; this altogether makes a hundred and thirty-two.

But if the instance is "ten less one, to be multiplied by ten less one," then ten times ten is a hundred; the negative one by ten is ten negative; the other negative one by ten is likewise ten negative, so that it becomes eighty; but the negative one by the negative one is one positive, and this makes the result eighty-one.

Or if the instance be "ten and two, to be multiplied by ten less one," then ten times ten is a hundred, and the negative one by ten is ten negative; the positive two by ten is twenty positive; this together is a hundred and ten; the positive two by the negative one gives two negative. This makes the product a hundred and eight.

I have explained this, that it might serve as an introduction to the multiplication of unknown sums, when numbers are added to them, or when numbers are subtracted from them, or when they are subtracted from numbers.

If the instance be "ten minus thing to be multiplied by ten minus thing," then ten times ten is a hundred; and minus thing by ten is minus ten things; and again, minus thing by ten is minus ten things. But minus thing multiplied by minus thing is a positive square. The product is therefore a hundred and a square, minus twenty things.

In like manner if the following question be proposed to you: "one dirham minus one-sixth to be multiplied by one dirham minus one-sixth"; that is to say, five sixths by themselves, the product is five and twenty parts of a dirham, which is divided into six and thirty parts, or two-thirds and one-sixth of a sixth. Computation: You multiply one dirham by one dirham, the product is one dirham; then one dirham by minus one-sixth, that is one-sixth negative; then, again, one dirham by minus one-sixth is one-sixth negative; so far, then, the result is two-thirds of a dirham; but there is still minus one-sixth to be multiplied by minus one-sixth, which is one-sixth of a sixth positive; the product is, therefore, two-thirds and one sixth of a sixth.

On legacies

On Capital, and Money lent: A man dies, leaving two sons behind him, and bequeathing one-third of his capital to a stranger. He leaves ten dirhams of property and a claim of ten dirhams upon one of the sons.

Computation: You call the sum which is taken out of the debt thing. Add this to the capital, which is ten dirhams. The sum is ten and thing. Subtract one-third of this, since he has bequeathed one-third of his property, that is, three dirhams and one-third plus one-third of thing. The remainder is six dirhams and two-thirds plus two-thirds of thing. Divide this between the two sons. The portion of each of them is three dirhams and one-third plus one-third of thing. This is equal to the thing which was sought for. Reduce it, by removing one-third from thing, on account of the other third of thing. There remain two-thirds of thing, equal to three dirhams and one-third. It is then only required that you complete the thing, by adding to it as much as one half of the same; accordingly, you add to three and one-third as much as one-half of them. This gives five dirhams, which is the thing that is taken out of the debts.

On another Species of Legacy: A man dies, leaving his mother, his wife, and two brothers and two sisters by the same father and mother with himself; and he bequeaths to a stranger one-ninth of his capital.

Computation: You constitute their shares by taking them out of forty-eight parts. You know that if you take one-ninth from any capital, eight-ninths of it will remain. Add now to the eight-ninths one-eighth of the same, and to the forty-eight also one-eighth of them, namely, six, in order to complete your capital. This gives fifty-four. The person to whom one-ninth is bequeathed receives six out of this, being one-ninth of the whole capital. The remaining forty-eight will be distributed among the heirs, proportionally to their legal shares.

Thābit ibn Qurra (836–901) was from Ḥarrān, in northern Mesopotamia. According to one account, the Banū Mūsā met him as a money changer in his hometown, but were so impressed with his linguistic capabilities that they brought him back to Baghdad to work with them in the House of Wisdom. Thābit's goal in this passage is to show how the procedures of the algebraists (such as al-Khwārizmī) for solving quadratic equations can be represented geometrically and justified by an appeal to Propositions II, 5 and II, 6 of Euclid's *Elements.* Book II of the *Elements* was a fundamental part of the toolkit of Greek and Islamic mathematics. It begins with ten theorems, each asserting that if a straight line is divided into two or more segments in certain ways then certain relationships hold between rectangles and squares formed from the line and these segments. And it ends with applications of these theorems to four advanced geometrical problems.

Particularly important are Propositions 5 and 6 of Book II, which deal with a straight line bisected at one point and (in the case of 5) divided arbitrarily at another point or (in the case of 6) extended by an arbitrary straight line. How these are used in Euclidean geometry is well illustrated by Euclid's use of Proposition 6 to prove (in Prop. 11) a crucial lemma for the construction of a regular pentagon and his use of Prop. 5 to construct (in Prop. 14) a square equal to a given polygon. Thābit's treatise extends the usefulness of these theorems by showing how they may be used to justify the known procedures for solving quadratic equations, a concept not found in the *Elements.*

The order in which Thābit lists the three basic forms of quadratic equations that have both square and linear terms is the same as the order found in al-Khwārizmī's *Algebra.* It is interesting, also, that in the case of $x^2 + c = bx$ he notes, as did al-Khwārizmī, that there are two solutions. Because of the interest in the exact terminology used in the early history of algebra we have tried to make the translation fairly literal. Since the first two proofs show how the two propositions of Euclid referred to above are used, we translate only them and the statement of the third type of equation. We have translated the following excerpt from the German translation by P. Luckey of the Arabic text in the manuscript Aya Sofya 2457,3[27] which we have also consulted in a number of places.

Thābit ibn Qurra on the geometric proof of the correctness of procedures for solving quadratic equations

In the name of God, the Merciful, the Compassionate! A letter by Abū al-Ḥasan Thābit ibn Qurra on the verification of problems in algebra by geometric proofs.

The basic forms, to which most problems of algebra are reducible, are three.

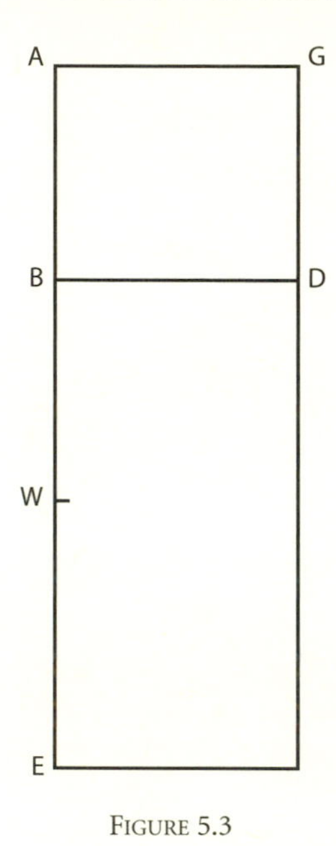

FIGURE 5.3

(1) The first basic form is *māl* and roots[28] equal a number. The way and means of the solution of same, by the sixth [proposition] of Book II of Euclid's book [*The Elements*], is as I now describe: We make [fig. 5.3] the *māl* into the square *ABGD*, we make in *BE* (an amount) of the multiples of the unit by which the lines are measured like the given amount of roots contained,[29] and we complete the area *DE*.

Then the root is obviously *AB* since the *māl* is the square *ABGD*, and this (squaring) is in the domain of the calculation. And the number is like the product of *AB* in the unit by which the lines are measured. Therefore the product of *AB* in the unit by which the lines are measured, is the root on the side [in the domain] of calculation and number. Now however in *BE* is [contained an amount] of these units like the given amount of roots; therefore, the product *AB* in *BE* equals the roots of the problem in the domain of calculation and number. But the product *AB* in *BE* is the area *DE*, since *AB* is like *BD*. Thus, in this way, the area *DE* is equal to the roots of the problem, so the whole area, *GE*, is like the *māl* together with the roots. Now, however, the *māl* together with the roots is like a known number. Thus the area *GE* is known and it is like the product *EA* in *AB*, since *AB* is like *AG*.

Thus the product *EA* in *AB* is known, and the line *BE* is known since the number of its units is known. Thus the matter is reduced to a given geometric problem, to wit: The line *BE* is known. To it is added *AB*, and thereby the product *EA* in *AB* is known. Now it was proved in proposition six of Book II of the *Elements* that, if the line *BE* is halved at the point *W*, the product *EA* in *AB*, together with the square of *BW*, is like the square of *AW*. But the product *EA* in *AB* is known and the square of *BW* is known. Thus, the

[27]We have compared Luckey's translation with the Arabic and have made a number of changes, not because Luckey's translation is wrong but simply because, due to the interest in the vocabulary of early algebra texts, we have decided to translate more literally than Luckey does in some places. Specifically we have translated one standard Arabic expression for "the product of *A* by *B*" as "the product of *A* in *B*," and we have translated *mithl* as "the like of" rather than "equal," or as "its like" rather than "itself."

[28]*Māl* in this context refers to the square of the root. (One ordinary meaning of this Arabic word is "asset," and Luckey translated the Arabic word *māl* by the German word for "asset.") "Roots" refers to what we write as *bx*. When Thābit wants to speak of just the coefficient *b* he refers to "the number of roots."

[29]I.e., make *BE* equal to as many units of length, as there are units in the coefficient of the linear term.

square of *AW* is known, and so *AW* is known.[30] And if *BW*, which is known, is subtracted from it, then *AB* is left as known, and that is the root. And if we multiply it in itself, then the square *ABGD* is known, i.e. the *māl*, and that is what we wanted to show.

This procedure agrees with the procedure of the algebraists in solving this problem. Namely, their taking one half of the number of the roots, is as if we take half of the line *BE*. That they multiply it in itself, is just as if we take the square of half of the line *BE*. That they add the number to the result obtained is like our adding the product *EA* in *AB*. So that, out of all this, the square of the sum of *AB* and half of the line [*BW* = *BE*/2] are put together. That they take the root of the result is as if we say: The sum of *AB* and half the line [*BE*] is known when its square is known.[31] That they subtract from this <half the number of the roots, so that they obtain the remainder, namely the root, is as if we take away half of *BE*>[32] so that the remainder results, as *AB* resulted for us. They multiply it in its like, and thus they determine the *māl*, [just] as we determined from *AB* its square, and that is the *māl*.[33]

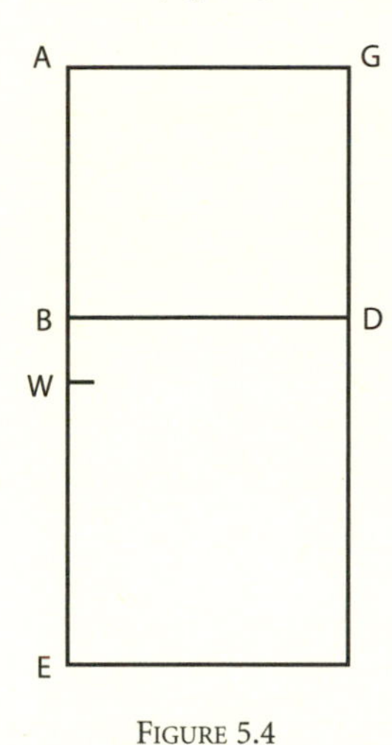

FIGURE 5.4

(2) The second fundamental form is that *māl* and number are equal to roots. The guiding precept in solving that is from the fifth [theorem] of Book II of the writings of Euclid, as I describe it: We make [see figs. 5.4 and 5.5] the *māl* the square *ABGD* and make (it so that there is) in *AE* [a number of] the multiples of the unit, by which the lines will be measured, equal to the given number of roots.

Then obviously *AE* is longer than *AB*, since the roots, which in the domain of calculation are the product *GA* in *AE*, are bigger than the *māl*.[34] We complete the area *GE* and prove, as was said, that it equals the roots according to the ways of the calculation. And when *BG* [the *māl*] is subtracted from it, *DE*, equal to the number, remains. Thus *DE* is known, and it is like the product *AB* in *BE*. And the line *AE* is [also] known. So the matter now amounts to the fact that the line *AE* is known and is divided at *B* so that the product *AB* in *BE* is known. Now it is proven in the fifth [proposition] of Book II of the writing of Euclid, that, if *AE* is halved at *W*, then the product *AB* in *BE* together with the square of *BW* is like the square on *AW*. But *AW* is known, and its square is known, and the product *AB* in *BE* is known. So the square of *BW* is known, and hence *BW* is known. And if it is (fig. 5.4) subtracted from *AW* or (fig. 5.5) added to it, *AB* results as known. And that is the root. And if we multiply it in its like, then *ABGD*, i.e., the *māl*, is known. And that is what we wanted to prove.

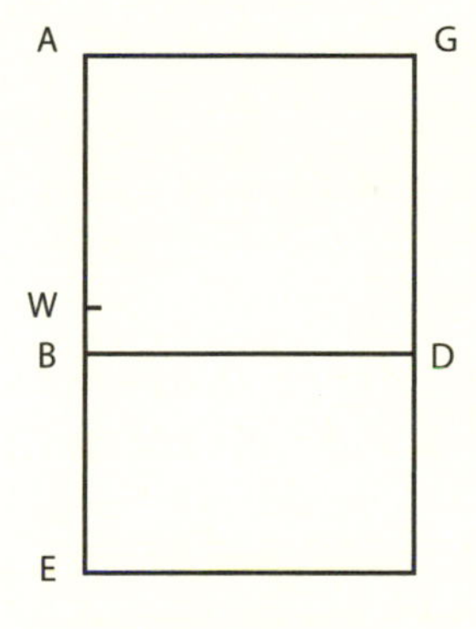

FIGURE 5.5

---

[30]This is a consequence of Prop. 55 of Euclid's *Data*.

[31]The Arabic says the sum is known "if it is a known square." The reference, again, is to Euclid's *Data*, Prop. 55.

[32]The words within the pointed brackets are based on Luckey's restoration of a portion of the text that has been destroyed.

[33]Computing the square of the unknown was part of the solution of the problem in Thābit's time.

[34]Since the roots are equal to *māl* plus a number.

This procedure too agrees with the procedure of the algebraists in the calculation of this problem. Therefore this allows both types of procedure, the [use of] augmentation and deduction in regards to the line *BD*.

(3) The third fundamental form (is): Number and roots are equal to *māl*.

Abū Kāmil ibn Aslam (c. 850–930), "the Egyptian calculator," used justifications similar to those of Thābit ibn Qurra in his own algebra text, a text similar to that of al-Khwārizmī but with many more complicated examples and geometrical applications. Of the two excerpts we present here, the first asks for three positive numbers x, y, and z so that $x+y+z=10$, $z^2=x^2+y^2$, and $xz=y^2$. To solve these he uses an ancient method known as "false position," in which he provisionally sets $x=1$ and, with this assumption, solves the last two equations for provisional values y and z. He then compares the value of the sum of the provisional values with the desired value, 10. Because the three equations are homogeneous he can now find the desired values by multiplying each of the provisional values by 10 divided by their sum.

The first half of the translated text sets up the problem and finds an expression for the value of the least of the three unknowns. The second half, beginning with the words "But, we know," works out the value of this unknown in simpler terms. A sometimes faulty English translation of the whole solution can be found in [Levey 1966, 186–192]. (Levey translated from a Hebrew version of the work by the fifteenth-century Jewish scholar Mordecai Finzi of Mantua.)

Our second problem from Abū Kāmil asks for the computation of the side of an equilateral pentagon inscribed in a square (in such a way that the one of the vertices of the pentagon coincides with a vertex of the square). Considerably later in the tenth century Abū Sahl al-Kūhī, in a selection we give in our section on geometry, constructed such a pentagon, but one with none of its vertices on a vertex of the square.

We have translated both problems from the Arabic text of Abū Kāmil's *Algebra* as found in a facsimile edition of the Arabic text from the Beyazid Library in Istanbul (Qara Mustafa Pasha, No. 379). In the first excerpt, Abū Kāmil's methods are sufficiently clear and his achievement sufficiently impressive from the first part of the solution, where he finds the value of the smallest of the three unknowns, that we have translated only that part. An exposition of Abū Kāmil's solution in modern mathematical symbols may be found in [Berggren 1986, 110–111].[35]

Abū Kāmil on the solution of three equations in three unknowns

And if it is said to you that you will divide ten *dirhams* into three parts so that you will multiply the smallest by its like and the middle by its like and it [the result of putting the two together] is like the largest [multiplied] by its like. And you multiply the smallest by the largest and it will be the like of the middle multiplied by its like. So, you omit for now [the condition about] the ten *dirhams*. Then we say that when you are faced with three different quantities [such that] if you multiply the smallest by its like and the middle by its like, it [the sum] is the like of [the product of] the biggest by its like, and when you multiply the smallest by the biggest it is the like of the middle [multiplied] by its like, then:

[35]Where it is wrongly stated that Abū Kāmil's work is a commentary on the *Algebra* of al-Khwārizmī.

Its rule [for solution] is that we make the smallest [quantity] a *dirham*, the middle a thing, and the largest a square (since, when you multiply the smallest by the largest, it is like the product of the middle by its like). Then we multiply the smallest by its like and the middle by its like and we put them together, so that it (the problem) will be: A square and one *dirham* is equal to a square square (which is like the product of the largest by itself).

And so we do as I described to you and the square will be half a *dirham* and the root of one and a fourth, and it is the largest [of the three quantities]. And the middle quantity is the root of this, and it is the root of the sum of half a *dirham* and the root of one and a fourth.[36] And the small quantity is a *dirham*. And you add the three quantities together and it is a *dirham* and a half and the root of one and a fourth and the root of the sum of half a *dirham* and the root of one and a fourth.

Then you return to [the condition about] the 10 *dirhams*.[37] And we say: We divide 10 *dirhams* by 1½, and the root of 1¼, and the root of the sum of ½ *dirham* and the root of 1¼. And so there results thing.[38] But, we know that whenever we multiply the quotient by the divisor the dividend results. And so we multiply thing by 1½ *dirhams* and the root of 1¼ and the root of the sum of ½ *dirham* and the root of 1¼. And so 1½ things and the root of 1¼ *māl*[39] and the root of the sum of ½ *māl* and the root of 1¼ *māl māl* will be equal to ten *dirhams*.

So subtract 1½ things and the root of 1¼ *māl* from 10 *dirhams* and there remain 10 *dirhams* less 1½ things[40] and less the root of 1¼ *māl* equal to the root of the sum of ½ *māl* and the root of 1¼ *māl māl*. And so multiply 10 *dirhams* less 1½ things[41] and less the root of 1¼ *māl* by its like, so 100 *dirhams* and 3½ *māl* and the root of 11¼ *māl māl* less 30 things and less the root of 500 *māl* will be equal to ½ *māl* and the root of 1¼ *māl māl*. And so complete the 100 *dirhams* by 30 things and the root of 500 *māl* and add the like to ½ *māl* and the root of *māl māl* and ¼ of *māl māl*. And throw away the half *māl* from the 3½ *māl*, and throw away the root of 1¼ of *māl māl* from the root of 11¼ *māl māl*. And so there remain 100 *dirhams* and 3 *māl* and the root of 5 *māl māl*, equal to 30 things and the root of 500 *māl*.

And so refer each of your terms to a [single] *māl*,[42] i.e., we multiply it by ¾ less the root of ¼ and ½ of ⅛. And so you multiply everything you have by ¾ less the root of ¼ and ½ of ⅛. And so *māl* and 75 *dirhams* less the root of 3125 *dirhams* will be equal to 10 things. Thus, halve the things, so it will be 5, and multiply it by its like, and it will be twenty-five. And throw away from it the 75 less the root of 3125. So there remains the root of 3125 less 50. And so the root of that is subtracted from 5, and what remains is the smallest of the three parts whose sum is 10 *dirhams*.

And if you want to know the largest part then make the largest of the three quantities a *dirham*, the smallest a thing, and the middle the root of the thing. Etc.

---

[36]We have modernized the text here and at other appearances of the same phrase. The Arabic could be more literally translated "it is half a *dirham* and the root of one and a fourth, its root taken."

[37]From now on we replace the text's verbal form of the numbers by our usual ciphers.

[38]This is now the value of the "thing" for the original problem.

[39]In this context *māl* refers to the square of the thing.

[40]The text says "one thing and a half."

[41]Again, text says "a thing and a half."

[42] We would say, "Multiply your equation by a number that makes the coefficient of x^2 equal to 1."

Abū Kāmil on Constructing an Equilateral Pentagon in a Given Square

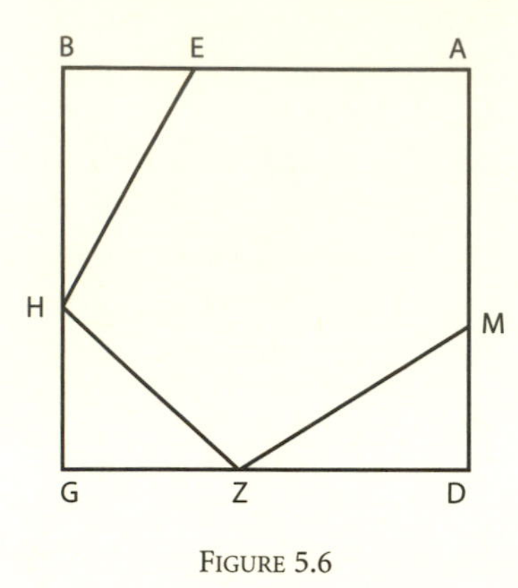

FIGURE 5.6

And if it is said to you: A quadrilateral[43] *ABGD* is given, whose sides are equal, each of whose angles is right, and each of whose sides is equal to 10, in which we shall inscribe an [equilateral] pentagon in this [way].

Let it be the pentagon *AEHZM* [fig. 5.6]. And to know each side of the pentagon you make one of the sides of the pentagon, say the line, *AE*, a thing.[44] Then remains the line *BE*, ten less a thing. And the line *GH* will be the root of one-half *māl*.[45] So there remains the line *HB* [equal to] ten less the root of one-half *māl*. And the line *EB* is ten less a thing. And so we multiply each of the two by its like and we put the two of them [the resulting squares] together and it will be two-hundred *dirhams* and *māl* and half *māl* less twenty things[46] and less the root of two-hundred *māl* equal to *māl*.[47] Solve this[48] along the lines I have shown you and *AE*, a side of the pentagon, is the root of the sum of 200 and the root of 320,000 subtracted from 20 and the root of 200.[49]

All we know about the life of Abū Bakr al-Karajī is that he worked in Baghdad around the year 1000 and that in the first decade of the eleventh century he dedicated a book on algebra to a vizier Fakhr al-Mulk. He appears to have been the first person to develop the algebra of expressions containing arbitrarily high powers of the unknown. We present here an excerpt of his development of the binomial coefficients and "Pascal's triangle." This excerpt is translated from the work of Al-Samaw'al ben Yahyā ben Yahūdā al-Maghribī (1125–1174), since al-Karajī's original work on the subject is no longer extant.

Al-Samaw'al on al-Karajī and binomial coefficients

Let us now recall a principle for knowing the necessary number of multiplications of these degrees by each other, for any number divided into two parts. Al-Karajī said that in order to succeed, we must place "one" on a table [fig. 5.7] and "one" below the first "one," move the [first] "one" into another column, add the [first] "one" to the one below it. We thus obtain "two," which we put below the [transferred] "one," and we place the [second] "one" below two. We have therefore "one," "two," and "one." This shows that for every number composed of two numbers, if

[43]The Arabic word used here (*murabba'*) often means simply "square," but Abū Kāmil's stipulation that it has equal sides and right angles suggests that he is using it in its wider (and literal) sense of a quadrilateral.

[44]Or, as we would say, "the unknown," x.

[45]This Arabic word means "possessions," and in an algebraic context usually signified the square of the unknown—as it does here. Since, by symmetry, the right triangle *HGZ* is isosceles the value for *GH* is clear.

[46]The Arabic says "parts" here.

[47]$x^2 = 200 + 1\frac{1}{2}x^2 - 20x - \sqrt{(200x^2)}$.

[48]Originally the Arabic term here signified removing common terms from the two sides of the equation as the x^2 on the left from the $1\frac{1}{2}x^2$ on the right. Now the sense has become, simply, "solve this."

[49]$AE = 20 + \sqrt{200} - \sqrt{200 + \sqrt{320{,}000}}$.

x	x^2	x^3	x^4	x^5	x^6	x^7	x^8	x^9	x^{10}	x^{11}	x^{12}
1	1	1	1	1	1	1	1	1	1	1	1
1	2	3	4	5	6	7	8	9	10	11	12
	1	3	6	10	15	21	28	36	45	55	66
		1	4	10	20	35	56	84	120	165	220
			1	5	15	35	70	126	210	330	495
				1	6	21	56	126	252	462	792
					1	7	28	84	210	462	924
						1	8	36	120	330	792
							1	9	45	165	495
								1	10	55	220
									1	11	66
										1	12
											1

FIGURE 5.7

we multiply each of them by itself once—since the two extremes are "one" and "one"—and if we multiply each one by the other twice—since the intermediary term is two—we obtain the square of this number. If we then transfer the "one" in the second column into another column, then add "one" [from the second column] to "two" [below it], we obtain "three" to be written under the "one" [in the third column], if we then add "two" (from the second column) to "one" below it, we have "three" which is written under "three," then we write "one" under this "three"; we thus obtain a third column whose numbers are "one," "three," "three," and "one." This teaches us that the cube of any number composed of two numbers is given by the sum of the cube of each of them and three times the product of each of them by the square of the other. If we transfer again "one" from the third column to another column, and if we add "one" [from the third column] to "three" below it, we have "four" which is written under "one"; if we then add "three" to the "three" below it, we obtain 6 which is written under "four"; if we then add the second "three" to the "one" below it, we have "four" which is written under "six," then we place "one" below "four"; the result is another column whose numbers pre "one," "four," 6, "four," and "one." This teaches us that the formation of the square-square of a number consisting of two numbers is given by the square-square of each of them—since we have "one" at each end—then by four times the product of each number by the cube of the other—since "four" follows "one" at either end—since the root multiplied by the cube is the square-square, and lastly by six times the product of the square of each by the square of the other—since the product of the square by the square is a square-square. Then if we transfer "one" from the fourth column into the fifth column, and we add "one" to "four" below it, "four" to "six," "six" to "four" and "four" to "one," then we write down the results under the transferred "one" in the aforesaid manner, and lastly write down last the remaining "one," we have a fifth column whose numbers are "one," 5, "ten," "ten," 5, and "one." This teaches us that for any number divided into two parts, its quadrato-cube is equal to the quadrato-cube of each part—since both ends have "one" and one—and five times the product of each one by the square-square of the other—since "five" succeeds at both ends on either side, and ten times the product of the square of each one by the cube of the other—since "ten" succeeds each five. Each of these terms belongs to the kind of quadrato-cube since the product of the root by the square-square and that of the cube by the square, both give the quadrato-cube; we can thus find the numbers of squares and cubes to the power desired.

In Book X of his *Elements* Euclid assumes that a straight line is given, and any straight line that is commensurable with it he calls *rational.* However, he also calls *rational* any straight line whose square is commensurable with the square of the given line. For example, if the given line is taken as the side of a square, then Euclid, in contrast to modern usage, also calls the diagonal of that square "rational." Lines other than these two kinds of rational lines he calls irrational. Among the irrational magnitudes that Euclid discusses in Book X are those he calls *medial.* These are lines arising as the sides of squares that are equal to a rectangle whose sides are commensurable in square only. Put differently, a medial is a mean proportional between two incommensurable lines whose squares are commensurable (such as lines of lengths 2 and $\sqrt{5}$). Finally (in X, 36) a straight line made from two segments that are commensurable in square only is called *binomial* if the two segments are added to one another and *apotome* if the smaller is taken away from the greater.

In the following selection (Chapter 6 of his *Wonderful Book on Ḥisāb*[50]) al-Karajī introduces the reader to Euclid's terminology and then takes the significant step of transferring the terminology to numbers. In Book X Euclid was concerned with certain classes of line segments as geometrical magnitudes. But al-Karajī showed how mathematicians could speak not just of irrational lines, etc., but of irrational numbers, and he discussed them with the terminology of algebra. In this work, then we see a widening of the concept of number and (especially in the following chapters) the application of algebra to the study of the widened domain of number. That al-Karajī worked in the late tenth century, about 13 centuries after Euclid wrote, gives one some sense of how long it took for new mathematical concepts to develop.

Al-Karajī on Book X of Euclid's Elements

Know that Euclid divided the individual lines into three categories: the first is the rational in length; the second is the rational in power[51] [only] (which reveals itself known in relation to its square;[52] and the third is the medial—which reveals itself in relation to the square square. And they are incommensurable with one other.[53]

---

[50]This Arabic term, meaning "calculation" or "reckoning," came to include not only all of what we call arithmetic but algebra and, as we see here, theoretical considerations relating to numbers.

[51]The phrase in the Arabic text here is the standard Arabic translation of the Greek *dunamei,* which means, literally, "in power" and—in mathematical terms—"with respect to squares," as al-Karajī explains in the following phrase.

[52]Unless otherwise specified, "square" translates the Arabic term *murabbaʿ.*

[53]That is to say, a line in one of the three categories is not commensurable with a line in either of the other categories.

Then he composed[54] two [types of] magnitudes out of the first two categories, calling each of them *binomial*, because each of their two parts maintains its own name. And the first type is that in which the greater of its two parts exceeds the smaller in power by a square whose side is commensurable with the first in length.[55] And it [the first type] is [itself] divided into three types: first, second and third. If its greater term is commensurable with a line rational in length this is the first; if the smaller is commensurable with a line rational in length this is the second; and if neither is commensurable with a line rational in length this is the third.

As for the second type of [binomial], it is that in which the square of the greater of its two terms exceeds the square of the smaller by a square whose side is not commensurable in length with the greater. And this, too, is divided into three types: fourth, fifth, and sixth. Thus the fourth is that [in which] the greater of its two terms is commensurable with a line rational in length. In the fifth the smaller of its two terms is commensurable with a line rational in length. And in the sixth neither of the two is commensurable with a line rational in length.

Next, he [Euclid], in each one of these six types [of binomial], took away the smaller of its two terms from the greater of the two, so there comes from that subtraction the six *apotomes*:[56] first, second, third, fourth, fifth, and sixth. And each of these magnitudes can be divided only by its two terms.[57]

Now I shall show you how these terms can be transferred to numbers, and [shall] add to them, because they are insufficient for *ḥisāb*, due to its extent. So, I say that the single magnitudes are endless [in multitude]. Thus, the first [of them] is the absolutely rational, such as five; the second, the rational in power, such as the root of ten; the third, that which is known in connection with its cube, such as the side[58] of twenty; the fourth, the medial, which makes itself known in connection with the square-square, as the root of the root of ten;[59] the fifth, the side of the square-cube;[60] then, the side of the cube-cube. And according to this [pattern] it will be divided without end.

And the binomial is also divided endlessly. So, as for the first one (the one where the square of the greater of its two terms exceeds the square of the smaller by a square whose side is commensurable in length with the greater), the greater of its two terms is known.[61] An example is ten and the root of seventy-five. And the

---

[54]Here and at the end of this selection "to compose" means to make one line segment out of two by putting them end-to-end to make a single straight line.

[55]That is to say, the square on the greater segment exceeds the square on the smaller segment by an amount equal to a square whose side has a common measure with the side of the greater square.

[56]The Greek word *apotome* refers to cutting off something from something else, and in this technical context means cutting off a smaller segment from a greater, instead of, as in the case of the binomial, extending the one by the other.

[57]He is saying that a magnitude made of two types of magnitudes cannot be made of a pair of magnitudes even one of which is a different type.

[58]"Side" refers to the side of a cube in this context, and so he is speaking of the cube root.

[59]Called "medial" because the fourth root of ten (for example) can be thought of as the mean proportional between $\sqrt{2}$ and $\sqrt{5}$.

[60]The medieval system was to express all exponents, as we call them, in a form representable as $2m + 3n$. Hence "square cube" means an exponent of 5 $(2 + 3)$, not $2 \cdot 3$ (this being "cube cube").

[61]When he starts talking about numbers he begins to use the term *maʿlūm* for known. This term is standard in arguments involving geometrical magnitudes being known (or "given" as the Greeks put it) in a context of geometrical analysis that derives from the sense of the word as used in Euclid's book *Data*.

second is that the smaller of its two terms is known, an example being the root of twelve and three, since nine, the square of three, is less than twelve by three, whose root is equal to half the root of twelve.[62] And the third [type] is that neither of the two of them [the constituent terms] is known, such as the root of twenty and the root of fifteen. And the fourth (that the square of the greater of the two terms exceeds <the square of> the smaller by a square whose side is incommensurable in length with the greater), the greater of its two terms will be known, such as six and the root of twenty-eight. And the fifth, the smaller of the two terms is known, such as the root of twenty-four and four. And the sixth, neither of the two is known, such as the root of twelve and the root of eight.

And we have mentioned the apotomes in which the smaller term is removed from the greater. And so [an example of] the first is three minus the root of eight; the second, the root of eighteen less four; the third, the root of 27 less the root of 24; the fourth, six less the root of 28; the fifth, the root of 20 less 4; and the sixth, the root of 12 less the root of 8.

And as for binomials composed of other than these two magnitudes, they are very many, such as the root of ten and the side of 15. And the ones made of three, four, five, or more single terms are boundless in multitude, such as the root of 10 and the side of 15 and the root of the root of 20, etc.

'Umar al-Khayyāmī (1048–1131), more familiarly known as Omar Khayyam, was born in what is now Iran, shortly after the area was conquered by the Seljuk Turks. He was able during most of his life to enjoy the support of the Seljuk rulers. In fact, he spent many years at the observatory in Isfahan at the head of a group working to reform the calendar. Around 1070, he wrote his great work on algebra, in which he classified the cubic equations and gave solutions to each type through the intersection of appropriately chosen conic sections whenever such solutions exist. We present here three excerpts from that book. In the first excerpt, from the introductory part of his *Algebra*, Omar states the prerequisites for understanding his work. He also reflects on the meaning of fourth powers in algebra, and shows his awareness that he is part of a growing tradition in which the future can bring solutions to problems that are now beyond one's grasp. In the second and third excerpts, Omar solves the case "a cube and number are equal to sides" ($x^3 + c = bx$) and the case "a cube and sides are equal to squares and numbers" ($x^3 + bx = ax^2 + c$).

The Algebra of Omar Khayyām

From the Introduction

You must realize that this treatise will be understood only by those who have mastered Euclid's *Elements* and *Data*[63] as well as the two [first] books of Apollonius's *Conics*. He who lacks knowledge of one of these treatises will not reach

---

[62]He is showing why the number he describes, $12 + \sqrt{3}$, is an example of the second binomial, the point being that $\sqrt{12/2}$ is obviously not commensurable with 12.

[63]Euclid's *Data* provides over 80 theorems stating that if certain aspects of a geometrical situation are given then so are other aspects. A valuable recent translation and study is [Taisbak 2003].

an understanding of this book. However, I have taken considerable pains to ensure that I refer to only these three works in this book.

Solutions in algebra, as one well knows, are done by equations, that is to say, by equating some of these degrees to the others. And, if the algebraist uses the "square-square" in problems of geometry it is only metaphorically, not properly, for it is impossible that the "square-square" be counted as a magnitude. What composes the magnitudes is, first, the one dimensional, i.e., the "root"—or, in relation to its square, the "side." Then [there is] the two dimensional, i.e., the "surface." The square in magnitudes is the squared surface. Finally, the three dimensions, i.e., the solid. The cube in magnitudes is the solid bounded by six squares. But, since there exists no other dimension, the square-square (and, *a fortiori*, anything higher than this) are not part of magnitude. And, if one speaks of the square-square in magnitudes, one says it for the number of their parts, when one measures them, but not for themselves, as measurable magnitudes, which is different.[64] Thus the square-square is not a magnitude, neither essentially nor by virtue of some property it might have. It is not like the odd and the even, which are properties of magnitudes by virtue of the number that divides up the continuity of these.[65]

The equations that one finds in the books of the algebraists between the four geometrical degrees, i.e., absolute numbers, sides, squares and cubes, are three equations between numbers, sides, and squares. As for us we are going to give methods by which one can determine the unknown between these four degrees which (as we said), are the only ones we can find among magnitudes, i.e., number, thing, square, and cube.

[...]

But, as for the proof concerning these kinds [of equations], if the subject of the question is simply a number, neither we nor any of the algebraists have been able to do it except in the three first degrees: number, thing, and *māl*. But perhaps someone else, who will come after us, will know [how to do] it.

And perhaps I shall indicate numerical demonstrations [for the first three degrees] of the sort that one finds in the book of Euclid.[66] For, know that proof of these methods by geometry does not replace proof by numbers if the subject is numbers and not simply magnitudes. Do you not see that Euclid proved things sought having to do with ratios of magnitudes in the fifth book [of the *Elements*], and then redid the proofs of those very same things sought dealing with ratios when the subject was number in the seventh book?

[...]

---

[64]I.e., one may think of a magnitude as measured by some part of it, say a centimeter, and so its fourth power would be measured by fourth powers of a centimeter, i.e. we could assign a number to it.

[65]Thus, a line segment may be said to be odd if, when measured, we find it to be 11 cm. long. But that is ignoring its continuity.

[66]Omar does, in fact, discuss numerical demonstrations of these cases, though often it is only to say, "When one understands the geometric demonstration one will see the numerical demonstration."

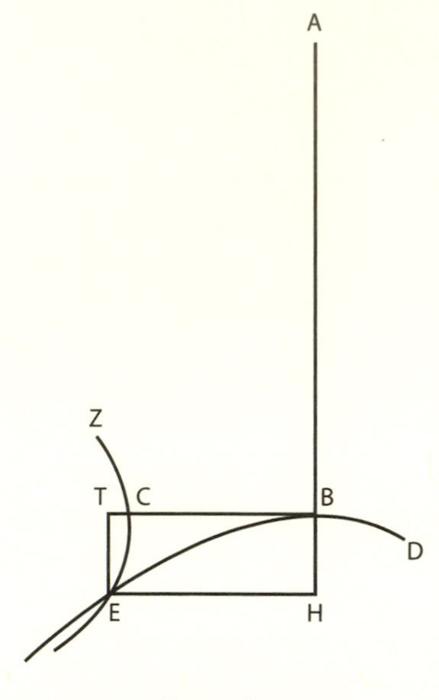

FIGURE 5.8

A cube and a number are equal to sides.

Let the line *AB* (fig. 5.8) be the side of a square equal to the number of the roots, and construct a solid having as its base the square of *AB* and equal to the given number, and let its height *BC* be perpendicular to *AB*. Describe a parabola having as its vertex the point *B* and its axis along the direction *AB* and its parameter *AB*.[67] This is, then, the curve *DBE*, whose position is known. Construct also a second conic, namely, a hyperbola whose vertex is the point *C* and whose axis is along the direction of *BC*. Each one of its two parameters, the perpendicular and the oblique, is equal to *BC*. It is the curve *ECZ*.[68] This hyperbola also is known in position, as was shown by Apollonius in the 58th proposition of his first book. The two conics will either meet or will not meet. If they do not meet, the problem is impossible of solution. If they do meet, they do it tangentially at a point or by intersection at two points.

Suppose they meet at a point and let it be at *E*, whose position is known. Then drop from it two perpendiculars *ET* and *EH* on the two lines *BT* and *BH*. The two perpendiculars are known unerringly in position and magnitude. The line *ET* is an ordinate of the hyperbola. Consequently, the square of *ET* is to the product of *BT* and *TC* as the perpendicular is to the oblique, as was demonstrated by Apollonius in the twentieth proposition of the first book. The two sides, the perpendicular and the oblique, are equal. Then the square *ET* is equal to the product of *BT* and *TC*, and *BT* to *TE* is as *TE* to *TC*. But the square of *EH*, which is equivalent to *BT*, is equal to the product of *BH* and *BA*, as was demonstrated in the second proposition of the first book of the treatise on conics. Consequently, *AB* is to *BT* as *BT* is to *BH*, and as *BH*, which is equal to *ET*, is to *TC*. The four lines, *AB*, *BT*, *ET*, *TC*, then, are in continuous proportion, and the square of *AB*, the first, is to the square of *BT*, the second, as *BT*, the second, is to *TC*, the fourth. The cube of *BT*, then, is equal to the solid whose base is the square *AB* and whose height is *CT*. Let the solid whose base is the square of *AB* and whose height is *BC*, which was made equal to the given number, be added to both. Then the cube *BT* plus the given number is equal to the solid whose base is the square of *AB* and whose height is *BT*, which represents the number of the sides of the cube.

Thus it is shown that this species includes different cases and among its problems are some that are impossible. The species has been solved by means of the properties of the two conics, the parabola and the hyperbola.

[...]

A cube and sides are equal to squares and numbers.

---

[67]In modern terms, the equation of the parabola is $x^2 = \sqrt{b}y$.

[68]In modern terms, the equation of the hyperbola is $y^2 = x^2 - (c/b)x$.

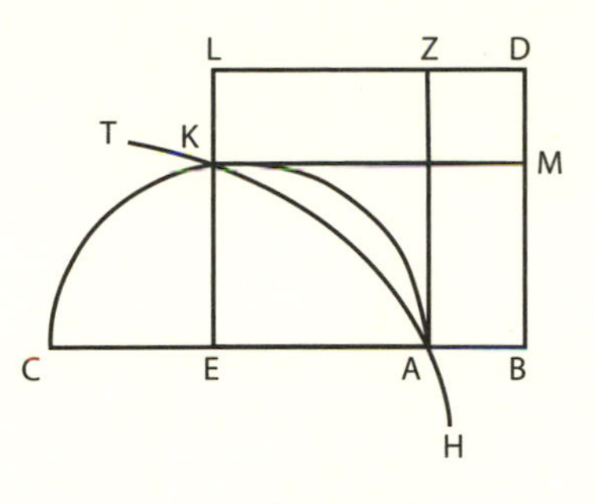

FIGURE 5.9

Let *BC* be equal to the given number of the squares and *BD* equal to the side of a square equal to the number of the sides and let *BD* be perpendicular to *BC*. Construct a solid equal to the given number and having as base the square of *BD* and call its height *S*. The line *S* will be either smaller than, equal to, or greater than *BC*. Let it first be smaller than *BC* (see fig. 5.9).

Take on *BC* the segment *BA* equal to *S* and complete the rectangle *AD*, and describe on *AC* as a diameter a circle *AKC* whose position will be known.[69] Pass through the point *A* a hyperbola with *BD* and *DZ* as asymptotes. It will be the conic *HAK*, which is of known position.[70] *HAK* intersects *AZ* which is tangent to the circle and consequently cuts the circle, because if it falls between the circle and *AZ* we could draw from the point *A* a tangent to the conic, as was expounded by Apollonius in the sixtieth proposition of the Second Book. Then this tangent will either fall between *AZ* and the circle (and that is impossible) or outside *AZ*. In that case *AZ* will be a straight line falling between the conic and the tangent, which is also impossible. Therefore the conic *TAH* does not fall between the circle and *AZ*. Hence, it intersects the latter and it necessarily cuts it in two points. Let *K* be the other point of intersection. The position of *K* will be known. Drop from this point two perpendiculars *KM* and *KE* on *BC* and *ED*. Both of them will be known in position and magnitude. Complete the rectangle *KD*. The rectangle *AD* will be equal to the rectangle *KD*. Subtract the common rectangle *MZ* and add to both the rectangle *AK*. Then *BK* will be equal to *AL*. The sides of these two rectangles, and in the same manner the squares of their sides, will be reciprocally proportional. But the square of *KE* is to the square of *EA* as *EC* is to *EA*. Consequently the square of *BD* is to the square of *BE* as *EC* is to *EA*. Then the solid whose base is the square of *BE* and whose height is *EC* is equal to the solid whose base is the square of *BD* and whose height is *EA*. Let the cube of *BE* be added to both. The solid whose base is the square of *BE* and whose height is *BC* will be equal to the cube of *BE* plus the solid whose base is the square of *BD* and whose height is *EA*. But the first solid is equal to the given number of the squares of the cube *BE*. Let the solid whose base is the square *BD* and whose height is *BA*, which we have made equal to the given number, be added to both solids. Then the cube of *BE* plus the solid whose base is the square of *BD* and whose height is *BE* (which is equal to the given number of the sides of the cube *BE*) will be equal to the given number of the squares of the same plus the given number, which is what was required.

If *S* is equal to *BC*, *BC* will be the required side of the cube. The cube of *BC* is equal to the given number of squares, and the solid whose height is *BC* and whose base is the square of *BD* is equal to the given number and is equal also to the given number of the sides of the cube *BC*. Consequently, the cube *BC* plus the given number of its sides is equal to the given number of its squares plus the given number. Also this case is included in the third species, because the given number of the sides of the cube *BC* is equal to the given number. Therefore, the cube of *BC* plus the given number is equal to the given number of the squares plus the given number of the sides of the cube.

---

[69]In modern terms, the equation of the circle is $y^2 + x^2 - (c/b + a)x - 2\sqrt{b}y + b + ac/b = 0$.

[70]In modern terms, the equation of the hyperbola is $yx - c/\sqrt{b} = 0$.

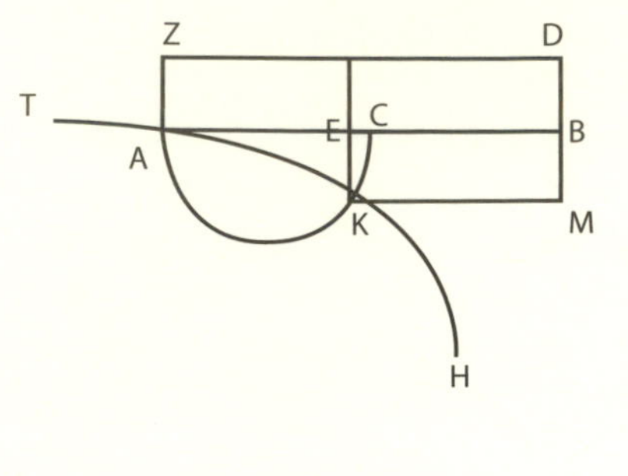

FIGURE 5.10

In case S is greater than BC (fig. 5.10), let BA be equal to S and on AC as a diameter describe a circle. Then the hyperbola that passes through the point A will cut the circle at the point A also at the point K, as we have previously shown. Drop from the point K two perpendiculars KE and KM, as we did in the previous figure. EB will be the side of the required cube, and the proof is similar to the former one. Subtract the common rectangle ED. Then the sides of the rectangles EM and EZ, and also their squares, will be reciprocally proportional, and the proof is the same as the above without any change.

It is evident that this species has varieties of cases and forms, and that one of these forms is included in the third species. It has no impossible problems. It has been solved by means of the properties of the circle and hyperbola.

V. Number Theory

The extract below consists of the introduction and the statements of the 10 propositions that, together with their proofs, make up Thābit ibn Qurra's treatise on amicable numbers. Thābit introduces this topic historically within the context of Pythagorean number theory, and the mathematical context of his treatise is the deductive approach found in the number theoretical books (VII–IX) of Euclid's *Elements.*

The idea of amicable numbers originated in Greek mathematics, and the fourth-century Greek writer Iamblichus defines the concept in his *Commentary on Nicomachos's Introduction to Arithmetic.* (Like Thābit after him, Iamblichus attributes the concept to the Pythagoreans.) Nicomachos's *Introduction to Arithmetic* was written in the first century CE and was intended for the general reader. Nicomachos was evidently successful in his aim of addressing a popular audience, for the book was translated from Greek into Syriac and (twice) into Arabic, the second time by Thābit ibn Qurra. It was, however, not from Nicomachos or Euclid that Thābit learned about amicable numbers, as he himself tells us, but from some other, unspecified, source.

Our translation is based on the French translation of Fr. Woepcke, which we have modified in places on the basis of the Arabic text that A. Saidan published in 1977.

Thābit ibn Qurra on an Easy Way of Finding Amicable Numbers

The way in which Pythagoras and the ancient philosophers of his school used numbers in their doctrine, the predilection they had for this case, and the manner that they used them as illustrations in most theories of the philosophy they wished to establish—these are things known widely among those dealing with the Greeks' works. Among the numbers that these philosophers used in this way, there were mainly two kinds they needed to find. One of these two kinds is well known: the numbers called *perfect.* The other is the numbers that they called *amicable*; numbers which they constructed and mentioned.

As for a *perfect* number, it is known that when we add all its parts,[71] their sum is exactly the number itself. The two kinds related to the perfect number are the abundant number and the deficient number. The *abundant* number is a number such that if we add all its divisors, this sum is larger than the number itself, and the *deficient* number is a number such that if we add all its divisors, this sum is smaller than the number itself. The difference between the number and the sum of all its divisors is called *excess* (when it is an abundant number) and *defect* (when it is a deficient number). As for *amicable* numbers, they are two numbers such that if one adds all the divisors of one of the two, the sum is equal to the other number, which is the *conjugate* of the one whose divisors were added.

Of these two kinds that we just mentioned, it is for the perfect numbers that Nicomachos described the method of finding them, without however giving its demonstration. Euclid, on the contrary, described the method that is used to find them and also carefully gave the demonstration in the arithmetic books of his treatise the *Elements*. He put this theory at the end of his [number theoretic] research, and as the highest degree he achieved, in such a way that some people believed that this theory was his highest goal and the highest level of the research included in these books.

As for amicable numbers, I found that neither of these authors mentioned them, nor did they pay any attention to them. But when I thought about the theory of these numbers, and when I found a proof for them, I did not want, since these numbers had been mentioned the way I just said, to give this demonstration without establishing it with a perfect precision. It is then I who will establish this theory after stating first certain propositions necessary to this subject, and which are the following:

1. No plane number[72] having as sides two prime numbers, is divisible by any number other than these two numbers.

2. Any plane number having as one of its sides a prime number, and as the other a composite number, is divided by its two sides, by each number which divides the composite number, and by each number which results from the multiplication of the prime side into each number which divides the composite number—but by no other number other than the ones just mentioned.

3. Any plane number having as sides two composite numbers is divided by the following numbers: its two sides; each number that divides its sides; each of its sides multiplied by each number which divides the other side; each number obtained through the multiplication of each number which divides one of the two sides by each number that divides the other side, and by no other number except these.

4. In any sequence of numbers following each other in double progression[73] whatever the number of terms, the largest of these numbers is greater than the sum of

---

[71]In *Elements* VII, def. 3, Euclid defines a "part" of a number as a number (less than the given number) that measures (or, as we should say, divides) it. And the definition of perfect number given here is that in *Elements* VII, def. 22.

[72]Thābit is using the terminology of Euclid's *Elements* VII, def. 16. We would describe these as numbers of the form pk, where p and k, whole numbers greater than 1, are what Euclid called the *sides* of such numbers.

[73]The terminology refers to a sequence of whole numbers in which each term after the first is double the preceding term—e.g., 6, 12, 24, 48.

the other numbers by a quantity equal to the smallest [term of the given sequence]; and the same thing happens, when the smallest one is the unit.[74]

5. When one adds a sequence of numbers following each other in double progression starting from the unit, and when one gets a certain sum, then, when one multiplies the greatest of the numbers added by a prime number other than two, the number obtained by this multiplication will be a *perfect number*, if the prime number is equal to the sum obtained [from the sequence].[75] But, if the prime number is smaller than this sum, the product will be an *abundant number*, and if the prime number is larger than the sum, the product will be a *deficient number*. And, the magnitude of its excess, if it is an abundant number—or of its defect, if it is a deficient number—is equal to the difference between the sum and the prime number mentioned above.

6. If one adds a sequence of numbers in double progression starting from the unit, and if one obtains a certain sum, then, if we multiply the largest of the numbers added by a plane number whose two sides are two different prime numbers, other than two, the resulting number will be abundant or deficient. If the plane number is smaller than the sum obtained added to the product of this sum by the sum of the two sides of the plane number, then the obtained number is an abundant number. And the magnitude of its excess is equal to the excess of the sum just mentioned over the plane number. Or, if that plane number is larger than the sum obtained added to the product of this sum by the sum of the two sides of the plane number, then the obtained number is a deficient number. And the magnitude of its defect is equal to the defect of the sum just mentioned relative to the plane number.

7. When one has any four numbers in double progression,[76] with the first one being the smallest, the solid number having for one of its sides the third number, as second side the sum of the third and fourth number, and as third side the sum of the third and second number, will be equal to the solid number having as one side the third number, as second side the fourth number and as third side the sum of the fourth and first number.

8. $4a(2a + 8a + 2 \cdot 4a) = 8a(8a + a)$.[77]

9. $8a(a + 8a - 1) = 4a[\{8a(8a + a) - 1\} - \{8a + 4a - 1\}(2a + 4a - 1)]$.

10. To find amicable numbers, as many as we want, let us take numbers in double progression starting with the unit, inclusively. Let these numbers be *a, b, g, d, e,* and let the sum of *a, b, g, d, e* added together, just as we did with the perfect numbers, be the number *z*.[78] Let us add to number *z* the last of the numbers whose sum we took, namely *e*, and let their sum be number *h*. Then let us subtract from number *z* the number before *e*, which is *d*, and let the remainder be the number *t*. Now, if each of the two numbers *h, t,* is a prime number other than two, it will be what we wish

---

[74]It was necessary to say this since the unit was not regarded as a number, but only as the generator of the numbers.

[75]This is simply *Elements* IX, 36, Euclid's famous sufficient condition for a number to be perfect.

[76]Thābit means here a four-term sequence of the form a, $2a$, $4a$, $8a$.

[77]This proposition and the next are summarized in modern notation.

[78]In what is called the *abjad* system, the letters of the Arabic alphabet represented numbers, and the letters used here appear in numerical order in that system. So the Arabic reader would not see these as just a jumble of letters. But they represent terms of an arbitrary double progression.

for. But, if not, we continue the sequence of numbers[79] that were added up, until we arrive at numbers [*h* and *t*] that are prime.

And so, now, let the two numbers *h*, *t* be prime numbers, other than two. Let us multiply them and let the result of the product be *k*. Let us multiply *k* by the last of the numbers that we summed, namely *e*; and let the result of the product be number *l*. This is one of the (two) numbers that we shall attend to, and we keep it in mind. Then let us add the number following number *e* in the sequence of numbers in double progression, which is the number *w*, to the one preceding the next-to-last number of those that one summed (the number *g*). Let the sum of these two numbers be number *m*. Then let the result of the multiplication of number *m* by number *w* be number *n*, from which we subtract "one." Let the remainder be the number *s*. If *s* is a prime number, then it is what we wish for. If not, we continue the sequence of numbers summed, up to where this number [*s*] becomes a prime number as well as the other [two] numbers [*h* and *t*].

So, let *s* be a prime number, let us multiply it by number *e*; and let the result of this multiplication be number *o*. I say that the numbers *l*, *o* are amicable numbers.

Abū Manṣūr al-Bāghdadī (d. 1027) was a jurist and theologian who wrote on geometry and arithmetic. (Such a seemingly wide combination of areas of serious interest would not have been unusual for a medieval Muslim scholar.) In his treatise on arithmetic, the *Completion on Reckoning*, he discussed a notion related to amicable numbers, that of "balanced numbers."

Al-Bāghdadī on Balanced Numbers

"Balanced" are any two numbers such that the sum of the aliquot parts of the one equals the sum of the aliquot parts of the other. Then, when we have a given number and wish to know two numbers such that [the sum of] the aliquot parts of each one equals this given number [we proceed as follows]: we subtract 1 from the given number; next, we divide the remainder into two prime numbers, and we divide it once again into two other prime numbers, and we continue to divide it similarly into two prime numbers, as long as this may be carried out. Next, we multiply the two parts from the first division one by the other, and we multiply the two parts from the second division one by the other, and we do the same with the two parts of the third division, of the fourth, and so on. The, the results from these multiplications are such that [the sums of] the aliquot parts of each are equal to that given number.

Example [with the number] 57: We wish to know two numbers such that [the sum of] the aliquot parts of each give 57. So, we subtract 1 from it, and there remains 56. We divide this into two prime parts, [say] 3 and 53, and we multiply the one by the other; the result is 159, and summing its aliquot parts gives 57. Next, we again divide 56 into two other prime parts, [say] 13 and 43, and we multiply the one by the other; then, [the sum of] the aliquot parts of the result is 57.

Hence any two numbers such that [the sum of] their aliquot parts is the same number are balanced. [But], whenever one subtracts from the [given] number 1 and the remainder is in no way divisible into two unidentical primes, or is so in only one way, the said rule will fail.

---

[79]He means the sequence of numbers "in double progression."

VI. Geometry

Islamic mathematicians were the true heirs of the Greeks in their careful treatment of geometrical questions. In particular, they were interested in geometrical constructions, both Euclidean constructions using straightedge and compass and more advanced constructions, using conic sections. In this section, we present several excerpts on constructions as well as excerpts dealing with other geometric questions, including centers of gravity, the volume of a paraboloid of revolution, and the theory of parallels. But, since Islamic mathematicians were also interested in the practical applications of geometry in such areas as architecture and building decoration, we also present some excerpts in dealing with practical geometry.

VI.a. Theoretical geometry

The first three excerpts are taken from the work of Ibrāhīm ibn Sinān (908–946), the grandson of Thābit ibn Qurra. In the first two selections Ibn Sinān makes reference to the method of analysis and synthesis, which goes back to ancient Greece and provides geometers with an approach to finding solutions to problems. For example, one might be given a circle and be required to inscribe a square in it. In analyzing this problem one would start not only with the center and radius—which one knows since the circle is given—but with the assumption that the problem has been solved, i.e., that one had constructed an inscribed square. The assumption that the problem has been solved then gives one some additional things to work with, in the case in question these being the vertices of the square we are trying to construct and the line segments forming its sides. One then proceeds to draw consequences from the enlarged set of givens. At some point in this chain of consequences, if the analysis is successful, one derives something that one can establish independently of the assumption that the problem has been solved, or a construction that one can do independently of that assumption. And, now that one has established a path, as it were, from the enlarged set of givens to something that can be established with only the original set of givens one shows that the path can be reversed so that, in the end, one shows how the thing sought can be derived from the original set of givens.

In the case of our example, the first consequence might be that the triangle formed by radii to two adjacent vertices of the inscribed square and the side joining them is an isosceles triangle, and the next might be that the angles in this triangle opposite the radii are equal. After further reasoning one deduces that the radii to two consecutive vertices of the square are perpendicular to each other. And this is something one can construct with only the given circle and the usual tools of geometry.

In the selection from his treatise on drawing the parabola Ibn Sinān's essential tool is something demonstrated within the proof of *Elements* II, 14, namely that if GD is a perpendicular to the diameter AGB of semicircle ADB, then a square whose side is GD is equal to the rectangle whose perpendicular sides are equal to AG and GB.

Ibrāhīm Ibn Sinān's Treatise on the Drawing of the Three Sections

Ibrāhīm Ibn Sinān said: in the propositions that we are going to mention here, we will not follow the way we followed in the thirteen treatises, because what we did in those was through analysis and synthesis but here we will use, in what we will present, demonstration only. [...]

Apollonius showed in his book *Conics* that the cone has sections, which he named. Among these, there are three sections, which are surrounded by convex lines that cannot be superposed on the circle: the parabola, the hyperbola, the ellipse. He then explained how to generate each one and what is to be found on them: diameters and parallel lines, and everything happening in each section.

When we found that it was difficult to draw these three sections with a compass or other instruments, we tried hard to draw numerous points to which man can add as many as he wants and such that these points will be on one of the three sections. Everything [that was] determined that way proved how these sections, along with others, are generated from the circle.

Let us start with the parabola, and let our intention be to find points whose number can be increased as much as wanted and that will be on the parabola. Let us draw a line on which we will put two points, namely *B* and *D* [fig. 5.11]. Let us draw from *D* the perpendicular *DE* and put a point which is outside *BD*, namely *G*; let us construct on *BG* half a circle which intersects *ED* at [point] *E*. It is clear that the square of *ED* is equal to the product of *BD* and *DG*. Likewise if we draw from the [point] *E* the line *EH* parallel to *BD* and from *G* a line parallel to *DE*, namely *GH*, then the area *GDHE* is a parallelogram, and the square of *GH* is then equal to the square of *DE*. Likewise, the square of *GH* is equal to the product of *GD* and *DB*. Let *DI* be in the prolongation of *DE* and equal to the line *DB*, it is then clear that the point *H* is on the parabola that goes through the point *D*, its axis being *DG* and its right side [parameter] *DI*. Indeed, we constructed, as Apollonius proved, a parabola that goes through *D* and such that its right side is *DI* and its diameter *DG* (and its extension). And the *ordinates* meet the line *DG* at right angles.

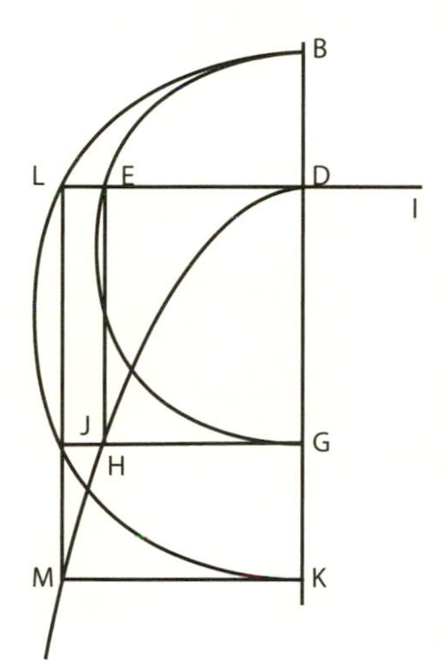

FIGURE 5.11

If one argues that it does not go through the point *H*, we reply, "Then let it go through point *J*." The square of *JG* will be equal to the product of *GD* and the right side *DI*. Indeed, *JG* is one *ordinate*—the square of *GH* was also one—hence *GH* would be equal to *GJ*; which is impossible. Consequently, the section goes through the point *H*, thus the section *ADHC*.

Likewise, if we put a point on the line *DG* and its prolongation and we construct on this point half a circle as the half circle *BLK* that meets the line *ED* at the point *L*, then the square of *DL* is equal to the product of *BD* and *DK*. Let us draw *LM* parallel to *DK* and *KM* parallel to *DL*; then because of the parallelism of the sides and the equality of those that are opposite, the square of *KM* is equal to the product of *KD* and *DB*, which is *DI*. Consequently, the point *M* is on the section *ADH*. Thus we found, through this construction, two points *H* and *M* that are on a parabola and we proved why it is so. However, we must proceed in this through a bare-bones construction so that from it the procedure will be certain.

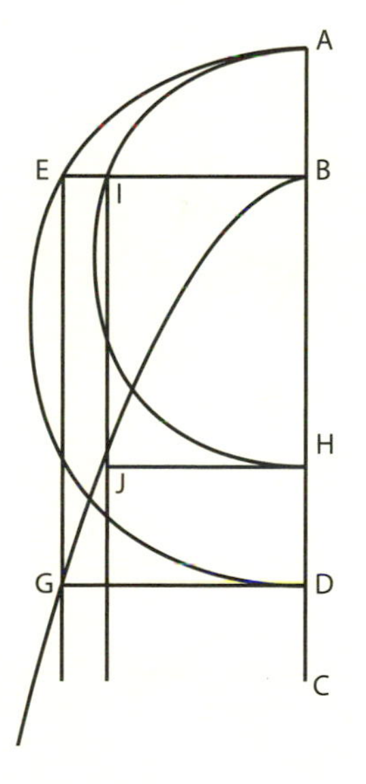

FIGURE 5.12

We then say: let us draw [the line] *ABC* and put on *BC* and its prolongation as many points as we want [fig. 5.12]. Let the point *D* be one of them. Let us construct on the line *DA* a half circle *AED*; let us draw *BE* perpendicular to *AB*; let us draw from *E* a line parallel to *AB* and from *D* a line parallel to *BE* which intersect at [the point] *G*. Likewise, let us put another point, namely *H*; let us construct on the line *ABH* a half circle *AIH* that intersects *BE* at *I*. Let us draw from *H* and *I* two lines following the same example—from *I* a line parallel to the line *ABH* and from *H* a line parallel to *BIE*—that intersect at *J*. We will always proceed this way.

It is then obvious that the points *J*, *B*, *G* are on a parabola with *BC* as an axis and the remaining diameters can be found if we draw from any point found on this section a line parallel to *BC*.

Ibrāhīm Ibn Sinān's Anthology of Problems

Problem 19

Further here is the analysis of another problem related to this art: *two circles AB and CD are given, [and] we want to draw a circle that is tangent to them and that is such that the straight line drawn between the two points of tangency is known.*

Let us suppose that [has been done], and let the circle be *BDJ*; let the center of the circle *AB* be point *E*, the center of circle *DC* be point *G*, the center of the circle .*BDJ* point *I*, and let the two points of tangency be *B* and *D* [fig. 5.13]. Line *EBI* is a straight line and line *GDI* is a straight line. Let line *BD* be equal to the known line; we extend it up to points *A* and *C*, we join *AE* and *GC*. The ratio between *AE* and *EB* is equal to the ratio between *ID* and *IB*; line *ABD* is a straight line, angle *E* is then equal to angle *I*, and, hence, *AE* is parallel to *DG*; likewise *BE* is parallel to *GC*.

If EB is equal to GD and BI is equal to ID: the ratio between *EB* and *BI* is then equal to the ratio between *DG* and *DI* and line *EG*, known, is parallel to *BD*, known; the ratio between the two is then known and the ratio between *EI* and *IB* is known; *EB* is known, *EI* is then known; likewise, *GI* is known. The two circles that we drew having as respective centers *E* and *G* and as respective distances *EI* and *GI* are known and their intersection point [i.e, that of *EI* and *BI*], *I*, is known.

But if the two circles are not equal: we need to use what we proved, the parallelism of line *EB* and line *GC*. Since they are parallel, without being equal, then line *BC* intersects line *EG*; it intersects it at *H*. The ratio between *GC*, [which is] known, and *EB*, [also] known, becomes equal to the ratio between *GH* and *EH*. But, *EG* is known, so point *H* is known. If we draw line *HK*, tangent to circle *AB*, it is known, and, hence, its square is known. And it is equal to the product of *BH* and *AH*.

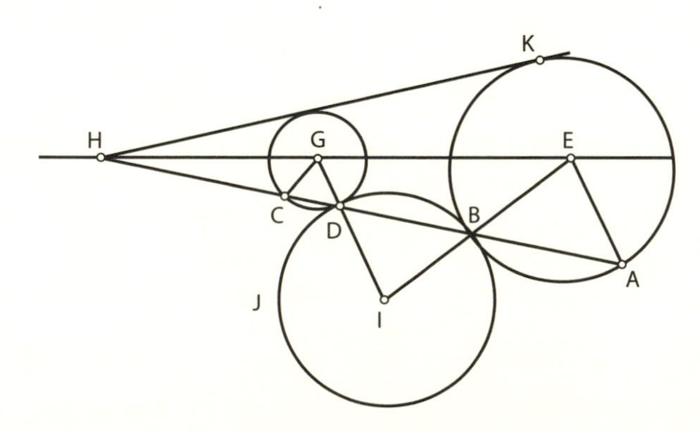

FIGURE 5.13

But the known ratio between *DG* and *EA* is equal to the ratio between *DH* and *HA*, [so] the ratio between *DH* and *HA* is then known. And the product of *BH* and *AH* is known so the product of *BH* and *DH* is known. Line *DB* is known, hence line *BH* is known, and point *H* is known, [so] the circle drawn with *H* as its center and *HB* as its radius is known. Let it be circle *LBM*, so this circle is then known. Circle *AB* is [also] known, [so] their intersection point, *B*, is then known. Line *DH* becomes known, hence point *D* is known. Consequently, lines *EBI* and *GDI* have given positions.

That is what we wanted to construct.

Problem 25

Let chords AB, AC and AD be positioned in circle ABCD; the chords are known, angle BAC is equal to angle CAD, [and] we want to know the diameter [fig. 5.14].

We join *BD*, intersecting *AC* at *E*. The ratio between *BA* and *AD* is then equal to the ratio between *BE* and *ED*, so the ratio between *ED* and *EB* is known. But, this ratio is the ratio between the product of *DE* and *EB* and the square of *EB*. And, the product of *DE* and *EB* is equal to the product of *AE* and *EC*, hence the ratio between the product of *AE* and *EC* and the square of *BE* is known.

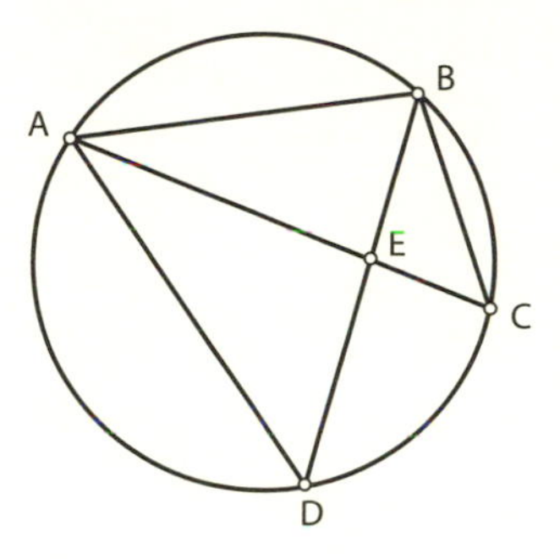

FIGURE 5.14

Besides, since angle *BAC* is equal to angle *CAD*, and since angle *CAD* is equal to angle *CBE*, for they are on the same segment of the circle, angle *CAB* is equal to angle *CBE*. And angle *ACB* is common to both triangles *ABC* and *ECB*, so the remaining angle *CEB* is equal to angle *ABC*. Consequently triangles *ABC* and *ECB* are similar; so the ratio between *BC* and *BE* is equal to the known ratio between *CA* and *AB*. And so the ratio between *BC* and *BE* is known; hence the ratio between the square of *BC* and the square of *EB* is known. But the ratio between the square of *BE* and the surface of *AE* by *EC* is known, so the ratio between the surface of *AE* by *EC* and the square of *BC* is known.

Besides, since triangle *ABC* is similar to triangle *ECB*, the square of *BC* is equal to the product of *AC* and *CE*. Hence the ratio between the product of *AC* and *CE* and the product of *AE* and *EC* is known. And this ratio is the ratio between *CA*, known, and *AE*, so *AE* is known.

Each of the two straight lines *AE* and *EC* is then known, the product of the two is then known and it is equal to the product of *BE* and *ED*. The ratio between *BE* and *ED* is known, hence each one of them is known. Hence triangle *ABD* became known for all its sides. The circle that is circumscribed about it then has known diameter, and this is what we wanted to do.

The next selection contains three extracts from the correspondence of two men active in the second half of the tenth century. The one, Abū Sahl al-Kūhī, came from Tabaristan, just south of the Caspian Sea, and was one of the most talented geometers in all of medieval Islam. He worked as a mathematician and astronomer during the last half of the tenth century and received support from at least three kings of the Būyid Dynasty, which ruled eastern Iraq and most of Iran at that time. The other correspondent, Abū Isḥāq al-Ṣābī, was a court official for two of the Būyid kings. (He was also a witness at the observations of the sun that Abū Sahl directed in Shiraz, and made an astrolabe for himself.) But he fell into disfavor with a third king, ʿAḍud al-Dawla, who became so angry with him that he wanted him trampled by elephants! (Cooler heads prevailed, however, and he died of natural causes in 994.)

We begin with Abū Sahl writing to al-Ṣābī of results he has obtained on the centers of gravity of plane and solid figures. (He knows of a treatise of Archimedes on the topic, but which of Archimedes' writings this was is uncertain.) He has, as he says, proved two beautiful lemmas (which he states) and five of six results on the centers of gravity of three plane figures (the triangle, the parabolic section, and a semicircle) and three solids (the cone, the paraboloid of revolution, and the hemisphere). And he has conjectured the sixth based on a very convincing pattern he perceived in the other five. Unfortunately, as al-Kūhī shows, the sixth result implies that the value of π is 3 1/9!

In al-Ṣābī's reply we see an enthusiastic amateur questioning, ever so politely, a result which he clearly believes to be wrong, and gently hinting that a value so widely at variance with Archimedes' bounds for π must prove his conjecture false. In the last selection, al-Kūhī questions the authenticity of Archimedes' treatise *On the Quadrature of the Circle* and questions the validity of mathematics that uses approximations. These three selections, taken together, provide us a rare glimpse of dialogue between two medieval scientists as they debate solutions to problems and the validity of approximation in mathematics.

Al-Kūhī and Al-Ṣābī on Centers of Gravity

From Al-Kūhī

As for the centers of gravity, there remains of them a slight thing until six consecutive chapters are finished, four of them which I have done here in Basra and two there in Baghdad. Then there will be done after that, God the Exalted willing, a chapter in which there are problems about centers of gravity and it will be the best of the chapters and biggest of them. Next chapters will be appended to this chapter about the matters of the centers of gravities, three or four [chapters about] liquid and non-liquid bodies. After all this [introductory detail I turn to] the first of these chapters, God willing. As for the four chapters which I did here, all of them point to an arrangement of deeds of the Creator, to Whom belong might and majesty, like the things that are in Archimedes' *Sphere and Cylinder*. Are we not astonished at the occurrence of the sphere's happening [to be] two thirds of the cylinder according to what he described and proved, and at the paraboloid, that it is its (the cylinder's) half as Thābit ibn Qurra proved, and at the cone that it is its third as the ancients made plain? And so we found in the matters of centers of gravity an arrangement more wonderful than that. Among them (our discoveries) is that if we rotate the semicircle *ABG*, whose center is *D*, along with the parabola whose axis is *BD*, and along with the rectilineal triangle *ABG* around the line *BD* perpendicular to the line *AG*, so that there results from the rotation of the semicircle a hemisphere and from the parabola the paraboloid and from the triangle a cone, then the cone is a solid for the triangle as the paraboloid is for the parabola and the hemisphere for the semicircle [fig. 5.15].

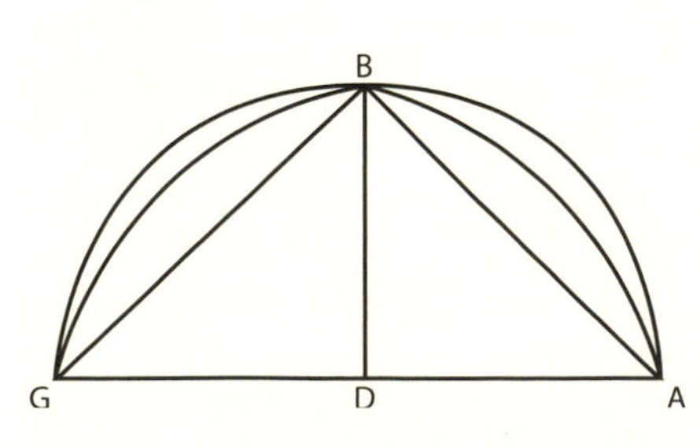

FIGURE 5.15

We found the arrangement for these solids, as regards centers of gravity, more wonderful than the corresponding arrangement for measurement. As for the centers of gravity of these solids, the center of gravity of the solid of the triangle, I mean the cone, falls according to the ratio of one to four [parts] of the diameter, and of the paraboloid according to the ratio of two to six, and of the [hemi]sphere according to the ratio of three to eight. Now the planar [figures]. As for the center of gravity of the triangle [it occurs] according to the ratio of one to three, and of the parabola according to the ratio of two to five, and of the semicircle according to the ratio of three to seven. And this is a chart for that:

Center of gravity

of the triangle according to one of three	(1 of 3)
and of the cone according to one of four	(1 of 4)
and of the parabola according to two of five	(2 of 5)
and of the paraboloid according to two of six	(2 of 6)
and of the semicircle according to three of seven	(3 of 7)
and of the hemisphere according to three of eight	(3 of 8)

This is the natural sequence in which we found the centers of gravity and we were amazed at the occurrence of this arrangement. Next, one theorem is a premise for the discovery of the center of gravity of a section of the circumference of the circle, and it has premises also. And it (the first-mentioned theorem) is that if there are two sectors of two concentric circles, and the ratio of the radius of one of the two to the radius of the other is the ratio of 3 to 2, and they are similar [sectors], then the center of gravity of the arc of the smaller of the two and the center of gravity of the surface of the larger of the two is one. An example of that: If the point *E*, is the center of two circles *AB*, *GD* and the line *EBD* is straight, and similarly the line *EAG*, and the ratio of the line *GE* to the line *EA* is equal to the ratio of three to two and the center of gravity of the arc *AB* is the point *Z*, then the point *Z* is the center of gravity of the surface *GED*, the sector, also [fig. 5.16].

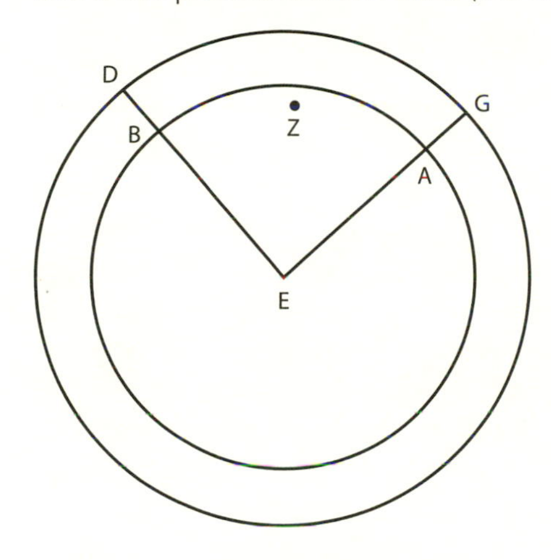

FIGURE 5.16

I proved that in the chapter whose premier theorem I sent in the writing that I wrote before that. In that chapter is also another theorem and it is the proof that the ratio of every arc to its chord in the circle is equal to the ratio of the radius of that circle to the line between the center of the circle and the center of gravity of the arc, and it is a good [and] very remarkable theorem since that straight line is always equal to an arc of the circumference of the circle. This is a wonderful thing that has not been mentioned. The example of that: If the arc *AEB* is part of the circumference of the circle whose center is *G* and whose radius is *GE*, and the center of gravity of the arc *AEB* is the point *D*, I say that the ratio of the arc *AEB* to its chord *AB* is always

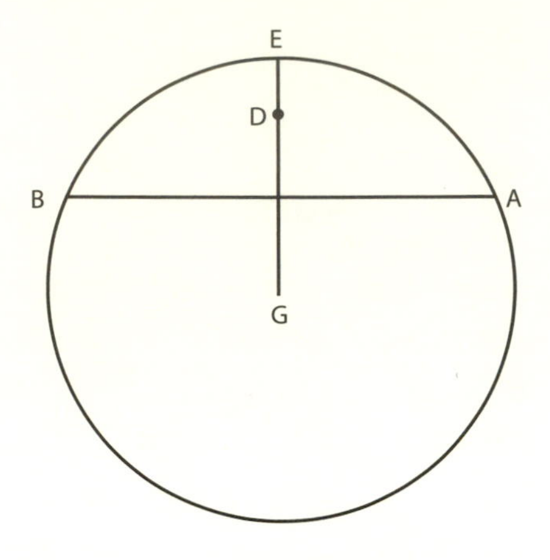

FIGURE 5.17

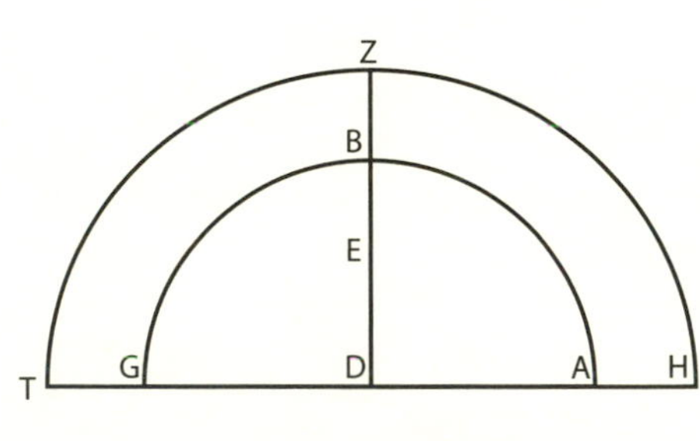

FIGURE 5.18

equal to the ratio of the radius, *EG*, of the circle to the line *GD*, which it is between the center of the circle *G* and the center of gravity of the arc *AEB*, i.e., the point *D*, and I proved that the straight line *GD* is always equal to a curved line from the circumference of the circle [fig. 5.17]. All of these things are from the totality of the theorems of the *Book of Centers of Gravity*.

As for the ratio of the diameter to the circumference [being] equal to the ratio of a number to a number it is not part of it (the totality), but when these facts from [the science of] the centers of gravity occurred to us we looked into the matter of the diameter [compared] with the circumference and we postulated the semicircle *ABG* of the circle whose center is *D*, and the line *DB* perpendicular to the diameter *AG*, and the point *E*, the center of gravity of the arc *ABG*, and we knew that the ratio of the arc *ABG* to the line *AG*, its chord, is equal to the ratio of the radius of the circle, the line *BD*, to the line *DE*, since we proved that concerning every section of the circumference of a circle so particularly for the semicircle [fig. 5.18].

Then we made the ratio of the line *DZ* to the line *DB* equal to the ratio of three to two, and we drew about the center *D* and with distance *DZ* the circle *HZT* so that the point *E* is the center of gravity of the surface of the semicircle *HZT* also, as we said. Since the ratio of the line *BD* to the line *DE* is equal to the ratio of the arc *ABG* to the line *AG* it is equal to the ratio of half the arc *ABG*, the arc *BG*, to half the line *AG*, the line *BD*, since the point *D* is the center of the circle, and the ratio of the arc *GB* to the line *BD* is equal to the ratio of the line *BD* to the line *DE*. Thus the product of the arc *BG* by the line *DE* is equal to the square of the line *BD*.

Furthermore, since the ratio of the line *ZD* to the line *DB* is as the ratio of three to two, the ratio of the square of the line *ZD* to the square of the line *DB* is as the ratio of nine to four. Also, the square of *BD* is equal to the product of the arc *BG* by the line *DE*, and [so] the ratio of the square of *ZD* to the product of the arc *BG* by the line *DE* is as the ratio of 9 to 4.

Further, the ratio of the product of the arc *BG* by the line *DE* to the product of the arc *BG* by the line *ZD* is as the ratio of four to nine and a third since the two of them are in the ratio of three to seven, and *ex aequali* the ratio of the square of the line *ZD* to the product of the arc *BG* by the line *ZD* will be as the ratio of nine to nine and a third. And the ratio of the product of the arc *BG* by the line *ZD* to the product of the arc *ZT* by the line *ZD* is as the ratio of the arc *BG* to the arc *ZT* since the line *ZD* is a common altitude to the two of them. Also the ratio of the arc *BG* to the arc *ZT* is as the ratio of the line *BD* to the line *ZD* since the two arcs *BG*, *ZT* are similar and *D* is the center of the circle.

Further, the ratio of the line *BD* to the line *DZ* is as the ratio of two to three, so the ratio of the product of the arc *BG* by the line *ZD* to the arc *ZT* by the line *DZ* is as the ratio of two to three, which is as the ratio of nine and a third to fourteen. *Ex aequali* the ratio of the square of the line *ZD* to the product of the line *ZD* by the arc *ZT* will be as the ratio of the line *ZD* to the arc *ZT*, and so the ratio of the line *ZD* to the arc *ZT* is as the ratio of nine to fourteen and the ratio of twice the arc *ZT*, the arc *HZT*, to twice *DZ*, the line *HT*, is as nine to fourteen.

But the line *HT* is the diameter of the circle and the arc *HZT* is half its circumference, so the ratio of the diameter to the whole circumference is as the ratio of nine to twenty-eight, and it is as the ratio of a number to a number. So the circumference turns out to be three likenesses of the diameter and a ninth.

Thus, when that occurred to us we looked into the work of Archimedes in which he says that the circumference of the circle is less than three likenesses of its diameter and ten parts of seventy parts, I mean the seventh, and this is conformable to our dictum, not contradictory to it, since the ninth is less than the seventh without doubt. However, he also says in it that it (the circumference) is greater than three likenesses and ten parts of seventy-one parts, and this is not conformable unless he says ninety-one parts in place of seventy-one parts so that it is conformable.[80] And according to us it is no more [serious] than that. We do not suspect any of the ancients [of being anything] but beautiful and good, so how much more, Archimedes, the leader in that.

From Al-Ṣābī

When I had finished my writing up to this place, his composition came to me from Abū-l Mufaḍḍal al-Anṣārī. I understood it and was reassured by his health that it showed. I praised God for it and asked Him for its continuance and its increase. I studied the discovery he mentioned he had deduced of the center of gravity of the triangle and its solid the cone, and the discovery of the center of the parabola and its solid and the discovery of the center [of gravity] of the semicircle and its solid, the hemisphere, and I marveled greatly at it and at the matter that appeared in it, like something natural in the necessity of that succession and arrangement that he explained and showed. And my excitement doubled at the magnificent gift in it, and by God, he never saw the like of himself and we cannot hope to see his like. It pains me that the [present] time and its people do not give him his due. Who will grant to me that some town will bring him and me together in the remainder of my life, so that I might occupy my time with him and with benefit from him?

Then I attended to all he mentioned about the discovery of the center of gravity of a section of the circumference of a circle and the proof that the ratio of every arc to its chord is as the ratio of half the diameter of the circle in which it is to the line joining the center of the circle and the center of gravity of that arc, and the

---

[80]In Arabic, the difference between the word for "ninth" and "seventh" is two dots above or one dot below words whose forms are otherwise identical, and confusion with these dots was common in manuscripts.

discovery of the ratio between the diameter of the circle and its circumference, that it is as the ratio of a number to a number, I mean the ratio of nine to twenty eight.

This is astonishing, but more astonishing is the difference between it and what Archimedes sets forth, and he (Abū Sahl) has mentioned that there are fundamental principles and premises for all that, on which he built. For this reason I am greatly in suspense that I might know their consequences and what precedes them, in order that I might share his certainty as well as the elimination of doubts and the objections of the adversaries. I hope that he will be so good and comply with my wishes for that, one by one, [by] the excellence of his grace, and his benefit to me will be complete.

Thus he knows, may God support him, that these things are magnificent and of great significance, and when the geometers hear of them they will wonder and desire to know the true state of affairs. Confidence in these things will not come about except with the security of the premises from doubts and objections, and so for this reason I ask that he send them to me, in order, from beginning to end.

From Al-Kūhī

He said, "More astonishing than that is the discrepancy between it and what Archimedes has set forth." I understand that, but the matter is not as he fancied, and [indeed] there never was disagreement between any of us and Archimedes. That cannot be, because disagreement between scholars, when it occurs, is in the things of which their knowledge is through opinion, dogma, and likelihood as it was between Aristotle and Galen and other physicists in the matters of the psyche, and the conditions of the faculties, and similar things. As for the things that refer to geometry and arithmetic they name "erroneous" that which is erroneous and [identify] negligence where negligence occurs, for they know that disagreement quickly disappears when they look into it.

Error in arithmetic, when it occurs, is not strange, nor does it indicate the inferiority of him who commits it. Do you not see that Ptolemy, in spite of his assertion of the superiority of Hipparchos, and his (Hipparchos) advancement and his accomplishment and his fairness and his preference for that which is true, says in his book *The Almagest* that there occurred in the calculation of Hipparchos an error, he does not intend by that to denigrate him, and how could he when he (Hipparchos) is in his opinion the most superior person?

Similarly [for] the calculation of the ratio of the diameter to the circumference by Archimedes. Although it is not clear to us that this calculation has missed the mark, in my opinion, it is [merely] attributed to Archimedes and does not befit him, so if we were to say it is not his work, it would be nearer [the truth] and it would be nearer to praising [him] than our saying that it was his, because the opinion [expressed therein] is not his opinion, the purpose is not his purpose.

Archimedes never had as his purpose any such thing as this, neither in *The Sphere and the Cylinder*, nor in *The Lemmas*, nor in other books of his. We never saw mention of this [book] anywhere in his writings, like the mention of the area of the parabola in the preface to the *Book of the Sphere and the Cylinder*, along with the mention of some of his [other] derivations. Neither did he use that in some theorem, since it is clear that that method does not lead to the truth at all. Rather it is an approximation and his purpose always is to discover knowledge of things exactly—not approximately, such as the discovery of the ratio between the square and the parabola, and between the circle and the spherical surface, and between the sphere and the cylinder and the cone, etc., an exact discovery and not approximate.

Also, this [method of] calculation [in *Measurement of the Circle*] although it is impossible that it is ever exact, is not [even] very fine, since its calculator did not revert to chords finer than the chord of four degrees less one-fourth, and this is very coarse compared to what is done in the *Almagest* since Ptolemy reverts to a chord approximating half a degree, which is much finer than this by quite a bit.

Because of [all] this I say this [work] is [merely] attributed to Archimedes and this calculation is, in my opinion, not like a work of Archimedes. And it is not a work of proficient calculators or astronomers either, so that if we were to attribute this derivation to one of our companions, he would not be pleased by it, much less glory in it.

What a difference between this way of measurement and the way of Archimedes, Thābit, and Ibrāhīm ibn Sinān, by which it appeared that the two [parabolic] sections *ABD*, *BGE* are a third of the triangle *ABG*, exactly rather than only approximately [fig. 5.19]! By the method of adding the triangles by calculation it is not possible that we are led to truth at all, since there are infinitely many triangles, and between the method which is only by approximation (and of which one does not accept that it is ever exact) and the method which is only exact, and which cannot be approximate at all, there is no analogy.

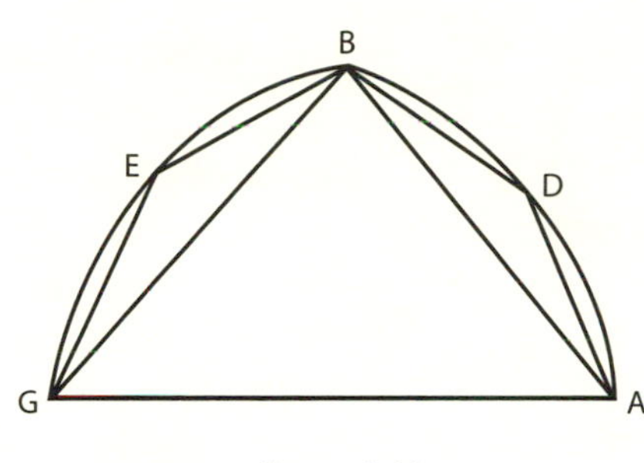

FIGURE 5.19

Al-Kūhī[81] found, in scientific instruments, a source of mathematical problems, and wrote on these in his treatise on the perfect compass (for drawing conic sections) and, in the selection below, on astrolabes. The planispheric astrolabe, of Greek origin, uses what is now called stereographic projection of a sphere onto the plane of a great circle to represent points on the sphere by points on the plane. In stereographic projection a fixed point *A* (see fig. 5.20) is chosen on the sphere and it is joined by a line to the point, *D*, of the sphere diametrically opposite it. A plane perpendicular to this line, and cutting the sphere, *EZ*, is then chosen. (In the design of the astrolabe the sphere represents the celestial, starry sphere, *A* represents the south pole (when the astrolabe is for use in the northern hemisphere), and the plane *EZ* is chosen to contain the center of the sphere, so it represents the equator plane.) If *B*, *G* are points other than *A* on the sphere they are projected to the points where the lines *AB* and *AG* cut the equator plane. (In fig. 5.20, *B* and *G* are both shown on the other side of the plane *EZ* from

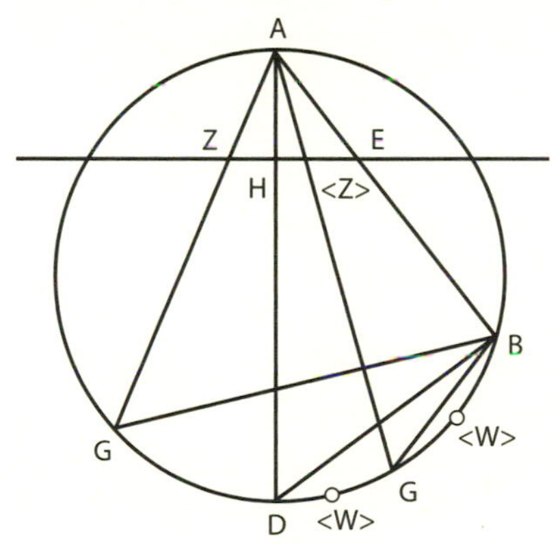

FIGURE 5.20

[81]On this author see our introduction to his correspondence on centers of gravity.

A, but they could equally well be on the same side, in which case the lines *AB* and *AG* would cut the equator plane outside of the sphere.) In this way, every point other than *A* corresponds to a unique point on the plane *EZ*, so that the sphere, less the point *A*, is mapped in a one-to-one fashion onto the plane.

In designing an astrolabe for use in the northern hemisphere one limits the mapping to points on the sphere north of the Tropic of Capricorn, since that circle marks the southern limit of the sun. Clearly the circle determined by plane *EZ* is mapped to itself and circles parallel to it are mapped onto circles concentric with the image of the plane *EZ*.

Al-Kūhī divides astrolabes into two types. One is three-dimensional, and he calls it "spherical" because it provides a direct model of the celestial sphere in the form of a spherical surface that can be rotated over an inner spherical surface. The others he calls "superficial" (i.e., having to do with surfaces other than the sphere). And he recognizes that the planispheric astrolabe is but one of many possible examples of this type. For example, he writes of astrolabes in which the sphere is mapped onto conical or cylindrical surfaces. (In this latter type, for example, a sphere is mapped onto a cylindrical surface tangent to the sphere along the whole of the equator by perpendiculars from the sphere onto the surface.)

The stars of the sphere are mapped onto one plate of the astrolabe, called (using the medieval Latin term) the *rete*, and the horizon and other celestial circles are mapped onto a second plate, called (in Latin, once again) the *tympan*. Both plates are circles of the same size, and—indeed—stereographic projection maps circles on the sphere onto circles (or, in case of circles passing though *A*, onto straight lines) on the plane and angles onto equal angles.

In the selection presented here al-Kūhī discusses the matters we have mentioned above and gives a proof that stereographic projection preserves circles. (That it also preserves angles is not mentioned—let alone proved—in the medieval literature.)

Al-Kūhī on the Astrolabe

In the name of God, the Merciful, the Compassionate. *Treatise on the Construction of the Astrolabe with Proof*, composed by Abū Sahl Wījan b. Rustam al-Kūhī.

Chapter One: On the Description of the Astrolabe and the Drawings on It

The astrolabe is an instrument on which is drawn the likeness of two surfaces, one of which moves with a circular motion on top of the other, while the other is fixed. If it is spherical, then [it is composed of] two spheres, and if it is superficial, then [it is composed of] two surfaces. The engravings inscribed on both are [derived] from astronomy to the degree that the operations require, to the extent that workmanship can master and that sense perception can achieve.

And the goal in making it is its beauty and its accuracy, to the extent that people at the time choose. Its accuracy [can be judged] by comparing it to true things. But as for beauty, it must have the proper measure, mass, thickness, thinness, smoothness and similar things, in addition to the surfaces, the lines and points engraved and the writing [inscribed on it]. As for accuracy, the surfaces and lines on them must be correct, and the placing of the lines and points on the surfaces correct also.

And the correct placing in the likeness [of the celestial sphere] used in the astrolabe is in two parts: the first is [of things whose position is] known exactly, and the other is [of things] known from observation. What is known exactly is on the fixed surface, and what is known by observation is on the movable surface. It is incumbent on the astrolabe maker, therefore, that he be cognizant of what is known exactly by the masters of this craft and of what is known by observation, and that he knows as much of what is found by observation as is necessary for this instrument, or he should refer in this matter to the observations of a competent observer, who will do the measuring for him. And so, if he intends to make a spherical astrolabe let him [simply] make a copy of what he (the observer) has measured for him, as we have said.

But if he wants to make a superficial one then, he needs to know [as well] the science of flattening the sphere. And the sphere is flattened onto different kinds of surfaces from different positions, but one of the two surfaces [derived] from it moves on the other with the movement of the sphere only if it is [flattened] on conical or cylindrical surfaces, or [on] something similar to them of those [figures] that have an axis (whose axis is the axis of the sphere) or [on] planes to which the axis of the sphere is perpendicular.[82]

In the case of conical or cylindrical surfaces, flattening circles that are on the sphere results in intersections of cone and cylinder, or of two cones or of two cylinders, for the sphere may be flattened in two ways, one of them cylindrical and the other conical. The cylindrical is that which, from circles on the sphere, forms cylinders, whose axes are parallel, [standing] on the surface onto which the sphere is flattened, and from the lines and points which are on that sphere forms surfaces and lines which are parallel to those axes [that fall] on that surface. And the conical is that which, from the circles that are on the sphere, forms cones with a common vertex, whose bases are on the surface onto which the sphere is flattened. And all the surfaces, lines and points that are on the sphere are opposite to all the surfaces, lines and points of that surface upon which the sphere is flattened, each one opposite to the other, and to one point, the vertex of the cones.

Now, if the flattening of the sphere is cylindrical, whose axis is parallel to the axis of the sphere, or conical, whose vertex is on the axis but not on the pole of the sphere, then two surfaces of the sphere coincide, one on the other, in that [image] surface. And circles on the sphere, other than those to which the axis of the sphere is perpendicular, fall not as circles in that surface, but as conic sections or something else. Further, if the flattening is onto anything other than a plane to which the axis of the sphere is perpendicular, then it is possible that all of the drawings on the sphere, or some of them, will not be flattened.

And if the flattening is cylindrical, whose axis is not parallel to the axis of the sphere, or conical, whose vertex is not on the axis, then its flattening takes other forms different from what we have described, but we have opted not to mention them because this is not our aim in this treatise.

On the other hand, if it is conical, whose vertex is on the pole of the sphere, and its flattening is on a plane to which the axis of the sphere is perpendicular, then none of these conditions obtains and every part of the sphere flattens. No two surfaces on the sphere will coincide, one upon the other, in that [image] plane and the circles

---

[82]Al-Kūhī is saying that the sphere should be projected onto a surface which has an axis of rotational symmetry coinciding with a diameter of the sphere.

which are on the sphere will in that plane not be conic sections, but circles or, if the flattening is of the circles that pass through that same pole, straight lines.

Now we want to prove that when the vertex of the cones is at the pole of the sphere then the flattening of the circles on the sphere results in circles or straight lines on the plane perpendicular to the axis of the sphere, and it will form straight lines from the circles that pass through that same pole.

Example: The circle *ABGD* is the circle that passes through the axis of the sphere, *AD*, and through the pole of the [given] circle on the sphere *W*, and the plane *EZ* is a plane to which the axis of the sphere, *AD*, is perpendicular. And let it be imagined that this plane is [seen] in thickness, and line *EZ* is the intersection of the two of the [planes *ABGD* and *EZ*]. We join lines *AB*, *AG*, *BD*, and *BG*, and triangle *ABG* will be perpendicular to the plane *EZ* and [to] the [given] circle, whose diameter is *BG*, because it is on the plane which passes through the axis of the sphere and the pole of that circle [fig. 5.20].

Now, because angle *ABD* is equal to angle *AHE*, since each one of them is a right angle, and angle *DAB* is common [to both] in this example, triangle *ABD* is similar to triangle *AHE*, and so angle *ADB* is equal to angle *AEH*. And angle *AGB* is equal to angle *ADB*, since both subtend arc *AB*. Hence angle *AGB* is equal to angle *AEH*, and so triangle *ABG* is similar to triangle *AZE*. But Apollonius has proved in the *Conics* [I, 5], that if this is so and one of the two planes to which the triangle is perpendicular is a circle, then the other plane will also be a circle. But the one of these two planes in the sphere is a circle, so, if we were to assume that it is in the cone whose vertex is point *A*, the remaining plane, *EZ*, in the cone will be a circle and the line *EZ* is the diameter of that circle. And Apollonius has proved also [*Conics* I, 9] that the planes other than these two planes, or those which are parallel to them, are not circles in the cone, but conic sections.

As for the circles that pass through that same pole [*A*], because that [pole] is on the plane upon which that circle lies and the surface upon which the sphere flattens is planar, and because the common section of two planes, which is the flattening of that circle, is a straight line, straight lines result from the circles that pass through that same pole. Therefore, the flattening of the circles that are on the sphere will be circles and straight lines on the plane perpendicular to the axis of the sphere, and that is what we wanted to show.

Abū al-Wafā'al-Būzjānī (940–997) served in the court of the Būyid dynasty in Baghdad in the second half of the tenth century. The selections presented here are from his *Book on the Geometrical Constructions Necessary to the Artisan*, which deals with the following kinds of construction problems (among others): constructing regular polygons that satisfy certain other conditions (e.g., that the side be a given segment); inscribing (or circumscribing) circles in (or around) various polygons, and conversely; inscribing a polygon such as a triangle in one of a different type, such as a square; dividing triangles, quadrilaterals and circles so that the parts have a given relationship to one another; dividing the surface of a sphere into regions of given shapes; decomposing a square into a given number of squares or constructing a square equal to a given number of squares.

We have prepared these selections primarily on the basis of H. Suter's German translation of an Arabic text of the treatise, but we have also used F. Woepcke's earlier French translation of a student's notes in Persian, and we have occasionally modified what we have found in these sources on the basis of copies of the relevant parts of an Arabic text that Prof. Jan Hogendijk has kindly sent to us. Like the Persian version, this latter Arabic version has no proofs. So our translation of the proof of the validity of the procedure in Problem 5[83] is from Suter's translation.

The statement of Problem 3 specifically requires that the construction be done with what is now called a "rusty compass," i.e. (as the author puts it), a compass with a given opening that does not change. This additional requirement is also as good as stated in Problem 4 as well. Note, too, that the last construction of the five chosen here is of the same type as the problem of constructing an equilateral pentagon in a given square, a problem whose solution by Abū Kāmil we presented earlier.

Abū al-Wafā'al-Būzjānī on Five Geometrical Constructions

1. If we have a certain angular magnitude and want to test whether its angle is a right angle or not then we may do it with the previous construction. For example, to test whether angle *BAG* is exactly a right angle or not we take [fig. 5.21] on line *AG* the point *D* and proceed as in the previous way: Let *Z* be a point where the two circles [around *A* and *D* with radius *AD*] intersect and draw *DZ* and extend it to *E* so that *EZ* is equal to *DZ*. Then we observe the point *E*. If it falls on the line *AB* then the angle *A* is a right angle and the angular magnitude is correct. If it falls outside of line *AB* [i.e. to the right of it] then angle *A* is an acute angle. And if it falls inside line *AB* then angle *A* is an obtuse angle.

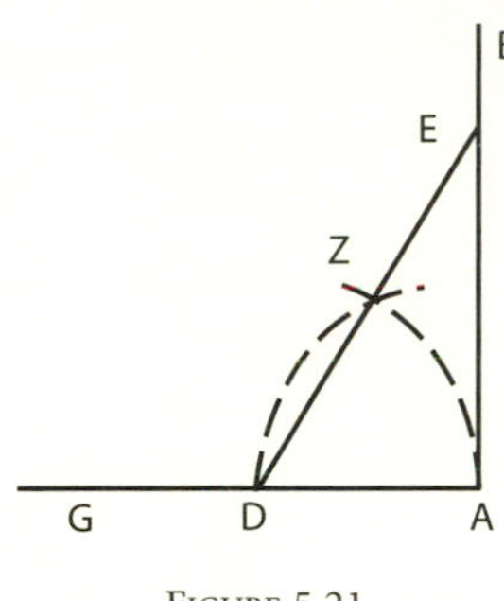

FIGURE 5.21

2. Another way of testing the correctness of an angular magnitude. The artisans proceed from a different consideration to test for the correctness of an angular magnitude, namely the following:

If they want to know [fig. 5.22] whether angle *A* of an angular magnitude is a right angle they measure off with the compass on *AE*, beginning with *A*, three equal parts (of an arbitrary magnitude), and they measure off four parts of the same magnitude on *AB*, starting at *A*. They connect the two endpoints [on lines *AE* and *AB*]. If the line joining them is equal to 5 such parts then angle *A* is exactly a right angle. But if it is greater, then angle *A* is obtuse, and if it is less, then it [angle *A*] is acute.

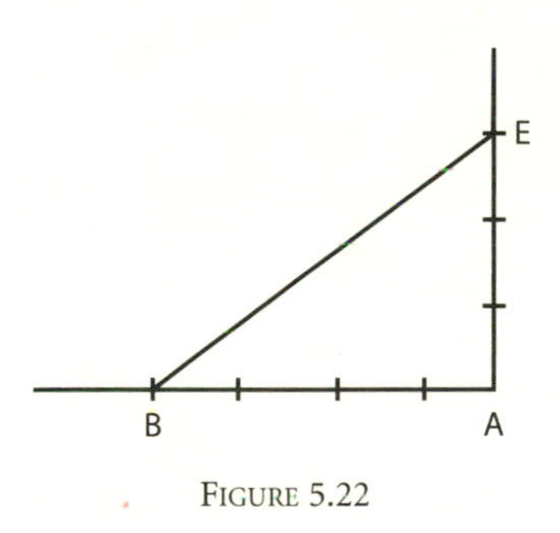

FIGURE 5.22

3. And if he [someone] says, how do we make on line *AB* an equilateral [and equiangular] pentagon on the condition that the compass opening is equal to *AB* and does not change. We erect [fig. 5.23] on line *AB* a line *BG* per-

[83]Our numbering is solely for convenience here and in no way reflects the numbers these problems have in the various manuscripts of the treatise.

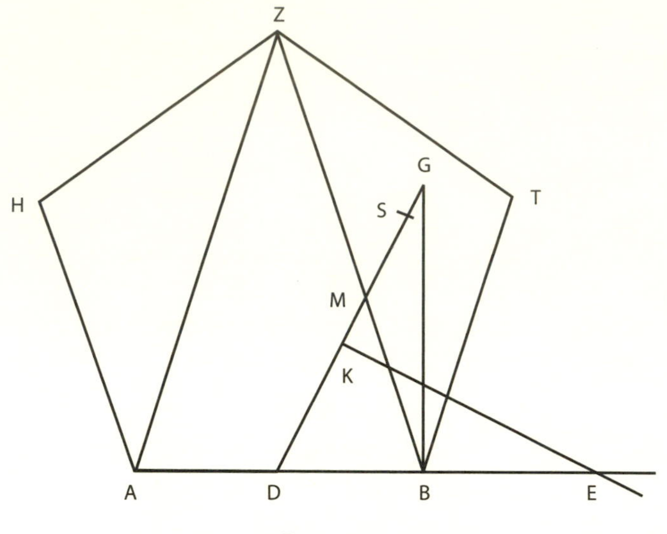

Figure 5.23

pendicular and equal to it, and we divide the line *AB* into two halves at point *D*. And we join *DG* and we make *DS* equal to *AB* and divide it into two halves at point *K*. And we produce from point *K* a perpendicular *KE* that meets the line *AB* at the point *E*. Then we make each one of the two points *A*, *E* a center, and with distance *AB* [we make] two arcs that cut each other at point *M*. We join *BM* and produce it in a straight line to point *Z* so that *MZ* is equal to *AB*. And we join *AZ*. And we make the two points *A*, *Z* centers and with distance *AB* we mark the point *H*. And we make the two points *Z*, *B* centers and with distance *AB* [we make] a mark *T*. And we join the lines *AH*, *HZ*, *ZT*, *TE*. Then the pentagon *ABTZH* will be equilateral and equiangular. And here is the drawing.

4. And if he says how do we make in a circle *ABGD* a quadrilateral with equal sides.[84] We open a compass whose magnitude is the magnitude of the radius of the circle *ABGD* [fig. 5.24], and we draw the diameter *AG*. Then we make point *A* a center and with the [given] opening of the compass we [draw a circle to] make two points *E*, *Z* [on the circle *ABGD*], and we join *EZ* [which intersects *AG* at *I*]. Then we make point *G* a center and with distance *AE*, two marks *H* and *T* [, and we join *HT*, which intersects *AG* at *K*]. And we join the two lines *KZ*, *IT* which cut each other at point *M*. And we join [a line] between the point *M* and the center [*S*], and we produce it in a straight line to two points *B*, *D*. And we join the lines *AB*, *BG*, *GD*, and *DA*. Then the quadrilateral *ABGD* will be equilateral. And this is the drawing.

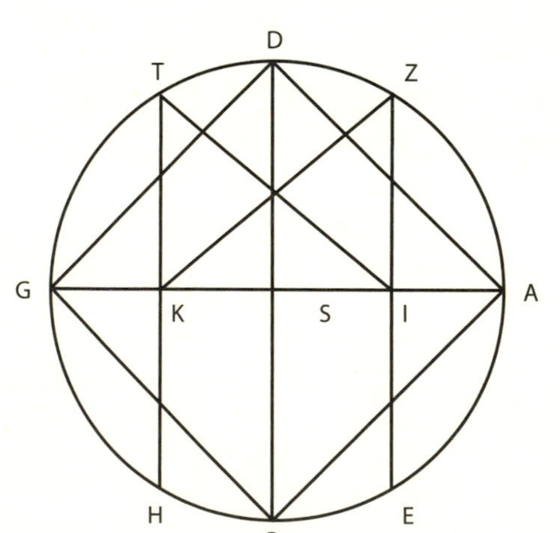

Figure 5.24

[84]Such an inscribed quadrilateral must be a square, of course.

5. To inscribe an equilateral triangle in a square so that its angles touch its sides. We make the square *ABGD* and we prolong the line *DG* [fig. 5.25] to *E* to make *DE* equal to *GD*. And on *GE* we construct the semicircle *EAG*. Then with *G* as center and distance *DG* [as radius] we mark *Z* [on the semicircle *EAD*], and with *E* as center and distance *EZ* we mark *H* [on line *DG*]. And we make *AT* equal to *GH*. Then we draw *BT*, *BH*, and *HT*. Then triangle *BTH* is equilateral and it was constructed in the square *ABGD*.

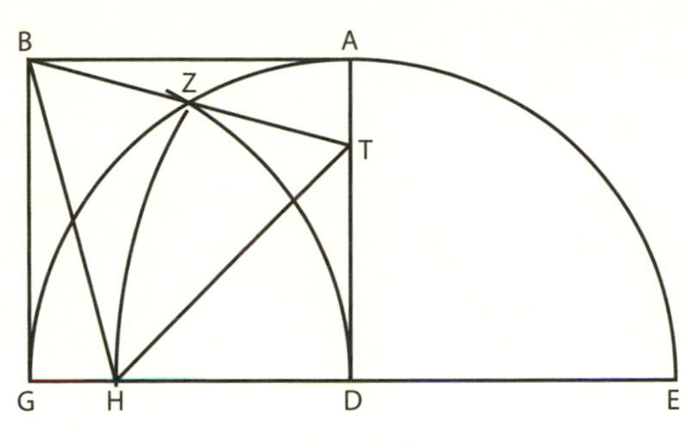

FIGURE 5.25

Proof: For this a lemma is necessary. Let the circle *ABG* be given [fig. 5.26], *AG* its diameter, and *GB* the side of the regular hexagon [inscribed in the circle]. One draws the line *AB*. It is the side of the regular triangle [inscribed in the circle]. One prolongs it to *D*, such that *AD* = *AG*. Since, now, arc *AB* is twice arc *BG*, angle *AGB* is twice angle *BAG*. The two together, however, are a right angle, so angle *AGB* is two-thirds a right angle, and *BAG* is one-third a right angle. Therefore [since the angles of triangle *AGD* are two right angles] there remain for angles *AGD* and *ADG* one and two-thirds of a right angle. However, since they are equally large, each of them is one-half and one-third of a right angle. Since angle *AGB* is two-thirds of a right angle, angle *BGD* is one-sixth of a right angle. Thus, it is proven that if, in a right-angled triangle [*DBG*], the two sides about the right angle are the side of a hexagon [of the circumcircle [of triangle *ABG*]] and the excess of the diameter [of the circumcircle] over the side of the inscribed regular triangle, then the smaller angle of this triangle [*DBG*] is one sixth of a right angle.

Now, in our previous figure [fig. 5.25], in triangle *BHG* the angle *G* is a right angle, formed by *BG* (i.e., the side of the hexagon in circle *EAG*) and *GH* (the

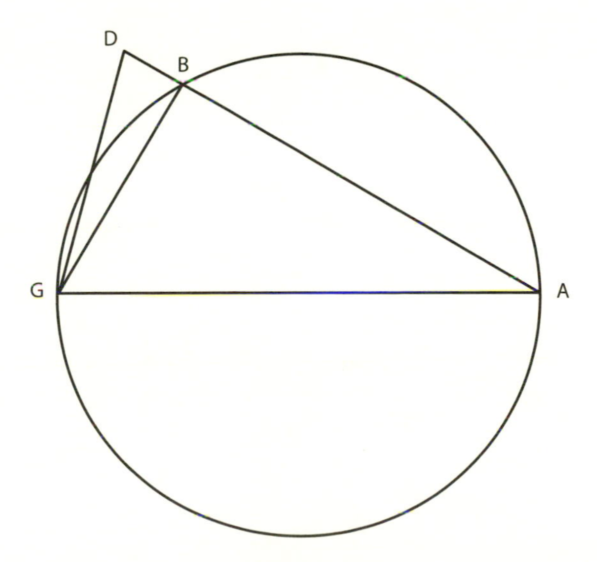

FIGURE 5.26

excess of diameter *EG* over *EZ*, which is the side of the triangle). Thus, by the lemma, angle *GBH* is equal to one-sixth of a right angle. Since the lines *AB* and *AT* are equal to the lines *BG* and *GH*, and the angles at *A* and *G* are each equal to a right angle, angle *ABT* is also equal to one-sixth of a right angle. Thus, angle *TBH* is equal to two-thirds of a right angle. However, the two angles *BTH* and *BHT* are equal to each other, so each of them is equal to two-thirds of a right angle. For that reason, the three sides, *BT*, *BH*, and *TH* are equally large and the triangle *TBH* is equilateral.

Aḥmad ibn Muḥammad ibn ʿAbd al-Jalīl al-Sijzī (late tenth century) was a younger contemporary of al-Kūhī. He was a prolific writer, with at least 50 surviving works to his credit, and an active correspondent. Although he was not of the first rank mathematically, he stressed the need to substitute constructions by conic sections for the ancient verging constructions, such as that used by Archimedes in his construction of the regular heptagon. (He called the latter constructions "moving geometry.") And his treatise on the construction of the astrolabe was an important source for al-Bīrūnī's treatise on the same subject. Al-Sijzī's work on the regular heptagon, which we quote from here, brought him into conflict with his contemporary, Abū al-Jūd, whose erroneous construction he exposed and with whom he exchanged mutual allegations of plagiarism.[85]

Al-Sijzī on the Construction of the Heptagon

In the name of God, the Merciful, the Compassionate. *The book on the Construction of the Heptagon and the Division of the Rectilineal Angle into Three Equal Parts*, by Aḥmad ibn Muḥammad ibn ʿAbd al-Jalīl al-Sijzī.

I am astonished that anybody who pursues and occupies himself with the art of geometry, even though he acquires it from the excellent Ancients, thinks that there are weakness and shortcomings in them; and especially when he is a beginner and a student, with so little knowledge of it that he imagines that he can achieve with very little effort things which he believes to be easy to handle and easily understood, although that was far beyond the understanding of those who are trained in this art and skilled in it. I wish I knew of any power, perspicacity, skill and profundity that would favor the opinion that the heptagon can be found by means of the lemmas of somebody who is reading some introductory book—I mean the *Book of Euclid on the Elements*, somebody who has neither skill nor training and finds fault with those who are prominent in this art. What makes it necessary to believe in the weakness of the excellent Archimedes, with his superiority in geometry over the rest of the geometers? He reached such a high level in geometry that the Greeks called him "the geometer Archimedes." None among the Ancients nor any of the later geometers were called by his name because of his excellence in geometry. He took great pains to find out useful things. By his power he completed the tools, the instruments, and the mechanical procedures. He established the lemmas for the heptagon and followed a path leading to success. By his power we have understood the heptagon just as Heron understood the machines by his [Archimedes'] power and his hard work in mathematical matters.

[85]For an account of this episode see [Hogendijk 1984, 242–69].

This being so, and in spite of his excellence, superiority and high level in the art of geometry, this evil erring man [Abū'l-Jūd] finds fault with him. He [Abū'l-Jūd] refers to the first [group] of his [Abū'l-Jūd's] corrupt, false lemmas, which are far from the path leading to success, and by means of which one cannot arrive at the construction of the heptagon, and [he refers] to the false argument with which he misled himself. He thought that he could mislead somebody, but, by God, [he] only [misled] those who have no mastery of geometry, or of the introduction to it.

Then, in addition to this, he accused Archimedes of things which are ignominious even for those who have a minimum of intelligence, not to speak of the geometers. He maintains that the lemma set forth by Archimedes is more difficult than what is sought, he says that his [Archimedes'] method is ugly, and he accuses him of improperly assuming [something].

It is a wonderful achievement, the proof which Archimedes discovered for the lemmas of the heptagon. But he did not write it down in his book to prevent anyone, such as this outcast, who is not worthy of it, from making use of it. When I had acquired instruction from the knowledge of Archimedes, the lemmas of Apollonius, and in particular from my contemporaries such as Al-'Alā' ibn Sahl, I was also eager [to know] of this noble, abstruse proposition, and the division of the rectilineal angle into three equal parts, which I achieved with very little effort by means of the first treatise of the Book of Apollonius *On Conics*.

Now I shall describe the affair. I shall quote the words of this person who misleads himself, so that it may serve as an education for beginners. I shall describe the wickedness in his words and the mistake in what he constructed. Then I shall follow it with the lemmas of the heptagon. I shall follow that with the construction of the heptagon. I shall finish the book with the division of the rectilineal angle into three equal parts. To God belongs success.

This is the beginning of his book and the arrangement of his lemmas. He [Abū al-Jūd] said: "Archimedes improperly assumed a lemma among the many lemmas he used as a basis for the division of the circle into seven equal parts. He neither explained its construction nor proved its correctness. Maybe it is more difficult to construct and more intricate to prove than that for which he used it as a basis. It is as follows: He said—that is, Archimedes said—Let us draw the diagonal *AZG* of the square *ABGD*. We extend *AB* towards *E*, without end. Let us draw from a point on *BE*, let it be *E*, a straight line to the angle of the square at point *D*, intersecting diagonal *AG* in point *Z* and side *BG* in point *H* such that triangle *BHE* outside the square becomes equal to triangle *GDZ* [fig. 5.27].

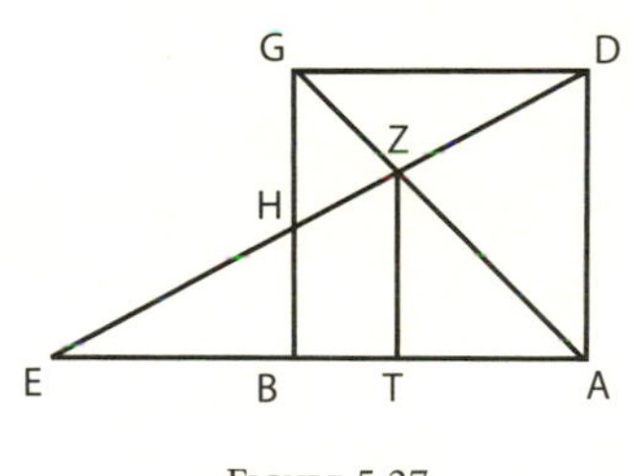

FIGURE 5.27

"Archimedes wished by means of what he improperly assumed therein, to draw the perpendicular *ZT* onto *AB*. Thus *AB* is divided in *T* in such a way that the product of *AT* and the whole *AE* becomes equal to the product of *AB* and *TE*, and the product of *AB* and *AT* [becomes] equal to the square of *BE*. But the division of *AB* with this condition is easier to construct and to prove than to draw line *ED* satisfying the condition Archimedes mentioned. Perhaps it is not possible [to draw *DE*] without dividing *AB* on the condition mentioned above. And perhaps it is more difficult to divide *AB* in that way than to divide the circle into seven equal parts.

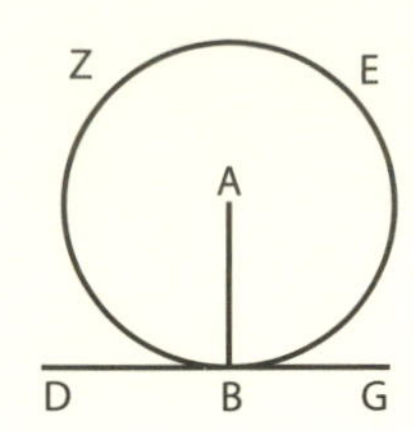

FIGURE 5.28

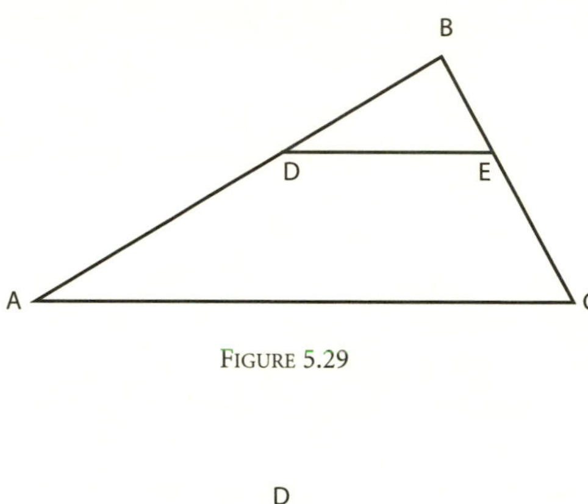

FIGURE 5.29

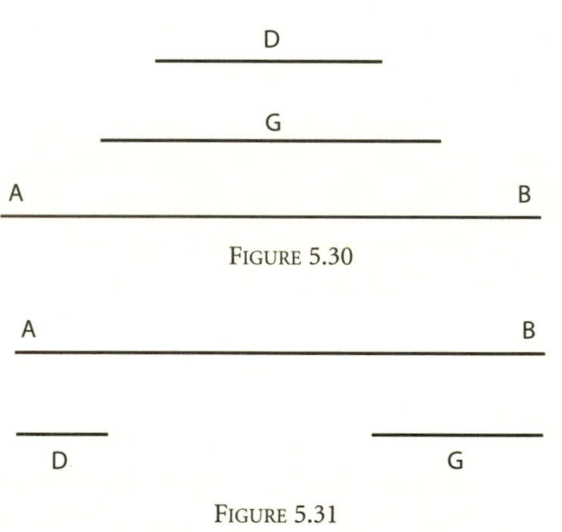

FIGURE 5.30

FIGURE 5.31

"In [the idea] that occurred to me in this chapter, I adhere to another principle, and easier way, a simpler construction and fewer and easier lemmas. The first of them is as follows:

"If a circle is drawn with (radius) a perpendicular onto a line, then it is tangent to the line to which the perpendicular is perpendicular. Line *AB* is perpendicular to *GD* at point *B*. We have drawn with radius *AB* the circle *BEZ* [fig. 5.28]. I say that it is tangent to line *GD*. The proof is easy.

"The second one. We wish to draw from one of the sides of an assumed triangle, for example side *AB* of triangle *ABG*, toward the second side *BG* a line equal to what it cuts off from it (*BG*) outside the small triangle and parallel to the third side *AG* [fig. 5.29].

"The third one. [Fig. 5.30] we wish to draw a line such that the ratio of it to a known line, for example line *AB*, is a known ratio, for example the ratio of *G* to *D*.

"The fourth one. [Fig. 5.31] We wish to divide a known line, for example line *AB*, into two parts such that the product of the whole line and one of the two parts is equal to the square of a line whose ratio to the other part is a known ratio, and let it be the ratio of *G* to *D*." [end of quotation of Abū'l-Jūd]

These are the lemmas which he [Abū'l-Jūd] used as a basis for his construction of the heptagon. Thereupon he used in the construction of the heptagon [the fact] that an assumed line can be divided into two parts such that the product of the whole line and one of the two parts is equal to the square of a line whose ratio to the other part is equal to the ratio of the whole line to the whole line and this part taken together. Thus he gave in the fourth proposition a ratio, but he used in the construction of the heptagon another ratio, different from what he presented in his lemma. He thought that the construction of that was possible by means of the fourth lemma. But the construction of that can only be achieved by means of conic sections, [that construction] which he does not know [since he does not know] the cone or its sections in geometry.

For by means of these lemmas, which are described in the books of the Ancients, [the books he despises], it is possible to construct the heptagon [for] the person who adds his own lemmas to them. But as regards his [Abū al-Jūd's] lemmas and similar lemmas: finding the hexagon in the circle was difficult—this is what the carpenters construct on the top of pots with one [fixed] compass-opening—not to speak of finding the heptagon. Thus this is his mistake and his cheating in the lemmas for the heptagon and its construction.

But now let us begin with what we promised concerning the heptagon and its lemmas, and the division of the rectilineal angle into three equal parts. We wish to divide line *AB* into two parts, for example in *G*, such that the ratio of [a] the line, the square of which is equal to *AB* by *BG* to [b] the line *AG* is equal to the ratio of [c] *AB* to [d] *AB* and *AG*, taken together. We extend *BA* in a straight line towards *D* such that *AD* is equal to *AB*. We apply to *AD* square *ADEZ*. We construct through point *A* a hyperbola *AHK* with asymptotes lines *EZ*, *ED*, as in the fourth [proposition] of the second [book] of the *Conics* of the excellent Apollonius, and in the first [proposition] in the translation of Iṣḥāq. We construct a parabola *BHL* with axis *BD* and parameter *AB*. We drop from the point of intersection *H* of the two conics the perpendicular *HG* onto line *AB* [fig. 5.32]. I say that we have divided line *BA* in point *G* as we desired.

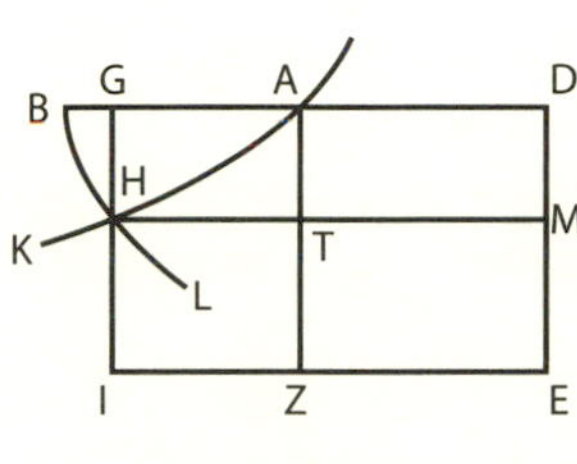

FIGURE 5.32

Proof of this: We extend *EZ* and *GH* in straight lines, to meet in *I*. We draw *HTM* parallel to *IE* and *ATZ* parallel to *GI*. Since rectangle *MI* is equal to square *ZD*, *IT* is equal to *TD*. We add rectangle *TG*; then rectangle *IA* is equal to rectangle *HD*. But rectangle *HD* is line *GH* by line *GD*, and *IA* is *GA* by *AZ*, that is *AB*. Thus *AB* by *AG* is equal to *GH* by *GD*. Thus the ratio of *GH* to *AG* is equal to the ratio of *AB* to *GD*. But the square of *GH* is equal to *AB* by *BG*, since *AB* is the parameter of parabola *BHL*. And *GD* is *AB* together with *GA*. Thus the ratio of [1] the line, the square of which is equal to *AB* by *BG* to [2] the line *GA* is equal to the ratio of [3] line *BA* to [4] *BA* and *AG* [taken together] as one line. Thus we have constructed what we wished, and that is what we wished to demonstrate.

Abū Sa'd al-'Alā' ibn Sahl set up this proposition, following the method of analysis. Our synthesis is part of his analysis.

This is another lemma. We wish to construct on line *AB* an isosceles triangle such that both of its angles at the base are three times the remaining angle. We divide *AB* into two parts in *G* such that *AB* by *AG* is equal to the square of a line *AD* such that the ratio of it to *GB* is equal to the ratio of *AB* to *AB* and *BG* [taken together] as one straight line, by means of the preceding construction. We draw *BD* such as to be equal to *AB* and to contain with *AD* an angle *D*. I say that triangle *ADB* is the required one, and that both of the angles *A* and *D* are three times angle *B* [fig. 5.33].

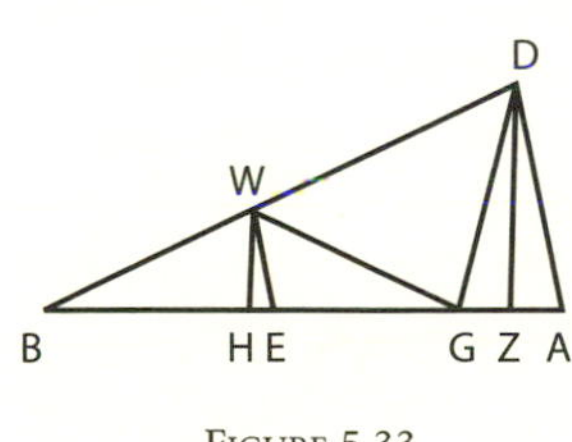

FIGURE 5.33

Proof of this: We draw *DG*. Since the ratio of *AD* to *GB* is equal to the ratio of *AB* to *AB* and *BG* [taken together] as one line, and since *AB* is less than *AB* and *BG* together, *AD* is less than *GB*. We remove from *GB* [line] *BE* equal to *AD*. We draw *EW* parallel to *AD*, and we draw *WH* and *DZ* perpendicular to *AB*. Since the product of *AB* and *AG* is equal to the square of *AD*, the ratio of *AB* to *AD* in triangle *ABD* is equal to the ratio of *AD* to *AG* in triangle *ADG*. And angle *A* is common to the two triangles. Thus triangle *ADG* is similar to triangle *ABD*. Thus line *DG* is equal to line *AD*. Thus *AZ* is equal to *ZG*.

Since *ZG* is half of *AG* and *GB* is half of twice *GB*, *ZB* is half of *AB* and *GB* taken together; and the ratio of *EB*, which is equal to *AD*, to half of *GB* is equal to the ratio of *AB* to half of *AB* and *GB* taken together, that is *ZB*. But the ratio of *WB*, which is equal to *EB*, to *HB* is equal to the ratio of *DB*, which is equal to *AB*, to *ZB*. Thus the ratio of *EB* to *HB* is equal to the ratio of *AB* to *ZB*, that is half of *AB* and *BG* [together]. But the ratio of *EB* to half of *GB* is also equal to the ratio of *AB* to *BZ*. Therefore half of *GB* is equal to *HB*. Thus line *WG* is equal to line *WB*, that is, *EB*. Thus the lines *AD*, *DG*, *WG*, *WB* are all equal. But angle *AGD* is equal to the two angles *GDW* and *DBG*, and angle *DWG* is twice angle *B*. So angle *AGD*, that is angle *GAD*, is three times angle *B*. Thus both of the angles *A* and *D* in triangle *ADB* are three times angle *B*. Thus we have constructed what we wished. That is what we wished to demonstrate.

We wish to construct in a circle *ABG* an equilateral heptagon. We construct an isosceles triangle *EZD* such that both of the angles *E* and *Z* are three times angle *D*. We inscribe in the circle *ABGH* a triangle *ABG* whose angles are equal to the angles of triangle *EZD*. We draw *BH* such that angle *GBH* is equal to angle *BAG*. We bisect angle *HBA* by means of line *BT* [fig. 5.34]. It is clear that the three angles *GBH, HBT, TBA* are equal. We treat angle *BGA* in the same way as angle *GBA*, and we draw lines *GI* and *GK*. Since the six angles at *B* and *G*[86] are equal to each other and to angle *A*, their arcs *GH, HT, TA, BI, IK, KA* are equal to each other and to arc *BG*. Thus we have constructed in the circle *ABGH* an equilateral heptagon. This is what we wished to construct.

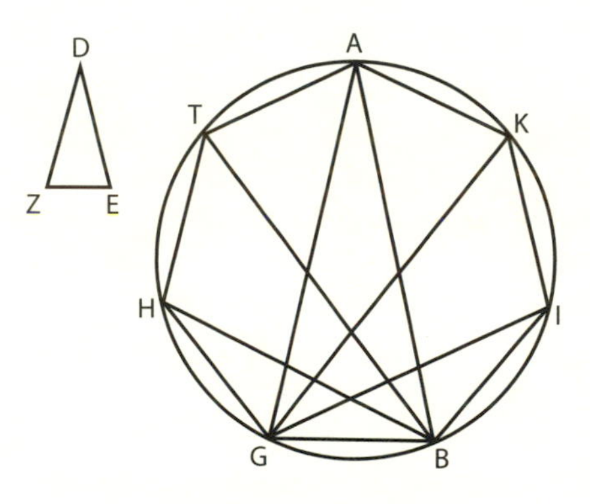

FIGURE 5.34

The following is a lemma for the division of the rectilineal angle into three equal parts. *AGB* is a given semicircle, and line *AZ* is known in position. The diameter [of the semicircle] is *AB*, the center is *D*. We wish to find on diameter *AB* a point *E* such that if a line *EG* is drawn from it to the circumference of semicircle *ABG*, parallel to line *AZ*, then the square of it, I mean of *EG*, is equal to line *BE* by *ED*.

Let us construct on diameter *DB* a hyperbola *DGH* with latus transversum *DB*, latus rectum equal to line *DB* and ordinates at angles equal to angle *ZAB*. Let it intersect the circumference of the semicircle in point *G*. We draw *GE* parallel to

[86] Al-Sijzī assumes the reader can figure out for himself which six of the eight angles he means.

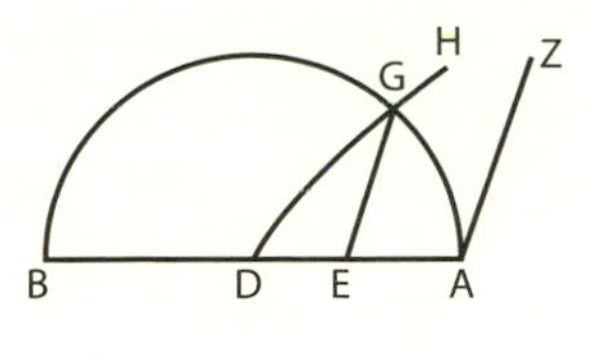

FIGURE 5.35

AZ. I say that *EB* by *ED* is equal to the square of *EG* [fig. 5.35].

Proof of that: The ratio of *EB* by *ED* to the square of *EG* is equal to the ratio of *DB* to the latus rectum. But *DB* is equal to its [the hyperbola's] latus rectum. Therefore *BE* by *ED* is equal to the square of *EG*. Thus we have constructed what we wished, and that is what we wished to demonstrate.

Let the division of the rectilineal angle into three equal parts be [made] easy by means of what we have presented. Let angle *BAG* be given. We wish to divide it into three equal parts. Let us extend *BA* in a straight line to *D*, to any extent we wish. We draw on diameter *AD* semicircle *AZD*. The center is *E*. We draw *HZ* parallel to *AG*, such that *HD* by *HE* is equal to the square of *HZ*, by means of the preceding construction. We join *ZD* and *ZE*, and we draw *AT* parallel to *EZ*. I say that angle *BAT* is twice angle *TAG* [fig. 5.36].

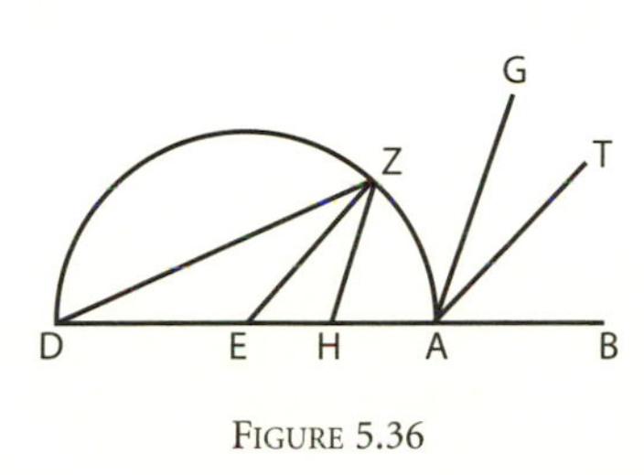

FIGURE 5.36

Proof of this: Since the product of *HD* and *HE* is equal to the square of *HZ*, the ratio of *HD* to *HZ* in triangle *HDZ* is equal to the ratio of *HZ* to *HE* in triangle *HZE*. Angle *H* is common to both triangles, so the two triangles are similar. Thus angle *HDZ* is equal to angle *HZE*. Therefore angle *HZE* is equal to angle *EZD*. But the exterior angle *HEZ* is equal to twice angle *EZD*, since *EZ* is equal to *ED*. Thus angle *HEZ* is twice angle *EZH*. But angle *AHZ*, which is equal to *BAG*, is equal to the [sum of the] two interior angles *HZE* and *ZEH* in the triangle. And angle *BAT* is equal to angle *HEZ*. Thus, by subtraction, angle *TAG* is equal to angle *EZH*. Thus angle *BAT* is twice angle *TAG*. We bisect angle *BAT*, and we have then divided angle *BAG* into three equal parts. That is what we wished to demonstrate.

Anonymous on the Nine-sided Figure

This anonymous treatise explains a solution to the problem of constructing a regular 9-gon in a circle. Because it uses a verging construction, which was the "moving geometry" that al-Sijzī found inadequate, it is likely that the construction dates from before, say, 950 CE. Both the initial question "What is the proof of the assertion etc." and the author's careful citation of a basic result from Book I of the *Elements* suggests to us that this treatise was composed for the use of a student. Constructing a nine-sided polygon is, of course, equivalent to trisecting an arc of the circle that is subtended by the side of an equilateral triangle in the circle. The Arabic *Book of Lemmas*, which most agree contains Archimedean material, contains a theorem that would reduce the trisection of an arc of a circle to a verging construction that reminds one of that in this treatise (see [Heath 1981, I: 240–241] for details).

What is the proof of the assertion of one who says, "Given the circle *ABG* whose center is *D* and whose quartering diameters are *AE*, *ZH*, if the two chords *AB*, *BG* are drawn in it subject to the condition that *AB* is equal to half its diameter and *BG*

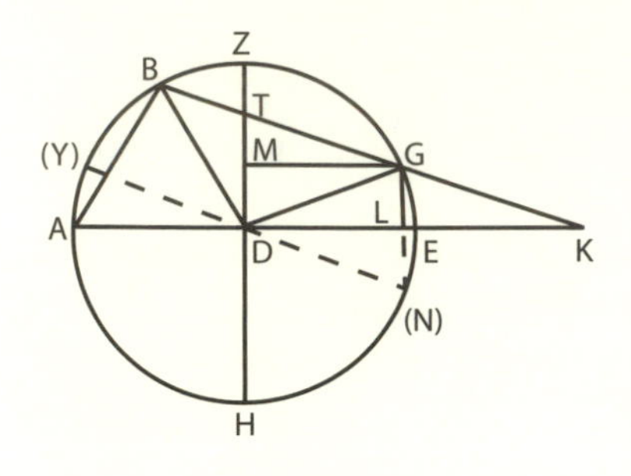

FIGURE 5.37

cuts the diameter at the point *T* and the circumference at the point *G*, and *TG* is equal to half the diameter, then I say that the line *TD* is always equal to the side of the regular nine-sided figure in it (the circle)"? [fig. 5.37] The reply is that that is true; what I claim about it is sound.

The proof of it [is as follows]: We produce the diameter *AE* and the chord *BG* in straight lines in the directions of *E*, *G* so that they meet; and I say first that their meeting is possible and the contrary is impossible, for if it were possible that the two were produced and did not meet, then we draw from the point *G* a perpendicular to the diameter, *GL*. Then either the lines *AE*, *BG* are parallels or their distance in the directions *E*, *G* is farther as they run side-by-side. If they are parallels, then *TG* is equal to *DL* because of the parallelism; but it was supposed equal to *DE*, i.e., equal to half the diameter, and that is a contradiction. Thus their distance in the directions *E*, *G* is wider than parallelism; and that is even more of a contradiction because of what we proved (since such a *GT* would be even smaller than the previous *GT*, and hence less than the radius).

Thus it is necessary that the two lines *EA*, *BG* meet if they are produced in straight lines in the directions *E*, *G*; so let them be produced and let their meeting be at the point *K* and draw *BD*, *DG* and produce *GM* parallel to *DK*. Then the ratio of *TM* to *MD* is as the ratio of *TG* to *GK*; and *TM* is equal to *MD*, since *TG* is equal to *GD*, and *GM* is perpendicular to *TD*. Thus *TG* is equal to *GK*, and because of that *DL* is equal to *LK*. Since the exterior angle *BGD* of the triangle *GDK* is equal to the two opposite interior angles *GDK*, *GKD*, as proved in the first book of *The Book of The Elements*, while the angle *BGD* is equal to the angle *GBD*, since *BD* is equal to *DG*, and the angle *GDK* is equal to the angle *GKD*, the angle *KBD* is equal to twice the angle *BKD*. Similarly the exterior angle *BDA* of the triangle *BDK* is equal to the two opposite interior angles *DBK*, *DKB*; (so) angle *DBK* is two-thirds of angle *BDA* and the angle *BKD* is one-third of angle *BDA*. However, triangle *ABD* is equilateral, since *AB* was supposed equal to half the diameter, so angle *BDA* is two-thirds of a right angle and thus angle *BKD*, i.e., angle *GDK* which is equal to it, is two-ninths of a right angle.

Certainly the sum of the angles around the center in any circle is four right angles; so it is necessary that the angle whose chord is the side of a regular nine-sided figure in any circle is four-ninths of a right angle. We have already proved that the angle *GDK* is two-ninths of a right angle and the line *GL* is half the chord of double the arc *GE*; (so) the line *GL* is half the side of the regular nine-sided figure in the circle *ABG*. Also certainly the line *TD* is double the line *GL*, since its ratio to it is as the ratio of *TK* to *GK*, and *TK* is double *GK* by what we proved. Thus *TD* is equal to the side of the regular nine-sided figure in the circle *ABG*; and that is what we wanted to prove. This is its figure. It is finished with praise to God and with His good success. His blessings be on His prophet Muḥammad and his family.

Abū ʿAlī al-Ḥasan ibn al-Ḥasan ibn al-Haytham (965–1040) was born in Basra, but spent most of his life in Egypt, after he was invited by the caliph al-Ḥākim to work there on a Nile control project. His most influential work was his *Optics*, which was translated into Latin in the thirteenth century and was studied and commented on in Europe for several centuries thereafter.

In the introduction to his treatise *On the Measurement*[87] *of the Paraboloid*, Ibn al-Haytham mentions two medieval Islamic mathematicians who found the volume of a paraboloid formed by rotating a section of a parabola about a diameter. These were Thābit ibn Qurra and Abū Sahl al-Kūhī.[88] Ibn al-Haytham emphasizes, however, that there are two kinds of solids formed from a parabola, the second kind being formed when a section of a parabola is rotated about an ordinate. He also mentions that al-Kūhī spoke only of the first of the two kinds of paraboloids but he did not mention that Thābit wrote of both kinds. (In fact Thābit gave quite an elaborate classification of all kinds, although he—like Abū Sahl—found the volume of the first kind only.)

Apparently, Ibn al-Haytham was the first to find the volume of the second kind of paraboloid, and we translate here his proof of the case when the parabola is rotated about an ordinate at right angles to its axis. (We translate, as well, some of the lemmas Ibn al-Haytham proves on the sums of powers of whole numbers.[89]) In addition, in a part that we have not translated, Ibn al-Haytham finds the volume of a paraboloid of the second type when the angle of the ordinates, angle AGB (in figure 5.41), is acute or obtuse.

Interestingly enough, Ibn al-Haytham uses the word *qubba* to describe the shape of the second type of paraboloid. The word *qubba* refers to a dome erected over a grave, and a glance at the section of the second type of paraboloid, shown in figure 5.41, will reveal why he chose this word.[90] (However, as Ibn al-Haytham tells us, his motive for writing the treatise was mathematical and not architectural.)

Medieval Islamic mathematicians were unaware that Archimedes had found the volume of a segment of the first kind of paraboloid in his treatise *On Conoids and Spheroids.* This was for the very good reason that this treatise was not translated into Arabic. Indeed, of Archimedes' treatises known today, the mathematicians of medieval Islam knew only *The Measurement of the Circle* and *On the Sphere and Cylinder, Books I and II*, the latter work dealing with the surface area and volume of the sphere. Nevertheless, Archimedes did not deal with the second kind of paraboloid.

Our translation is based on the German translation of H. Suter, which we have checked against (and sometimes modified according to) the edition of the Arabic text and French translation published by R. Rashed. Figure 5.41 is Suter's figure 7, which we have re-lettered. (The two letters x and y refer to a later part of the argument which we have not translated.)

Ibn al-Haytham on the Measurement of the Paraboloid

For every statement and every treatise the speaker or the author has an impetus, which motivates him to do it. When we had examined the book by Abū al-Ḥasan Thābit ibn Qurra on the measurement of the paraboloid, we found that he proceeds in a roundabout

---

[87]This word, whether in its Greek or Arabic version, was the usual one the ancients and medievals used for what we would call "finding the volume (or area) of a solid (or plane) figure."

[88]Al-Kūhī was an older contemporary of Ibn al-Haytham, whom he may possibly have met.

[89]The author thanks the editor, Victor Katz, for kindly offering to do this.

[90]See [Dold-Samplonius 1993].

way and that in his exposition he chose a long-drawn-out and difficult method. Then we came across the treatise by Abū Sahl Wījan ibn Rustam al-Kūhī on the same topic, and we found it treated [the subject] somewhat too lightly and too tersely. He also mentioned in it, that the impetus that made him to compose his treatise was his study of the book by Abū al-Ḥasan Thābit ibn Qurra on this topic, which [study] let him recognize his [Thābit's] methods as difficult and far removed from his own.

But when we studied the treatise by Abū Sahl we found that even though he made it easy, he taught in it only the measurement of one kind of parabolic body. There exist, however, two kinds, as we shall find later. The one is fairly easy to treat, but the other one is considerably more difficult, and we found that Abū Sahl in his treatise restricts himself to the calculation of the easier kind and only mentions the second one.

Since we found these two treatises [to be] as we described here, this persuaded us to compose the present work. In this we intend to include the measurement of both types of these bodies, and to treat thoroughly all the concepts pertaining to their measurement. And we shall—for everything we mention and prove—choose the shortest path by which it can be executed (without affecting the clarity of the presentation) and the simplest means by which it can be made clear (without damaging the thorough treatment of the proofs).

Now we begin with our presentation, and God is the helper in that which pleases him.

If we take in the plane of a figure an arbitrary straight line, and around this straight line let the figure revolve until it returns to its original position, the figure forms a solid of revolution. Every piece of a parabola forms, when it is rotated around a fixed line in its plane, a solid, called a paraboloid. One of the chosen lines in the plane of the parabola can be parallel to the axis of the parabola, or the axis itself, or it can cut the axis. In the latter case, it can cut the axis itself or the axis extended. In this case, the line cuts the parabola in two points and is therefore an ordinate to a diameter of the parabola. (This has all been explained by Apollonius in his work on the conic sections.)

The collection of straight lines that can be drawn in the plane of a parabola is divided into two types, either diameters or ordinates of the parabola. Thus all paraboloids, that are formed by the rotation of a parabola around a line in its plane, are divided into two classes: the first consists of those formed by the rotation of a parabola around a diameter; the others are those formed by rotation of the parabola around an ordinate. We will now undertake the measurement of both types. To accomplish this goal, we must set out some lemmas. Concerning the class formed by rotation around a diameter, we do not require these, since this class, as we have said earlier, is easy to handle. However, the other class, formed by rotation of the parabola around an ordinate, is the more difficult case and requires some number theoretical lemmas. These are

Lemma 1: If one has a sequence of natural numbers, beginning with one, and one takes half the largest and half of one, adds these halves and multiplies this sum by the largest number, one has the sum of all the given numbers.[91]

[91]In modern notation, this lemma may be written as $\sum_{i=1}^{n} i = \left(\frac{1}{2}n + \frac{1}{2}\right)n$.

Lemma 2: One has again the same sequence of numbers; one takes the third part of the largest and the third part of one, adds these parts, and multiplies the sum by the largest numbers. Then one adds to the largest number the half of one and multiplies this sum by the former product, one has the sum of the squares of the given numbers.[92]

Lemma 3: One is again given the same sequence of numbers. One takes the fourth part of the largest and the fourth part of one, adds these parts, and multiplies the sum by the largest number. One then adds one to the largest number, multiplies this sum by the largest number. And multiplies this product by the former product. One then has the sum of the cubes of the given numbers.[93]

Lemma 4: One is again given the same sequence of numbers. One takes the fifth part of the largest and the fifth part of one, adds these parts, and multiplies the sum by the largest number; then one adds to the largest number half of one, and multiplies this sum by the former product. Now one adds one to the largest number, multiplies this sum by the largest number, subtracts from the product the third part of one, and multiplies this result with the previous product. One then has the sum of the fourth powers of the given numbers.[94]

We will now prove these lemmas.

Proof of 1: Let *ab*, *gd*, *ez*, *ht* be the sequence of given numbers,[95] and let $ab = 1$; then I say that if half of *ht* is added to half of *ab*, and this sum is multiplied by *ht*, the result will be the sum of the numbers *ab*, *gd*, *ez*, *ht* (fig. 5.38).

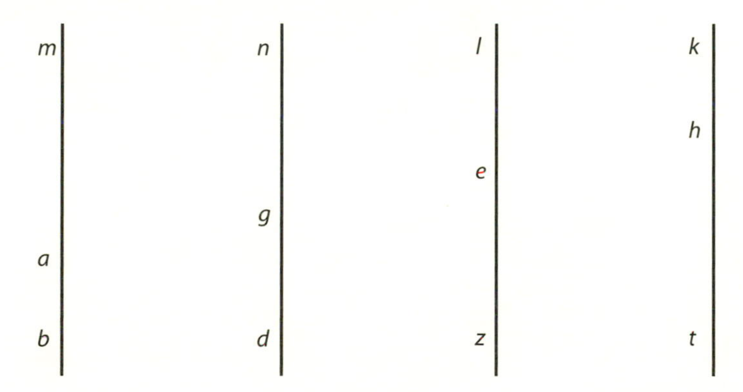

FIGURE 5.38

We adjoin to these numbers the same numbers, but in the reverse order; these are *kh*, *le*, *ng*, *ma*, and let $kh = 1$, with each one greater than the previous by 1. Since now *ht* is greater than *ez* by 1 and $kh = 1$, then *kt* is greater than *ez* by 2, so $le = 2$

---

[92]In modern notation, this lemma may be written as $\sum_{i=1}^{n} i^2 = n\left(\frac{1}{3}n + \frac{1}{3}\right)\left(n + \frac{1}{2}\right)$.

[93]In modern notation, this lemma may be written as $\sum_{i=1}^{n} i^3 = \left(\frac{1}{4}n + \frac{1}{4}\right)nn(n+1)$.

[94]In modern notation, this lemma may be written as $\sum_{i=1}^{n} i^4 = n\left(\frac{1}{5}n + \frac{1}{5}\right)\left(n + \frac{1}{2}\right)\left((n+1)n - \frac{1}{3}\right)$.

[95]The representation of whole numbers as line segments goes back (at least) to the arithmetic books (7–9) of Euclid's *Elements*.

and $lz = kt$. Since ht is 2 greater than gd, then kt is 3 greater than gd, but $ng = 3$, so $nd = kt$. Therefore, since $mb = kt$, the numbers mb, nd, lz, kt are all equal. The number of consecutive numbers beginning with 1 is however equal to the number of ones in the last number [ht], and is also equal to the number of equal numbers mb, ..., kt. If one multiplies the number kt by the number of ones in ht, the result is the sum of the numbers mb, ..., kt, and the numbers ab, gd, ez, ht are the same as the numbers kh, le, ng, ma. Then the sum of both rows of numbers is double the sum of the numbers ab, gd, ez, ht; and the sum of these numbers is half the sum of the numbers mb, nd, lz, kt, therefore equal to half the product of kt and the number of ones in ht. Now the half of kt is equal to the half of ht plus the half of 1, since $kt = ht + 1$. Thus we will have the sum of all the numbers ab, gd, ez, ht, when we add the half of the largest [ht] to the half of one and multiply this sum by the largest number. And this will always be true, whatever the number of numbers is. Thus we also have that the sum of a sequence of natural numbers beginning with 1 is half of the square of the largest number plus half of the largest number itself.

[We omit the proofs of 2 and 3, except that we note that at the end of each of those proofs, Ibn al-Haytham restates the results in "polynomial" form, as follows.]

Restatement of 2: "We also have that the sum of the squares of a sequence of natural numbers beginning with one is equal to the third of the cube of the largest number plus half the square of the largest number plus one sixth this largest number itself."

Restatement of 3: "We also have that the sum of the cubes of a sequence of natural numbers beginning with one is equal to one fourth of the fourth power of the largest number plus one half of its cube, plus one fourth of its square."

Proof of 4: Let now ab, bg, gd, de be a sequence of cubes (beginning with the cube of 1) and let bz, gh, dt, ek be the natural numbers themselves (fig. 5.39).

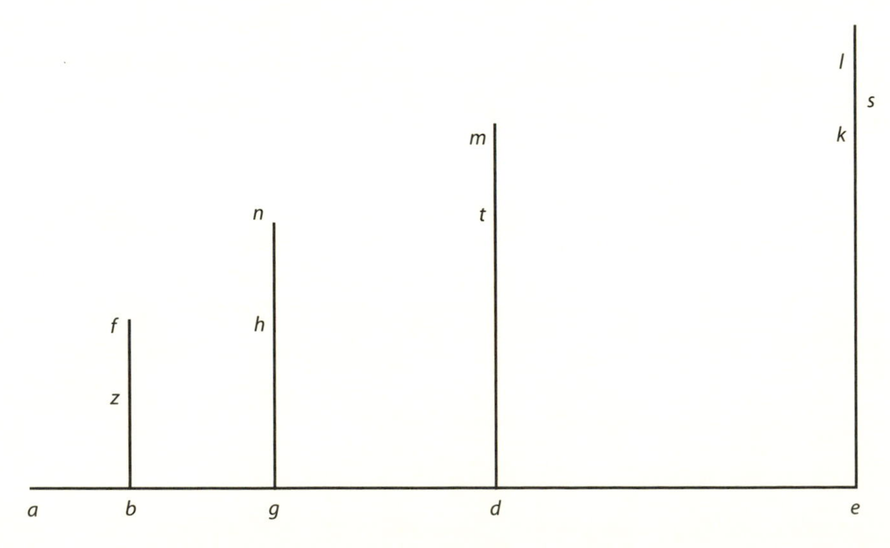

FIGURE 5.39

So $de \cdot ek = ek^4$, $gd \cdot dt = dt^4$, etc. We add unity to each of the numbers in the sequence of natural numbers as before; thus $ae \cdot el = ae \cdot ek + ae \cdot kl$; but $ae \cdot kl = ae$ itself [because $kl = 1$] and $ae \cdot ek = de \cdot ek + ad \cdot ek = ek^4 + ad \cdot dm$, since $de = ek^3$ and $dm = ek$.[96] Therefore, $ae \cdot el = ae + ek^4 + ad \cdot dm$. In the same way, we have $ad \cdot dm = ad + dt^4 + ag \cdot gn$, etc. Therefore, we have $ae \cdot el = ek^4 + dt^4 + gh^4 + bz^4 + ae + ad + ag + ab$. But ae is the sum of a sequence of cubes of integers, so [by Lemma 3] is equal to $\frac{1}{4}ek^4 + \frac{1}{2}ek^3 + \frac{1}{4}ek^2$; similarly, $ad = \frac{1}{4}dt^4 + \frac{1}{2}dt^3 + \frac{1}{4}dt^2$; $ag = \frac{1}{4}gh^4 + \frac{1}{2}gh^3 + \frac{1}{4}gh^2$; and $ab = \frac{1}{4}bz^4 + \frac{1}{2}bz^3 + \frac{1}{4}bz^2$. Thus $ae \cdot el$ equals the sum of the fourth powers of the sequence of consecutive integers, whose largest is ek, plus a fourth of this sum, plus half the sum of the cubes of these integers, plus a fourth of the sum of the squares of the same integers. Therefore, $(4/5)ae \cdot el$ equals the sum of the fourth powers of the sequence of integers, plus 2/5 of the sum of their cubes, plus 1/5 of the sum of their squares.

Now since $sl = 1/2$, $(4/5)ae \cdot sl = (2/5)ae = 2/5$ of the sum of the cubes of the sequence of integers, which, if we subtract this from the previous equality, leaves $(4/5)ae \cdot es$ equal to the sum of the fourth powers of the sequence of consecutive integers plus 1/5 of the sum of their squares. But ae is [by Lemma 3] equal to $(1/4el \cdot ek) \cdot (el \cdot ek)$,[97] so $(4/5)ae \cdot es = (1/5)el \cdot ek \cdot el \cdot ek \cdot es$ equals the sum of the fourth powers of the sequence of consecutive integers plus 1/5 of the sum of their squares. But, by Lemma 2, we have $(1/3)el \cdot ek \cdot es$ equals the sum of the squares of the sequence of consecutive integers, whose greatest is ek, and $(1/5)el \cdot ek \cdot es$ equals 3/5 of the sum of the squares of the sequence of consecutive integers; so 3/5 of the sum of these squares multiplied by $el \cdot ek$ is equal to the sum of the fourth powers of the sequence of consecutive integers plus 1/5 of the sum of their squares; but 1/5 of the sum of the squares is (1/3)(3/5) of the sum of these squares. Therefore, 3/5 of the sum of the squares multiplied by $(el \cdot ek - 1/3)$, is equal to the sum of the fourth powers of the sequence of consecutive integers. But 3/5 of the sum of the squares multiplied by $(el \cdot ek - 1/3)$ is equal to $(1/5)el \cdot ek \cdot es \cdot (el \cdot ek - 1/3)$, so this quantity is equal to the sum of the fourth powers of the sequences of consecutive integers.

If one now takes from any sequence of natural numbers, beginning with 1, one fifth of the largest, adds this to one fifth of unity, multiplies this sum ($= (1/5)el$) by the largest number ($= ek$), multiplies the result by the largest number increased by 1/2 ($= es$), then multiplies the new result by 1/3 less than the product of the largest number and the number 1 greater ($= el \cdot ek - 1/3$), one then has the sum of the fourth powers of the sequence of consecutive integers.

Lemma 5: Let *ab*, *gd*, *ez*, *ht*, *kl* be the squares of the sequence of natural numbers and let each of the quantities *mb*, *nd*, *fz*, *ot* be equal to *kl*, then I say that the sum of the squares of *am*, *gn*, *ef*, *ho* is less than 8/15 of the sum of the squares of *mb*, *nd*, *fz*, *ot*, *kl*, and greater than 8/15 of the sum of the squares of *mb*, *nd*, *fz*, *ot*,

[96] Ibn al-Haytham's statement that "*de* is the cube of *ek*" shows that, despite the geometrical representation of these quantities by line segments, they are numbers, pure and simple.

[97] Products such as this Ibn al-Haytham describes as "the product of one-fourth of *el* by *ek*, which one then multiplies by the product of *el* by *ek*." We shall, however, write them in the modern fashion in what follows.

and that the sum of the squares of *am, gn, ef, ho, kl* is greater than 8/15 of the sum of the squares of *mb, nd, fz, ot, kl* (fig. 5.40).

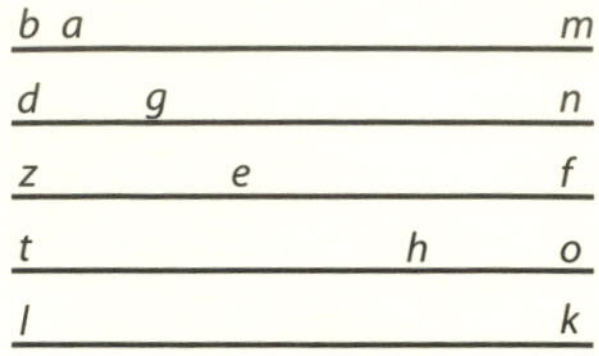

FIGURE 5.40

[We omit the proof of this lemma (based on lemmas 1–4) since it is long and tedious, but sheds no additional light on Ibn al-Haytham's thinking. (It is not difficult to work out an equivalent modern proof.) After the proof, Ibn al-Haytham continues:]

From this proof, we get the following: Let an arbitrary number of equal line segments be given, and equally many squares of the natural numbers beginning with 1, and from the first line segment let a part be cut such that the ratio of the entire segment to the cut off part is equal to the ratio of the greatest square to the square of unity, in our case thus equal to *mb*: *ab*; further from the second segment let a part be cut such that the ratio of the entire segment to the cut off part is equal to the ratio of the greatest square to the square of the number following unity, in our case thus equal to *nd*: *gd*; and continue in this way up to the last line segment (*kl*), from which nothing more can be cut because it corresponds to the greatest square. Then the sum of the squares of the line segments that remain from the originally given equal line segments is less than 8/15 the sum of the squares of the original line segments, including the final segment. Furthermore, the sum of the squares of the line segments that remain together with the square of an uncut segment is greater than 8/15 of the sum of the squares of the original line segments, including the final segment.

[Ibn al-Haytham now calculates the volume of a paraboloid formed by rotating a parabola around a diameter. He then continues:]

What we have now shown refers to the calculation of [the] one kind of paraboloids, which are created by the rotating the parabola around a diameter. We next want to consider the second kind, which results from rotating a segment of parabola about an ordinate.

Let *ABG* [fig. 5.41] be a segment of a parabola, *BG* its diameter, *AG* the ordinate, and [let] the angle *AGB* [be] a right angle. Through *B* draw *BE* parallel to *AG*, and in the same way *AE* parallel to *GB*. Then let the rectangle *AGBE* be rotated about the fixed straight line *AG*, so in this way the [right] circular cylinder *BZ* is generated, with base a circle of radius *BG* [and *AG* as height]. And, by rotating the segment of parabola *ABG*, the paraboloid *BAD* is generated, with base a circle of radius *BG* [and height the same as the cylinder]. I now state that the paraboloid *BAD* is 1/3 plus 1/5 of cylinder *BZ*.[98]

[98]Arabic writers frequently expressed fractions as the sum of unit fractions, and Ibn al-Haytham expresses these unit fractions verbally. The sum here is one we would write as 8/15 (= 5/15 + 3/15), and we shall write it thus in what follows.

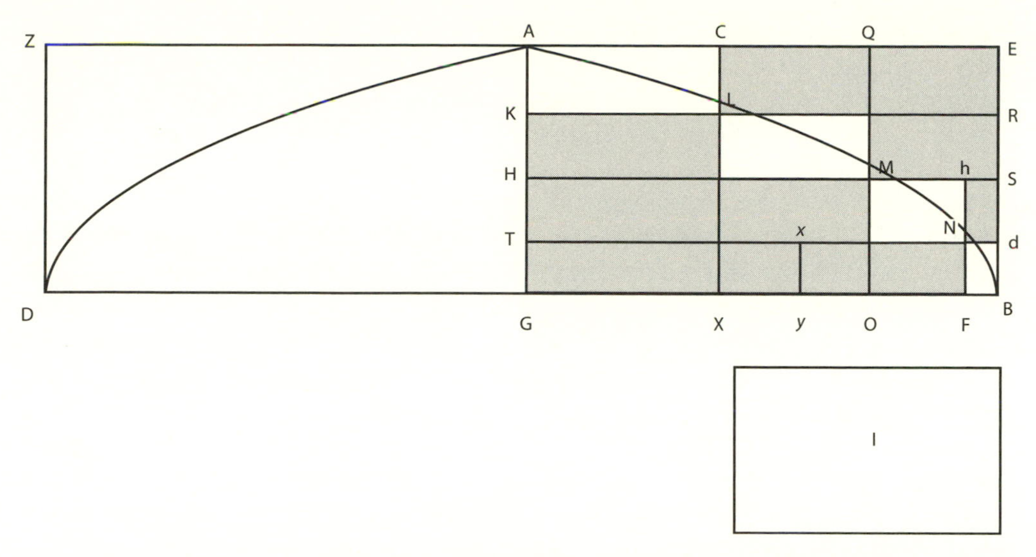

Figure 5.41

The proof consists in assuming that it is not 8/15 of the cylinder, so it must be either bigger or smaller than this. First, let it be bigger than 8/15 of the cylinder and [let] the excess over 8/15 of the cylinder be the solid *I*. We now halve *AG* in the point *H*, draw *HMS* parallel [to] *BG*, and through the point *M* the line *QMO* parallel to *BE* and *AG*. Now, since *QM* is equal to *MO*—*AH* being equal to *HG*—the area *EM* is equal to the area *MB*, and [similarly] the area *AM* is equal to the area *MG*. Thus, if the area *BA* rotates about the straight line *AG* until it returns to its initial position, then the solids of revolution that are created by rotating the surfaces *AM*, *MG* are equal to each other, and the two surfaces of revolution created by the two surfaces *EM* and *MB* are also equal to each other. The two surfaces of revolution created by rotating the surfaces *ME* and *MG* are therefore, added together, [equal to] half of cylinder *BZ*.[99]

Next, we halve *AH* at point *K*, and draw through *K* the line *KLR* parallel to *HS*, and through *L* the straight line *CLX* parallel to *AG*. And in the same way we also halve *HG* at point *T*, draw *TNd* parallel [to] *GB*, and through the point *N* [draw] the straight line *hNF* parallel [to] *HG*. Then the solids of rotation that are created by rotating the areas *QL* and *LH* are together equal to the half of the cylinder that is created by rotating the area *AM*. And in the same way, the solids of rotation that are created by rotating the areas *NS* and *NO*, are, together, half the solid of rotation arising from rotating the area *SO*. Thus the four solids of rotation that arise from rotating the areas *NS*, *NO*, *QL*, and *LH* are, together, half of the two solids of rotation arising by rotating the areas *SO* and *AM*.

But if one subtracts from the whole cylinder, *BZ*, the two solids of rotation arising by rotating the areas *EM* and *MG*, and which together are half of the cylinder *BZ*, then the two solids of rotation remain that arise through rotating the areas *SO* and

[99]We have translated this paragraph as literally as is consistent with the English language. In the following we shall omit designations such as "the point," "the area," "the line" since the context makes it clear what sort of geometrical object is being referred to.

AM, and which together are (therefore) equal to half the cylinder *BZ*. And if one subtracts from the two last-named solids of rotation the four solids of rotation, which arise from rotating the areas *QL*, *LH*, *NS*, *NO*, and which together are half of those two, then the four solids of rotation that arise through rotating the areas *BN*, *NM*, *ML*, and *LA* remain.

Were one now further to halve each of the [four] parts of *AG* and to draw from each of the points of bisection lines parallel to the line *BG*, and if one draws through the places where these lines intersect the section *AB* lines parallel to *AG*, one then has the solids of revolution arising from rotating the surfaces, and which [solids] the pairs compose half of the solids of revolution in which they are found,[100] as we have shown previously.

If, now, one has two unequal magnitudes and subtracts from the bigger one its half, from the remainder again the [i.e. its] half, and always continues, then one must necessarily reach a magnitude that is smaller than the smaller of the two (original) magnitudes, as was shown.[101] Thus if from the large cylinder *BZ* the halves [of the cylinder and successive remainders] are repeatedly removed, in this way then, necessarily, a quantity finally remains that is smaller than the solid *I*. We assume now, that the process is so far advanced, that the remaining four solids of rotation, which arise from rotating the areas *BN*, *NM*, *ML*, and *LA* are together smaller than the solid *I*. In that case, the part of this solid of revolution that lies inside the paraboloid is smaller still than the solid *I*. And since [by assumption] the paraboloid *BAD* is bigger than 8/15 of the cylinder *BZ*, (namely, by the quantity *I*) and since the parts of the small solids of rotation remaining inside the paraboloid are [together] smaller than the solid *I*, then the paraboloid, less the part of the solids of rotation inside [it], is [still] bigger than 8/15 of the cylinder *BZ*, i.e. the solid of slices[102] that has as base the circle with radius *FG* and as top the circle with radius *KL*, is bigger than 8/15 of the cylinder *BZ*.

Now, since *ABG* is a segment of a parabola with diameter [i.e., axis] *BG* and ordinate *AG*, $AG^2 = BG \cdot p$ and similarly $LX^2 = BX \cdot p$, $MO^2 = BO \cdot p$ and $NF^2 = BF \cdot p$.[103] Thus, one has the proportions: $AG^2 : LX^2 = BG : BX$, $LX^2 : MO^2 = BX : BO$, $MO^2 : NF^2 = BO : BF$. Thus, $BG : BX : BO : BF = AG^2 : LX^2 : MO^2 : NF^2$. Since, now, $NF = GT$, $MO = HG$, and $HG = 2GT$, it follows that $MO = 2NF$; and, since $AK = KH = HT = TG$, it follows that $KG = 3GT$. Thus, also $LX = 3NF$. And in the same way $AG = 4GT = 4NF$. Thus, with the measure that makes *NF* [equal to] one, $MO = 2$, $LX = 3$ and $AG = 4$; so, $NF : MO : LX : AG = 1 : 2 : 3 : 4$. So the ratios of *NF*, *MO*, *LX*, and *AG* are, one to another, equal to the ratios of the consecutive [natural] numbers, starting from 1. And had one divided *AG* into more than four parts, all these lines would again be to each other as the natural numbers.

From this follows now that NF^2, MO^2, LX^2, and AG^2 are to each other as the squares of the natural numbers, so one also has: $BF : BO : BX : BG = 1^2 : 2^2 : 3^2 : 4^2$. But $BF = Nd$, $BO = MS$, $BX = LR$, $BG = AE$, thus: $Nd : MS : LR : AE = 1^2 : 2^2 : 3^2 : 4^2$; and for the same reason $Td = HS = KR = AE = GB$.

[100]He is saying in general terms here what he said earlier as, for example, "two solids of rotation remain, arising through rotating the areas *SO* and *AM*, and which together are therefore equal to half of the cylinder *BZ*."

[101]Euclid showed this in *Elements* X, 1.

[102]The Arabic text has the word *manshūr* here, and one meaning of the verbal root of this Arabic participle is "to saw off," which nicely catches the image of the body made up of short cylinders of different heights, sawed off of cylinders of decreasing radii and placed one next to the other. (I am indebted to J. Hogendijk for this suggestion.)

[103]Ibn al-Haytham describes the parabola by the words "the square of *AG* is like the product of *BG* in the right side" (i.e., the parameter), and similarly for the other ordinates.

Now, we have proven in the introductory lemmas,[104] that, if one has a sequence of equal line [segments], from all but one of which certain lines are cut that are to one another as the squares of the consecutive natural numbers, then the sum of the squares of the remaining lines is smaller than 8/15 the sum of the squares of the lines equal to each other and to the largest line. And the sum of the squares of the remaining lines plus the square of the [one] undivided segment is larger than 8/15 the sum of the squares of the lines equal to each other. Thus, the squares of *NT*, *MH*, and *LK* are less than 8/15 the [sum of the] squares of lines *dT*, *SH*, *RK*, and *AE*; and the squares of *NT*, *MH*, *LK*, and *AE* are larger than 8/15 of [the sum of the squares of] *dT*, *SH*, *RK*, and *AE*.

But the ratio between the squares of these lines, one to another, is equal to the ratio, one to another, of the circles[105] that have these lines as radii, so the circles, whose radii are the lines *TN*, *HM*, and *KL* are, together, smaller than 8/15 the sum of the circles whose radii are *Td*, *SH*, *RK*, and *AE*. And the circles whose radii are the distances *TN*, *HM*, *KL*, and *EA* are together bigger than 8/15 of the circles whose radii are *Td*, *SH*, *RK*, and *EA*.

Taking now for all circles a common height, *AK*, the small cylinders, whose bases are the circles with radii *TN*, *HM*, *KL* and whose heights all equal *AK*, are together smaller than 8/15 the sum of the cylinders, whose bases are the circles with radii *Td*, *SH*, *RK*, and *AE*, and whose heights all equal *AK*. But the cylinders, whose bases are the circles with radii *TN*, *HM*, *KL* and whose heights all equal *AK*, form the solid of slices whose base has the radius *GF* (=*TN*) and whose top surface has radius *KL*. And the cylinders, whose bases are the circles with the radii *Td*, *SH*, *RK*, and *AE* and whose heights all equal *AK*, form together the cylinder whose base has the radius *BG* (=*AE*) and whose height equals *AG*, *i.e.*, the cylinder *BZ*. Thus the body is smaller than 8/15 of the cylinder *BZ*. But we found previously, that it is larger than 8/15 the cylinder *BZ*. This is a contradiction, so the paraboloid is not, as we assumed, bigger than 8/15 of the cylinder *BZ*.

I claim now, that neither can it be smaller than 8/15 of the cylinder.

[...]

The following two extracts concern a drawing tool known in medieval Islam as "the perfect (or 'complete') compass." This was a compass in which [see fig. 5.45] one arm stays in a fixed position relative to the drawing plane, so that the other arm, as it rotates around the first, describes the surface of a cone, *C*, whose axis is the first arm. At the end of the second arm a stylus slides freely up and down. This allows the tip of the stylus to describe the curve resulting when the cone, *C*, is cut by the plane represented by the paper.

In Eutocius of Ascalon's commentary on Book II of Archimedes' *Sphere and Cylinder* one finds the statement that Isidore of Miletus (sixth century CE) invented a compass with which one could draw parabolas, so it appears that some form of the perfect compass was known in late antiquity. The passages we have translated here show that Muslim writers knew of the earlier discovery of this instrument[106] but had no treatise on it at hand.

The first selection on the perfect compass is part of a treatise in which the tenth-century geometer al-Sijzī describes its general form and functioning. We have based our translation on the French text by Franz Woepcke.

[104] The result is stated in Lemma 5.

[105] Recall that for ancient and medieval mathematicians, the circle was an area.

[106] Possibly via Eutocius's commentary, part of which, at least, was translated into Arabic.

Al-Sijzī on the Perfect Compass

If we consider *GD* [fig. 5.42] as an axis, and we rotate side *GB* around axis *GD*, imagining that line *GB* gets longer and shorter to stay constantly against the surface of the given plane, if the angle at the top of the compass, which is angle *DGB*, stays invariable, if axis *GD* rotates around itself, and if the plane is of unbounded extent [then, if the drawing takes place on the plane going through line *ZH*, parallel to the circular base AB] and which is perpendicular to the plane of the triangle, the drawing resulting from this motion will produce the circumference of a circle. If the drawing takes place on the plane going through line *ZL* and which is perpendicular to the plane of the triangle, an ellipse will be described. If the drawing takes place on the plane going through line *ZT* and which is perpendicular to the plane of the triangle, a parabola will be described. If the drawing takes place on the plane going through line *OKZ* and which is perpendicular to the plane of the triangle, a hyperbola will be described. And, finally, if the two sides of the triangle formed in the cone are extended towards the vertex of the cone in order to generate another cone, and if the plane on the side of *Z* is extended indefinitely outside the cone up to when it intersects the other cone, the two opposite branches of a hyperbola will be described.

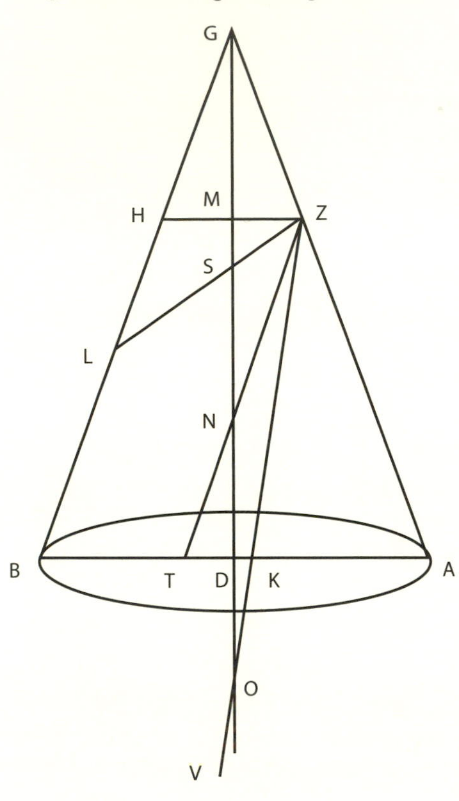

FIGURE 5.42

It is obvious that, if angle *GMH* is right, the drawing resulting from the motion of side *GB* produces on the plane going through *ZH* [and perpendicular to the plane of the axial triangle][107] a circle. If the two angles *BGS* and *GSL* are smaller than two right angles, the drawing resulting from the motion of line *GL* on the plane going through *ZL* produces the ellipse. If the two angles *BGN* and *GNT* are equal to two right angles, the rotation of line *GB* generates, on the plane going through line *ZNT*, the parabola. And if the two angles *BGO* and *GOF* are larger than two right angles, the drawing resulting from the rotation of line *GB* produces, on the plane going through *ZKO*, the hyperbola.

Now, let us assume that the straight line *GD* is the axis of the [perfect] compass, imagining that line *GB* moves in the tube, going in and out, so that a stylus, replacing the other arm of the [usual] compass, gets longer and shorter, as we described in our *Treatise on the construction of the conic compass*.[108] Then, if the position of

[107]Al-Sijzī omits this phrase, which is (technically speaking) necessary, in this and the following three cases. He assumed, no doubt correctly, that anyone knowing enough mathematics to read his treatise would not need to be reminded of this requirement.

[108]This treatise is not listed in Sezgin, Vol. 5 (1974).

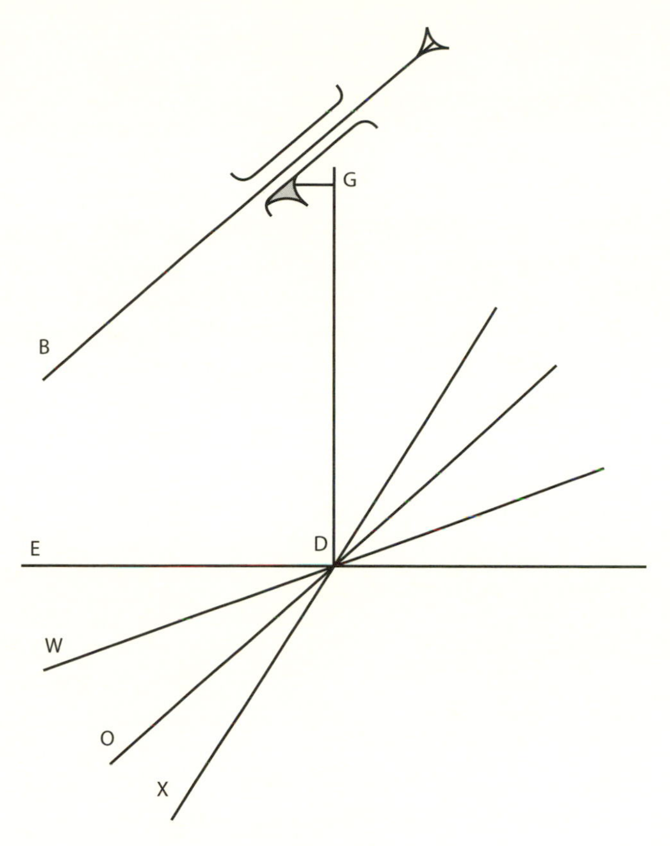

FIGURE 5.43

the axis *GD* relative to the plane is one of the positions with respect to which we just discussed, [in relation to] how they determine the angle at the end of the arm of the compass as well as the angle formed at the endpoint *G* (between arms *GD* and *GB*, of the compass) the compass draws the above-mentioned sections. And this is the figure of the conic compass [fig. 5.43].

This second selection on the perfect compass was written in the late twelfth century by the otherwise unknown Muḥammad b. al-Ḥusayn[109] and was dedicated to the famous Kurdish general, Saladin, who defeated the Crusaders in the twelfth century. The author knew from a treatise of al-Bīrūnī that Abū Sahl al-Kūhī had written on the instrument in the tenth century, but he was, he says, unable to get a copy of al-Kūhī's treatise. With the help of a colleague, one Abū al-Maʿalī b. Yūnus, he was able to compose this treatise. From this treatise we have chosen some of the introductory material and the construction of a parabola by the perfect compass (including the introductory lemmas).

Although such instruments were probably made and used, one finds in Arabic manuscripts that most conic sections are represented by circular arcs or even a connected path of line segments. This selection, like the previous one, was translated from Arabic into French by Franz Woepcke. We have based our translation on the French text, which we have checked (and occasionally altered) against the Arabic.

---

[109]According to de Slane's identification, quoted on p. 2 of Jules Mohl's preface to Woepcke's treatise.

Muḥammad b. al-Ḥusayn on the Perfect Compass

In the name of the lenient and the merciful God! Glory to God, Lord of the worlds. May his blessing be upon his best creation, Mohammed the prophet, and on his most noble family!

After your humble servant[110] wrote *Al-Ishārah al-Naṣīriyya* and he decided to put it into the royal library—may the God on high make it flourish!—he desired to add to it something that could show the goodness of its fruit and the assurance of its results. The author had promised, in his work, to make the perfect compass known and to put in order his memoir [on this instrument], some of whose elements had been put in order and the principles of its propositions established before the author left his place of residence to go to the imperial residence. He hastened to keep his promise, immediately after he completed his *Kitāb al-Ishārah*, and had given, by the way of an exact diagram, a good representation of the shape of the perfect compass. He explains in this memoir the utility of this instrument, the help it can give, how to build and how to use it, generally and in detail. In all that, his goal was to pay homage to our Master, the glorious Lord, The Victorious King, who reunited the Moslems and who tamed the Crusaders,[111] the salvation (*salāḥ*) of the world and of religion (*al-dīn*), the Sultan of Islam and of Moslems, Abū al-Muẓaffar Yūsuf ibn Ayyūb,[112] who gave a new life to the empire of the Commander of the Believers (the Calif of Baghdad).

[We omit several lines that continue this fulsome praise.]

Beginning of the discourse on the perfect compass

The construction of three conic sections is one of the most important problems in geometry, one of the matters that, in practice, give rise to the most elegant procedures and in theory, give rise to the most ingenious and most useful considerations. Apollonius, the great geometer, developed, in his work entitled *The Conics*, most of the marvelous characteristics and wonderful properties of these three curves. By penetrating their inner nature through the most diverse speculations, he filled intellects with admiration, and he amazed spirits through discoveries and particular stratagems based on the theory of conics. These are used [among other things] for the description of the *muqanṭar*s[113] on the plane astrolabe by conic projection, as well as for drawing hour lines on sundials, whether parallel to the horizon or on walls perpendicular to the horizon. And this is because the *muqanṭars* and the tips of shadows always project themselves on these surfaces as hyperbolas, ellipses or parabolas, depending on the latitudes and horizons in the astrolabe, and according to their positioning [relative to the horizon] in horizontal or vertical sundials.

---

[110]So the author of the treatise refers to himself. Henceforth, simply "the author."

[111]Literally, "the cross-worshippers."

[112]This is the famous Saladin (as the name is represented in English), who lived from 1138 to 1193. The Arabic words of his title, Salah al-Din, gave rise to the English form of his name, and the author of the treatise neatly weaves these words into the opening encomium by calling him "the salvation ... of religion" (salah ... al-din).

[113]This Arabic word denotes the circles parallel to the horizon on the celestial sphere, so they are circles of constant altitude.

[We omit some lines on the problems of drawing smooth curves by pointwise constructions.]

The ancients had invented, for the description of conic sections, an instrument that they called a "perfect compass," because, with this instrument, all kinds of curved and straight lines could be drawn. But no writing of the inventor reached us, nor by any other of the ancients who would have shed light on the manner of constructing this instrument or of using it.

We however met, in the work of Abū al-Rayḥān al-Bīrūnī, titled *Complete account on possible ways of constructing the astrolabe*, a passage where this author mentions Shaykh Abū Sahl Wījan ibn Rustam al-Kūhī, and that in a treatise the latter had written on how to construct the instrument mentioned and how to use it. Abū al-Rayḥān also says that al-Kūhī had founded his methods of describing the conic sections by means of this instrument on theorems that he had set out in a book entitled *Division of Lines According to Ratios of Surfaces*. But this book has not come down to us [either].

That is what gave me a great desire to discover the demonstrations of the methods that Abū al-Rayḥān had mentioned about Abū Sahl (despite the lack of his treatise), either the very demonstrations of al-Kūhī, or others leading to those, [and] through reasoning proper to deduce one from the other.

[...]

I will now set out the description of the perfect compass, its form and how to construct it. Then I will give lemmas to furnish the proof that the lines drawn with this instrument are conic sections, I will give the reader an ordered work to fulfill the desire of both the learned and the artisan, and gather into a complete discussion all demonstrations useful to the problem. A few other propositions might be inserted toward the end (of the work) because they result from these theorems even though they are there only incidentally, being foreign to the proper goal of the work.

[...]

Description of the perfect compass and how to construct it

Description of the perfect compass—Suppose a straight line stands on a plane surface at a given point of this surface, one of whose endpoints is fixed to this point, and that it can move above this surface in two opposite directions. And [suppose also that] through a point of this line (the one with the fixed endpoint) is drawn another line which can move three different ways: first a motion corresponding to that of the line with a fixed endpoint,[114] second a rotational motion around that [first] line, third a linear motion of the second line itself along its extension in both directions. Then, when an instrument includes these 3 motions, it is called a perfect compass. For example, line *AB* [fig. 5.44] will be the base of the compass and positioned in the plane in which is found the center of the compass. From *G* on line *AB* will be drawn line *GD* which moves in the plane going through *A, G, D, B,* toward both sides of *A* and *B*. Through *D* on line *GD* will be drawn line *EDZ* which has three motions: a motion corresponding

[114]The reader should think of the arm of an ordinary compass that carries the stylus or pencil.

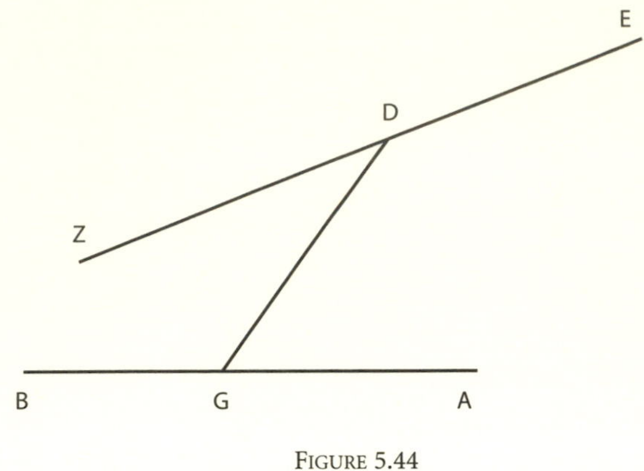

FIGURE 5.44

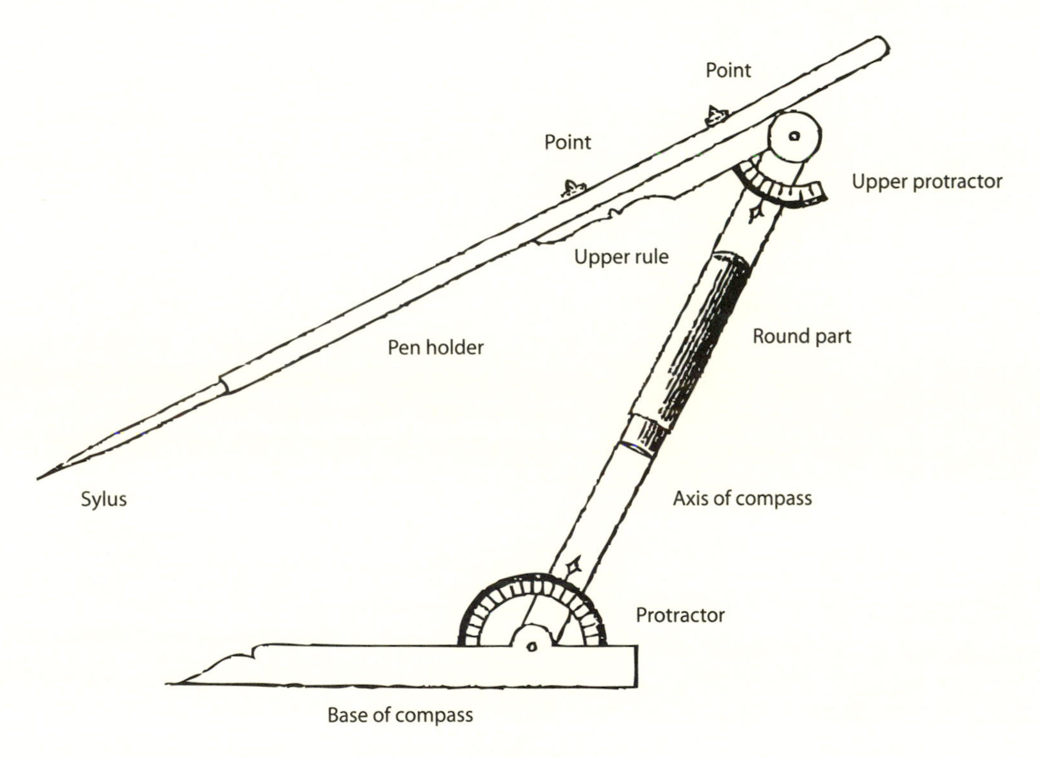

FIGURE 5.45

to the motion of line *GD*, another one around *G*, and a third one according to which line *EDZ* moves itself along its extension towards both sides of *E* and *Z*.

[We omit most of the description of the instrument, since we include a picture of it [fig. 5.45], and give only the last two paragraphs. The description is so detailed, however, that one assumes the author had actually seen such instruments and perhaps constructed them himself. We should add that when he speaks of the "axis" (labeled as such in fig. 5.45) of the perfect compass he is referring to the arm around which it rotates. An ordinary compass cannot have such an arm since the other arm does not have adjustable length.]

In the two angles that the compass axis makes with the base and the transversal ruler, are constructed two brass quarters of a circle[115] with their circumferences divided into as many parts as possible. These quarters of a circle must be fixed to measure the magnitude of the angles without disrupting the motion of the compass axis, as seen in fig. 5.45. The fitting of the two joints, the one found near the base of the compass and the one found near the pen, must be very exact, and extreme care is needed to execute all the work of the instrument with the highest cleverness and exactness.

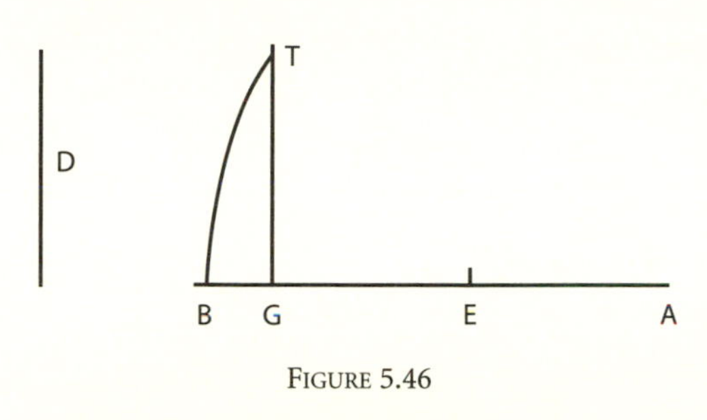

FIGURE 5.46

Proposition I

Problem: We wish to apply on the given straight line *AG* [fig. 5.46] a rectangle equal to the square of the given line *D* and larger than *AG* by a square.[116]

Solution: We halve *AG* at point *E*; we draw from point *G*, perpendicularly to *AG*, a straight line [*GT*] equal to straight line *D*. Taking point *E* as the center, we describe, with a radius equal to distance *ET*, a circle which intersects the extension of *AG* at point *B*.

Proof: *AB* times *BG* plus the square of *EG* is [*Elements* II, 6] equal to the square of *EB*, which is the sum of the two squares of *EG* and of *GT*. Removing the square of *EG* common to these two sums, the rectangle *AB* times *BG* is left equal to the square of *GT*, which is the square of *D*. That is what was to be proved.

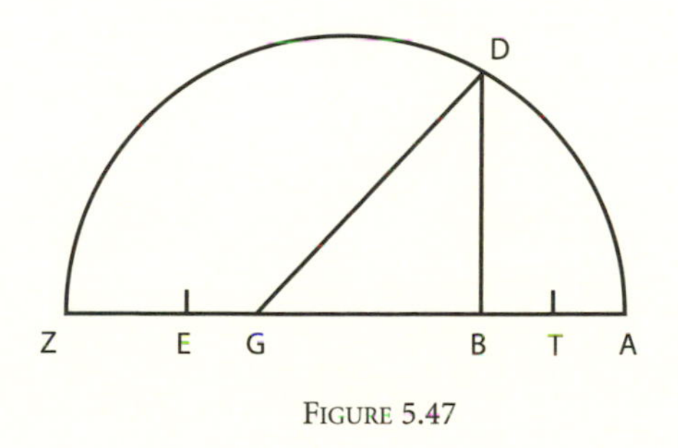

FIGURE 5.47

Proposition II

Problem: Given two straight lines *AB*, *BG* [fig. 5.47] joined [end-to-end] in a straight line, and a line perpendicular to *AG* being drawn from *B* and indefinitely extended, we wish to extend *AG* by an additional quantity such that a semicircle drawn on line *AG*, increased by the addition, as diameter, intersects the perpendicular at a point such that the line joined between this point and point *G* is equal to straight line *BG* increased by the addition.

Solution: We apply [by Prop. 1] on the double of *BG* a rectangle equal to the square of half of *AB* and exceeding the straight line that is the double of *BG* by a square. Let the side of the square be *GE*. We extend *BE* up to *Z* making *ZE* equal to half of *AB*, and we describe on *AZ* the semicircle *ADZ*. I say that line *DG* is equal to line *BZ*.

[We omit the demonstration of the validity of this solution.]

---

[115]In fig. 5.45 one of the two is shown as a semicircle.

[116]On "application of areas" see Heath 1981, i, 150–53.

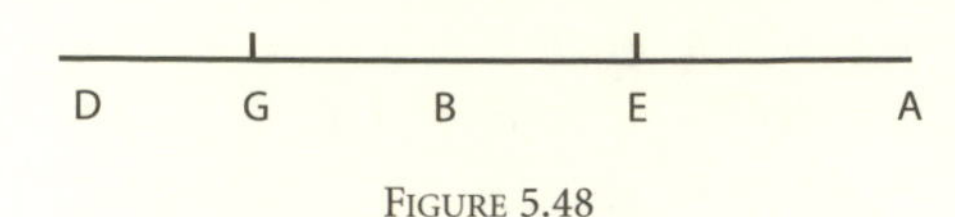

FIGURE 5.48

Proposition III

Problem: From the previous proposition, it follows that, given two straight lines, such as *AB*, *BG* [fig. 5.48] put end-to-end in a straight line, we can divide one of them, *AB*, at a point [*E*] such that the segment between this point and point *G* is to *GB* as the segment between this point and *B* is to the segment between the [same] point and point *A*.

Solution: We extend *BG* in the direction of *G* by the same amount as previously.[117] Let this extension be *GD*. We then take away [a segment *AE* equal to] *GD* from *BA*, to obtain *BE*. I say that the desired proportion is found, since *EG* is to *AB* as *EB* is to *EA*.

Proposition IV

Problem: Describe a parabola, given that: the *latus rectum* of the parabola is equal to the straight line *O* [fig. 5.49], the axis of the perfect compass with which one draws is equal to line *S*, and the axis of the parabola is the prolongation of a given line, *AB*.

Solution: We make segment *AG* equal to a quarter of *latus rectum O*, and segment *GB* equal to half of the axis *S*. We construct at point *G* a perpendicular, indefinitely extended, and we extend *AB* by the amount previously mentioned.[118] Let this extension be *BE* and let the semicircle *ADE* be the one making *DB* equal to *GE*.

We describe on *GE* [as diameter] a semicircle *GZE*, we make *G* center, and we mark with a compass opening equal to distance *GB* a mark *Z*. We join *GZ* and *ZE*;

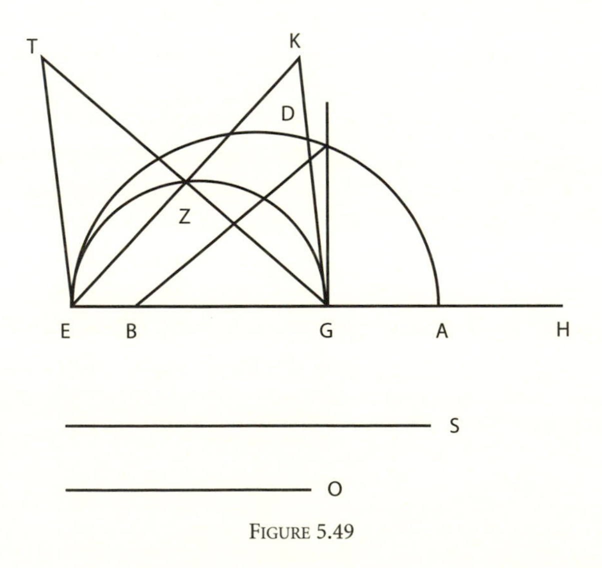

FIGURE 5.49

[117]This is an addition *GD* that satisfies the same conditions as *GZ* in fig. 5.47.

[118]I.e. by *GZ* in fig. 5.47. The straight lines *AG*, *GB* in fig. 5.49 correspond to *AB* and *BG* in fig. 5.47.

we extend *GZ* in a straight line to point *T*, we make *ZT* equal to *GZ*, and we join *TE*. Then the two angles [formed at points] *G* and *T* will be equal[119] and *GT* will be equal to the magnitude of the axis of the perfect compass.

I say that, if the straight line *TE* is put on the given line, if the two angles [at *T* and *G*] are preserved, and if the straight line *GT* is rotated around itself, the stylus of the perfect compass draws on the plane a parabola with point *E* as its vertex and *O* as its *latus rectum*.

Proof: The square of straight line *GE* is equal to the square of *DB*, which is the sum of the squares of *GB* and *GD*, i.e. of the squares of *GZ* and *GD*. But the square of *GE*, which is *BD*, is [also] equal to the sum of the squares of *GZ* and *ZE*. Removing on each side the square of *GZ*, there remains the square of *ZE* equal to the square of *GD*, which is equal to the rectangle *GE* by *GA*. Consequently *AG* is to *ZE* as *ZE* is to *GE*; hence, the double of *AG*, which is *HG* (i.e. half of the *latus rectum*) is to *ZE* as the double of *ZE*, which is *KE*, is to *GE*. Hence the whole *latus rectum* is to *KE* as *KE* is to *GE*; so, the square of *KE* is to the square of *GE* (which is the rectangle *GE* by *GK*) as the *latus rectum* is to *GE*. Thus the axis, *GT*, rotating around itself, generates by means of triangle *GZE* a cone with *EGK* as the axial triangle and which is intersected by the plane perpendicular to the plane of the axial triangle, the common intersection [*ET*] of the plane of the [axial] triangle and of the intersecting plane being parallel to one of the sides of the axial triangle [namely *KG*]. But the ratio of the square of the base, *KE*, of the triangle to the area [formed by the product of] one of the two sides of the triangle, *GE*, in the other side, *GK*, is as a certain straight line, *O*, is to a certain [other] straight line, *GE*. So, the straight line *O* is the *latus rectum* of a parabola with point *E* as its vertex and with *ET* as its axis, as was proved in the eleventh proposition of the first book of Apollonius's *Conics*.[120] And that is what was to be proved.

Abū al-ʿIzz Ismāʿīl ibn al-Razzāz al-Jazarī, whose nickname was Badīʿ al-Zamān ("The Prodigy of his Age"), was born in upper Mesopotamia, probably around 1150. He served a family of Artuqid princes who ruled in Diyar Bakir and were, by his time, vassals of the great Kurdish general Saladin and his successors. Al-Jazarī completed his *Book of Knowledge of Ingenious Mechanical Devices*, from which the present selection has been taken, in about 1205. It is a unique collection of descriptions of how to construct such devices as water clocks, "vessels suitable for drinking sessions," a mechanized slave who would pour water over a worshipper's hands for ritual ablutions, and fountains that would display different patterns of water at different times of the day. It is part of a tradition of writing about practical mechanics that goes back at least to Heron of Alexandria, who, both in his *On the Art of Constructing Automata* and in his *Mechanics*, wrote about similar topics in the first century CE. (The latter work was translated into Arabic in the mid-ninth century.)

---

[119]This is because *EZ* is perpendicular to *GZ*, extending *ZT* = *GZ* makes *EZ* the perpendicular bisector of *TG*. So *EG* = *DT*, and the angles at *G* and *T* are equal.

[120]*Conics* I, 11 proves, roughly speaking, that if a plane cuts a cone so that the diameter of the section it makes is parallel to a side of the axial triangle of the cone, then, if an appropriate length is taken as parameter, the curve bounding the section satisfies the equation of a parabola with that parameter.

Al-Jazarī on Constructing Geometrical Instruments

Chapter 2 of Category VI

It is an instrument by means of which the center-point of three points of unknown position, lying on the surface of a sphere, may be precisely determined; also [for three points] on a horizontal surface, provided that they are not in a straight line; the other angles in [general] use, acute and obtuse, may be also determined with it. It is divided into three sections.

Section 1

On the purpose of the instrument, and its construction

I say that 2/3 of a circle, 1/3 of a circle, an arc of a circle, or more or less than these, can pass through any three points lying on the surface of a sphere. When I mentioned this, some people denied it, and demanded that I should fix one foot of the dividers on the center point and pass the other leg through the three unknown points [implying that] there was an error in that.

The straight line joining two of the points is divided at its center by a mark, upon which a line is erected making two equal angles with the first line. Similarly a line is drawn between the second and third points and on its center a mark is made, upon which a line is erected making two equal angles with the first line. Where the lines cross is the center-point.

I made an instrument for that to facilitate the determination of the required center-point, and the determination of all the angles in [general] use, acute and obtuse. This is [the method]: we take a brass ruler of [adequate] thickness, about 3 spans[121] long: in the center of the ruler is a projection in the shape of a semi-circle. I have shown a picture of this [fig. 5.50]: the ends of the ruler are marked *j*, *h* [on the left end] and the center point of the semi-circle *y*.

On center-point *y* a semi-circle *z* is drawn, a short distance inside the perimeter of the [first] semi-circle. The edge of the ruler is verified [for straightness]. Then from center point *y* on the second drawing [i.e., referring to fig. 5.52] a line *n*[122] is drawn which divides the semi-circle into two halves, and which finishes at the ruler's edge,

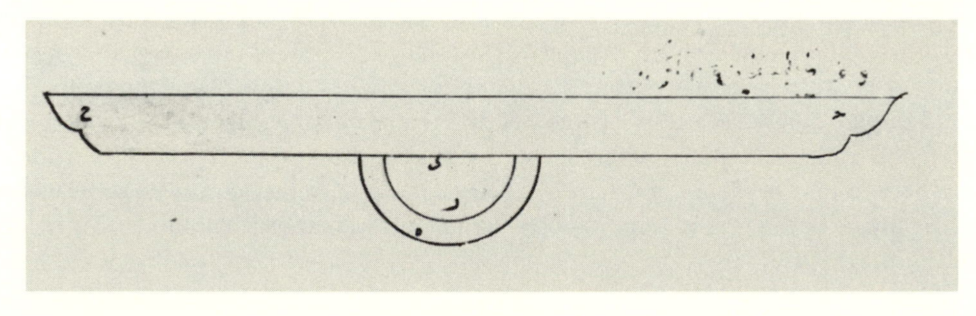

FIGURE 5.50

[121]A span is 25 cm. Henceforth we abbreviate it as "sp."

[122]The line *n*, which did not reproduce well in the diagram here, is a vertical line running up the middle of the vertical black ruler.

dividing its length into two halves. The intersection of the semi-circle with the end of the line is marked *f*. Then the face of the ruler from line *n* to the end *j* of the ruler is divided into as many divisions as possible, and from line *n* to end *h* it is divided in the same way as from line *n* to end *j*. Between every five divisions a "5" is written on the ruler.

Then an alidade is made, about 1½ sp. long, and as wide as five divisions of the ruler. Near one of its ends, a projection is fitted, a very small semi-circle is drawn on the projection, and a narrow hole is made in its point-point. In the end of the alidade nearest the semi-circle a hole is made—the distance between this hole and the hole in the center-point of the semi-circle equals the distance between center-point *y* and semi-circle *z* on the ruler. This is a picture of the alidade [fig. 5.51]: on the pierced end of the alidade is *t*, on the hole in the center of the semi-circle is *p* and on the other end is *q*.

In the two holes [i.e., holes *y* and *p*] a nail is inserted, and its ends are hammered gently until they bend down over the ruler and the alidade, in such a manner that when the alidade is moved over the ruler it rotates truly. Then the alidade is moved over the ruler until its edge with the semi-circle on it is on line *n* of the ruler, and the alidade is at right angles to the ruler. At this juncture the end of the drill is placed in hole *p* and a hole is made on circle *z* and through line *n*. The drill is lifted out and a nail is inserted tightly in the hole. On its end is a protruding piece for lifting it out and for inserting it. This is the complete picture [fig. 5.52].

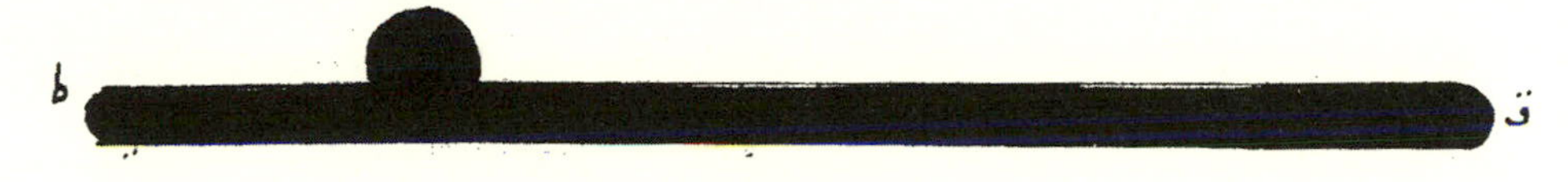

FIGURE 5.51

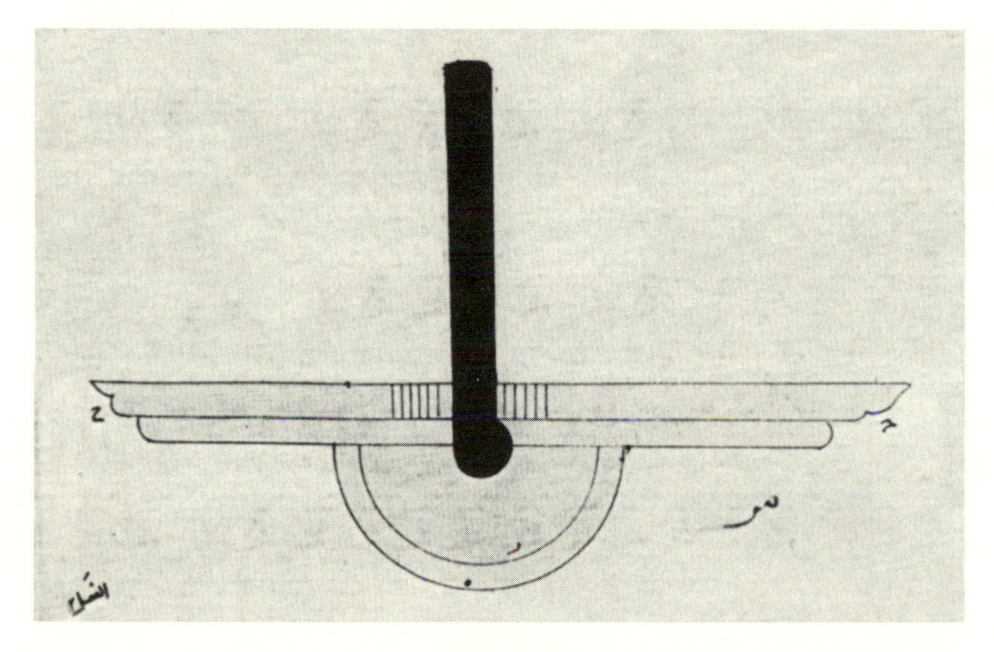

FIGURE 5.52

Section 2

Use of this instrument for determining the center-point: either for determining the center-point of three points of unknown position in the surface of a sphere, with precision; or for determining the center point of three points of unknown position on a horizontal surface, with the exception of one case, i.e., when the points are in a straight line.

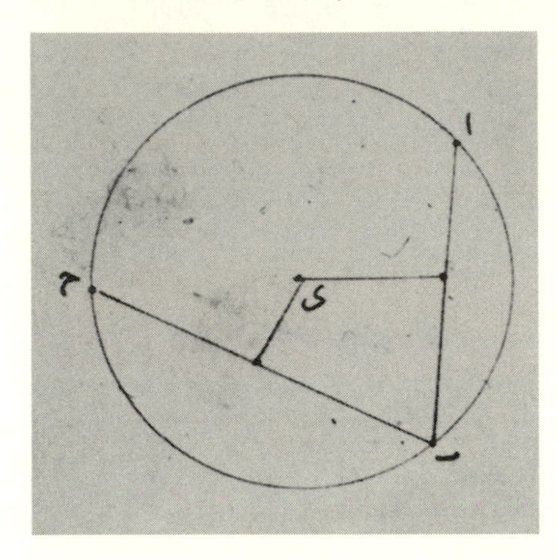

FIGURE 5.53

Consider the points *a*, *b*, *j* to be on the surface of a sphere or on a horizontal surface [fig. 5.53].[123]

The ruler is laid on points *a*, *b* and the alidade is near the center of *ab* and towards the space between the three points. The divisions are to the right and to the left of the alidade. The ruler is moved until the edge of the alidade is exactly in the center of the two points, whereupon a line is drawn along the edge of the alidade. Then the ruler is moved to points *b*, *j* and adjusted to bring the edge of the alidade exactly in the center of points *b*, *j*, whereupon a line is drawn along the edge of the alidade, cutting the first line. The intersection is the center-point of the three points. This is the picture of the three points [fig. 5.53]: through them are two lines drawn along the edge of the alidade, intersecting in *y*. If one foot of the dividers is placed on the intersection, the other foot will pass through the three points.

Section 3

Use of this instrument for setting out different angles

When the alidade is at right angles to the ruler, right angles can be set out with it, without altering it. Then one takes a flat board, the edge of which has been made true with the ruler. The angle of an equilateral triangle is drawn on it, the edge of the board being one of its sides. Then the ruler is placed against the edge of the board. The nail is removed from the holes in the alidade and the ruler. The alidade and the ruler are moved until the edge of the alidade is aligned with the side of the triangle. Then the drill is placed in the hole at the end of the alidade and a hole is made through the ruler's semi-circle. The nail is inserted in this hole, and the alidade on the ruler is now showing two angles, i.e. the angle of an equilateral triangle. From this one can obtain a sextuple acute angle and a sextuple obtuse angle.

Then an acute quintuple angle is marked on the edge of the board, the edge of the ruler is placed against the edge of the board, the nail is lifted from the hole of the triple [angle] and the ruler and the alidade are moved until the alidade is on the side of the quintuple [angle]. The drill is placed in the hole in the alidade and a hole is made through the ruler's [semi-]circle. The nail is inserted in it, and the alidade on

[123] The points *a*, *b*, *j* are the three points in clockwise order on the circumference of the circle in fig. 5.53. The point *y* is the center.

the ruler is now showing two angles—a quintuple acute angle and a quintuple obtuse angle. [One proceeds] in this way for whatever angles are chosen. On every hole on the ruler's semi-circle one writes the name of its angle. To the end of the nail the end of a light chain is attached, the other end of which is attached to a staple on the semi-circle, so it [i.e., the nail] is permanently retained.

That is what I wished to describe clearly.

Naṣīr al-Dīn al-Ṭūsī (1201–1274), from Ṭūs in what is now Iran, completed his formal education in Nishapur, then a major center of learning, and soon gained a great reputation as a scholar. During the early part of his adult life, he worked for the Ismaʿīlī rulers, but after their dynasty was defeated by the Mongol leader Hūlāgū in 1256, he transferred his allegiance to the new ruler. During the final two decades of his life, in fact, al-Ṭūsī worked at the observatory at Maragha at the head of a large group of astronomers.

In this selection from al-Ṭūsī's geometrical work, he investigates the theory of parallels and attempts to prove Euclid's parallel postulate. We give here the statements of seven propositions, the final one being the parallel postulate, but only give his proofs of two of them.

Al-Ṭūsī on the Theory of Parallels

As for the approaches by means of which I investigated this after studying the words of these scholars, we will set forth our discourse in seven propositions, two of which are taken from al-Khayyām's propositions; they are our second and fourth which are his first and fourth. Let the beginning of the book the *Elements*, the twenty-eight propositions that do not include a doubtful postulate remain unchanged, and then we will add these propositions.

Proposition I: The shortest one of the lines drawn from a point to any line whose ends are not bounded, called the distance from that point to that line, is the perpendicular dropped from the point to the line.

Proposition II: If one erects two equal perpendiculars to a straight line and joins their ends by means of a straight line then they form equal angles [with the latter].

Proposition III: If one erects two equal perpendiculars and joins their ends by means of a straight line then the angles formed by them are right angles.

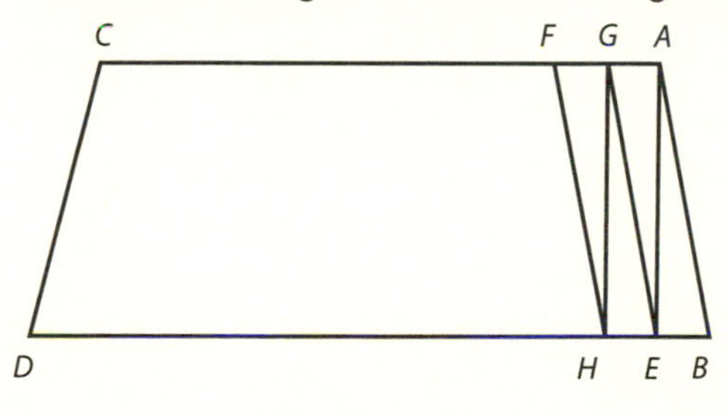

FIGURE 5.54

Example: Equal perpendiculars *AB* and *CD* are erected on the line *ED*, and their ends are connected by means of the line *AC*. I claim that the equal angles *BAC* and *DCA* are right angles.

Proof: If they are not right angles then they are obtuse or acute. First we assume that they are obtuse and we erect in the first drawing (fig. 5.54) at the point *A* a perpendicular *AE* to the line *AC* as proved in proposition II. It necessarily falls between the lines *AB* and *CD* and, as proved in proposition 16,[124] the angle *AED*, an exterior angle of the triangle *ABE*, is greater than the interior right angle. Therefore

[124]The proposition numbers in the proof are from propositions in Book I of Euclid's *Elements*.

it is also obtuse. Now we erect at the point *E* a perpendicular *EG* to the line *ED*. It falls between the lines *AE* and *CD*, and the angle *EGC*, an exterior angle of the triangle *EAG*, is greater than the interior right angle *A*. Therefore it is also obtuse. Further, we erect at the point *G* a perpendicular *GH*, again to the line *AC*, and in this order we will indefinitely continue to erect perpendiculars. Then the perpendiculars drawn from points located on the line *AC* at a right angle to the line *BD*, [that is], the perpendiculars *AB*, *GE*, *FH*, successively increase in length. The shortest of them is the perpendicular *AB* which subtends the acute angle *AEB* in the triangle *AEB* and is therefore shorter than *AE* which subtends the right angle *ABE*; this follows from proposition 19. *AE*, which subtends the acute angle *AGE* in the triangle *AEG*, is shorter than *GE* which subtends the right angle *EAG*. Therefore *AB* is also shorter than *GE*. In just this way it is shown that *GE* is shorter than *FH* and *FG* is shorter than that which follows it. Thus the perpendiculars closer to *AB* will be shorter and the distances between the points that are the feet of the perpendiculars dropped from the points of the line *AC* to the line *ED* successively increase in the direction of *C*, so that the lines *AC* and *ED* diverge in the direction of *C* and converge in the direction of *A*. But the angle *DCA* is also obtuse, for it is equal to the angle *BAC* by the previous proposition. Thus we prove, as before, that the lines *CA* and *DB* diverge in the direction of *A* and converge in the direction of *C*. But this is absurd. Hence the angles *BAC* and *DCA* are not obtuse.

If, on the other hand, these angles are acute, then in the second drawing (fig. 5.55) we will drop from the point *B* a perpendicular *BE* to the line *AC*, as proved in proposition 12. It necessarily falls between the lines *AB* and *CD*, for the angle *A* is acute and it is impossible that it falls outside these lines. In the right triangle *AEB* the angle *ABE* is acute, so that the angle *EBD* which, together with the angle *ABE*, forms the right angle *ABD* is also acute. Next we drop from the point *E* a perpendicular *EG* to the line *BD*. It falls between the lines *AB* and *CD* and the angle *GEC* is acute. Then we drop from the point *G* a perpendicular *GH* again to the line *AC* and in this order we will indefinitely continue to drop perpendiculars. Then the perpendiculars drawn from the points on the line *AC* at right angles to the line *BD*, the perpendiculars *AB*, *EG*, *HF*, successively decrease in length. The longest one of them is the perpendicular *AB*. In this way it is shown that the lines *AC* and *BD* converge in the direction of *C* and diverge in the direction of *A*. But the angle *DCA* is also acute, for it is equal to the angle *BAC* by the previous proposition. Hence, as before, it is shown that the lines *CA* and *DB* converge in the direction of *A* and diverge in the direction of *C*. But this is absurd. Hence the angles *BAC* and *DCA* are not acute, and since they are not obtuse either, they are right angles. And this is what we wished to prove.

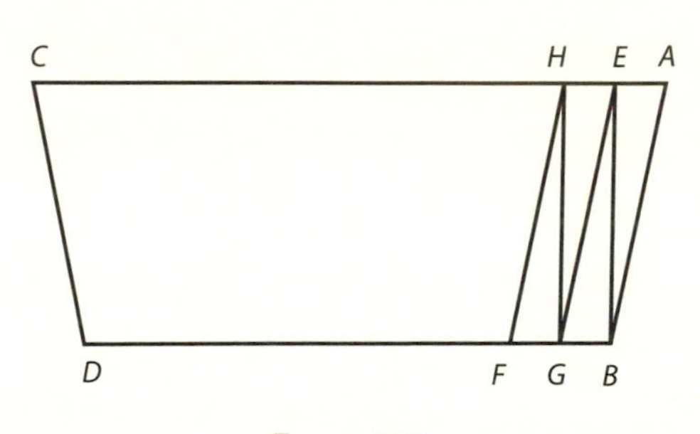

FIGURE 5.55

Proposition IV: Every two opposite sides of a right-angled quadrilateral are equal.

Proposition V: If a straight line falls on two perpendiculars erected in an arbitrary way on another straight line, then the alternate angles are equal, each exterior angle is equal to the [corresponding] interior opposite angle, and interior angles on the same side are equal to two right angles.

[In proposition VI Nasīr al-Dīn offers a proof of the parallel postulate for the case when the transversal is perpendicular to one of the two given lines and oblique to the other.]

Proposition VII: If a straight line falling on two straight lines makes the interior angles on the same side [together] less than two right angles, the two straight lines, if produced indefinitely, meet on that side.

Example: The line *AB* falls on the lines *CD* and *EG* and forms angles *CHF* and *EFH* that are together less than two right angles (fig. 5.56).[125] I claim that the lines *CD* and *EG*, if extended in the direction of *C*, will meet.

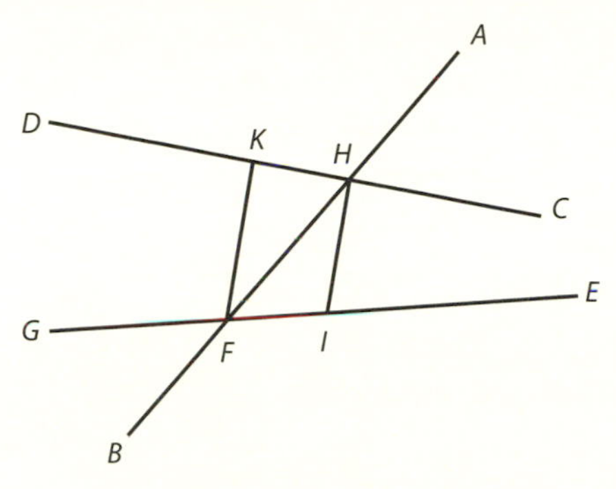

FIGURE 5.56

Proof: If one of the angles *CHF* and *EFH* is a right angle, then the other angle is necessarily acute. Therefore one of the lines *CD* and *EF* intersects the line *AB* at an angle that is not a right angle and the other is perpendicular to it, so that if we extend them, they will meet on the side of the acute [angle], as was proved in the previous proposition.

If one of the angles is obtuse, let it be the angle *CHF*, we erect at *H* a perpendicular *HI* to the line *CD*, as shown in proposition 11, and from *F* we drop to it [*DC*] a perpendicular *FK* as shown in proposition 12.[126] Then we claim that, since the angles *CHF* and *EFH* are together less than two right angles and *CHI* is a right angle, the angles *IHF* and *HFI* are together less than a right angle. But, as was shown in the fifth of these propositions, the angles *IHF* and *HFK*, as alternate angles formed by the line *AB* falling on two perpendiculars *IH* and *FK*, are equal. Hence the whole angle *KFI* is less than a right angle, that is, it is acute. Therefore the lines *KF* and *EF* intersect at an angle that is not a right angle and the line *HK* is perpendicular to one of them, namely to the line *KF*. Therefore, as shown in the previous proposition, the lines *CK* and *EF* will meet if extended in the direction of *C* and *E*.

If both angles are acute then we will erect at the point *F* a perpendicular *FK* to the line *GE*, as shown in proposition 11, and drop to it from the point *H* a perpendicular *HI*, as shown in proposition 12. Then *EFK* is a right angle, and the angles *KFH* and *FHI* are equal as alternate angles formed by the line *AB* falling on two perpendiculars *HI* and *KF*, as proved in the fifth of these propositions. Hence the angles *FHI* and *HFI* are together equal to a right angle. Since, by assumption, the angles *EFH* and *CHF* are less than two right angles, the angle *IHC* is less than a right angle, that is, it is acute. Therefore the lines *IH* and *CH* intersect at an angle that is not a right angle and *EI* is perpendicular to one of them, namely to *IH*. Therefore [the lines] *CD* and *EG* intersect when extended in the direction of *C* and *E*, as shown in the previous proposition. And this is what had to be proved.

---

[125]We have redrawn this figure since the one in our source is incorrect.

[126]The numbers 11 and 12 refer to the propositions with those numbers in *Elements* I.

VI.b. Practical geometry

In this work, Abū Sahl al-Kūhī, who worked in the latter part of the tenth century, solves a group of problems concerning a tower built on an island in the sea. The problems he solves are those of determining any two of the following three quantities from a knowledge of the third: the height of the tower, the surface area of the part of the sea visible from the top of the tower, or the angular excess of the part of the sky, imagined as a great sphere, above the horizon over that below the horizon. Our translation, of the part where he finds the surface area from the height, is from the Arabic manuscript Mashhad 5414 (Riad 184).

Al-Kūhī on What is Seen of Sea and Sky

[Al-Kūhī begins with a dedication to a court official, a vizier whom he names only as "The Professor". He then continues:]

The benefits of the science of what this epistle contains are many, and especially to one who wants to build on an island of the sea a guide [tower] in order to lead sailors by it, and by the light spread from its top by night, and the smoke [spread] by day. And that will be a reason for their being saved from foundering and by that he will seek nearness to God, Mighty and Majestic, and recompense and reward [for good deeds] in the next world, and remembrance and praise in this. And that is that he knows by this science how far away [from its base] the top of anything elevated above the surface of the sea is seen, and he will begin with making what is possible and he will not seek what is impossible.

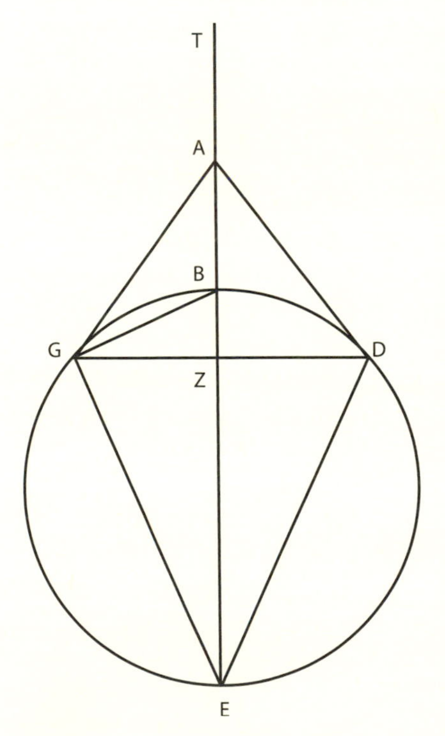

FIGURE 5.57

[...]

If the thing of known height above the surface of the water of the sea is AB (fig 5.57) and the magnitude of the surface of the water of the sea seen from the point A is GBD, then between the eye and the surface of the water of the sea there is a cone, whose vertex is at the point A and whose surface is tangent to the sphere of the water of the sea along a circle (for the water of the sea is spherical).[127] And the diameter of that circle is GD. So I say that if the height of AB is known then the surface that the circle whose diameter is GD cuts from the surface of the water of the sea is known.

Proof: The intersection of the conical surface and the plane that passes through the point A

[127]The sphericity of the earth was known to educated Greeks from the time of Plato on, and consequently to the educated in medieval Islam as well.

and the center of the earth are the two lines *AG*, *AD*, and they are tangent to the circumference of the circle that is the intersection of that same plane and the surface of the sphere of the water of the sea at the two points *G*, *D*, which circle is *GBDE*. And its diameter, *EB*, is known since the diameter of the earth is known by the observation of certain observers.[128] And the line *BA* is known, so the line *EA* is known. And hence the product of the line *EA* in *AB* is known. And it is equal to the square of the line *GA* (since it is tangent to the circle). And so the square of the line *AG* is known. But the square of the line *AG* is equal to the two squares [taken together] of the two lines *GZ*, *ZA*, so the squares of the two lines *GZ* and *ZA* [taken together] are known.

Now we make the line *AT* equal to line *AB*, and [so] the square of the line *ZA* is equal to the product of the line *TZ* by *ZB* with the square of the line *AB* (since line *TB* is bisected and the line *BZ* is added to it[129]). And so the square of the line *GZ*, with the product of the line *TZ* by line *ZB*, and with the square of the line *AB* is known. And the square of the known line *AB* is cast away, and so there remains known the square of the line *GZ* with the area *TZ* in *ZB*. And the square of line *GZ* is equal to the product of line *EZ* by line *ZB* since line *EB* is a diameter of the circle,[130] and so the product of line *EZ* in line *ZB* with the product of line *TZ* in line *ZB*, i.e., the product of line *ET* in line *BZ*, is known. And line *ET* is known so line *ZB* is known, and so its product in line *ZE* is known since *ZE*, the line remaining from the diameter of the circle is known. And the product of the line *BZ* in line *ZE* is equal to the square of line *ZG*. And so the square of line *ZG* is known. And the square of line *ZB* is known. And we join line *GB*, and so the sum of the two squares of the two lines *GZ*, *ZB* is equal to the square of line *GB*. And so the square of line *GB* is known. And so the circle whose radius is line *GB* is known,[131] and it is equal to the surface cut off from the spherical surface of the sea by the circle whose diameter is *GD*, as Archimedes proved in the book *Sphere and Cylinder*.[132] And so *GBD*, the spherical surface of the water is known, and it is the magnitude of what is seen of the surface of the water of the sea from the point that is the top of the elevated thing. And so if the height of that thing from the surface of the water of the sea is known then the magnitude of the surface of the water of the sea which is seen from the head of that thing elevated from it is known. And that is what we wanted to prove.

Abū al-Wafā' al-Būzjānī tells us in his book, *On the Geometric Constructions Necessary for the Artisan*, that he collaborated with artisans on geometry. In this passage, he notes that, although artisans frequently used the technique of cutting up squares of material and rearranging them into other designs, they sometimes made elementary geometric errors.

---

[128]Among the "observers" to whom Abū Sahl might have been referring were the two teams of astronomers and surveyors who, according to a report by al-Bīrūnī (which we have reproduced in this volume), the Caliph al-Ma'mūn commissioned to find the circumference of the earth.

[129]The reference here is to *Elements* II, 6.

[130]The reference here is to *Elements* III, 35.

[131]Euclid's *Data*, a work Abū Sahl knew well, states in Prop. 55 that if a square is given then its side is given. For the Arabic authors, the Greek "given" became the Arabic "known," and since Abū Sahl showed that GB^2 is known he deduced from *Data*, 55 that *GB* is known.

[132]Abū Sahl is referring to Archimedes' *Sphere and Cylinder*, Book I, Prop. 42.

Abū al-Wafā' on the Geometry of Artisans

A number of geometers and artisans have erred in the matter of these squares and their assembling. The geometers [have erred] because they have little practice in constructing, and the artisans [have erred] because they lack knowledge of proofs. The reason is that, since the geometers do not have experience in construction. it is difficult for them to approximate [i.e., construct in practice]—in the way required by the artisan—what is known to be correct by proofs by means of lines [i.e., diagrams].

The aim of an artisan is what is approximated by the construction, and [for him] correctness is apparent by what he perceives through his senses and by inspection. He is not concerned with the proofs by means of lines. If, for a geometer, the proof of something is established by way of imagination, he is not concerned with the [apparent] correctness or incorrectness of something by inspection. But we do not doubt that everything that the artisan sees is taken from what the geometer had worked out previously and what had been proved to he correct. Therefore, the artisan and the surveyor take the choice parts [literally, the cream] of the thing, and they do not think about the methods by which correctness is established. Thus occur the errors and mistakes.

The geometer knows the correctness of what he wants by means of proofs, since he is the one who has derived the notions on which the artisan and the surveyor base their work. However, it is difficult for him to transform what he has proved into a [practical] construction, since he has no experience with the practical work of the artisan and the surveyor. If the skillful among these geometers are asked about something in dividing the figures or multiplying the lines, they are confused and need a long time to think. Sometimes they are successful. and it is easy for them; but sometimes it is difficult for them and they do not find its construction.

I was present at some meetings in which a group of geometers and artisans participated. They were asked about the construction of a square from three squares. A geometer easily constructed a line such that the square of it is equal to the three squares, but none of the artisans was satisfied with what he had done. The artisan wants to divide those squares into pieces from which one square can be assembled, as we have described for two squares and five squares.

The artisans proposed a number of methods, some of which can be proved and others of which are incorrect, but the methods which cannot be proved resemble the truth in appearance, so someone who looks at them may think that they are correct. We shall present these methods so that the correct ones may be distinguished from the false ones. and someone who looks into this subject will not make a mistake by accepting a false method, God willing.

One of the artisans placed one of the squares in the middle and bisected the second by means of the diagonal and placed it [the parts] on both sides of the [first] square. He drew from the center of the third square two straight lines to two of its angles, not on one diagonal. and he drew a line from it [the center] to the midpoint of the side opposite the triangle which is produced by two lines. Then the square is divided into two trapezia and a triangle. Then he put three triangles below the first square and the two trapezia above it, and he joined the two longer sides [of the trapezia] in the middle. Thus he obtained a square as in this figure (fig. 5.58):

Abū al-Wafā' said: But this figure which he constructed is fanciful, and someone who has no experience in the art or in geometry may consider it correct, but if he is

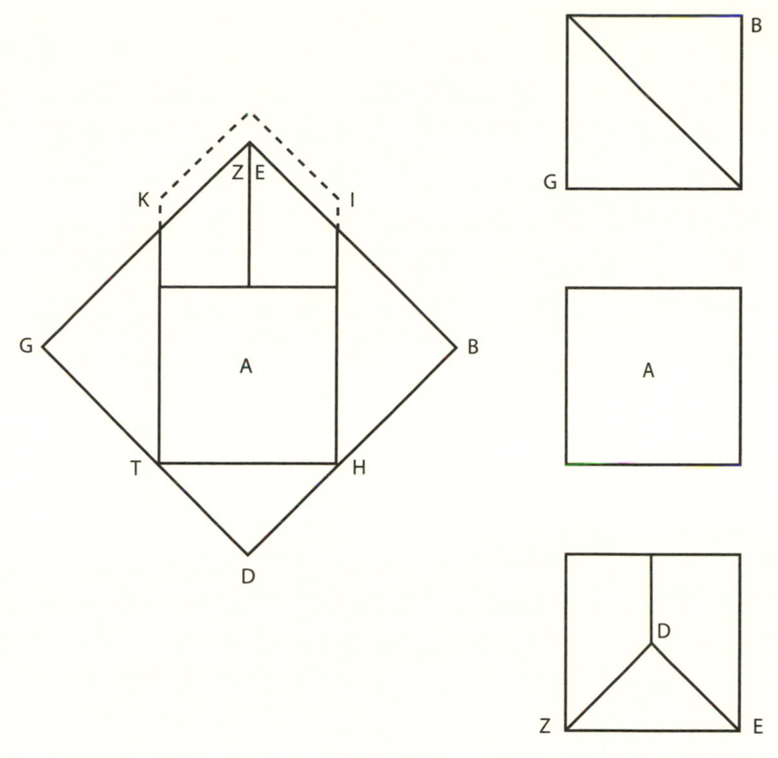

FIGURE 5.58

informed about it he knows that it is false. He may imagine it to be correct because of the correctness of the angles and the equality of the sides. The angles of the square are correct, each of them is a right angle, and the sides are equal, and because of this it is imagined to be correct. The reason is as follows. Each of the angles of the [three] triangles, namely *G*, *B*, and *D* which are [also] angles of the square, is a right angle; and the fourth angle is composed of two angles each of which is half a right angle, for they are the angles of the trapezia. The sides are straight and equal, since each of these sides is composed of a side of one of the squares plus half its diagonal that are equal.[133] It is also clear that they are straight [lines] when assembled, because the sums of the angles at the meeting points of the lines are all equal to two right angles. For the three angles at point *H* are equal to two right angles, since they are one angle of a square and two angles of a triangle, each of which is equal to half a right angle. The same [is true] for angle *T*. Angle *I* consists of two angles; one of them is the angle of the triangle, that is half a right angle, and the other is the angle of the trapezium, that is one right angle plus half a right angle. The same [is true] for the two angles at point *K*. Since the angles are right angles and the sides are straight and equal, everybody imagines that it [i.e., the figure] is a square constructed from three squares.

But they do not notice the place where the error and mistake enters their argument. This is clear if we know that each of the sides of this square is equal to the side of one of the [unit] squares plus half of its diagonal. It is not possible that the side of the square

---

[133]He means any two sides of the square are equal, as are any two half-diagonals. So, the sum of a side and a half diagonal is equal to the sum of another side and half diagonal.

composed from three squares has this magnitude, since it must be greater. The reason is as follows. If we make the side of each [unit] square approximately ten ells, to make it easy for the student, the side of the square composed of three squares is by approximation seventeen and one-third ells. But the side of this square [which has been constructed] is seventeen and one-fourth ells, and there is a big difference between them.

Again, since we bisected [the unit] square *BG* and placed each half of it at the side of the square *A*, [the constructed] square *BG* is cut by two lines, *HI* and *TK*, that [the latter] does not coincide with it [the former] because of two things. First, the diagonal of square *BG* is irrational but line *HI* is rational since it is equal to the side of square *BG* plus half of it. Second, it is less than that because the diagonal of square *BG* is by approximation fourteen and one-seventh, and side *HI* is fifteen. Thus the incorrectness of this [method of] division and assembling has become clear.

Some people have divided these squares in another way, which is even more clearly incorrect than the first division. This [method] is as follows. In the middle of the diagonal of two of the squares, a segment is cut off equal to one of their sides. From the two endpoints of [this segment of] the diagonal, four triangles are cut off. Thus, two squares become four irregular pentagons and four triangles. Then each pentagon is placed at one side of the third square. At its four angles one obtains [empty] places for four triangles. Then the remaining triangles are transported to these places, and one obtains a square from three squares as in this figure (fig. 5.59):

The division of the squares by the correct method, as required by proof, will be clearer according to the method that we now mention. We bisect two of the squares

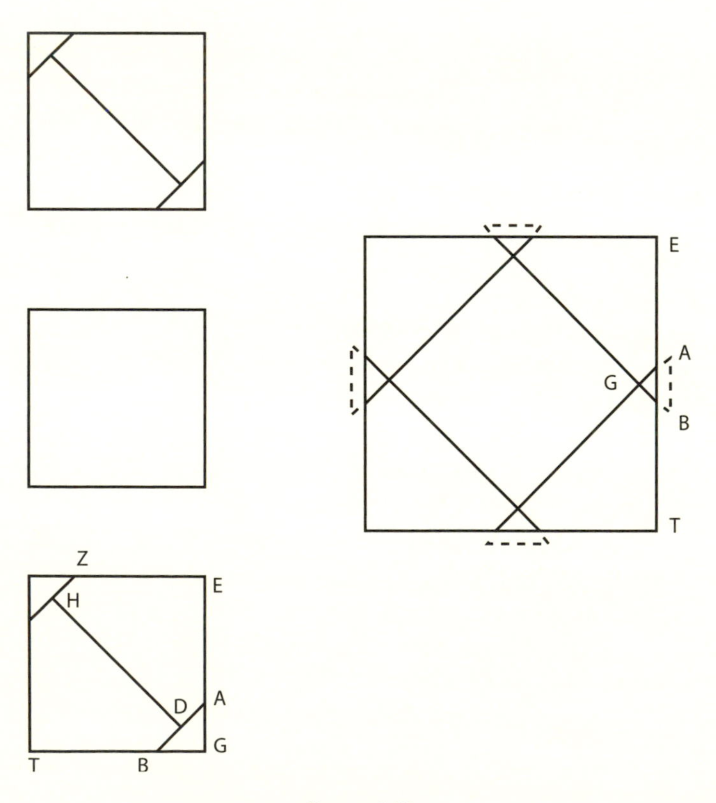

FIGURE 5.59

along their diagonals. Each of those is applied to one side of the third square; we place one of the angles of the triangle which is half a right angle at one angle of the square, and the hypotenuse of it [the triangle] at the side [of the square]. Thus, part of the triangle sticks out at the other angle [of the square]. Then we join the right angles of the triangles by means of straight lines. That becomes the side of the desired square. From each big triangle, a small triangle is cut off for us [by a straight line], and we transfer it to the [empty] position of the triangle that is produced at the other angle.

Example of this: If we want to construct a square from three equal squares *ABGD*, *EWZH*, *TIKL*, we bisect two of the squares at their diagonals, by means of lines *AG*, *EH*, and we transport [them] to the sides of the [third] square. Then we join the right angles of the triangles by lines *BZ*, *ZW*, *WD*, *DB*. On either side [of the straight line], a small triangle has now been produced from the sides of the [two big] triangles. That [empty position of the triangle] is equal to the triangle which has been cut off from the big triangle. Thus triangle *BGM* is equal to triangle *MZH*, since angle *G* is half a right angle, angle *H* is half a right angle, the two opposite angles of the triangles at *M* are equal, and side *BG* is equal to side *ZH*. Therefore, the remaining sides of the triangles [*BGM*, *MZH*] and the triangles are equal. Thus, if we take triangle *BGM* and put it in the position of triangle *MZH*, line *BZ* is the side of the square constructed from three squares. This is a correct method, easier than what was constructed [by the artisans], and the proof of it has been established. This is the figure for it (fig. 5.60):

If the geometer is asked for a construction of a square equal to any number of squares, he will find for you the line that is equal in square to these squares, and he

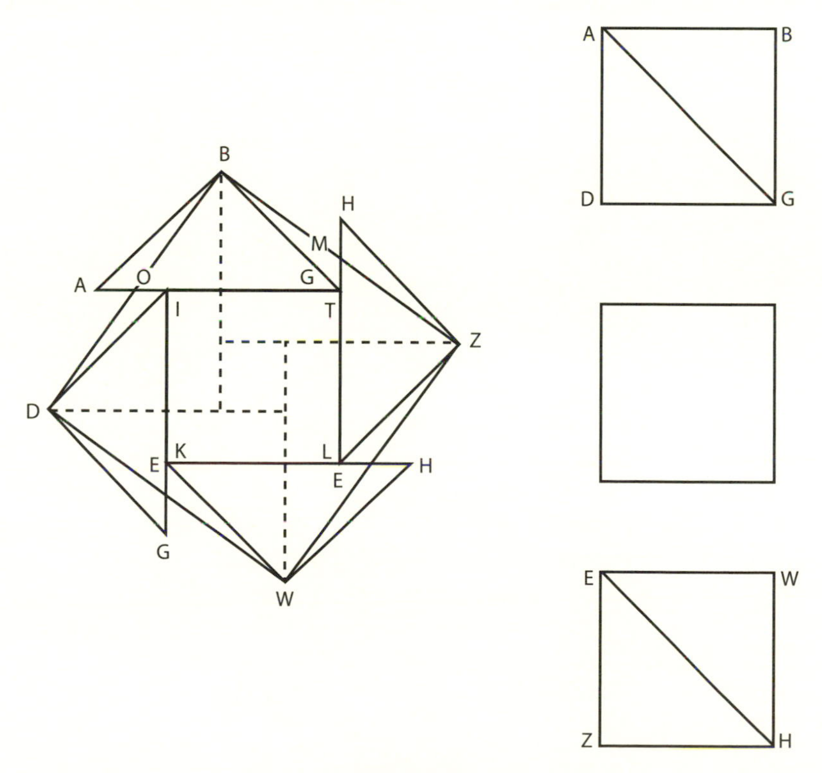

FIGURE 5.60

will not be concerned with the way in which the squares have to be cut. That is to say, if he is asked for the construction of a square from three [equal] squares, he will draw the diagonal of one of the squares, and erect at one of the endpoints of the diagonal a line perpendicular to it and equal to the side of the square. He will join its endpoint with the [other] endpoint of the diagonal by means of a straight line. Then that is the side of the square composed of three [equal] squares. Example of this: If we want to construct one square equal to three squares, each of which is equal to square *ABGD*, we draw diagonal *AD* (fig. 5.61). Then *AD* is the side of the square composed of two squares. Then we erect at point *A* of line *AD* perpendicular *AE* equal to line *AG*, and we join *ED*. The line *ED* is the side of the square equal to three squares, each of which is equal to square *ABGD*.

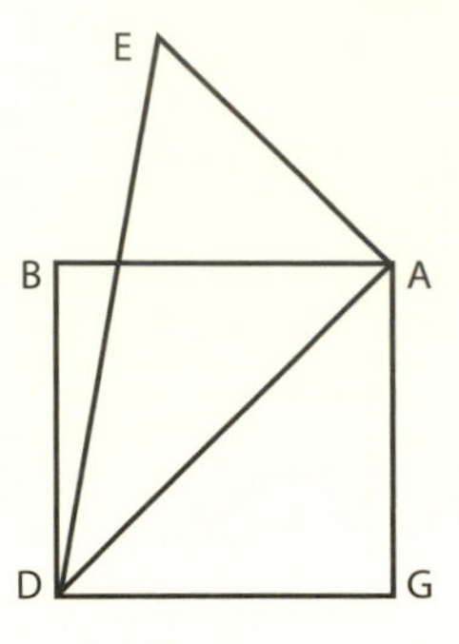

FIGURE 5.61

If the geometer has obtained this line, he is not concerned any more with the way in which the squares have to be cut. He will say that if a square is constructed on line *ED*, it is equal to three squares. And [we can construct] similarly, if we want the square to be equal to more than three squares or less than three squares.

A Persian manuscript found in the Bibliothèque Nationale in Paris, titled *On Inscribing Similar and Congruent Figures*, contains an extensive set of geometric problems.[134] These demand either constructions (some of them with a compass opening given in advance) or the dissection of polygons (not necessarily convex) into polygonal pieces that can be reassembled to form other polygons. The unknown author gives no proofs for any of the solutions, nor does he usually distinguish approximate solutions from exact ones.

The following three problems are chosen from those that deal with dissections. The reader will see that the author gives, for each problem, a procedure for doing the thing demanded and illustrates it with appropriate figures. These figures carry a lot of information because in the case of dissection problems it is they, and not the text, that show how the figures of one piece can be rearranged to form another. And it is they, not the text, that make it clear what part of the diagram the letters in the text refer to.

The problem of dissecting polygons to form other polygons is an ancient one. A number of proofs of equality of figures in Book I of Euclid's *Elements* are proofs by dissection. And Archimedes wrote a treatise, *The Stomachion*, a word referring to a game that was based on a geometric puzzle of arranging a given set of flat, ivory pieces in the form of polygons so as to make various shapes.[135] About 2100 years later David Hilbert posed the following problem: If Π and Π' are two polygonal shapes of equal areas, is it possible to dissect Π and Π' into finite sets of polygons, say $\{R, S, \ldots, W\}$ and $\{R', S', \ldots, W'\}$, respectively, so that each set has the same number of members and the members may be paired, say $R \leftrightarrow R', \ldots, W \leftrightarrow W'$, so that corresponding polygons are congruent?[136]

[134]Dr. Jan Hogendijk has made a preliminary study of the manuscript and has generously given his permission to use his translation and the figures he has drawn on the basis of those in the manuscript.

[135]Fragments of the treatise exist in Greek and in Arabic.

[136]Max Dehn showed that this was indeed true, although the answer to the corresponding question for solids is "No."

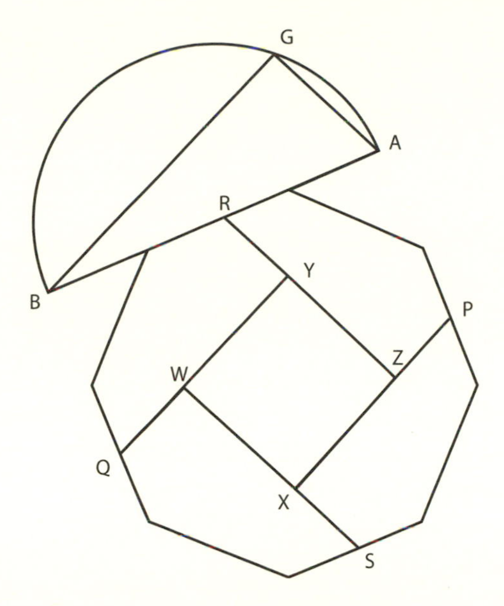

FIGURE 5.62

Geometrical Dissections from a Persian Manuscript

Problem 3: Each time when we want to find the side of a square whose area is equal to the area of an octagon, it is necessary that if the square of the side of the octagon is subtracted from the square of the [diameter of] the circle inscribed in the octagon the root of the difference is the side of the square (fig. 5.62).[137]

Problem 5: If it is said: We want to make for a hexagon a square such that the pieces of the hexagon fit exactly in the square, we say: The method for this is that we make on the extension of line *AB* line *AD* equal to line *AG* (fig. 5.63). We bisect line *DB* at point *E*. With compass-opening *EB* we describe arc *BZD*. Again, on the straight line *AG* we draw line *AZ* to end at the circumference of the circle. Line *AZ* is the side of the square that is equal to the hexagon. Then we put on line *AB* line *AH* equal to line *AZ*, then point *H* is known. The rest is easy.[138] God knows best.

Problem 7: The way to divide a hexagon into triangles [to make one equilateral triangle] is as follows: If triangle *ADZ* [in hexagon *ABZEDG*] is known, divide each angle in the triangle into four equal parts by lines *AT*, *DI*, *ZH* and extend these lines until they meet each other at points *H*, *T*, and *I* (fig. 5.64). Then triangle *HTI* is found. Divide this triangle into six parts by perpendiculars *IL*, *HN*, *TM*. From all these pieces compose one triangle (fig. 5.65). God knows best.

---

[137]In the figure *R*, *P*, *S* and *Q* are the midpoints of four sides of the regular octagon. The semicircle has *R* as center and diameter, *AB*, equal to that of the incircle of the octagon. The side *AG* of the right triangle *AGB* is equal to a side of the octagon. The lines *RZ* and *WS* are parallel to the side *AG* of the right triangle and lines *QY* and *PX* are parallel to the other side, *BG*, of the triangle. The octagon is thus dissected into a square and four pieces that border it, five pieces which can be rearranged to form a square whose side is equal to *GB*, which one may show is equal to *RZ*.

[138]*BH* is bisected (at *W*), *AH* is bisected at *Y*. Then GA_1 is divided at W_1, H_1 and Y_1 so that $GW_1 = BW$, $W_1H_1 = WH$, and Y_1 is the midpoint of A_1H_1. Finally perpendiculars are drawn to *AB* at *H* and *Y*, and to GA_1 at Y_1 and H_1 as indicated in the figure. The accompanying square shows how the resulting 8 pieces can be reassembled to form a square.

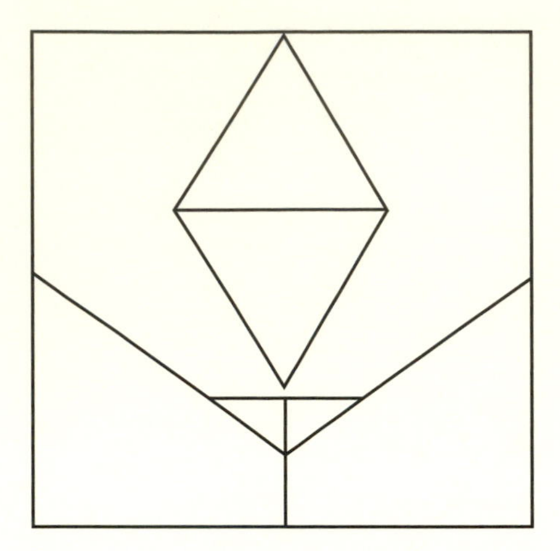

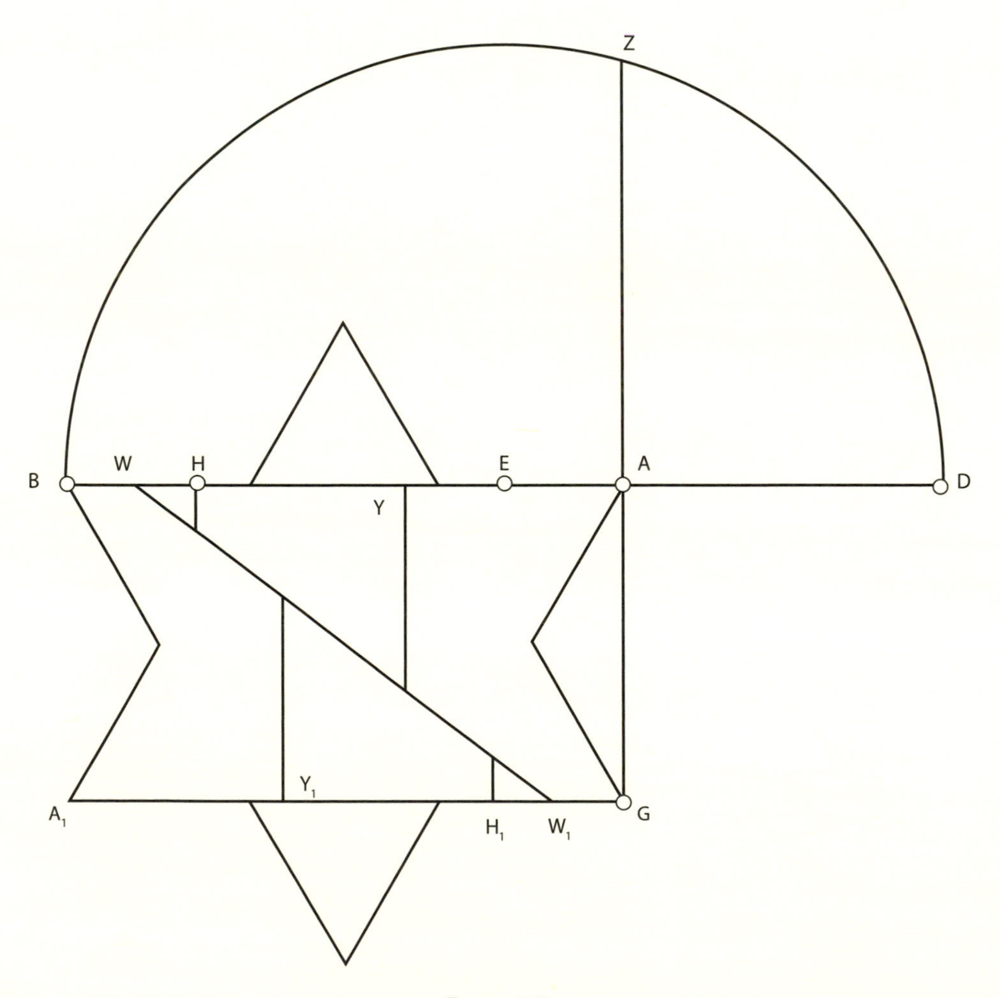

FIGURE 5.63

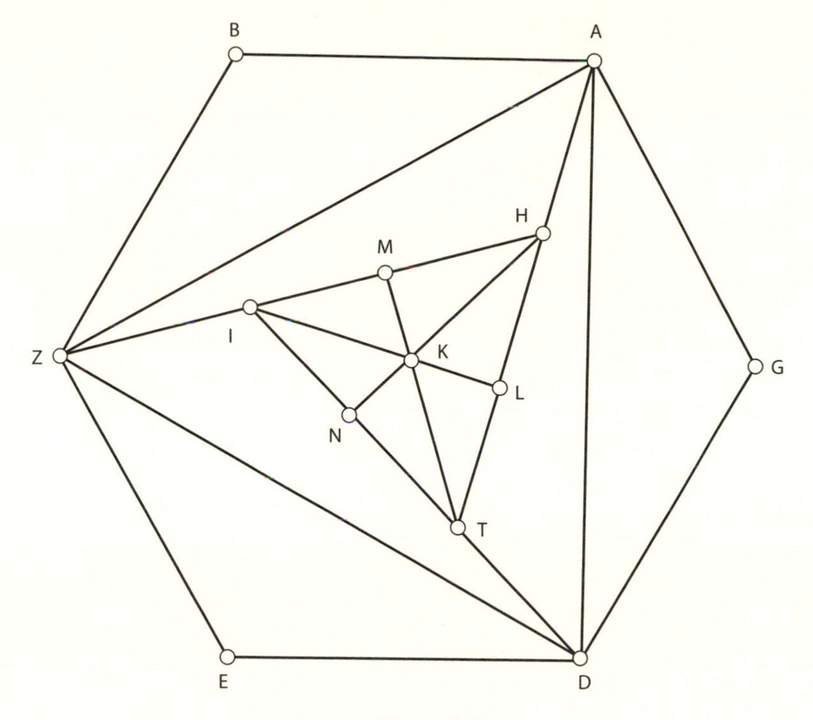

Figure 5.64

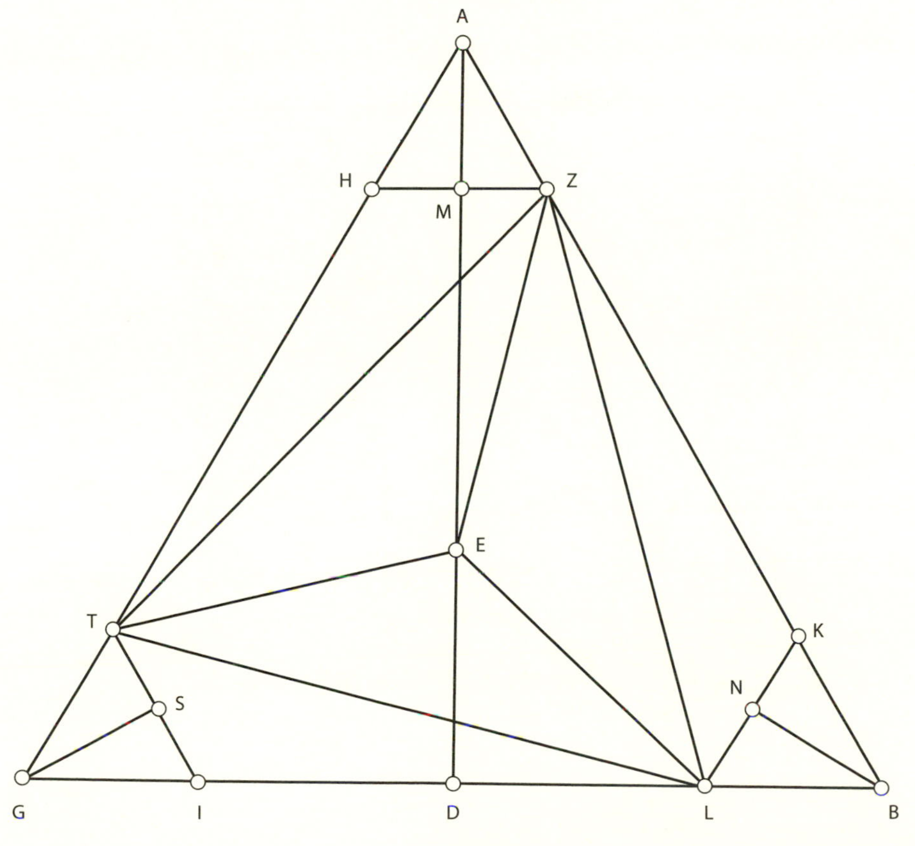

Figure 5.65

The two patterns shown here are taken from a pattern book published by Gülru Necipoglu and now found in the Topkapi Museum in Istanbul.[139] Both show designs of quarter vaults for *muqarnas*, the architectural structures that look like clusters of stalactites on the ceilings of Islamic mosques and *madrasas*. The first pattern is extendible to a semicircular vault only, but the second may be extended to a full circular vault. The mathematics of these structures attracted the attention of mathematicians of the caliber of al-Kāshī, and considerable mathematical investigations must have been necessary for many of the patterns. The Topkapi scroll, however, contains only the patterns, but in Necipoglu's edition she has supplied for each drawing inked-in construction lines that in some cases are found ruled into the paper of the original with a sharp point and in other cases are absent.

The first figure is built on a fan pattern radiating out from the upper left corner (fig. 5.66). The vertices of the eight non-convex hexagons surrounding this corner (drawn in black) determine 15 radii of the quadrant centered on that corner whose circumference passes through the black points inside the hexagons. Note the pattern of five-pointed stars running along the circumference of that quadrant and the heptagonal star a bit in from the lower right corner. Its center is the center of the squinch.[140] Red and black dots mark surfaces in different planes.

The second pattern (fig 5.67) again shows a quarter vault of a *muqarna*, one in which the six-pointed star in the lower right occupies the squinch corner of the vault. Five equi-spaced radii (none of them shown) of the quarter circle whose center is at the upper left corner and whose circumference passes through the bases of the three red figures divide the quarter circle into six equal parts. So the whole vault would have 24-fold rotational symmetry about the center point. Again, the regions dotted more or less densely are in different planes.

Two Patterns for Muqarnas

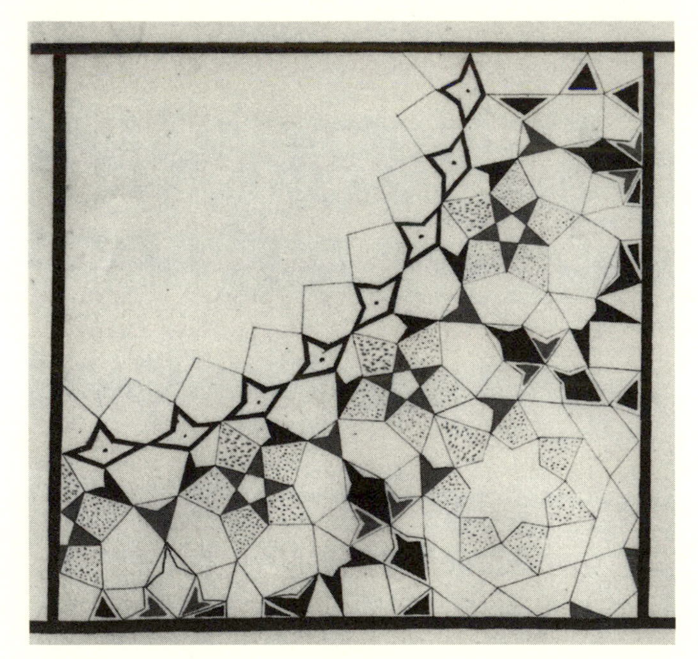

FIGURE 5.66

---

[139]Our descriptions of the two figures are shortened versions of those given by Necipoglu.

[140]A squinch is an interior corner support in a building.

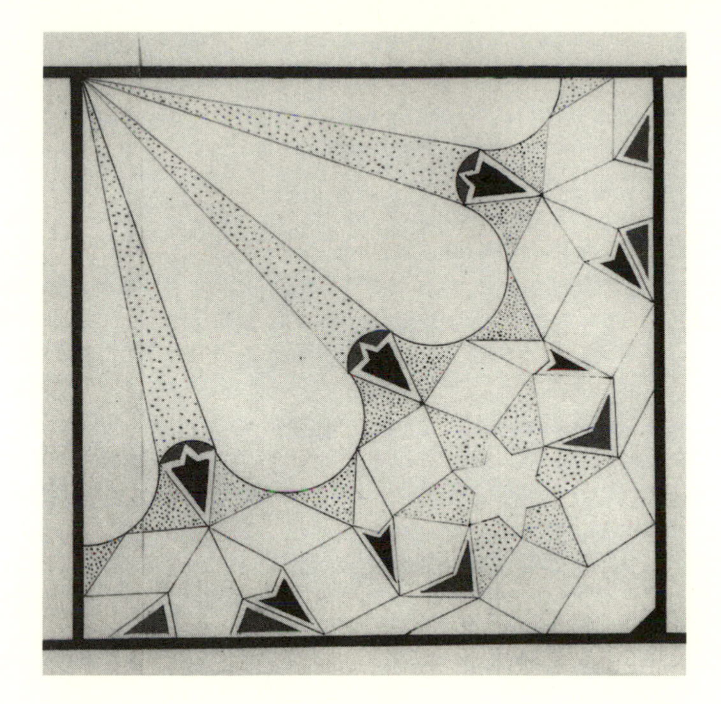

FIGURE 5.67

VII. Trigonometry

Abū al-Wafā' was one of the Islamic mathematicians responsible for simplifying and extending the spherical trigonometry they learned from Greek sources. We present two selections from his work. In the first selection, translated from the Arabic text of al-Bīrūnī's *Keys of Astronomy* published by Dr. M.-Th. Debarnot, al-Bīrūnī states and proves the theorem sometimes known as "The Rule of Four Quantities" on the basis of what he found in Abū al-Wafā''s astronomical handbook called *The Almagest.* Abū al-Wafā' was an older contemporary of al-Bīrūnī, and his *Almagest* should not, of course, to be confused with the work of the same name by the Alexandrian astronomer Claudius Ptolemy written in the second century of our era. The title was probably meant as a tribute to Ptolemy and a bit of an "advertisement" for his own work. Bīrūnī's name for the theorem, "The Rule that Dispenses," is a short way of saying "The rule that dispenses with the need for using the sector figure of Menelaus." In fact, in his *Spherics,* Menelaus proved a more general theorem than this, and Ptolemy, in his *Almagest,* used a special case of the rule. However, although the theorem was known in the Hellenistic world, medieval Islamic mathematicians gave their own proofs of it.

The geometrical term "inclination" in the statement of the theorem is defined later, in the proof. And the reader must remember that the medieval sine is equal to the modern sine multiplied by the radius, so it is a line segment and not a ratio. Of course, when one is considering ratios of sines, the ratio is the same whether one uses the medieval sine or the modern one.

Abū al-Wafā' on the Rule of Four Quantities

Demonstration of the "Theorem that Dispenses" as presented by Abū al-Wafā' in his *Almagest.*

If two arcs of great circles intersect on the surface of a sphere at an angle less than a right angle[141] and if one takes on one of them any points, then the ratios to each other of the sines of the arcs contained between these points and the common point [of the two intersecting arcs] are equal to the ratios of the sines of their inclinations to each other.

Thus, let *AB* and *AG*, two arcs of great circles on the surface of a sphere, cut each other at point *A* at angle *BAG* (less than a right [angle]) [fig 5.68].[142] On the circle *AB*, points *B* and *D* are assumed [given]. The inclination of arc *AB* to the circle *BG* is arc *AG* and that of arc *AD* [to the circle *BG* is] the arc *DE*, (where these two arcs are [chosen so they are each] perpendicular to circle *AG*. I say that the ratio of the sine of arc *AD* to the sine of arc *AB* is equal to the ratio of the sine of arc *DE* to the sine of arc *BG*.

The proof of that is that we suppose that the center of the sphere is *Z* and we join the lines *AZ*, *GZ*, and *EZ*. And we produce from the points *B* and *D* the [lines] *BH* and *DT* perpendicular to the lines *GZ* and *EZ* [respectively], and so they are perpendicular to the plane *AGZ*.[143] And we produce from them [also] *BY* and *DK*, perpendiculars to the line *AZ*, and we join the lines *HY*, *TK*.

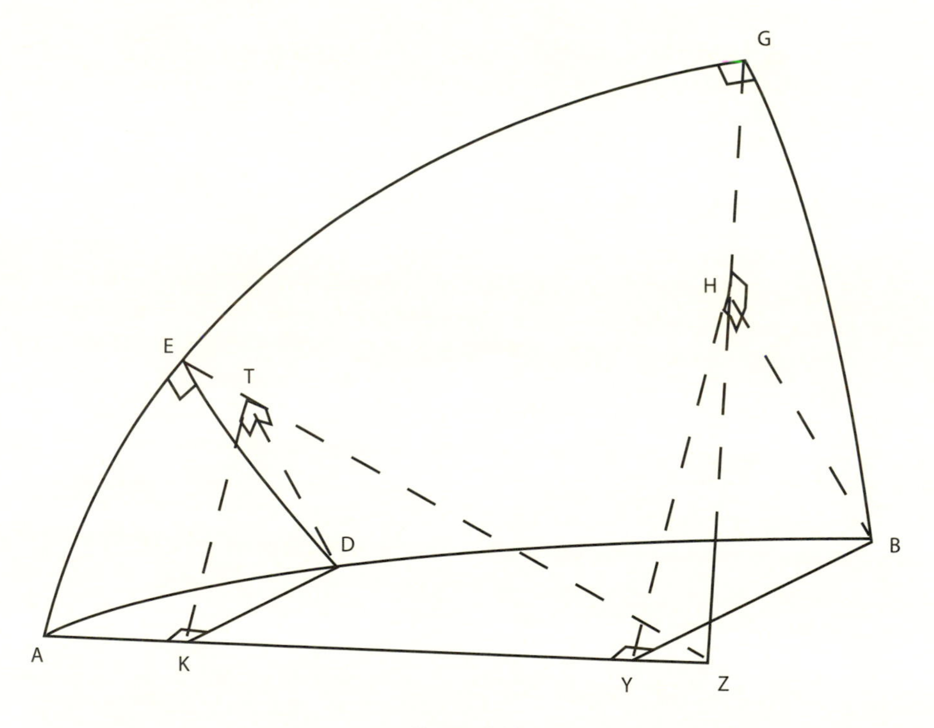

FIGURE 5.68

[141]The angle of intersection of two arcs on a sphere is, in modern language, the angle between the tangent vectors to those two arcs which are in the tangent plane to the sphere at the point of intersection. The ancients, however, defined this angle as the measure of the arc of the great circle joining the two points on the intersecting arcs that are located 90° from the point of intersection and in the same direction from it.

[142]In the figure, the spherical surface *ABG* has its convexity facing the reader, and the arcs *BG* and *DE* are, as Biruni states explicitly a bit further on, perpendicular to the arc *AEG*. *Z* is the center of the sphere. The reader needs to imagine the pairs of lines, *BH*, *BY* and *DT*, *DK* as going into the sphere from points, *D* and *B*, on the surface *ABG*.

[143]By construction, line *BH* is perpendicular to line *GZ*, the intersection of two perpendicular planes, *GBZ* and *GAZ*. It follows from *Elements* XI, Def. 4, that *BH* is perpendicular to the plane *GAZ*. Similarly, *DT* is perpendicular to the same plane.

Now, since the two lines *BH*, *BY* are parallel to the two lines *DT*, *DK*,[144] the angles *YBH*, *KDT* are equal [*Elements* XI, 10]. And the angles *YHB*, *KTD* are right [angles]. And so the two triangles *YHB*, *TKD* are similar. And the ratio of *DK*, the sine of arc *AD*, to *BY*, the sine of arc *AB*, is as the ratio of *DT*, the sine of arc *DE*, to *BH*, the sine of arc *BG*. And that is what we wanted to prove.

A fundamental result of both plane and spherical trigonometry that the mathematicians of medieval Islam discovered is the sine theorem, which asserts that, in the case of spherical triangles, the ratio of the sines of any two angles is equal to the ratio of the sines of the great circle arcs forming the sides opposite the angles. In the case of plane triangles the ratio of sines of any two angles is simply equal to the ratio of the two opposite sides.[145]

The result was discovered almost simultaneously in the late tenth century by Prince Abū Naṣr ibn ʻIrāq and Abū al-Wafā' al-Būzjānī, and considerable controversy ensued over who deserved the credit for being the first.

Abū al-Wafā''s proof relies on the Rule of Four Quantities, whose proof is given in the previous selection. That Rule says that if two right spherical triangles share a common acute angle,[146] then the ratio of the sines of the arcs opposite that angle is equal to the ratio of the sines of the corresponding arcs opposite the right angles. We now present Abū al-Wafā''s proof of the sine theorem, from his astronomical handbook, the *Almagest*. Our translation is from the French translation by Carra de Vaux.

Abū al-Wafā' on the Sine Theorem

Let the triangle *ABG* on the surface of the sphere be formed by the arcs of great circles, *AB*, *BG*, *AG*. I say that the ratio of the sine of *B* to the sine of *C* is equal to the ratio of the sin of arc *AC* to the sine of arc *AB*.

We draw the arcs of great circles *DE*, *EZ* having (respectively) *B* and *G* for poles, and we prolong arcs *AB*, *BG*, *GA* until they meet arcs *DE* and *EZ* at points *D*, *T*, *H*, *Z*. The angles at *Z*, *D* are both right angles,[147] and consequently the point *E* is the pole of circle *DGZ* (fig. 5.69).

Through the two points *E*, *A* we pass the arc *AEI* of a great circle. The angle at *I* will be right. There results the proportion that the sine of *DT* is to the sine of *AI* as the sine of *TB* is to the sine of *AB*, according to the first theorem of our discourse [the Rule of Four Quantities]. And, also, the sine of *ZH* is to the sine of *AI* as the sine of *GH* is to the sine of *GA*. From which, taking account of equal quantities, the sine of *DT* (which is equal to the sine of angle *B*) is to the sine of *ZH* (which is equal to the sine of angle *G*) as the sine of *GA* is to the sine of *AB*.

And that is what we wanted to prove.

---

[144] *BH* and *DT*, being both perpendicular to the plane *GAZ*, are parallel by *Elements* XI, 6. And since *BY* and *DK* are two lines in the plane *BAZ* which are perpendicular to line *AZ* they are parallel as well.

[145] For Naṣīr al-Dīn al-Ṭūsī's proof of the theorem for plane triangles, see the excerpt from his trigonometry work later in this section. For more details on how he uses it to solve plane triangles systematically, see [Berggren 1986, 138–141].

[146] Or have vertically opposite angles.

[147] For *G* is a pole of *EZ*, *B* is a pole of *DT*, and it was proved by Theodosius in his *Spherics* that a great circle arc (e.g., *GBZ* or *BGD*) joining the pole (e.g. *G* or *B*) of a great circle (e.g. *ZE* or *DT*) to a point on the circumference of that great circle) is perpendicular to it at that point.

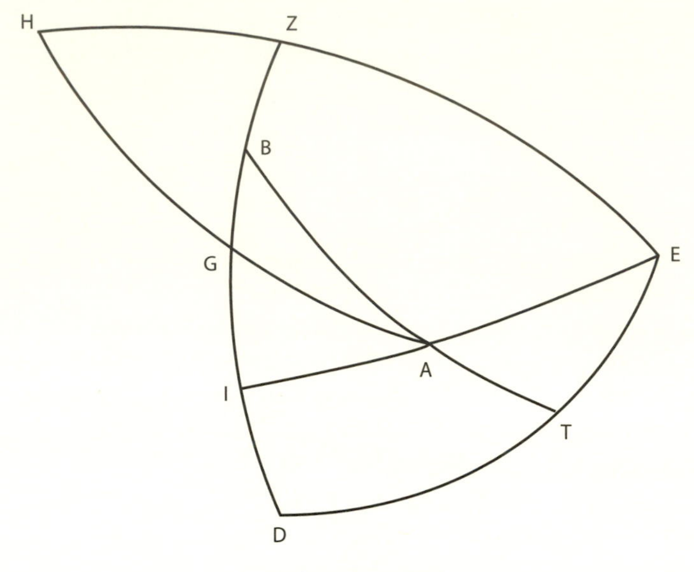

FIGURE 5.69

Abū al-Wafā' used trigonometry to solve many astronomical problems, including determining time by the altitude of the sun. On any given day of the year the sun makes a circle, called its day circle, around the heavens and parallel to the equator. After sunrise, the altitude of the sun above the horizon[148] (measured in degrees) increases from sunrise until noon, and then decreases until sunset. The question arises if it is possible to tell the time of day (provided one knows whether it is morning or afternoon) from the altitude of the sun (observed by, say, an astrolabe or a quadrant). It turns out that this is, indeed, the case if one knows, for example, the latitude of one's locality and the declination of the sun on the day in question.[149]

The solution to the problem given here is from the Abū al-Wafā''s treatise *On the Rotation of the Celestial Sphere.* (The "rotation" referred to in the title is the number of degrees the sphere has rotated since sunrise, i.e., the arc of the sun's day circle between the sun and the eastern horizon.) Abū al-Wafā' begins by stating a rule for calculating this rotation known in seventh-century India, whence it made its way (via Iran) to Baghdad. And there the famous astronomer Ḥabash al-Ḥāsib used it in the mid-ninth century.[150] He tells us that some astronomers believed that the rule gives only approximate results; but, he says, he has found a proof—indeed several proofs—of it. This concern with rigorous demonstration of the rules is one of the features that distinguished medieval Islamic from Indian mathematical sciences.

What follows is our translation of the part of Abū al-Wafā''s treatise which treats a method for solving the problem that, because it is based on explicit use of the transversal theorem,

---

[148]This is the arc, measured in degrees, of the great circle between the sun and the horizon that passes through the zenith. (This circle, for obvious reasons, is called an altitude circle.)

[149]Standard astronomical works would have contained tables displaying such data or rules for computing it.

[150]More specifically, the rule is found in the *Khandakadyaka*, an astronomical work by the Hindu scholar Brahmagupta (fl. 650 CE), whence it made its way to Baghdad via a Persian work known as the *Royal Astronomical Tables.*

provides both a procedure for calculating the arc and a proof of the correctness of that procedure.[151] The author begins by considering the simpler case when the sun is on the equator, but the case for the days of the year other than these two does not need this special case so we translate only the case for when the sun is in a northern or southern sign.

Abū al-Wafā' on Telling Time by the Altitude of the Sun

Let the sun be in the northern or southern signs [of the ecliptic], and we make the circle *AGB* the meridian, *ADB* half the circle of the horizon, *GD* a quarter of the equator, the point *Z* the center of the sun, and the zenith the point *E* (fig. 5.70).[152]

We join the points *E*, *Z* with an arc [of a great circle], *EZT*. Then *ZT* will be the arc of the altitude [of the sun], and it is [assumed] known.[153] And since two arcs, *KL*[154] and *HE*, intersect between two arcs *KG* and *HG*,[155] the ratio of the sine of arc *KG* to the sine of arc *GE* is composed of the ratio of the sine of arc *KL* to the sine of arc *LZ* and of the ratio of the sine of arc *HZ* to the sine of arc *HE*.[156] But arc *KG* is equal to arc *KL*[157] [and so] the ratio of the sine of arc *LZ* to the sine of arc *GE* is as the ratio of the sine of arc *HZ* to the sine of arc *HE*. And the arc *LZ* is known since it is the

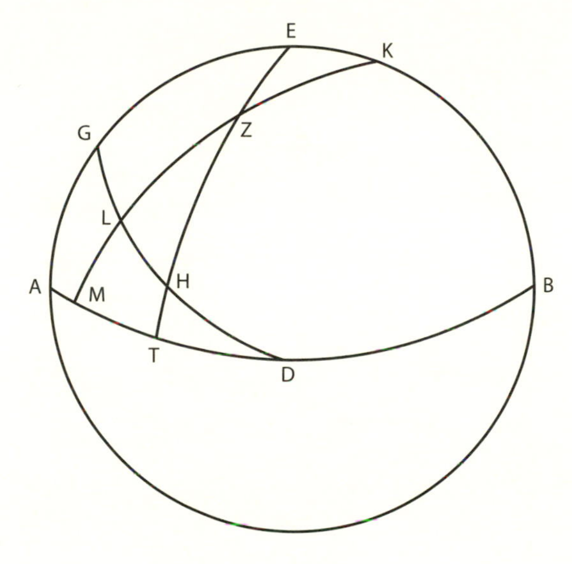

FIGURE 5.70

[151]A concise description of the procedure is available in [Lorch 2001, 398–399]. A summary of all of al-Būzjānī's proofs of the rule, with remarks about medieval Islamic trigonometric functions, may be found in [Nadir 1960, 460–463], reprinted in [King and Kennedy 1983, 280–283].

[152]The published Arabic text has a number of figures each consisting of a bounding circle, representing the horizon, and various straight lines within the circle. We have redrawn, instead, a diagram, from [Lorch 2001, 398], in which circular arcs on the sphere are represented by circular arcs in the diagram. The figure (5.70) shows the celestial sphere viewed from the outside.

[153]The arc could be determined by observation with an astrolabe.

[154]*K*, not identified in the treatise, represents the north pole of the celestial sphere *ABE* and *L* is the intersection of the equator with the meridian of that sphere that passes through the sun.

[155]This identifies the four circular arcs making up the transversal figure.

[156]This is a straightforward application of the transversal theorem, the first of three such applications.

[157]Both arcs are quadrants of great circles.

declination of the degree of the sun.[158] And *GE* is known since it is the latitude of the locality.[159] [And so] *HZ* is known since *ZE*, the excess of the two arcs *HE* and *HZ*, is known.[160] There remains[161] arc *HT*, known.

And also, the ratio of the sine of arc *EA* to the sine of arc *GA* is composed of the ratio of the sine of arc *ET* to the sine of arc *TH* and the ratio of the sine of arc *DH* to the sine of arc *DG*. [So,] because of what we premised, arc *DH* is known, and thus the arc *HG* is known.

And also, because the ratio of the sine of arc *KE* to the sine of arc *GE* is composed of the ratio of the sine of arc *KZ* to the sine of arc *ZL* and the ratio of the sine of arc *HL* to the sine of arc *HG*, arc *LH* is known and arc *GL* is known, [which is] the complement of the revolution to the half of the day arc. [Hence, the revolution itself is known.]

The accuracy of a sine table depended, in general, on the accuracy of the calculation of the sine of 1°. But even with an accurate sine table given for each degree or each half degree, in order to get results accurate enough for astronomy, it was necessary to interpolate in the tables to get additional values. One method of interpolation, used by Ptolemy and adopted by some Islamic mathematicians, was linear interpolation. But since this method was often not as accurate as desired, other methods were developed. Here we present a method of non-linear interpolation described by the Egyptian astronomer Abū al-Ḥasan ʿAlī ibn Yūnus (c. 930–1009). It is a second-order scheme which, if imagined in terms of modern analytic geometry, calculates the value of $\sin(\theta^\circ\ \tau')$ by replacing the graph of the sine function between θ and $(\theta + 1)^\circ$ by the graph of a parabola through the three points θ°, $(\theta + 1/2)^\circ$, and $(\theta + 1)^\circ$.

Ibn Yūnus on Nonlinear Interpolation in Sine Tables

Discussion of an idea which occurred to me for finding Sines of arcs after one has found the Sines for each half degree of arc. It is a good method, and is exact. Sines of arcs found by linear interpolation are always less than the actual values. By the method I use, I find the difference between the Sine of an arc found by linear interpolation and the actual value, and I add it to the value found by linear interpolation: the result will be the actual value. If you want to find the Sine of an arc other than the half-degrees whose Sines are given [in the table], and [you want to find] the difference between the Sine found by linear interpolation and the actual value, then first find the Sine of the arc in question by linear interpolation, using the Sines obtained rigorously for each degree and computing to thirds or fourths according to the degree of accuracy you desire. Also find the Sine of the half degree which is in between the two integral degrees less and greater than the arc in question, using

[158] This would be determined for the given date by consulting a table.

[159] Since it is the arc of the meridian of the locality between the zenith and the equator.

[160] The argument has shown that both the ratio of the sines and the difference of the arcs *HZ* and *HE* are known. Abū al-Wafāʾ now makes implicit use of a lemma found in Ptolemy's *Almagest*, that, in such a case, the arcs are known. Here and in the following, the statement that something is known means that it is either known by assumption or can be calculated on the basis of what is already known and demonstrated mathematical relations (e.g., the transversal theorem).

[161] Once one has removed from the quadrant *TE*, the known arcs *HZ* and *ZE*.

the Sines calculated rigorously for each degree. Note the Sine of that half degree by linear interpolation, and note the difference between it and the Sine of the half degree given in the *Zīj*, which was calculated rigorously. Multiply the difference by four, and the result will be the base for interpolation, so keep it in mind. Then look at the minutes of the arc in question over the integer degrees, and subtract them from 60'. Multiply the remainder by the minutes which you subtracted, and then multiply the product by the interpolation base: the product will be the difference between the Sine of the degrees and minutes (of the arc in question found) by linear interpolation and its actual Sine, so add it to the value of the Sine which you found first by linear interpolation. The result will be the Sine of the degrees and minutes of arc, which you wanted to find exactly. Success is with God.

In his *Book of the Determination of Coordinates of Localities*, the polymath Abū al-Rayḥān al-Bīrūnī (973–1055) gives a masterful exposition of the methods of mathematical geography. As a context for the exposition he sets himself the task of (quite literally) putting the capital city of his patron, Maḥmūd of Ghazna,[162] "on the map." In a systematic exposition, with a number of delightful digressions, he shows how to find the circumference of the earth, how to determine the latitude and longitude of localities, and how to find the distance between two localities. And since Moslems must pray in the direction of Mecca, he gives several methods for determining accurately that direction, known as the *qibla*. We present here several excerpts from al-Bīrūnī's text, known in Arabic as the *Kitāb taḥdīd al-amākin* ("Determination of the Coordinates of Localities").

Al-Bīrūnī's Book of the Determination at Coordinates of Localities

Chapter 5

As to the observations that were made by al-Ma'mūn['s astronomers], they were started after he had read in some Greek books that one degree of the meridian is equivalent to five hundred *stadia*, where a *stadium* is the standard measure of length which was used by the Greeks for measuring distances. However, he found that its actual length was not satisfactorily known to the translators, to enable them to identify it with local standards of length. Then, according to Ḥabash, who obtained his information from Khālid al-Marwarūdhī, al-Ma'mūn ordered a group of learned astronomers, and expert carpenters and workers in brass, to prepare the required instruments and to select a locality for a geodetic survey. They chose a spot in the plain of Sinjār, which is in the neighborhood of Mosul, nineteen *farsakhs* from the town itself, and forty-three *farsakhs* from Samarra. They liked its level ground, and transported their instruments to it. They chose a site and observed with their instruments the sun's meridian altitude. Then they departed in two parties.

Khālid, with the first party of surveyors and artisans, headed in the direction of the north pole; and 'Alī ibn 'Īsā, the maker of astrolabes, and Aḥmad al-Buḥtarī, the surveyor, with the second party, headed in the direction of the south pole. Each party observed the meridian altitude of the sun until they found that the change in its

---

[162]Ghazna is a town in eastern Afghanistan, ancient enough to have been mentioned by the second-century (CE) Alexandrian geographer Claudius Ptolemy in his *Geography*.

meridian altitude had amounted to one degree, apart from the change due to variation in the declination. While proceeding on their paths, they measured the distances they had traversed, and planted arrows at different stages of their paths [to mark their courses]. While on their way back, they verified, by a second survey, their former estimates of the lengths of the courses they had followed, until both parties met at the place whence they had departed. They found that one degree of a terrestrial meridian is equivalent to fifty-six miles. He (Ḥabash) claimed that he had heard Khālid dictating that number to Judge Yaḥyā ibn Aktham. So he heard of that achievement from Khālid himself. Also, a similar narrative was told by Abū Ḥāmid al-Ṣaghānī, who obtained his information from Thābit ibn Qurra. But it is said that al-Farghānī has reported an extra two thirds of a mile, in addition to the [above-]mentioned number of miles.

I have found all the other narratives in agreement about these two thirds. However, I cannot ascribe that omission to an oversight in the manuscript of the *Kitāb al-ab'ād wal-ajrām*, because Ḥabash has derived on that basis the circumference of the earth, its diameter, and all other distances. When I examined those, I found that they were based on that assumption of fifty-six miles. So it is preferable to ascribe the difference in the two narratives to the two parties [who participated independently in that expedition]. That difference is a puzzle; it is an incentive for a fresh examination and observations. Who is prepared to help me in this [project]? It requires a strong command over a vast tract of land and extreme caution is needed from the dangerous treacheries of those spread over it. I once chose for this project the localities between Dahistān, in the vicinity of Jurjān and the land of the Ghuzz (Turks), but the findings were not encouraging, and then the patrons who financed the project lost interest in it.

[...]

Here is another method for the determination of the circumference of the earth. It does not require walking in deserts.

We climb a high mountain close to the seashore, or close to a large level desert. If we find that sea, or that desert, to be lying to the east whence the sun rises, or to the west wherein it sets, we observe it until half its disc is screened from our eyes, and we measure its angle of dip at that instant by a ring provided with an alidade, like the ring *ABGD* [fig. 5.71].

When the position of the alidade is along *HZ*, the amount of the dip is given by arc *BZ*, and its complement by arc *ZG*. If the level plane does not happen to lie in one of those two [above-] mentioned directions, we set the ring vertically downward, and we look with one eye through the two sights of the alidade until we see the spot of contact between the sky and the earth. The alidade would then take up a position like the first one, and the line of sight would coincide with the line *HEZT*, the direction of the alidade. We join *T* to the center of the earth, which is point *K*. Then we survey the perpendicular (height) of the mountain, which is *EL*. We drop the perpendicular *ZM*. The two triangles *EZM* and *EKT* are similar, and the ratio of *EZ*, the total sine, to *ZM*, the cosine of the dip angle, is as the ratio of *EK* to *KT*. If we apply [the rule of] *invertendo et dividendo* [to the proportion], then the ratio of *EZ*

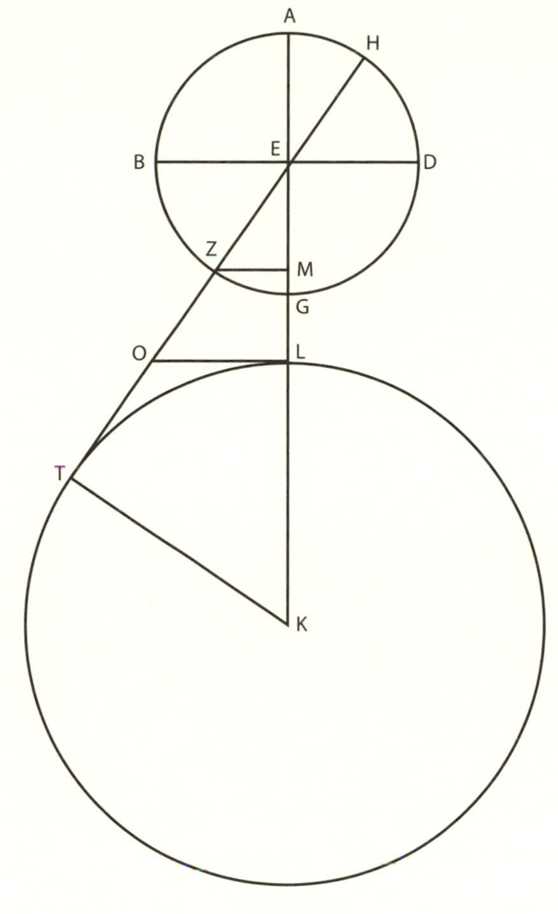

FIGURE 5.71

to the difference between it and *ZM*, which is equal to the versed sine of are *BZ*, is as the ratio of *EK* to the difference between *EK* and *KT*, which is *EL*. Hence *EK* is known. But *EL* is known. Therefore *LK* is known, in the units of length in which *EL* was surveyed. And if the radius of the earth is known, then its circumference would also be known.

Again, we draw *LO* tangent to the earth at point *L*. Since angle *E* is known, the ratio of *EL* to *LO* is as the ratio of the sine of angle *EOL*, the dip angle, to the sine of angle *GEL*, the complement of the dip angle. Hence *LO* is known, and it is equal to *OT*. And *EO* is known, therefore *ET* is known. But the ratio of *ET* to *KT* is as the ratio of the cosine of the dip angle to the sine of the dip angle. Therefore the sides of the triangle *KTE* are known.

By the same method al-Ma'mūn's astronomer has derived the circumference of the earth. Abū al-Tayyib Sanad bin 'Alī has narrated that he was in the company of al-Ma'mūn when he made his campaign against the Byzantines, and that on his way he passed by a high mountain close to the sea. Then al-Ma'mūn summoned him to his presence and ordered him to climb that mountain, and to measure at its summit the dip of the sun. He executed the order, and derived the circumference of the

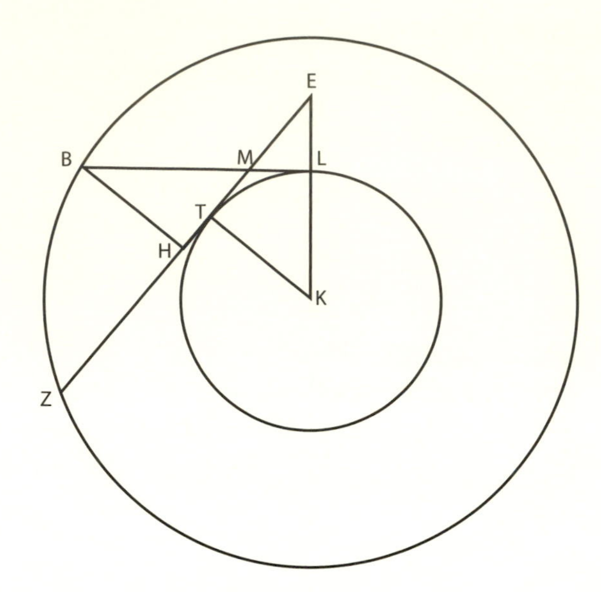

FIGURE 5.72

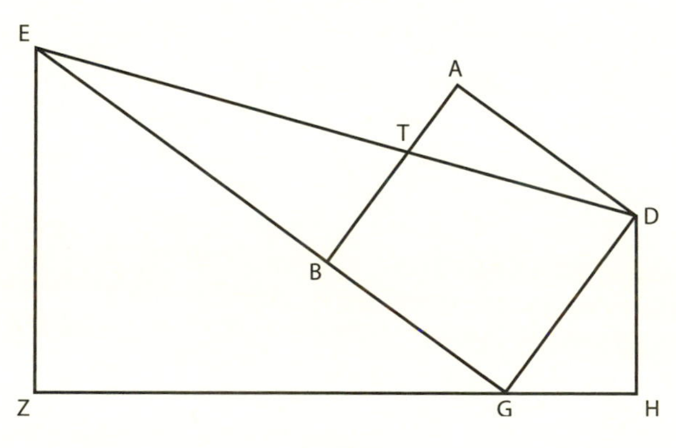

FIGURE 5.73

earth as follows: Let *LT* [fig. 5.72] represent the circle of the earth, with center at *K*, *LE* the perpendicular [height] of the mountain, and *LB* [the tangent line at *L*] in the plane of the apparent horizon. We draw *EZ*, a tangent line to the earth, touching it at point *T*. Then arc *BZ* is a measure of the dip in the altitude circle. We join *KT*, and we drop *BH* perpendicular to *EZ*. Then *BH* is the sine of the dip, because *M* can be regarded as the center and *MB* as the radius. Hence *MH*, the cosine of the dip, is known, and *MB* is the total sine. So the sides of the triangle *BMH* are known, and it is similar to the triangle *ETK*. Therefore the ratio of *MB* to *MH* is as the ratio of *EK* to *KT*. Hence by [the rule of] *invertendo et dividendo*, the ratio of *MB* to the difference between *MB* and *MH* is equal to the ratio of *EK* to *EL*. So [*EK* is known and hence] *LK* is known, and this is what we wished to find.

As to the determination of the perpendicular (height) of the mountain, it is made by one of the methods for the determination of distances. Let us construct for it a plate, with right-angled corners and of square shape, one cubit by one cubit, like the right-angled square *ABGD* [fig. 5.73].

We divide each of the two sides *AB* and *AD* into any number of parts we desire, provided that the divisions are equal in number and in magnitude. We fix at the two comers *B* and *G* two pegs normal to the surface of the square, and at corner *D* (*G*) we set an alidade provided with two visors, or with a sharp edge and two pegs, whose length is equal to the diagonal of the square. Then let *EZ* represent the required perpendicular of the mountain, and *ZG* the plane of the horizon. We set the instrument perpendicular to it. Then we look from corner *G*, and we adjust the instrument by raising or lowering it until we see the tips of both pegs at *G*, and *B*, in line with the summit of the mountain, point *E*. We fix the instrument in that position. From *D* we drop a stone. Suppose that it falls at point *H*. We mark the distance between *G* and *H*, where the stone has fallen, in terms of the divisions on the side of the instrument. Then we turn to *D*; and we adjust the alidade by raising or lowering it, until we see the summit *E*, through the visors, in line with the tips of the two pegs. This takes place as if the alidade were mounted at *T*. Since the triangles *DAT* and *EGD* are similar, the ratio of *TA* to *AD* is as the ratio of *DG* to *GE*. So we multiply the number of divisions in *AD* by the cubit *DG*, and we divide the product by the number of divisions in *AT*. The quotient obtained thereby is a measure of *GE*, expressed in cubits. But the ratio of *GE* to *EZ* is as the ratio of *DG* to *GH*, because the sum of the two angles *DGH* and *EGZ* is a right angle, and the sum of the two angles *EGZ* and *GEZ* is a right angle, and if we subtract the common angle *EGZ*, there remains angle *DGH* equal to angle *GEZ* and, therefore, angle *GDH* is equal to angle *EGZ*. So we multiply *EG* by *GH*, and we divide the product by the number of divisions in *DG*, the side of the square. The quotient obtained thereby is the required height, *EZ*.

When I happened to be living in the fort of Nandana in the land of India, I observed from an adjacent high mountain standing west of the fort, a large plain lying south of the mountain. It occurred to me that I should examine this method [from the previous section] there. So, from the top of the mountain, I made an empirical measurement of the contact between the earth and the blue sky. I found that the line of sight had dipped below the reference line by the amount 0;34°. Then I measured the perpendicular of the mountain and found it to be 652;3,18 cubits, where the cubit is a standard of length used in that region for measuring cloth. Let it be represented in the figure by *EL* [fig. 5.74].

Because angle *T* is a right angle, and angle *K* is 0;34°, the dip angle, and angle *E* is 89;26°, the complement of the dip angle, the angles of the triangle *ETK* are known. Therefore its sides are known in the scale in which *EK* represents the total sine. In this scale, *TK* is 59;59,49 and the difference between it and the total sine is 0;0,11, which is the measure of the perpendicular *EL*. But it is known in cubits, and the ratio of its cubits to the cubits of *LK* is as the ratio of 0;0,11 to 59;59,49. The product of 652;3,18, the cubits of *EL*, times 59;59,49, the parts of *LK*, is 39,121;18,27,42. If it (the product) is divided by 0;0,11, the parts of *EL*, the quotient would be 12,803,337;2,9 cubits, the radius of the earth. So its circumference is 80,478,118;30,39, and the share (or equivalent) of one degree out of three hundred and sixty is 223,550;19,45 cubits. If this amount is divided by four thousand, the quotient is 55;53,15 miles per degree, and that is not far from Ḥabash's version. God is the true Helper!

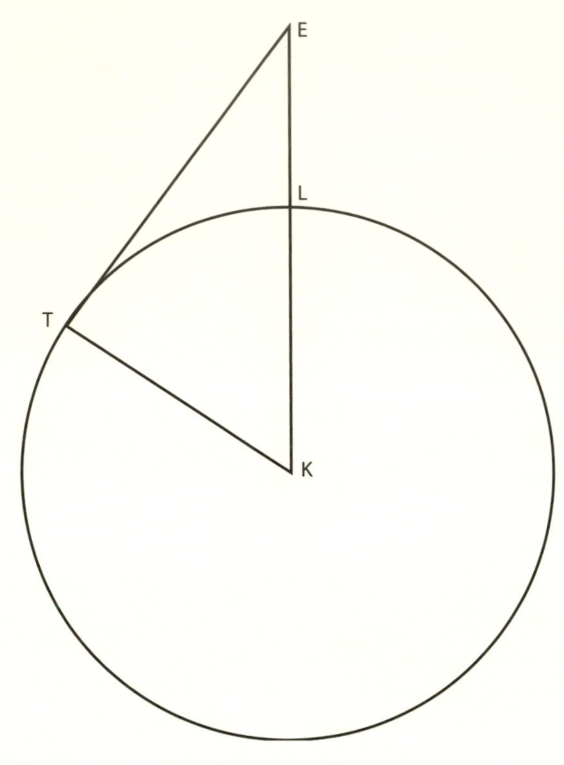

FIGURE 5.74

Chapter 6

[The next excerpt is devoted to finding the distance between two localities, given their geographical coordinates. In figure 5.75, E is the north pole and A and B are two localities whose geographical coordinates are known, EH and ET meridians through these localities, HT an arc of the equator, BD and AZ parallels of latitude through two localities, and AB is an arc of the great circle joining the two localities. The question is: What is the distance AB? Since al-Bīrūnī assumes as given the circumference of the earth the question is answered once we know the measure of arc AB in degrees.

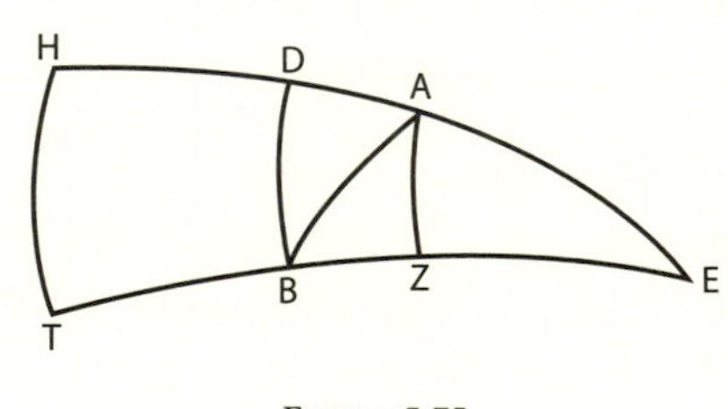

FIGURE 5.75

The argument here is straightforward, using only:

1. The similarity of the isosceles plane triangles EAZ, EDB, and EHT, which he uses to establish the first group of proportions, and
2. A result known as Ptolemy's Theorem which states that if A, Z, B, and D are consecutive vertices of a quadrilateral inscribed in a circle then $AZ \cdot DB + AD \cdot ZB = AB \cdot ZD$.

The reader also needs to understand al-Bīrūnī's reference to the chord of the arc of AZ and of arc BD as "the chord of the difference between the two longitudes, in its (respective) circle of latitude." His point here is to remind the reader that the lengths of the two chords are taken not relative to the radius of the sphere but to the radii of the respective circles of latitude.]

If two towns are on the same meridian, that is, if the two longitudes are the same but the two latitudes are different, then the difference between the two latitudes is equal to the displacement between the two towns, measured on the circle. Hence, if it (the difference in latitude) is multiplied by the amount of the surveyed degree mentioned heretofore, the product obtained is a measure of the distance between the two towns.

But if both towns are on the same parallel, that is, if the two latitudes are the same but the two longitudes are different, the displacement between them is the arc of the great circle passing through both of them, and it is not the arc of the parallel of latitude between them. The chord of the displacement is the chord of the parallel of latitude between them and its ratio to the chord between the two longitudes is equal to the ratio of the cosine of their common latitude to the total sine. Hence, if we multiply the chord between the two longitudes by the cosine of their latitude and we divide the product by the total sine, the quotient obtained is the chord of the displacement. If we multiply the displacement by the amount surveyed of one degree, the product obtained is a measure of the distance.

In case the two longitudes as well as the two latitudes are different, let *A* (fig. 5.75) be one of the two towns, and *B* the other. We draw *AB*, the displacement arc between them. Let *E* be the north pole of the equator, *EAH* the meridian of town *A*, and *EBT* the meridian of town *B*. With *E* as pole, we draw with (polar) distance *EA* the parallel of latitude *AZ*, and with distance *EB* the parallel *ED*. Then the points *A*, *D*, *B*, and *Z* are concyclic, because the chords *AD* and *BZ* are equal, and the chords *AZ* and *ED* are parallel. But each of the two ratios, the sine of *EA*, the complement of the latitude, to the chord of *AZ*, and the sine of *EB* to the chord of *ED*, is as the ratio of the sine of *EH*, a quadrant, to the [chord] of *HT*, (the difference) between the two longitudes. Hence, if we multiply the cosine of each of the two latitudes by the chord (of the difference) between the two longitudes, and we divide the product by the total sine, the quotient (in each case) is the chord (of the difference) between the two longitudes, in its (respective) parallel of latitude. And the sum of the product of chord *AZ* times chord *BD* plus the product of chord *AD* times the equal chord *BZ* is equal to the product of chord *AB* times the equal chord *ZD*. So, if we multiply the quotients of the two divisions, the one by the other, and if we multiply the chord of the difference in latitude by itself, and if we add the two products, and then extract the (square) root of the sum, the root obtained thereby is a measure of the chord of the displacement *AB*. If we multiply the displacement by the amount of the surveyed degree, the product will be a measure of the distance.

Then I say that between each pair of towns, these four things are interrelated: their two latitudes, their difference in longitude, and the distance between them. Whenever three of these are known, the fourth is deducible from them. The number of combinations is three. The first is made of the two latitudes with the longitudinal difference. The distance is deducible from this combination, and this has already been discussed. The second is made of the two latitudes with the distance. The longitudinal difference is deducible from this combination. The third is made of the distance, the longitudinal difference, and one of the two latitudes. The other latitude is deducible from it. These are the two main objects we have been pursuing in our study right from the beginning.

Now, let us evaluate the longitudes of some towns, or their latitudes, from the correct data known to us about one of them; or from data correctly derived for another town, so that we can derive the unknown data for the remaining towns. We make Baghdad, the City of Peace, the reference base for measuring longitudes, because astronomical observations are made there, and it is the seat of the Caliphate and the source of royalty and princes. The difference (of longitude) between it and Alexandria is known, because Baghdad is in the vicinity of Babylon; and Babylon was an ancient city long before the Deluge, and it existed after it, down to the time of Alexander, in its present location.

As to the towns of known latitudes, which I take as reference bases in my worked examples, they are: Baghdad, Shiraz, and Sijistan, then Rayy, Nishapur, and Jurjaniya in Khwārizm, and Balkh. Also other towns are added for corroborative evidence. If they do not follow their normal courses, I keep measuring one of them against another until my mind is fairly satisfied with the longitudes obtained and then I will gradually proceed to the intended base at Ghazna, because my observation and operations are based at it. It is known that by taking them in pairs, some of them occupy extreme or mean positions, and that some are related to others in simple or complex relationships. The following examples will serve as guides to the calculator and as aids for formulation and investigation. I do not feel secure against the occurrence of slips in the calculations, because of the intensity of my worries. God is the best Guide to the truth!

Chapter 10

That evaluation [of the longitudinal difference between two towns] tallies with what I had determined by observation. I had arranged with Abū al-Wafā' Muḥammad ibn Muḥammad al-Būzjānī, he being in Baghdad and I in the city of Khwārizm, for a joint observation by the two of us of a lunar eclipse which we observed in the year three hundred eighty-seven of the Hijira. The difference in time between the two observations showed that there is a difference of approximately one equinoctial hour between their two meridians. Moreover, I observed several lunar eclipses, and the result in each case was close to this amount, differing from it only by an amount of insignificant order.

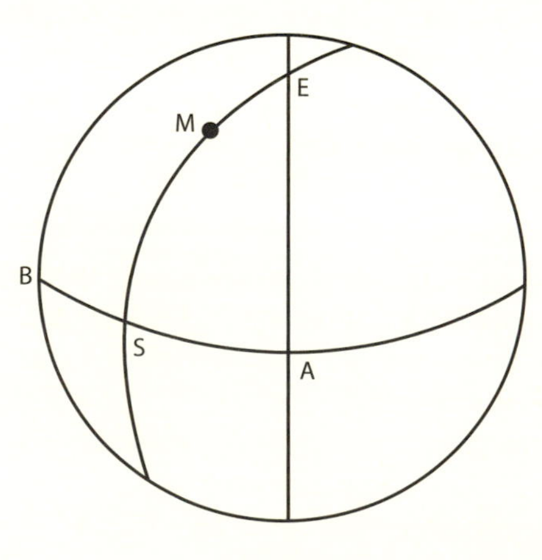

FIGURE 5.76

Chapter 23

[The following section explains how to apply mathematics to find the direction of Mecca relative to the meridian of one's locality. Figure 5.76 shows the celestial sphere viewed from the outside. The zenith of a locality is the point on the sphere directly above the locality, and the meridian of a locality is the great circle on the celestial sphere passing through the north pole and the zenith of the locality. Let E be the zenith of one's own locality and M the zenith of Mecca. Imagine that the great circle *ASB* is the horizon of one's locality, with *A* its south point and *B* its west point. The great circle *EA* is one's meridian[163] and the great circle arc *EMS* passes through the zenith of the locality and that of Mecca,[164] so the arc *AS* of the horizon represents the angle through which one must turn (westwards, in the case shown here) to be facing Mecca.

Muslims must do this five times a day, when they say the obligatory prayers. Bīrūnī states the importance of knowing this direction, called in Arabic the *qibla*, as follows:

> If the investigation of distances between towns, and the mapping of the habitable world, so that the relative positions of towns become known, serve none of our needs except the need for correcting the direction of the *qibla* we should find it our duty to pay all our attention and energy for that investigation. The faith of Islam has spread over most parts of the earth, and its kingdom has extended to the farthest west; and every Muslim has to perform his prayers and to propagate the call of Islam for prayer in the direction of the *qibla*.

It is typical of al-Bīrūnī's broad tolerance that he also mentions that this method is useful for Jews who wish to face Jerusalem. In this chapter of his work Bīrūnī explains a number of methods for finding the *qibla* of one's locality. The one we give here he calls "the method of the *zīj*es," a *zīj* being an astronomical handbook with tables, directions for their use, and a varying amount of theory. Since the diagrams for this method and an earlier one are nearly the same, Bīrūnī refers to an earlier diagram to set up the situation for the method of the *zīj*es, but we have taken just the diagram (fig. 5.77) for the method of the *zij*es itself. (We have, however, taken Bīrūnī's initial description of the various circles and arcs on the sphere in fig. 5.77 from the earlier method, since it uses the same lettering.) As the reader will discover, the procedure for finding the *qibla* is based on multiple applications of the Rule of Four Quantities to the many right triangles in fig 5.77.]

We represent the horizon of Ghazna by circle *ABG*, whose pole is *E*, and the circle of its meridian by *AEG*, whose pole is the west point *B*, because Mecca is west of it (Ghazna). Let *BH* be a quadrant of the equator whose pole is *T*. We draw *TL* the circle of the meridian of Mecca, then arc *HL* is a measure of the longitudinal difference, and, on circle *TL*, we take the arc *LM* equal to the latitude of Mecca. Then *M* is the zenith of the people of Mecca. Through the two points *E* and *M* we draw a great circle; then this circle gives the boundary of the azimuth of the *qibla*. Let this circle intersect the horizon at point *S*, then *S* is the azimuth of the *qibla*. The displacement of *S* from point *A*, which is the south point at Ghazna, is given by arc *AS*, and from the west point *B* by arc *SB*.

[...]

Let us reproduce the first figure which was given to illustrate the method described in the *zijes* for the determination of the azimuth of the *qibla*, and let us draw in it the great

---

[163]The great circle is represented by a straight line since we imagine ourselves looking at the meridian of the locality from outside the locality and "edge-on," as it were.

[164]*EMS* is called "the altitude circle of Mecca for the locality," since the arc *MS* represents the altitude of the zenith of Mecca relative to one's own horizon.

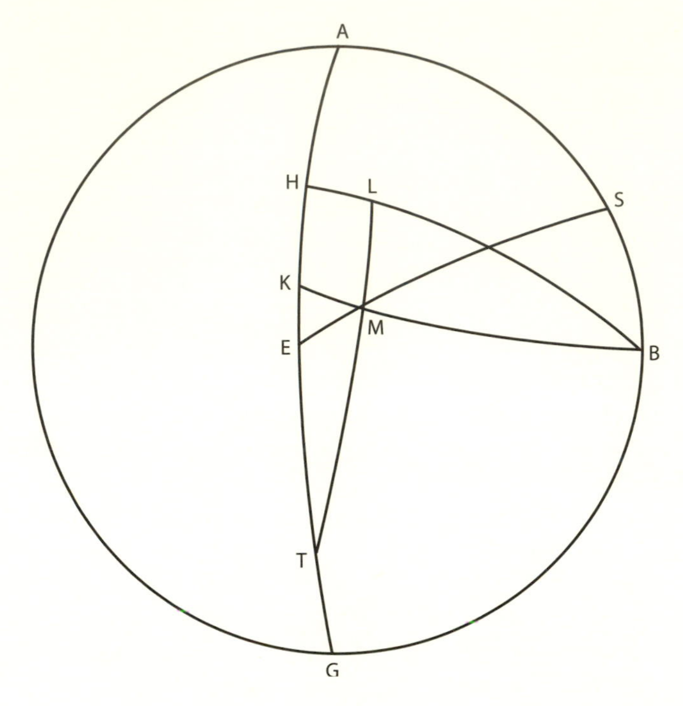

FIGURE 5.77

circle arc *BMK* (fig. 5.77). Then the ratio of the sine of *TM*, the complement of the latitude of Mecca, to the sine of *MK* is as the ratio of the sine of the quadrant *TL* to the sine of *LH*, the longitudinal difference. Therefore, arc *MK*, the modified longitude is known. Also the ratio of the sine of *BM*, the complement of the modified longitude, to the sine of *ML*, the latitude of Mecca, is as the ratio of the sine of the quadrant *BK* to the sine of *KH*, where *KH* is called the modified latitude. So arc *KH* is known. Hence *KE*, which is equal to the difference between *KH* and *HE*, the latitude of the town, is known, and its complement *KA* is known. Further, the ratio of the sine of *BM*, the complement of the modified longitude, to the sine of *MS*, the altitude of Mecca, is as the ratio of the sine of the quadrant *BK* to the sine of *KA*, the complement of that difference. Hence *MS* is known, and its complement *ME* is the distance between the town and Mecca. Also, the ratio of the sine of *ME* to the sine of *MK*, the modified longitude, is as the ratio of the sine of the quadrant *ES* to the sine of *SA*, the displacement of the azimuth of the *qibla* from the meridian line.

We repeat the example. The sine of the modified longitude, which was obtained above, is 25;38,17. Its arc sine is 25;17,47°, and its complement is 64;42,13° and the sine of this is 54;14,48. We multiplied the sine of the latitude of Mecca by the total sine and obtained the product 1329;8,0; then we divided this product by the cosine of the modified longitude, and obtained the quotient 24;30,6. The arc (sine) of this quotient, 24;6,7°, is the modified latitude, and the difference between this and the latitude of Ghazna is 9;28,53°. The complement of this difference is 80;31,7°; we multiplied its sine, which is 59;10,49, by the cosine of the modified longitude and obtained the product 3210;19,58,5,12, then we divided this product by the total sine and obtained the quotient 53;30,19, whose arc (sine) is 63;5,40°. The complement of this arc is 26;54,20°; it is the direct (great circle) distance between Ghazna and Mecca, which is equivalent to 1524;38,53 miles or 508;12,58 *farsakhs*. We divided the product of the sine of the modified longitude times the total sine by 27;9,4, the sine of the distance, and obtained the quotient 56;39,23. The arc (sine) of this quotient is 70;46,56°; it is the measure of the displacement of the azimuth (line) of the *qibla* from the meridian line.

The first systematic trigonometry text, independent of an astronomical work, was written by Naṣīr al-Dīn al-Ṭūsī in the middle of the thirteenth century. First written in Persian and then translated by its author into Arabic, it was entitled *The Complete Quadrilateral* and included methods for solving all types of plane and spherical triangles. Most of these methods had been worked out earlier, but al-Ṭūsī organized all of them into a work in five books. We include here several significant extracts from al-Ṭūsī's work, translated from the published Arabic text as well as from a somewhat free French translation by Alexander Carathéodory published in (then) Constantinople in 1894.

Naṣīr al-Dīn al-Ṭūsī on the Sector Figure

Book I: On the composition of ratios

[The selection from Book I deals with the notion, fundamental to the rest of the work, of composite ratio. This notion appears in the spurious Def. VI, 5 of the *Elements* and T. L. Heath translates the corresponding Greek term as "compound ratio" (a term some authors still use). Heath's translation of this definition reads: "A ratio is said to be compounded of ratios, when the sizes of the ratios, multiplied together, make some (?ratio, or size)." (The bracketed portion is also Heath's.) The Arabic word corresponding to Heath's "multiplied" we have translated as "being repeated" since it has that general sense; however, it refers to a range of ideas, from a piece of material being folded repeatedly to something being multiplied. We have, however, rejected the translation "multiplied" because the specific word for multiplication appears later in the selection; so, it seems, Naṣīr al-Dīn intended some distinction between the two ideas.

R. Lorch (2001) has published the Arabic text and English translation of a much earlier treatise on composite ratios, written in the late ninth century by Thābit ibn Qurra.

Following our usual custom, we have, for the first part, not used symbolism foreign to the Arabic text; but, having made the text's usage clear, we shift to a limited use of modern symbolism. For typographic reasons we use the non-standard symbol "∘" to denote composition of ratios. The idea behind the figures is sufficiently illustrated by the first to omit the subsequent ones.]

General rule

Just as, in the measurement of discontinuous quantity, one uses certain properties essential to continuous quantity, such as divisibility *ad infinitum*, so, too, in the measure of continuous quantity one requires a certain property essential to discontinuous quantity: the supposition that it is composed of discrete or discontinuous units that one imagines to measure those continuous magnitudes. Examining [further] the particulars of what one of these two studies borrows from the other does not [, however,] belong to our subject.

Note on the definition of composition and decomposition of ratios

The beginning of Book VI of Euclid's *Elements* states that a ratio is said to be composed of ratios when, the values of these [other] ratios being repeated one by the other, there results a certain ratio. And we say that the ratios are decomposed into ratios when, divided by each other, they create other ratios.

Having laid these foundations, I now say that:

Proposition 1: [Given] any three magnitudes that can form ratios with one another, i.e., are all of the same kind, the ratio between any of these three and a second one is composed of the ratio of the first to the third and of the ratio of the third to the second.

FIGURE 5.78

Let the three magnitudes A, B, G be homogeneous.[165] I say that the ratio of A to B is composed of the ratio of A to G and of the ratio of G to B, [$A/B = A/G \circ G/B$] and that $A/G = A/B \circ B/G$ and $B/G = B/A \circ A/G$, etc. It suffices to prove the correctness of the first of these to conclude the correctness of the others.

Proof: Let us postulate a unit with which we measure these magnitudes [A, B and G], and that this unit measures E as A measures G [$1/E = A/G$][166]; $1/D = G/B$; and $1/H$ as A/B.

E will then be the magnitude of A/G; D the magnitude of G/B, and H the magnitude of A/B. This is because any ratio has the same name as the number that the unit measures the same way as the first term of the ratio measures the second, and because the number that has the *same name* as the ratio is its magnitude. But, we said that the composition of a ratio and another is the repetition of the value of one of the two by the value of the other. But the repetition of a number by another is nothing but the multiplication of one of these numbers by the other. So H is also the product of the multiplication of E by D. (Indeed, $A/G = 1/E$; and, inversely, $G/A = E/1$. But, $A/B = 1/H$, so, *ex æquali*,[167] one gets $G/B = E/H$. And since $G/B = 1/D$, one has $1/D = E/H$, which gives $D \times E = H \times 1$. But the product of any number by the unit is the number itself, so $H = D \times E$.) Hence, $A/B = A/G \circ G/B$. *QED*.

Proposition 2: Conversely, given three homogeneous magnitudes, if one composes the ratio of any one of these magnitudes to a second with the ratio of this second to the third, one gets as a result the ratio of the first to the third.

---

[165]The Arabic for "homogeneous" is "of one kind," i.e., all of them line segments, or areas, or volumes, or angles, etc.

[166]This is the only interpretation of the text that the rest of the proof allows, even though it seems counterintuitive to express the unit measuring E by the ratio "$1:E$" rather than "$E:1$."

[167]This refers to *Elements* V, 22, according to which, as Heath explains in his translation of the *Elements*, Vol. II, p. 136, given the sequence of proportions ($a/b = A/B$; $b/c = B/C$, etc. through, say, $k/l = K/L$, it follows that $a/l = A/L$.

Let three magnitudes be A, B, G. Let $A/B = 1/E$; $B/G = 1/D$; and let H result from the repetition of E by D. I say that H is the magnitude of A/G.

Proof: For the magnitude E, $E \times 1 = E$; and $E \times D = H$. So the ratios of the rectangles E and H is as the ratio of their sides, 1 and D. But $1/D = B/G$, so $E/H = B/G$. And since $1/E = A/B$, one obtains, *ex æquali*, $1/H = A/G$, which proves that H, the result of [the multiplication of] E by D, is the magnitude of A/G. *QED*.

Or, put differently, E is the magnitude of A/B and D is the magnitude of B/G, and H is the result of the product of D, E. But the unit counts the multiplier as the multiplicand counts the product. And the unit counts E as D [counts] H. But the unit counts E as A counts B. And so $D/H = A/B$. But $1/D = B/G$, so by perturbed proportion, $1/H = A/G$. Hence the magnitude of A/G is H itself, which results from the product of E by D, i.e. from the product (A/B) by (B/G). Hence A/G is what results from the product of A/B by B/G. *QED*

Proposition 3: The preceding is still true when dealing with more than three magnitudes.

Let four homogeneous magnitudes be A, B, G, D; I say that A/D is composed of the ratios A/B, B/G, and G/D.

Proof: It was established earlier that $A/G = A/B \circ B/G$; but A, G, D being also three homogeneous magnitudes, one will have $A/D = A/G \circ G/D$. And, since $A/G = A/B \circ B/G$, one will have $A/D = A/B \circ B/G \circ G/D$. It will be likewise for the converse. Observe that the number of ratios formed this way with a certain number of magnitudes is always less by one unit than the number of magnitudes that make the terms, provided that these magnitudes are common [i.e., the consequent of one is used as the antecedent of another].

Book II: On the sector figure in the plane and its ratios

Chapter 2: On the parts of this figure [the sector figure] and theorems relative to the ratios that one finds there

[In Book II Naṣīr al-Dīn introduces the plane sector figure (fig. 5.79 below) as a way of leading up to the spherical sector figure. This allows him both to introduce terminology that will be basic to the spherical sector figure and to develop the theory of the plane figure, to which the proof of the spherical figure reduces.

With reference to fig. 5.79 below there are two basic theorems:

Theorem I. $\frac{BD}{DA} = \frac{BZ}{ZE} \circ \frac{EG}{GA}$, and

Theorem II. $\frac{BA}{AD} = \frac{BE}{EZ} \circ \frac{ZG}{GD}$,

where, as above, the small circle denotes composition of the two ratios on either side of it. Naṣīr al-Dīn proves these by drawing auxiliary lines. For example, in the case of I, he draws a line through A parallel to DG and extends BE to meet that line at H. The proof then follows by the following short argument, for which the reader may easily supply the reasons.

$$\frac{BD}{DA} = \frac{BZ}{ZH} = \frac{BZ}{ZE} \circ \frac{ZE}{ZH} = \frac{BZ}{ZE} \circ \frac{EG}{GA}. \ \square$$]

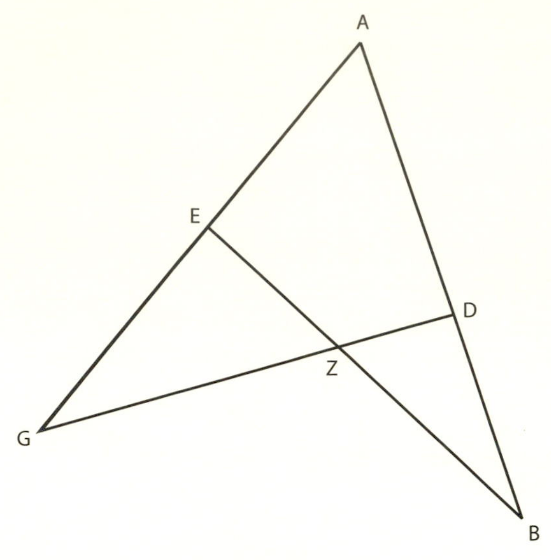

FIGURE 5.79

The different points of view from which this figure may be considered all revert easily to one inasmuch as this figure is composed of two outer lines and two inner lines, which cut each other in the manner here indicated.

And in the following we keep the same labeling of points as here.

[The columns and their segments]

I call *columns* the four lines, *AB*, *AG*, *GD*, *BE*, neither parallel nor juxtaposed, that constitute the [sector] figure. These columns have six intersections: *A*, *B*, *G*, *D*, *E*, *Z*. Each column contains three line [segments] determined by the three points [on it]. Thus, for *AB* one has the three lines *AB*, *AD*, and *DB* … .[168] Thus, in all there are twelve lines and four triangles: *ABE*, *ADG*, *BDZ*, *GZE*, which have as sides the twelve lines in question. The four columns, taken in pairs, give six pairs: *AB* and *BE*; *AB* and *AG*; *AB* and *GD*, *BE* and *AG*; *BE* and *GD*; *AG* and *GD*. And between every pair of them there are two [of the twelve] lines, and all of them are the twelve lines.[169]

[Lines that partner or avoid each other]

And each of these twelve lines is *partner* with five lines and *avoids* six lines. And a partner [of a line] is one that enters [with it] into a composite ratio or [into] a simple ratio that is part of a composite, by virtue of being an antecedent or a consequent [of the composite or constituent simple ratios].[170] And one that avoids it is one that does not enter into such a ratio [with it].

---

[168] Al-Ṭūsī lists all four cases.

[169] For example, between *AG* and *BE* are the two lines *AB* and *GZ*.

[170] In the proportion $a/b = c/d \circ e/f$ the ratio a/b is "the composed ratio" and, e.g., e/f is a simple ratio into which f "enters with" e.

Two lines are partners in one of three cases: when one of the two lines is on the prolongation of the other;[171] when they form the two sides of an angle of a triangle; when they fall between two columns. Each line has two lines partnered with it by the first criterion, two by the second, and only one by the third. So these are the five lines partnered with the one line. As for the remaining six lines [of the total of twelve] they are the ones that avoid it.

And we designate the three [possible sorts of] partners [as partners of] the first, second and third [kinds]. Thus line *AD*, for example, is a partner of the first kind with *AB* and *DB*; of the second kind with the lines *AG* and *DG*; and of the third kind with the line *EZ*. And it avoids the six other lines, *AE*, *EG*, etc. A line is a partner of the third kind with only one other line, but it is virtually equivalent to a double partnership, as we shall explain below.[172]

[Inactive parts of the figure]

Moreover, each ratio composed of two others has six terms [in the whole proportion], as was established in Book I. So, each time a ratio occurs in this figure, six [of the 12] lines are involved, and six others remain inactive. In this case, three of the latter six are always found on the same prolongation and constitute what one calls the *inactive column*, and three others bound a triangle, which we call the *inactive triangle*.

[Ordered and mixed-up ratios]

As for the six lines that are the terms of those three ratios, the two terms of each of the three ratios [occurring in the proportion] necessarily fall in one of the three kinds of partnership. Then, if the two terms of [each of] the three ratios are related by the same kind of partnership [of the three listed above], we say that the [three] ratios are *ordered*. And in the other case we say the ratios are *mixed up*. It remains to observe that the three lines that enter into the same domain of a composite ratio are never partners.

These basics having been set out, we shall call a proposition one of the *first, second,* or *third kind* if it has for its object a composite ratio of which the two terms [of each of its three ratios] are between themselves in a partnership of the same kind (first, second, or third). And each of these propositions has many cases, some of them ordered and others mixed-up.

The foundation of the theory is the ordered proposition of the first kind.[173] The other cases are only supplementary, as we shall explain with God's help.

---

[171]With the one word "prolongation" al-Ṭūsī includes two cases. The first case (Theorem I of the introduction above), in which the two terms (here *BD* and *DA*) of the composite ratio *d* have only an endpoint in common, he calls a case of *explicit* ratio and the second case (Theorem II of the introduction above) where one term of the compound ratio contains the other he calls the *implicit* case.

[172]The explanation, which appears in a later section, is not relevant to the main points here and is omitted.

[173]By definition, this is a statement of a composite ratio in which one segment in each of the three ratios occurring is a prolongation of the other segment in that ratio.

Book III Preliminaries on the spherical sector figure and on what is necessary to make use of it

Chapter 2: On calculating the sides and angles of a triangle

[The author first calculates the sides and angles of plane triangles by the method of "arcs and their chords." He then gives a different method, that of "arcs and their sines," where he uses the law of sines for plane triangles. Thus, he begins with a proof of that law.]

In calculating [in triangles] by arcs and their sines, the fundamental notion is that the ratio of the sides is is equal to the ratio of the sines of the angles opposite to those sides: given a triangle *ABC*, I say that $AB : AC = \text{sine(angle } ACB) : \text{sine(angle } ABC)$.[174]

[In fig. 5.80, in the diagram on the left, triangle *ABC* has an obtuse angle at *B*, while in the diagram on the right, all the angles are acute. The same proof works in either case.]

Proof: Prolong *BC* so that $CE = 60$ [the radius of the circle in which sines are being calculated]. With center *C* and radius equal to *CE*, describe the arc *ED* and prolong *CA* until it meets that arc at *D*. From *D* drop the perpendicular *DF* on *CE*; then $DF = \text{sine}(ACB)$. Also prolong *BC* so that $BH = 60$. With center *B* and radius equal to *BH*, describe the arc *HT* until it meets at point *T* the prolongation of *AB*. Draw the perpendicular *TK*, which is equal to the sine of angle *ABC*. From *A* draw *AL* perpendicular to *CE*. Then $AB: AL = TB$ (radius)$: TK$ (because of the similarity of triangles *ABL*, *TBK*) and, because of the similarity of triangles *ALC*, *DFC*, we have $AL : AC = DF : DC$ (radius). Therefore, by the rules of proportions, $AB : AC = DF$ (sin *ACB*) $: TK$ (sin *ABC*).

Chapter 3: Rules whose knowledge is very useful in the theory of the plane sector figure

[Naṣīr al-Dīn al-Ṭūsī devotes Chapter 3 of Book III of *The Complete Quadrilateral* to the topic of determining two arcs whose sum or difference is known. The additional condition that makes the problem solvable is that one also assumes knowledge of the ratio of the sines of the two known arcs.[175] The result of this problem is used in a number of places in the work, and the problem was sufficiently important that al-Ṭūsī follows the theoretical arguments that the individual arcs are known with a description of the computational procedures needed to compute their values.]

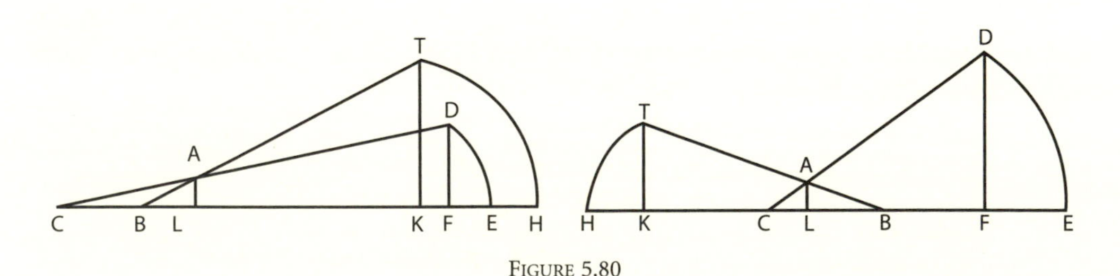

FIGURE 5.80

[174] Modern symbols, including ":" and "=," are used in this proof to help the reader.

[175] "Sine" refers to the medieval sine of an arc: the length of half the chord of double the arc.

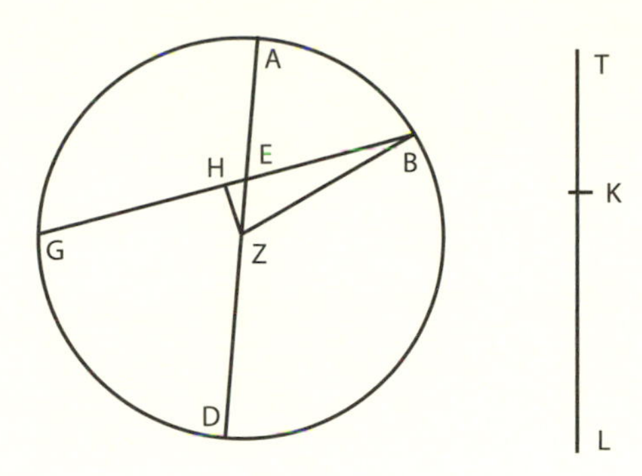

FIGURE 5.81

If two different arcs of a circle that, together, are less than a semicircle, join at a point, and their sum is known, and if the ratio of the sine of one of them to the sine of the other is known, then each of them is known.

Let there be in circle *ABG* two arcs, *AB* and *AG*, which join at *A*, and let their sum, *BAG*, which is less than a semicircle, be known [fig. 5.81]. Let the ratio of the sine of arc *AB* to the sine of arc *AG* be known. I say, then, that each of the two arcs, *AB* and *AG*, is known.

Its Proof: The chord *BG* and diameter *AD* are produced, and so they meet at [say] *E*. From the center, *Z*, the perpendicular *ZH* is produced onto *BG*, and *BZ* is joined. Because arc *BAG* is known, the chord *BG* is known. And because sin(*BA*)/sin(*AG*) is known *BE*/*EG* is known.[176] So, let it be as the ratio of *TK* to *KL*, and *BG*/*BE* = *TL*/*TK*, so *BE* is known.[177] and [hence] *EG* is known. And *BH*, half of *BG*, is known, and so *EH* is known. And *ZH*, the sine of the complement of half the arc *BG*,[178] is known. And in the right triangle *EZH* the two sides *EH* and *HZ* containing the right angle are known, so angle *EZH* is known. And angle *BZH*, which is in magnitude half of arc *BAG*, is known. So the remaining angle, *BZA*, is known, and it is the magnitude of arc *BA*, [which is therefore] known. And [so] the remaining arc *AG* is known as well. And that is what we wanted to demonstrate.

And also, if one arc in a circle falls on another, not equal to it, and they begin at the same point, and if each of them is less than half the circumference, and if the excess of the one of them over the other is known, and if the ratio of the sine of one of them to the sine of the other is known then each of the arcs is known.

So, let there be in circle *ABG* two arcs, *AB* and *AG*, which begin at *A*, and let the difference between them, *BG*, be known [fig. 5.82]. Let the ratio of the sine of [arc] *AB* to the sine of [arc *AG*] be known. I say that each of the arcs, *AB* and *AG*, is known.

Its Proof: We join the diameter *AI* and we produce it, and we produce chord *BG* until the two of them meet at *E*. From the center, *Z*, we produce the perpendicular

[176]Naṣīr al-Dīn demonstrates this in Chapter 1 of Book III. If *BT* and *GI* are perpendiculars from *B* and *G* onto the diameter, *AZD*, they are the sines of arcs *AG* and *GZ*, respectively. The right triangles *BTE* and *GIE* are similar and the ratio of corresponding sides, the sine of arc *AB* to the sine of arc *AG*, is known. So the ratio of two other corresponding sides, *BE* and *EG*, is also known.

[177]Because both the ratio *BE*/*BG* and *BG* are known.

[178]This is the cosine of one-half arc(*BG*).

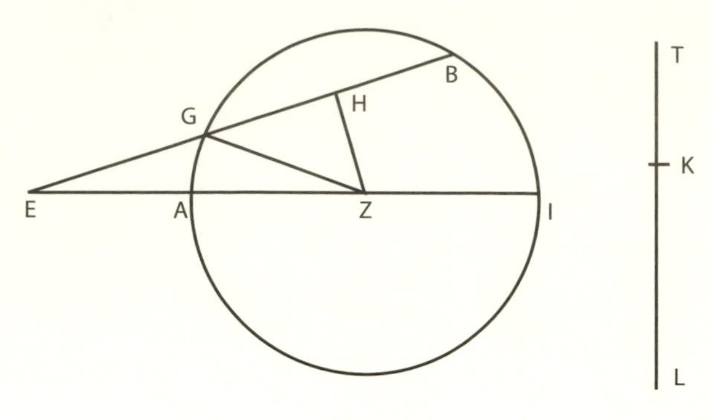

FIGURE 5.82

ZH onto *BG*, and we join *GZ*. And because *BG*, the chord of the excess, is known, half of it, *HG* is known. And because sin(*AB*)/sin(*AG*) is known, *BE*/*EG* is known. And, let it be as the ratio of *TL* to *KL*. And *BG*/*GE* = *TK* /*KL*, so *BE* is known and *GE* is known. And *HG* is known, so *HE* is known. And *HZ*, the sine of the complement of half [arc] *BG*, is known. And so in the right triangle *HZE* the two sides *HZ* and *HE* are known, so angle *HZE* is known. And angle *HZG*, which is in magnitude half of arc *BG*, is known. And so angle *GZA* is known. And it is in magnitude the arc *AG*, and so arc *AG* is known. And arc *AB* is known. And that is what we wanted.

And so it is evident that if the sine of *AB* is larger than the sine of *AG* then the intersection is in the direction of *A*, and if it is less than it then the intersection is in the direction of *I*. And if it is equal to it then the chord is parallel to the diameter.

Book IV On the spherical sector figure and the ratios found in it

Chapter 1: What the spherical sector figure consists of and indications regarding the statements of the ratios found in it

If four great circles of the surface of a sphere intersect so that no more than two intersect in the same point,[179] then there result twelve points of intersection[180] and each circle will be divided into six arcs,[181] each one of which will constitute a side of a figure. [This way] there will be twenty-four arcs altogether, and the surface of the sphere will be divided into fourteen parts; six quadrilaterals.[182] Each arc is a side of both a quadrilateral and a triangle, and each angle is vertically opposite an angle of another figure of the same kind.[183] Each quadrilateral will, in this way, have its four sides equal to four sides of four triangles, and its four angles equal to four angles of four [other] quadrilaterals, but each triangle will have its sides equal to sides of the quadrilaterals and its angles equal to angles of triangles. And

[179]In fact, as Naṣīr al-Dīn and his readers well knew, any two great circles intersect in exactly two points.

[180]The points *A, B, G, D, E, Z, H, K, L, M, N, T* in fig. 5.83 are on the circles *ABKZ, BGZN, AGKN,* and *DETL*.

[181]Any one of the circles intersects each of the three others in two points, having no member in common with any of the other resulting pairs.

[182]The six quadrilaterals are labeled 1, 3, 5, 9, 12, 13 in our figure, and the eight triangles are 2, 4, 6, 8, 10, 11, and 14, (which is represented by the area outside the boundary *NKZ*). The areas are not, of course, numbered in the figure in the text.

[183]The French translator points out that "It is seen on the picture that the angles formed around point *A*, for example, belong two-by-two, to two triangles *AGB, ADE* and to two quadrilaterals *ABMD, AGHE*."

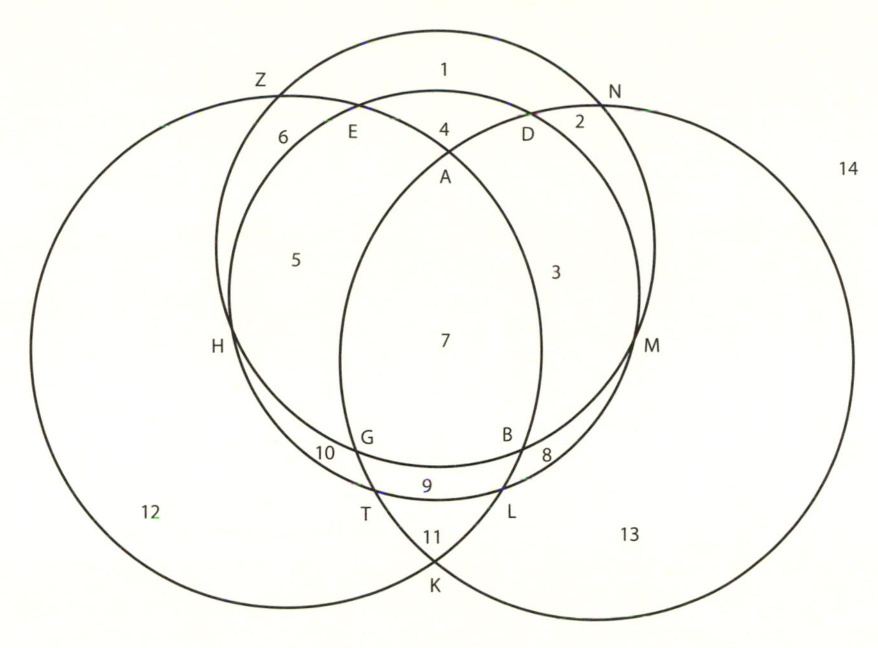

FIGURE 5.83

so the six quadrilaterals will meet each other at their angles, but they meet the triangles on their sides. And similarly for the triangles.[184] And here is the figure for that.[185]

[The author at this point lists the six arcs of each circle as well as the six quadrilaterals and the eight triangles that make up the configuration.]

Now we say that each of these quadrilaterals, together with the two triangles positioned on two of its adjacent sides, has the form of the plane *sector figure*[186] because it has four *columns*[187] mutually intersecting at six points from which there result twelve arcs, three on each column,[188] and four triangles, each made of three lines [i.e. arcs], as was explained. And by the totality of what was said about the figure, the figure formed by a quadrilateral and the two triangles is what we call a *spherical sector [figure]*.[189]

For example, if we consider quadrilateral *AEHG* with its two triangles *ABG*, *GHT*, positioned on its adjacent sides *AG*, *GH*, it looks like the figure [5.84]. It gives us a

---

[184]The French translator: "Each quadrilateral this way will have its four sides equal to four sides of four triangles, and its four angles equal to four angles of four [other] quadrilaterals, but each triangle will have its sides equal to sides of the quadrilaterals and its angles equal to angles of triangles."

[185]Fig. 5.83 represents the four great circles on the sphere as four circles on the plane. The points of intersection are denoted by letters and the regions bounded by the circular arcs are denoted by numerals. Note that region 14, though it is infinite on the plane, represents a finite region on the sphere.

[186]The same figure is sometimes called "the complete quadrilateral."

[187]For this terminology, see Chapter 2 of Book II, on the plane sector figure.

[188]The Arabic actually says "every three of them on a column," which is not accurate.

[189]The Arabic calls it "the spherical sector."

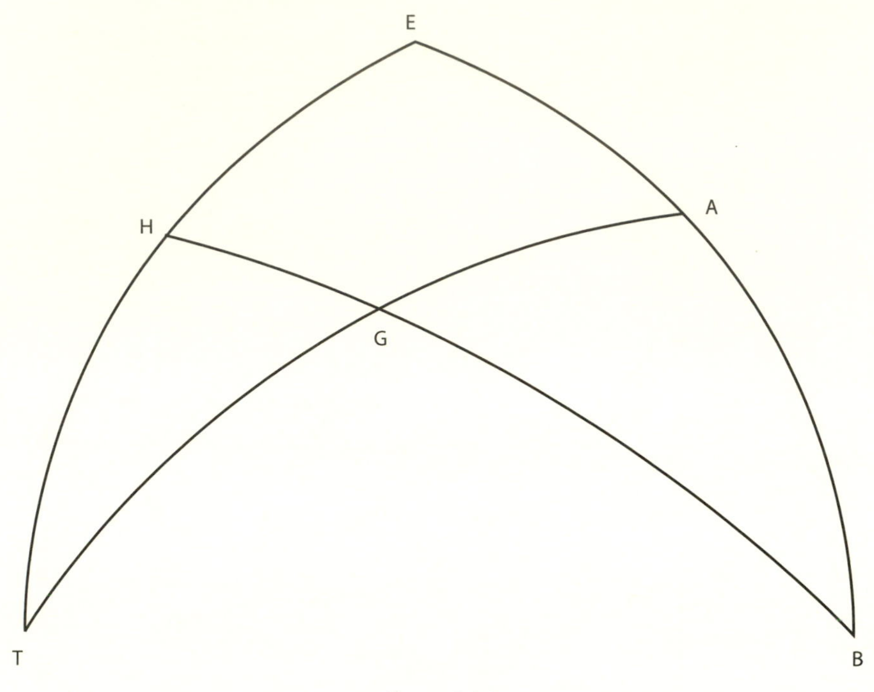

FIGURE 5.84

complete spherical sector figure, and it is itself like the plane sector [figure], except for one thing, that figure is composed of straight lines in a plane and this figure is composed of arcs of great circles.

And since every quadrilateral has four sides and on every side there is a triangle, every quadrilateral this way appears in four sector [figures]. For example, the quadrilateral *AEGH* taken with triangles *ABG*, *GHT* is, alone, as sector figure, and with *ABG*, *ADE* a second, and with *ADE*, *EHZ* a third, and with *EHZ*, *HGT* a fourth. And, since the intersection of four circles gives, as it is seen, six quadrilaterals, there will be altogether twenty-four figures of complete spherical quadrilaterals on the surface of the sphere.

But, each sector [figure] is corresponding and equal to another [fig. 5.85]. The figure *MNZEAD*, for example, is corresponding and equal to the figure [*KL*]*BGHT*, since column *MNZ* of the first sector is equal to column *HGB* of the second for the reason that *MNZH* is a semicircle because any two great circles on the surface of a sphere necessarily bisect each other, as Theodosius proved in Book I, Theorem 12 [of his *Spherics*]. And also *ZHGB* is a semicircle, so if we remove the common [component] *ZH* there remains column *MDZ* equal to column *HGB* in the two sectors. And in the same way we prove that column *MDE* is equal to *HTL* and column *AEZ* to column *KLB*, and column *ADN* to column *KTG*. And also in the same way we prove that the two sides of every corresponding pair of triangles, and every two of their quadrilaterals are equal. And [we also prove that] every two corresponding angles are equal, and from that it is proved that twelve sectors of the twenty-four mentioned correspond to twelve others. And the ratios occurring between the lines of the plane sectors, which comprise the three propositions

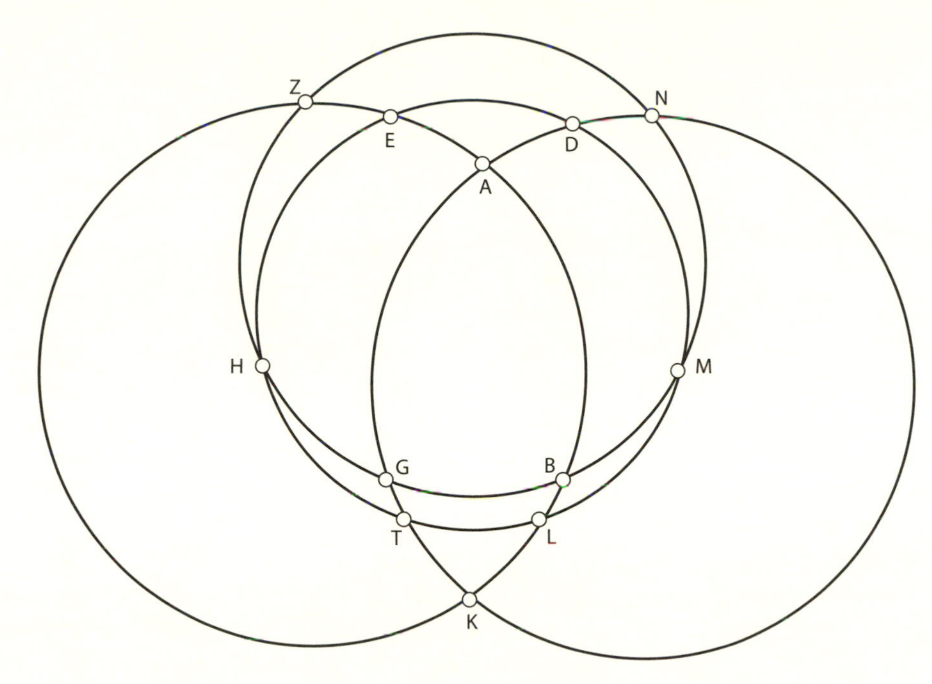

FIGURE 5.85

there [i.e., in Book II of this work], hold here between the sines of the arcs of the spherical sector without varying a thing, and it is unnecessary to refer to them. So let us now prove them.

Chapter 2: General indication about the demonstration—and demonstration of the proposition of the first kind, known as the "disjunctive ratio" of Ptolemy

We shall, first of all, apply the demonstration to the ordered proposition of the first kind,[190] and we shall use what we say about it to speak concerning the proof of what he [Ptolemy] omitted of the ratios occurring in the three propositions.

When we want to prove that the ratio of the sines of two arcs of the spherical sector figure is composed of two ratios provided by the sines of four other arcs, such that these six arcs generate, among their sines, an ordered ratio of the first kind, we must first determine the inactive column and triangle, as we have done previously.[191] Afterward, we join the vertices of the angles of the inactive triangle by lines. In other words, we draw the chords of the arcs that constitute the triangle, and also three straight lines that, starting from the center of the sphere, end up at the three points positioned on the inactive column, lines that will be radii. These will necessarily intersect the three chords of the inactive triangle[192] in such a way that each

---

[190]For an explanation of this terminology see the selection from Chapter 2 of Book II and especially the last sentence of that chapter.

[191]I.e., for the plane case, in Chapter 2. To make this general description concrete in the present, spherical, case see fig. 5.83 of this section, where *ABE* is the inactive triangle and *DZG* the inactive column.

[192]This is because the inactive column contains three points on the sides of the inactive triangle.

radius intersects the chord of the arc on which is found the point where it (*the radius*) ends.

These three intersections of the radii with the chords will be positioned [both] in the plane of the triangle determined by the three chords of the inactive triangle and in the plane of the great circle to which belongs the arc of the inactive column. In other words, the three points considered will be positioned on the intersection of these two planes. But Euclid proves in his *Elements* that the intersection of two planes can only be a straight line, so the three points will also be positioned on a same straight line. And this straight line will [then] form, with the three chords of the inactive triangle, a plane sector figure, whose ratios and properties will be used to prove those of the spherical sector figure, using what we established in Book III. (If it should happen that one of these three chords is parallel to one of the radii taken into consideration, there would be a composite ratio formed either of a ratio equal to the composite ratio itself and a ratio of equality,[193] or of another ratio and its inverse, as we shall prove.)

Let the spherical sector figure be determined by the 6 points *ABGDEZ*, as it was with the plane sector [fig. 5.86]. And let us first speak of the case known as the disjunction of Ptolemy, which is the same as the ordered proposition of the first kind. And we say that the ratio of the sine of arc *BD* to the sine of arc *DA* is composed of the ratio of the sine of arc *BZ* to the sine of arc *ZE* and of the ratio of the sine of arc *EG* to the sine of arc *GA*.[194]

Here arc *DZG* will be the inactive column; *ABE*, the inactive triangle.[195] Let us join the straight lines *BA*, *BE*, *AE*, the chords of triangle *ABE*. And let *H* be the center of the sphere. From it we produce the radii *HD*, *HZ*, *HG* to the three points *D*, *Z*, *G* of the inactive column—each of these three radii being necessarily in the plane of

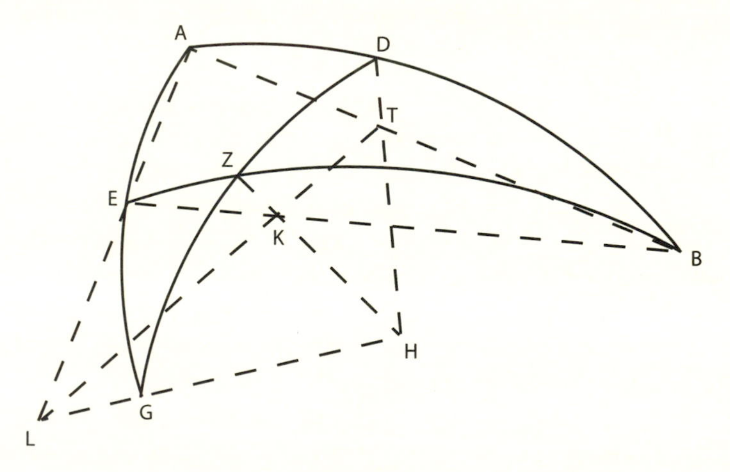

FIGURE 5.86

[193]This is a ratio in which the antecedent is equal to the consequent, i.e., in modern terms, a ratio equal to 1.

[194]The medieval sine of an arc was a line equal to half the chord subtending twice the arc. In modern terms, if the arc is θ and Sin θ is the medieval sine of θ, then Sin $\theta = (r/2)\text{chord}(2\theta)$. The radius, r, was often taken to be 60.

[195]The proposition says that *BD*/*DA* is composed of *BZ*/*ZE* and *EG*/*GA*. (Note that all six terms appearing are partners of the first kind.) The remaining six terms, which avoid these, are *BA*, *BE*, *EA*, *GZ*, *ZD*, and *DG*. The first three contain the triangle *ABE* and the remaining three are segments of the line *GD*.

the [respective] circle that provides an arc of the inactive triangle, in which [plane] will be also found respectively the chords of those same arcs. Consequently, because each of *HD* and chord *BA* being in the plane of the circle *BDA* they will intersect each other, say at point *T*. Similarly, *HZ* and chord *BE*, which are in the plane of circle *BZE*, will meet each other, say at point *K*. And because radius *HG* and chord *AE* are in the plane of circle *AEG*, they will, being extended, either intersect each other or be parallel.[196]

Let us suppose first that they intersect. In this case, these two lines will either intersect on the same side as *GE* or on the same side as *EA*.[197] First, let them intersect each other in the direction of *GE*, at, say, point *L*. Then points *T*, *K*, *L* fall in the plane of triangle *BAE*, formed by the chords of the inactive triangle, because of their all being on its sides, and [also fall] in the plane of circle *DZG*, which is its inactive column, because they are on the radii passing through the points on it. So they are all on the intersection [of the two planes]. And, because of their being planes, this intersection is a straight line. And so the line *TKL* is straight, and there results from it and from the three chords the plane sector figure *BTAELK*. From this we have that the ratio of *BT* to *TA* is composed of the ratio of *BK* to *KE* and the ratio of *EL* to *LA*. But the ratio of *BT* to *TA* is like the ratio of the sine of arc *BD* to the sine of arc *DA*, the ratio of *BK* to *KE* is like the ratio of the sine of arc *BZ* to the sine of arc *ZE*, and the ratio of *EL* to *LA* is like the ratio of the sine of arc *EG* to the sine of arc *GA*, as we established in Book III. Hence replacing the ratios[198] BT/TA, BK/KE, EL/LA by their equals, one gets $\sin(BD)/\sin(DA) = \sin(BZ)/\sin(ZE) \circ \sin(EG)/\sin(GA)$, which was to be proved.

But if the intersection of chord *AE* and radius *HG* takes place on the side *AE*,[199] then we have recourse to another sector [figure] on the surface of the sphere. The proof will be special to that sector and the objective will be attained by applying [the result of] that proof to this postulated sector. And the way to do it is to produce each of the arcs *GEA*, *GZD* until they meet on the other side at point *M*. Then *GEM* and *GZM* will be semicircles, according to what was established in Prop. 12 of Book I of Theodosius's *Spherics* (that great circles on a sphere bisect each other)[200] [fig. 5.87].

Now, when we have also extended the radius *GH* beyond the point *M* it will meet the chord *EA* at *L*, and so the three points *K*, *T*, *L* will again be found on a straight line, *KTL*, because of their being on the plane of triangle *ABE* (the chords of the arcs of the inactive triangle) and in the plane of the inactive circle *DZG*.[201] And so there results the plane sector *BKEALT* in which $BT/TA = BK/KE \circ EL/LA$. But

[196]Since *G* is not on the arc *AE*, *HG* does not necessarily cut the arc (and hence its chord), so it may be parallel to the chord.

[197]I.e., the intersection, which is outside of arc *EA*, will be either closer to *E* or closer to *A*.

[198]Having given the reader some feeling for what a text virtually devoid of symbolism is like we now use modern symbolism (with the sign "∘" for composition of ratios) to make the argument easier to follow.

[199]Both the Arabic text and the French translation have, erroneously, *AH* instead of *AE* here.

[200]This is Theorem I, 11 in the Greek text. It is not unusual for Arabic versions of Greek works to differ from the extant Greek version in the numbering of propositions. This sometimes happened because the translators used a different version of the Greek text than the one that came down to us. Other times it is the result of medieval editors revising the work to a greater or lesser extent.

[201]He means, of course, "the inactive column, which is an arc of the circle *DZG*." The full argument here is that *K*, *T*, *L* are in the plane of the great circle *GZDM* since they are on radii (in one case extended) of that circle. But *K* and *T* are also in the plane of the inactive triangle, *EBA*, and so is *L*, which is on *EA*.

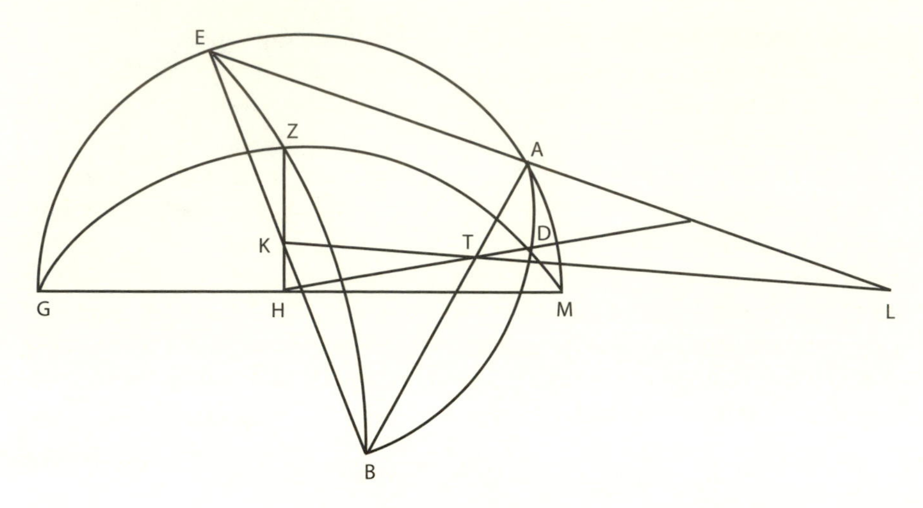

FIGURE 5.87

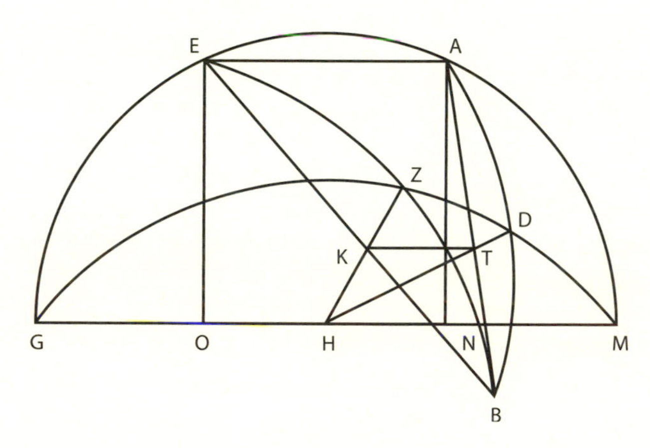

FIGURE 5.88

$BT/BA = \sin(BD)/\sin(DA)$, $BK/KE = \sin(BZ)/\sin(ZE)$ and $EL/LA = \sin(EM)/\sin(MA)$. Hence, in the special sector with the demonstration, i.e. the sector *MEBD*, $\sin(BD)/\sin(DA) = \sin(BZ)/\sin(ZE) \circ \sin(EM)/\sin(MA)$. But, the sector postulated [here] is not the sector *BAGZ*.[202] And $\sin(EM) = \sin(EG)$, *EM* and *EG* being supplementary, and, similarly, $\sin(AM) = \sin(AG)$. Hence, in the original postulated case, one will have $\sin(BD)/\sin(DA) = \sin(BZ)/\sin(ZE) \circ \sin(EG)/\sin(GA)$. And this is what was to be proved.

But if the chord *AE* is parallel to the radius *HG*, let us complete the half-circumferences *MAG*, *MZG* as was done [above] and join diameter *MG*[203] [fig. 5.88]. Then we say that *TK* and the diameter *MG*, which are in the plane of circle *MDG*, are parallel; for, if not, they would meet at some point, *L*. Then the three

---

[202]The text has *BAZG*.

[203]Most of the notation here is tacitly taken over from previous cases.

points L, A, E would be both in the plane of triangle BAE and in the plane of circle MAG. So LAE would be a straight line, and thus the two lines EA, GM, would meet each other at L. But we supposed that they are parallel, which is a contradiction. Hence TK will be parallel to diameter MG. And AE is parallel to it [also]. And so line KT is parallel to AE.[204]

Hence, in triangle BAE, $BT/TA = BK/KE$. But, $\sin(BD)/\sin(DA) = BT/TA$ and $\sin(BZ)/\sin(ZE) = BK/KE$, hence $\sin(BD)/\sin(DA) = \sin(BZ)/\sin(ZE)$. On the other hand, $\sin(AG) = \sin(GE)$. This is because the sine of arc AG is perpendicular [to GM], and AN [which] is the sine of arc GE is perpendicular [to] EO [also]. And [so] the two of them [i.e. the two sines] are parallels. And they fall between the parallels AE and MG, so they are equal. And so $\sin(BD)/\sin(DA) = \sin(BZ)/\sin(ZE) \circ \sin(GE)/\sin(GA)$ (the first term being equal to the second term and the last ratio being one of equality[205]). *QED*.

And this is how the ratio considered is composed of the two aforementioned ratios in all cases, which is what we wanted to show.

In all cases that lead back to the proposition of the first kind, every time that arc DZ is the inactive column, and triangle BAE the inactive triangle, the figure will be as was just said and the proof will be done consequently. And also every time arc BZE will be the inactive column and triangle GDA the inactive triangle, the difference will consist [only] in a simple arrangement of the sides, but the figure and the demonstration stay exactly the same, so we need not go into details on this subject.

Book V: Explanation of the methods of using the sector figure in the study of arcs of great circles which meet on the surface of a sphere
Chapter 6: On tangents, cotangents and the tangent theorem

[Naṣīr al-Dīn begins with the following:]

According to Abū al-Rayḥān [al-Bīrūnī], the priority in the discovery of this theorem belongs incontestably to Abū al-Wafā' al-Būzjānī. Its enunciation is that in a right-angled triangle [whose sides are] formed by arcs of great circles the ratio of the sine of one of the two sides of the right [angle] to the sine of the right angle is as the ratio of the tangent[206] of the other side of the right [angle] to the tangent of the angle it subtends. But, before embarking on a proof of it, I first say that the tangent of an arc is what two diameters through its extremities cut off from the perpendicular, drawn from one of the two extremities, to the diameter that passes through that extremity. And that perpendicular is parallel to the sine of that arc if it [the sine] is [the one] perpendicular to that [same] diameter.[207]

---

[204]In the medieval style, Naṣīr al-Dīn inserts here an alternate proof of what has just been proved, as follows: And, in a different way: If TK is not parallel to chord AE and diameter MG, let us produce from point T, in the plane of triangle ABE, line TS parallel to AE, and, in the plane of circle MZG, TF parallel to MG. Then TF and AE, being parallel to MG, will also be parallel to each other. And also TF and TS, both parallel to AE, will be parallel to each other. But they meet at T. And this is a contradiction so TK is parallel to chord AE.

[205]That is, a ratio in which the antecedent and consequent are equal—in modern terms, a ratio equal to 1.

[206]The term in ordinary Arabic means, literally, "shadow."

[207]Since an arc has two extremities it has two sines (both equal).

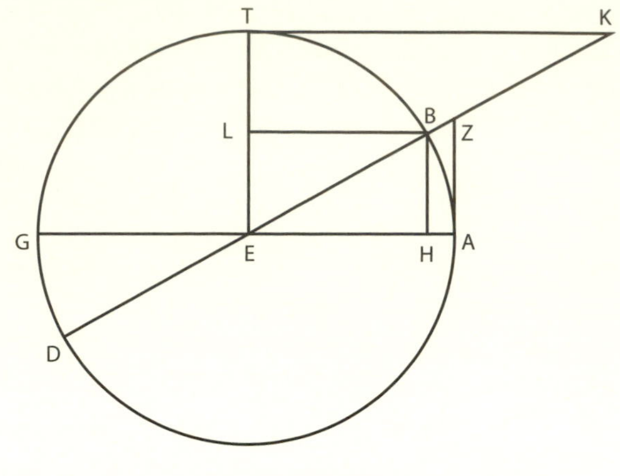

FIGURE 5.89

So let us draw [fig. 5.89] circle *ABGD*, whose center is *E*, and let us cut off from it some arc, say *AB*. And we produce two diameters that pass through the points *A* and *B*, *AG* and *BD*. And from *A* we produce *AZ* perpendicular to *AG* until it meets the diameter *DB* at *Z*. Then *AZ* is the tangent of arc *AB*, and it is parallel to its [the arc's] sine, *BH*. And we also draw from the center [the line] *ET* perpendicular to *AG*, and the perpendicular *TK* from the point *T*. Then *TK* is the tangent of arc *TB*, and *BL* is its sine. And the two of them are [in other terms] the tangent of the [complement of the] arc *AB* and its sine. And what we have called the tangent the astronomers call the first, or reversed, tangent of the arc *AB*, which [arc] is analogous to an arc of altitude.[208] And they call *TK* the second, or level, tangent of arc *AB*. And they call *ZE* the diameter of the first tangent and *KE* the diameter of the second tangent.

They make use of the diameter [of circle *ABGD*] to measure the first tangent in the same way that they use it to measure sines and chords.[209] Sometimes they divide the second tangent into twelve parts, which they call digits, or even into nine or six. They give to the half the name of feet. The first tangent of any arc is the second tangent of its complement, and vice versa. The ratio of a tangent to the radius is equal to the ratio of the sine of an arc to the sine of its complement; the ratio of the tangent to the diameter of the tangent is equal to the ratio of the sine to the radius, because of the similarity of triangles *ZAE* and *BHE*.

[After some further preliminaries Naṣīr al-Dīn now states and proves a crucial lemma:]

If [fig. 5.90] two planes cut each other at an obtuse or acute angle and if, taking a point on one of these planes, we construct a perpendicular to this plane, and [also] from this point we draw a line perpendicular to the line of intersection of the two planes, [then] the line that joins the intersection of the first perpendicular with the

[208]Al-Ṭūsī is thinking of arc *AB* measuring the altitude of a star at *B* for an observer at *E* relative to the horizon *EA*.

[209]That is to say, they divide the diameter into a convenient number of units (e.g., 120 if one is using a sexagesimal system) and use those units to measure the tangent.

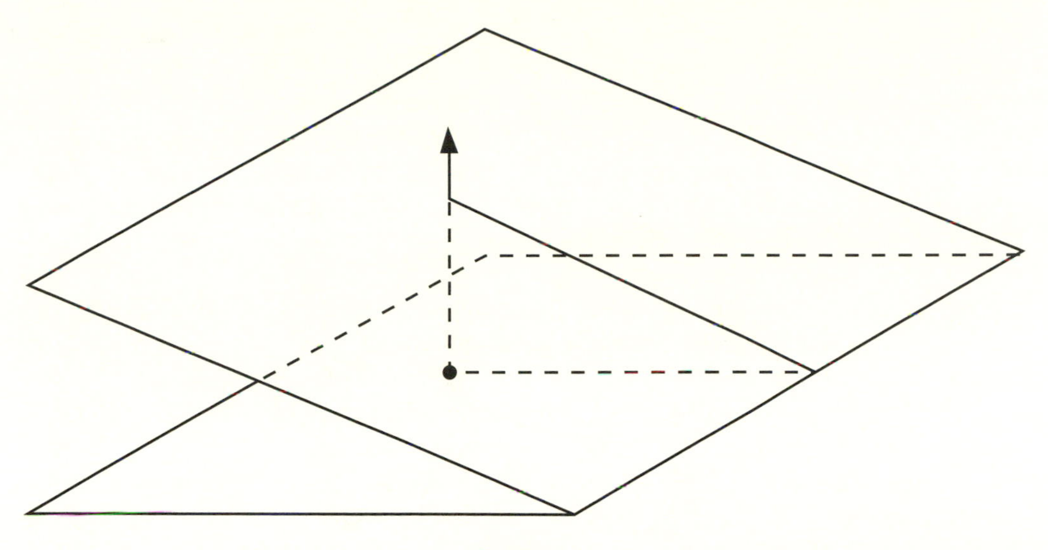

FIGURE 5.90

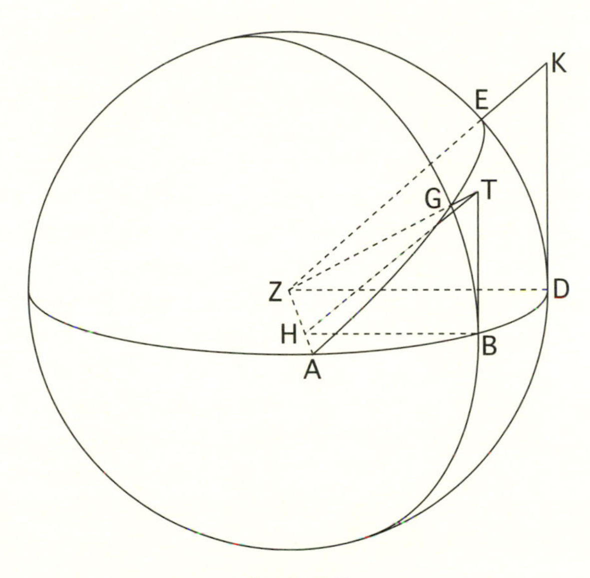

FIGURE 5.91

second plane to the foot of the perpendicular [we drew] to the line of intersection will also be perpendicular to the line of intersection.[210]

[Now he is ready to state and prove the tangent law:]

Let there be triangle *ABG* [formed] from arcs of great circles [fig. 5.91], one of its angles, *B*, a right angle, and angle *A* an acute [angle]. Then I say that the ratio of the sine of arc *AB* to the whole sine (which is the sine of angle *B*) is as the ratio of the tangent of [arc] *BG* to the tangent of angle *A*.

[210]This is a partial converse of *Elements* XI, 11, and Naṣīr al-Dīn's proof uses the Pythagorean theorem to show that a certain triangle in the plane on which the first perpendicular was constructed is a right triangle.

Proof: Extend the two sides *AB*, *AG* until they become complete quadrants, *AD*, *AE*, and let the great circle arc, *DE*, pass through the two of them [*D* and *E*], which [arc] will measure angle *A*. From points *B*, *D* produce two perpendiculars *BT*, *DK* on the plane of circle *ABD* until they end at the plane of circle *AGE* at the two points *T*, *K*. And so these two perpendiculars will, without doubt, be in the two planes of the two circles *BG* and *DE*, each in the plane of its circle, because these two circles make right angles with the plane of circle *ABD* and the perpendiculars are drawn from the intersection [points].

And we draw from the center [of the sphere], *Z*, two radii, *ZE* and *ZG*, and we extend them to points *K* and *T*; we also produce the radius *AZ*, the common section of the two circles *AD*, *AE*; we draw from *B* the perpendicular, *BH*, to *AZ*; and we join the [half] diameter *ZD*, which is also perpendicular to *AZ*, because of *AD* being a quadrant. And we produce *TH*.

Then the two triangles *TBH* and *KDZ* will be similar because of the parallelism of *BH* and *DZ*, which are both in the plane of circle *ABD* and perpendicular to *AZ*, and the parallelism of *DK* and *BT*, two perpendiculars to one plane (the plane of the circle *ABD*).

And [so] *TH* and *KZ* are also parallel to each other according to the lemma we stated earlier.[211]

[There follows an alternate proof of the parallelism, which we omit.]

In any case, having proved the parallelism of the corresponding sides of the two similar triangles, *TBH* and *KDZ*, the ratio of *BH*, the sine of arc *AB*, to *DZ*, the radius (i.e., the sine of the right angle), is as the ratio of *BT*, the tangent of arc *BG*, to *DK*, the tangent of arc *DE* (i.e., the tangent of angle *A*). And that is what we wanted to prove.

From this it follows that if we suppose another arc, for example arc *LM*, perpendicular to circle *AD*, we may also show that the ratio of the sine of *AL* to the sine of *AD* is also as the ratio of the tangent of *LM* to the tangent of *DE*. And so the ratio of the sine of *AB* to the sine of *AL* is as the ratio of the tangent of *BG* to the tangent of *LM*. So the ratio of the sines, the one to the other, is as the ratio of the tangents of the widths[212] of those arcs, the one to the other. Hence, alternately, the ratio of the sine of any arc to the tangent of its width is as the ratio of the maximum sine to the tangent of angle *A*. So that in two right triangles with an acute angle equal, even if the triangles are not superposable (as in the case with triangles *ABG*, *ADE* of the figures where the two acute angles, *A*, are equal and angles *B* and *E* are right), the principle may be established as it was above.

Chapter 7: Complementary developments on the manner by which one determines the unknowns from the knowns in spherical triangles

[In the first part of this chapter, al-Ṭūsī discusses methods for solving right triangles. In general, he gives two methods for each case, one using the Rule of Four Quantities and one using the law of tangents. He then proceeds to methods for solving non-right triangles.]

---

[211]For, according to the lemma, applied twice, they will be perpendicular to *AZ*, the line of intersection of the planes *KZA* and *DZA*.

[212]Arcs *BG* and *LM* are called "widths" of arcs *AB* and *AL* evidently because they make right angles with the arcs at *B* and *L*, respectively.

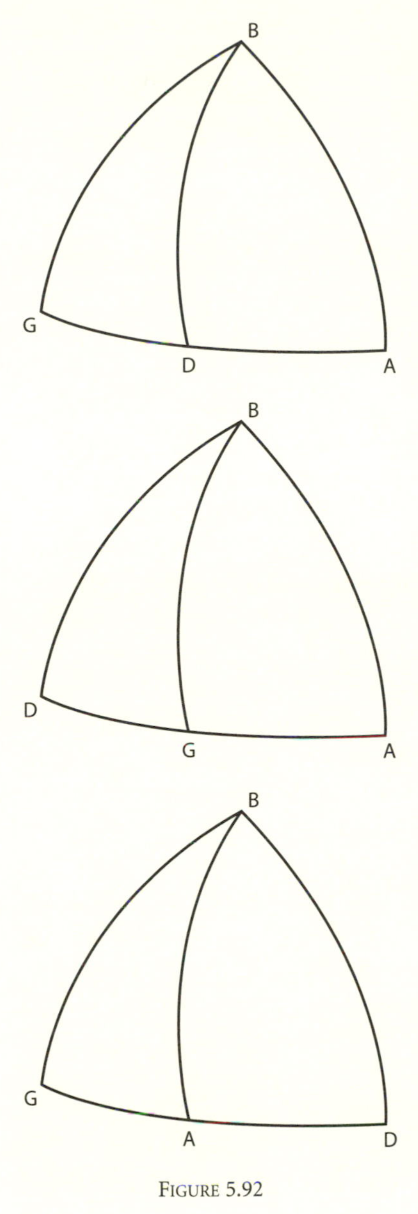

FIGURE 5.92

Regarding acute-angle or obtuse-angle triangles, we start from the same point [as in our previous discussion of right triangles] namely: that three things in each one must be known in order to find another unknown, by way of ratio, according to what we already said. The three knowns may be: two sides and one angle; two angles and one side; three sides; three angles. Each of the first and second cases is subdivided into two others, depending on whether, in the first, the known angle is between the two sides or opposite one of them[213] or, in the second, depending on whether the known side is between the two known angles, or opposite one of them.

And so there are six cases to be examined for these triangles, namely:

I. Let the known be two sides and the angle between them. Say, in triangle *ABG*, one knows the sides *AB*, *AG*, and angle *A*. Let us draw an arc of a great circle through one of the unknown angles and through the pole of the side opposite it,[214] say *BD* [fig. 5.92], so the angles *D* [arising in this way] are right. In acute angle triangles, point *D* will be inside the triangle, and the same will happen in the triangles with an obtuse angle at *B*. But, if the obtuse angle is *A* or *G*, point *D* will be outside the triangle and on the same side as the obtuse angle.

Now in [the right] triangle *ABD*, *AB* and angle *A* are known, so the remaining sides and angles of this triangle will be found, as in Case Four of [the section dealing with] right angled triangles. Then in triangle *BDG*, the sides *BD* and *GD* are known and so the remaining side and the other angles become known, according to Case Two [of the right angled triangle section]. So the angles *B*, *G* and side *BG* become known, using either the Rule of Four Quantities[215] or the tangent law.

---

[213]Since al-Ṭūsī is dealing with spherical triangles, the knowledge of two angles does not determine the third, so there are indeed two subcases in Case 2 of the list.

[214]A pole of a great circle is one of the two points on a sphere where the perpendicular to the great circle through its center intersects the surface of the sphere. The pole of an arc (of a great circle—since those are the arcs we are dealing with here) is a pole of the great circle containing the arc.

[215]This is a special case of the law of sines for spherical triangles, one sometimes referred to as "the theorem that frees [one from the use of the complete quadrilateral of Ptolemy's *Almagest*]." It states that if *ADE* and *ABG* are spherical triangles with right angles at *D* and *B* and a common acute angle at *A* then the ratio of the sines of arcs *DE* and *BG* is as the ratio of the sines of arcs *EA* and *GA*.

II. Let two sides and one angle not between them be known ....

[One method is to use the law of sines to find a second angle, then draw a perpendicular from the unknown angle to the unknown side and use laws of right spherical triangles.]

III. Let two angles and the side adjacent to them be known ....

[One method is to draw a perpendicular from one of the known angles to the opposite side and use the the laws of right spherical triangles twice to determine the third angle. Then use the law of sines to complete the solution.]

IV. Let two angles and a side not adjacent to both angles be known ....

[One method is to use the law of sines to find a second side, then draw a perpendicular from the unknown angles to the unknown side and use laws of right spherical triangles.]

V. Let the three sides, but none of the angles, be known and the triangle be *ABG*. And we make *AB*, *AG* quadrants at points *D*, *E* and we produce *DE* and we complete the quadrilateral [fig. 5.93]. And we say that, *AB* and *AG* being known, *BD* and *GE* will also be (known). And they are the inclinations of arcs *ZB*, *ZG*, the angles *D* and *E* being right. Hence the ratio of their sines will be equal to the ratio of the sines of their [associated] arcs *ZB*, *ZG*. And so it [the latter ratio] is known. But *BG* is also known, and so, according to what was established in Book III, each of the two arcs *ZB*, *ZG* will be known.[216] And in the right triangles *ZBD*, *ZGE*, two sides will be known, and so *ZD*, *ZE* will be known. Consequently, arc *DE*, i.e., angle *A*, will be known. It will be likewise for the two remaining angles.

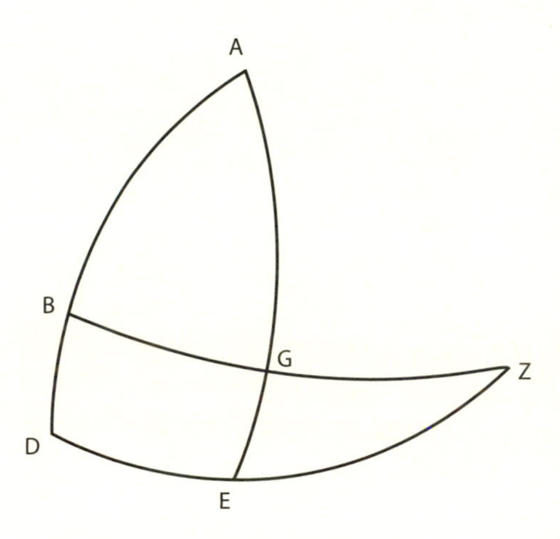

FIGURE 5.93

If one of the sides, say *AG*, should be a quadrant, we would make *AB* a quadrant [the later ratio] also [by adding or removing an appropriate arc *BD*], and we would draw arc *DG* [fig. 5.94]. And because *AB* is known and *AD* is a quadrant, we would have *BD* known. And in triangle *BDG* the two sides *BD*, *BG* would be known, and in triangle *ADG*, the two sides *DG*, *AG*, would be known. And the angles of triangle *ABG* will [then] become known.

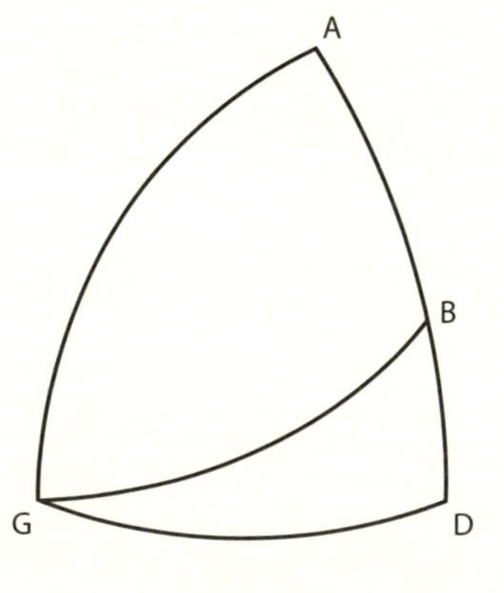

FIGURE 5.94

VI. Let the known be all of the angles of

[216]See the selection from Chapter 3 of Book III for al-Ṭūsī's demonstration of this result.

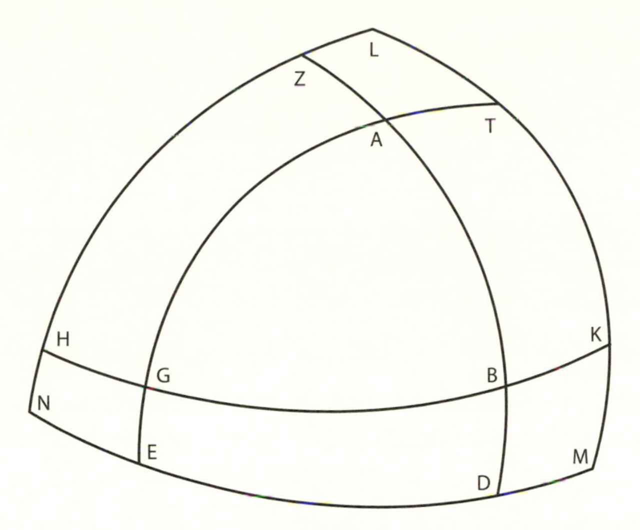

FIGURE 5.95

the triangle, but none of the sides, and [let] the triangle [be] *ABG*. We extend *AB*, *AG* to where *AD*, *AE* become quadrants; likewise *BA*, *BG* up to where *BZ*, *BH* become quadrants; and *GA*, *GB* up to where *GK*, *GT* are also quadrants [fig. 5.95].

Let us draw the great circle arcs *DE*, *ZH*, *TK*. The points of intersection will be *L*, *M*, *N*, and this way triangle *LMN* will be formed, with its sides being arcs of great circles. Angles *A*, *B*, *G*, being known, *DE*, *ZH*, *TK* will also be known. And *K* and *H* being right angles, *L* will be the pole of *KH*. Likewise *M* will be the pole of *TE*, and *N* the pole of *ZD*. Now *TL*, *KM* being each a complement of arc *KT*, *LM* will be known, and similarly for *LN* and *MN*. And so, the sides of triangle *LMN* will be determined. Consequently its angles also, according to the previous case. And so, arcs *KH*, *DZ*, *ET* will be known. But, each of the arcs *KG*, *BH* being a quadrant, the supplement of *HK* will be equal to *BG*. And so, *BG* will be known, and we may say the same for *AB*, *AG*. And thus, the sides of triangles *ABG* will be known.

If one of the sides is a quadrant or larger ....

In regard to the calculation details, I suppressed them for these last six cases, because I wanted to be brief and also because the calculations are not frequent in practice. Anyhow, those who have understood what has been said up to here will have no difficulty extracting the necessary calculations.

Jamshīd al-Kāshī, known for his detailed work on calculation, developed a method of interpolation, here excerpted from his astronomical treatise the *Zīj-i Khāqānī*. The method is designed to save labor in computing the true longitudes of slow-moving planets by doing the full computation only for every ten days and computing the longitudes for the intervening days of the interval by the method described here. In modern terms one is putting a parabola through the three points corresponding to the beginning day of the interval, the day previous to the one for which one is computing, and the last day of the interval. In the passage, the Persian-Arabic word *buht* (derived from the Sanskrit *bhukti*) means longitudinal speed, here in degrees per day.

Al-Kāshī on Interpolation Using Second Order Differences

... That ten days, or whatever it is which occurs between the two longitude [computations], we have called the increment. So we divide the motion of the planet in each increment by the number of that increment, so that the daily *buht* results, and that we call the mean *buht*. With that *buht* we advance the true longitude of the planet in that time, so that the true longitude for each day results. But if the mean *buht* of one increment is much different from the mean *buht* of another increment, one must operate with the difference arc. That is thus: that we add one to the number of the increment, and we take half of that amount and that we call the preserved [number]. And we obtain the difference between the preceding *buht*, i.e., the *buht* of the last day of the past increment and the mean *buht*, and we divide it by the preserved [amount]. The quotient we call the equation of the *buht*, and if the number of the increment is even we divide the double of the difference by the double of the preserved [amount] so that it will be easier. Then, if the [new] mean *buht* is more than the preceding [day's] *buht*, we add the equation of the *buht* to the previous *buht*, and [this is continued] one time after another; otherwise we subtract it one time after another, so that the distance and the increments of the successive *buhts* differing by that amount result, and the measure of the correctness of this operation is this, that the sum of the *buht* of the first day of the increment and of the last day of the increment should be equal to the double of the mean *buht*. Also, if the number of the period is an odd integer, the *buht* of the middle day should be the mean *buht* exactly. Then, if the planet is in forward motion, we proceed with the days of the increment with those *buhts* in such fashion that we add every *buht* to the true longitude of the preceding day, so that the longitude(s) for the day(s), one after another, result. But if the planet is in retrograde motion we subtract each *buht* from the true longitude of the previous day that the [longitudes of the] succeeding days will result.

VIII. Combinatorics

Although combinatorics was studied in India early, it was Islamic mathematicians, along with medieval Jewish mathematicians, who were the first to give detailed discussions with arguments. Magic squares were also a subject of interest. The first excerpt here is from Ibn al-Haytham on a certain construction of such squares.

Ibn al-Haytham's Method of Constructing Magic Squares of Odd Orders

One draws the square to be made magic.[217] Then one cuts the four sides into two [equal] parts and one joins with a line the midpoints of each pair of contiguous sides; then one joins the intersections of the lines in each of the two directions (parallel to the sides). This way another square appears inside the square, one similar to the square that one tries to fill, as is shown in the figure.[218]

---

[217]The account begins as if the first square were to be the magic one, but—by the end—it is the smaller square, constructed inside the larger, that is magic.

[218]The figures are missing in the manuscript, which often happened when manuscripts were copied by hand. Sometimes the person doing the figures simply never got around to it.

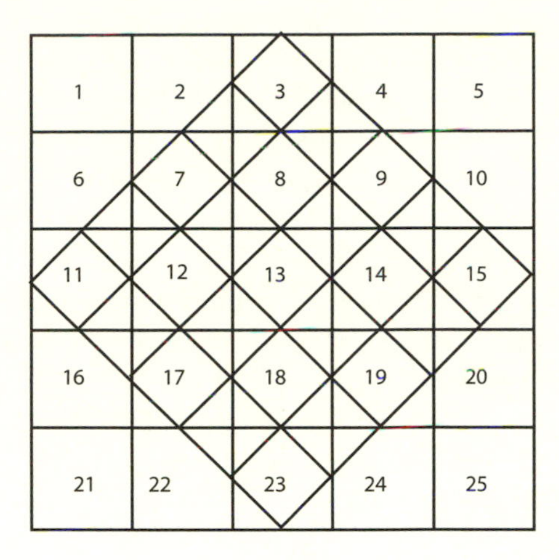

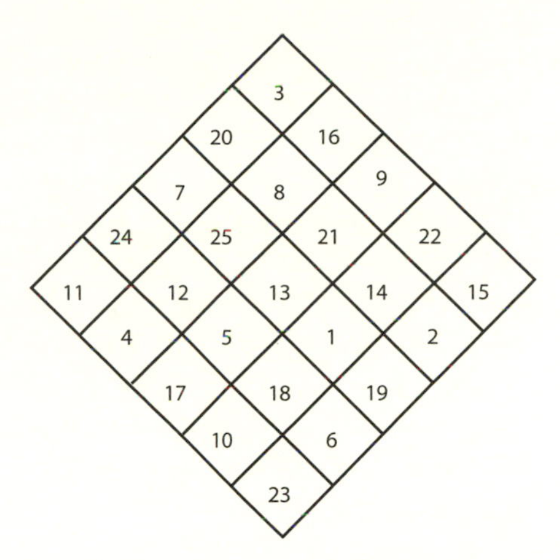

FIGURE 5.96

Then one fills the cells of the larger square—which is the one drawn first—with successive numbers, starting with the one [number] that is proper. Starting this way from 1 for the natural order, one fills up the boxes one after the other, row after row, up to the last one of the horizontal rows, as shown in the figure in which the start is "one"—because from now on one will deal with the natural order, the other cases [of numbers in an arithmetic progression] being similar. Then [at this stage—shown in the left-hand part of fig. 5.96], some cells of the small square are filled and the remaining are empty. One completes these in the following manner: one goes toward each empty cell, and puts there, out of the numbers situated in the larger square [and] outside the smaller, the one that is in the furthest cell diagonally opposite to it [right-hand part of fig. 5.96]. This being done, the magic will be completed in the smaller square—because it is the one that was supposed to be filled, when the first one, the larger, was only filled to be used as an auxiliary.

In this excerpt, al-Bīrūnī discusses arrangements of certain astronomical data.

Al-Bīruni on Number of Possibilities for Eclipse Data

This shows that there are seven possibilities for the time of an eclipse. If the time reported by one time recorder is combined with that of the other, since each time has seven possibilities, the total number of time combinations in pairs is equal to the sum of the first seven natural numbers from one to seven, and is given by seven times half of seven plus one, and that is twenty-eight. Every combination of two times can be interchanged between the two towns, so the number [of permutations] is fifty-six. In every one of these [arrangements], the latitudes of the two towns may both be known, or both of them may be unknown, or one of them may be known but the other is unknown. But if one is unknown and the other is known, an interchange can be arranged, and thus there are four possible arrangements for every combination.

Hence the total number of arrangements is two hundred and fifty-six. This result was worked out by distribution, and there was no need for deriving it by induction. However, the logical distribution has led Abū Zakarīya Yaḥyā ibn 'Adīy to declare that someone's dictum: [transliterated, a character for each Arabic letter:] 'NN ALQĀ'M GHYR ALQĀ'D[219] can be arranged in sixteen thousand, three hundred and eighty-four different ways. Then it was realized that he had made a slip in multiplication, and it is said that it is eighteen thousand, four hundred and thirty-two. This was increased by Abū al-Qāsim al-Ḥasūlī, who claimed that the number of arrangements is twenty-five thousand eighty-eight. Both results were increased by Abū Sahl 'Īsā ibn Yaḥyā, the Christian, who stated in a letter to me that the number of arrangements is one hundred and twenty-eight thousand thousand thousand and four hundred and fifty thousand thousand and five hundred and sixty thousand [i.e., 128,450,560,000]. Also, he wrote to me recently and claimed that he had obtained some extra arrangements which make the total a multiple of this number, and has promised to send me his work on that problem.

Aḥmad ibn Mun'im probably lived at the Almohade court in Marrakech (now in Morocco) during the reign of Muḥammed al-Nāṣir (1199–1213). It was at this time that the dynasty lost a major battle in Spain to a coalition of Christian kings, thereby losing much of its extensive Spanish domains. Ibn Mun'im's work is the earliest extant Islamic discussion of methods for deriving the basic combinatorial formulas.

Ibn Mun'im on Denumerating the Words Such That a Person Cannot Express Himself without One of Them

[Introduction]

We have wanted to describe the manner of proceeding to denumerate the words that a person cannot speak without one of them. Al-Khalīl—may Allah be merciful to him—has indicated only the number of configurations of the word in which the letters do not repeat. But this section contains [the study of] words whose letters repeat, as well as the number of quinqueliteral or sextiliteral words composed of letters of the alphabet and of which the letters are all distinct, or of which one, two or all of them are repeated, as well as the denumeration of all of this.

We agree, in this example, that the number of letters of the alphabet is twenty-eight, that the longest word, taking account of affixes and repetitions, has ten letters (like *Arisṭūṭālīs*,[220] that following a letter can be three vowels and a *sukun*, that one does not begin with a *sukūn*, and two *sukūn* cannot occur in succession. Someone might object, saying that it happens that more than three vowels follow a letter, as in the case of vocalic inflection and others, that certain non-Arabs begin [words] with a *sukun*, but that we have ignored all this because of the inability of our language to express it, that the non-Arabs speak with other letters, even if

[219]The "GH" in "GHYR" represents one Arabic letter.

[220]This is the name of the Greek philosopher, Aristotle. The Arabic form of this name has ten letters, the "i" not appearing in the Arabic version.

they are not in the speech of the Arabs, like the *kāf* which is similar to the *qāf*, the *jīm* which is similar to the *shīn*, as well as others, and that it happens that the number of letters of a word may be more than ten, as when Allah says *layastakhlafnahum* ("let us appoint them as successors"), the supporting letter equivalent to two letters.

But I would respond by saying that our goal, in fact, is the description of a method by which it is possible to denumerate words, and the description will be identical, even if the number of letters and vowels should increase to whatever value. If we have supposed that the number of letters is twenty-eight, that the longest word, taking account of affixes and repetitions, has ten letters (like *Arisṭūṭālīs*), that more than three vowels and a *sukūn* cannot follow one and the same letter, it is [only] to illustrate the procedure of the method we propose. The procedure having been acquired, you will follow [the same process] whether the number of letters and vowels is larger or smaller.

May Allah inspire us with the truth. There is no other Master than he.

[A. Establishing general rules]

[Problem 1] Preliminary proposition for that which we want to demonstrate:

Ten colors of silk being given, with which we want to make pom-poms [respectively], of one, two, three, etc. colors, we want to know what is the number of pom-poms of each type, the colors of each pom-pom being known, or what is the number of all of the pom-poms taken together, account being taken of the different numbers of colors of the pom-poms. We arrange the colors, one after another in a line, according to the size of the page, as in the example.

Sum	The example written in a table										
1	1	Line of pom-poms of ten colors									
10	9	1	Line of pom-poms of nine colors								
45	36	8	1	Line of pom-poms of eight colors							
120	84	28	7	1	Line of pom-poms of seven colors						
210	126	56	21	6	1	Line of pom-poms of six colors					
252	126	70	35	15	5	1	Line of pom-poms of five colors				
210	84	56	35	20	10	4	1	Four colors			
120	36	28	21	15	10	6	3	1	Three colors		
45	9	8	7	6	5	4	3	2	1	Two colors	
10	1	1	1	1	1	1	1	1	1	1	One color
All	10th	9th	8th	7th	6th	5th	4th	3rd	2nd	1st	

FIGURE 5.97

Then, if you think about the problem, you note that the pom-poms of two colors are obtained by combining the second color with the first, the third color with the first and with the second, the fourth color with the first, second and third, the fifth color with the first, second, third, and fourth, etc. according to this order [of enumeration], until one ends up with combinations of the tenth color with each of the colors that precede it.

In general it is by combining each of the colors with those that precede it in the numeration, and according to this order of enumeration, that the number of combinations of each color with each of the colors will be determined.

You write "1" in the first case of the second line opposite the second color, and it is the pom-pom made up of the [combination] of the second color with the first. You write "2" in the second case of the second line, similarly opposite the third color, and these are the two pom-poms made up of the [combination] of the third color with the first and the second. You write "3" in this line, similarly opposite the fourth color, and this is the number of pom-poms made up of the [combination] of the fourth color with the first, second and third. In the same way, you write "4" in this line opposite the fifth color, and it is the number of pom-poms obtained by the combination of the fifth color with the first, second, third and fourth. And in this way you finish the second line when you write "9" at its end opposite the tenth color, and this nine is the number of pom-poms obtained by combining the tenth color with the first, second and [so on] to the ninth. The result is that in the second line the numbers follow from one to nine and it is the number of pom-poms composed of ten colors. The number of pom-poms of two colors is then equal to the sum of the successive whole numbers from one to the number that is one less than the number of colors.

As for determining the number of pom-poms of three colors it is obtained by combining the third color with the first and the second, then by combining the fourth color with each pair of colors among the three colors preceding which are the first, the second, and the third, then by the combination of the fifth color with each pair of colors among the four colors preceding, then by the combination of the sixth color with each pair among the five colors preceding, and so on, until [the combination of] the tenth color, with each pair of colors among the nine colors preceding.

But each pair of colors is a pom-pom of the second line. For this reason, we write: one, in the first case of the third line opposite the third color, and this will be the pom-pom composed of the first, second and third color; then, we write, in the next case, which is opposite the fourth color, the number of pom-poms obtained by the combination of the fourth color with each pair among the colors preceding, and it is equal to the number of pom-poms of two colors composed of colors preceding the fourth color, and it is also equal to the sum of the content of the two first cases of the second line, and it is three. We then write three in the second case of the third line. And we write in the third case of the third line—this case being that opposite the fifth color—[the number] of pom-poms [obtained] by the combination of the fifth color with the pairs of colors preceding the fifth color. And it is also the sum of the content of the three first cases of the second line. And it is six. We [therefore] write six in the third case of the third line. Thus, according to this order, one shows that the content of the following case—the fourth from the third line—is equal to the sum of the four cases of the second line, and it is 10; and the content of the following case, the fifth, is equal to the sum of the five cases of the second line, the [so on] up to the end of the third line. The sum of the cases of the third line is then equal to the set of pom-poms of three colors each, [obtained] beginning with the [given] colors.

As for the knowledge of the number of pom-poms of four colors, it is obtained by the combination of the fourth color with the three preceding ones, by the combination of the fifth color with the set of triplets among the colors preceding the fifth, then by the combination of the sixth color with the set of triplets among the colors preceding the sixth, and so on up to the tenth color [combined] with the set of triplets among the colors preceding the tenth. But, each triplet is a pom-pom of the third line. Because of this, then, we write, in the first case of the fourth line 'one', which is the pom-pom composed of the four first colors, and we write in the following case which is facing the fifth color, the [number of] pom-poms obtained by the combination of the fifth color with each triplet among the colors preceding the fifth, which is also equal to the sum of the contents of the two first cases of the third line, and it is four. It appears in the same way that you must write, in the third case of the fourth line, that which is equivalent to the sum of the three first cases if the third line, and it is ten. One proceeds thus for the construction of the set of the fourth line, beginning from the third line, [construction] which is identical to that of the third line beginning with the second line, and that of the second line beginning with the first.

One proceeds in the same way for the construction of the fifth line beginning with the fourth. It is analogous to the other construction of the fourth line beginning from the third, to that of the sixth line beginning from the fifth, of the seventh beginning with the sixth, of the eighth beginning from the seventh, of the ninth beginning with the eighth, and of the tenth beginning with the ninth. But, in our example here, the tenth line has only one case containing only one pom-pom of ten colors.

May God inspire us.

[Table of the combinations]

Then, if you reflect on the particularities of this table and on that which appears here as a surprising harmony, it will show you extraordinary symmetries and surprising properties whose discussion would take a long time. We have renounced this, counting on the reflection of the student and, equally, desiring to abandon excess and to make a choice.

May God aid us.

[How to proceed with the tables]

If you have some colors of silk and you want [to know] how many pompoms they can make so that each pom-pom has a known number of colors, you enter in the table vertically, by the color whose number is equal to the number of colors of your silk. Then you enter equally by the line corresponding to the number of colors of each pom-pom and you count the number of the case where the line and the column meet with the number of cases that are, in this line, to the right of this case, [and you do this] until the one. The sum that you get is the number of pom-poms.

There is a better and simpler method, which consists of entering in the table vertically, beginning from the color of which the number exceeds by one the number of colors of your silk, and entering by the line of pom-poms of which the number of colors in each exceeds by one the number of colors of your pom-poms. The number of the case where the line and column meet is then the number of pom-poms that you seek. If the number of colors you have is larger than ten, you add [columns] to the table until the number of its colors is equal to that of your colors.

[Problem II]

Problem: We want to know a canonical procedure to determine the number of permutations of the letters of a word of a known number of letters and in which no letter is repeated.

[He lists the two possible ways for a word with two letters and then, to find the number for a word of three letters, lists the three possible places that a third letter could be placed in a word of two letters. He then carries on this argument, letter by letter, to show that the answer for five letters is 120. He then says:] And one will demonstrate it thus however large [the number] may be.

One has thus shown, with this, that if you have a word of which the number of letters is known and of which no letter is repeated, and you want to know the number of permutations of the letters of this word, you multiply one by two, and the result by three, and then this result by four, and then this result by five, and so on, each result being multiplied by the next number in the sequence of whole numbers, until one reaches the product by the number equal to the number of letters in the word. The result is equal to the number of permutations of letters of this word. And this is what we wanted to demonstrate.

[Problem III]

We want to know the number of permutations of letters of a word of which the number of letters is known and of which one, two, or more letters are repeated a known number of times.

The method, when only one letter is repeated, consists in determining the number of permutations of [the letters of] a word [having the same number of letters but] in which no letter repeats and [hence] where the number of letters is equal to the number of letters of the given word with their repetitions. You divide the result you obtain by the number of permutations of the letters of a word whose number of letters is equal to the number of repetitions of the letter. The result you obtain will be the number of permutations of the letters of the given word.

The proof of that is that when only one letter is repeated in a word, to each position of these letters represented there would correspond, if they were different, the permutation of the letters of a word of which the number of letters would be equal to the number of repetitions of this letter.

If two, three, or more letters are repeated, the procedure consists of determining the number of permutations of the letters of another word in which no letter is repeated but whose number of letters is equal to the number of letters of the given work, with their repetitions. You keep the result you have obtained, and then you count the number of repetitions of just one of the letters that are repeated as if they were distinct letter of a word. In the same way, the number of repetitions of the second letter repeated will be the number of letters of a second word. And if it contains a third letter that repeats you count, equally, the number of its repetitions as being the number of letters of another word. Then, you multiply the numbers of permutations of these words, the one by the other, and you divide, by this result, that which you have saved. The result is the number of permutations of the letters of the given word. The proof of this is analogous to that which has proceeded in the demonstration concerning the repetition of only one letter.

[For the next problem we remark that the Arabic language is built up on what are called "skeletons" of three consonants (e.g., k-t-b, which carries the basic meaning of "writing"). One then may attach at most one vowel to each consonant (e.g., *ku-ti-ba* = "it was written"). If a vowel is not attached to a letter then the letter has what is called in Arabic a *sukūn*. An example is the verb ba-r-hana. (A word cannot begin with a *sukūn* and there cannot be two consecutive *sukūn*s.) In order to avoid a discussion of the details of Arabic word-formation we simply note here that Ibn Mun'im solves next Problem IV:

We want to know the number of configurations of a word of which the number of letters is known, taking account of the vowels and of the *sukūns*, which succeed one another on the letters [i.e. consonants], but taking no account of the permutations of the letters. His solution to this problem he follows by a table which records three possible words of one letter,[221] 12 of two letters, 45 of three letters, and so on, up to 507,627 for a word of 10 letters.]

Having shown this, we go back to our problem: To know the number of words composed of letters of the alphabet if the smallest has three letters and the largest ten.

We describe the procedure, to start, when in the word no letter is repeated. The method here consists of considering the number of letters of the alphabet as being colors of silk and to say: "How many pom-poms do they make so that the number of colors of each pom-pom is, for example, three?" You obtain a result, which you keep. Since you have supposed that the number of colors of a pom-pom is three you put, on a line, the sequence of whole numbers from one to three, thus: 1, 2, 3. Then you begin by multiplying one by two, then the product by three, which follows on the line. And if the following were another [number], you would multiply the product by it. Then you multiply the result by the number of configurations of the word, according the vowels and the *sukuns*. Then you keep the result and you multiply it by the number of pom-poms of silk that are, each, of three colors (as we have supposed in this example). The result is the number of words of the alphabet and each of which has three letters. You proceed in the same manner for more or less than three letters. Then you sum the number of groups of letters of the whole,[222] and it will be the number of words composed of letters of the alphabet, having [at least] three letters and not more than ten, and in which no letter is repeated.

---

[221]The three possible words are *wa/wi/wu*, of which only one (*wa*) corresponds to a word ("and").

[222]That is to say, one sums over the different possible numbers of letters in the words (from 3 to 10 in this case).

IX. On Mathematics

In this excerpt, al-Kūhī discuss the value of mathematics.

Al-Kūhī on the Certainty Deriving from Mathematics

As for this luminous object which is called the shooting star, whether it is really a star, a ray of a star which appears and then vanishes, a fire which burns for a moment and then dies out, a flame of fire which exists there [i.e., in the sky] continuously, an object in which fire is kindled and then extinguished, or a fire which results from the speed of the movement of something else other than fire, all these [explanations] have been presented by a group of people whom Galen had continuously criticized, both them and their knowledge, because they spoke much about everything without any proof.

But the other group, whom neither Galen nor anyone else could criticize, neither them or their knowledge, because they depended on proofs in all their sciences and books, those were the mathematicians. Among them were those who spoke about the knowledge of the science of distances and sizes of [heavenly] bodies and things dependent on the science of observations and the stars. And they wrote many books about these subjects, [scholars] like Archimedes, Aristarchos, Timocharis, and Ptolemy, beside other ancients of whom books or anecdotes have reached us. [Nevertheless], they did not mention in their books how to know the magnitude of the distance between the center of the Earth and the position of any shooting star. And this must be because of one of the following:

Either they did not mention it, or they did mention it but they did not record it in books, or they recorded it but those books in which it was recorded did not reach us, neither our region nor to us personally.

But I think only the best of them and judge them favorably. For what I know of their innovations, their deep and meticulous endeavors in pursuing the obscure sciences, no other person in this our time knows.

In this excerpt, al-Sijzī discusses how one learns mathematics. In particular, he shows how one learns how to develop proofs.

Al-Sijzī on Heuristics

In the name of God, the merciful, the compassionate Lord, help us. The book of Ahmad ibn Muhammad ibn 'Abd-al-Jalīl al-Sijzī on *Making Easy the Ways of Deriving Geometrical Figures.*

We want to enumerate, in this book of ours, the rules which will make it easy for the researcher who knows and masters them to derive whatever geometrical constructions he wants. We will mention the methods and ways which will make the mind of the researcher who follows them competent in the different aspects of deriving figures.

Some people think that there is no way of learning the rules for deriving [new propositions] even with much research, practice, study, and lessons in the elements of geometry unless a man has an innate natural talent which enables him to discover figures, because study and practice are insufficient. But this is not the case. There are people who have a natural aptitude and an excellent ability for deriving figures, but who do not have much knowledge and who do not work hard to study these things. But there are also people who work hard, who study the elements and the methods, but who do not have an excellent natural ability. If a man has an inborn natural talent and if he works hard to study and practice, then he will be first-rate and outstanding. If he does not have a perfect ability, but if he works hard and studies, then he can also become outstanding by means of study. As for someone who has the ability but does not study the elements, and does not devote himself to the constructions of geometry, he will not benefit from it in any way. Since this is as we said, if someone thinks that discovery in geometry proceeds only by means of innate ability and not by study, then he thinks nonsense.

The first thing which is required of a beginner in this art is that he knows the theorems which come after the axioms. This is [also] considered to be part of our aim: the figures which we want to discover. But [here] our aim is the methods [of discovery] which can be found by means of the theorems, not only by means of the axioms, which are preliminaries to the theorems.

It is necessary for someone who wants to learn this art, to thoroughly master the theorems which Euclid presented in his *Elements*. For between mastering the thing and the thing itself there is a very deep gap. It is necessary that he has a thorough idea of their genera and their special properties, so if he needs to look for their properties, he is well prepared to find them. If he has to do any research, then it is necessary for him to study and visualize in his imagination the preliminaries and theorems that are of that genus, or that have [something] in common with it. For example: if we want to derive a figure of the genus of the triangle, then we have to visualize all the properties belonging to triangles, and the theorems which Euclid mentioned, and the angles, arcs, sides, and parallel lines which are involved in the properties of the triangle, so that this may be easy for him [i.e., the researcher], and that he may be prepared for deriving them.

There are figures which share one or more special properties, and there are figures which have nothing in common, and there are figures which are more closely or more distantly related, according to configuration, proportionality, and genus. If we want to find some figure by means of preliminaries, and we mean by a "preliminary" the figure which comes before it and is a basis for the derivation of it, and if we cannot derive it by means of this preliminary, then we must try to seek [it] by means of a preliminary related to that preliminary—if our search on the basis of that preliminary can be successful. It is a consequence of this affair that if any figure can be derived from some preliminary, then it can also be derived from preliminaries which are related to it in the way we mentioned, or from some of them, according to the extent of the relationship.

Among the special properties of figures is [the fact] that some (figures) can easily be derived by means of many different preliminaries and in many ways, and some can be found by means of only one preliminary, and there are some for which no preliminary exists, even though that figure can be imagined, or its correctness is described in nature. This is a consequence of the close relation to the special properties of the preliminaries, or the difference between it [the proposition] and them.

Sometimes there appears to the researcher a method by means of which it was easy for him to derive many difficult figures. This [method] is the *transformation*. We shall explain it and give an example of it, God Willing.

Another method is easy for the researcher if he follows it: He assumes the desired aim as if it were already constructed, if the aim is a construction, or he assumes that it is true, if the aim is the investigation of a special property. Then he unravels [analyzes] it by means of a succession of preliminaries, or by means of [mutually] linked preliminaries, until he ends up with the correct and true preliminaries, or with false preliminaries. If he ends up with true preliminaries, the desired thing can be found as a consequence. If he ends up with false preliminaries, the impossibility of the desired thing follows. This is called analysis by inversion. This method is of more general use than the other methods. We shall give an example of it below, God willing.

Synthesis is the inverse of analysis. That is, synthesis is: following the road [i.e., reasoning] toward the result, by means of the preliminaries. Analysis is following it towards the preliminaries which produce the desired thing.

It is the business of geometry that the unknown becomes constructed or known. Here it [the unknown] is necessarily constructions or special properties.

First the researcher must think about the question and the things which are desired. There are questions of which the essence is possible in nature, but not for us, or the investigation of them is impossible for us, because of the lack of preliminaries for them. Such is the quadrature of the circle.

These are the methods for discovery in this art. We will enumerate them separately, so that the researcher may visualize them with his mind and master them, if God Most High wills so and grants good success.

First, cleverness and intelligence, and bearing in mind the conditions which the proper order [of the problem] makes necessary.

The second is the profound mastery of the [relevant] theorems and preliminaries.

The third is: following of the methods of them [these theorems and preliminaries] in a profound and successful way, so that you rely not only on the theorems and preliminaries and constructions and their arrangement which we mentioned, but you must combine with that [your own] cleverness and guesswork and tricks. The pivotal factor in this art is the application of tricks, and not only [your own] intelligence, but also the thought of the experienced [mathematicians], the skilled and those who use tricks.

The fourth is: information about the common features [of figures], their differences, and their special properties. In this particular approach, the special properties, the resemblance and the opposition are [considered by themselves] without enumeration of the theorems and preliminaries.

The fifth is the use of transformation.

The sixth is the use of analysis.

The seventh is the use of tricks, such as Heron used.

Since we have presented and mentioned these things in a loose manner, let us now bring for each of them examples, so that the researcher learns their true nature.

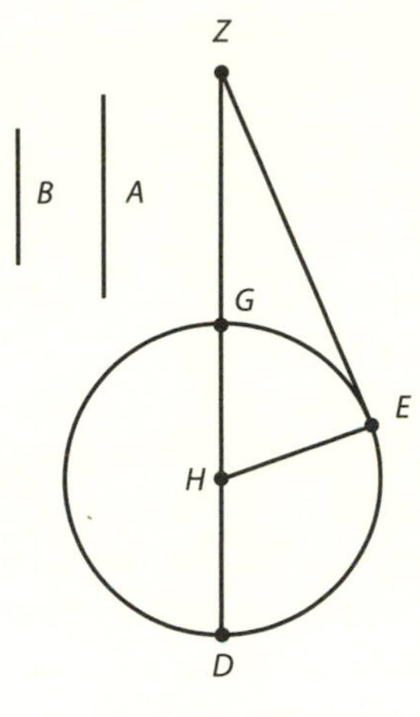

Figure 5.98

[Example 1, concerning "transformation"]

A question on the construction of a figure: How do we find two lines proportional to two assumed lines, in such a way that one of them is tangent to an assumed circle and the other meets the circle in such a way that its [rectilinear] extension inside the circle passes through its center?

Thus we assume that the figure has been constructed, by way of analysis, so that we may look for its preliminaries. For example [fig. 5.98] we assume that the ratio is the ratio of *A* to *B*, that the circle is circle *GD*, that *ZE*, *ZG* are in the ratio of *A* to *B*, and that they are the desired lines—as if it [the figure] were already constructed and as if the construction were available to us, as we mentioned before—in such a way that if *ZG* is extended inside the circle toward *D*, *GD* is a diameter of it.

Then we ask from what construction and what preliminaries its construction could have been found.

Since point *Z*, lines *ZG*, *ZE*, and the point of contact of [circle] *DG* at point *E* are all unknown to us, and again, since the situation of the inclination of angle *Z* is also unknown, the figure is difficult to derive. This guess is what I have called before: information on the level of easiness or the difficulty of them [the propositions]. If there are many unknowns in the figure, it is difficult to find by means of known things. [This is] especially [true] if its configuration is such that the components of the figure do not have a relationship as we have mentioned. In this figure neither lines *ZG*, *ZE* and the circumference of the circle are closely related, nor angle *Z* and arc *GE*. Then we also use guesswork and intelligence, and we undertake its construction by means of transformation; for as we mentioned before, it makes the derivation of difficult figures easy.

So we say: How can we place lines *ZG*, *ZE* in such a configuration that if a circle is drawn, it is tangent to *ZE* and meets *ZG*? This is only possible for us by the assumption of angle *Z* and the knowledge of it [i.e., its magnitude]. Thus we must seek the knowledge of angle *Z*, but we can only know it if we seek something else of the same kind, namely angles. Then, how can we seek [this] from the combination of lines *ZG*, *ZE* or *ZE*, *ZH* or *ZE*, *ZD*, for in this figure it is not possible for us [to seek this] from the combination of any other line. Here guesswork and intelligence must be used. If we join *E* and *G* [by a straight line], it is sometimes difficult to find it [the angle], and sometimes it is impossible to find it in this way, because the angles which are produced here are also not known in this figure by means of these preliminaries. Therefore we join *E* and *H*. Here we have found that angle *E* of the three angles is known. Then it is necessary that we find the shape of triangle *ZEH* from the combination of the lines and the angles.

We now seek, after having found them [i.e. these insights] here, [the solution of] another problem. If we have found this [new] problem, by means of it the [original] problem will be solved. It [the new problem] is [as follows]: the configuration of triangle *ZEH* is restricted [i.e., determined] by the fact that it is a right-angled triangle

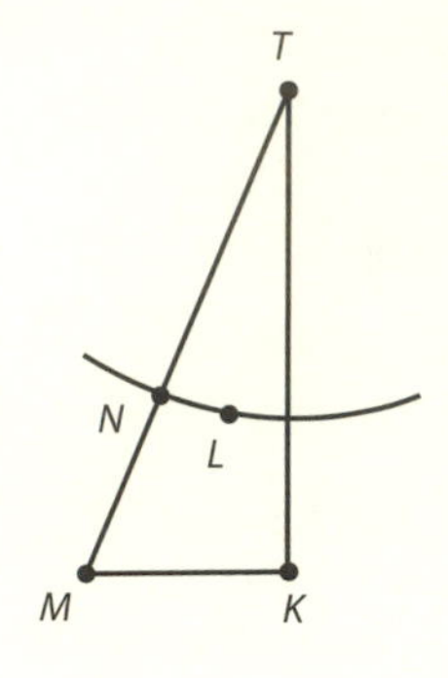

FIGURE 5.99

such that the ratio of one side to the hypotenuse minus the other side is as an assumed ratio. Thus our first question has been reduced to this question, by means of this method which we have now followed, in order that it produces what the [first] question requires. Thus we assume that the triangle is constructed in the usual way [fig. 5.99]: triangle *TKM* is right-angled, its right angle is angle *K*, but *NM* is equal to *KM*, and the ratio of *TN* to *TK* is as the ratio of *B* to *A*. Here skill and intelligence must be used, for each time when we seek [the solution of] an original [i.e. new] problem, we must use intelligence and guesswork, not learning.

We must ask how we assume *TK* such that the ratio of *TK* to *TN* is as the ratio of *A* to *B*. We extend *KM* indefinitely, and then we draw *TL* in our imagination [in such a way that] if you extend it to line *KM*, the difference between the moving line *TL* and its extension to line *KM* is equal to the line between point *K* and the junction point on line *KM*. So here is a problem of two unknowns.

So we construct a circle with center *T* and radius *TL* because we imagined line *TL* moving on point *T*, so that we can be sure that the endpoint *L* of line *TL* in the imaginary motion falls always on the circumference of the circle. But the shape of the triangle is placed in front of us so we can see the figure with our eyes at the time of construction in the correct configuration. We seek the center of a circle which [center] is common to lines *TM*, *KM*. So here guesswork and intelligence must be used successfully.

Again, we can only do this by means of an extra construction. So we imagine this construction: how do we extend *TN* to *S* in such a configuration that line *KM* bisects it and the whole [segment] *NS* is twice *KM*. So we transfer the problem to another figure, and it is this [fig. 5.100].

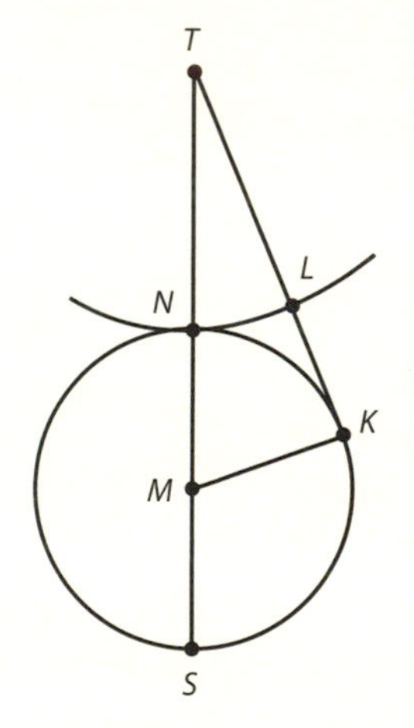

FIGURE 5.100

Then we do some thinking, and we visualize that the aim has been achieved, as we usually do. That is to say: we assume *TNS* such that *NS* is twice *KM*, and *NM* is equal to *KM*. We draw with center *M* and radius *MK* circle *KS*. Then it is clear that *TK* is tangent to the circle. But [now] we use guesswork and intelligence. If this is so, we must look for a special property of this figure caused by the tangency, [a property] which Euclid established in the *Elements*. The simplest special property of this figure is: the square of *TK* is equal to *TS* times *TN*. Thus we have found from this special property in this construction an aid [i.e., an auxiliary construction]: we make *NS* the line such that *TK* is equal in square to *TS* times *TN*. If we do this, the rest of the construction is easy. That is to say, we have found lines *TN* and *TK* and *TS*, so what is left for us is that we find the configuration of *TS* such that *KM* bisects *NS*. So we first bisect *NS* at *M* and in our imagination we move *TS* around point *T* [until] *KM* bisects *NS*. That is easy to construct by drawing with point *T* as center and *TM* as radius a circle which *KM* intersects at point *M*. We draw *TMS* and we draw arc *SKN*. So we have constructed this figure as we wanted.

Then we transform it to the assumed circle by similarity and ratio, and we prove it, and that is what we wanted to explain.

X. Appendices

X.a. Sources

Note that when a reference indicates that the source has been retranslated from a French or German translation, that translation has always been compared with the original Arabic and modified where necessary.

Appropriation of the ancient heritage

1. Banū Mūsā on Conics, G. J. Toomer, *Apollonius' Conics Books V–VII: The Arabic Translation of the Lost Greek Original in the Version of the Banū Mūsā* (New York: Springer-Verlag, 1990), 620–628.

2. Ibn al-Haytham's Preface to his Completion of the Conics, J. P. Hogendijk, *Ibn al-Haytham's Completion of the Conics* (New York: Springer-Verlag, 1985), 134–138.

Arithmetic

3. Al-Khwārizmī on Number, John Crossley and Alan Henry, "Thus Spake al-Khwārizmī: A Translation of the Text of Cambridge University Library Ms. Ii.vi.5," *Historia Mathematica* 17 (1990): 103–132, pp. 110–113, 120–122.

4. Al-Uqlīdisī on Arithmetic, A. Saidan, *The Arithmetic of al-Uqlīdisī* (Boston: Reidel, 1978), 74, 110, 114, 247, 258–259, 313–314, 343.

5. Kūshyār ibn Labbān on Arithmetic, M. Levey and M. Petruck, *Kushyar ibn Labbān Principles of Hindu Reckoning* (Madison: University of Wisconsin Press, 1966), 50–68.

6. Al-Kāshī on extracting a fifth root (1426), translated by J. L. Berggren from Jamshid al Kāshī's *Miftāh al-hisāb*, ed. and notes by Nabulsi Nader (Damascus: University of Damascus Press, 1977), 81–84.

Algebra

7. Al-Khwārizmī's algebra, F. Rosen, *The Algebra of Mohammad ben Musa* (London: Oriental Translation Fund, 1831), 3–4, 5–12, 21–24, 87–88. (Reprinted Hildesheim: Georg Olms Verlag, 1986.)

8. Thābit ibn Qurra on a geometric proof of a solution method for quadratic equations, translated from P. Luckey, "Tābit b. Qurra über den geometricschen Richtigkeitsnachweis der Auflösung der quadratischen Gleichungen," *Berichte über die Verhandlungen der Sachsischen Akademie der Wissenschaften zu Leipzig, Math-phys Klasse* 93, (1941): 93–114.

9. Abū Kāmil's algebra, translated from the Arabic text (Beyazid Library, Istanbul, Qara Mustafa Pasha Collection, No. 379), facsimile edition published by the Institute for the History of Arabic-Islamic Science (ed. F. Sezgin), Series C, Vol. 24, Frankfurt, 1986.

10. Al-Karajī on binomial coefficients, R. Rashed, *The Development of Arabic Mathematics* (Boston: Kluwer, 1994), 66–67.

11. Al-Karajī on irrational numbers, translated from the Arabic text in the edition of A. Anbouba (Beirut, 1964). (The author thanks Fr. John Ayoub, of Surrey, British Columbia, for his assistance in parts of the translation.)

12. Omar Khayyam's algebra, D. Kasir, *The Algebra of Omar Khayyam* (New York: Columbia University Press, 1931), 76–77, 99–102.

Number theory

13. Thābit ibn Qurra on amicable numbers, translated from F. Woepcke, "Notice sur une théorie ajoutée par Thâbit Ben Korrah à l'arithmétique spéculative des Grecs," *Journal Asiatique* 4th ser. 20 (1852): 420–429.

14. Abū Manṣūr al-Bāghdadī on balanced numbers, J. Sesiano, "Two Problems of Number Theory in Islamic Times," *Archive for History of Exact Sciences* 41 (1991): 235–238.

Theoretical geometry

15. Ibrāhīm ibn Sinān ibn Thābit ibn Qurra on the drawing of the three conic sections, R. Rashed and H. Bellosta (eds.), *Ibrāhīm ibn Sinān Logique et géométrie au Xe siècle* (Leiden: Brill, 2000), 264–274.

16. Ibrāhīm ibn Sinān ibn Thābit ibn Qurra's anthology of problems, R. Rashed and H. Bellosta (eds.), *Ibrāhīm ibn Sinān Logique et géométrie au Xe siècle* (Leiden: Brill, 2000), 664–666, 700–702.

17. Abū Sahl al-Kūhī and Abū Isḥāq al-Ṣābī on centers of gravity, J. L. Berggren, "The Correspondence of Abū Sahl al-Kūhī and Abū Isḥāq al-Ṣābī," *Journal for the History of Arabic Science* 7 (1983): 39–124.

18. Abū Sahl al-Kūhī on the astrolabe, J. L. Berggren, "Abū Sahl al-Kūhī's Treatise on the Construction of the Astrolabe with Proof: Text, Translation and Commentary," *Physis* 31 (1994): 141–252.

19. Abū al-Wafā' al-Būzjānī on five geometrical constructions, translated from H. Suter, "Das Buch der geometrischen Konstruktionen der Abu'l-Wefa," *Abhandlungen zur Geschichte der Naturwissenschaften und der Medizin* 4 (1922): 94–109.

20. Aḥmad ibn Muḥammad al-Sijzī on the construction of the heptagon, J. P. Hogendijk, "Greek and Arabic Constructions of the Regular Heptagon," *Archive for History of Exact Sciences* 30 (1984): 197–330, pp. 305–316.

21. Anonymous on the nonagon, J. L. Berggren, "An Anonymous Treatise on the Regular Nonagon," *Journal for the History of Arabic Science* 5 (1981): 37–41.

22. Ibn al-Haytham on the measurement of the paraboloid, translated from H. Suter, "Die Abhandlung über die Ausmessung des Paraboloides von el-Ḥasan b. el-Ḥasan b. el-Haitham," *Bibliotheca Mathematica* 3rd ser. 12 (1911–12): 289–322.

23. Al-Sijzī on the perfect compass, translated from F. Woepcke, "Trois traits arabes sur le Compas Parfait," *Notices et Extraits des Manuscrits de la Bibliothèque Imperial et autres Bibliothèques* 22 (1874): 18–32.

24. Muḥammad ibn al-Ḥusayn on the perfect compass, translated from F. Woepcke, "Trois traits arabes sur le Compas Parfait," *Notices et Extraits des Manuscrits de la Bibliothèque Imperial et autres Bibliothèques* 22 (1874): 18–32.

25. Badīʿ al-Zamān Abū al-ʿIzz Ismaʿīl b. al-Razzāz al-Jazarī on constructing geometrical instruments, Donald Hill, *The Book of Knowledge of Ingenious Mechanical Devices by Ibn al-Razzāz al-Jazarī* (Boston: Reidel, 1974), 196–198.

26. Naṣīr al-Dīn al-Ṭūsī on the theory of parallels, B. A. Rosenfeld, *A History of Non-Euclidean Geometry: Evolution of the Concept of a Geometric Space* (New York: Springer-Verlag, 1988), 74–80.

Practical geometry

27. Al-Kūhī on what is seen of sea and sky, translation by J. L. Berggren from the Arabic manuscript 5412 (Riād. 184) of the Holy Shrine Library (Mashad, Iran).

28. Abū al-Wafā' al Būzjanī on the geometry of artisans, Alpay Özdural, "Mathematics and Arts: Connections between Theory and Practice in the Medieval Islamic World," *Historia Mathematica* 27 (2000): 171–201, pp. 176–183.

29. Persian manuscript on geometrical dissections, translated by Jan Hogendijk from BN MS Persan 169.

30. Two patterns for Muqarnas, from G. Necipoglu, *Topkapi Scroll-Geometry and Ornament in Islamic Architecture* (Getty Center for the History of Art and the Humanities, Santa Monica, 1995), 332–333.

Trigonometry

31. Abū al-Wafā' al-Būzjānī's spherical trigonometry from his *Almagest*, translated from M.-Th. Debarnot, *Al-Bīrūnī, Les clefs de l'astronomie: La trigonométrie sphérique chez les Arabes de l'est à la fin du* X^{e} *siècle*, Institut Français de Damas, Damascus, 1985 and B. Carra de Vaux, "L'Almageste d'Abu'lwefa Albuzdjani," *Journal Asiatique*, 8th ser. 19 (1892): 408–471, pp. 423–424.

32. Abū al-Wafā' al-Būzjānī on telling time by the altitude of the sun, from *On the Rotation of the Celestial Sphere*, translated by J. L. Berggren from the Arabic text of Abū al-Wafā' al-Būzjānī, *Risāla … Fī iqāmat al-burhān ʿalā al-dā'ir min al-falak …* printed in *Al-rasā'il al-mutafariqa fī al-hay'a li 'l-mutaqadimīn wa-muʿāsirī al-Bīrūnī* (Hyderabad, 1948), 11–12.

33. Abū l-Ḥasan ibn Yūnus on nonlinear interpolation in sine tables, David A. King, "The Astronomical Works of Ibn Yūnus, PhD dessertation, Yale University, 1972, 82.

34. Al-Bīrūnī on geodesy, Jamil Ali, *The Determination of the Coordinates of Cities: Al-Bīrūnī's Taḥdīd al-Amākin* (Beirut: American University of Beirut, 1967), 178–180, 183–189, 192–193, 200–201, 214–215, 243–245, 253–255.

35. Naṣīr al-Dīn al-Ṭusī's *The Complete Quadrilateral*, translated from Alexandre Pacha Caratheodory, *Traité du quadrilatère attribué à Nassiruddin-el-Toussy* (Constantinople, 1891).

36. Al-Kashī on interpolation using second order differences, E. S. Kennedy, "A Medieval Interpolation Scheme Using Second Order Differences," reprinted in D. A. King and M. H. Kennedy (eds), *Studies in the Islamic Exact Sciences by E. S. Kennedy, Colleagues, and Former Students* (Beirut: American University of Beirut, 1983), 522–525. Original publication in *A Locust's Leg: Studies in Honor of S. H. Taqizadeh* (London), pp. 117–120.

Combinatorics

37. Ibn al-Haytham on constructing magic squares, translated from J. Sesiano, "Herstellungsverfahren magischer Quadrate aus islamischer Zeit (I)," *Sudhoffs Archiv* 64/2 (1980): 187–196.

38. Al-Bīrūnī on the number of possibilities for eclipse data, Jamil Ali, *The Determination of the Coordinates of Cities: Al-Bīrūnī's Taḥdīd al-Amākin* (Beirut: American University of Beirut, 1967), 132–133.

39. Aḥmad ibn Mun'im on counting the words, translated from Ahmed Djebbar, *L'Analyse combinatoire au Maghreb: L'Exemple d'Ibn Mun'im (XIIe–XIIIe s.)* (Orsay: Université de Paris Sud, 1985).

On mathematics

40. Al-Kūhī on the certainty deriving from mathematics, J. L. Berggren, "The Correspondence of Abū Sahl al-Kūhī and Abū Isḥāq al–Ṣābī: A Translation with commentaries," *Journal for the History of Arabic Science* 7 (1983): 39–124.

41. Aḥmad ibn Muḥammad al-Sijzī on heuristics, Jan P. Hogendijk, *Al-Sijzī's Treatise on Geometrical Problem Solving* (Teheran: Fatemi Publishing Co.), 1–10.

X.b. References

Note: A comprehensive bibliography of material on Islamic mathematics, both original sources and commentaries, can be found in [Lewis 2000].

Berggren, J. Lennart. 1986, 2003. *Episodes in the Mathematics of Medieval Islam.* New York: Springer-Verlag.

Clagett, Marshall. 1964. *Archimedes in the Middle Ages.* Vol. 1, Madison: University of Wisconsin Press.

Dakhel, Abdul-Kader. 1960. "Al-Kāshī on Root Extraction." In W. A. Hijab and E. S. Kennedy, eds., *Sources and Studies in the History of the Exact Sciences*, Vol. 2. Beirut: American University of Beirut.

Dold-Simplonius, Y. 1993. "The Volume of Domes in Arabic Mathematics." In M. Folkerts and J. P. Hogendijk, eds., *Vestigia Mathematica.* Amsterdam: Rodopi, pp. 93–106.

Heath, Thomas L. 1981. *A History of Greek Mathematics.* New York: Dover.

Hogendijk, Jan. 1984. "Greek and Arabic Constructions of the Regular Heptagon." *Archive for History of Exact Sciences* 30: 197–330.

Levey, Martin. 1966. *The Algebra of Abū Kāmil in a Commentary by Mordecai Finzi.* Madison: University of Wisconsin Press.

Lewis, Albert. 2000. *The History of Mathematics from Antiquity to the Present: A Selective Annotated Bibliography.* Rev. ed. (first edition edited by Joseph Dauben). Providence: American Mathematical Society.

Lorch, Richard. 2001. *Thābit ibn Qurra: The Sector Figure and Related Texts.* Islamic Mathematics and Astronomy, 108. Frankfurt: Institute for the History of Arabic Islamic Science.

Nadir, Nadi. 1960. "Abū al-Wafā' on the Solar Altitude." *Mathematics Teacher* 53: 460–463.

Taisbak, Christian M. 2003. *Euclid's Data: The Importance of Being Given.* Copenhagen: Museum Tusculanum Press.

About the Contributors

VICTOR J. KATZ received his Ph.D. in mathematics from Brandeis University in 1968 and is now Professor Emeritus of Mathematics at the University of the District of Columbia. He has long been interested in the history of mathematics and, in particular, in its use in teaching. His well-regarded textbook, *A History of Mathematics: An Introduction*, won the Watson Davis Prize of the History of Science Society in 1995, a prize awarded annually to the best book on the history of science aimed at undergraduates. Professor Katz has published many articles on the history of mathematics and its use in teaching and has edited two recent books dealing with this subject, *Learn from the Masters* (1994) and *Using History to Teach Mathematics* (2000). He has also directed two NSF-sponsored projects that helped college teachers learn the history of mathematics and how to use it in teaching, and also involved secondary school teachers in writing materials using history in the teaching of various topics in the high school curriculum. These materials, *Historical Modules for the Teaching and Learning of Mathematics*, have now been published on a CD by the MAA. Currently, Professor Katz is the editor of *Convergence: Where Mathematics, History and Teaching Interact*, an online magazine of the MAA that publishes articles and reviews designed to help teachers of grades 9 through 14 mathematics use history in the classroom.

ANNETTE IMHAUSEN studied mathematics (including history of mathematics) and Egyptology at the Universities of Mainz, Heidelberg and Berlin. She carried out her doctoral research in history of mathematics under the supervision of David Rowe (Mainz University) and Jim Ritter (Paris University) and received her doctoral degree summa cum laude in 2000. Her doctoral dissertation was awarded the dissertation prize of Mainz University and published in 2003. Imhausen has been a postdoctoral fellow at the Dibner Institute for the History of Science and Technology at MIT, a visiting fellow at the Department for History of Science at Harvard University (2000–2002), and a research fellow (2002–2005) at Trinity Hall, Cambridge (England). In 2001 she organized, in collaboration with Dr. John Steele and Christopher Walker, the international conference "Under One Sky: Astronomy and Mathematics in the Ancient Near East" (held at the British Museum, London). Since 2001 she has been an honorary fellow of the Centre for the History of Mathematical Sciences (Open University, England). She currently holds a Junior professorship at Mainz University, where she is teaching courses in history of mathematics and Egyptology. Her research interests focus on Egyptian science, its cultural and social context as well as influences from neighboring cultures. Her next research projects will explore tradition, transmission, and development of Egyptian mathematics in the Graeco-Roman period and the development of number and metrological systems at the beginning of Egyptian history (around 3000 BCE).

ELEANOR ROBSON is a Lecturer in the Department of History and Philosophy of Science at the University of Cambridge and a Fellow of All Souls College, Oxford. She holds a BSc (Hons) in mathematics from the University of Warwick and a Ph.D. in Oriental Studies from Wolfson College, Oxford. Her research centers on the social history of literacy and numeracy in ancient Iraq, from Sumerian literature to Hellenistic astronomy. She is a council member of the British School of Archaeology in Iraq and assistant editor of *The BSHM Bulletin*, the journal of the British Society for the History of Mathematics. Her publications include: *Mesopotamian Mathematics, 2100–1600 BC* (Clarendon Press, 1999); with M. Campbell-Kelly et al., *The History of Mathematical Tables from Sumer to Spreadsheets* (Oxford University Press, 2003); with J. A. Black et al., *The Literature of Ancient Sumer* (Oxford University Press, 2004); and, with W. L. Treadwell and C. Gosden, *Who Owns Objects? The Ethics and Politics of Collecting Archaeological Artefacts* (Oxbow, 2006). She is consultant editor for the children's book *Eyewitness: Mesopotamia* (Dorling Kindersley, 2006). Current book projects include *Mathematics in Ancient Iraq: A Social History* (Princeton University Press) and, with J. A. Stedall and T. Archibald, *The Oxford Companion to the History of Mathematics* (Oxford University Press).

JOSEPH W. DAUBEN graduated with honors in mathematics from Claremont McKenna College and received his MA (1968) and Ph.D. (1972) from Harvard University in History of Science. Since 1972 he has been a member of the Department of History at Herbert H. Lehman College of the City University of New York, where he is now Distinguished Professor of History and History of Science. He is also a member of the Ph.D. Program in History at the Graduate Center of the City University of New York, where he directs the Master of Arts Program in Liberal Studies. A former editor of *Historia Mathematica* and Chairman of the International Commission on History of Mathematics, Professor Dauben is an honorary member of the Institute for History of Natural Science of the Chinese Academy of Sciences. In 2006 he was Zhu Kezhen Visiting Professor of History of Science at the IHNS in Beijing, and is currently completing a critical edition in English of the "Ten Classics of Ancient Chinese Mathematics." His publications include biographies of Georg Cantor and Abraham Robinson (both translated into Chinese). He has also edited, with Christoph Scriba, *Writing the History of History of Mathematics.* His most recent publication is a translation of the oldest yet known mathematical work from ancient China, the *Suan shu shu* (Book on Numbers and Computations), in the *Archive for History of Exact Sciences.*

KIM PLOFKER received her doctoral degree from the History of Mathematics Department at Brown University, where she studied with the late David Pingree. She specializes in the history of Indian and Islamic mathematics and astronomy, and also works on early modern exact sciences in Latin. She has held various postdoctoral positions at Brown, at the Dibner Institute for the History of Science and Technology at MIT and, most recently, at the Mathematics Institute of the University of Utrecht and the International Institute for Asian Studies in Leiden. In 1994–1995 and 2003–2004 she received grants from the American Institute for Indian Studies to research Sanskrit scientific manuscripts in India. Her book on the history of mathematics in India from 500 BCE to 1800 CE will be published shortly by Princeton University Press.

J. LENNART BERGGREN is Emeritus Professor in the Department of Mathematics at Simon Fraser University, Canada, where he taught mathematics and its history for forty years. During this time he held visiting positions in the Mathematics Institute at the University of Warwick, and in the History of Science Departments at Yale and Harvard Universities. He has published fifty five refereed papers and six books on the history of mathematics. His special interest is the history of mathematical sciences in ancient Greece and medieval Islam—including cartography, spherical astronomy, and such mathematical instruments as sundials and astrolabes. Among his books are *Episodes in the Mathematics of Medieval Islam* (1986), *Euclid's "Phænomena"* (with Robert Thomas, 1996; reprinted by the AMS, 2006), and *Pi: A Sourcebook* (3rd edition, 2004). He is a member of the Canadian Society for the History and Philosophy of Science (of which he has served three two-year terms as president).

INDEX

Note: In general, birth and death dates are approximate. All dates are CE unless noted as BCE.